McGraw-Hill

Dictionary of
Environmental
Science

McGraw-Hill

New York   Chicago   San Francisco   Lisbon   London   Madrid
Mexico City   Milan   New Delhi   San Juan   Seoul   Singapore
Sydney   Toronto

The McGraw·Hill Companies

Materials in this dictionary are derived from the McGRAW-HILL DICTIONARY OF SCIENTIFIC AND TECHNICAL TERMS, Sixth Edition, copyright © 2003 by The McGraw-Hill Companies, Inc., and the McGRAW-HILL ENCYCLOPEDIA OF SCIENCE & TECHNOLOGY, Ninth Edition, copyright © 2002 by The McGraw-Hill Companies, Inc. All rights reserved.

2 3 4 5 6 7 8 9 0   DOC/DOC   0 9 8 7 6 5 4 3

ISBN 0-07-142177-7

 This book is printed on recycled, acid-free paper containing a minimum of 50% recycled, de-inked fiber.

This book was set in Helvetica Bold and Novarese Book by TechBooks, Fairfax, Virginia. It was printed and bound by RR Donnelley, The Lakeside Press.

McGraw-Hill books are available at special quantity discounts to use as premiums and sales promotions, or for use in corporate training programs. For more information, please write to the Director of Special Sales, Professional Publishing, McGraw-Hill, Two Penn Plaza, New York, NY 10121-2298. Or contact your local bookstore.

Library of Congress Cataloging-in-Publication Data

McGraw-Hill dictionary of environmental science.
      p.   cm.
   Includes index.
   ISBN 0-07-142177-7
   1. Environmental sciences—Dictionaries.   I. Title: Dictionary of environmental science.

GE10.M378   2003
628—dc21   2003051208

On the cover: Wind turbines convert wind energy to electricity.

Contents

Preface

The McGraw-Hill Dictionary of Environmental Science provides a compendium of 8,800 terms that are relevant to environmental science and related fields of science and technology. The coverage includes terminology from more than 30 disciplines, including agriculture, botany, chemical engineering, civil engineering, climatology, ecology, forestry, genetics and evolution, geochemistry, geography and mapping, meteorology, microbiology, mycology, oceanography, petroleum and mining engineering, plant pathology, systematics, and zoology.

The definitions are derived from the McGraw-Hill Dictionary of Scientific and Technical Terms, 6th edition (2003). The pronunciation of each term is provided along with synonyms, acronyms, and abbreviations where appropriate. A guide to the use of the Dictionary is included, explaining the alphabetical organization of terms, the format of the book, cross referencing, and how synonyms, variant spellings, abbreviations, and similar information are handled. A pronunciation key is also provided to assist the reader. An appendix provides definitions and conversion tables for commonly used scientific units as well as charts and listings of useful environmental data.

Many of the terms used in environmental science are often found in specialized dictionaries and glossaries; the McGraw-Hill Dictionary of Environmental Science, however, aims to provide the user with the convenience of a single, comprehensive reference. It is the editors' hope that it will serve the needs of scientists, engineers, students, teachers, librarians, and writers for high-quality information, and that it will contribute to scientific literacy and communication.

Mark D. Licker
Publisher

Staff

Mark D. Licker, Publisher—Science

Elizabeth Geller, Managing Editor
Jonathan Weil, Senior Staff Editor
David Blumel, Staff Editor
Alyssa Rappaport, Staff Editor
Charles Wagner, Digital Content Manager
Renee Taylor, Editorial Assistant

Roger Kasunic, Vice President—Editing, Design, and Production

Joe Faulk, Editing Manager
Frank Kotowski, Jr., Senior Editing Supervisor

Ron Lane, Art Director

Thomas G. Kowalczyk, Production Manager
Pamela A. Pelton, Senior Production Supervisor

Henry F. Beechhold, Pronunciation Editor
Professor Emeritus of English
Former Chairman, Linguistics Program
The College of New Jersey
Trenton, New Jersey

How to Use the Dictionary

ALPHABETIZATION.   The terms in the McGraw-Hill Dictionary of Environmental Science are alphabetized on a letter-by-letter basis; word spacing, hyphen, comma, solidus, and apostrophe in a term are ignored in the sequencing. For example, an ordering of terms would be:

Animalia	**apple scab disease**
animal kingdom	**Darwinism**
apple-cedar rust	**Darwin's theory**

FORMAT.   The basic format for a defining entry provides the term in boldface, the field in small capitals, and the single definition in lightface:

> **term**   [FIELD] Definition.

A term may be followed by multiple definitions, each introduced by a boldface number:

> **term**   [FIELD] **1.** Definition. **2.** Definition. **3.** Definition.

A term may have difinitions in two or more fields:

> **term**   [ECOL] Definition.   [GEN] Definition.

A simple cross-reference entry appears as:

> **term**   *See* another term.

A cross reference may also appear in combination with definitions:

> **term**   [ECOL] Definition.   [GEN] *See* another term.

CROSS REFERENCING.   A cross-reference entry directs the user to the defining entry. For example, the user looking up "aiophyllous" finds:

> **aiophyllous**   *See* evergreen.

The user then turns to the "E" terms for the definition. Cross references are also made from variant spellings, acronyms, abbreviations, and symbols.

> **aestivation**   *See* estivation.
> **ED$_{50}$**   *See* effective dose 50.
> **PVC**   *See* polyvinyl chloride.

ALSO KNOWN AS ..., etc.   A definition may conclude with a mention of a synonym of the term, a variant spelling, an abbreviation for the term, or other such information, introduced by "Also known as . . . ," "Also spelled . . . ," "Abbreviated . . . ," "Symbolized . . . ," "Derived from . . . ," When a term has more than one definition, the positioning of any of these phrases conveys the extent of applicability. For example:

term [ECOL] **1.** Definition. Also known as synonym. **2.** Definition. Symbolized T.

In the above arrangement, "Also known as. . ." applies only to the first definition; "Symbolized. . ." applies only to the second definition.

term [ECOL] **1.** Definition. **2.** Definition. [GEN] Definition. Also known as synonym.

In the above arrangement, "Also known as . . ." applies only to the second field.

term [ECOL] Also known as synonym. **1.** Definition. **2.** Definition. [GEN] Definition.

In the above arrangement, "Also known as . . ." applies only to both definitions in the first field.

term Also known as synonym. [ECOL] **1.** Definition. **2.** Definition. [GEN] Definition.

In the above arrangement, "Also known as . . ." applies to all definitions in both fields.

CHEMICAL FORMULAS. Chemistry definitions may include either an empirical formula (say, for acephate, $C_4H_{10}NO_3PS$) or a line formula (for sodium propionate, CH_3CH_2COONa), whichever is appropriate.

Fields and Their Scope

[AGR] **agriculture**—The production of plants and animals useful to humans, involving soil cultivation and the breeding and management of crops and livestock.

[BIOL] **biology**—The science of living organisms, including such fields as anatomy, biochemistry, biophysics, cell and molecular biology, and physiology.

[BOT] **botany**—That branch of biology dealing with the structure, function, diversity, evolution, reproduction, and utilization of plants and their interactions within the environment.

[CHEM] **chemistry**—The scientific study of the properties, composition, and structure of matter, the changes in the structure and composition of matter, and accompanying energy changes; includes the fields of analytical chemistry, inorganic chemistry, organic chemistry, physical chemistry, and spectroscopy.

[CHEM ENG] **chemical engineering**—A branch of engineering which involves the design of chemical products and processes for a wide range of engineering fields, including petroleum, materials science, agricultural, energy, environmental, pharmaceutical, and biomedical.

[CIV ENG] **civil engineering**—The planning, design, construction, and maintenance of fixed structures and ground facilities for industry, for transportation, for use and control of water, for occupancy, and for harbor facilities.

[CLIMATOL] **climatology**—That branch of meteorology concerned with the mean physical state of the atmosphere together with its statistical variations in both space and time as reflected in the weather behavior over a period of many years.

[ECOL] **ecology**—The study of the interrelationships between organisms and their environment.

[ENG] **engineering**—The art and science by which the properties of matter and the sources of power in nature are made useful to humans, for example, in structures, machines, processes, and products; subfields include aerospace engineering, building construction, design engineering, food engineering, industrial engineering, mechanical engineering, mechanics, and metallurgy.

[FOR] **forestry**—The science of developing, cultivating, and managing forest lands for wood, forage, water, wildlife, and recreation; the management of growing timber.

[GEN] **genetics and evolution**—The branches of biological science concerned with biological inheritance, that is, with the causes of the resemblances and differences among related individuals (genetics); and the processes and history of biological change in populations of organisms by which descendants come to differ from their ancestors (evolution).

[GEOCHEM] **geochemistry**—The field that encompasses the investigation of the chemical composition of the earth, other planets, and the solar system and universe as a whole, as well as the chemical processes that occur within them.

[GEOGR] **geography and mapping**—The science that deals with the description of land, sea, and air and the distribution of plant and animal life, including humans (geography); and the creation of representations indicating the relative size and shape of areas including such features (mapping).

[GEOL] **geology**—The science of the earth, its history, and its life as recorded in the rocks; includes the study of the geologic features of an area, such as the geometry of rock formations, weathering and erosion, and sedimentation, as well the structure and origins of the rocks (petrology) and minerals (mineralogy) themselves.

[GEOPHYS] **geophysics**—The branch of geology in which the principles and practices of physics are used to study the earth and its environment, that is, earth, air, and (by extension) space.

[HYD] **hydrology**—The science dealing with all aspects of the waters on earth, including their occurrence, circulation, and distribution; their chemical and physical properties; and their reaction with the environment, including their relation to living things.

[MED] **medicine**—The study of the causes, effects, and treatment of human diseases, including the subfields of immunology (the study of the native or acquired resistance of higher animal forms and humans to infection with microorganisms); pathology (the study of disease, including the biochemical and microbiological examination of bodily substances and the study of structural abnormalities of cells, tissues, and organs); and pharmacology (the study of the action of drugs and other chemical substances on biological systems).

[METEOROL] **meteorology**—The science concerned primarily with the observation of the atmosphere and its phenomena, including temperature, density, winds, clouds, and precipitation.

[MICROBIO] **microbiology**—The study of organisms of microscopic size, such as bacteria, viruses, and protozoa.

[MYCOL] **mycology**—The branch of biological science concerned with the study of fungi.

[OCEANOGR] **oceanography**—The science of the sea, including physical oceanography (the study of the physical properties of seawater and its motion in waves, tides, and currents), marine chemistry, marine geology, and marine biology.

[PETR MIN] **petroleum and mining engineering**—Branches of engineering concerned with the search for and extraction from the earth of oil, gas, and liquifiable hydrocarbons (petroleum engineering), and of coal and mineral resources (mining engineering), and the processing of these products for use.

[PHYS] **physics**—The science concerned with those aspects of nature that can be understood in terms of elementary principles and laws, including the subfields of acoustics, astrophysics, electromagnetism, fluid mechanics, nuclear physics, nucleonics, optics, plasma physics, and thermodynamics.

[PL PATH] **plant pathology**—The branch of botany concerned with diseases of plants.

[SCI TECH] **science and technology**—The logical study of natural phenomena and application of this knowledge for practical purposes, and the general terms and concepts used in such endeavors.

[STAT] **statistics**—The science dealing with the collection, analysis, interpretation, and presentation of masses of numerical data.

[SYST] **systematics**—The science of animal and plant classification.

[VET MED] **veterinary medicine**—The branch of medical practice which treats the diseases and injuries of animals.

[ZOO] **zoology**—The science that deals with the taxonomy, behavior, and morphology of animal life, usually divided into vertebrate and invertebrate zoology.

Pronunciation Key

Vowels

a	as in b**a**t, th**a**t
ā	as in b**ai**t, cr**a**te
ä	as in b**o**ther, f**a**ther
e	as in b**e**t, n**e**t
ē	as in b**ee**t, tr**ea**t
i	as in b**i**t, sk**i**t
ī	as in b**i**te, l**igh**t
ō	as in b**oa**t, n**o**te
ȯ	as in b**ough**t, t**au**t
u̇	as in b**oo**k, p**u**ll
ü	as in b**oo**t, p**oo**l
ə	as in b**u**t, sof**a**
au̇	as in cr**ow**d, p**ow**er
ȯi	as in b**oi**l, sp**oi**l
yə	as in form**u**la, spectac**u**lar
yü	as in f**ue**l, m**u**le

Semivowels/Semiconsonants

w	as in **w**ind, t**w**in
y	as in **y**et, on**i**on

Stress (Accent)

ˈ   precedes syllable with primary stress

ˌ   precedes syllable with secondary stress

ˌ̧   precedes syllable with variable or indeterminate primary/ secondary stress

Consonants

b	as in **b**i**b**, dri**bb**le
ch	as in **ch**arge, stre**tch**
d	as in **d**og, ba**d**
f	as in **f**ix, sa**f**e
g	as in **g**ood, si**g**nal
h	as in **h**and, be**h**ind
j	as in **j**oint, di**g**it
k	as in **c**ast, bri**ck**
k̲	as in Ba**ch** (used rarely)
l	as in **l**oud, be**ll**
m	as in **m**ild, su**mm**er
n	as in **n**ew, de**n**t
n̲	indicates nasalization of preceding vowel
ŋ	as in ri**ng**, si**ng**le
p	as in **p**ier, sli**p**
r	as in **r**ed, sca**r**
s	as in **s**ign, po**s**t
sh	as in **s**ugar, **sh**oe
t	as in **t**imid, ca**t**
th	as in **th**in, brea**th**
t̲h̲	as in **th**en, brea**the**
v	as in **v**eil, wea**v**e
z	as in **z**oo, crui**s**e
zh	as in bei**g**e, trea**s**ure

Syllabication

·   Indicates syllable boundary when following syllable is unstressed

aapamoor [ECOL] A moor with elevated areas or mounds supporting dwarf shrubs and sphagnum, interspersed with low areas containing sedges and sphagnum, thus forming a mosaic. { 'äp·ə‚mür }

abandoned channel *See* oxbow. { ə'ban·dənd 'chan·əl }

abatement [ENG] A decrease in the amount of a substance or other quantity, such as atmospheric pollution. { ə'bāt·mənt }

abiocoen [ECOL] A nonbiotic habitat. { 'ā‚bī·ō‚sēn }

abiogenesis [BIOL] The origin of life from nonliving matter, as occurred with the appearance of the first lifeform on earth. Also the discredited idea of spontaneous generation. { ¦ā‚bī·ō'jen·ə·sis }

abioseston [OCEANOGR] A general term for dead organic matter floating in ocean water. { ¦ā‚bī·ō'ses·tən }

abiotic [BIOL] Referring to the absence of living organisms. { ¦a‚bī'äd·ik }

abiotic environment [ECOL] All physical and nonliving chemical factors, such as soil, water, and atmosphere, which influence living organisms. { ¦a‚bī'äd·ik in'vī·rən‚mənt }

abiotic substance [ECOL] Any fundamental chemical element or compound in the environment. { ¦a‚bī'äd·ik 'səb·stəns }

ablation [HYD] The reduction in volume of a glacier due to melting and evaporation. { ə'blā·shən }

ablation area [HYD] The section in a glacier or snowfield where ablation exceeds accumulation. { ə'blā·shən 'er·ē·ə }

ablation cone [HYD] A debris-covered cone of ice, firn, or snow formed by differential ablation. { ə'blā·shən kōn }

ablation factor [HYD] The rate at which a snow or ice surface wastes away. { ə'blā·shən 'fak·tər }

ablation form [HYD] A feature on a snow or ice surface caused by melting or evaporation. { ə'blā·shən fȯrm }

abrade [GEOL] To wear away by abrasion or friction. { ə'brād }

Abraham's tree [METEOROL] The popular name given to a form of cirrus radiatus clouds, consisting of an assemblage of long feathers and plumes of cirrus that seems to radiate from a single point on the horizon. { 'ā·brə‚hamz 'trē }

abrasion [GEOL] Wearing away of sedimentary rock chiefly by currents of water laden with sand and other rock debris and by glaciers. { ə'brā·zhən }

abrasion platform [GEOL] An uplifted marine peneplain or plain, according to the smoothness of the surface produced by wave erosion, which is of large area. { ə'brā·zhən 'plat·fȯrm }

abrin [BIOL] A highly poisonous protein found in the seeds of *Abrus precatorius*, the rosary pea. { 'a·brin }

abs *See* absolute.

abscission [BOT] A physiological process promoted by abscisic acid whereby plants shed a part, such as a leaf, flower, seed, or fruit. { ab'sizh·ən }

absolute [METEOROL] Referring to the highest or lowest recorded value of a meteorological element, whether at a single station or over an area, during a given period. Abbreviated abs. { ,ab·sə'lüt }

absolute drought [METEOROL] In Britain, a period of at least 15 consecutive days during which no measurable daily precipitation has fallen. { 'ab·sə,lüt ,draut }

absolute instability [METEOROL] The state of a column of air in the atmosphere when it has a superadiabatic lapse rate of temperature, that is, greater than the dry-adiabatic lapse rate. Also known as autoconvective instability; mechanical instability. { 'ab·sə ,lüt ,in·stə'bil·ə·dē }

absolute stability [METEOROL] The state of a column of air in the atmosphere when its lapse rate of temperature is less than the saturation-adiabatic lapse rate. { 'ab·sə ,lüt stə'bil·ə·dē }

absorb [CHEM] To take up a substance in bulk. [PHYS] To take up energy from radiation. { əb'sòrb }

absorbed-dose rate [PHYS] The absorbed dose of ionizing radiation imparted at a given location per unit of time (second, minute, hour, or day). { əb'sòrbd 'dōs ,rāt }

absorber [ENG] The surface on a solar collector that absorbs the solar radiation. { əb'sòr·bər }

absorber plate [ENG] A part of a flat-plate solar collector that provides a surface for absorbing incident solar radiation. { əb'sòr·bər ,plāt }

absorption [HYD] Entrance of surface water into the lithosphere. { əb'sòrp·shən }

absorption spectrum [CHEM] A plot of how much radiation a sample absorbs over a range of wavelengths; the spectrum can be a plot of either absorbance or transmittance versus wavelength, frequency, or wavenumber. { əb'sòrp·shən ,spek·trəm }

abstraction [HYD] **1.** The draining of water from a stream by another having more rapid corroding action. **2.** The part of precipitation that does not become direct runoff. { ab'strak·shən }

abundance [GEOCHEM] The relative amount of a given element among other elements. { ə'bən·dəns }

abyssal [OCEANOGR] Pertaining to the abyssal zone. { ə'bis·əl }

abyssal-benthic [OCEANOGR] Pertaining to the bottom of the abyssal zone. { ə'bis·əl 'ben·thik }

abyssal floor [GEOL] The ocean floor, or bottom of the abyssal zone. { ə'bis·əl 'flòr }

abyssal plain [GEOL] A flat, almost level area occupying the deepest parts of many of the ocean basins. { ə'bis·əl 'plān }

abyssal zone [OCEANOGR] The biogeographic realm of the great depths of the ocean beyond the limits of the continental shelf, generally below 1000 meters. { ə'bis·əl 'zōn }

abyssopelagic [OCEANOGR] Pertaining to the open waters of the abyssal zone. { ə'bis·ō·pə'la·jik }

Ac *See* altocumulus cloud.

acanthocheilonemiasis [MED] A parasitic infection of humans caused by the filarial nematode *Acanthocheilonema perstans*. { ə‚kan·thə‚kī·lə·ne'mī·ə·səs }

acaricide [AGR] A pesticide used to destroy mites on domestic animals, crops, and humans. Also known as miticide. { ə'kar·ə‚sīd }

Acaridiae [ZOO] A group of pale, weakly sclerotized mites in the suborder Sarcoptiformes, including serious pests of stored food products and skin parasites of warm-blooded vertebrates. { ‚a·kə'rid·ē‚ē }

acarophily [ECOL] A symbiotic relationship between plants and mites. { ¦a·kə·rȯ¦fil·ē }

acarpous [BOT] Not producing fruit. { ā'kär·pəs }

accelerated erosion [GEOL] Soil erosion that occurs more rapidly than soil horizons can form from the parent regolith. { ak'sel·ər‚ā·dəd i'rō·zhən }

accessory cloud [METEOROL] A cloud form that is dependent, for its formation and continuation, upon the existence of one of the major cloud genera; may be an appendage of the parent cloud or an immediately adjacent cloudy mass. { ak'ses·ə·rē ‚klau̇d }

accessory element *See* trace element. { ak'ses·ə·rē 'el·ə·mənt }

accessory pigments [BIOL] Light-absorbing pigments, including carotenoids and phycobilins, which complement chlorophyll in plants, algae, and bacteria by trapping light energy for photosynthesis. { ak¦ses·ə·rē 'pig·məns }

accident [HYD] An interruption in a river that interferes with, or sometimes stops, the normal development of the river system. { 'ak·sə‚dent }

accidental species [ECOL] Species that are not characteristic of a particular habitat type and occur there only by chance. { ¦ak·sə¦den·təl 'spē·shēz }

acclimated microorganism [ECOL] Any microorganism that is able to adapt to environmental changes such as a change in temperature, or a change in the quantity of oxygen or other gases. { ə'klīm·əd·əd ‚mī·krō'ȯr·gə·niz·əm }

acclimation *See* acclimatization. { ‚ak·lə'mā·shən }

acclimatization [BIOL] Physiological, emotional, and behavioral adjustment by an individual to changes in the environment. [GEN] Adaptation of a species or population to a changed environment over several generations. Also known as acclimation. { ə‚klī·mə·tə'zā·shən }

accordant drainage [HYD] Flow of surface water that follows the dip of the strata over which it flows. Also known as concordant drainage. { ə¦kȯrd·ənt 'drān·ij }

accretion [METEOROL] The growth of a precipitation particle by the collision of a frozen particle (ice crystal or snowflake) with a supercooled liquid droplet which freezes upon contact. { ə'krē·shən }

accretionary ridge [GEOL] A beach ridge located inland from the modern beach, indicating that the coast has been built seaward. { ə'krē·shən‚er·ē ‚rij }

accretion tectonics [GEOL] The bringing together, or suturing, of terranes; regarded by many geologists as an important mechanism of continental growth. Also known as accretion. { ə'krē·shən tek'tän·iks }

accumulated dose [MED] The total amount of radiation absorbed by an organism as a result of exposure to radiation. { ə'kyü·myə‚lād·əd 'dōs }

accumulated temperature [METEOROL] A value based on the integrated product of the number of degrees that air temperature rises above a given threshold value and the number of days in the period during which this excess is maintained. { ə'kyü·myə ‚lād·əd 'tem·prə·chər }

accumulation [HYD] The quantity of snow or other solid form of water added to a glacier or snowfield mainly by snowfall. { ə·kyü·myə'lā·shən }

accumulation area [HYD] The portion of a glacier above the firn line, where the accumulation exceeds ablation. Also known as firn field; zone of accumulation. { ə·kyü·myə'lā·shən 'er·ē·ə }

accumulation zone [GEOL] The area where the bulk of the snow contributing to an avalanche was originally deposited. { ə·kyü·myə'lā·shən ˌzōn }

accumulator plant [BOT] A plant or tree that grows in a metal-bearing soil and accumulates an abnormal content of the metal. { ə'kyü·myəˌlād·ər ˌplant }

accustomization [ENG] The process of learning the techniques of living with a minimum of discomfort in an extreme or new environment. { əˌkəs·tə·mə'zā·shən }

acephate [CHEM] $C_4H_{10}NO_3PS$ A white solid with a melting point of 72–80°C; very soluble in water; used as an insecticide for a wide range of aphids and foliage pests. { 'as·ə·fāt }

acephatemet [CHEM] $CH_3OCH_3SPONH_2$ A white, crystalline solid with a melting point of 39–41°C; limited solubility in water; used as an insecticide to control cutworms and borers on vegetables. { as·ə'fāt·mət }

acervate [BIOL] Growing in heaps or dense clusters. { 'a·sərˌvāt }

acetoclastis [MICROBIO] The process, carried out by some methanogens, of splitting acetate into methane and carbon dioxide. { ˌa·sə·tō'klas·təs }

acetogenic bacteria [BIOL] Anaerobic bacteria capable of reducing carbon dioxide to acetic acid or converting sugars into acetate. { ˌa·sə·tō¦jen·ik bak'tir·ē·ə }

acetone cyanohydrin [CHEM] $(CH_3)_2COHCN$ A colorless liquid obtained from condensation of acetone with hydrocyanic acid; used as an insecticide or as an organic chemical intermediate. { 'as·əˌtōn sī,ə·nō'hīd·rən }

acetylacetone [CHEM] $CH_3COCH_2OCCH_3$ A colorless liquid with a pleasant odor and a boiling point of 140.5°C; soluble in water; used as a solvent, lubricant additive, paint drier, and pesticide. { ə¦sed·əl'as·əˌtōn }

acetyl benzoyl peroxide [CHEM] $C_6H_5CO·O_2·OCCH_3$ White crystals with a melting point of 36.6°C; moderately soluble in ether, chloroform, carbon tetrachloride, and water; used as a germicide and disinfectant. { ə'sed·əl 'ben·zȯil pə'räkˌsīd }

acetyl-CoA pathway [BIOL] A pathway of autotrophic carbon dioxide fixation. { a¦sed·əl ˌkō¦ā'pathˌwā }

acheb [ECOL] Short-lived vegetation regions of the Sahara composed principally of mustards (Cruciferae) and grasses (Gramineae). { ə'cheb }

acicular ice [HYD] Fresh-water ice composed of many long crystals and layered hollow tubes of varying shape containing air bubbles. Also known as fibrous ice; satin ice. { ə'sik·yə·lər 'īs }

acid clay [GEOL] A type of clay that gives off hydrogen ions when it dissolves in water. { 'as·əd 'klā }

acid gases [CHEM ENG] The hydrogen sulfide and carbon dioxide found in natural and refinery gases which, when combined with moisture, form corrosive acids; known as sour gases when hydrogen sulfide and mercaptans are present. { 'as·əd 'gas·əz }

acidity coefficient [GEOCHEM] The ratio of the oxygen content of the bases in a rock to the oxygen content in the silica. Also known as oxygen ratio. { ə'sid·ə·tē ˌkō·ə'fish·ənt }

acidophile [BIOL] **1.** Any substance, tissue, or organism having an affinity for acid stains. **2.** An organism having a preference for an acid environment. { ə'sid·əˌfil }

acidotrophic [BIOL] Having an acid nutrient requirement. { əˌsid·ə'trōf·ik }

acid pickle [CHEM ENG] Industrial waste water that is the spent liquor from a chemical process used to clean metal surfaces. { 'as·əd 'pik·əl }

acid precipitation [METEOROL] Rain or snow with a pH of less than 5.6. { 'as·əd prə ˌsip·ə'tā·shən }

acid rain [METEOROL] Precipitation in the form of water drops that incorporates anthropogenic acids and acid materials. { ˈas·əd 'rān }

acid soil [GEOL] A soil with pH less than 7; results from presence of exchangeable hydrogen and aluminum ions. { 'as·əd 'sȯil }

acid soot [ENG] Carbon particles that have absorbed acid fumes as a by-product of combustion; hydrochloric acid absorbed on carbon particulates is frequently the cause of metal corrosion in incineration. { 'as·əd ˌsüt }

acidulous water [HYD] Mineral water either with dissolved carbonic acid or dissolved sulfur compounds such as sulfates. { ə'sij·ə·ləs 'wȯd·ər }

acid-water pollution [ENG] Industrial wastewaters that are acidic; usually appears in effluent from the manufacture of chemicals, batteries, artificial and natural fiber, fermentation processes (beer), and mining. { 'as·əd 'wȯd·ər pə'lü·shən }

acorn disease [PL PATH] A virus disease of citrus plants characterized by malformation of the fruit, which is somewhat acorn-shaped. { 'ā,kȯrn diz,ēz }

acoustic absorption See sound absorption. { ə'küs·tik əb'sȯrp·shən }

acoustical door [ENG] A solid door with gasketing along the top and sides, and usually an automatic door bottom, designed to reduce noise transmission. { ə'küs·tə·kəl 'dȯr }

acoustic noise [PHYS] Noise in the acoustic spectrum; usually measured in decibels. { ə'küs·tik ˌnȯiz }

acoustic shielding [PHYS] A sound barrier that prevents the transmission of acoustic energy. { ə'küs·tik 'shēld·iŋ }

acquired [BIOL] Not present at birth, but developed by an individual in response to the environment and not subject to hereditary transmission. { ə'kwīrd }

acquired immune deficiency syndrome [MED] A disease that is caused by the human immunodeficiency virus (HIV) and compromises the competency of the immune system; characterized by persistent lymphadenopathy, opportunistic infections, and malignancies. HIV infection is transmitted by sexual intercourse, by blood and blood products, and perinatally from infected mother to child (prepartum, intrapartum, and postpartum via breast milk). { ə'kwīrd əˌmyün dəˌfish·ən·sē 'sin,drōm }

acre-foot [HYD] The volume of water required to cover 1 acre to a depth of 1 foot, hence 43,560 cubic feet; a convenient unit for measuring irrigation water, runoff volume, and reservoir capacity. { 'ā·kər 'fút }

acre-foot per day [HYD] The United States unit of volume rate of water flow. Abbreviated acre-ft/d. { 'ā·kər 'fút pər 'dā }

acre-ft/d See acre-foot per day.

acre-in. See acre-inch.

acre-inch [HYD] A unit of volume used in the United States for water flow, equal to 3630 cubic feet. Abbreviated acre-in. { 'ā·kər 'inch }

acre-yield [GEOL] The average amount of oil, gas, or water taken from one acre of a reservoir. { 'ā·kər ˌyēld }

acrodomatia [ECOL] Specialized structures on certain plants adapted to shelter mites; relationship is presumably symbiotic. { ˌak·rə·də'māsh·ē·ə }

acrodynia [MED] A childhood syndrome associated with mercury ingestion and characterized by periods of irritability alternating with apathy, anorexia, pink itching hands and feet, photophobia, sweating, tachycardia, hypertension, and hypotonia. { ˌak·rōˈdin·ē·ə }

acrolein [CHEM] $CH_2=CHCHO$ A colorless to yellow liquid with a pungent odor and a boiling point of 52.7°C; soluble in water, alcohol, and ether; used in organic synthesis, pharmaceuticals manufacture, and as an herbicide and tear gas. { əˈkrōl·ē·ən }

acrylamide [CHEM] $CH_2CHCONH_2$ Colorless, odorless crystals with a melting point of 84.5°C; soluble in water, alcohol, and acetone; used in organic synthesis, polymerization, sewage treatment, ore processing, and permanent press fabric; a probable human carcinogen. { əˈkril·əˌmīd }

actinochemistry [CHEM] A branch of chemistry concerned with chemical reactions produced by light or other radiation. { ˌak·tə·nōˈkem·ə·strē }

Actinomyces [MICROBIO] The type genus of the family Actinomycetaceae; anaerobic to facultatively anaerobic; includes human and animal pathogens. { ˌak·tə·nōˈmī·sēs }

actinomycosis [MED] An infectious bacterial disease caused by *Actinomyces bovis* in cattle, hogs, and occasionally in humans. Also known as lumpy jaw. { ˌak·tə·nōˌmī ˈkō·səs }

activate [PHYS] To induce radioactivity through bombardment by neutrons or by other types of radiation. { ˈak·təˌvāt }

activated sludge [CIV ENG] A semiliquid mass removed from the liquid flow of sewage and subjected to aeration and aerobic microbial action; the end product is dark to golden brown, partially decomposed, granular, and flocculent, and has an earthy odor when fresh. { ˈak·təˌvād·əd ˈsləj }

activated-sludge process [CIV ENG] A sewage treatment process in which the sludge in the secondary stage is put into aeration tanks to facilitate aerobic decomposition by microorganisms; the sludge and supernatant liquor are separated in a settling tank; the supernatant liquor or effluent is further treated by chlorination or oxidation. { ˈak·təˌvād·əd ˌsləj ˈprä·ˌsəs }

activation [CHEM] Treatment of a substance by heat, radiation, or activating reagent to produce a more complete or rapid chemical or physical change. [ENG] The process of inducing radioactivity by bombardment with neutrons or with other types of radiation. { ˌak·təˈvā·shən }

active front [METEOROL] A front, or portion thereof, which produces appreciable cloudiness and, usually, precipitation. { ˈak·tiv frənt }

active glacier [HYD] A glacier in which some of the ice is flowing. { ˈak·tiv ˈglā·shər }

active immunity [MED] Disease resistance in an individual due to antibody production after exposure to a microbial antigen following disease, inapparent infection, or inoculation. { ˈak·tiv imˈyü·nət·ē }

active layer [GEOL] That part of the soil which is within the suprapermafrost layer and which usually freezes in winter and thaws in summer. Also known as frost zone. { ˈak·tiv ˈlā·ər }

active permafrost [GEOL] Permanently frozen ground (permafrost) which, after thawing by artificial or unusual natural means, reverts to permafrost under normal climatic conditions. { ˈak·tiv ˈpər·məˌfröst }

active sludge [CIV ENG] A sludge rich in destructive bacteria used to break down raw sewage. { ˈak·tiv ˈsləj }

active solar system [ENG] A solar heating or cooling system that operates by mechanical means, such as motors, pumps, or valves. { ˈak·tiv ˈsō·lər ˌsis·təm }

activity [PHYS] The intensity of a radioactive source. Also known as radioactivity. { ,ak'tiv·əd·ē }

actual elevation [METEOROL] The vertical distance above mean sea level of the ground at the meteorological station. { 'ak·chə·wəl ,el·ə'vā·shən }

actual pressure [METEOROL] The atmospheric pressure at the level of the barometer (elevation of ivory point), as obtained from the observed reading after applying the necessary corrections for temperature, gravity, and instrumental errors. { 'ak·chə·wəl 'presh·ər }

acute radiation syndrome [MED] A complex of symptoms involving the intestinal tract, blood-forming organs, and skin following whole-body irradiation. { ə'kyüt 'rād·ē'a·shən 'sin,drōm }

acute rhinitis [MED] Inflammation of the nasal mucous membrane due to either infection or allergy. { ə'kyüt rī'nīd·əs }

acute yellow atrophy [MED] Rapid liver destruction following viral hepatitis, toxic chemicals, or other agents. { ə'kyüt 'yel·ō'a·trə·fē }

adaptation [BIOL] The occurrence of physiological changes in an individual exposed to changed conditions; for example, tanning of the skin in sunshine, or increased red blood cell counts at high altitudes. [GEN] Adjustment to new or altered environmental conditions by changes in genotype (natural selection) or phenotype. { ,a,dap'tā·shən }

adaptive disease [MED] The physiologic changes impairing an organism's health as the result of exposure to an unfamiliar environment. { ə'dap·tiv di,zēz }

adaptive divergence [GEN] Divergence of new forms from a common ancestral form due to adaptation to different environmental conditions. { ə'dap·tiv də'vər·jəns }

adaptive mutations [GEN] Mutations conferring an advantage in a selective environment which arise after nongrowing or slowly growing cells are exposed to the selective environment. { ə¦dap·tiv myü'tā·shənz }

adaptive value [GEN] The property of a given genotype that confers fitness to an organism in a given environment. { ə'dap·tiv 'val·yü }

Adenoviridae [MICROBIO] A family of double-stranded DNA viruses with icosahedral symmetry; usually found in the respiratory tract of the host species and often associated with respiratory diseases. Also known as adenovirus. { ,ad·ən·ō'vīr·ə,dē }

adenovirus See Adenoviridae. { ¦ad·ən,o'vī·rəs }

adequate contact [MED] The degree of contact required between an infectious and a susceptible individual to cause infection of the latter. { 'ad·ə·kwət 'kän,takt }

adfluvial [BIOL] Migrating between lakes and rivers or streams. { ad'flü·vē·əl }

adfreezing [HYD] The process by which one object adheres to another by the binding action of ice; applied to permafrost studies. { ,ad'frēz·iŋ }

adiabat [METEOROL] The relatively constant rate (5.5°F/100 feet or 10°C/kilometer) at which a mass of air cools as it rises. { 'ad·ē·ə,bat }

adiabatic [PHYS] Referring to any change in which there is no gain or loss of heat. { ¦ad·ē·ə¦bad·ik }

adiabatic atmosphere [METEOROL] A model atmosphere characterized by a dry-adiabatic lapse rate throughout its vertical extent. { ¦ad·ē·ə¦bad·ik 'at·mə,sfir }

adiabatic chart See Stuve chart. { ¦ad·ē·ə¦bad·ik 'chärt }

adiabatic condensation pressure See condensation pressure. { ¦ad·ē·ə¦bad·ik ,kän ,den'sā·shən ,presh·ər }

adiabatic condensation temperature See condensation temperature. { ¦ad·ē·ə¦bad·ik ‚kän‚den'sā·shən 'tem·prə·chər }

adiabatic equilibrium [METEOROL] A vertical distribution of temperature and pressure in an atmosphere in hydrostatic equilibrium such that an air parcel displaced adiabatically will continue to possess the same temperature and pressure as its surroundings, so that no restoring force acts on a parcel displaced vertically. Also known as convective equilibrium. { ¦ad·ē·ə¦bad·ik ‚ē·kwə'lib·rē·əm }

adiabatic equivalent temperature See equivalent temperature. { ¦ad·ē·ə¦bad·ik i'kwiv·ə·lənt 'tem·prə‚chər }

adiabatic lapse rate See dry adiabatic lapse rate. { ¦ad·ē·ə¦bad·ik 'laps ‚rāt }

adiabatic rate See dry adiabatic lapse rate. { ¦ad·ē·ə¦bad·ik 'rāt }

adiabatic saturation pressure See condensation pressure. { ¦ad·ē·ə¦bad·ik ‚sach·ə'rā·shən ‚presh·ər }

adiabatic saturation temperature See condensation temperature. { ¦ad·ē·ə¦bad·ik ‚sach·ə'rā·shən ‚tem·prə·chər }

adiabatic system [SCI TECH] A body or system whose condition is altered without gaining heat from or losing heat to the surroundings. { ¦ad·ē·ə¦bad·ik 'sis·təm }

adjacent sea [GEOGR] A sea connected with the oceans but semienclosed by land; examples are the Caribbean Sea and North Polar Sea. { ə'jās·ənt 'sē }

adjusted stream [HYD] A stream which flows mostly parallel to the strike and as little as necessary in other courses. { ə'jəs·təd 'strēm }

adlittoral [OCEANOGR] Of, pertaining to, or occurring in shallow waters adjacent to a shore. { ‚ad'lid·ə·rəl }

adobe [GEOL] Heavy-textured clay soil found in the southwestern United States and in Mexico. { ə'dō·bē }

adolescent coast [GEOL] A type of shoreline characterized by low but nearly continuous sea cliffs. { ‚ad·əl'es·ənt ‚kōst }

adolescent river [HYD] A river with a graded bed and a well-cut channel that reaches base level at its mouth, its waterfalls and lakes of the youthful stage having been destroyed. { ‚ad·əl'es·ənt 'riv·ər }

adolescent stream [HYD] A stream characterized by a well-cut, smoothly graded channel that may reach base level at its mouth. { ‚ad·əl'es·ənt 'strēm }

adret [ECOL] The sunny (usually south) face of a mountain featuring high timber and snow lines. { 'ad·rət }

advance [HYD] The forward movement of a glacier. { əd'vans }

advanced sewage treatment See tertiary sewage treatment. { əd¦vanst 'sü·ij ‚trēt·mənt }

advection [METEOROL] The process of transport of an atmospheric property solely by the mass motion of the atmosphere. [OCEANOGR] The process of transport of water, or of an acqueous property, solely by the mass motion of the oceans, most typically via horizontal currents. { ‚ad'vek·shən }

advectional inversion [METEOROL] An inverted temperature gradient in the air resulting from a horizontal inflow of colder air into an area. { ad'vek·shən·əl in'vər·zhən }

advection fog [METEOROL] A type of fog caused by the horizontal movement of moist air over a cold surface and the consequent cooling of that air to below its dew point. { ‚ad'vek·shən ‚fäg }

advective hypothesis [METEOROL] The assumption that local temperature changes are the result only of horizontal or isobaric advection. { ‚ad'vek·tiv hī'päth·ə·səs }

advective thunderstorm [METEOROL] A thunderstorm resulting from static instability produced by advection of relatively colder air at high levels or relatively warmer air at low levels or by a combination of both conditions. { ‚ad'vek·tiv 'thən·dər‚störm }

adventitious [BIOL] Acquired spontaneously or accidentally, not by heredity. Also known as adventive. { ‚ad·ven'tish·əs }

adventitious root [BOT] A root that arises from any plant part other than the primary root (radicle) or its branches. { ‚ad·ven'tish·əs 'rüt }

adventive [BIOL] **1.** An organism that is introduced accidentally and is imperfectly naturalized; not native. **2.** See adventitious. { ad'ven·tiv }

aelophilous [BOT] Describing a plant whose disseminules are dispersed by wind. { ‚ē'lä·fə·ləs }

aeolian See eolian. { ē'ōl·ē·ən }

aeration [ENG] **1.** Exposing to the action of air. **2.** Causing air to bubble through. **3.** Introducing air into a solution by spraying, stirring, or similar method. **4.** Supplying or infusing with air, as in sand or soil. { e'rā·shən }

aerator [ENG] **1.** One who aerates. **2.** Equipment used for aeration. **3.** Any device for supplying air or gas under pressure, as for fumigating, welding, or ventilating. **4.** Equipment used to inject compressed air into sewage in the treatment process. { 'e‚rād·ər }

aerial [BIOL] Of, in, or belonging to the air or atmosphere. { 'e·rē·əl }

aerial mapping [GEOGR] The making of planimetric and contoured maps and charts on the basis of photographs of the ground surface from an aircraft, spacecraft, or rocket. Also known as aerocartography. { 'e·rē·əl 'map·iŋ }

aerial root [BOT] A root exposed to the air, usually anchoring the plant to a tree, and often functioning in photosynthesis. { 'e·rē·əl 'rüt }

aeroallergen [MED] Any airborne particulate matter that can induce allergic responses in sensitive persons. { ‚e·rō'al·ər·jən }

aerobe [BIOL] An organism that requires air or free oxygen to maintain its life processes. { 'e‚rōb }

aerobic-anaerobic interface [CIV ENG] That point in bacterial action in the body of a sewage sludge or compost heap where both aerobic and anaerobic microorganisms participate, and the decomposition of the material goes no further. { e'rōb·ik 'an·ə ‚rōb·ik 'in·tər‚fās }

aerobic-anaerobic lagoon [CIV ENG] A pond in which the solids from a sewage plant are placed in the lower layer; the solids are partially decomposed by anaerobic bacteria, while air or oxygen is bubbled through the upper layer to create an aerobic condition. { e'rōb·ik 'an·ə‚rōb·ik lə'gün }

aerobic bacteria [MICROBIO] Any bacteria requiring free oxygen for the metabolic breakdown of materials. { e'rōb·ik ‚bak'tir·ē·ə }

aerobic digestion [CHEM ENG] Digestion of matter suspended or dissolved in waste by microorganisms under favorable conditions of oxygenation. { e'rōb·ik də'jes·chən }

aerobic lagoon [CIV ENG] An aerated pond in which sewage solids are placed, and are decomposed by aerobic bacteria. Also known as aerobic pond. { e'rō·bik lə'gün }

aerobic pond See aerobic lagoon. { e'rō·bik 'pänd }

aerobic process [BIOL] A process requiring the presence of oxygen. { e'rōb·ik 'präs·əs }

aerobiology [BIOL] The study of the atmospheric dispersal of airborne fungus spores, pollen grains, and microorganisms; and, more broadly, of airborne propagules of algae and protozoans, minute insects such as aphids, and pollution gases and particles which exert specific biologic effects. { ,e·rō,bī'äl·ə·jē }

aerobioscope [MICROBIO] An apparatus for collecting and determining the bacterial content of a sample of air. { ,e·rō'bi·ə,skōp }

aerobiosis [BIOL] Life existing in air or oxygen. { ,e·rō,bi'ō·səs }

aerocartography See aerial mapping. { ,e·rō,kär'täg·rə·fē }

aerochlorination [CIV ENG] Treatment of sewage with compressed air and chlorine gas to remove fatty substances. { ,e·rō,klȯr·ə'nā·shən }

AERO code [METEOROL] An international code used to encode for transmission, in words five numerical digits long, synoptic weather observations of particular interest to aviation operations. { 'e·rō 'kōd }

aerofilter [CIV ENG] A filter bed for sewage treatment consisting of coarse material and operated at high speed, often with recirculation. { 'e·rō,fil·tər }

aerogenerator [ENG] A generator that is driven by the wind, designed to utilize wind power on a commercial scale. { ,e·rō'jen·ə,rād·ər }

aerogeography [GEOGR] The geographic study of earth features by means of aerial observations and aerial photography. { ,e·rō·jē'äg·rə·fē }

aerography [METEOROL] **1.** The study of the air or atmosphere. **2.** The practice of weather observation, map plotting, and maintaining records. See descriptive meteorology. { e'räg·rə·fē }

aerological days [METEOROL] Specified days on which additional upper-air observations are made; an outgrowth of the International Polar Year. { ,e·rə'lä·jə·kəl 'dāz }

aerological diagram [METEOROL] A diagram of atmospheric thermodynamics plotted from upper-atmospheric soundings; usually contains various reference lines such as isobars and isotherms. { ,e·rə,lä·jə·kəl 'dī·ə,gram }

aerology [METEOROL] The study of the free atmosphere throughout its vertical extent, as distinguished from studies confined to the layer of the atmosphere near the earth's surface. { e'rä·lə·jē }

aeronautical climatology [METEOROL] The application of the data and techniques of climatology to aviation meteorological problems. { e·rə'nȯd·ə·kəl ,klī·mə'täl·ə·je }

aeronautical meteorology [METEOROL] The study of the effects of weather upon aviation. { e·rə'nȯd·ə·kəl ,mēd·ē·ə'räl·ə·jē }

aerophyte See epiphyte. { 'e·rō,fīt }

aeroplankton [ECOL] Small airborne organisms such as insects. { ¦e·rō'plaŋk·tən }

aeroponics [AGR] The practice of growing plants without soil while suspended in air; a nutrient and water solution is sprayed on the roots and allowed to drain off to be discarded or recycled. { ,er·ə'pän·iks }

aerosol [METEOROL] A small droplet or particle suspended in the atmosphere and formed from both natural and anthropogenic sources. { 'e·rə,sȯl }

aerosol propellant [ENG] Compressed gas or vapor in a container which, upon release of pressure and expansion through a valve, carries another substance from the container; used for cosmetics, household cleaners, and so on; examples are butanes, propane, nitrogen, fluorocarbons, and carbon dioxide. { 'e·rə,sȯl prə'pel·ənt }

aerospace See airspace. { ¦e·rō¦spās }

aerotaxis [BIOL] The movement of an organism, especially aerobic and anaerobic bacteria, with reference to the direction of oxygen or air. { ˌe·rōˈtak·səs }

aerotolerant [MICROBIO] Able to survive in the presence of oxygen. { ˌe·rōˈtäl·ə·rənt }

aerotropism [BOT] A response in which the growth direction of a plant component changes due to modifications in oxygen tension. { ˌe·rōˈtrō,piz·əm }

aestivation *See* estivation.

afforestation [FOR] Establishment of a new forest by seeding or planting on non-forested land. { a,fär·əˈstā·shən }

aflatoxicosis [MED] Aflatoxin poisoning. { əˌflād·ō,täk·səˈkō·səs }

aflatoxin [BIOL] The toxin produced by some strains of the fungus *Aspergillus flavus*, the most potent carcinogen yet discovered. { ˌaf·ləˈtäk·sin }

A frame [OCEANOGR] An A-shaped frame used for outboard suspension of oceanographic gear on a research vessel. { ˈā ,frām }

Africa [GEOGR] The second largest continent, with an area of 11,700,000 square miles (30,420,000 square kilometers); bisected midway by the Equator, above and below which it shows symmetry of climate and vegetation zones. { ˈaf·ri·kə }

African swine fever *See* hog cholera. { ˈaf·ri·kən ˈswīn ,fēv·ər }

afterglow [METEOROL] A broad, high arch of radiance or glow seen occasionally in the western sky above the highest clouds in deepening twilight, caused by the scattering effect of very fine particles of dust suspended in the upper atmosphere. { ˈaf·tər,glō }

afterripening [BOT] A period of dormancy after a seed is shed during which the synthetic machinery of the seed is prepared for germination and growth. { ˈaf·tər,rī·pən·iŋ }

agar [BOT] A gelatinous product extracted from certain red algae and used chiefly as a gelling agent in culture media. { ˈäg·ər }

agarophyte [BOT] Any seaweed that yields agar. { əˈgar·ə,fīt }

Agassiz trawl [OCEANOGR] A dredge consisting of a net attached to an iron frame with a hoop at each end that is used to collect organisms, particularly invertebrates, living on the ocean bottom. { ˈag·ə·sē ˈtrōl }

Agassiz Valleys [GEOL] Undersea valleys in the Gulf of Mexico between Cuba and Key West. { ˈag·ə·sē ˈval·ēz }

agatized wood *See* silicified wood. { ˈag·ə·tīzd ˈwu̇d }

age [BIOL] Period of time from origin or birth to a later time designated or understood; length of existence. [GEOL] **1.** Any one of the named epochs in the history of the earth marked by specific phases of physical conditions or organic evolution, such as the Age of Mammals. **2.** One of the smaller subdivisions of the epoch as geologic time, corresponding to the stage or the formation, such as the Lockport Age in the Niagara Epoch. { āj }

aged [GEOL] Of a ground configuration, having been reduced to base level. { ˈā·jəd }

age determination [GEOL] Identification of the geologic age of a biological or geological specimen by using the methods of dendrochronology or radiometric dating. { ˈāj di,tər·məˈnā·shən }

age distribution [ECOL] The proportions of a population falling into different age groups.. { ˈāj dis·trəˈbyü·shən }

aged shore [GEOL] A shore long established at a constant level and adjusted to the waves and currents of the sea. { ˈā·jəd ˈshȯr }

age ratio [GEOL] The ratio of the amount of daughter to parent isotope in a mineral being dated radiometrically. { 'āj ˌrā·shō }

agglomeration [METEOROL] The process in which particles grow by collision with and assimilation of cloud particles or other precipitation particles. Also known as coagulation. { əˌgläm·ə'rā·shən }

aggradation [HYD] A process of shifting equilibrium of stream deposition, with upbuilding approximately at grade. { ˌag·rə'dā·shən }

aggraded valley plain *See* alluvial plain. { ə'grād·əd 'val·ē 'plān }

aggregate [BOT] Referring to fruit formed in a cluster, from a single flower, such as raspberry, or from several flowers, such as pineapple. [GEOL] A collection of soil grains or particles gathered into a mass. { 'ag·rə·gət }

aggregate fruit [BOT] A type of fruit composed of a number of small fruitlets all derived from the ovaries of a single flower. { 'ag·rə·gət 'früt }

aggregation [BIOL] A grouping or clustering of separate organisms. { ˌag·rə'gā·shən }

aggressive carbon dioxide [CHEM ENG] The carbon dioxide dissolved in water in excess of the amount required to precipitate a specified concentration of calcium ions as calcium carbonate; used as a measure of the corrosivity and scaling properties of water. { ə'gres·iv 'kär·bən dī'äkˌsīd }

aggressive water [HYD] Any of the waters which force their way into place. { ə'gres·iv 'wȯd·ər }

agrestal [ECOL] Growing wild in the fields. { ə'grest·əl }

agribiotechnology [AGR] Biotechnology applied to agriculture. { ˌag·rəˌbī·ō,tek'näl·ə·jē }

agricere [GEOL] A waxy or resinous organic coating on soil particles. { 'ag·rəˌsir }

agricultural chemicals [AGR] Fertilizers, soil conditioners, fungicides, insecticides, weed killers, and other chemicals used to increase farm crop productivity and quality. { ¦ag·rə¦kəl·chə·rəl 'kem·ə·kəls }

agricultural chemistry [AGR] The science of chemical compositions and changes involved in the production, protection, and use of crops and livestock; includes all the life processes through which food and fiber are obtained for humans and animals, and control of these processes to increase yields, improve quality, and reduce costs. { ¦ag·rə¦kəl·chə·rəl 'kem·ə·strē }

agricultural climatology [AGR] In general, the study of climate as to its effect on crops; it includes, for example, the relation of growth rate and crop yields to the various climatic factors and hence the optimum and limiting climates for any given crop. Also known as agroclimatology. { ¦ag·rə¦kəl·chə·rəl ˌklī·mə'täl·ə·jē }

agricultural engineering [AGR] A discipline concerned with developing and improving the means for providing food and fiber for human needs. { ¦ag·rə¦kəl·chə·rəl ˌen·jə'nir·iŋ }

agricultural geography [GEOGR] A branch of geography that deals with areas of land cultivation and the effect of such cultivation on the physical landscape. { ¦ag·ri¦kəl·chə·rəl jē'ag·rə·fē }

agricultural geology [GEOL] A branch of geology that deals with the nature and distribution of soils, the occurrence of mineral fertilizers, and the behavior of underground water. { ¦ag·rə¦kəl·chə·rəl jē'äl·ə·jē }

agricultural meteorology [AGR] The study and application of relationships between meteorology and agriculture, involving problems such as timing the planting of crops. Also known as agrometeorology. { ¦ag·rə¦kəl·chə·rəl ˌmēd·ē·ə'räl·ə·jē }

agricultural science |AGR| A discipline dealing with the selection, breeding, and management of crops and domestic animals for more economical production. { ¦ag·rə ¦kəl·chə·rəl 'sī·əns }

agricultural wastes [AGR] Those liquid or solid wastes that result from agricultural practices, such as cattle manure, crop residue (for example, corn stalks), pesticides, and fertilizers. { ¦ag·rə¦kəl·chə·rəl 'wāsts }

agriculture |BIOL| The production of plants and animals useful to humans, involving soil cultivation and the breeding and management of crops and livestock. { 'ag·rə ˌkəl·chər }

Agrobacterium |MICROBIO| A genus of bacteria in the family Rhizobiaceae; cells do not fix free nitrogen, and three of the four species are plant pathogens, producing galls and hairy root. { ¦ag·rō͵bak'tir·e·əm }

agroclimatology See agricultural climatology. { ¦ag·rō'klī·mə'täl·ə·je }

agroecology [ECOL] The ecology of agricultural ecosystems. { ¦ag·rō·e'käl·ə·je }

agroecosystem [ECOL] Any ecosystem involving cultivated plants. { ¦ag·rō'ek·ō ˌsis·təm }

agroenvironment |AGR| The soil and climate of a region as they affect agriculture. { ¦ag·rō͵en'vī·rən·mənt }

agroforestry |AGR| The practice of growing trees in association with agricultural crops or animals to provide both ecological and economic benefits. { ¦ag· rō'fär·əs·trē }

agrometeorology See agricultural meteorology. { ¦ag·rō'mēd·ē·ə'räl·ə·je }

agronomy |AGR| The principles and procedures of soil management and of field crop and special-purpose plant improvement, management, and production. { ə'grän·ə·mē }

agrophilous |ECOL| Having a natural habitat in grain fields. { ə'gräf·ə·ləs }

agrostology |BOT| A division of systematic botany concerned with the study of grasses. { ¦ag·rə¦stä·lə·je }

agrotechnology |AGR| An innovative technology designed to render agricultural production more efficient and profitable. { ¦ag·rō͵tek'näl·ə·je }

Agulhas Current |OCEANOGR| A fast current flowing in a southwestward direction along the southeastern coast of Africa. { ə'gəl·əs 'kər·ənt }

ahermatypic |ZOO| Non-reef-building, as applied to corals. { ¦ā͵hər·mə¦tip·ik }

AIDS See acquired immune deficiency syndrome. { ādz }

aiguille |GEOL| The needle-top of the summit of certain glaciated mountains, such as near Mont Blanc. { ͵ā'gwēl }

aimless drainage |HYD| Drainage without a well-developed system, as in areas of glacial drift or karst topography. { 'ām·ləs 'drān·ij }

aiophyllous See evergreen. { ͵ī·ō'fil·əs }

air composition |METEOROL| The kinds and amounts of the constituent substances of air, the amounts being expressed as percentages of the total volume or mass. { 'er ͵käm·pə'zish·ən }

aircraft ceiling |METEOROL| After United States weather observing practice, the ceiling classification applied when the reported ceiling value has been determined by a pilot while in flight within 1.5 nautical miles (2.8 kilometers) of any runway of the airport. { 'er͵kraft 'sēl·iŋ }

13

aircraft noise |PHYS| Effective sound output of the various sources of noise associated with aircraft operation, such as propeller and engine exhaust, jet noise, and sonic boom. { 'er₁kraft ₁nȯiz }

aircraft thermometry [METEOROL] The science of temperature measurement from aircraft. { 'er₁kraft thər'mäm·ə·trē }

aircraft weather reconnaissance [METEOROL] The making of detailed weather observations or investigations from aircraft in flight. { 'er₁kraft 'we<u>th</u>·ər ri₁kän·ə₁səns }

air drainage [METEOROL] General term for gravity-induced, downslope flow of relatively cold air. { 'er 'drān·ij }

air hoar [HYD] Hoarfrost growing on objects above the ground or snow. { 'er ₁hȯr }

air layering [BOT] A method of vegetative propagation, usually of a wounded part, in which the branch or shoot is enclosed in a moist medium until roots develop, and then it is severed and cultivated as an independent plant. { 'er ₁lā·ər·iŋ }

airlight [METEOROL] In determinations of visual range, light from sun and sky which is scattered into the eyes of an observer by atmospheric suspensoids (and, to slight extent, by air molecules) lying in the observer's cone of vision. { 'er₁līt }

air mass [METEOROL] An extensive body of the atmosphere which approximates horizontal homogeneity in its weather characteristics, particularly with reference to temperature and moisture distribution. { 'er ₁mas }

air-mass analysis [METEOROL] In general, the theory and practice of synoptic surface-chart analysis by the so-called Norwegian methods, which involve the concepts of the polar front and of the broad-scale air masses which it separates. { 'er₁mas ə'nal·ə·səs }

air-mass climatology [CLIMATOL] The representation of the climate of a region by the frequency and characteristics of the air masses under which it lies; basically, a type of synoptic climatology. { 'er ₁mas klīm·ə'täl·ə·jē }

air-mass precipitation [METEOROL] Any precipitation that can be attributed only to moisture and temperature distribution within an air mass when that air mass is not, at that location, being influenced by a front or by orographic lifting. { 'er ₁mas pri ₁sip·ə'tā·shən }

air-mass shower [METEOROL] A shower that is produced by local convection within an unstable air mass; the most common type of air-mass precipitation. { 'er ₁mas 'shau·ər }

air-mass source region [METEOROL] An extensive area of the earth's surface over which bodies of air frequently remain for a sufficient time to acquire characteristic temperature and moisture properties imparted by that surface. { 'er ₁mas 'sȯrs ₁rē·jən }

air parcel [METEOROL] An imaginary body of air to which may be assigned any or all of the basic dynamic and thermodynamic properties of atmospheric air. { 'er ₁pär·səl }

air pocket [METEOROL] An expression used in the early days of aviation for a downdraft; such downdrafts were thought to be pockets in which there was insufficient air to support the plane. { 'er ₁päk·ət }

air pollution [ECOL] The presence in the outdoor atmosphere of one or more contaminants such as dust, fumes, gas, mist, odor, smoke, or vapor in quantities and of characteristics and duration such as to be injurious to human, plant, or animal life or to property, or to interfere unreasonably with the comfortable enjoyment of life and property. { ¦er pə'lü·shən }

air-pollution control [ENG] A practical means of treating polluting sources to maintain a desired degree of air cleanliness. { ¦er pə'lü·shən kən₁trōl }

air sac [ZOO] In birds, any of the small vesicles that are connected with the respiratory system and located in bones and muscles to increase buoyancy. { 'er ˌsak }

air sampling [ENG] The collection and analysis of samples of air to measure the amounts of various pollutants or other substances in the air, or the air's radioactivity. { 'er ˌsam·pliŋ }

airshed [METEOROL] The air supply in a given region. { 'er,shed }

air sounding [METEOROL] The act of measuring atmospheric phenomena or determining atmospheric conditions at altitude, especially by means of apparatus carried by balloons or rockets. { 'er ˌsaund·iŋ }

airspace [METEOROL] **1.** Of or pertaining to both the earth's atmosphere and space. Also known as aerospace. **2.** The portion of the atmosphere above a particular land area, especially a nation or other political subdivision. { 'er,spās }

air spora [BIOL] Airborne fungus spores, pollen grains, and microorganisms. { 'er ˌspór·ə }

air temperature [METEOROL] **1.** The temperature of the atmosphere which represents the average kinetic energy of the molecular motion in a small region and is defined in terms of a standard or calibrated thermometer in thermal equilibrium with the air. **2.** The temperature that the air outside of an aircraft is assumed to have as indicated on a cockpit instrument. { 'er ˌtem·prə·chər }

air toxics *See* hazardous air pollutants. { 'er ˌtäk·siks }

air trap [CIV ENG] A U-shaped pipe filled with water that prevents the escape of foul air or gas from such systems as drains and sewers. { 'er ˌtrap }

air turbulence [METEOROL] Highly irregular atmospheric motion characterized by rapid changes in wind speed and direction and by the presence, usually, of up and down currents. { 'er ˌtər·byə·ləns }

airwave [METEOROL] A wavelike oscillation in the pattern of wind flow aloft, usually with reference to the stronger portion of the westerly current. { 'er,wāv }

airways code *See* United States airways code. { 'er,wāz ˌkōd }

airways forecast *See* aviation weather forecast. { 'er,wāz ˌfòr,kast }

airways observation *See* aviation weather observation. { 'er,wāz ˌäb·zər'vā·shən }

aktological [GEOL] Nearshore shallow-water areas, conditions, sediments, or life. { ˌak·tə'läj·ə·kəl }

Alaska Current [OCEANOGR] A current that flows northwestward and westward along the coasts of Canada and Alaska to the Aleutian Islands. { ə'las·kə 'kər·ənt }

Alberta low [METEOROL] A low centered on the eastern slope of the Canadian Rockies in the province of Alberta, Canada. { al'bərt·ə 'lō }

Alboll [GEOL] A suborder of the soil order Mollisol with distinct horizons, wet for some part of the year; occurs mostly on upland flats and in shallow depressions. { 'al,ból }

alburnum *See* sapwood. { al'bər·nəm }

alcohol [CHEM] Any member of a class of organic compounds in which a hydrogen atom of a hydrocarbon has been replaced by a hydroxy (−OH) group. { 'al·kə,hòl }

alcove [GEOL] A large niche formed by a stream in a face of horizontal strata. { 'al ˌkōv }

aldicarb [CHEM] $C_7H_{14}N_2O_2S$ A colorless, crystalline compound with a melting point of 100°C; used as an insecticide, miticide, and nematicide to treat soil for cotton, sugarbeets, potatoes, peanuts, and ornamentals. { 'al·də,kärb }

Aldrin |CHEM| $C_{12}H_8Cl_6$ Trade name for a water-insoluble, white, crystalline compound, consisting mainly of chlorinated dimethanonaphthalene; used as a pesticide. { 'al·drən }

aletophyte |ECOL| A weedy plant growing on the roadside or in fields where natural vegetation has been disrupted by humans. { ə'lēd·ə,fīt }

Aleutian Current |OCEANOGR| A current setting southwestward along the southern coasts of the Aleutian Islands. { ə'lü·shən ,kər·ənt }

Aleutian low |METEOROL| The low-pressure center located near the Aleutian Islands on mean charts of sea-level pressure; represents one of the main centers of action in the atmospheric circulation of the Northern Hemisphere. { ə'lü·shən ,lō }

Aleyrodidae |ZOO| The whiteflies, a family of homopteran insects included in the series Sternorrhyncha; economically important as plant pests. { ,al·ə'räd·ə,dē }

alfalfa |BOT| *Medicago sativa*. A herbaceous perennial legume in the order Rosales, characterized by a deep taproot. Also known as lucerne. { al'fal·fə }

Alfalfa mosaic virus group *See* Alfamovirus. { al¦fal·fə mō¦zā·ik 'vī·rəs ,grüp }

Alfamovirus |MICROBIO| A genus of plant viruses in the family Bromoviridae that is characterized by virions which are either bacilliform or ellipsoidal and contain single-stranded ribonucleic acid genomes; alfalfa mosaic virus is the type species. Also known as Alfalfa mosaic virus group. { al'fam·ə,vī·rəs }

Alfisol |GEOL| An order of soils with gray to brown surface horizons, a medium-to-high base supply, and horizons of clay accumulation. { 'al·fə,sōl }

algae |BOT| General name for the chlorophyll-bearing organisms in the plant subkingdom Thallobionta. { 'al·jē }

algae bloom |ECOL| A heavy growth of algae in and on a body of water as a result of high phosphate concentration from farm fertilizers and detergents. { 'al·jē ,blüm }

algae wash |ECOL| A shoreline drift consisting almost entirely of filamentous algae. { 'al·jē ,wash }

algal |BOT| Of or pertaining to algae. |GEOL| Formed from or by algae. { 'al·gəl }

algal coal |GEOL| Coal formed mainly from algal remains. { 'al·gəl ,kōl }

algal limestone |GEOL| A type of limestone either formed from the remains of calcium-secreting algae or formed when algae bind together the fragments of other lime-secreting organisms. { 'al·gəl 'līm,stōn }

algal reef |GEOL| An organic reef which has been formed largely of algal remains and in which algae are or were the main lime-secreting organisms. { 'al·gəl ,rēf }

algal ridge |GEOL| Elevated margin of a windward coral reef built by actively growing calcareous algae. { 'al·gəl ,rij }

algal rim |GEOL| Low rim built by actively growing calcareous algae on the lagoonal side of a leeward reef or on the windward side of a patch reef in a lagoon. { 'al·gəl ,rim }

algal structure |GEOL| A deposit, most frequently calcareous, with banding, irregular concentric structures, crusts, and pseudo-pisolites or pseudo-concretionary forms resulting from organic, colonial secretion and precipitation. { ¦al·gəl ¦strək·chər }

algicide |AGR| A chemical used to kill algae. { 'al·jə,sīd }

algin |BOT| A hydrophilic polysaccharide extracted from brown algae, such as giant kelp. { 'al·jən }

alginate |BOT| An algal polysaccharide that is a major constituent of the cell walls of brown algae. { 'al·jə,nāt }

algology |BOT| The study of algae. Also known as phycology. { al'gäl·ə·jē }

algophage *See* cyanophage. { 'al·gə,fāj }

alimentation *See* accumulation. { ,al·ə·mən'tā·shən }

alkali chlorosis [PL PATH] Yellowing of plant foliage due to excess amounts of soluble salts in the soil. { 'al·kə,lī klə'rō·səs }

alkali disease [VET MED] **1.** Botulism of ducks. **2.** Trembles of cattle. { 'al·kə,lī diz,ēz }

alkali flat [GEOL] A level lakelike plain formed by the evaporation of water in a depression and deposition of its fine sediment and dissolved minerals. { 'al·kə,lī ,flat }

alkali lake [HYD] A lake with large quantities of dissolved sodium and potassium carbonates as well as sodium chloride. { 'al·kə,lī 'lāk }

alkaline soil [GEOL] Soil containing soluble salts of magnesium, sodium, or the like, and having a pH value between 7.3 and 8.5. { 'al·kə,līn 'sȯil }

alkaliphile [BIOL] An organism that prefers or is able to withstand an alkaline environment (pH value above 9). { 'al·kə·lə,fīl }

alkali soil [GEOL] A soil, with salts injurious to plant life, having a pH value of 8.5 or higher. { 'al·kə,lī ,sȯil }

alkenones [GEOL] Long-chain (37–39 carbon atoms) di-, tri-, and tetraunsaturated methyl and ethyl ketones produced by certain phytoplankton (coccolithophorids), which biosynthetically control the degree of unsaturation (number of carbon-carbon double bonds) in response to the water temperature; the survival of this temperature signal in marine sediment sequences provides a temporal record of sea surface temperatures that reflect past climates. { 'al·kə,nōnz }

alkylbenzene sulfonates [CHEM] Widely used nonbiodegradable detergents, commonly dodecylbenzene or tridecylbenzene sulfonates. { ¦al·kəl¦ben,zēn 'səl·fə,nāts }

Alleghenian life zone [ECOL] A biome that includes the eastern mixed coniferous and deciduous forests of New England. { ¦al·ə¦gān·ē·ən 'līf ,zōn }

allele [GEN] One of the alternate forms of a gene at a gene locus on a chromosome. Also known as allelomorph. { ə'lēl }

allelomorph *See* allele. { ə'lē·lə,mȯrf }

allelopathy [ECOL] The harmful effect of one plant or microorganism on another owing to the release of secondary metabolic products into the environment. { ,a·lə'läp·ə·thē }

allelotoxin [ECOL] A toxic compound released in an allelopathic process. { ə¦lē·lō ¦täk·sən }

allergen [MED] Any antigen, such as pollen, a drug, or food, that induces an allergic state in humans or animals. { 'al·ər,jen }

allergic dermatitis [MED] Inflammation of the skin following contact of an allergen with sensitized tissue. { ə'lərj·ik 'dər·mə'tīd·əs }

allergic reaction *See* allergy. { ə'lərj·ik rē'ak·shən }

allergic rhinitis *See* hay fever. { ə'lərj·ik rī'nīd·əs }

allergy [MED] A type of antigen-antibody reaction marked by an exaggerated physiologic response to a substance that causes no symptoms in nonsensitive individuals. Also known as allergic reaction. { 'al·ər·jē }

allethrin [CHEM] An insecticide, a synthetic pyrethroid, more effective than pyrethrin. { 'al·ə·thrən }

allidochlor [CHEM] $C_8H_{12}NOCl$ An amber liquid having slight solubility in water; used as a preemergence herbicide for vegetable crops, soybeans, sorghum, and ornamentals. { ə'lid·ə,klȯr }

allochoric |BOT| Describing a species that inhabits two or more closely related communities, such as forest and grassland, in the same region. { ‚a·lə'kȯr·ik }

allochthonous |ECOL| Materials that come from outside the system, such as plant material in the sediment of a lake that did not originate in the lake. { ə'läk·thə·nəs }

allochthonous coal |GEOL| A type of coal arising from accumulations of plant debris moved from their place of growth and deposited elsewhere. { ə'läk·thə·nəs ‚kōl }

allochthonous stream |HYD| A stream flowing in a channel that it did not form. { ə'läk·thə·nəs ‚strēm }

allogenic |ECOL| Caused by external factors, as in reference to the change in habitat of a natural community resulting from drought. { ¦a·lə¦jen·ik }

Alloionematoidea |ZOO| A superfamily of parasitic nematodes belonging to the order Rhabditida, having either no lips or six small amalgamated lips, and a rhabditiform esophagus with a weakly developed valve in the posterior bulb. { ə‚lȯi·ō‚nem·ə'tȯid·ē·ə }

allometry |BIOL| **1.** The quantitative relation between a part and the whole or another part as the organism increases in size. Also known as heterauxesis; heterogony. **2.** The quantitative relation between the size of a part and the whole or another part, in a series of related organisms that differ in size. { ə'läm·ə·trē }

allopatric |ECOL| Referring to populations or species that are geographically separated from one another. { ¦a·lō¦pa·trik }

allopatric speciation |ECOL| Differentiation of populations in geographical isolation to the point where they are recognized as separate species. { ¦al·ō¦pa·trik ‚spē·sē'ā·shən }

allopelagic |ECOL| Relating to organisms living at various depths in the sea in response to influences other than temperature. { ¦a·lō·pə¦laj·ik }

alluvial |GEOL| **1.** Of a placer, or its associated valuable mineral, formed by the action of running water. **2.** Pertaining to or consisting of alluvium, or deposited by running water. { ə'lüv·ē·əl }

alluvial deposit See alluvium. { ə'lüv·ē·əl di'päz·ət }

alluvial fan |GEOL| A fan-shaped deposit formed by a stream either where it issues from a narrow mountain valley onto a plain or broad valley, or where a tributary stream joins a main stream. { ə'lüv·ē·əl 'fan }

alluvial plain |GEOL| A plain formed from the deposition of alluvium usually adjacent to a river that periodically overflows. Also known as aggraded valley plain; river plain; wash plain; waste plain. { ə'lüv·ē·əl 'plān }

alluvial soil |GEOL| A soil deposit developed on floodplain and delta deposits. { ə'lüv·ē·əl 'sȯil }

alluvial valley |GEOL| A valley filled with a stream deposit. { ə'lüv·ē·əl 'val·ē }

alluviation |GEOL| The deposition of sediment by a river. { ə‚lüv·ē'ā·shən }

alluvion See alluvium. { ə'lüv·ē·ən }

alluvium |GEOL| The detrital materials that are eroded, transported, and deposited by streams; an important constituent of continental shelf deposits. Also known as alluvial deposit; alluvion. { ə'lüv·ē·əm }

allylacetone |CHEM| $CH_2CHCH_2CH_2COCH_3$ A colorless liquid, soluble in water and organic solvents; used in pharmaceutical synthesis, perfumes, fungicides, and insecticides. { ‚al·əl'as·ə‚tōn }

allyl isothiocyanate |CHEM| $CH_2CH:CH_2NCS$ A pungent, colorless to pale-yellow liquid; soluble in alcohol, slightly soluble in water; irritating odor; boiling point 152°C;

used as a fumigant and as a poison gas. Also known as mustard oil. { 'al·əl ¦ī·sō,thī·ō 'sī·ə,nāt }

allyxycarb |CHEM| $C_{16}H_{22}N_2O_2$ A yellow, crystalline compound used as an insecticide for fruit orchards, vegetable crops, rice, and citrus. { ə'liks·ə,karb }

alm |ECOL| A meadow in alpine or subalpine mountain regions. { älm }

alongshore current See littoral current. { ə'lȯŋ,shȯr 'kər·ənt }

alpenglow |METEOROL| A reappearance of sunset colors on a mountain summit after the original mountain colors have faded into shadow; also, a similar phenomenon preceding the regular coloration at sunrise. { 'al·pən,glō }

alpestrine |ECOL| Referring to organisms that live at high elevation but below the timberline. Also known as subalpine. { ˌal'pes·trən }

alpine |ECOL| Any plant native to mountain peaks or boreal regions. { 'al,pīn }

alpine glacier |HYD| A glacier lying on or occupying a depression in mountainous terrain. Also known as mountain glacier. { 'al,pīn 'glā·shər }

alpine tundra |ECOL| Large, flat or gently sloping, treeless tracts of land above the timberline. { 'al,pīn 'tən,dra }

alternation of generations See metagenesis. { ˌȯl·tər'nā·shən əv ˌjen·ə'rā·shənz }

altithermal soil |GEOL| Soil recording a period of rising or high temperature. { ¦al·tə ¦thər·məl 'sȯil }

altitudinal vegetation zone |ECOL| A geographical band of physiognomically similar vegetation correlated with vertical and horizontal gradients of environmental conditions. { ¦al·tə¦tüd·ən·əl ˌvej·ə'tā·shən ˌzōn }

altocumulus cloud |METEOROL| A principal cloud type, white or gray or both white and gray in color; occurs as a layer or patch with a waved aspect, the elements of which appear as laminae, rounded masses, or rolls; frequently appears at different levels in a given sky. Abbreviated Ac. { ¦al·tō¦kyüm·yə·ləs 'klaud }

altostratus cloud |METEOROL| A principal cloud type in the form of a gray or bluish (never white) sheet or layer of striated, fibrous, or uniform appearance; very often totally covers the sky and may cover an area of several thousand square miles; vertical extent may be from several hundred to thousands of meters. Abbreviated As. { ¦al·tō ¦strat·əs 'klaud }

alvar |ECOL| Dwarfed vegetation characteristic of certain Scandinavian steppelike communities with a limestone base. { 'al,vär }

amanthophilous |BOT| Of plants having a habitat in sandy plains or hills. { ˌa·mən'thä·fə·ləs }

amatoxin |BIOL| Any of a group of toxic peptides that selectively inhibit ribonucleic acid polymerase in mammalian cells; produced by the mushroom *Amanita phalloides*. { ¦am·ə¦täk·sən }

ambient |ENG| Surrounding; especially, of or pertaining to the environment about a flying aircraft or other body but undisturbed or unaffected by it, as in ambient air or ambient temperature. { 'am·bē·ənt }

ambient noise |PHYS| The pervasive noise associated with a given environment, being usually a composite of sounds from sources both near and distant. { 'am·bē·ənt 'nȯiz }

amensalism |ECOL| A type of interaction that is neutral to one species but harmful to a second species. { ā'men·sə,liz·əm }

American boreal faunal region [ECOL] A zoogeographic region comprising marine littoral animal communities of the coastal waters off east-central North America. { ə'mer·ə·kən 'bȯr·ē·əl 'fȯn·əl ˌrē·jən }

American spotted fever See Rocky Mountain spotted fever. { ə'mer·ə·kən ˌspäd·əd 'fēv·ər }

Ames test [CHEM] A bioassay that uses a set of histidine auxotrophic mutants of *Salmonella typhimurium* for detecting mutagenic and possibly carcinogenic compounds. { 'āmz ˌtest }

ametoecious [ECOL] Of a parasite that remains with the same host. { ˌam·ə'tēsh·əs }

amictic lake [HYD] A lake in which there is no thermal stratification and no overturn. These lakes occur in the arctic. { ə'mik·tik 'lāk }

aminocarb [CHEM] $C_{11}H_{16}N_2O_2$ A tan, crystalline compound with a melting point of 93–94°C; slightly soluble in water; used as an insecticide for control of forest insects and pests of cotton, tomatoes, tobacco, and fruit crops. { ə'mē·nō,kärb }

3-amino-2,5-dichlorobenzoic acid [CHEM] $C_7H_5O_2\text{-}NCl_2$ A white solid with a melting point of 200–201°C; solubility in water is 700 parts per million at 20°C; used as a preemergence herbicide for soybeans, corn, and sweet potatoes. { ¦thrē ə¦mē·nō ¦tü ¦fīv dī,klȯr·ə,ben¦zō·ik 'as·əd }

2-aminopropane See isopropylamine. { ¦tü ə,mē·nō'prō,pān }

aminotriazole [CHEM] $C_2H_4N_4$ Crystals with a melting point of 159°C; soluble in water, methanol, chloroform, and ethanol; used as an herbicide, cotton plant defoliant, and growth regulator for annual grasses and broadleaf and aquatic weeds. Abbreviated ATA. { ə¦mē·nō'trī·ə,zól }

ammocolous [ECOL] Describing plants having a habitat in dry sand. { ə'mä·kə·ləs }

ammonifiers [ECOL] Fungi, or actinomycetous bacteria, that participate in the ammonification part of the nitrogen cycle and release ammonia (NH_3) by decomposition of organic matter. { ə'män·ə,fī·ərz }

ammonium acid fluoride See ammonium bifluoride. { ə'mōn·yəm 'as·əd 'flùr,īd }

ammonium bifluoride [CHEM] $NH_4F\cdot HF$ A salt that crystallizes in the orthorhombic system and is soluble in water; prepared in the form of white flakes from ammonia treated with hydrogen fluoride; used in solution as a fungicide and wood preservative. Also known as ammonium acid fluoride; ammonium hydrogen fluoride. { ə'mōn·yəm bī'flùr,īd }

ammonium hydrogen fluoride See ammonium bifluoride. { ə'mōn·yəm 'hi·drə·jən 'flùr ,īd }

ammonium sulfamate [CHEM] $NH_4OSO_2NH_2$ White crystals with a melting point of 130°C; soluble in water; used for flameproofing textiles, in electroplating, and as an herbicide to control woody plant species. { ə'mōn·yəm 'səl·fə,māt }

ammonotelic [BIOL] Pertaining to the excretion of nitrogen primarily as ammonium ion, [NH_4^+]. { ə¦mä·nō'tēl·ik }

amoeboid glacier [HYD] A glacier connected with its snowfield for a portion of the year only. { ə'mē,bȯid 'glā·shər }

amorphous frost [HYD] Hoar frost which possesses no apparent simple crystalline structure; opposite of crystalline frost. { ə'mȯr·fəs 'frȯst }

amorphous peat [GEOL] Peat composed of fine grains of organic matter; it is plastic like wet, heavy soil, with all original plant structures destroyed by decomposition of cellulosic matter. { ə'mȯr·fəs 'pēt }

amorphous sky [METEOROL] A sky characterized by an abundance of fractus clouds, usually accompanied by precipitation falling from a higher, overcast cloud layer. { ə'mȯr·fəs 'skī }

amorphous snow [HYD] A type of snow with irregular crystalline structure. { ə'mȯr·fəs 'snō }

amphibious [BIOL] Capable of living both on dry or moist land and in water. { ˌam'fib·ē·əs }

amphicarpic [BOT] Having two types of fruit, differing either in form or ripening time. { ¦am·fəˌkär·pik }

amphicryptophyte [BOT] A marsh plant with amphibious vegetative organs. { ¦am·fə ¦krip·tə¦fīt }

amphidromic [OCEANOGR] Of or pertaining to progression of a tide wave or bulge around a point or center of little or no tide. { ¦am·fəˌdräm·ik }

amphimorphic [GEOL] A rock or mineral formed by two geologic processes. { ˌam·fə'mȯr·fik }

amphisarca [BOT] An indehiscent fruit characterized by many cells and seeds, pulpy flesh, and a hard rind; melon is an example. { ˌam·fə'sär·kə }

amphitheater [GEOGR] A valley or gulch having an oval or circular floor and formed by glacial action. { 'am·fəˌthē·ə·tər }

amphotericin [MICROBIO] An amphoteric antifungal antibiotic produced by *Streptomyces nodosus* and having two components, A and B. { ˌam·fə'ter·ə·sən }

AMV See Alfalfa mosaic virus.

anabasine [CHEM] A colorless, liquid alkaloid extracted from the plants *Anabasis aphylla* and *Nicotiana glauca*; boiling point is 105°C; soluble in alcohol and ether; used as an insecticide. { ə'na·bəˌsēn }

anabatic wind [METEOROL] An upslope wind; usually applied only when the wind is blowing up a hill or mountain as the result of a local surface heating, and apart from the effects of the larger-scale circulation. { ¦an·ə¦bad·ik 'wind }

anabranch [HYD] A diverging branch of a stream or river that loses itself in sandy soil or rejoins the main flow downstream. { 'an·əˌbranch }

anacardium gum See cashew gum. { ˌan·ə'kärd·ē·əm 'gəm }

anaerobe [BIOL] An organism that does not require air or free oxygen to maintain its life processes. { 'an·əˌrōb }

anaerobic bacteria [MICROBIO] Any bacteria that can survive in the partial or complete absence of air; two types are facultative and obligate. { ¦an·ə¦rōb·ik ˌbak'tir·ē·ə }

anaerobic condition [BIOL] The absence of oxygen, preventing normal life for organisms that depend on oxygen. { ¦an·ə¦rōb·ik kən'dish·ən }

anaerobic process [SCI TECH] A process from which air or oxygen not in chemical combination is excluded. { ¦an·ə¦rōb·ik 'präs·əs }

anaerobic sediment [GEOL] A highly organic sediment formed in the absence or near absence of oxygen in water that is rich in hydrogen sulfide. { ¦an·ə¦rōb·ik 'sed·ə·mənt }

anaerobiosis [BIOL] A mode of life carried on in the absence of molecular oxygen. { ˌan·ə·ˌrō'bī·ə·səs }

anaerophyte [ECOL] A plant that does not need free oxygen for respiration. { ə'ner·ə ˌfīt }

anafront [METEOROL] A front at which the warm air is ascending the frontal surface up to high altitudes. { 'an·əˌfrənt }

anagyrine |CHEM| $C_{15}H_{20}N_2O$ A toxic alkaloid found in several species of *Lupinus* in the western United States; acute poisoning produces nervousness, depression, loss of muscular control, convulsions, and coma. { ˌan·ə'jī͵rēn }

analog |METEOROL| A past large-scale synoptic weather pattern which resembles a given (usually current) situation in its essential characteristics. { 'an·əl͵äg }

analysis |METEOROL| A detailed study in synoptic meteorology of the state of the atmosphere based on actual observations, usually including a separation of the entity into its component patterns and involving the drawing of families of isopleths for various elements. { ə'nal·ə·səs }

anastatic water |HYD| That part of the subterranean water in the capillary fringe between the zone of aeration and the zone of saturation in the soil. { ¦an·ə¦stad·ik 'wȯd·ər }

anautogenous insect |ZOO| Any insect in which the adult female must feed before producing eggs. { ¦ā͵nȯ¦täj·ə·nəs 'in͵sekt }

anchored dune |GEOL| A sand dune stabilized by growth of vegetation. { 'aŋ·kərd 'dün }

anchor ice |HYD| Ice formed beneath the surface of water, as in a lake or stream, and attached to the bottom or to submerged objects. Also known as bottom ice; ground ice. { 'aŋ·kər ͵īs }

anchor station |OCEANOGR| An anchoring site by a research vessel for the purpose of making a set of scientific observations. { 'aŋ·kər ͵stā·shən }

anchor stone |GEOL| A rock or pebble that has marine plants attached to it. { 'aŋ·kər ͵stōn }

androecious |BOT| Pertaining to plants that have only male flowers. { an'drē·shəs }

androecium |BOT| The aggregate of stamens in a flower. { ˌan'drēsh·ē·əm }

androphile |ECOL| An organism, such as a mosquito, showing a preference for humans as opposed to animals. { 'an·drō͵fīl }

anemochory |ECOL| Wind dispersal of plant and animal disseminules. { ə'nēm·ə ͵kȯr·ē }

anemoclastic |GEOL| Referring to rock that was broken by wind erosion and rounded by wind action. { ¦a·nə·mō¦klas·tik }

anemology |METEOROL| Scientific investigation of winds. { ˌan·ə'mäl·ə·jē }

anemometry |METEOROL| The study of measuring and recording the direction and speed (or force) of the wind, including its vertical component. { ˌan·ə'mäm·ə·trē }

anemophilous |BOT| Pollinated by wind-carried pollen. { ¦an·ə¦mäf·ə·ləs }

anemotaxis |BIOL| Orientation movement of a free-living organism in response to wind. { ˌan·ə·mō'tak·səs }

anemotropism |BIOL| Orientation response of a sessile organism to air currents and wind. { ˌan·ə'mä·trə͵piz·əm }

aneroid |ENG| Containing no liquid or using no liquid. *See* aneroid barometer. { 'an·ə ͵rȯid }

aneroid barograph |ENG| An aneroid barometer arranged so that the deflection of the aneroid capsule actuates a pen which graphs a record on a rotating drum. Also known as aneroidograph; barograph; barometrograph. { 'an·ə͵rȯid 'bar·ə͵graf }

aneroid barometer |ENG| A barometer which utilizes an aneroid capsule. Also known as aneroid. { 'an·ə͵rȯid bə'räm·əd·ər }

aneroidograph *See* aneroid barograph. { 'an·ə͵rȯid·ə·graf }

22

angiosperm [BOT] The common name for members of the plant division Magnoliophyta; flowering plants characterized by the production of seeds that are enclosed in an ovary. { 'an·jē·ō‚spərm }

angle of current [HYD] In stream gaging, the angular difference between 90° and the angle made by the current with a measuring section. { 'aŋ·gəl əv 'kər·ənt }

angular spreading [OCEANOGR] The lateral extension of ocean waves as they move out of the wave-generating area as swell. { 'aŋ·gyə·lər 'spred·iŋ }

angular-spreading factor [OCEANOGR] The ratio of the actual wave energy present at a point to that which would have been present in the absence of angular spreading. { 'aŋ·gyə·lər 'spred·iŋ ‚fak·tər }

anhydrous ferric chloride *See* ferric chloride. { an'hī·drəs ‚fer·ik 'klȯr‚īd }

anhydrous hydrogen chloride [CHEM] HCl Hazardous, toxic, colorless gas used in polymerization, isomerization, alkylation, nitration, and chlorination reactions; becomes hydrochloric acid in aqueous solutions. { an'hī·drəs ‚hī·drə·jən 'klȯr‚īd }

animal [ZOO] A multicellular eukaryote that typically ingests its food, has the ability to move from place to place, and reproduces sexually. { 'an·ə·məl }

animal community [ECOL] An aggregation of animal species held together in a continuous or discontinuous geographic area by ties to the same physical environment, mainly vegetation. { 'an·ə·məl kə'myü·nəd·ē }

animal ecology [ECOL] A study of the relationships of animals to their environment and each other. { 'an·ə·məl i'käl·ə·jē }

Animalia [SYST] The animal kingdom. { ‚an·ə'māl·yə }

animal kingdom [SYST] The worldwide array of animal life, constituting a major division of living organisms. { 'an·ə·məl ‚kiŋ·dəm }

animal virus [MICROBIO] A small infectious agent able to propagate only within living animal cells. { 'an·ə·məl 'vī·rəs }

Annelida [ZOO] A diverse phylum comprising the multisegmented wormlike animals. { ə'nel·ə·də }

annual flood [HYD] The highest flow at a point on a stream during any particular calendar year or water year. { 'an·yə·wəl 'fləd }

annual growth ring *See* annual ring. { 'an·yə·wəl 'grōth ‚riŋ }

annual inequality [OCEANOGR] Seasonal variation in water level or tidal current speed, more or less periodic, due chiefly to meteorological causes. { 'an·yə·wəl ‚in·i'kwäl·əd·ē }

annual plant [BOT] A plant that completes its growth in one growing season and therefore must be planted annually. { 'an·yə·wəl 'plant }

annual ring [BOT] A line appearing on tree cross sections marking the end of a growing season and showing the volume of wood added during the year. Also known as annual growth ring. { 'an·yə·wəl 'riŋ }

annual storage [HYD] The capacity of a reservoir that can handle a watershed's annual runoff but cannot carry over any portion of the water for longer than the year. { 'an·yə·wəl 'stȯr·ij }

annular drainage pattern [HYD] A ringlike pattern subsequent in origin and associated with maturely dissected dome or basin structures. { 'an·yə·lər 'drān·ij ‚pad·ərn }

anomaly [OCEANOGR] The difference between conditions actually observed at a serial station and those that would have existed had the water all been of a given arbitrary temperature and salinity. { ə'näm·ə·lē }

anoxic zone [OCEANOGR] An oxygen-depleted region in a marine environment. { a'nak·sik ‚zōn }

antagonism [BIOL] **1.** Mutual opposition as seen between organisms, muscles, physiologic actions, and drugs. **2.** Opposing action between drugs and disease or drugs and functions. { an'tag·ə‚niz·əm }

Antarctica [GEOGR] A continent roughly centered on the South Pole and surrounded by an ocean consisting of the southern parts of the Atlantic, Pacific, and Indian oceans. { ‚ant'ärd·ik·ə }

antarctic air [METEOROL] A type of air whose characteristics are developed in an antarctic region. { ‚ant'ärd·ik 'er }

antarctic anticyclone [METEOROL] The glacial anticyclone which has been said to overlie the continent of Antarctica; analogous to the Greenland anticyclone. { ‚ant'ärd·ik ‚ant·i'sī‚klōn }

Antarctic Circumpolar Current [OCEANOGR] The ocean current flowing from west to east through all the oceans around the Antarctic Continent. Also known as West Wind Drift. { ‚ant'ärd·ik ‚sər·kəm'pōl·ər 'kər·ənt }

Antarctic Convergence [OCEANOGR] The oceanic polar front indicating the boundary between the subantarctic and subtropical waters. Also known as Southern Polar Front. { ‚ant'ärd·ik kən'vər·jəns }

Antarctic faunal region [ECOL] A zoogeographic region describing both the marine littoral and terrestrial animal communities on and around Antarctica. { ‚ant'ärd·ik 'fȯn·əl ‚rē·jən }

antarctic front [METEOROL] The semipermanent, semicontinuous front between the antarctic air of the Antarctic continent and the polar air of the southern oceans; generally comparable to the arctic front of the Northern Hemisphere. { ‚ant'ärd·ik 'frənt }

Antarctic Intermediate Water [OCEANOGR] A water mass in the Southern Hemisphere, formed at the surface near the Antarctic Convergence between 45° and 55°S; it can be traced in the North Atlantic to about 25°N. { ‚ant'ärd·ik in·tər'mēd·ē·ət 'wȯd·ər }

Antarctic Ocean [GEOGR] A circumpolar ocean belt including those portions of the Atlantic, Pacific, and Indian oceans which reach the Antarctic continent and are bounded on the north by the Subtropical Convergence; not recognized as a separate ocean. { ‚ant'ärd·ik ‚ō·shən }

Antarctic ozone hole [METEOROL] In the spring, the depletion of stratospheric ozone over the Antarctic region, typically south of 55° latitude, the formation of the hole is explained by the activation of chlorine and the catalytic destruction of O_3, it occurs during September, when the polar regions are sunlit but the air is still cold and isolated from midlatitude air by a strong polar vortex. Also known as ozone hole. { ant¦ärt·ik 'ō‚zōn ‚hōl }

Antarctic vortex *See* polar vortex. { ant¦ärt·ik 'vȯr‚teks }

Antarctic Zone [GEOGR] The region between the Antarctic Circle (66°32'S) and the South Pole. { ‚ant'ärd·ik ‚zōn }

antecedent precipitation index [METEOROL] A weighted summation of daily precipitation amounts; used as an index of soil moisture. { ‚ant·ə'sēd·ənt pri‚sip·ə'tā·shən 'in‚deks }

antecedent stream [HYD] A stream that has retained its early course in spite of geologic changes since its course was assumed. { ‚ant·ə'sēd·ənt 'strēm }

anthelminthic [MED] A chemical substance used to destroy tapeworms in domestic animals. Also spelled anthelmintic. { ¦an·thel¦min·thik }

anthelmintic *See* anthelminthic. { ¦an·thel¦min·tik }

anther |BOT| The pollen-producing structure of a flower. { 'an·thər }

antheridium |BOT| **1.** The sex organ that produces male gametes in cryptogams. **2.** A minute structure within the pollen grain of seed plants. { ‚an·thə'rid·ē·əm }

antheriferous |BOT| Anther-bearing. { ‚an·thə'rif·ə·rəs }

anther smut |MYCOL| U*stilago violacea.* A smut fungus that attacks certain plants and forms spores in the anthers. { 'an·thər ‚smət }

anthesis |BOT| The flowering period in plants. { an'thē·səs }

anthracitization |GEOCHEM| The natural process by which bituminous coal is transformed into anthracite coal. { ‚an·thrə‚sīd·ə'zā·shən }

anthracosilicosis |MED| Chronic lung inflammation caused by inhalation of carbon and silicon particles. { ‚an·thrə·kə‚sil·ə'kō·səs }

anthracosis |MED| The accumulation of inhaled black coal dust particles in the lung accompanied by chronic inflammation. Also known as blacklung. { 'an·thrə'kō·səs }

anthraquinone pigments |BIOL| Coloring materials which occur in plants, fungi, lichens, and insects; consists of about 50 derivatives of the parent compound, anthraquinone. { ‚an·thrə·kwi'nōn 'pig·məns }

anthrax |VET MED| An acute, infectious bacterial disease of sheep and cattle caused by *Bacillus anthracis*; transmissible to humans. Also known as splenic fever; wool-sorter's disease. { 'an‚thraks }

anthraxylon |GEOL| The vitreous-appearing components of coal that are derived from the woody tissues of plants. { an'thrak·sə‚län }

anthropochory |ECOL| Dispersal of plant and animal disseminules by humans. { ¦an·thrə·pə¦kór·ē }

anthropogenic |ECOL| Referring to environmental alterations resulting from the presence or activities of humans. { ¦an·thrə·pə¦jen·ik }

anthropogeography *See* human geography. { ¦an·thrə·pō·jē·'äg·rə·fē }

anthropogeomorphology |GEOL| The study of the effects of humans on the physical landscape, such as in the development and operation of an open-pit mine. { ¦an·thrə·pō‚jē·ə·mór'fäl·ə·jē }

anthroposphere |ECOL| That aspect of the biosphere that has been modified by the activities of humankind. Also known as noosphere. { an'thrä·pə‚sfir }

antibacterial agent |MICROBIO| A synthetic or natural compound which inhibits the growth and division of bacteria. { ¦an·tē‚bak'tir·ē·əl 'ā·jənt }

antibiosis |ECOL| Antagonistic association between two organisms in which one is adversely affected. { ¦an·tē‚bī'ō·səs }

antibiotic |MICROBIO| A chemical substance, produced by microorganisms and synthetically, that has the capacity in dilute solutions to inhibit the growth of, and even to destroy, bacteria and other microorganisms. { ¦an·tē‚bī'äd·ik }

antiboreal faunal region |ECOL| A zoogeographic region including marine littoral faunal communities at the southern end of South America. { 'an·tē‚bór·ē·əl 'fón·əl ‚rē·jən }

anticarcinogen |MED| Any substance which is antagonistic to the action of a carcinogen. { 'an·tē‚kär'sin·ə·jən }

anticryptic |ECOL| Pertaining to protective coloration that makes an animal resemble its surroundings so that it is inconspicuous to its prey. { ¦an·tē¦krip·tik }

anticyclone [METEOROL] High-pressure atmospheric closed circulation whose relative direction of rotation is clockwise in the Northern Hemisphere, counterclockwise in the Southern Hemisphere, and undefined at the Equator. Also known as high-pressure area. { ˌan·tē'sīˌklōn }

anticyclonic [METEOROL] Referring to a rotation about the local vertical that is clockwise in the Northern Hemisphere, counterclockwise in the Southern Hemisphere, undefined at the Equator. { ¦an·tē,sī¦klän·ik }

antidip stream [HYD] A stream that flows in a direction opposite to the general dip of the strata. { ˌant·ē'dip ¦strēm }

antiestuarine circulation [OCEANOGR] In an estuary, the inflow of low-salinity surface water over a deeper outflowing (seaward), dense, high-salinity water layer. { ˌan·tē ¦es·chə·wə,rēn ˌser·kyü'lā·shən }

antifreeze proteins [BIOL] Proteins that decrease the nonequilibrium freezing point of water without significantly affecting the melting point by directly binding to the surface of an ice crystal, thereby disrupting its normal structure and growth pattern and inhibiting further ice growth; found in a number of fish, insects, and plants. { 'an·ti,frēz ¦prō,tēnz }

antigen [MED] A substance which reacts with the products of specific humoral or cellular immunity, even those induced by related heterologous immunogens. { 'an·tə·jən }

antigenic drift [MED] Minor change of an antigen on the surface of a pathogenic microorganism. { ˌan·tə¦jen·ik 'drift }

antigenic variation [MED] Alteration of an antigen on the surface of a microorganism; may enable a pathogenic microorganism to evade destruction by the host's immune system. { ˌan·tə¦jen·ik ˌver·ē'ā·shən }

Antilles Current [OCEANOGR] A current formed by part of the North Equatorial Current that flows along the northern side of the Greater Antilles. { an'til·ēz ¦kər·ənt }

antimalarial [MED] **1.** A drug, such as quinacrine, that prevents or suppresses malaria. **2.** Acting against malaria. { ˌan·tē·mə'ler·ē·əl }

antimicrobial agent [MICROBIO] A chemical compound that either destroys or inhibits the growth of microscopic and submicroscopic organisms. { ˌan·tē,mī'krōb·ē·əl ˌā·jənt }

antioxidant [CHEM] A substance that, when present at a lower concentration than that of the oxidizable substrate, significantly inhibits or delays oxidative processes, while being itself oxidized. Antioxidants are used in polymers to prevent degradation, and in foods, beverages, and cosmetic products to inhibit deterioration and spoilage. { ˌan·tē'äk·sə·dənt }

antiseptic [MICROBIO] A substance used to destroy or prevent the growth of infectious microorganisms on or in the human or animal body. { ¦an·tə¦sep·tik }

antismallpox vaccine *See* smallpox vaccine. { ¦an·tē¦smȯl,päks ˌvak'sēn }

antitoxin [MED] An antibody elaborated by the body in response to a bacterial toxin that will combine with and generally neutralize the toxin. { ˌan·tē'täk·sən }

antitrades [METEOROL] A deep layer of westerly winds in the troposphere above the surface trade winds of the tropics. { 'an·tē,trādz }

antivenin [MED] An immune serum that neutralizes the venoms of certain poisonous snakes and black widow spiders. { ¦an·tē¦ven·ən }

antivernalization [BOT] Delayed flowering in plants due to treatment with heat. { ˌan·tē,vərn·əl·ə'zā·shən }

anvil *See* incus. { 'an·vəl }

anvil cloud [METEOROL] The popular name given to a cumulonimbus capillatus cloud, a thunderhead whose upper portion spreads in the form of an anvil with a fibrous or smooth aspect; it also refers to such an upper portion alone when it persists beyond the parent cloud. { 'an·vəl ˌklaůd }

Ao horizon [GEOL] That portion of the A horizon of a soil profile which is composed of pure humus. { ¦ā¦ōhə'rīz·ən }

Aoo horizon [GEOL] Uppermost portion of the A horizon of a soil profile which consists of undecomposed vegetable litter. { ¦ā¦ō¦ōhə'rīz·ən }

apandrous [BOT] Lacking male organs or having nonfunctional male organs. { ˌa'pan·drəs }

apatetic [ECOL] Pertaining to the imitative protective coloration of an animal subject to being preyed upon. { ¦a·pə¦ted·ik }

apetalous [BOT] Lacking petals. { ˌā'ped·əl·əs }

aphid [ZOO] The common name applied to the soft-bodied insects of the family Aphididae; they are phytophagous plant pests and vectors for plant viruses and fungal parasites. { ā·fəd }

aphotic zone [OCEANOGR] The deeper part of the ocean where sunlight is absent. { a'fäd·ik ˌzōn }

aphyllous [BOT] Lacking foliage leaves. { ā'fil·əs }

aphytic zone [ECOL] The part of a lake floor that lacks plants because it is too deep for adequate light penetration. { ā'fid·ik ˌzōn }

apical bud *See* terminal bud. { 'ap·i·kəl ˌbəd }

apob [METEOROL] An observation of pressure, temperature, and relative humidity taken aloft by means of an aerometeorograph; a type of aircraft sounding. { 'āˌpäb }

apogean tidal currents [OCEANOGR] Tidal currents of decreased speed occurring at the time of apogean tides. { ¦ap·ə¦jē·ən ¦tīd·əl ¦kər·əns }

apogean tides [OCEANOGR] Tides of decreased range occurring when the moon is near apogee. { ¦ap·ə¦jē·ən 'tīdz }

apogeny [BOT] Loss of the function of reproduction. { ə'päj·ə·nē }

apogeotropism [BOT] Negative geotropism; growth up or away from the soil. { ¦a·pō ˌjē·ō'trä,piz·əm }

aposematic coloration [ECOL] Warning coloration that is used to discourage potential predators; usually the animal is poisonous or unpalatable. { ¦ap·ə·sə¦mad·ik }

apostatic selection [ECOL] Predation on the most abundant forms in a population, which gives a selective advantage to rare forms. { ¦ap·ə¦stad·ik sə'lek·shən }

apparent cohesion [GEOL] In soil mechanics, the resistance of particles to being pulled apart due to the surface tension of the moisture film surrounding each particle. Also known as film cohesion. { ə'pa·rənt ˌkō'hē·zhən }

apparent shoreline [GEOGR] The outer edge of marine vegetation (marsh, mangrove, cypress) delineated on photogrammetric surveys where the actual shoreline is obscured. { ə'pa·rənt 'shȯr,līn }

apparent water table *See* perched water table. { ə'pa·rənt 'wȯd·ər ˌtā·bəl }

apple-cedar rust [PL PATH] A disease of apples and Eastern red cedars that is caused by the fungus *Gymnosporangium juniperi-virginianae*; on cedar branches, it manifests itself as brown round galls that do not cause injury, and on apple leaves, as yellow spots that later turn brown and result in cupping and curling of the leaf. { 'ap·əl ¦sēd·ər ˌrəst }

apple pox   *See* blister canker.   { 'ap·əl ˌpäks }

apple scab disease   [PL PATH] A plant disease caused by the fungus *Venturia inaequalis* that may cause premature defoliation, June drop of young fruits, and unsightly blemishes on ripe apples.   { 'ap·əl ¦skab diz‚ēz }

applied climatology   [CLIMATOL] The scientific analysis of climatic data in the light of a useful application for an operational purpose.   { ə'plīd ˌklīm·ə'täl·ə·jē }

applied ecology   [ECOL] Using ecological science to benefit humans or advance human goals.   { ə'plīd i'käl·ə·jē }

applied meteorology   [METEOROL] The application of current weather data, analyses, or forecasts to specific practical problems.   { ə'plīd ˌmēd·ē·ə'räl·ə·jē }

apposition beach   [GEOL] One of a series of parallel beaches formed on the seaward side of an older beach.   { ˌap·ə'zish·ən ˌbēch }

apron   *See* ram.   { 'ā·prən }

aquaculture   *See* aquiculture.   { 'ak·wə‚kəl·chər }

aquatic   [BIOL] Living or growing in, on, or near water; having a water habitat.   { ə'kwäd·ik }

aqueous   [SCI TECH] Relating to or made with water.   { 'āk·wē·əs }

aqueous desert   [ECOL] A marine bottom environment with little or no macroscopic invertebrate shelled life.   { 'āk·wē·əs 'dez·ərt }

aquiclude   [GEOL] A porous formation that absorbs water slowly but will not transmit it fast enough to furnish an appreciable supply for a well or spring.   { 'ak·wə‚klüd }

aquiculture   [BIOL] The controlled cultivation of fresh-water animals and plants for food. Also known as mariculture.   { 'ak·wə‚kəl·chər }

aquifer   [HYD] A subsurface zone that yields economically important amounts of water to wells.   { 'ak·wə·fər }

Aquult   [GEOL] A suborder of the soil order Ultisol; seasonally wet, it is saturated with water a significant part of the year unless drained; surface horizon of the soil profile is dark and varies in thickness, grading to gray in the deeper portions; it occurs in depressions or on wide upland flats from which water drains very slowly.   { 'ak·wəlt }

arboreal   [BOT] Relating to or resembling a tree. Also known as arboreous.   { är'bór·ē·əl }

arboreous   [BOT] **1.** Wooded. **2.** *See* arboreal.   { är'bór·ē·əs }

arboretum   [BOT] An area where trees and shrubs are cultivated for educational and scientific purposes.   { ˌär·bə'rēd·əm }

arboriculture   [BOT] The cultivation of ornamental trees and shrubs.   { ¦är·bə·rə ¦kəl·chər }

arbovirus   [MICROBIO] Small, arthropod-borne animal viruses that are unstable at room temperature and inactivated by sodium deoxycholate; cause several types of encephalitis.   { 'är·bə‚vī·rəs }

arborvitae   [BOT] Any of the ornamental trees, sometimes called the tree of life, in the genus *Thuja* of the order Pinales.   { ¦ar·bər¦vīd·ē }

arbuscular mycorrhizae   *See* vesicular-arbuscular mycorrhizal fungi.   { är'bə·skyül·ər ˌmīk·ə'rīz·ē }

archibenthic zone   [OCEANOGR] The biogeographic realm of the ocean extending from a depth of about 665 feet to 2625–3610 feet (200 meters to 800–1100 meters).   { ¦är·kē ¦ben·thik ˌzōn }

archibole   *See* positive element.   { 'är·kē₁bōl }

archipelago   [GEOGR] **1.** A large group of islands.   **2.** A sea that has a large group of islands within it.   { ₁är·kə'pel·ə₁gō }

arctic air   [METEOROL] An air mass whose characteristics are developed mostly in winter over arctic surfaces of ice and snow.   { ¦ärd·ik *or* 'ärk·tik 'er }

arctic-alpine   [ECOL] Of or pertaining to areas above the timberline in mountainous regions.   { ¦ärd·ik ¦al₁pīn }

arctic climate   *See* polar climate.   { ¦ärd·ik 'klī·mət }

arctic desert   *See* polar desert.   { ¦ärd·ik 'dez·ərt }

arctic front   [METEOROL] The semipermanent, semicontinuous front between the deep, cold arctic air and the shallower, basically less cold polar air of northern latitudes.   { ¦ärd·ik 'frənt }

arctic mist   [METEOROL] A mist of ice crystals; a very light ice fog.   { 'ärd·ik 'mist }

Arctic Ocean   [GEOGR] The north polar ocean lying between North America, Greenland, and Asia.   { 'ärd·ik 'ō·shən }

Arctic Oscillation   [METEOROL] Atmospheric pressure fluctuations (positive and negative phases) between the polar and middle latitudes (above 45° North) that strengthen and weaken the winds circulating counterclockwise from the surface to the lower stratosphere around the Arctic and, as a result, modulate the severity of the winter weather over most Northern Hemisphere middle and high latitudes. Also known as the Northern Hemisphere annular mode.   { ¦ärd·ik ₁äs·ə'lā·shən }

arctic sea smoke   [METEOROL] Steam fog; but often specifically applied to steam fog rising from small areas of open water within sea ice.   { 'ärd·ik ¦sē ₁smōk }

arctic tree line   [ECOL] The northern limit of tree growth; the sinuous boundary between tundra and boreal forest.   { 'ärd·ik ¦trē ₁līn }

Arctic Zone   [GEOGR] The area north of the Arctic Circle (66°32′N).   { ¦ärd·ik ¦zōn }

arcus   [METEOROL] A dense and horizontal roll-shaped accessory cloud, with more or less tattered edges, situated on the lower front part of the main cloud.   { 'ar·kəs }

ARDC model atmosphere   *See* standard atmosphere.   { ¦ā¦är¦dē¦sē ₁mäd·əl 'at·mə ₁sfēr }

area drain   [CIV ENG] A receptacle designed to collect surface or rain water from an open area.   { 'er·ē·ə ¦drān }

area forecast   [METEOROL] A weather forecast for a specified geographic area; usually applied to a form of aviation weather forecast. Also known as regional forecast.   { 'er·ē·ə 'fȯr₁kast }

area landfill   [CIV ENG] A sanitary landfill operation that takes care of the solid waste of more than one municipality in a region.   { 'er·ē·ə 'land₁fil }

areal geology   [GEOL] Geologic features of the area over which a rock or sediment unit occurs.   { 'er·ē·əl jē'äl·ə·jē }

areg   [ECOL] A sand desert.   { 'a₁reg }

areography   [ECOL] Descriptive biogeography.   { ₁ar·ē'äg·rə·fē }

ARFOR   [METEOROL] A code word used internationally to indicate an area forecast; usually applied to an aviation weather forecast.   { 'är₁fȯr }

ARFOT   [METEOROL] A code word used internationally to indicate an area forecast with units in the English system; usually applied to an aviation weather forecast.   { 'är₁fōt }

arhythmicity [BIOL] A condition that is characterized by the absence of an expected behavioral or physiologic rhythm. { ¦ā‚rith'mis·əd·ē }

arid biogeographic zone [ECOL] Any region of the world that supports relatively little vegetation due to lack of water. { 'ar·əd ¦bī·ō‚gē·ō'graf·ik ‚zōn }

arid climate [CLIMATOL] Any extremely dry climate. { 'ar·əd 'klī·mət }

arid erosion [GEOL] Erosion or wearing away of rock that occurs in arid regions, due largely to the wind. { 'ar·əd i'rō·zhən }

Aridisol [GEOL] A soil order characterized by pedogenic horizons; low in organic matter and nitrogen and high in calcium, magnesium, and more soluble elements; usually dry. { a'rid·ə‚sȯl }

aridity [CLIMATOL] The degree to which a climate lacks effective, life-promoting moisture. { ə'rid·əd·ē }

aridity coefficient [CLIMATOL] A function of precipitation and temperature designed by W. Gorczynski to represent the relative lack of effective moisture (the aridity) of a place. { ə'rid·əd·ē ‚kō·ə'fish·ənt }

aridity index [CLIMATOL] An index of the degree of water deficiency below water need at any given station; a measure of aridity. { ə'rid·əd·ē ‚in‚deks }

Arid Transition life zone [ECOL] The zone of climate and biotic communities occurring in the chaparrals and steppes from the Rocky Mountain forest margin to California. { 'ar·əd trans'ish·ən 'līf ‚zōn }

arid zone See equatorial dry zone. { 'ar·əd ‚zōn }

arm [OCEANOGR] A long, narrow inlet of water extending from another body of water. { ärm }

ARMET [METEOROL] An international code word used to indicate an area forecast with units in the metric system. { 'är‚met }

armyworm [ZOO] Any of the larvae of certain species of noctuid moths composing the family Phalaenidae; economically important pests of corn and other grasses. { 'är·mē ‚wərm }

arrested decay [GEOL] A stage in coal formation where biochemical action ceases. { ə'res·təd di'kā }

arroyo [GEOL] Small, deep gully produced by flash flooding in arid and semiarid regions of the southwestern United States. { ə'rȯi·ō }

arsenic acid [CHEM] $H_3AsO_4 \cdot \frac{1}{2}H_2O$ White, poisonous crystals, soluble in water and alcohol; used in manufacturing insecticides, glass, and arsenates and as a defoliant. Also known as orthoarsenic acid. { är'sen·ik 'as·əd }

artesian aquifer [HYD] An aquifer that is bounded above and below by impermeable beds and that contains artesian water. Also known as confined aquifer. { är'tē·zhən 'ak·wə·fər }

artesian basin [HYD] A geologic structural feature or combination of such features in which water is confined under artesian pressure. { är'tē·zhən 'bās·ən }

artesian leakage [HYD] The slow percolation of water from artesian formations into the confining materials of a less permeable, but not strictly impermeable, character. { är'tē·zhən 'lēk·ij }

artesian pressure [HYD] Confining internal pressure of ground water in an artesian aquifer; it is significantly greater than atmospheric pressure, causing ground water to rise above its natural level in the aquifer. { är'tē·zhən 'presh·ər }

artesian spring [HYD] A spring whose water issues under artesian pressure, generally through some fissure or other opening in the confining bed that overlies the aquifer. Also known as fissure spring. { är'tē·zhən 'spriŋ }

artesian water [HYD] Ground water that is under sufficient pressure to rise above the level at which it encounters a well, but which does not necessarily rise to or above the surface of the ground. { är'tē·zhən 'wȯd·ər }

artesian well [HYD] A well in which the water rises above the top of the water-bearing bed. { är'tē·zhən 'wel }

arthropod [ZOO] Any invertebrate (of the phylum Arthropoda) with a hard exoskeleton, segmented body, and jointed legs (for example, insects, arachnids, myriapods, and crustaceans). { 'arth·rō͵päd }

artificial malachite *See* copper carbonate. { ͺärd·ə͵fish·əl 'mal·ə͵kīt }

arviculture [AGR] The cultivation of field crops. { 'är·və͵kəl·chər }

aryloxy compound [CHEM] One of a group of compounds useful as organic weed killers, such as 2,4-dichlorophenoxyacetic acid (2,4-D). { ͺar·əl͵äk·sē ͵käm͵paủnd }

As *See* altostratus cloud.

asbestos-cement pipe [CIV ENG] A concrete pipe made of a mixture of portland cement and asbestos fiber and highly resistant to corrosion; used in drainage systems, waterworks systems, and gas lines. { as'bes·təs si͵ment 'pīp }

ascariasis [MED] Any parasitic infection of humans or domestic mammals caused by species of *Ascaris*. { ͵as·kə'rī·ə·səs }

A selection [ECOL] Selection that favors species adapted to consistently adverse environments. { 'āsi͵lek·shən }

ash [BOT] **1.** A tree of the genus *Fraxinus*, deciduous trees of the olive family (Oleaceae) characterized by opposite, pinnate leaflets. **2.** Any of various Australian trees having wood of great toughness and strength; used for tool handles and in work requiring flexibility. [GEOL] Volcanic dust and particles less than 4 millimeters in diameter. { ash }

Asia [GEOGR] The largest continent, comprising the major portion of the broad east-west extent of the Northern Hemisphere landmasses. { 'āzh·ə }

Asian flu [MED] An acute viral respiratory infection of humans caused by influenza A-2 virus. { 'āzh·ən 'flü }

aspergillic acid [BIOL] $C_{12}H_{20}O_2N_2$ A diketopiperazine-like antifungal antibiotic produced by certain strains of *Aspergillus flavus*. { ͵as·pər͵jil·ik 'as·əd }

Aspergillus [MYCOL] A genus of fungi including several species of common molds and some human and plant pathogens. { ͵as·pər'jil·əs }

assemblage [ECOL] A group of organisms sharing a common habitat by chance. { ə'sem·blij }

assimilation [BIOL] The conversion of nutritive materials into protoplasm. { ə͵sim·ə'lā·shən }

assimilative nitrate reduction [MICROBIO] The reduction of nitrates by some aerobic bacteria for purposes of assimilation. { ə͵sim·ə'lād·iv 'nī͵trāt ri͵dək·shən }

assimilative sulfate reduction [MICROBIO] The reduction of sulfates by certain obligate anaerobic bacteria for purposes of assimilation. { ə͵sim·ə'lād·iv 'səl͵fāt ri͵dək·shən }

association [ECOL] Major segment of a biome formed by a climax community, such as an oak-hickory forest of the deciduous forest biome. { ə͵sō·sē'ā·shən }

asthenosphere |GEOL| That portion of the upper mantle beneath the rigid lithosphere which is plastic enough for rock flowage to occur; extends from a depth of 30–60 miles (50–100 kilometers) to about 240 miles (400 kilometers) and is seismically equivalent to the low velocity zone. { as'then·ə,sfir }

asthma |MED| A pulmonary disease marked by labored breathing, wheezing, and coughing; cause may be emotional stress, chemical irritation, or exposure to an allergen. { 'az·mə }

astronomical tide |OCEANOGR| An equilibrium tide due to attractions of the sun and moon. { ,as·trə'näm·ə·kəl 'tīd }

ATA See aminotriazole.

Atlantic Ocean |GEOGR| The large body of water separating the continents of North and South America from Europe and Africa and extending from the Arctic Ocean to the continent of Antarctica. { ət'lan·tik 'ō·shən }

atmidometer See atmometer. { ,at·mə'däm·əd·ər }

atmometer |ENG| The general name for an instrument which measures the evaporation rate of water into the atmosphere. Also known as atmidometer; evaporation gage; evaporimeter. { ət'mäm·əd·ər }

atmophile element [METEOROL] **1.** Any of the most typical elements of the atmosphere (hydrogen, carbon, nitrogen, oxygen, iodine, mercury, and inert gases). **2.** Any of the elements which either occur in the uncombined state or, as volatile compounds, concentrate in the gaseous primordial atmosphere. { 'at·mō,fīl 'el·ə·mənt }

atmosphere [METEOROL] The gaseous envelope surrounding a planet or celestial body. { 'at·mə,sfir }

atmospheric boundary layer See surface boundary layer. { ¦at·mə¦sfir·ik 'baun·drē ,lā·ər }

atmospheric cell [METEOROL] An air parcel that exhibits a specific type of motion within its boundaries, such as the vertical circular motion of the Hadley cell. { ¦at·mə ,sfir·ik 'sel }

atmospheric chemistry [METEOROL] The study of the production, transport, modification, and removal of atmospheric constituents in the troposphere and stratosphere. { ¦at·mə¦sfir·ik 'kem·ə·strē }

atmospheric composition [METEOROL] The chemical abundance in the earth's atmosphere of its constituents, including nitrogen, oxygen, argon, carbon dioxide, water vapor, ozone, neon, helium, krypton, methane, hydrogen, and nitrous oxide. { ¦at·mə ¦sfir·ik ,käm·pə'zish·ən }

atmospheric condensation [METEOROL] The transformation of water in the air from a vapor phase to dew, fog, or cloud. { ¦at·mə¦sfir·ik ,kän·dən'sā·shən }

atmospheric density [METEOROL] The ratio of the mass of a portion of the atmosphere to the volume it occupies. { ¦at·mə¦sfir·ik 'den·səd·ē }

atmospheric diffusion [METEOROL] The exchange of fluid parcels between regions in the atmosphere in the apparently random motions of a scale too small to be treated by equations of motion. { ¦at·mə¦sfir·ik di'fyü·zhən }

atmospheric disturbance [METEOROL] Any agitation or disruption of the atmospheric steady state. { ¦at·mə¦sfir·ik dis'tər·bəns }

atmospheric evaporation |HYD| The exchange of water between the earth's oceans, lakes, rivers, ice, snow, and soil and the atmosphere. { ¦at·mə¦sfir·ik i,vap·ə'rā·shən }

atmospheric gas [METEOROL] One of the constituents of air, which is a gaseous mixture primarily of nitrogen, oxygen, argon, carbon dioxide, water vapor, ozone, neon, helium, krypton, methane, hydrogen, and nitrous oxide. { ¦at·mə¦sfir·ik 'gas }

atmospheric general circulation [METEOROL] The statistical mean global flow pattern of the atmosphere. { ¦at·mə¦sfir·ik ¦jen·rəl sərk·yə'lā·shən }

atmospheric interference [GEOPHYS] Electromagnetic radiation, caused by natural electrical disturbances in the atmosphere, which interferes with radio systems. Also known as atmospherics; sferics; strays. { ¦at·mə¦sfir·ik ˌin·tər'fir·əns }

atmospheric lapse rate See environmental lapse rate. { ¦at·mə¦sfir·ik 'laps ˌrāt }

atmospheric layer See atmospheric shell. { ¦at·mə¦sfir·ik 'lā·ər }

atmospheric pressure [PHYS] The pressure at any point in an atmosphere due solely to the weight of the atmospheric gases above the point concerned. Also known as barometric pressure. { ¦at·mə¦sfir·ik 'presh·ər }

atmospheric radiation [GEOPHYS] Infrared radiation emitted by or being propagated through the atmosphere. { ¦at·mə¦sfir·ik ˌrād·ē·ē'ā·shən }

atmospheric region See atmospheric shell. { ¦at·mə¦sfir·ik 'rē·jən }

atmospherics See atmospheric interference. { ¦at·mə¦sfir·iks }

atmospheric shell [METEOROL] Any one of a number of strata or layers of the earth's atmosphere; temperature distribution is the most common criterion used for denoting the various shells. Also known as atmospheric layer; atmospheric region. { ¦at·mə ¦sfir·ik 'shel }

atmospheric sounding [METEOROL] A measurement of atmospheric conditions aloft, above the effective range of surface weather observations. { ¦at·mə¦sfir·ik 'saùnd·iŋ }

atmospheric structure [METEOROL] Atmospheric characteristics, including wind direction and velocity, altitude, air density, and velocity of sound. { ¦at·mə¦sfir·ik 'strək·chər }

atmospheric suspensoids [METEOROL] Moderately finely divided particles suspended in the atmosphere; dust is an example. { ¦at·mə¦sfir·ik sə'spen,soidz }

atmospheric turbulence [METEOROL] Apparently random fluctuations of the atmosphere that often constitute major deformations of its state of fluid flow. { ¦at·mə ¦sfir·ik 'tər·byə·ləns }

atoll [GEOGR] A ring-shaped coral reef that surrounds a lagoon without projecting land area and that is surrounded by open sea. { 'a,tól }

atomic fallout See fallout. { ə'täm·ik 'fól,aùt }

atomic fission See fission. { ə'täm·ik 'fish·ən }

atomic power plant See nuclear power plant. { ə'täm·ik 'paù·ər ,plant }

ATP synthase [BIOL] An enzyme that catalyzes the conversion of phosphate and adenosine diphosphate into adenosine triphosphate during oxidative phosphorylation in mitochondria and bacteria or phosphorylation in chloroplasts. { ¦ā¦tē¦pē 'sin,thās }

atrazine [CHEM] $C_8H_{14}ClN_5$ A white crystalline compound widely used as a photosynthesis-inhibiting herbicide for weeds. { 'a·trə,zēn }

attached groundwater [HYD] The portion of subsurface water adhering to pore walls in the soil. { ə'tacht 'graùnd,wód·ər }

attenuation [BOT] Tapering, sometimes to a long point. [MICROBIO] Weakening or reduction of the virulence of a microorganism. { ə,ten·yə'wā·shən }

attrition [GEOL] The act of wearing and smoothing of rock surfaces by the flow of water charged with sand and gravel, by the passage of sand drifts, or by the movement of glaciers. { ə'trish·ən }

attritus [GEOL] **1.** Visible-to-ultramicroscopic particles of vegetable matter produced by microscopic and other organisms in vegetable deposits, particularly in swamps and bogs. **2.** The dull gray to nearly black, frequently striped portion of material that makes up the bulk of some coals and alternate bands of bright anthraxylon in well-banded coals. { ə'trīd·əs }

aufwuch [ECOL] A plant or animal organism which is attached or clings to surfaces of leaves or stems of rooted plants above the bottom sediments of freshwater ecosystems. { 'ȯf,wək }

aulophyte [ECOL] A nonparasitic plant that lives in the cavity of another plant for shelter. { 'ȯl·ə,fīt }

Australia [GEOGR] An island continent of 2,941,526 square miles (7,618,517 square kilometers), with low elevation and moderate relief, situated in the southern Pacific. { ȯ'strāl·yə }

Australian faunal region [ECOL] A zoogeographic region that includes the terrestrial animal communities of Australia and all surrounding islands except those of Asia. { ȯ'strāl·yən 'fȯn·əl ,rē·jən }

Austroriparian life zone [ECOL] The zone in which occurs the climate and biotic communities of the southeastern coniferous forests of North America. { ¦ȯs·trō,rī 'per·ē·ən 'līf ,zōn }

autecology See autoecology. { ,ȯd·i'käl·ə·jē }

autoallelopathy [PL PATH] Inhibition of a species by self-produced toxins. { ¦ȯd·ō ,a·lə'lä·pə·thē }

autocarpy [BOT] Production of fruit by self-fertilization. { 'ȯd·ō,kärp·ē }

autochthonous [ECOL] Pertaining to organisms or organic sediments that are indigenous to a given ecosystem. { ȯ'täk·thə·nas }

autochthonous microorganism [MICROBIO] An indigenous form of soil microorganisms, responsible for chemical processes that occur in the soil under normal conditions. { ȯ'täk·thə·nas ,mī·krō'ȯr·gə,niz·əm }

autochthonous stream [HYD] A stream flowing in its original channel. { ȯ'täk·thə·nas 'strēm }

autoconsequent falls [HYD] Waterfalls in streams carrying a heavy load of calcium carbonate in solution which develop at particular sites along the stream course where warming, evaporation, and other factors cause part of the solution load to be precipitated. { ¦ȯd·ō'kän·sə·kwənt 'fȯlz }

autoconsequent stream [HYD] A stream in the process of building a fan or an alluvial plain, the course of which is guided by the slopes of the alluvium the stream itself has deposited. { ¦ȯd·ō'kän·sə·kwənt 'strēm }

autoconvection [METEOROL] The phenomenon of the spontaneous initiation of convection in an atmospheric layer in which the lapse rate is equal to or greater than the autoconvective lapse rate. { ¦ȯd·ō·kən'vek·shən }

autoconvective instability See absolute instability. { ¦ȯd·ō·kən'vek·tiv ,in·stə'bil·əd·ē }

autodeme [ECOL] A plant population in which most individuals are self-fertilized. { 'ȯd·ō,dēm }

autoecious [BIOL] See autoicous. [MYCOL] Referring to a parasitic fungus that completes its entire life cycle on a single host. { ȯ'tēsh·əs }

autoecology [ECOL] The study of how a particular species responds to the environment. Also spelled autecology. { ,ȯd·ōi'käl·ə·jē }

autogenetic drainage [HYD] A self-established drainage system developed solely by headwater erosion. { ¦ȯd·ō·jə¦ned·ik 'drān·ij }

autogenetic topography [GEOL] Conformation of land due to the physical action of rain and streams. { ¦ȯd·ō·jə¦ned·ik tə'päg·rə·fē }

autogenous insect [ZOO] Any insect in which adult females can produce eggs without first feeding. { ȯ'täj·ə·nəs 'in‚sekt }

autohemorrhage [ZOO] Voluntary exudation or ejection of nauseous or poisonous blood by certain insects as a defense against predators. { ¦ȯd·ō'hem·rij }

autoicous [BOT] Having male and female organs on the same plant but on different branches. Also spelled autoecious. { ȯ'tȯi·kəs }

autolysis [GEOCHEM] Return of a substance to solution, as of phosphate removed from seawater by plankton and returned when these organisms die and decay. { ȯ'täl·ə·səs }

automatic weather station [METEOROL] A weather station at which the services of an observer are not required; usually equipped with telemetric apparatus. { ¦ȯd·ə¦mad·ik 'weth·ər ‚stā·shən }

autoradiography [ENG] A technique for detecting radioactivity in a specimen by producing an image on a photographic film or plate. Also known as radioautography. { ¦ȯd·ō‚rād·ē'äg·rə·fē }

autospore [BOT] In algae, a nonmotile asexual reproductive cell or a nonmotile spore that is a miniature of the cell that produces it. { 'ȯd·ȯ‚spȯr }

autotroph [BIOL] An organism capable of synthesizing organic nutrients directly from simple inorganic substances, such as carbon dioxide and inorganic nitrogen. { 'ȯd·ō ‚träf }

autotrophic ecosystem [ECOL] An ecosystem that has primary producers as a principal component, and sunlight as the major initial energy source. { ¦ȯd·ə‚trȯf·ik 'ek·ō ‚sis·təm }

autotrophic succession [ECOL] A type of ecological succession that involves organisms that can utilize renewable resources. { ¦ȯd·ə‚trō·fik sək'sesh·ən }

autumn ice [OCEANOGR] Sea ice in early stage of formation; comparatively salty, and crystalline in appearance. { ¦ȯd·əm 'īs }

available moisture [HYD] Moisture in soil that is available for use by plants. { ə'vāl· ə·bəl 'mȯis·chər }

avalanche [HYD] A mass of snow or ice moving rapidly down a mountain slope or cliff. { 'av·ə‚lanch }

avalanche wind [METEOROL] The rush of air produced in front of an avalanche of dry snow or in front of a landslide. { 'av·ə‚lanch ‚wind }

average limit of ice [OCEANOGR] The average seaward extent of ice formation during a normal winter. { 'av·rij 'lim·ət əv 'īs }

avian leukosis [VET MED] A disease complex in fowl probably caused by viruses and characterized by autonomous proliferation of blood-forming cells. { 'av·ē·ən lü'kō· səs }

avian pneumoencephalitis See Newcastle disease. { 'av·ē·ən ¦nü·mō·in‚sef·ə'līd·əs }

avian pseudoplague See Newcastle disease. { 'av·ē·ən 'süd·ō‚plāg }

avian tuberculosis [VET MED] A tuberculosis-like mycobacterial disease of fowl caused by *Mycobacterium avium*. { 'av·ē·ən tə‚bər·kyə'lō·səs }

aviation weather forecast [METEOROL] A forecast of weather elements of particular interest to aviation, such as the ceiling, visibility, upper winds, icing, turbulence, and types of precipitation or storms. Also known as airways forecast. { ‚ā·vē'ā·shən ¦weth·ər ‚fȯr‚kast }

aviation weather observation [METEOROL] An evaluation, according to set procedure, of those weather elements which are most important for aircraft operations. Also known as airways observation. { ‚ā·vē'ā·shən ¦weth·ər ‚äb·zər'vā·shən }

avicolous [ECOL] Living on birds, as of certain insects. { ā'vik·ə·ləs }

avifauna [ZOO] **1.** Birds, collectively. **2.** Birds characterizing a period, region, or environment. { ¦ā·və¦fón·ə }

avulsion [HYD] A sudden change in the course of a stream by which a portion of land is cut off, as where a stream cuts across and forms an oxbow. { ə'vəl·shən }

axial stream [HYD] The chief stream of an intermontane valley, the course of which is along the deepest part of the valley and is parallel to its longer dimension. { 'ak·sē·əl 'strēm }

azonal soil [GEOL] Any group of soils without well-developed profile characteristics, owing to their youth, conditions of parent material, or relief that prevents development of normal soil-profile characteristics. Also known as immature soil. { 'ā‚zōn·əl 'sóil }

Azores high [METEOROL] The semipermanent subtropical high over the North Atlantic Ocean, especially when it is located over the eastern part of the ocean; when in the western part of the Atlantic, it becomes the Bermuda high. { 'ā‚zórz 'hī }

Azotobacteraceae [MICROBIO] A family of large, bluntly rod-shaped, gram-negative, aerobic bacteria capable of fixing molecular nitrogen. { ə¦zōd·ə‚bak·tə'rās·ē‚ē }

B

babesiasis [VET MED] A tick-borne protozoan disease of mammals other than humans caused by species of *Babesia*. { ‚bab·ə'zī·ə·səs }

baccate [BOT] **1.** Bearing berries. **2.** Having pulp like a berry. { 'bak‚āt }

Bacillariophyceae [BOT] The diatoms, a class of algae in the division Chrysophyta. { ‚bas·ə‚ler·ē·ə'fīs·ē‚ē }

Bacillariophyta [BOT] An equivalent name for the Bacillariophyceae. { ‚bas·ə‚ler·ē'ä·fəd·ə }

bacillary dysentery [MED] A highly infectious bacterial disease of humans, localized in the bowels; caused by *Shigella*. { 'bas·ə‚ler·ē 'dis·ən‚ter·ē }

bacillary white diarrhea *See* pullorum disease. { 'bas·ə‚ler·ē ¦wīt ‚di·ə'rē·ə }

Bacillus anthracis [MICROBIO] A gram-positive, rod-shaped, endospore-forming bacterium that is the causative agent of anthrax; its spores can remain viable for many years in soil, water, and animal hides and products. { bə¦sil·əs ‚an'thrak·əs }

Bacillus cereus [MICROBIO] A spore-forming bacterium that often survives cooking and grows to large numbers in improperly refrigerated foods; it produces both a diarrheal toxin and an emetic toxin in the gastrointestinal tract following its ingestion via contaminated meats, dried foods, and rice. { bə‚sil·əs 'sir·ē·əs }

back beach *See* backshore. { 'bak ‚bēch }

backcross [GEN] A cross between an F_1 heterozygote and an individual of P_1 genotype. { 'bak‚krȯs }

back-door cold front [METEOROL] A front which leads a cold air mass toward the south and southwest along the Atlantic seaboard of the United States. { 'bak ‚dȯr 'kōld ‚frənt }

backflooding [HYD] A reversal of flow of water at the water table resulting from changes in precipitation. { 'bak‚fləd·iŋ }

background radiation [PHYS] The radiation in humans' natural environment, including cosmic rays and radiation from the naturally radioactive elements. Also known as natural radiation. { 'bak‚graὐnd ‚rād·ē'ä·shən }

backing [METEOROL] **1.** Internationally, a change in wind direction in a counterclockwise sense (for example, south to east) in either hemisphere of the earth. **2.** In United States usage, a change in wind direction in a counterclockwise sense in the Northern Hemisphere, clockwise in the Southern Hemisphere. { 'bak·iŋ }

backmarsh [ECOL] Marshland formed in poorly drained areas of an alluvial floodplain. { 'bak‚märsh }

back reef [GEOGR] The area between a reef and the land. { 'bak ‚rēf }

back rush [OCEANOGR] Return of water seaward after the uprush of the waves. { 'bak ‚rəsh }

backshore [GEOL] The upper shore zone that is beyond the advance of the usual waves and tides. Also known as back beach; backshore beach. { 'bak,shȯr }

backshore beach See backshore. { 'bak,shȯr ,bēch }

backshore terrace See berm. { 'bak,shȯr 'ter·əs }

back siphonage [CIV ENG] The flowing back of used, contaminated, or polluted water from a plumbing fixture or vessel into the pipe which feeds it; caused by reduced pressure in the pipe. { 'bak ¦sī·fən·ij }

backswamp depression [ECOL] A low swamp found adjacent to river levees. { 'bak ,swamp di'presh·ən }

backwash [OCEANOGR] **1.** Water or waves thrown back by an obstruction such as a ship or breakwater. **2.** The seaward return of water after a rush of waves onto the beach foreshore. { 'bak,wäsh }

backwater [HYD] **1.** A series of connected lagoons, or a creek parallel to a coast, narrowly separated from the sea and connected to it by barred outlets. **2.** Accumulation of water resulting from and held back by an obstruction. **3.** Water reversed in its course by an obstruction. { 'bak,wȯd·ər }

bacteria [MICROBIO] Extremely small, relatively simple prokaryotic microorganisms traditionally classified with the fungi as Schizomycetes. { bak'tir·ē·ə }

bacterial blight [PL PATH] Any blight disease of plants caused by bacteria, including common bacterial blight, halo blight, and fuscous blight. { bak'tir·ē·əl 'blīt }

bacterial brown spot [PL PATH] A bacterial blight disease of plants caused by *Pseudomonas syringae*; marked by water-soaked reddish-brown spots or cankers. Also known as bacterial canker. { bak'tir·ē·əl ¦braůn ,spät }

bacterial canker See bacterial brown spot. { bak'tir·ē·əl 'kaŋ·kər }

bacterial encephalitis [MED] Inflammation of the brain caused by primary or secondary bacterial infection. { bak'tir·ē·əl in,sef·ə'līd·əs }

bacterial infection [MED] Establishment of an infective bacterial agent in or on the body of a host. { bak'tir·ē·əl in,sef·ə'līd·əs }

bacterial leaf spot [PL PATH] A bacterial disease of plants characterized by spotty discolorations on the leaves; examples are angular leaf spot and leaf blotch. { bak'tir·ē·əl 'lēf ,spät }

bacterial methanogenesis See methanogenesis. { bak'tir·ē·əl 'meth·ə·nō'jen·ə·səs }

bacterial photosynthesis [MICROBIO] Use of light energy to synthesize organic compounds in green and purple bacteria. { bak'tir·ē·əl ,fōd·ō'sin·thə·səs }

bacterial soft rot [PL PATH] A bacterial disease of plants marked by disintegration of tissues. { bak'tir·ē·əl ¦sȯft ,rät }

bacterial speck [PL PATH] A bacterial disease of plants characterized by small lesions on plant parts. { bak'tir·ē·əl 'spek }

bacterial spot [PL PATH] Any bacterial disease of plants marked by spotting of the infected part. { bak'tir·ē·əl 'spät }

bacterial wilt disease [PL PATH] A common bacterial disease of cucumber and muskmelon, caused by *Erwinia tracheiphila*, characterized by wilting and shriveling of the leaves and stems. { bak'tir·ē·əl ¦wilt di,zēz }

bacteriocyte [ZOO] A modified fat cell found in certain insects that contains bacterium-shaped rods believed to be symbiotic bacteria. { bak'tir·ē·ə,sīt }

bacteriological warfare [MICROBIO] Warfare conducted with pathogenic microorganisms as offensive weapons; a type of biological warfare. { bak¦tir·ē·ə¦läj·ə·kəl 'wȯr ,fer }

bacteriophage [MICROBIO] Any of the viruses that infect bacterial cells; each has a narrow host range. Also known as phage. { bak'tir·ē·ə‚fāj }

bacteriosis [PL PATH] Any bacterial disease of plants. { bak‚tir·ē'ō·səs }

bacteriostasis [MICROBIO] Inhibition of bacterial growth and metabolism. { bak ‚tir·ē·ō'stā·səs }

bacteriostatic agent [MICROBIO] A substance that inhibits the growth of bacteria. { ‚bak¦tir·ē·ō¦stad·ik 'ā·jənt }

bacteriotoxin [MICROBIO] **1.** Any toxin that destroys or inhibits growth of bacteria. **2.** A toxin produced by bacteria. { bak‚tir·ē·ō'täk·sən }

badge meter *See* film badge. { 'baj ‚mēd·ər }

badlands [GEOGR] An erosive physiographic feature in semiarid regions characterized by sharp-edged, sinuous ridges separated by steep-sided, narrow, winding gullies. { 'bad‚lanz }

baffling wind [METEOROL] A wind that is shifting so that nautical movement by sailing vessels is impeded. { ¦baf·liŋ 'wind }

baguio [METEOROL] A tropical cyclone that occurs in the Philippines. { bäg'yō }

bai [METEOROL] A yellow mist prevalent in China and Japan in spring and fall, when the loose surface of the interior of China is churned up by the wind, and clouds of sand rise to a great height and are carried eastward, where they collect moisture and fall as a yellow mist. { bī }

Bakanae disease [PL PATH] A fungus disease of rice in Japan, caused by *Gibberella fujikurae*; a foot rot disease. { bə'kä·nē di‚zēz }

balance equation [METEOROL] A diagnostic equation expressing a balance between the pressure field and the horizontal field of motion of the atmosphere. { 'bal·əns i'kwä·zhən }

balantidiasis [MED] An intestinal infection of humans caused by the protozoan *Balantidium coli*. { ‚bal·ən·tə'dī·ə·səs }

bald [GEOGR] An elevated grassy, treeless area, as on the top of a mountain. { bȯld }

Bali wind [METEOROL] A strong east wind at the eastern end of Java. { ¦bäl·ē ¦wind }

ball ice [OCEANOGR] Numerous floating spheres of sea ice having diameters of 1–2 inches (2.5–5 centimeters), generally in belts similar to slush which forms at the same time. { 'bȯl ‚īs }

ballistic separator [CIV ENG] A device that takes out noncompostable material like stones, glass, metal, and rubber, from solid waste by passing the waste over a rotor that has impellers to fling the material in the air; the lighter organic (compostable) material travels a shorter distance than the heavier (noncompostable) material. { bə'lis·tik 'sep·ə‚rād·ər }

Baltic Sea [GEOGR] An intracontinental, Mediterranean-type sea, connected with the North Sea and surrounded by Sweden, Denmark, Germany, Poland, the Baltic States, and Finland. { 'bȯl·tik 'sē }

banana freckle [PL PATH] A fungus disease of the banana caused by *Macrophoma musae*, producing brown or black spots on the fruit and leaves. { bə'nan·ə ‚frek·əl }

banco [HYD] A meander or oxbow lake separated from a river by a change in its course. { 'baŋ·kō }

banded peat [GEOL] Peat formed of alternate layers of vegetable debris. { 'ban·dəd 'pēt }

Bang's disease *See* contagious abortion. { 'baŋz diz'ēz }

bank [OCEANOGR] A relatively flat-topped raised portion of the sea floor occurring at shallow depth and characteristically on the continental shelf or near an island. { baŋk }

bankfull stage [HYD] The flow stage of a river in which the stream completely fills its channel and the elevation of the water surface coincides with the bank margins. { ¦baŋk ¦fúl ˌstāj }

bank-inset reef [GEOL] A coral reef situated on island or continental shelves well inside the outer edges. { 'baŋk 'inˌset ˌrēf }

bank reef [GEOL] A reef which rises at a distance back from the outer margin of rimless shoals. { 'baŋk ˌrēf }

bank storage [HYD] Water absorbed in the permeable bed and banks of a lake, reservoir, or stream. { 'baŋk ˌstór·ij }

banner cloud [METEOROL] A cloud plume often observed to extend downwind from isolated mountain peaks, even on otherwise cloud-free days. Also known as cloud banner. { 'ban·ər ˌklaúd }

barban [CHEM] $C_{11}H_9O_2NCl_2$ A white, crystalline compound with a melting point of 75–76°C; used as a postemergence herbicide of wild oats in barley, flax, lentil, mustard, and peas. { 'bärˌban }

bar beach [GEOL] A straight beach of offshore bars that are separated by shallow bodies of water from the mainland. { 'bär ˌbēch }

barbed tributary [HYD] A tributary that enters the main stream in an upstream direction instead of pointing downstream. { 'bärbd 'trib·yəˌter·ē }

baric topography See height pattern. { 'bar·ik tə'päg·rə·fē }

baric wind law See Buys-Ballot's law. { 'bar·ik ¦wind ˌló }

baring See overburden. { 'ba·riŋ }

barium [CHEM] A chemical element, symbol Ba, with atomic number 56 and atomic weight of 137.34. { 'bar·ē·əm }

barium carbonate [CHEM] $BaCO_3$ A white powder with a melting point of 174°C; soluble in acids (except sulfuric acid); used in rodenticides, ceramic flux, optical glass, and television picture tubes. { 'bar·ē·əm 'kär·bə·nət }

barium chloride [CHEM] $BaCl_2$ A toxic salt obtained as colorless, water-soluble cubic crystals, melting at 963°C; used as a rat poison, in metal surface treatment, and as a laboratory reagent. { 'bar·ē·əm 'klórˌīd }

barium fluosilicate [CHEM] $BaSiF_6H$ A white, crystalline powder; insoluble in water; used in ceramics and insecticides. Also known as barium silicofluoride. { 'bar·ē·əm ˌflü·ə'sil·əˌkāt }

barium permanganate [CHEM] $Ba(MnO_4)_2$ Brownish-violet, toxic crystals; soluble in water; used as a disinfectant. { 'bar·ē·əm pər'maŋ·gəˌnāt }

barium silicofluoride See barium fluosilicate. { 'bar·ē·əm 'sil·ə·kō'flúrˌīd }

bark [BOT] The tissues external to the cambium in a stem or root. { bärk }

barker [ENG] A machine, used mainly in pulp mills, which removes the bark from logs. [FOR] **1.** A worker who subjects logs and pulpwood to water pressure in a stream barker or tumbling in a drum barker, in order to free them from bark and dirt. Also known as power barker. **2.** A worker who prepares or shovels bark for tanning. { 'bär·kər }

barley [BOT] A plant of the genus *Hordeum* in the order Cyperales that is cultivated as a grain crop; the seed is used to manufacture malt beverages and as a cereal. { 'bär·lē }

barley scald |PL PATH| A fungus disease of barley caused by *Rhynchosporium secalis* and characterized by bluish-green to yellow blotches and blighting of the foliage. { 'bär·lē ‚skȯld }

barley smut |PL PATH| **1.** A loose smut disease of barley caused by *Ustilago nuda.* **2.** A covered smut disease of barley caused by U. *hordei.* { 'bär·lē ‚smət }

barley stripe |PL PATH| A fungus disease of barley characterized by light green or yellow stripes on the leaves; incited by the diffusible toxin of *Helminthosporium gramineum.* { 'bär·lē ‚strīp }

baroclinic model [METEOROL] A concept of stratification in the atmosphere, involving surfaces of constant pressure intersecting surfaces of constant density. { ¦bar·ə¦klin·ik 'mäd·əl }

baroduric bacteria [MICROBIO] Bacteria that can tolerate conditions of high hydrostatic pressure. { ¦bar·ə¦dùr·ik bak'tir·ē·ə }

barograph See aneroid barograph. { 'bar·ə‚graf }

barometer [ENG] An absolute pressure gage specifically designed to measure atmospheric pressure. { bə'räm·əd·ər }

barometer elevation [METEOROL] The vertical distance above mean sea level of the ivory point (zero point) of a weather station's mercurial barometer; frequently the same as station elevation. Also known as elevation of ivory point. { bə'räm·əd·ər el·ə'vā·shən }

barometric pressure See atmospheric pressure. { bar·ə'met·rik 'presh·ər }

barometric tendency See pressure tendency. { bar·ə'met·rik 'ten·dən·sē }

barometric wave [METEOROL] Any wave in the atmospheric pressure field; the term is usually reserved for short-period variations not associated with cyclonic-scale motions or with atmospheric tides. { bar·ə'met·rik 'wāv }

barometrograph See aneroid barograph. { bar·ə'me·trə‚graf }

barophile [MICROBIO] An organism that thrives under conditions of high hydrostatic pressure. { 'bar·ə‚fīl }

barotaxis [BIOL] Orientation movement of an organism in response to pressure changes. { ¦bar·ə¦tak·səs }

barotropic disturbance [METEOROL] **1.** A wave disturbance in a two-dimensional nondivergent flow; the driving mechanism lies in the variation of either vorticity of the basic current or the variation of the vorticity of the earth about the local vertical. **2.** An atmospheric wave of cyclonic scale in which troughs and ridges are approximately vertical. { ‚bar·ə'träp·ik dis'tər·bəns }

barotropic model [METEOROL] Any of a number of model atmospheres in which some of the following conditions exist throughout the motion: coincidence of pressure and temperature surfaces, absence of vertical wind shear, absence of vertical motions, absence of horizontal velocity divergence, and conservation of the vertical component of absolute vorticity. { ‚bar·ə'träp·ik 'mäd·əl }

barranca [GEOL] A hole or deep break made by heavy rain; a ravine. { bə'raŋ·kə }

barrens [GEOGR] An area that because of adverse environmental conditions is relatively devoid of vegetation compared with adjacent areas. { 'bar·ənz }

barricade shield [ENG] A type of movable shield made of a material designed to absorb ionizing radiation, for protection from radiation. { 'bar·ə‚kād ‚shēld }

barrier [ECOL] Any physical or biological factor that restricts the migration or free movement of individuals or populations. { 'bar·ē·ər }

barrier basin [GEOL] A basin formed by natural damming, for example, by landslides or moraines. { 'bar·ē·ər ,bās·ən }

barrier beach [GEOL] A single, long, narrow ridge of sand which rises slightly above the level of high tide and lies parallel to the shore, from which it is separated by a lagoon. Also known as offshore beach. { 'bar·ē·ər ,bēch }

barrier chain [GEOL] A series of barrier spits, barrier islands, and barrier beaches extending along a coastline. { 'bar·ē·ər ,chān }

barrier ice *See* shelf ice. { 'bar·ē·ər ,īs }

barrier island [GEOL] An elongate accumulation of sediment formed in the shallow coastal zone and separated from the mainland by some combination of coastal bays and their associated marshes and tidal flats; barrier islands are typically several times longer than their width and are interrupted by tidal inlets. { 'bar·ē·ər ,ī·lənd }

barrier lagoon [GEOGR] A shallow body of water that separates the shore and a barrier reef. { 'barṁēṁər lə'gün }

barrier lake [HYD] A small body of water that lies in a basin, retained there by a natural dam or barrier. { 'bar·ē·ər ,lāk }

barrier marsh [ECOL] A type of marsh that restricts or prevents invasion of the area beyond it by new species of animals. { 'bar·ē·ər ,märsh }

barrier reef [GEOL] A coral reef that runs parallel to the coast of an island or continent, from which it is separated by a lagoon. { 'bar·ē·ər ,rēf }

barrier shield [ENG] A wall or enclosure made of a material designed to absorb ionizing radiation, shielding the operator from an area where radioactive material is being used or processed by remote-control equipment. { 'bar·ē·ər ,shēld }

barrier theory of cyclones [METEOROL] A theory of cyclone development, proposed by F. M. Exner, which states that a slow-moving mass of cold air in the path of rapidly eastward-moving warmer air will bring about the formation of low pressure on the lee side of the cold air; analogous to the formation of a dynamic trough on the lee side of an orographic barrier. Also known as drop theory. { 'bar·ē·ər ,thē·ə·rē əv 'sī,klōnz }

Bartonella [MICROBIO] A genus of the family Bartonellaceae; parasites in or on red blood cells and within fixed tissue cells; found in humans and in the arthropod genus *Phlebotomus*. { ,bärt·ən'el·ə }

Bartonellaceae [MICROBIO] A family of the order Rickettsiales; rod-shaped, coccoid, ring- or disk-shaped cells; parasites of human and other vertebrate red blood cells. { ,bärt·ən,e'lās·ē,ē }

basal [BIOL] Of, pertaining to, or located at the base. { 'bā·səl }

basal groundwater [HYD] A large body of groundwater that floats on and is in hydrodynamic equilibrium with sea water. { 'bā·səl 'graund,wod·ər }

basal rot [PL PATH] Any rot that affects the basal parts of a plant, especially bulbs. { 'bā·səl 'rät }

basal water table [HYD] The water table of basal groundwater. { 'bā·səl 'wod·ər ,tā·bəl }

base [CHEM] Any chemical species, ionic or molecular, capable of accepting or receiving a proton (hydrogen ion) from another substance; the other substance acts as an acid in giving of the proton. Also known as Brønsted base. { bās }

base exchange [GEOCHEM] Replacement of certain ions by others in clay. { 'bās iks'chānj }

base flow [HYD] The flow of water entering stream channels from groundwater sources in the drainage of large lakes. { 'bās ,flō }

base-leveled plain [GEOL] Any land surface changed almost to a plain by subaerial erosion. Also known as peneplain. { 'bās ‚lev·əld 'plān }

basidiocarp [MYCOL] The fruiting body of a fungus in the class Basidiomycetes. { bə'sid·ē·ə‚kärp }

Basidiomycetes [MYCOL] A class of fungi in the subdivision Eumycetes; important as food and as causal agents of plant diseases. { bə‚sid·ē·ō‚mī'sēd‚ēz }

basidium [MYCOL] A cell, usually terminal, occurring in Basidiomycetes and producing spores (basidiospores) by nuclear fusion followed by meiosis. { bə'sid·ē·əm }

basin [OCEANOGR] Deep portion of sea surrounded by shallower regions. { 'bās·ən }

basin accounting See hydrologic accounting. { 'bās·ən ə'kaůnt·iŋ }

basin cultivation [AGR] A type of cultivation in which small basins are enclosed by low earthen ridges to check runoff from heavy rains, thus conserving soil moisture and minimizing soil erosion. { 'bā·sən ‚kəl·tə'vā·shən }

basin length [GEOL] Length in a straight line from the mouth of a stream to the farthest point on the drainage divide of its basin. { 'bās·ən ‚leŋkth }

basin peat See local peat. { 'bās·ən ‚pēt }

basin swamp [ECOL] A fresh-water swamp at the margin of a small calm lake, or near a large lake protected by shallow water or a barrier. { 'bās·ən ‚swämp }

basophilous [ECOL] Of plants, growing best in alkaline soils. { bə'säf·ə·ləs }

bast See phloem. { bast }

bast fiber [BOT] Any fiber stripped from the inner bark of plants, such as flax, hemp, jute, and ramie; used in textile and paper manufacturing. { 'bast ‚fī·bər }

Batesian mimicry [ECOL] Resemblance of an innocuous species to one that is distasteful to predators. { 'bāt·sē·ən 'mim·ə·krē }

bathyal zone [OCEANOGR] The biogeographic realm of the ocean depths between 100 and 1000 fathoms (180 and 1800 meters). { 'bath·ē·əl ‚zōn }

bathymetric chart [GEOGR] A topographic map of the floor of the ocean. { ¦bath·ə'me·trik 'chärt }

bathymetry [ENG] The science of measuring ocean depths in order to determine the sea floor topography. { bə'thim·ə·trē }

bathypelagic zone [OCEANOGR] The biogeographic realm of the ocean lying between depths of 500 and 2550 fathoms (900 and 3700 meters). { ¦bath·ə·pə'laj·ik 'zōn }

battery reefs See Kimberley reefs. { 'bad·ə·rē ‚rēfs }

bay [BOT] *Laurus nobilis.* An evergreen tree of the laurel family. [GEOGR] **1.** A body of water, smaller than a gulf and larger than a cove in a recess in the shoreline. **2.** A narrow neck of water leading from the sea between two headlands. { bā }

bay ice [OCEANOGR] Sea ice that is young and flat but sufficiently thick to impede navigation. { ¦bā‚īs }

bayou [HYD] A small, sluggish secondary stream or lake that exists often in an abandoned channel or a river delta. { 'bī‚yü }

beach [GEOL] The zone of unconsolidated material that extends landward from the low-water line to where there is marked change in material or physiographic form or to the line of permanent vegetation. { bēch }

beach cycle [GEOL] Periodic retreat and outbuilding of beaches resulting from waves and tides. { bēch ‚sī·kəl }

beach nourishment |GEOL| The replenishment of a beach, either naturally (such as by littoral transport) or artificially (such as by deposition of dredged materials). { 'nər·ish·mənt }

beach scarp |GEOL| A nearly vertical slope along the beach caused by wave erosion. { 'bēch ,skärp }

beaded lake *See* paternoster lake. { 'bēd·əd ¦lāk }

bean |BOT| The common name for various leguminous plants used as food for humans and livestock; important commercial beans are true beans (*Phaseolus*) and California blackeye (*Vigna sinensis*). { bēn }

bean anthracnose |PL PATH| A fungus disease of the bean caused by *Colletotrichum lindemuthianum*, producing pink to brown lesions on the pod and seed and dark discolorations on the veins on the lower surface of the leaf. { 'bēn an'thrak,nōs }

bean blight |PL PATH| A bacterial disease of the bean caused by *Xanthomonas phaseoli*, producing water-soaked lesions that become yellowish-brown spots on all plant parts. { bēn ,blīt }

Beaufort force |METEOROL| A number denoting the speed (or so-called strength) of the wind according to the Beaufort wind scale. Also known as Beaufort number. { 'bō ,fərt ,fòrs }

Beaufort number *See* Beaufort force. { 'bō,fərt ,nəm·bər }

Beaufort wind scale |METEOROL| A system of code numbers from 0 to 12 classifying wind speeds into groups from 0–1 mile per hour or 0–1.6 kilometers per hour (Beaufort 0) to those over 75 miles per hour or 121 kilometers per hour (Beaufort 12). { 'bō·fərt 'wind ,skāl }

bed |HYD| The bottom of a channel for the passage of water. { bed }

bed load |GEOL| Particles of sand, gravel, or soil carried by the natural flow of a stream on or immediately above its bed. Also known as bottom load. { 'bed ,lōd }

bedrock |GEOL| General term applied to the solid rock underlying soil or any other unconsolidated surficial cover. { 'bed,räk }

beetle |ZOO| The common name given to members of the insect order Coleoptera. { 'bēd·əl }

behavioral ecology |ECOL| The branch of ecology that focuses on the evolutionary causes of variation in behavior among populations and species. { bi'hāv·yə·rəl ē'käl·ə·jē }

behavioral isolation |ECOL| An isolating mechanism in which two sympatric species do not mate because of differences in courtship behavior. Also known as ethological isolation. { bi'hāv·yə·rəl ī·sə'lā·shən }

behavioral toxicology |MED| The study of behavioral abnormalities induced by exogenous agents such as drugs, chemicals in the general environment, and chemicals encountered in the workplace. { bə¦hāv·yə·rəl ,täk·sə'käl·ə·jē }

beheaded stream |HYD| A water course whose upper portion, through erosion, has been cut off and captured by another water course. { bi'hed·əd ¦strēm }

belt |HYD| A long area or strip of pack ice, with a width of 1 kilometer (0.6 mile) to more than 100 kilometers (60 miles). { belt }

belt of soil moisture *See* belt of soil water. { ¦belt əv 'sòil ,mòis·chər }

belt of soil water |GEOL| The upper subdivision of the zone of aeration limited above by the land surface and below by the intermediate belt; this zone contains plant roots and water available for plant growth. Also known as belt of soil moisture; discrete film zone; soil-water belt; soil-water zone; zone of soil water. { ¦belt əv 'sòil ,wòd·ər }

bend [GEOL] **1.** A curve or turn occurring in a stream course, bed, or channel which has not yet become a meander. **2.** The land area partly encircled by a bend or meander. { bend }

bending [OCEANOGR] The first stage in the formation of pressure ice caused by the action of current, wind, tide, or air temperature changes. { 'ben·diŋ }

benequinox [CHEM] $C_{13}H_{11}N_3O_2$ A yellow-brown powder that decomposes at 195°C; used as a fungicide for grain seeds and seedlings. { ben'ē·kwə,näks }

Benguela Current [OCEANOGR] A strong current flowing northward along the southwestern coast of Africa. { ben'gwel·ə¦kər·ənt }

benign [MED] Of no danger to life or health. { bə'nīn }

benomyl [CHEM] $C_{14}H_{18}N_4O_3$ Methyl-l-butylcarbamoyl-2-benzimidazole carbamate; a fungicide used to control plant disease. { 'ben·ə,mil }

bensulide [CHEM] $C_{14}H_{24}O_4NPS_3$ An S-(O,O-diisopropyl phosphorodithioate) ester of N-(2-mercaptoethyl)-benzenesulfonamide; an amber liquid slightly soluble in water; melting point is 34.4°C; used as a preemergent herbicide for annual grasses and for broadleaf weeds in lawns and vegetable and cotton crops. { 'ben·sə,līd }

benthic [OCEANOGR] Of, pertaining to, or living on the bottom or at the greatest depths of a large body of water. Also known as benthonic. { 'ben·thik }

benthiocarb [CHEM] $C_{12}H_{16}NOCl$ An amber liquid with a boiling point of 126–129°C; slightly soluble in water; used as an herbicide to control aquatic weeds in rice crops. { ben'thī·ō,kärb }

benthonic See benthic. { ben'thän·ik }

benthos [OCEANOGR] The floor or deepest part of a sea or ocean. { 'ben,thäs }

benzalkonium chloride [CHEM] $C_6H_5CH_2(CH_3)_2NRCl$ A yellow-white powder soluble in water; used as a fungicide and bactericide; the R is a mixture of alkyls from C_8H_{17} to $C_{18}H_{37}$. { ,benz·əl'kōn·ē·əm 'klór,īd }

benzene [CHEM] C_6H_6 A colorless, liquid, flammable, aromatic hydrocarbon that boils at 80.1°C and freezes at 5.4–5.5°C; used to manufacture styrene and phenol. Also known as benzol. { 'ben,zēn }

benzol See benzene. { 'ben,zól }

benzomate [CHEM] $C_{18}H_{18}O_5N$ A white solid that melts at 71.5–73°C; used as a wettable powder as a miticide. { 'ben·zə,māt }

benzopyrene [CHEM] $C_{20}H_{12}$ A five-ring aromatic hydrocarbon found in coal tar, in cigarette smoke, and as a product of incomplete combustion; yellow crystals with a melting point of 179°C; soluble in benzene, toluene, and xylene. { 'ben·zō¦pī,rēn }

4-benzothienyl-N-methylcarbamate [CHEM] $C_{10}H_9NO_2S$ A white powder compound with a melting point of 128°C; used as an insecticide for crop insects. { ¦fór ¦ben·zō 'thī·ə,nil ¦en ¦meth·əl'kär·bə,māt }

benzotrifluoride [CHEM] Colorless liquid, boiling point 102.1°C; used for dyes and pharmaceuticals, as solvent and vulcanizing agent, in insecticides. { ben·zō,trī'flúr ,īd }

benzoylpropethyl [CHEM] $C_{18}H_{17}Cl_2NO_3$ An off-white, crystalline compound with a melting point of 72°C; used as a preemergence herbicide for control of wild oats. { ¦ben·zə·wəl¦prō·pə·thəl }

benzthiazuron [CHEM] $C_9H_9N_3SO$ A white powder that decomposes at 287°C; slightly soluble in water; used as a preemergent herbicide for sugarbeets and fodder beet crops. { ,benz,thī'az·yə,rän }

benzyl penicillin sodium [MICROBIO] $C_{16}H_{17}N_2NaO_4S$ Crystals obtained from a methanol-ethyl acetate acidified extract of fermentation broth of *Penicillium chrysogenum*; used as an antimicrobial in human and animal disease. { 'ben·zəl ˌpen·ə'sil·ən 'sōd·ē·əm }

3,4-benzpyrene [CHEM] $C_{20}H_{12}$ A polycyclic hydrocarbon; a chemical carcinogen that will cause skin cancer in many species when applied in low dosage. { ¦thrē ¦fȯr ˌbenz'pī ˌrēn }

Beranek scale [PHYS] A scale which measures the subjective loudness of a noise; noises are arranged into six arbitrary categories: very quiet, quiet, moderately quiet, noisy, very noisy, and intolerably noisy. { bə'ran·ik ˌskāl }

Bergeron-Findeisen theory [METEOROL] The theoretical explanation that precipitation particles form within a mixed cloud (composed of both ice crystals and liquid water drops) because the equilibrium vapor pressure of water vapor with respect to ice is less than that with respect to liquid water at the same temperature. Also known as ice-crystal theory; Wedener-Bergeron process. { 'berzh·əˌrän ¦finˌdīz·ən ˌthē·ə·rē }

Bergmann's rule [ECOL] The principle that in a wide-ranging species of warm-blooded animals the average body size increases in populations living in colder environments. { 'bərg·mənz ˌrül }

bergschrund [HYD] A type of crevice in a glacier; formed when ice and snow break away from a rock face. { 'berk ˌshrȯnt }

bergy-bit *See* growler. { 'bərg·ē ˌbit }

Bering Sea [GEOGR] A body of water north of the Pacific Ocean, bounded by Siberia, Alaska, and the Aleutian Islands. { 'ber·iŋ 'sē }

berm [GEOL] **1.** A narrow terrace which originates from the interruption of an erosion cycle with rejuvenation of a stream in the mature stage of its development and renewed dissection. **2.** A horizontal portion of a beach or backshore formed by deposit of material as a result of wave action. Also known as backshore terrace; coastal berm. { bərm }

Bermuda high [METEOROL] The semipermanent subtropical high of the North Atlantic Ocean, especially when it is located in the western part of that ocean area. { bər'myüd·ə ¦hī }

berry [BOT] A usually small, simple, fleshy or pulpy fruit, such as a strawberry, grape, tomato, or banana. { 'ber·ē }

beta [PHYS] The amount of reactivity of a nuclear reactor corresponding to the delayed neutron fraction. { 'bād·ə }

betrunked river [GEOL] A river that is shorn of its lower course as a result of submergence of the land margin by the sea. { bē'trəŋkt 'riv·ər }

B horizon [GEOL] The zone of accumulation in soil below the A horizon (zone of leaching). Also known as illuvial horizon; subsoil; zone of accumulation; zone of illuviation. { 'bē hə'rīz·ən }

bichloride of mercury *See* mercuric chloride. { bī'klȯrˌīd əv 'mər·kyə·rē }

biennial plant [BOT] A plant that requires two growing seasons to complete its life cycle. { bī'en·ē·əl ¦plant }

bifenox [CHEM] $C_{14}H_9Cl_2NO_5$ A tan, crystalline compound with a melting point of 84–86°C; insoluble in water; used as a preemergence herbicide for weed control in soybeans, corn, and sorghum, and as a pre- and postemergence herbicide in rice and small greens. { bī'fenˌäks }

bight [OCEANOGR] An indentation in shelf ice, fast ice, or a floe. { bīt }

bilateral   [BIOL] Of or relating to both right and left sides of an area, organ, or organism. { bī'lad·ə·rəl }

billow cloud   [METEOROL] Broad, nearly parallel lines of cloud oriented normal to the wind direction, with cloud bases near an inversion surface. Also known as undulatus. { 'bil·ō,klaůd }

bioacoustics   [BIOL] The study of the relation between living organisms and sound. { ¦bī·ō·ə'kü·stiks }

bioactivity   [BIOL] The effect that a substance has on a living organism or tissue after interaction.   { ,bī·ō·ak'tiv·əd·ē }

biobubble   [ECOL] A model concept of the ecosphere in which all living things are considered as particles held together by nonliving forces.   { 'bī·ō,bəb·əl }

biocenology   [ECOL] The study of natural communities and of interactions among the members of these communities.   { ,bī·ō·sə'näl·ə·jē }

biocenose   See biotic community.   { ,bī·ō'sē,nōs }

biochemical deposit   [GEOL] A precipitated deposit formed directly or indirectly from vital activities of organisms, such as bacterial iron ore and limestone.   { ¦bī·ō'kem·ə·kəl di'päz·ət }

biochemical engineering   [BIOL] The application of chemical engineering principles to conceive, design, develop, operate, or utilize processes and products based on biological and biochemical phenomena; this field is included in a wide range of industries, such as health care, agriculture, food, enzymes, chemicals, waste treatment, and energy.   { ,bī·ō¦kem·i·kəl ,en·jə'nir·iŋ }

biochemical fuel cell   [ENG] An electrochemical power generator in which the fuel source is bioorganic matter; air is the oxidant at the cathode, and microorganisms catalyze the oxidation of the bioorganic matter at the anode.   { ¦bī·ō'kem·ə·kəl 'fyül ,sel }

biochemical oxygen demand   [MICROBIO] The amount of dissolved oxygen required to meet the metabolic needs of aerobic microorganisms in water rich in organic matter, such as sewage. Abbreviated BOD. Also known as biological oxygen demand.   { ¦bī·ō 'kem·ə·kəl 'äk·sə·jən di'mand }

biochemical oxygen demand test   [MICROBIO] A standard laboratory procedure for measuring biochemical oxygen demand; standard measurement is made for 5 days at 20°C. Abbreviated BOD test.   { ¦bī·ō'kem·ə·kəl 'äk·sə·jən di'mand ,test }

biochronology   [GEOL] The relative age dating of rock units based on their fossil content.   { ,bī·ō·krə'näl·ə·jē }

biocide   See pesticide.   { 'bī·ə,sīd }

bioclimatic law   [ECOL] The law which states that phenological events are altered by about 4 days for each 5° change of latitude northward or longitude eastward; events are accelerated in spring and retreat in autumn.   { ¦bī·ō,klī'mad·ik 'lȯ }

bioclimatograph   [ECOL] A climatograph showing the relation between climatic conditions and some living organisms.   { ¦bī·ō,klī'mad·ə,graf }

bioclimatology   [ECOL] The study of the effects of the natural environment on living organisms.   { ¦bī·ō,klī·mə'täl·ə·jē }

biocoenosis   [ECOL] A group of organisms that live closely together and form a natural ecologic unit.   { ,bī·ō·sə'nō·səs }

biocontrol   See biological control.   { ,bī·ō·kən'trōl }

bioconversion   [BIOL] The process of converting biomass to a source of usable energy. { ,bī·o·kən'vər·zhən }

biocycle   [ECOL] A group of similar biotopes composing a major division of the biosphere; there are three biocycles: terrestrial, marine, and fresh-water.   { 'bī·ō ‚sī·kəl }

biodegradation   [ECOL] The destruction of organic compounds by microorganisms.   { ‚bī·ō‚deg·rə'dā·shən }

biodeterioration   [BOT] Decay of wood or other material caused by fungi, bacteria, insects, or marine boring organisms.   { ‚bī·ō·di‚tir·ē·ər'ā·shən }

biodistribution kinetics   [BIOL] A mathematical description of the in vivo distribution of a radionuclide present in various organs as a function of time following its administration.   { ‚bī·ō‚dis·trə¦byü·shən ki'ned·iks }

biodiversity   [ECOL] All aspects of biological diversity, especially species richness, genetic variation, and the complexity of ecosystems.   { ‚bī·ō·di'vər·sə·dē }

biodynamic   [BIOL] Of or pertaining to the dynamic relation between an organism and its environment.   { ‚bi·ō·dī'nam·ik }

bioerosion   [OCEANOGR] The process by which animals, through drilling, grazing, and burrowing, erode hard substances such as rocks and coral reefs.   { ‚bī·ō·i'rōzh·ən }

bioethics   [BIOL] A discipline concerned with the application of ethics to biological problems, especially in the field of medicine.   { ‚bī·ō'eth·iks }

biofilm   [MICROBIO] A microbial (bacterial, fungal, algal) community, enveloped by the extracellular biopolymer which these microbial cells produce, that adheres to the interface of a liquid and a surface.   { 'bī·ō‚film }

biofilter   [ENG] An emission control device that uses microorganisms to destroy volatile organic compounds and hazardous air pollutants.   { 'bī·ō‚fil·tər }

biofog   [METEOROL] A type of steam fog caused by contact between extremely cold air and the warm, moist air surrounding human or animal bodies or generated by human activity.   { 'bī·ō‚fäg }

biogas   [BIOL] A mixture of methane and carbon dioxide generated from the bacterial decomposition of animal and vegetable wastes.   { 'bī·ō‚gas }

biogenic reef   [GEOL] A mass consisting of the hard parts of organisms, or of a biogenically constructed frame enclosing detrital particles, in a body of water; most biogenic reefs are made of corals or associated organisms.   { ¦bī·ō¦jen·ik 'rēf }

biogenic sediment   [GEOL] A deposit resulting from the physiological activities of organisms.   { ¦bī·ō¦jen·ik 'sed·ə·mənt }

biogeochemical cycle   [GEOCHEM] The chemical interactions that exist between the atmosphere, hydrosphere, lithosphere, and biosphere.   { ‚bī·ō‚jē·ō'kem·ə·kəl 'sīkəl }

biogeochemical prospecting   [GEOCHEM] A prospecting technique for subsurface ore deposits based on interpretation of the growth of certain plants which reflect subsoil concentrations of some elements.   { ‚bī·ō‚jē·ō'kem·ə·kəl 'präs‚pek·tiŋ }

biogeochemistry   [GEOCHEM] A branch of geochemistry that is concerned with biologic materials and their relation to earth chemicals in an area.   { ‚bī·ō‚jē·ō'kem·ə·strē }

biogeographic realm   [ECOL] Any of the divisions of the landmasses of the world according to their distinctive floras and faunas.   { ‚bī·ō‚jē·ə‚graf·ik 'relm }

biogeography   [ECOL] The science concerned with the geographical distribution of animal and plant life.   { ¦bī·ō·jē¦äg·rə·fē }

biogeosphere   [ECOL] The region of the earth extending from the surface of the upper crust to the maximum depth at which organic life exists.   { ‚bī·ō'¦jē·ə‚sfir }

biohazard   [BIOL] Any biological agent or condition that presents a hazard to life.   { 'bī·ō ‚haz·ərd }

bioherm |GEOL| A circumscribed mass of rock exclusively or mainly constructed by marine sedimentary organisms such as corals, algae, and stromatoporoids. Also known as organic mound.   { 'bī·ō,hərm }

biohermal limestone |GEOL| Reefs or reeflike mounds of carbonate that accumulated much in the same fashion as modern reefs and atolls of the Pacific Ocean.   { ¦bī·ō ¦hər·məl 'līm,stōn }

biohydrology |ECOL| Study of the interactions between water, plants, and animals, including the effects of water on biota as well as the physical and chemical changes in water or its environment produced by biota.   { ¦bī·ō,hī'dräl·ə·jē }

biolite |GEOL| A concretion formed of concentric layers through the action of living organisms.   { 'bī·ō,līt }

biological |MED| A biological product used to induce immunity to various infectious diseases or noxious substances of biological origin.   { ¦bī·ə¦läj·ə·kəl }

biological agent |MICROBIO| Any of the viruses, microorganisms, and toxic substances derived from living organisms and used as offensive weapons to produce death or disease in humans, animals, and growing plants.   { ¦bī·ə¦läj·ə·kəl 'ā·jənt }

biological balance |ECOL| Dynamic equilibrium that exists among members of a stable natural community.   { ¦bī·ə¦läj·ə·kəl 'bal·əns }

biological control |ECOL| Natural or applied regulation of populations of pest organisms, especially insects, through the role or use of natural enemies. Also known as biocontrol.   { ¦bī·ə¦läj·ə·kəl kən'trōl }

biological indicator |BIOL| An organism that can be used to determine the concentration of a chemical in the environment.   { ,bī·ə¦läj·ə·kəl 'in·də,kād·ər }

biological invasion |ECOL| The process by which species (or genetically distinct populations), with no historical record in an area, breach biogeographic barriers and extend their range.   { ,bi·ə¦läj·i·kəl in'vā·zhən }

biological magnification |ECOL| The increasing concentration of toxins from pesticides, herbicides, and various types of waste in living organisms that accompanies cycling of nutrients through the trophic levels of food webs.   { ,bī·ə¦läj·ə·kəl ,mag·nə·fə'kā·shən }

biological oceanography |OCEANOGR| The study of the flora and fauna of oceans in relation to the marine environment.   { ¦bī·ə¦läj·ə·kəl ,ō·shə'näg·rə·fē }

biological oil-spill control |ECOL| The use of cultures of microorganisms capable of living on oil as a means of degrading an oil slick biologically.   { ¦bī·ə¦läj·ə·kəl 'óil ,spil kən'trōl }

biological oxygen demand   *See* biochemical oxygen demand **2.**   { ¦bī·ə¦läj·ə·kəl 'äk·sə·jən di'mand }

biological productivity |ECOL| The quantity of organic matter or its equivalent in dry matter, carbon, or energy content which is accumulated during a given period of time.   { ¦bī·ə¦läj·ə·kəl prə,dək'tiv·əd·ē }

biological warfare |MICROBIO| Abbreviated BW. **1.** Employment of living microorganisms, toxic biological products, and plant growth regulators to produce death or injury in humans, animals, or plants. **2.** Defense against such action.   { ¦bī·ə¦läj·ə·kəl 'wór ,fer }

biological weathering   *See* organic weathering.   { ¦bī·ə¦läj·ə·kəl 'weth·ə·riŋ }

biologic weathering   *See* organic weathering.   { ¦bī·ə¦läj·ik 'weth·ə·riŋ }

biology |SCI TECH| A division of the natural sciences concerned with the study of life and living organisms.   { bī'äl·ə·jē }

bioluminescence [BIOL] The emission of visible light by living organisms. { ˌbī·ō ˌlü·mə'nes·əns }

biolysis [BIOL] **1.** Death and the following tissue disintegration. **2.** Decomposition of organic materials, such as sewage, by living organisms. { bī'äl·ə·səs }

biomarkers [GEOL] Complex organic compounds found in oil, bitumen, rocks, and sediments that are linked with and distinctive of a particular source (such as algae, bacteria, or vascular plants); they are useful dating indicators in stratigraphy and molecular paleontology. Also known as chemical fossils; molecular fossils. { 'bī·ō ˌmär·kərz }

biomass [ECOL] The dry weight of living matter, including stored food, present in a species population and expressed in terms of a given area or volume of the habitat. { 'bī·ō,mas }

biome [ECOL] A complex biotic community covering a large geographic area and characterized by the distinctive life-forms of important climax species. { 'bī,ōm }

biometeorology [BIOL] The study of the relationship between living organisms and atmospheric phenomena. { ¦bī·ō,mēd·ē·ə'räl·ə·jē }

biometer [BIOL] An instrument which is used to measure minute amounts of carbon dioxide given off by the functioning tissue of an organism. { bī'äm·ə·tər }

biomining [MICROBIO] The use of microorganisms to recover metals of value, such as gold, silver, and copper, from sulfide minerals. { 'bī·ō,mīn·iŋ }

bionomics See ecology. { ˌbī·ō'näm·iks }

biophage See macroconsumer. { 'bī·ō,fāj }

biophile [BIOL] Any element concentrated or found in the bodies of living organisms and organic matter; examples are carbon, nitrogen, and oxygen. { 'bī·ə,fīl }

bioprospecting [MED] The search for new pharmaceutical (and sometimes nutritional or agricultural) products from natural sources, such as plants, microorganisms, and sometimes animals. { ˌbī·ō'prä·spek·tiŋ }

biorefinery [BIOL] A large, integrated processing facility that produces chemicals and biochemicals from plant matter, wood waste, and waste paper. { ¦bī·ō·ri¦fīn·rē }

bioregion [ECOL] A region with borders that are naturally defined by topographic systems (such as mountains, rivers, and oceans) and ecological systems (such as deserts, rainforests, and tundras). { 'bī·ō,rē·jən }

bioregionalism [ECOL] An environmentalist movement to make political boundaries coincide with bioregions. { ˌbī·ō'rē·jən·əl,iz·əm }

bioremediation [ECOL] The use of a biological process (via plants or microorganisms) to clean up a polluted environmental area (such as an oil spill). { ˌbī·ō·ri,mē·dē'ā·shən }

biorhythm [BIOL] A biologically inherent cyclic variation or recurrence of an event or state, such as a sleep cycle or circadian rhythm. { 'bī·ō,rith·əm }

biosafety [BIOL] The establishment and maintenance of safe conditions in a biological research laboratory to ensure that pathogenic microbes are contained (and not released to workers or the environment). { ¦bī·ō¦sāf·tē }

bioscience [BIOL] The study of the nature, behavior, and uses of living organisms as applied to biology. { ¦bī·ō¦sī·əns }

biosolid [CIV ENG] A recyclable, primarily organic solid material produced by wastewater treatment processes. { ¦bī·ō,säl·əd }

biosphere [ECOL] The life zone of the earth, including the lower part of the atmosphere, the hydrosphere, soil, and the lithosphere to a depth of about 1.2 miles (2 kilometers). { 'bī·ə,sfir }

biostabilizer [CIV ENG] A component in mechanized composting systems; consists of a drum in which moistened solid waste is comminuted and tumbled for about 5 days until the aeration and biodegradation turns the waste into a fine dark compost. { ,bī·ō 'stāb·əl,īz·ər }

biostasy [ECOL] Maximum development of organisms when, during tectonic repose, residual soils form extensively on the land and calcium carbonate deposition is widespread in the sea. { bī'äs·tə·sē }

biostromal limestone [GEOL] Biogenic carbonate accumulations that are laterally uniform in thickness, in contrast to the moundlike nature of bioherms. { ¦bī·ə¦strō·məl 'līm,stōn }

biostrome [GEOL] A bedded structure or layer composed of calcite and dolomitized calcarenitic fossil fragments distributed over the sea bottom as fine lentils, independent of or in association with bioherms or other areas of organic growth. { 'bī·ə,strōm }

biota [BIOL] **1.** Animal and plant life characterizing a given region. **2.** Flora and fauna, collectively. { bī'ōd·ə }

biotechnology [GEN] The use of advanced genetic techniques to construct novel microbial, plant, and animal strains or obtain site-directed mutants to improve the quantity or quality of products or obtain other desired phenotypes. { ¦bī·ō·tek'näl·ə·jē }

biotic [BIOL] **1.** Of or pertaining to life and living organisms. **2.** Induced by the actions of living organisms. { bī'äd·ik }

biotic community [ECOL] An aggregation of organisms characterized by a distinctive combination of both animal and plant species in a particular habitat. Also known as biocenose. { bī'äd·ik kə'myün·əd·ē }

biotic district [ECOL] A subdivision of a biotic province. { bī'äd·ik 'dis·trikt }

biotic environment [ECOL] That environment comprising living organisms, which interact with each other and their abiotic environment. { bī'äd·ik in'vī·ərn·mənt }

biotic isolation [ECOL] The occurrence of organisms in isolation from others of their species. { bī'äd·ik ,i·sə'lā·shən }

biotic potential [ECOL] The maximum possible growth rate of living things under ideal conditions. { bī'äd·ik pə'ten·chəl }

biotic province [ECOL] A community, according to some systems of classification, occupying an area where similarity of climate, physiography, and soils leads to the recurrence of similar combinations of organisms. { bī'äd·ik 'präv·əns }

biotope [ECOL] An area of uniform environmental conditions and biota. { 'bī·ə,tōp }

biotron [ENG] A test chamber used for biological research within which the environmental conditions can be completely controlled, thus allowing observations of the effect of variations in environment on living organisms. { 'bī·ə,trän }

bioturbation [GEOL] The disruption of marine sedimentary structures by the activities of benthic organisms. { ¦bī·o·tər'bā·shən }

biparasitic [ECOL] Parasitic upon or in a parasite. { ¦bī,par·ə'sid·ik }

biphasic [BOT] Possessing both a sporophyte and a gametophyte generation in the life cycle. { bī'fāz·ik }

bipotential [BIOL] Having the potential to develop in either of two mutually exclusive directions. { ¦bī·pə'ten·chəl }

birth [BIOL] The emergence of a new individual from the body of its parent. { bərth }

birth rate [BIOL] The ratio between the number of live births and a specified number of organisms in a population over a given period of time. { bərth ‚rāt }

2,2-bis(*para*-chlorophenyl)-1,1-dichloroethane [CHEM] $C_{14}H_{10}Cl_4$ A colorless, crystalline compound with a melting point of 109–111°C; insoluble in water; used as an insecticide on fruits and vegetables. Also known as DDD; TDE. { ¦tü ¦tü ‚bis 'par·ə ‚klȯr·ə'fen·əl ¦wən ¦wən di‚klȯrō'e‚thän }

bise [METEOROL] A cold, dry wind which blows from a northerly direction in the winter over the mountainous districts of southern Europe. Also spelled bize. { bēz }

Bishop's ring [METEOROL] A faint, broad, reddish-brown corona occasionally seen in dust clouds, especially those which result from violent volcanic eruptions. { 'bish·əps 'riŋ }

bismuth subsalicylate [CHEM] $Bi(C_7H_5O)_3Bi_2O_3$ A white powder that is insoluble in ethanol and water; used in medicine and as a fungicide for tobacco crops. { 'biz·məth ‚səb·sə'lis·ə‚lāt }

bitter lake [HYD] A lake rich in alkaline carbonates and sulfates. { ¦bid·ər 'lāk }

bitter rot [PL PATH] A fungus disease of apples, grapes, and other fruit caused by *Glomerella cingulata*. { 'bid·ər ‚rät }

bituminization *See* coalification. { bī‚tü·mə·nə'zā·shən }

bize *See* bise. { bēz }

black [CHEM] Fine particles of impure carbon that are made by the incomplete burning of carbon compounds, such as natural gas, naphthas, acetylene, bones, ivory, and vegetables. { blak }

black alkali [GEOL] A deposit of sodium carbonate that has formed on or near the surface in arid to semiarid areas. { ¦blak 'al·kə‚lī }

black band disease [ZOO] A coral reef disease that is characterized by a thick black band of tissue that advances rapidly across infected corals, leaving empty coral skeletons behind. { ¦blak¦band di‚zēz }

black blight [PL PATH] Any of several diseases of tropical plants caused by superficial sooty molds. { 'blak ‚blīt }

black chaff [PL PATH] A bacterial disease of wheat caused by *Xanthomonas translucens undulosa* and characterized by dark, longitudinal stripes on the chaff. { 'blak ¦chaf }

black cyanide *See* calcium cyanide. { ¦blak 'sī·ə‚nīd }

black death *See* plague. { ¦blak 'deth }

blackfire [PL PATH] A bacterial disease of tobacco caused by *Pseudomonas angulata* and characterized by angular leaf spots which gradually darken and may fall out, leaving ragged holes. { 'blak‚fīr }

black frost [HYD] A dry freeze with respect to its effects upon vegetation, that is, the internal freezing of vegetation unaccompanied by the protective formation of hoarfrost. Also known as hard frost. { ¦blak 'frȯst }

blackhead disease [PL PATH] **1.** A parasitic disease of the banana caused by eelworms of the family Tylenchidae. **2.** A rot disease of the banana rootstock that is caused by the fungus *Thielaviopsis paradoxa*. { 'blak‚hed di'zēz }

black ice [HYD] A type of ice forming on lake or salt water; compact, and dark in appearance because of its transparency. { 'blak ‚īs }

blacklung   *See* anthracosis.   { 'blak₁ləŋ }

black mold   [MYCOL] Any dark fungus belonging to the order Mucorales.   [PL PATH] A fungus disease of rose grafts and onion bulbs marked by black appearance due to the mold.   { 'blak ,mōld }

black ring   [PL PATH] **1.** A virus disease of cabbage and other members of the family Cruciferae characterized by dark necrotic and often sunken rings on the surface of the leaf.   **2.** A virus disease of the tomato characterized in the early stage by small black rings on young leaves.   { 'blak ,riŋ }

black root   [PL PATH] Any plant disease characterized by black discolorations of the roots.   { 'blak ,rüt }

black root rot   [PL PATH] **1.** Any of several plant diseases characterized by dark lesions of the root.   **2.** A fungus disease of the apple caused by *Xylaria mali.*   **3.** A fungus disease of tobacco and other plants caused by *Thielaviopsis basicola.*   { 'blak 'rüt ,rät }

black rot   [PL PATH] Any fungal or bacterial disease of plants characterized by dark brown discoloration and decay of a plant part.   { 'blak ,rät }

black sand   [GEOL] Heavy, dark, sandlike minerals found on beaches and in stream beds; usually magnetite and ilmenite and sometimes gold, platinum, and monazite are present.   { ¦blak 'sand }

Black Sea   [GEOGR] A large inland sea, area 163,400 square miles (423,000 square kilometers), bounded on the north and east by the Commonwealth of Independent States (former U.S.S.R.) on the south and southwest by Turkey, and on the west by Bulgaria and Rumania.   { ¦blak 'sē }

black smoker   *See* hydrothermal vent.   { ¦blak 'smōk·ər }

black snow   [HYD] Snow that falls through a particulate-laden atmosphere.   { ¦blak 'snō }

black spot   [PL PATH] Any bacterial or fungal disease of plants characterized by black spots on a plant part.   { 'blak ,spät }

black stem   [PL PATH] Any of several fungal diseases of plants characterized by blackening of the stem.   { 'blak ,stem }

bladder   [BIOL] Any saclike structure in humans and animals, such as a swimbladder or urinary bladder, that contains a gas or functions as a receptacle for fluid.   [GEOL] *See* vesicle.   { 'blad·ər }

blanket sand   [GEOL] A relatively thin body of sand or sandstone covering a large area. Also known as sheet sand.   { 'blaŋ·kət ,sand }

blasticidin-S   [CHEM] A compound with a melting point of 235–236°C; soluble in water; used as a fungicide for rice crops.   { ,blas'tis·ə·dən 'es }

blight   [PL PATH] Any plant disease or injury that results in general withering and death of the plant without rotting.   { blīt }

blind drainage   *See* closed drainage.   { ¦blīnd ¦drā·nij }

blind rollers   [OCEANOGR] Long, high swells which have increased in height, almost to the breaking point, as they pass over shoals or run in shoaling water. Also known as blind seas.   { ¦blīnd 'rō·lərz }

blind seas   *See* blind rollers.   { ¦blīnd 'sēz }

blind seed   [PL PATH] A fungus disease of forage grasses caused by *Phealea temulenta,* resulting in abortion of the seed.   { 'blīnd ,sēd }

blink   [METEOROL] A brightening of the base of a cloud layer, caused by the reflection of light from a snow- or ice-covered surface.   { bliŋk }

blister blight   [PL PATH] **1.** A fungus disease of the tea plant caused by *Exobasidium vexans* and characterized by blisterlike lesions on the leaves.   **2.** A rust disease of Scotch pine caused by *Cronartium asclepiadeum* and characterized by blisterlike lesions on the twigs.   { 'blis·tər ˌblīt }

blister canker   [PL PATH] A fungus disease of the apple tree caused by *Nummularia discreta* and characterized by rough, black cankers on the trunk and large branches. Also known as apple pox.   { 'blis·tər ˌkaŋ·kər }

blister spot   [PL PATH] A bacterial disease of the apple caused by *Pseudomonas papulans* and characterized by dark-brown blisters on the fruit and cankers on the branches.   { 'blis·tər ˌspät }

blizzard   [METEOROL] A severe weather condition characterized by low temperatures and by strong winds bearing a great amount of snow (mostly fine, dry snow picked up from the ground).   { 'bliz·ərd }

blocking   [METEOROL] Large-scale obstruction of the normal west-to-east progress of migratory cyclones and anticyclones.   { 'bläk·iŋ }

blocky iceberg   [OCEANOGR] An iceberg with steep, precipitous side and with a horizontal or nearly horizontal upper surface.   { ¦bläk·ē 'īs,bərg }

blood poisoning   *See* septicemia.   { 'bləd ˌpoiz·ən·iŋ }

blood rain   [METEOROL] Rain of a reddish color caused by dust particles containing iron oxide that were picked up by the raindrops during descent.   { 'bləd ˌrān }

bloom   [ECOL] A colored area on the surface of bodies of water caused by heavy planktonic growth.   { blüm }

blossom   [GEOL] The oxidized or decomposed outcrop of a vein or coal bed.   { 'bläs·əm }

blowball   [BOT] A fluffy seed ball, as of the dandelion.   { 'blō,bòl }

blowdown   [METEOROL] A wind storm that causes trees or structures to be blown down.   { 'blō,daùn }

blowhole   [GEOL] A longitudinal tunnel opening in a sea cliff, on the upland side away from shore; columns of sea spray are thrown up through the opening, usually during storms.   { 'blō,hōl }

blowing dust   [METEOROL] Dust picked up locally from the surface of the earth and blown about in clouds or sheets.   { ¦blō·iŋ 'dəst }

blowing sand   [METEOROL] Sand picked up from the surface of the earth by the wind and blown about in clouds or sheets.   { ¦blō·iŋ ¦sand }

blowing snow   [METEOROL] Snow lifted from the surface of the earth by the wind to a height of 6 feet (1.8 meters) or more (higher than drifting snow) and blown about in such quantities that horizontal visibility is restricted.   { ¦blō·iŋ ¦snō }

blowing spray   [METEOROL] Spray lifted from the sea surface by the wind and blown about in such quantities that horizontal visibility is restricted.   { ¦blō·iŋ ¦sprā }

blowout   [HYD] A bubbling spring which bursts from the ground behind a river levee when water at flood stage is forced under the levee through pervious layers of sand or silt. Also known as sand boil.   { 'blō,aùt }

bluegrass   [BOT] The common name for several species of perennial pasture and lawn grasses in the genus *Poa* of the order Cyperales.   { 'blü,gras }

blue-green algae   *See* cyanobacteria.   { ¦blü¦grēn 'al·jē }

blue-green algal virus   *See* cyanophage.   { ¦blü¦grēn ¦al·gəl 'vī·rəs }

blue ice   [HYD] Pure ice in the form of large, single crystals that is blue owing to the scattering of light by the ice molecules; the purer the ice, the deeper the blue.   { ¦blü 'īs }

blue mold [MYCOL] Any fungus of the genus *Penicillium*. { 'blü ,mōld }

blue mud [GEOL] A combination of terrigenous and deep-sea sediments having a bluish gray color due to the presence of organic matter and finely divided iron sulfides. { 'blü ,məd }

blue-sky scale *See* Linke scale. { ¦blü ¦skī 'skāl }

bluff [GEOGR] **1.** A steep, high bank. **2.** A broad-faced cliff. { bləf }

BOD *See* biochemical oxygen demand.

BOD test *See* biochemical oxygen demand test. { ¦bē¦ō'dē ,test }

body [GEOGR] A separate entity or mass of water, such as an ocean or a lake. [GEOL] An ore body, or pocket of mineral deposit. { 'bäd·ē }

bog [ECOL] A plant community that develops and grows in areas with permanently waterlogged peat substrates. Also known as moor; quagmire. { bäg }

bog moss [ECOL] Moss of the genus *Sphagnum* occurring as the characteristic vegetation of bogs. { 'bäg ,mòs }

boiler plate [HYD] A crusty, frozen surface of snow. { 'bòil·ər ,plāt }

boiling spring [HYD] **1.** A spring which emits water at a high temperature or at boiling point. **2.** A spring located at the head of an interior valley and rising from the bottom of a residual clay basin. **3.** A rapidly flowing spring that develops strong vertical eddies. { 'bòil·iŋ ,spriŋ }

boil smut [PL PATH] A fungus disease of corn caused by *Ustilago maydis*, characterized by galls containing black spores. { 'bòil ,smət }

bole [FOR] The main stem of a tree of substantial diameter; capable of yielding timber, veneer logs, and large poles. [GEOL] Any of various red, yellow, or brown earthy clays consisting chiefly of hydrous aluminum silicates. Also known as bolus; terra miraculosa. { bōl }

boll [BOT] A pod or capsule (pericarp), as of cotton and flax. { bōl }

boll rot [PL PATH] A fungus rot of cotton bolls caused by *Glomerella gossypii* and *Xanthomonas malvacearum*. { 'bōl ,rät }

bollseye *See* sodium cacodylate. { 'bōlz,ī }

boll weevil [ZOO] A beetle, *Anthonomus grandis*, of the order Coleoptera; larvae destroy cotton plants and are the most important pests in agriculture. { 'bōl ,wē·vəl }

bolt [FOR] A short section of tree trunk. { bōlt }

bolus *See* bole. { 'bō·ləs }

bombykol [BIOL] The first pheromone to be characterized chemically; it is an unsaturated straight-chain alcohol secreted in microgram amounts by females of the silkworm moth (*Bombyx mori*) and is capable of attracting male silkworm moths at large distances. { 'bäm·bə,kòl }

book louse [ZOO] A common name for a number of insects belonging to the order Psocoptera; important pests in herbaria, museums, and libraries. { 'bùk ,laùs }

bora [METEOROL] A fall wind whose source is so cold that when the air reaches the lowlands or coast the dynamic warming is insufficient to raise the air temperature to the normal level for the region; hence it appears as a cold wind. { 'bòr·ə }

bora fog [METEOROL] A dense fog caused when the bora lifts a spray of small drops from the surface of the sea. { 'bòr·ə ,fäg }

Bordeaux mixture [AGR] A fungicide made from a mixture of lime, copper sulfate, and water. { bòr'dō ,miks·chər }

Bordetella [MICROBIO] A genus of gram-negative, aerobic bacteria of uncertain affiliation; minute coccobacilli, parasitic and pathogenic in the respiratory tract of mammals. { ˌbȯr·də'tel·ə }

Bordetella avium [MICROBIO] A nonsporulating, gram-negative coccobacillus that causes respiratory infections in birds. { ˌbȯr·də,tel·ə 'ā·vē·əm }

Bordetella bronchiseptica [MICROBIO] An aerobic, gram-negative bacterium that is a pathogen in many domestic and wild mammals, including horses, swine, dogs, and rodents, and may cause a variety of respiratory diseases in them. { ˌbȯr·də,tel·ə ˌbraŋ·ki'sep·ti·kə }

bore [OCEANOGR] **1.** A high, breaking wave of water, advancing rapidly up an estuary. Also known as eager; mascaret; tidal bore. **2.** A submarine sand ridge, in very shallow water, whose crest may rise to intertidal level. { bȯr }

boreal [ECOL] Of or relating to northern geographic regions. { 'bȯr·ē·əl }

boreal forest *See* taiga. { 'bȯr·ē·əl 'fär·əst }

Boreal life zone [ECOL] The zone comprising the climate and biotic communities between the Arctic and Transitional zones. { 'bȯr·ē·əl 'līf ˌzōn }

borer [ZOO] Any insect or other invertebrate that burrows into wood, rock, or other substances. { 'bȯr·ər }

boron-10 [PHYS] A nonradioactive isotope of boron with a mass number of 10; it is a good absorber for slow neutrons, simultaneously emitting high-energy alpha particles, and is used as a radiation shield in Geiger counters. { 'bȯ,rän 'ten }

Borrelia [MICROBIO] A genus of bacteria in the family Spirochaetaceae; helical cells with uneven coils and parallel fibrils coiled around the cell body for locomotion; many species cause relapsing fever in humans. { bə'rel·ē·ə }

Borrelia anserina [MICROBIO] A motile, helical bacterial pathogen propagated by ticks of the genus *Argas* that causes borreliosis in geese, ducks, turkeys, pheasants, chickens, and other birds. { bə,rel·ē·ə an'ser·ə·nə }

Borrelia burgdorferi [MICROBIO] A gram-negative, helically shaped bacterium that is the causative agent of Lyme disease. { bə,rēl·yə ˌbərg'dȯr·fə·rē }

bosporus [GEOGR] A strait connecting two seas or a lake and a sea. { 'bäs·pə·rəs }

bosque *See* temperate and cold scrub. { 'bäsk *or* 'bä·skā }

botanical garden [BOT] An institution for the culture of plants collected chiefly for scientific and educational purposes. { bə'tan·ə·kəl 'gär·dən }

botany [BIOL] A branch of the biological sciences which embraces the study of plants and plant life. { 'bät·ən·ē }

bottom [GEOL] The bed of a body of running or still water. *See* root. { 'bäd·əm }

bottom fauna *See* benthos. { 'bäd·əm ˌfȯn·ə }

bottom flow [HYD] A density current that is denser than any section of the surrounding water and that flows along the bottom of the body of water. Also known as underflow. { 'bäd·əm ˌflō }

bottom ice *See* anchor ice. { 'bäd·əm ˌīs }

bottomland [GEOL] A lowland formed by alluvial deposit about a lake basin or a stream. { 'bäd·əm,land }

bottom load *See* bed load. { 'bäd·əm ˌlōd }

bottom rot [PL PATH] **1.** A fungus disease of lettuce, caused by *Pellicularia filamentosa*, that spreads from the base upward. **2.** A fungus disease of tree trunks caused by pore fungi. { 'bäd·əm ˌrät }

bottom water |HYD| Water lying beneath oil or gas in productive formations. |OCEANOGR| The water mass at the deepest part of a water column in the ocean. { 'bäd·əm ˌwȯd·ər }

botulin |MICROBIO| The neurogenic toxin which is produced by *Clostridium botulinum* and C. *parabotulinum* and causes botulism. Also known as botulinus toxin. { 'bäch·ə·lən }

botulinus |MICROBIO| A bacterium that causes botulism. { 'bäch·ə'lī·nəs }

botulinus toxin *See* botulin. { 'bäch·ə'lī·nəs 'täk·sən }

botulism |MED| Food poisoning due to intoxication by the exotoxin of *Clostridium botulinum* and C. *parabotulinum*. { 'bäch·əˌliz·əm }

boturon |CHEM| $C_{12}H_{13}N_2OCl$ A white solid with a melting point of 145–146°C; used as pre- and postemergence herbicide in cereals, orchards, and vineyards. Also known as butyron. { 'bäch·əˌrän }

bough |BOT| A main branch on a tree. { baů }

boulder clay *See* till. { 'bōl·dər ˌklā }

boundary layer |METEOROL| The lower portion of the atmosphere, extending to a height of approximately 1.2 miles (2 kilometers). { 'baůn·drē ˌlā·ər }

bourne |HYD| A small intermittent stream in a dry valley. { bůrn }

box canyon |GEOGR| A canyon with steep rock sides and a zigzag course, that is usually closed upstream. { 'bäks ˌkan·yən }

brace root *See* prop root. { 'brās ˌrüt }

brachiate |BOT| Possessing widely divergent branches. |ZOO| Having arms. { 'bra·kē,āt }

brackish |HYD| **1.** Of water, having salinity values ranging from approximately 0.50 to 17.00 parts per thousand. **2.** Of water, having less salt than sea water, but undrinkable. { 'brak·ish }

bract |BOT| A modified leaf associated with plant reproductive structures. { brakt }

braided stream |HYD| A stream flowing in several channels that divide and reunite. { 'brād·əd ˌstrēm }

brain coral |ZOO| A reef-building coral resembling the human cerebrum in appearance. { 'brān ˌkär·əl }

branch |HYD| A small stream that merges into another, generally bigger, stream. { branch }

branching adaptation *See* divergent adaptation. { 'branch·iŋ ˌad,ap'tā·shən }

Branhamella |MICROBIO| A genus of bacteria in the family Neisseriaceae; cocci occur in pairs with flattened adjacent sides; parasites of mammalian mucous membranes. { ˌbran·ə'mel·ə }

Brazil Current |OCEANOGR| The warm ocean current that flows southward along the Brazilian coast below Natal; the western boundary current in the South Atlantic Ocean. { brə'zil ˌkər·ənt }

break |METEOROL| **1.** A sudden change in the weather; usually applied to the end of an extended period of unusually hot, cold, wet, or dry weather. **2.** A hole or gap in a layer of clouds. { brāk }

break-bone fever *See* Dengue fever. { 'brāk,bōn ˌfē·vər }

breaker |OCEANOGR| A wave breaking on a shore, over a reef, or other mass in a body of water. { 'brā·kər }

breaker depth

breaker depth [OCEANOGR] The still-water depth measured at the point where a wave breaks. Also known as breaking depth. { 'brā·kər ˌdepth }

breaking depth See breaker depth. { 'brāk·iŋ ˌdepth }

breaks in overcast [METEOROL] In United States weather observing practice, a condition wherein the cloud cover is more than 0.9 but less than 1.0. { ¦brāks in 'ō·vərˌkast }

breakup [HYD] The spring melting of snow, ice, and frozen ground; specifically, the destruction of the ice cover on rivers during the spring thaw. { 'brāk‚əp }

breathing apparatus [ENG] An appliance that enables a person to function in irrespirable or poisonous gases or fluids; contains a supply of oxygen and a regenerator which removes the carbon dioxide exhaled. { 'brēth·iŋ ap·ə'rad·əs }

breed [AGR] A group of animals that have a common origin and possess characteristics that are not common to other individuals of the same species. { brēd }

breeding [AGR] The application of genetic principles to the improvement of farm animals and of cultivated plants. { 'brēd·iŋ }

breeze [METEOROL] **1.** A light, gentle, moderate, fresh wind. **2.** In the Beaufort scale, a wind speed ranging from 4 to 31 miles (6.4 to 49.6 kilometers) per hour. { brēz }

brevitoxin [BIOL] One of several ichthyotoxins produced by the dinoflagellate *Ptychodiscus brevis*. { ‚brev·ə'täk·sən }

bridled pressure plate [METEOROL] An instrument for measuring air velocity in which the pressure on a plate exposed to the wind is balanced by the force of a spring, and the deflection of the plate is measured by an inductance-type transducer. { ¦brīd·əld 'presh·ər ˌplāt }

brine [OCEANOGR] Sea water containing a higher concentration of dissolved salt than that of the ordinary ocean. { brīn }

brine spring [HYD] A salt-water spring. { 'brīn ˌspriŋ }

broadleaf tree [BOT] Any deciduous or evergreen tree having broad, flat leaves. { 'bròd ˌlēf ˌtrē }

broad-spectrum antibiotic [MICROBIO] An antibiotic that is effective against both gram-negative and gram-positive bacterial species. { ¦bròd ¦spek·trəm ˌant·i·bī'äd·ik }

broken [METEOROL] Descriptive of a sky cover of from 0.6 to 0.9 (expressed to the nearest tenth). { 'brō·kən }

broken belt [OCEANOGR] The transition zone between open water and consolidated ice. { ¦brō·kən 'belt }

broken stream [HYD] A stream that repeatedly disappears and reappears, such as occurs in an arid region. { ¦brō·kən 'strēm }

broken water [OCEANOGR] Water having a surface covered with ripples or eddies, and usually surrounded by calm water. { ¦brō·kən 'wòd·ər }

bromadiolone [CHEM] $C_{30}H_{23}BrO_4$ A rodenticide. { ‚brō·mə'dī·əˌlōn }

bromate [CHEM] **1.** BrO_3^- A negative ion derived from bromic acid, $HBrO_3$ **2.** A salt of bromic acid. **3.** $C_9H_9ClO_3$ A light brown solid with a melting point of 118–119°C; used as a herbicide to control weeds in crops such as flax, cereals, and legumes. { 'brō ˌmāt }

bromethalin [CHEM] $C_{14}H_7Br_3F_3N_3O_4$ A rodenticide. { ‚brō·mə'thal·ən }

bromochloroprene [CHEM] $CHCl=CHCH_2Br$ A compound used as a nematicide and soil fumigant. { ‚brō·mō'klòr·əˌprēn }

bromocyclen [CHEM] $C_8H_5BrCl_6$ A compound used as an insecticide for wheat crops. { ˌbrō·mō'sī·klən }

***O*-(4-bromo-2,5-dichlorophenyl) *O*-methyl phenylphosphorothioate** *See* leptophos. { ¦ō 'fȯr ¦brō·mō 'tü ¦fīv dī¦klȯr·ō·'fen·əl ¦ō meth·əl fen·əl·fäs·fə'rō·thī·ō·āt }

bromofenoxim [CHEM] $C_{13}H_7N_3O_6Br_2$ A cream-colored powder with melting point 196–197°C; slightly soluble in water; used as herbicide to control weeds in cereal crops. { ¦brō·mō·fə'näk·səm }

bromophos [CHEM] $C_8H_8SPBrCl_2O_3$ A yellow, crystalline compound with a melting point of 54°C; used as an insecticide and miticide for livestock, household insects, flies, and lice. { 'brō·mə,fäs }

bromoxynil [CHEM] $C_7H_3OBr_2N$ A colorless solid with a melting point of 194–195°C; slightly soluble in water; used as a herbicide in wheat, barley, oats, rye, and seeded turf. { ˌbrō'mäk·sə·nil }

bronchial asthma [MED] Asthma usually due to hypersensitivity to an inhaled or ingested allergen. { 'bräŋ·kē·əl 'az·mə }

Brønsted base *See* base. { 'brən·steth ˌbās }

brood [ZOO] **1.** The young of animals. **2.** To incubate eggs or cover the young for warmth. **3.** An animal kept for breeding. { brüd }

brood parasitism [ECOL] A type of social parasitism among birds characterized by a bird of one species laying and abandoning its eggs in the nest of a bird of another species. { ¦brüd ˌpar·ə·sə,tiz·əm }

brown algae [BOT] The common name for members of the Phaeophyta. { ¦braün ¦al·jē }

brown blight [PL PATH] A virus disease of lettuce characterized by spots and streaks on the leaves, reduction in leaf size, and gradual browning of the foliage, beginning at the base. { 'braün ¦blīt }

brown blotch [PL PATH] **1.** A bacterial disease of mushrooms caused by *Pseudomonas tolaasi* and characterized by brown blotchy discolorations. **2.** A fungus disease of the pear characterized by brown blotches on the fruit. { 'braün ¦bläch }

brown coal *See* lignite. { ¦braün ¦kōl }

browning [PL PATH] Any plant disorder or disease marked by brown discoloration of a part. Also known as stem break. { 'braü·niŋ }

brown leaf rust [PL PATH] A fungus disease of rye caused by *Puccinia dispersa*. { ¦braün 'lēf ˌrəst }

brown root rot [PL PATH] **1.** A fungus disease of plants of the pea, cucumber, and potato families caused by *Thielavia basicola* and characterized by blackish discoloration and decay of the roots and stem base. **2.** A disease of tobacco and other plants comparable to the fungus disease but believed to be caused by nematodes. { ¦braün 'rüt ˌrät }

brown rot [PL PATH] Any fungus or bacterial plant disease characterized by browning and tissue decay. { ¦braün ¦rät }

brown seaweed [BOT] A common name for the larger algae of the division Phaeophyta. { ¦braün 'sē,wēd }

brown smoke [ENG] Smoke with less particulates than black smoke; comes from burning fossil fuel, usually fuel oil. { ¦braün ¦smōk }

brown snow [METEOROL] Snow intermixed with dust particles. { ¦braün ¦snō }

brown spot [PL PATH] Any fungus disease of plants, especially Indian corn, characterized by brown leaf spots. { 'braün ˌspät }

browse |BIOL| **1.** Twigs, shoots, and leaves eaten by livestock and other grazing animals. **2.** To feed on this vegetation. { braúz }

Brucella |MICROBIO| A genus of gram-negative, aerobic bacteria of uncertain affiliation; single, nonmotile coccobacilli or short rods, all of which are parasites and pathogens of mammals. { brü'sel·ə }

brucellosis *See* contagious abortion. { ‚brü·sə'lō·səs }

Brückner cycle |CLIMATOL| An alternation of relatively cool-damp and warm-dry periods, forming an apparent cycle of about 35 years. { 'brük·nər ‚sī·kəl }

brush *See* tropical scrub. { brəsh }

brush fire |FOR| A fire involving growth that is heavier than grass but less than full tree size. { 'brəsh ‚fīr }

Bryales |BOT| An order of the subclass Bryidae; consists of mosses which often grow in disturbed places. { brī'ā·lēz }

Bryidae |BOT| A subclass of the class Bryopsida; includes most genera of the true mosses. { 'brī·ə‚dē }

bryology |BOT| The study of bryophytes. { brī'äl·ə·jē }

Bryophyta |BOT| A small phylum of the plant kingdom, including mosses, liverworts, and hornworts, characterized by the lack of true roots, stems, and leaves. { brī'ä·fə·də }

Bryopsida |BOT| The mosses, a class of small green plants in the phylum Bryophyta. Also known as Musci. { brī'äp·sə·də }

bubonic plague *See* plague. { bü¦ban·ik 'plāg }

buckwheat |AGR| A herbaceous and erect annual belonging to the Polygonaceae family; its dry seed or grain is used as a source of food and animal feed. { 'bək‚wēt }

bud |BOT| An embryonic shoot containing the growing stem tip surrounded by young leaves or flowers or both and frequently enclosed by bud scales. { bəd }

budding |BIOL| A form of asexual reproduction in which a new individual arises as an outgrowth of an older individual. Also known as gemmation. |BOT| A method of vegetative propagation in which a single bud is grafted laterally onto a stock. |MICROBIO| A form of virus release from the cell in which replication has occurred, common to all enveloped animal viruses; the cell membrane closes around the virus and the particle exits from the cell. { 'bəd·iŋ }

budget year |METEOROL| The 1-year period beginning with the start of the accumulation season at the firn line of a glacier or ice cap and extending through the following summer's ablation season. { 'bəj·ət ‚yir }

bud rot |PL PATH| Any plant disease or symptom involving bud decay. { 'bəd ‚rät }

buffer |ECOL| An animal that is introduced to serve as food for other animals to reduce the losses of more desirable animals. { 'bəf·ər }

buildup |ECOL| A significant increase in a natural population, usually as a result of progressive changes in ecological relations. { 'bil‚dəp }

bulb |BOT| A short, subterranean stem with many overlapping fleshy leaf bases or scales, such as in the onion and tulip. { bəlb }

bulb glacier |HYD| A glacier formed at the foot of a mountain and out into an open slope; the glacier ends spread out into an ice fan. { 'bəlb ‚glā·shər }

bull's-eye rot |PL PATH| A fungus disease of apples caused by either *Neofabraea malicorticis* or *Gloeosporium perennans* and characterized by spots resembling eyes on the fruit. { 'bůlz ‚ī ‚rät }

bunt [PL PATH] A fungus disease of wheat caused by two *Tilletia* species and characterized by grain replacement with fishy-smelling smut spores. { bənt }

Buprestoidea [ZOO] A superfamily of coleopteran insects in the suborder Polyphaga including many serious pests of fruit trees. { ˌbyü·presˈtȯid·ē·ə }

burn off [METEOROL] With reference to fog or low stratus cloud layers, to dissipate by daytime heating from the sun. { ¦bərn ˈȯf }

burnt lime *See* calcium oxide. { ¦bərnt ˈlīm }

burr [BOT] **1.** A rough or prickly envelope on a fruit. **2.** A fruit so characterized. { bər }

burr ball *See* lake ball. { ˈbər ˌbȯl }

burst slug detector [ENG] A radiation detector used for detecting small leaks in a fuel element of a nuclear reactor by measuring the radiation from short-lived fission products that escape into the coolant. { ¦bərst ˈsləg diˈtek·tər }

butanol [CHEM] Any one of four isomeric alcohols having the formula C_4H_9OH; colorless, toxic liquids soluble in most organic liquids. Also known as butyl alcohol. { ˈbyüt·ənˌȯl }

butte [GEOGR] A detached hill or ridge which rises abruptly. { byüt }

butyl alcohol *See* butanol. { ¦byüd·əl ˈal·kəˌhȯl }

butylate [CHEM] $C_{11}H_{23}NOS$ A colorless liquid used as an herbicide for preplant control of weeds in corn. { ˈbyüd·əlˌāt }

N-*sec*-butyl-4-*tert*-butyl-2,6-dinitroaniline [CHEM] $C_{14}H_{21}N_3O_4$ Orange crystals with a melting point of 60–61°C; solubility in water is 1.0 part per million at 24°C; used as a preemergence herbicide. { ¦en ¦sek ˈbyüd·əl ¦fȯr ¦tərt ˈbyüd·əl ¦tü ¦siks ˌdī̇ˌnī-trōˈan·ə ˌlēn }

butyron *See* boturon. { byüˌtəˈrän }

Buys-Ballot's law [METEOROL] A law describing the relationship of the horizontal wind direction in the atmosphere to the pressure distribution: if one stands with one's back to the wind, the pressure to the left is lower than to the right in the Northern Hemisphere; in the Southern Hemisphere the relation is reversed. Also known as baric wind law. { ˈbīz bəˈläts ˌlȯ }

BW *See* biological warfare.

C

c *See* centi-.

caballing |OCEANOGR| The mixing of two water masses of identical in situ densities but different in situ temperatures and salinities, such that the resulting mixture is denser than its components and therefore sinks. { kə'bal·iŋ }

cabbage yellows |PL PATH| A fungus disease of cabbage caused by *Fusarium conglutinans* and characterized by yellowing and dwarfing. { 'kab·ij ˌyel·ōz }

cable *See* cable length. { 'kā·bəl }

cable length |OCEANOGR| A unit of distance, originally equal to the length of a ship's anchor cable, now variously considered to be 600 feet (183 meters), 608 feet (185.3 meters, one-tenth of a British nautical mile), or 720 feet or 120 fathoms (219.5 meters). Also known as cable. { 'kā·bəl ˌlengkth }

cacodylic acid |CHEM| $(CH_3)_2AsOOH$ Colorless crystals that melt at 200°C; soluble in alcohol and water; used as a herbicide. { ˌkak·əˌdil·ik 'as·əd }

cactus |BOT| The common name for any member of the family Cactaceae, a group characterized by a fleshy habit, spines and bristles, and large, brightly colored, solitary flowers. { 'kak·təs }

cadmium chlorate |CHEM| $CdClO_3$ White crystals, soluble in water; a highly toxic material. { 'kad·mē·əm 'klòr,āt }

calcareous |SCI TECH| Resembling, containing, or composed of calcium carbonate. { kal'ker·ē·əs }

calcareous algae |BOT| Algae that grow on limestone or in soil impregnated with lime. { kal'ker·ē·əs 'al·jē }

calcareous ooze |GEOL| A fine-grained pelagic sediment containing undissolved sand- or silt-sized calcareous skeletal remains of small marine organisms mixed with amorphous clay-sized material. { kal'ker·ē·əs 'üz }

calcareous soil |GEOL| A soil containing accumulations of calcium and magnesium carbonate. { kal'ker·ē·əs 'sòil }

calcic |SCI TECH| Derived from or containing calcium. { 'kal·sik }

calcicole |BOT| Requiring soil rich in calcium carbonate for optimum growth. { 'kal·sə ˌkōl }

calciferous |BIOL| Containing or producing calcium or calcium carbonate. { kal 'sif·ə·rəs }

calcification |GEOCHEM| Any process of soil formation in which the soil colloids are saturated to a high degree with exchangeable calcium, thus rendering them relatively immobile and nearly neutral in reaction. { ˌkal·sə·fə'kā·shən }

calcifuge |ECOL| A plant that grows in an acid medium that is poor in calcareous matter. { 'kal·sə,fyüj }

calcium [CHEM] A chemical element, symbol Ca, atomic number 20, atomic weight 40.08; used in metallurgy as an alloying agent for aluminum-bearing metal, as an aid in removing bismuth from lead, and as a deoxidizer in steel manufacture, and also used as a cathode coating in some types of photo tubes. { 'kal·sē·əm }

calcium arsenate [CHEM] $Ca_3(AsO_4)_2$ An arsenic compound used as an insecticide to control cotton pests. { 'kal·sē·əm 'ärs·ən‚āt }

calcium arsenite [CHEM] $Ca_3(AsO_3)_2$ White granules that are soluble in water; used as an insecticide. { 'kal·sē·əm 'ärs·ən‚īt }

calcium cyanide [CHEM] $Ca(CN)_2$ In pure form, a white powder that gives off hydrogen cyanide in air at normal humidity; prepared commercially in impure black or gray flakes; used as an insecticide and rodenticide. Also known as black cyanide. { 'kal·sē·əm 'sī·ə‚nīd }

calcium hypochlorite [CHEM] $Ca(OCl)_2·4H_2O$ A white powder, used as a bleaching agent and disinfectant for swimming pools. { 'kal·sē·əm hī·pō'klȯr‚īt }

calcium orthoarsenate [CHEM] $Ca_3(AsO_4)_2$ A white powder, insoluble in water; used as a preemergence insecticide and herbicide for turf. { 'kal·se·əm ¦ȯr·thō'ärs·ən‚āt }

calcium oxide [CHEM] CaO A caustic white solid sparingly soluble in water; the commercial form is prepared by roasting calcium carbonate limestone in kilns until all the carbon dioxide is driven off; used as a refractory, in pulp and paper manufacture, and as a flux in manufacture of steel. Also known as burnt lime; calx; caustic lime. { 'kal·se·əm 'äk‚sīd }

calf See calved ice. { kaf }

California Current [OCEANOGR] The ocean current flowing southward along the western coast of the United States to northern Baja California. { ¦kal·ə¦fȯr·nyə 'kər·ənt }

calling song [ZOO] A high-intensity insect sound which may play a role in habitat selection among certain species. { 'kȯl·iŋ ‚sȯŋ }

callus [BOT] A hard tissue that forms over a damaged plant surface. { 'kal·əs }

calm [METEOROL] The absence of apparent motion of the air; in the Beaufort wind scale, smoke is observed to rise vertically, or the surface of the sea is smooth and mirrorlike; in U.S. weather observing practice, the wind has a speed under 1 mile per hour or 1 knot (1.6 kilometers per hour). { käm }

calm belt [METEOROL] A belt of latitude in which the winds are generally light and variable; the principal calm belts are the horse latitudes (the calms of Cancer and of Capricorn) and the doldrums. { 'käm ‚belt }

calms of Cancer [METEOROL] One of the two light, variable winds and calms which occur in the centers of the subtropical high-pressure belts over the oceans; their usual position is about latitude 30°N, the horse latitudes. { ¦kämz əv 'kan·sər }

calms of Capricorn [METEOROL] One of the two light, variable winds and calms which occur in the centers of the subtropical high-pressure belts over the oceans; their usual position is about latitude 30°S, the horse latitudes. { ¦kämz əv 'kap·ri‚kȯrn }

calved ice [OCEANOGR] A piece of ice floating in a body of water after breaking off from a mass of land ice or an iceberg. Also known as calf. { ¦kavd 'īs }

Calvin-Benson cycle See Calvin cycle. { ¦kal·vən 'ben·sən ‚sī·kəl }

Calvin cycle [BIOL] A metabolic process during photosynthesis that uses light indirectly to convert carbon dioxide to sugar in the stroma of chloroplasts. Also known as Calvin-Benson cycle; carbon fixation cycle. { 'kal·vən ‚sī·kəl }

calx See calcium oxide. { kalks }

calyx [BOT] The outermost whorl of a flower; composed of sepals. { 'kā‚liks }

camanchaca *See* garúa. { kä·män'chä·kə }

cambium [BOT] A layer of cells between the phloem and xylem of most vascular plants that is responsible for secondary growth and for generating new cells. { 'kam·bē·əm }

camouflage [ECOL] An organism's use of color, form, or behavior to blend into its surroundings and thus go undetected by predators. { 'kam·ə,fläzh }

camphene [CHEM] $C_{10}H_{16}$ A bicyclic terpene used as raw material in the synthesis of insecticides such as toxaphene and camphor. { 'kam,fēn }

campos [ECOL] The savanna of South America. { 'käm,pōs }

Campylobacter jejune [MICROBIO] A microaerophilic pathogen associated with raw meats and unpasteurized milk; ingestion of a small amount can cause diarrhea, cramps, and nausea. { kam¦pī·lə,bak·tər jə'jü·nē }

Canadian life zone [ECOL] The zone comprising the climate and biotic communities of the portion of the Boreal life zone exclusive of the Hudsonian and Arctic-Alpine zones. { kə'nād·ē·ən 'līf ,zōn }

canal [CIV ENG] An artificial open waterway used for transportation, waterpower, or irrigation. [ENG] A water-filled trench or conduit associated with a nuclear reactor, used for removing and sometimes storing radioactive objects taken from the reactor; the water acts as a shield against radiation. [GEOGR] A long, narrow arm of the sea extending far inland, between islands or between islands and the mainland. { kə'nal }

canaliculate [BIOL] Having small channels, canals, or grooves. { ,kan·əl'ik·yə,lāt }

Canary Current [OCEANOGR] The prevailing southward flow of water along the northwestern coast of Africa. { kə'ner·ē ,kər·ənt }

canary-pox virus [MICROBIO] An avian poxvirus that causes canary pox, a disease closely related to fowl pox. { kə'ner·ē ,päks ,vī·rəs }

cancer [MED] A group of diseases characterized by the uncontrolled growth of abnormal cells. { 'kan·sər }

Candida [MYCOL] A genus of yeastlike, pathogenic imperfect fungi that produce very small mycelia. { 'kan·də·də }

candidiasis [MED] A fungus infection of the skin, lungs, mucous membranes, and viscera of humans caused by a species of *Candida*, usually C. *albicans*. Also known as moniliasis. { ,kan·də'dī·ə·səs }

cane blight [PL PATH] A fungus disease affecting the canes of several bush fruits, such as currants and raspberries; caused by several species of fungi. { 'kān ,blīt }

canker [PL PATH] An area of necrosis on a woody stem resulting in shrinkage and cracking followed by the formation of callus, ultimately killing the stem. { 'kaŋ·kər }

cankerworm [ZOO] Any of several lepidopteran insect larvae in the family Geometridae which cause severe plant damage by feeding on buds and foliage. { 'kaŋ·kər,wərm }

canopy [FOR] The uppermost branching and spreading layer of a forest. { 'kan·ə·pē }

canthariasis [MED] Infection or disease caused by coleopteran insects or their larvae. { kan·thə'rī·ə·səs }

canyon [GEOGR] A chasm, gorge, or ravine cut in the surface of the earth by running water; the sides are steep and form cliffs. { 'kan·yən }

canyon wind [METEOROL] **1.** The mountain wind of a canyon; that is, the nighttime down-canyon flow of air caused by cooling at the canyon walls. **2.** Any wind modified by being forced to flow through a canyon or gorge; its speed may be increased as a jet-effect wind, and its direction is rigidly controlled. Also known as gorge wind. { 'kan·yən ¦wind }

capacity of the wind   [GEOL] The total weight of airborne particles (soil and rock) of given size, shape, and specific gravity, which can be carried in 1 cubic mile (4.17 cubic kilometers) of wind blowing at a given speed.   { kə'pas·əd·ē əv thə ¦wind }

cap cloud   [METEOROL] An approximately stationary cloud, or standing cloud, on or hovering above an isolated mountain peak; formed by the cooling and condensation of humid air forced up over the peak. Also known as cloud cap.   { 'kap ¦klaúd }

cape   [GEOGR] A prominent point of land jutting into a body of water. Also known as head; headland; mull; naze; ness; point; promontory.   { kāp }

cape doctor   [METEOROL] The strong southeast wind which blows on the South African coast.   { 'kāp ¦däk·tər }

Cape Horn Current   [OCEANOGR] That part of the west wind drift flowing eastward in the immediate vicinity of Cape Horn, and then curving northeastward to continue as the Falkland Current.   { ¦kāp ¦hórn 'kər·ənt }

capillary   [GEOL] A fissure or a crack in a formation which provides a route for flow of water or hydrocarbons.   { 'kap·ə¦ler·ē }

capillary fringe   [HYD] The lower subdivision of the zone of aeration that overlies the zone of saturation and in which the pressure of water in the interstices is lower than atmospheric.   { 'kap·ə¦ler·ē ¦frinj }

capillary migration   [HYD] Movement of water produced by the force of molecular attraction between rock material and the water.   { 'kap·ə¦ler·ē mī'grā·shən }

capillary ripple   *See* capillary wave.   { 'kap·ə¦ler·ē ¦rip·əl }

capillary water   [HYD] Soil water held by capillarity as a continuous film around soil particles and in interstices between particles above the phreatic line.   { 'kap·ə¦ler·ē ¦wód·ər }

capillary wave   [PHYS] A water wave of less than 1.7 centimeters. Also known as capillary ripple; ripple.   { 'kap·ə¦ler·ē ¦wāv }

capitate   [BIOL] Enlarged and swollen at the tip.   [BOT] Forming a head, as certain flowers of the Compositae.   { 'kap·ə¦tāt }

capped column   [HYD] A form of ice crystal consisting of a hexagonal column with plate or stellar crystals (so-called caps) at its ends and sometimes at intermediate positions; the caps are perpendicular to the column.   { 'kapt 'käl·əm }

capsaicin   [CHEM] $C_{18}H_{27}O_3N$ A toxic material extracted from the capsicum fruit.   { kap'sā·ə·sən }

capsid   [MICROBIO] In a virus, the protein shell surrounding the nucleic acid and its associated protein core. Also known as protein coat.   { 'kap·səd }

capsule   [BIOL] A membranous structure enclosing a body part or organ.   [BOT] A closed structure bearing seeds or spores; it is dehiscent at maturity.   [MED] A soluble shell in which drugs are enclosed for oral administration.   [MICROBIO] A thick, mucous envelope, composed of polypeptide or carbohydrate, surrounding certain microorganisms.   { 'kap·səl }

captan   [CHEM] $C_9H_8O_2NSCl_3$ A buff to white solid with a melting point of 175°C; used as a fungicide for diseases of fruits, vegetables, and flowers.   { 'kap¦tan }

capture   [HYD] The natural diversion of the headwaters of one stream into the channel of another stream having greater erosional activity and flowing at a lower level. Also known as piracy; river capture; river piracy; robbery; stream capture; stream piracy; stream robbery.   { 'kap·chər }

Carabidae   [ZOO] The ground beetles, a family of predatory coleopteran insects in the suborder Adephaga.   { kə'rab·ə¦dē }

carbamide   See urea.   { 'kär·bə,mīd }

carbaryl   [CHEM] $C_{12}H_{11}NO_2$ A colorless, crystalline compound with a melting point of 142°C; used as an insecticide for crops, forests, lawns, poultry, and pets.   { 'kär·bə ,ril }

carbide nuclear fuel   [ENG] A nuclear reactor fuel which is mixed with carbon compounds and a metal to give structural strength and oxidation resistance.   { 'kär ,bīd ¦nü·klē·ər 'fyül }

carbofuran   [CHEM] $C_{12}H_{15}NO_3$ A white solid with a melting point of 150–152°C; soluble in water; used as an insecticide, miticide, and nematicide in many crops.   { ,kär·bō 'fyúr,än }

carbohydrate   [BIOL] Any of the group of organic compounds composed of carbon, hydrogen, and oxygen, including sugars, starches, and celluloses.   { ,kär·bō'hī,drāt }

carbolic acid   See phenol.   { kär'bäl·ik 'as·əd }

carbonaceous   [SCI TECH] Relating to or composed of carbon.   { kär·bə'nā·shəs }

carbonate   [CHEM] **1.** An ester or salt of carbonic acid.   **2.** A compound containing the carbonate ($CO_3{}^{2-}$) ion.   **3.** Containing carbonates.   { 'kär·bə·nət }

carbonate cycle   [GEOCHEM] The cycling of carbon, as calcium carbonate, between organisms and the surface of the Earth.   { 'kär·bə·nət ,sī·kəl }

carbonate reservoir   [GEOL] An underground oil or gas trap formed in reefs, clastic limestones, chemical limestones, or dolomite.   { 'kär·bə·nət 'rez·əv,wär }

carbonate spring   [HYD] A type of spring containing dissolved carbon dioxide gas.   { 'kär·bə·nət 'spriŋ }

carbonation   [GEOCHEM] A process of chemical weathering whereby minerals that contain soda, lime, potash, or basic oxides are changed to carbonates by the carbonic acid in air or water.   { ,kär·bə'nā·shən }

carbon black   [CHEM] **1.** An amorphous form of carbon produced commercially by thermal or oxidative decomposition of hydrocarbons and used principally in rubber goods, pigments, and printer's ink.   **2.** See gas black.   { 'kär·bon ¦blak }

carbon cycle   [GEOCHEM] The cycle of carbon in the biosphere, in which plants convert carbon dioxide to organic compounds that are consumed by plants and animals, and the carbon is returned to the biosphere in the form of inorganic compounds by processes of respiration and decay.   { 'kär·bon ,sī·kəl }

carbon fixation   [BIOL] During photosynthesis, the process by which plants convert carbon dioxide from the air into organic molecules.   { 'kär·bon fik¦sā·shən }

carbon fixation cycle   See Calvin cycle.   { 'kär·bon fik'sā·shən ,sī·kəl }

carbon-14   [PHYS] A naturally occurring radioisotope of carbon having a mass number of 14 and half-life of 5780 years; used in radiocarbon dating and in the elucidation of the metabolic path of carbon in photosynthesis. Also known as radiocarbon.   { 'kär·bon 'fór,tēn }

carbon-14 dating   [ENG] Determining the approximate age of organic material associated with archeological or fossil artifacts by measuring the rate of radiation of the carbon-14 isotope. Also known as radioactive carbon dating; radiocarbon dating.   { ¦kär·bon ¦fór,tēn 'dād·iŋ }

carbonic acid   [CHEM] H_2CO_3 The acid formed by combination of carbon dioxide and water.   { kär'bän·ik 'as·əd }

carbonification   See coalification.   { kär,bän·ə·fə'kā·shən }

carbon isotope ratio   [GEOL] Ratio of carbon-12 to either of the less common isotopes, carbon-13 or carbon-14, or the reciprocal of one of these ratios; if not specified, the ratio refers to carbon-12/carbon-13. Also known as carbon ratio. { ¦kär·bən 'is·ə,tōp ,rā·shō }

carbonization [GEOCHEM] **1.** In the coalification process, the accumulation of residual carbon by changes in organic material and their decomposition products. **2.** Deposition of a thin film of carbon by slow decay of organic matter underwater. **3.** A process of converting a carbonaceous material to carbon by removal of other components. { ,kär·bə·nə'zā·shən }

carbon monoxide [CHEM] CO A colorless, odorless gas resulting from the incomplete oxidation of carbon; found, for example, in mines and automobile exhaust; poisonous to animals. { ¦kär·bən mə'näk,sīd }

carbon-nitrogen-phosphorus ratio [OCEANOGR] The relatively constant relationship between the concentrations of carbon, nitrogen, and phosphorus in plankton, and nitrogen and phosphorus in sea water, owing to removal of the elements by the organisms in the same proportions in which the elements occur and their return upon decomposition of the dead organisms. { ¦kär·bən ¦nī·trə·jən ¦fäs·fə·rəs ,rā·shō }

carbon number [CHEM] The number of carbon atoms in a material under analysis; plotted against chromatographic retention volume for compound identification. { 'kär·bən ,nəm·bər }

carbon ratio [GEOL] **1.** The ratio of fixed carbon to fixed carbon plus volatile hydrocarbons in a coal. **2.** *See* carbon isotope ratio. { 'kär·bən ,rā·shō }

carbon-12 [PHYS] A stable isotope of carbon with mass number of 12, forming about 98.9% of natural carbon; used as the basis of the newer scale of atomic masses, having an atomic mass of exactly 12u (relative nuclidic mass unit) by definition. { 'kär·bən 'twelv }

carbon-13 [PHYS] A heavy isotope of carbon having a mass number of 13. { 'kär·bən 'thər,tēn }

carbophenothion [CHEM] $C_{11}H_{16}ClO_2PS_3$ An amber liquid used to control pests on fruits, nuts, vegetables, and fiber crops. { ¦kär·bō¦fēn·ō'thī,än }

carcinogen [MED] Any agent that incites development of a carcinoma or any other sort of malignancy. { kär'sin·ə·jən }

carcinoma [MED] A malignant epithelial tumor. { ,kärs·ən'ō·mə }

cardinal winds [METEOROL] Winds from the four cardinal points of the compass, that is, north, east, south, and west winds. { 'kärd·nəl ,winz }

cardiovascular toxicity [MED] The adverse effects on the heart or blood systems which result from exposure to toxic chemicals. { ¦kärd·ē·ō¦vas·kyə·lər tak'sis·əd·ē }

Cardiovirus [MICROBIO] A genus of viruses of the family Picornaviridae; consists of strains of encephalomyocarditis virus and mouse encephalomyelitis. { 'kär·dē·ō ,vī·rəs }

Caribbean Current [OCEANOGR] A water current flowing westward through the Caribbean Sea. { kar·ə'bē·ən 'kər·ənt }

Caribbean Sea [GEOGR] One of the largest and deepest enclosed basins in the world, surrounded by Central and South America and the West Indian island chains. { kar·ə'bē·ən 'sē }

carnivore *See* secondary consumer. { 'kär·nə,vȯr }

carnivorous [BIOL] Eating flesh or, as in plants, subsisting on nutrients obtained from the breakdown of animal tissue. { kär'niv·ə·rəs }

carnivorous plant *See* insectivorous plant. { kär'niv·ə·rəs 'plant }

Carolina Bays [GEOGR] Shallow, marshy, often ovate depressions on the coastal plain of the mideastern and southeastern United States of unknown origin. { ˌkar·ə'lī·nə 'bāz }

Carolinian life zone [ECOL] A zone comprising the climate and biotic communities of the oak savannas of eastern North America. { ¦kar·ə¦lin·ē·ən 'līf ˌzōn }

carotene [BIOL] $C_{40}H_{56}$ Any of several red, crystalline, carotenoid hydrocarbon pigments occurring widely in nature, convertible in the animal body to vitamin A, and characterized by preferential solubility in petroleum ether. Also known as carotin. { 'kar·əˌtēn }

carotin See carotene. { 'kar·ə,tin }

carpel [BOT] The basic specialized leaf of the female reproductive structure in angiosperms; a megasporophyll. { 'kär·pəl }

carpology [BOT] The study of the morphology of fruit and seeds. { kär'päl·ə·jē }

carrageen [BOT] *Chondrus crispus.* A cartilaginous red algae harvested in the northern Atlantic as a source of carrageenan. Also known as Irish moss; pearl moss. { 'kar·ə ˌgēn }

carrageenan [CHEM] A polysaccharide derived from the red seaweed (Rhodophyceae) and used chiefly as an emulsifying, gelling, and stabilizing agent and as a viscosity builder in foods, cosmetics, and pharmaceuticals. Also spelled carrageenin. { ˌkar·ə'gē·nən }

carrageenin See carrageenan. { ˌkar·ə'gē·nən }

carrier [CHEM] A substance that, when associated with a trace of another substance, will carry the trace with it through a chemical or physical process. [MED] A person who harbors and eliminates an infectious agent and so transmits it to others, but who may not show signs of the disease. { 'kar·ē·ər }

carrion [ECOL] Dead, decaying animal flesh used as a source of food by scavengers. { 'kär·ē·ən }

carrying capacity [ECOL] The maximum population size that the environment can support without deterioration. { 'kar·ē·iŋ kə'pas·əd·ē }

carry-over [HYD] The portion of the stream flow during any month or year derived from precipitation in previous months or years. { 'kar·ē ˌō·vər }

caryopsis [BOT] A small, dry, indehiscent fruit having a single seed with such a thin, closely adherent pericarp that a single body, a grain, is formed. { ˌkar·ē'äp·səs }

cascade [HYD] A small waterfall or series of falls descending over rocks. { ka'skād }

cascading glacier [HYD] A glacier broken by numerous crevasses because of passing over a steep irregular bed, giving the appearance of a cascading stream. { ka'skād·iŋ 'glā·shər }

cashew gum [BOT] A gum obtained from the bark of the cashew tree; hard, yellowish-brown substance used for inks, insecticides, pharmaceuticals, varnishes, and bookbinders' gum. Also known as anacardium gum. { 'kash·ü ˌgəm }

cask See coffin. { kask }

casket See coffin. { 'kas·kət }

caste [ZOO] One of the levels of mature social insects in a colony that carry out a specific function; examples are workers and soldiers. { kast }

casual carrier [MED] A person who carries an infectious microorganism but never manifests the disease. { 'kazh·ə·wəl 'kar·ē·ər }

catalytic converter [CHEM ENG] A device that is fitted to the exhaust system of an automotive vehicle and contains a catalyst capable of converting potentially polluting exhaust gases into harmless or less harmful products. { ¦kad·ə¦lid·ik kən'vərd·ər }

cataract [HYD] A waterfall of considerable volume with the vertical fall concentrated in one sheer drop. { 'kad·ə,rakt }

catarobic [ECOL] Pertaining to a body of water characterized by the slow decomposition of organic matter, and oxygen utilization which is insufficient to prevent the activity of aerobic organisms. { ¦kad·ə¦rō·bik }

catarrhal jaundice See infectious hepatitis. { kə'tär·əl 'jȯn·dəs }

catch basin [CIV ENG] **1.** A basin at the point where a street gutter empties into a sewer, built to catch matter that would not easily pass through the sewer. **2.** A well or reservoir into which surface water may drain off. { 'kach ,bā·sən }

catch crop [AGR] A rapidly growing plant that can be intercropped between rows of the main crop; often used as a green manure. { 'kach ,kräp }

catchment area See drainage basin. { 'kach·mənt ,er·ē·ə }

catchment glacier See snowdrift glacier. { 'kach·mənt ,glā·shər }

caterpillar fungus See Cordyceps sinensis. { 'kat·ər,pil·ər ,fəŋ·gəs }

cat's paw [METEOROL] A puff of wind; a light breeze affecting a small area, as one that causes patches of ripples on the surface of water. { 'kats ,pȯ }

caudex [BOT] The main axis of a plant, including stem and roots. { 'kȯ,deks }

caulescent [BOT] Having an aboveground stem. { kȯ'les·ənt }

cauliflower disease [PL PATH] **1.** A disease of the strawberry plant caused by the eelworm and manifested as clustered, puckered, and malformed leaves. **2.** A bacterial disease of the strawberry and some other plants caused by *Corynebacterium fascians*. { 'kȯl·ə,flaü·ər di,zēz }

caulocarpic [BOT] Having stems that bear flowers and fruit every year. { ¦kȯl·ō¦kär·pik }

caustic barley See sabadilla. { 'kȯ·stik 'bär·lē }

caustic lime See calcium oxide. { 'kȯ·stik'līm }

cave [GEOL] A natural, hollow chamber or series of chambers and galleries beneath the earth's surface, or in the side of a mountain or hill, with an opening to the surface. { kāv }

cavern [GEOL] An underground chamber or series of chambers of indefinite extent carved out by rock springs in limestone. { 'kav·ərn }

cavernicolous [BIOL] Inhabiting caverns. { ¦kav·ər¦nik·ə·ləs }

CA virus See croup-associated virus. { sē'ā ,vī·rəs }

cay [GEOL] **1.** A flat coral island. **2.** A flat mound of sand built up on a reef slightly above high tide. **3.** A small, low coastal islet or emergent reef composed largely of sand or coral. { kā }

cay sandstone [GEOL] Firmly cemented or friable coral sand formed near the base of coral reef cays. { ¦kā'san,stōn }

Cc See cirrocumulus cloud.

cecidium [PL PATH] Plant gall produced either by insects in ovipositing or by fungi as a consequence of infection. { sə'sid·ē·əm }

ceiling [METEOROL] In the United States, the height ascribed to the lowest layer of clouds or of obscuring phenomena when it is reported as broken, overcast, or obscuration and not classified as thin or partial. { 'sē·liŋ }

ceiling classification [METEOROL] In aviation weather observations, a description or explanation of the manner in which the height of the ceiling is determined. { 'sē·liŋ ˌklas·ə·fə'kā·shən }

cell [BIOL] The typically microscopic functional and structural unit of all living organisms, consisting of a nucleus, cytoplasm, and a limiting membrane. { sel }

cellular [BIOL] Characterized by, consisting of, or pertaining to cells. { 'sel·yə·lər }

cellular convection [METEOROL] An organized, convective, fluid motion characterized by the presence of distinct convection cells or convective units, usually with upward motion (away from the heat source) in the central portions of the cell, and sinking or downward flow in the cell's outer regions. { 'sel·yə·lər kən'vek·shən }

cement [ZOO] Any of the various adhesive secretions, produced by certain invertebrates, that harden on exposure to air or water and are used to bind objects. { si'ment }

census [STAT] A complete counting of a population, as opposed to a partial counting or sampling. { 'sen·səs }

center jump [METEOROL] The formation of a second low-pressure center within an already well-developed low-pressure center; the latter diminishes in magnitude as the center of activity shifts or appears to jump to the new center. { 'sen·tər ¦jəmp }

center of action [METEOROL] A semipermanent high or low atmospheric pressure system at the surface of the earth; fluctuations in the intensity, position, orientation, shape, or size of such a center are associated with widespread weather changes. { 'sen·tər əv 'ak·shən }

centi- [SCI TECH] A prefix representing 10^{-2}, which is 0.01 or one-hundredth. Abbreviated c. { 'sen·tē *or* 'sent·ə }

central pressure [METEOROL] At any given instant, the atmospheric pressure at the center of a high or low; the highest pressure in a high, the lowest pressure in a low. { 'sen·trəl 'presh·ər }

central water [OCEANOGR] Upper water mass associated with the central region of oceanic gyre. { 'sen·trəl 'wȯd·ər }

cephalosporin [MICROBIO] Any of a group of antibiotics produced by strains of the imperfect fungus *Cephalosporium*. { ˌsef·ə·lə'spȯr·ən }

cereal [BOT] Any member of the grass family (Graminae) which produces edible, starchy grains usable as food by humans and livestock. Also known as grain. { 'sir·ē·əl }

cerium [CHEM] A chemical element, symbol Ce, atomic number 58, atomic weight 140.12; a rare-earth metal, used as a getter in the metal industry, as an opacifier and polisher in the glass industry, in carbon-arc lighting, and as a liquid-liquid extraction agent to remove fission products from spent uranium fuel. { 'sir·ē·əm }

cesium-134 [PHYS] An isotope of cesium, atomic mass number of 134; emits negative beta particles and has a half-life of 2.06 years. { 'sē·zē·əm ˌwən,thərd·ē'fȯr }

cesium-137 [PHYS] An isotope of cesium with atomic mass number of 137; emits negative beta particles and has a half-life of 30 years; offers promise as an encapsulated radiation source for therapeutic and other purposes. Also known as radiocesium. { 'sē·zē·əm ˌwən,thərd·ē'sev·ən }

cesspit *See* cesspool. { 'ses,pit }

cesspool [CIV ENG] An underground tank for raw sewage collection; used where there is no sewage system. Also known as cesspit. { 'ses,pül }

cevedilla *See* sabadilla. { ˌsev·ə'dil·ə }

CFC *See* chlorofluorocarbon.

chain |GEOL| A series of interconnected or related natural features, such as lakes, islands, or seamounts, arranged in a longitudinal sequence. { chān }

Chalcidoidea |ZOO| A superfamily of hymenopteran insects in the suborder Apocrita, including primarily insect parasites. { ˌkal·səˈdȯidˌē·ə }

chamaephyte |ECOL| Any perennial plant whose winter buds are within 10 inches (25 centimeters) of the soil surface. { ˈkam·əˌfīt }

Chandler wobble |GEOPHYS| A movement in the earth's axis of rotation, the period of motion being about 14 months. Also known as Eulerian nutation. { ˈchand·lər ˌwäb·əl }

change chart |METEOROL| A chart indicating the amount and direction of change of some meteorological element during a specified time interval; for example, a height-change chart or pressure-change chart. Also known as tendency chart. { ˈchānj ˌchärt }

change of tide |OCEANOGR| A reversal of the direction of motion (rising or falling) of a tide, or in the set of a tidal current. Also known as turn of the tide. { ˈchānj əv ˌtīd }

channel |HYD| The deeper portion of a waterway carrying the main current. { ˈchan·əl }

channel black See gas black. { ˈchan·əl ˌblak }

channel control |HYD| A condition whereby the stage of a stream is controlled only by discharge and the general configuration of the stream channel, that is, the contours of its bed, banks, and floodplains. { ˈchan·əl kənˈtrōl }

channel fill |GEOL| Accumulations of sand and detritus in a stream channel where the transporting capacity of the water is insufficient to remove the material as rapidly as it is delivered. { ˈchan·əl ˌfil }

channel morphology See river morphology. { ˈchan·əl ˌmȯrˈfäl·ə·jē }

channel net |HYD| Stream channel pattern within a drainage basin. { ˈchan·əl ˌnet }

channel order See stream order. { ˈchan·əl ˌȯrd·ər }

channel pattern |HYD| The configuration of a limited reach of a river channel as seen in plan view from an airplane. { ˈchan·əl ˌpad·ərn }

channel segment See stream segment. { ˈchan·əl ˌseg·mənt }

chaparral |ECOL| A vegetation formation characterized by woody plants of low stature, impenetrable because of tough, rigid, interlacing branches, which have simple, waxy, evergreen, thick leaves. { ˈshap·əˌral }

char See charcoal. { ˈchär }

character convergence |ECOL| An evolutionary process whereby two species interact so that one converges toward the other with respect to one or more traits. { ˈkar·ik·tər kənˌvər·jəns }

character displacement |ECOL| An evolutionary outcome of competition in which two species living in the same area evolve differences in morphology or other characteristics that lessen competition for food resources. { ˈkar·ik·tər disˈplās·mənt }

character progression |ECOL| The geographic gradation of expression of specific characters over the range of distribution of a race or species. { ˈkar·ik·tər prəˌgresh·ən }

charcoal |BOT| A porous solid product containing 85–98% carbon and produced by heating carbonaceous materials such as cellulose, wood, or peat at 500–600°C in the absence of air. Also known as char. |PETR MIN| **1.** The residue obtained from the carbonization of a noncoking coal, such as subbituminous coal, lignite, or anthracite. **2.** See low-temperature coke. { ˈchärˌkōl }

charcoal rot |PL PATH| A fungus disease of potato, corn, and other plants caused by *Macrophomina phaseoli*; tissues of the root and lower stem are destroyed and blackened. { ˈchärˌkōl ˌrät }

Charophyta |BOT| A group of aquatic plants, ranging in size from a few inches to several feet in height, that live entirely submerged in water. { kə'räf·əd·ə }

chart datum *See* datum plane. { 'chärt ,dad·əm }

charted depth |OCEANOGR| The vertical distance from a tidal datum to the ocean bottom. { 'char·təd 'depth }

chasmophyte |ECOL| A plant that grows in rock crevices. { 'kaz·mə,fīt }

chelerythrine |CHEM| $C_{21}H_{17}O_4H$ A poisonous, crystalline alkaloid, slightly soluble in alcohol; it is derived from the seeds of the herb celandine (*Chelidonium majus*) and has narcotic properties. { ,kel·ə'rī,thrēn }

chemical-cartridge respirator |PETR MIN| An air purification device worn by miners that removes small quantities of toxic gases or vapors from the inspired air; the cartridge contains chemicals which operate by processes of oxidation, absorption, or chemical reaction. { 'kem·i·kəl ,kär·trij 'res·pə,rād·ər }

chemical denudation |GEOL| Wasting of the land surface by water transport of soluble materials into the sea. { 'kem·i·kəl ,dē·nü'dā·shən }

chemical dosimeter |ENG| A dosimeter in which the accumulated radiation-exposure dose is indicated by color changes accompanying chemical reactions induced by the radiation. { 'kem·i·kəl dō'sim·əd·ər }

chemical ecology |ECOL| The study of ecological interactions mediated by the chemicals that organisms produce. { ¦kem·i·kəl ē'käl·ə·jē }

chemical element *See* element. { 'kem·i·kəl 'el·ə·mənt }

chemical fossils *See* biomarkers. { ¦kem·i·kəl 'fäs·əlz }

chemical operations *See* chemical warfare. { 'kem·i·kəl ,äp·ə'rā·shənz }

chemical precipitates |GEOL| A sediment formed from precipitated materials as distinguished from detrital particles that have been transported and deposited. { 'kem·i·kəl pri'sip·ə,tāts }

chemical reservoir |GEOL| An underground oil or gas trap formed in limestones or dolomites deposited in quiescent geologic environments. { 'kem·i·kəl 'rez·əv,wär }

chemical symbol |CHEM| A notation for one of the chemical elements, consisting of letters; for example Ne, O, C, and Na represent neon, oxygen, carbon, and sodium. { 'kem·i·kəl 'sim·bəl }

chemical warfare |ENG| The employment of chemical compounds to produce casualties or destroy crops. Also known as chemical operations. { 'kem·i·kəl 'wȯr,fer }

chemical weathering |GEOCHEM| A weathering process whereby rocks and minerals are transformed into new, fairly stable chemical combinations by such chemical reactions as hydrolysis, oxidation, ion exchange, and solution. Also known as decay; decomposition. { 'kem·i·kəl 'weth·ə·riŋ }

chemistry |SCI TECH| The scientific study of the properties, composition, and structure of matter, the changes in structure and composition of matter, and accompanying energy changes. { 'kem·ə·strē }

chemoautotroph |MICROBIO| Any of a number of autotrophic bacteria and protozoans which do not carry out photosynthesis. { ,kē·mō,ȯd·ə'träf·ik }

chemocline |HYD| The transition in a meromictic lake between the mixolimnion layer (at the top) and the monimolimnion layer (at the bottom). { 'kē·mə,klīn }

chemoheterotroph |BIOL| An organism that derives energy and carbon from the oxidation of preformed organic compounds. { ¦kē·mō'hed·ə·rə,träf }

chemoorganotroph [BIOL] An organism that requires an organic source of carbon and metabolic energy. { ¦kē·mō‚ȯr′gan·ə‚träf }

chemosphere [METEOROL] The vaguely defined region of the upper atmosphere in which photochemical reactions take place; generally considered to include the stratosphere (or the top thereof) and the mesosphere, and sometimes the lower part of the thermosphere. { 'kē·mō‚sfir }

chemosynthesis [BIOL] The synthesis of organic compounds from carbon dioxide by microorganisms using energy derived from chemical reactions. { ‚kē·mō′sin·thə·səs }

chemotaxis [BIOL] The orientation or movement of a motile organism with reference to a chemical agent. { ‚kē·mō′tak·səs }

chemotaxonomy [BOT] The classification of plants based on natural products. { ‚kē·mō‚tak′sän·ə·mē }

chemotropism [BIOL] Orientation response of a sessile organism with reference to chemical stimuli. { ‚kē·mō′trō‚piz·əm }

Chernozem [GEOL] One of the major groups of zonal soils, developed typically in temperate to cool, subhumid climate; the Chernozem soils in modern classification include Borolls, Ustolls, Udolls, and Xerolls. Also spelled Tchernozem. { ¦chər·nəz ¦yȯm }

cherry leaf spot [PL PATH] A fungus disease of the cherry caused by *Coccomyces hiemalis*; spotting and chlorosis of the leaves occurs, with consequent retardation of tree and fruit development. { 'cher·ē ‚lēf ‚spät }

chersophyte [ECOL] A plant that grows in dry wastelands. { 'kərz·ə‚fīt }

chickenpox [MED] A mild, highly infectious viral disease of humans caused by a herpesvirus and characterized by vesicular rash. Also known as varicella. { 'chik·ən ‚päks }

chimney cloud [METEOROL] A cumulus cloud in the tropics that has much greater vertical than horizontal extent. { 'chim‚nē ‚klaůd }

chimopelagic [ECOL] Pertaining to, belonging to, or being marine organisms living at great depths throughout most of the year; during the winter they move to the surface. { ¦kī·mō·pə′laj·ik }

chinook [METEOROL] The foehn on the eastern side of the Rocky Mountains. { shə′nůk }

chionophile [ECOL] Having a preference for snow. { ‚kī′än·ə‚fīl }

chipboard [FOR] A low-density paper board made from mixed waste paper and used where strength and quality are needed. { 'chip‚bȯrd }

chisel [AGR] A strong, heavy tool with curved points used for tilling; drawn by a tractor, it stirs the soil at an appreciable depth without turning it. { 'chiz·əl }

chitin [BIOL] A white or colorless amorphous polysaccharide that forms a base for the hard outer integuments of crustaceans, insects, and other invertebrates. { 'kīt·ən }

Chitral fever *See* phlebotomus fever. { 'chi·trəl ‚fē·vər }

Chlamydiales [MICROBIO] An order of coccoid, gram-negative bacteria that are obligate, intracellular parasites of vertebrates. { klə‚mid·ē′ā·lēz }

chlamydospore [MYCOL] A thick-walled, unicellular resting spore developed from vegetative hyphae in almost all parasitic fungi. { klə′mid·ə‚spȯr }

chloralosane *See* chloralose. { ‚klȯr·ə′lō‚sān }

chloralose [CHEM] $C_8H_{11}O_6Cl_3$ A crystalline compound with a melting point of 178°C; used as a repellent for birds. Also known as glucochloralose. { 'klȯr·ə‚lōs }

α-chloralose [CHEM] $C_8H_{11}O_6Cl_3$ Needlelike crystals with a melting point of 87°C; soluble in glacial acetic acid and ether; used on seed grains as a bird repellent and as a hypnotic for animals. Also known as chloralosane; glucochloral. { 'al·fə ¦klȯr·ə,lōs }

chloramine T [CHEM] $CH_3C_6H_4SO_2NClNa·3H_2O$ A white, crystalline powder that decomposes slowly in air, freeing chlorine; used as an antiseptic, a germicide, and an oxidizing agent and chlorinating agent. { 'klȯr·ə,mēn 'tē }

chloranil [CHEM] $C_6Cl_4O_2$ Yellow leaflets melting at 290°C; soluble in organic solvents; made from phenol by treatment with potassium chloride and hydrochloric acid; used as an agricultural fungicide and as an oxidizing agent in the manufacture of dyes. { klȯr'an·əl }

chlorbenside [CHEM] $C_{13}H_{10}SCl_2$ White crystals with a melting point of 72°C; used as a miticide for spider mites on fruit trees and ornamentals. { klȯr'ben,sīd }

chlorbromuron [CHEM] $C_9H_{10}ONBrCl$ A white solid with a melting point of 94–96°C; used as a pre- and postemergence herbicide for annual grass and for broadleaf weeds on crops, soybeans, and Irish potatoes. { ,klȯr·brə'myú·rən }

chlordan *See* chlordane. { 'klȯr,dan }

chlordane [CHEM] $C_{10}H_6Cl_8$ A volatile liquid insecticide; a chlorinated hexahydromethanoindene. Also spelled chlordan. { 'klȯr,dān }

chlordimeform [CHEM] $C_{10}H_{13}ClN_2$ A tan-colored solid, melting point 35°C; used as a miticide and insecticide for fruits, vegetables, and cotton. { ,klȯr'dī·mə,fȯrm }

chlorenchyma [BOT] Chlorophyll-containing tissue in parts of higher plants, as in leaves. { klȯr'eŋ·kə·mə }

chlorendic acid [CHEM] $C_9H_4Cl_6O_4$ White, fine crystals used in fire-resistant polyester resins and as an intermediate for dyes, fungicides, and insecticides. { klȯr'en·dik 'as·əd }

chlorfenpropmethyl [CHEM] $C_{10}H_{10}OCl_2$ A colorless to brown liquid used as a postemergence herbicide of wild oats, cereals, fodder beets, sugarbeets, and peas. { ¦klȯr·fən,präp'meth·əl }

chlorfensulfide [CHEM] $C_{12}H_6Cl_4N_2S$ A yellow, crystalline compound with a melting point of 123.5–124°C; used as a miticide for citrus. { ,klȯr·fən'səl,fīd }

chlorfenvinphos [CHEM] $C_{12}H_{14}Cl_3O_4P$ An amber liquid with a boiling point of 168–170°C; used as an insecticide for ticks, flies, lice, and mites on cattle. { ,klȯr·fən'vin ,fäs }

chlorine [CHEM] A chemical element, symbol Cl, atomic number 17, atomic weight 35.453; used in manufacture of solvents, insecticides, and many non-chlorine-containing compounds, and to bleach paper and pulp. { 'klȯr,ēn }

chlorine dioxide [CHEM] ClO_2 A green gas used to bleach cellulose and to treat water. { 'klȯr,ēn dī'äk,sīd }

chlorine war gas [ENG] Chlorine gas packaged to be released against enemy troops; greenish yellow, toxic, and gaseous at normal temperatures and pressures. { 'klȯr,ēn 'wȯr ,gas }

chlorine water [CHEM] A clear, yellowish liquid used as a deodorizer, antiseptic, and disinfectant. { 'klȯr,ēn ,wȯd·ər }

chlorinity [OCEANOGR] A measure of the chloride and other halogen content, by mass, of sea water. { klə'rin·əd·ē }

chlormephos [CHEM] $C_5H_{12}O_2S_2ClP$ A liquid used as an insecticide for soil. { 'klȯr·mə ,fäs }

chloroacetic acid [CHEM] $ClCH_2COOH$ White or colorless, deliquescent crystals that are soluble in water, ether, chloroform, benzene, and alcohol; used as an herbicide and in the manufacture of dyes and other organic molecules. { ¦klȯr·ə¦sēd·ik 'as·əd }

chloroacetonitrile [CHEM] $ClCH_2CN$ A colorless liquid with a pungent odor; soluble in hydrocarbons and alcohols; used as a fumigant. { ¦klȯr·ō₁as·ə·'tän·ə·trəl }

2-chloroallyl diethyldithiocarbamate *See* sulfallate. { ¦tü ¦klȯr·ō·al·əl 'dī₁eth·əl'dī₁thī·ō 'kär·bə₁māt }

chlorobenzilate [CHEM] $C_{16}H_{14}Cl_2O_3$ A yellow-brown, viscous liquid with a melting point of 35–37°C; used as a miticide in agriculture and horticulture. { ₁klȯr·ō'ben·zə ₁lāt }

chloroethene *See* vinyl chloride. { ₁klȯr·ō'eth₁ēn }

chloroethylene *See* vinyl chloride. { ₁klȯr·ō'eth·ə₁lēn }

chlorofluorocarbon [CHEM] A compound consisting of chlorine, fluorine, and carbon; has the potential to destroy ozone in the stratosphere. Abbreviated CFC. Also known as fluorochlorocarbon. { ¦klȯr·ə¦flür·ə₁kär·bən }

chlorofluoromethane [CHEM] A compound consisting of chlorine, fluorine, and carbon, has the potential to destroy ozone in the stratosphere. Abbreviated CFC. Also known as fluorochlorocarbon (FCC). { ₁klȯr·ə₁flür·ə'meth₁ān }

chloroform [CHEM] $CHCl_3$ A colorless, sweet-smelling, nonflammable liquid; used at one time as an anesthetic. Also known as trichloromethane. { 'klȯr·ə₁fȯrm }

chloromethane [CHEM] CH_3Cl A colorless, noncorrosive, liquefiable gas which condenses to a colorless liquid; used as a refrigerant, and as a catalyst carrier in manufacture of butyl rubber. Also known as methyl chloride. { ¦klȯr·ō'meth₁ān }

chlorophyll [BIOL] The generic name for any of several oil-soluble green tetrapyrrole plant pigments which function as photoreceptors of light energy for photosynthesis. { 'klȯr·ə₁fil }

chlorophyll a [BIOL] $C_{55}H_{72}O_5N_4Mg$ A magnesium chelate of dihydroporphyrin that is esterified with phytol and has a cyclopentanone ring; occurs in all higher plants and algae. { 'klȯr·ə₁fil 'ā }

chlorophyllase [BIOL] An enzyme that splits or hydrolyzes chlorophyll. { 'klȯr·ə·fə ₁lās }

chlorophyll b [BIOL] $C_{55}H_{70}O_6N_4Mg$ An ester similar to chlorophyll *a* but with a $-CHO$ substituted for a $-CH_3$; occurs in small amounts in all green plants and algae. { 'klȯr· ə₁fil 'bē }

Chlorophyta [BOT] The green algae, a highly diversified plant division characterized by chloroplasts, having chlorophyll *a* and *b* as the predominating pigments. { klō'räf·ə·də }

chloropicrin [CHEM] CCl_3NO_2 A colorless liquid with a sweet odor whose vapor is very irritating to the lungs and causes vomiting, coughing, and crying; used as a soil fumigant. Also known as nitrochloroform; trichloronitromethane. { ₁klȯr·ō'pik·rən }

chloroplast [BOT] A type of cell plastid occurring in the green parts of plants, containing chlorophyll pigments, and functioning in photosynthesis and protein synthesis. { 'klȯr·ə₁plast }

chlorosis [MED] A form of macrocytic anemia in young females characterized by marked reduction in hemoglobin and a greenish skin color. [PL PATH] A disease condition of green plants seen as yellowing of green parts of the plant. { klə'rō·səs }

chlorosity [OCEANOGR] The chlorine and bromide content of one liter of sea water; equals the chlorinity of the sample times its density at 20°C. { klə'räs·əd·ē }

chlorothalonil |CHEM| $C_8Cl_4N_2$ Colorless crystals with a melting point of 250–251°C; used as a fungicide for crops, turf, and ornamental flowers. { ‚klȯr·ə'thal·ə·nəl }

chlorothymol |CHEM| $CH_3C_6H_2(OH)(C_3H_7)Cl$ White crystals melting at 59–61°C; soluble in benzene alcohol, insoluble in water; used as a bactericide. { ‚klȯr·ə'thī‚mȯl }

4-chloro-3,5-xylenol |CHEM| $ClC_6H_2(CH_3)_2OH$ Crystals with a melting point of 115.5°C; soluble in water, 95% alcohol, benzene, terpenes, ether, and alkali hydroxides; used as an antiseptic and germicide and to stop mildew; used in humans as a topical and urinary antiseptic and as a topical antiseptic in animals. { ¦fȯr ¦klȯr·ō ¦thrē ¦fīv 'zī·lə ‚nȯl }

chlorthiamid |CHEM| $C_7H_5Cl_2NS$ An off-white, crystalline compound with a melting point of 151–152°C; used as a herbicide for selective weed control in industrial sites. { klȯr'thī·ə‚mid }

chocolate spot [PL PATH] A fungus disease of legumes caused by species of *Botrytis* and characterized by brown spots on leaves and stems, with withering of shoots. { 'chäk·lət ‚spät }

cholera |MED| 1. An acute, infectious bacterial disease of humans caused by *Vibrio comma*; characterized by diarrhea, delirium, stupor, and coma. 2. Any condition characterized by profuse vomiting and diarrhea. { 'käl·ə·rə }

cholera vibrio |MICROBIO| *Vibrio comma*, the bacterium that causes cholera. { 'käl·ə·rə 'vib·rē‚ō }

choppy sea |OCEANOGR| In popular usage, short, rough, irregular wave motion on a sea surface. { ¦chäp·ē ¦sē }

chronic carrier |MED| A person who harbors and transmits an infectious agent for an indefinite period. { 'krän·ik 'kar·ē·ər }

chronology |SCI TECH| The arrangement of data in order of time of appearance. { krə'näl·ə·jē }

Chrysophyceae |BOT| Golden-brown algae making up a class of fresh- and salt-water unicellular forms in the division Chrysophyta. { ‚kris·ō'fīs·ē‚ē }

Chrysophyta |BOT| The golden-brown algae, a division of plants with a predominance of carotene and xanthophyll pigments in addition to chlorophyll. { krə'säf·ə·də }

chute |HYD| A short channel across a narrow land area which bypasses a bend in a river; formed by the river's breaking through the land. { shüt }

chylophyllous |BOT| Having succulent or fleshy leaves. { ¦kīl·ō¦fil·əs }

Ci *See* cirrus cloud.

CI *See* temperature-humidity index.

ciguatoxin |BIOL| A toxin produced by the benthic dinoflagellate *Gambierdiscus toxicus*. { ¦sēg·wə¦täk·sən }

Ciidae |ZOO| The minute, tree-fungus beetles, a family of coleopteran insects in the superfamily Cucujoidea. { 'sī·ə‚dē }

circadian rhythm |BIOL| A self-sustained cycle of physiological changes that occurs over an approximately 24-hour cycle, generally synchronized to light-dark cycles in an organism's environment. { sər'kād·ē·ən 'rith·əm }

circular vortex |METEOROL| An atmospheric flow in parallel planes in which streamlines and other isopleths are concentric circles about a common axis; an atmospheric model of easterly and westerly winds is a circular vortex about the earth's polar axis. { 'sər·kyə·lər 'vȯr‚teks }

circulation |METEOROL| For an air mass, in the line integral of the tangential component of the velocity field about a closed curve. |OCEANOGR| A water current flow occurring within a large area, usually in a closed circular pattern. { ˌsər·kyə·'lā·shən }

circulation flux [METEOROL] Flux due to mean atmospheric motion as opposed to eddy flux; the dominant flux in low latitudes. { ˌsər·kyə·'lā·shən ˌfləks }

circulation index [METEOROL] A measure of the magnitude of one of several aspects of large-scale atmospheric circulation patterns; indices most frequently measured represent the strength of the zonal (east-west) or meridional (north-south) components of the wind, at the surface or at upper levels, usually averaged spatially and often averaged in time. { ˌsər·kyə·'lā·shən 'in,deks }

circulation pattern [METEOROL] The general geometric configuration of atmospheric circulation usually applied, in synoptic meteorology, to the large-scale features of synoptic charts and mean charts. { ˌsər·kyə·'lā·shən ˌpad·ərn }

circumboreal distribution [ECOL] The distribution of a Northern Hemisphere organism whose habitat includes North American, European, and Asian stations. { ¦sər·kəm'bȯr·ē·əl ˌdis·trə'byü·shən }

circumpolar [GEOGR] Located around one of the polar regions of earth. { ¦sər·kəm 'pō·lər }

circumpolar westerlies *See* westerlies. { ¦sər·kəm'pō·lər 'wes·tər,lēz }

circumpolar whirl *See* polar vortex. { ¦sər·kəm'pō·lər 'wərl }

cirque [GEOL] A steep elliptic to elongated enclave high on mountains in calcareous districts, usually forming the blunt end of a valley. Also known as corrie; cwm. { sərk }

cirque lake [HYD] A small body of water occupying a cirque. { 'sərk ˌlāk }

cirriform [METEOROL] Descriptive of clouds composed of small particles, mostly ice crystals, which are fairly widely dispersed, usually resulting in relative transparency and whiteness and often producing halo phenomena not observed with other cloud forms. { 'sir·ə,fȯrm }

cirrocumulus cloud [METEOROL] A principal cloud type, appearing as a thin, white path of cloud without shadows, composed of very small elements in the form of grains, ripples, and so on. Abbreviated Cc. { ¦sir·ō'kyü·myə·ləs ¦klaůd }

cirrostratus cloud [METEOROL] A principal cloud type, appearing as a whitish veil, usually fibrous but sometimes smooth, which may totally cover the sky and often produces halo phenomena, either partial or complete. Abbreviated Cs. { ¦sir·ō 'strad·əs ¦klaůd }

cirrus [ZOO] A tendrillike animal appendage. { 'sir·əs }

cirrus cloud [METEOROL] A principal cloud type composed of detached cirriform elements in the form of white, delicate filaments, of white (or mostly white) patches, or narrow bands. Abbreviated Ci. { 'sir·əs ¦klaůd }

citronella oil [AGR] A yellowish oil distilled from the leaves of either of two grasses, *Cymbopogon nardus* or *C. winterianus*; used as an insect repellent. Also known as Java citronella oil. { ˌsi·trə'nel·ə 'ȯil }

citrus anthracnose [PL PATH] A fungus disease of citrus plants caused by *Colletotrichum gloeosporioides* and characterized by tip blight, stains on the leaves, and spots, stains, or rot on the fruit. { 'si·trəs ˌan'thrak,nōs }

citrus blast [PL PATH] A bacterial disease of citrus trees caused by *Pseudomonas syringae* and marked by drying and browning of foliage and twigs and black pitting of the fruit. { 'si·trəs ˌblast }

citrus canker [PL PATH] A bacterial disease of citrus plants caused by *Xanthomonas citri* and producing lesions on twigs, foliage, and fruit. { 'si·trəs ˌkaŋ·kər }

citrus fruit |BOT| Any of the edible fruits having a pulpy endocarp and a firm exocarp that are produced by plants of the genus *Citrus* and related genera. { 'si·trəs ,früt }

citrus gummosis |PL PATH| A disease of citrus trees caused by the fungus *Phytophthora citrophthora*, characterized by the formation of narrow cracks in the bark which exude a pale yellow gum; infection is favored by excessive moisture. { ,si·trəs gə'mō·səs }

citrus scab |PL PATH| A fungus disease of citrus plants caused by *Sphaceloma rosarum*, producing scablike lesions on all plant parts. { 'si·trəs ,skab }

cladogenesis |GEN| The splitting of a single taxon into two new taxa. { ,klad·ə'jen·ə·səs }

cladogenic adaptation *See* divergent adaptation. { ¦klad·ə¦jen·ik ,ad,ap'tā·shən }

clan |ECOL| A very small community, perhaps a few square yards in area, in climax formation, and dominated by one species. { klan }

clarke |GEOCHEM| A unit of the average abundance of an element in the earth's crust, expressed as a percentage. Also known as crustal abundance. { klärk }

class |SYST| A taxonomic category ranking above the order and below the phylum or division. { klas }

classification |SYST| A systematic arrangement of plants and animals into categories based on a definite plan, considering evolutionary, physiologic, cytogenetic, and other relationships. { ,klas·ə·fə'kā·shən }

classify |SCI TECH| To sort into groups that have common properties. { 'klas·ə·fī }

clast |GEOL| An individual grain, fragment, or constituent of detrital sediment or sedimentary rock produced by physical breakdown of a larger mass. { klast }

clastation *See* weathering. { kla'stā·shən }

clay |GEOL| **1.** A natural, earthy, fine-grained material which develops plasticity when mixed with a limited amount of water; composed primarily of silica, alumina, and water, often with iron, alkalies, and alkaline earths. **2.** The fraction of an earthy material containing the smallest particles, that is, finer than 3 micrometers. { klā }

clay soil |GEOL| A fine-grained inorganic soil which forms hard lumps when dry and becomes sticky when wet. { ¦klā ¦sȯil }

clear |METEOROL| **1.** After United States weather observing practice, the state of the sky when it is cloudless or when the sky cover is less than 0.1 (to the nearest tenth). **2.** To change from a stormy or cloudy weather condition to one of no precipitation and decreased cloudiness. { klir }

clear-cutting |FOR| Felling and removing all trees in a forest area. { 'klir ,kəd·iŋ }

clear ice |HYD| Generally, a layer or mass of ice which is relatively transparent because of its homogeneous structure and small number and size of air pockets. { ¦klir ¦īs }

cliff |GEOGR| A high, steep, perpendicular or overhanging face of a rock; a precipice. { klif }

climagram *See* climatic diagram. { 'klī·mə,gram }

climagraph *See* climatic diagram. { 'klī·mə,graf }

climate |CLIMATOL| The long-term manifestations of weather. { 'klī·mət }

climate change |METEOROL| Any change in global temperatures and precipitation over time due to natural variability or to human activity. { 'klī·mət ,chānj }

climate control |CLIMATOL| Schemes for artificially altering or controlling the climate of a region. { 'klī·mət kən'trōl }

climate model |CLIMATOL| A mathematical representation of the earth's climate system capable of simulating its behavior under present and altered conditions. { 'klī·mət ‚mäd·əl }

climatic change |CLIMATOL| The long-term fluctuation in rainfall, temperature, and other aspects of the earth's climate. { klī'mad·ik 'chānj }

climatic classification |CLIMATOL| The division of the earth's climates into a system of contiguous regions, each one of which is defined by relative homogeneity of the climate elements. { klī'mad·ik ‚klas·ə·fə'kā·shən }

climatic climax |ECOL| A climax community viewed, by some authorities, as controlled by climate. { klī'mad·ik 'klī‚maks }

climatic controls |CLIMATOL| The relatively permanent factors which govern the general nature of the climate of a portion of the earth, including solar radiation, distribution of land and water masses, elevation and large-scale topography, and ocean currents. { klī'mad·ik kən'trōlz }

climatic cycle |CLIMATOL| A long-period oscillation of climate which recurs with some regularity, but which is not strictly periodic. Also known as climatic oscillation. { klī'mad·ik 'sī·kəl }

climatic diagram |CLIMATOL| A graphic presentation of climatic data; generally limited to a plot of the simultaneous variations of two climatic elements, usually through an annual cycle. Also known as climagram; climagraph; climatograph; climogram; climograph. { klī'mad·ik 'dī·ə‚gram }

climatic divide |CLIMATOL| A boundary between regions having different types of climate. { klī'mad·ik də'vīd }

climatic factor |CLIMATOL| Climatic control, but regarded as including more local influences; thus city smoke and the extent of the builtup metropolitan area are climatic factors, but not climatic controls. { klī'mad·ik 'fak·tər }

climatic forecast |CLIMATOL| A forecast of the future climate of a region; that is, a forecast of general weather conditions to be expected over a period of years. { klī'mad·ik 'fȯr‚kast }

climatic oscillation *See* climatic cycle. { klī'mad·ik ‚äs·ə'lā·shən }

climatic prediction |METEOROL| The description of the future state of the climate, that is, the average or expected atmospheric and earth-surface conditions, for example, temperature, precipitation, humidity, winds, and their range of variability. Seasonal and interannual climate predictions, made many months in advance, provide useful information for planners and policy makers. { klī'mad·ik prə'dik·shən }

climatic province |CLIMATOL| A region of the earth's surface characterized by an essentially homogeneous climate. { klī'mad·ik 'prä·vəns }

climatic snow line |METEOROL| The altitude above which a flat surface (fully exposed to sun, wind, and precipitation) would experience a net accumulation of snow over an extended period of time; below this altitude, ablation would predominate. { klī'mad·ik 'snō ‚līn }

climatic zone |CLIMATOL| A belt of the earth's surface within which the climate is generally homogeneous in some respect; an elemental region of a simple climatic classification. { klī'mad·ik ‚zōn }

climatograph *See* climatic diagram. { klī'mad·ə‚graf }

climatography |CLIMATOL| A quantitative description of climate, particularly with reference to the tables and charts which show the characteristic values of climatic elements at a station or over an area. { ‚klī·mə'täg·rə·fē }

climatological forecast [METEOROL] A weather forecast based upon the climate of a region instead of upon the dynamic implications of current weather, with consideration given to such synoptic weather features as cyclones and anticyclones, fronts, and the jet stream. { ˌklī·məd·əl'äj·ə·kəl 'fȯrˌkast }

climatological station elevation [CLIMATOL] The elevation above mean sea level chosen as the reference datum level for all climatological records of atmospheric pressure in a given locality. { ˌklī·məd·əl'äj·ə·kəl 'stā·shən ˌel·ə'vā·shən }

climatological station pressure [CLIMATOL] The atmospheric pressure computed for the level of the climatological station elevation, used to give all climatic records a common reference; it may or may not be the same as station pressure. { ˌklī·məd·əl'äj·ə·kəl 'stā·shən ˌpresh·ər }

climatological substation [CLIMATOL] A weather-observing station operated (by an unpaid volunteer) for the purpose of recording climatological observations. { ˌklī·məd·əl'äj·ə·kəl 'səbˌstā·shən }

climatology [METEOROL] That branch of meteorology concerned with the mean physical state of the atmosphere together with its statistical variations in both space and time as reflected in the weather behavior over a period of many years. { ˌklī·mə'täl·ə·jē }

climatopathology [MED] The study of disease in relation to the effects of the natural environment. { ¦klī·mə·tō·pə'thäl·ə·jē }

climatophysiology [BIOL] The study of the interaction of the natural environment with physiologic factors. { ¦klī·mə·tōˌfiz·ē'äl·ə·jē }

climax community [ECOL] The final stage in ecological succession in which a relatively constant environment is reached and species composition no longer changes in a directional fashion, but fluctuates about some mean, or average, community composition. { 'klīˌmaks kəˌmyü·nə·dē }

climax plant formation [ECOL] A mature, stable plant population in a climax community. { 'klīˌmaks 'plant fȯr'mā·shən }

climbing bog [ECOL] An elevated boggy area on a swamp margin, usually occurring where there is a short summer and considerable rainfall. { 'klīm·iŋ 'bäg }

climbing stem [BOT] A long, slender stem that climbs up a support or along the tops of other plants by using spines, adventitious roots, or tendrils for attachment. { 'klīm·iŋ 'stem }

climogram See climatic diagram. { 'klī·məˌgram }

climograph See climatic diagram. { 'klī·məˌgraf }

cline [BIOL] A graded series of morphological or physiological characters exhibited by a natural group (as a species) of related organisms, generally along a line of environmental or geographic transition. { klīn }

clinical microbiology [MED] The adaptation of microbiological techniques to the study of the etiological agents of infectious disease. { ˌklin·ə·kəl ˌmī·krō·bī'äl·ə·jē }

clod [AGR] A compact mass of soil, ranging from about 0.2 to 10 inches (0.5 to 25 centimeters) in size, which is produced by plowing and digging of excessively wet or dry soil. { kläd }

clonorchiasis [MED] A parasitic infection of humans and other fish-eating mammals which is caused by the trematode *Opisthorchis* (*Clonorchis*) *sinensis*, which is usually found in the bile ducts. { ˌklōn·ȯr'kī·ə·səs }

closed drainage [HYD] Drainage in which the surface flow of water collects in sinks or lakes having no surface outlet. Also known as blind drainage. { ¦klōzd 'drā·nij }

81

closed ecological system [ECOL] An ecosystem that is self-contained and does not gain organisms by immigration or lose them by emigration. { ¦klōzd ek·ə'läj·ə·kəl ‚sis·təm }

closed high [METEOROL] A high that may be completely encircled by an isobar or contour line. { ¦klōzd 'hī }

closed lake [HYD] A lake that does not have a surface effluent and that loses water by evaporation or by seepage. { ¦klōzd 'lāk }

closed low [METEOROL] A low that may be completely encircled by an isobar or contour line, that is, an isobar or contour line of any value, not necessarily restricted to those arbitrarily chosen for the analysis of the chart. { ¦klōzd 'lō }

closed sea [OCEANOGR] **1.** That part of the ocean enclosed by headlands, within narrow straits, or within other landforms. **2.** That part of the ocean within the territorial jurisdiction of a country. { ¦klōzd 'sē }

Closterovirus [MICROBIO] A genus of plant viruses belonging to the family Closteroviridae that has a wide host range and is transmitted primarily by aphids; beet yellows virus is the type species. { ¦klä·stə·rə'vī·rəs }

Clostridium perfringens [MICROBIO] A spore-forming, toxin-producing bacterium that can contaminate meat left at room temperature. The ingested cells release toxin in the digestive tract, resulting in cramps and diarrhea. { klä‚strid·ē·əm pər'frin·jənz }

Clostridium tetani [MED] A spore-forming bacterium that produces a powerful toxin, tetanospasmin, that blocks inhibitory synapses in the central nervous system and thus causes the severe muscle spasms characteristic of tetanus. { klä‚strid·ē·əm 'tet·ən‚ī }

clothing monitor [ENG] An instrument designed for monitoring radioactive contamination on clothing. { 'klō·thiŋ ‚män·əd·ər }

cloud [SCI TECH] Any suspension of particulate matter, such as dust or smoke, dense enough to be seen. { klaud }

cloud absorption [GEOPHYS] The absorption of electromagnetic radiation by the waterdrops and water vapor within a cloud. { 'klaud əb'sȯrp·shən }

cloudage See cloud cover. { 'klau·dij }

cloud band [METEOROL] A broad band of clouds, about 10 to 100 or more miles (16 to 160 kilometers) wide, and varying in length from a few tens of miles to hundreds of miles. { 'klaud ‚band }

cloud bank [METEOROL] A fairly well-defined mass of cloud observed at a distance; covers an appreciable portion of the horizon sky, but does not extend overhead. { 'klaud ‚baŋk }

cloud banner See banner cloud. { 'klaud ‚ban·ər }

cloud bar [METEOROL] **1.** A heavy bank of clouds that appears on the horizon with the approach of an intense tropical cyclone (hurricane or typhoon); it is the outer edge of the central cloud mass of the storm. **2.** Any long, narrow, unbroken line of cloud, such as a crest cloud or an element of billow cloud. { 'klaud ‚bär }

cloud base [METEOROL] For a given cloud or cloud layer, that lowest level in the atmosphere at which the air contains a perceptible quantity of cloud particles. { 'klaud ‚bās }

cloudburst [METEOROL] In popular terminology, any sudden and heavy fall of rain, usually of the shower type, and with a fall rate equal to or greater than 100 millimeters (3.94 inches) per hour. Also known as rain gush; rain gust. { 'klaud‚bərst }

cloud cap See cap cloud. { 'klaud ‚kap }

cloud classification [METEOROL] **1.** A scheme of distinguishing and grouping clouds according to their appearance and, where possible, to their process of formation. **2.** A scheme of classifying clouds according to their altitudes: high, middle, or low clouds. **3.** A scheme of classifying clouds according to their particulate composition: water clouds, ice-crystal clouds, or mixed clouds. { 'klaȯd ˌklas·ə·fə'kā·shən }

cloud cover [METEOROL] That portion of the sky cover which is attributed to clouds, usually measured in tenths of sky covered. Also known as cloudage; cloudiness. { 'klaȯd ˌkəv·ər }

cloud crest See crest cloud. { 'klaȯd ˌkrest }

cloud deck [METEOROL] The upper surface of a cloud. { 'klaȯd ˌdek }

cloud droplet [METEOROL] A particle of liquid water from a few micrometers to tens of micrometers in diameter, formed by condensation of atmospheric water vapor and suspended in the atmosphere with other drops to form a cloud. { 'klaȯd ˌdräp·lət }

cloud echo [METEOROL] The radar target signal returned from clouds alone, as detected by cloud detection radars or other very-short-wavelength equipment. { 'klaȯd ˌek·ō }

cloud forest See temperate rainforest. { 'klaȯd ˌfär·əst }

cloud formation [METEOROL] **1.** The process by which various types of clouds are formed, generally involving adiabatic cooling of ascending moist air. **2.** A particular arrangement of clouds in the sky, or a striking development of a particular cloud. { 'klaȯd fȯr'mā·shən }

cloud height [METEOROL] The absolute altitude of the base of a cloud. { 'klaȯd ˌhīt }

cloudiness See cloud cover. { 'klaȯd·ē·nəs }

cloud layer [METEOROL] An array of clouds, not necessarily all of the same type, whose bases are at approximately the same level; may be either continuous or composed of detached elements. { 'klaȯd ˌlā·ər }

cloud level [METEOROL] **1.** A layer in the atmosphere in which are found certain cloud genera; three levels are usually defined: high, middle, and low. **2.** At a particular time, the layer in the atmosphere bounded by the limits of the bases and tops of an existing cloud form. { 'klaȯd ˌlev·əl }

cloud modification [METEOROL] Any process by which the natural course of development of a cloud is altered by artificial means. { 'klaȯd ˌmäd·ə·fə'kā·shən }

cloud particle [METEOROL] A particle of water, either a drop of liquid water or an ice crystal, comprising a cloud. { 'klaȯd ˌpärd·ə·kəl }

cloud-phase chart [METEOROL] A chart designed to indicate and distinguish super-cooled water clouds from ice-crystal clouds. { 'klaȯd ˌfāz ˌchärt }

cloud physics [METEOROL] The study of the physical and dynamical processes governing the structure and development of clouds and the release from them of snow, rain, and hail. { 'klaȯd ˌfiz·iks }

cloud seeding [METEOROL] Any technique carried out with the intent of adding to a cloud certain particles that will alter its natural development. { 'klaȯd ˌsēd·iŋ }

cloud symbol [METEOROL] One of a set of specified ideograms that represent the various cloud types of greatest significance or those most commonly observed, and entered on a weather map as part of a station model. { 'klaȯd ˌsim·bəl }

cloud system [METEOROL] An array of clouds and precipitation associated with a cyclonic-scale feature of atmospheric circulation, and displaying typical patterns and continuity. Also known as nephsystem. { 'klaȯd ˌsis·təm }

cloud top [METEOROL] The highest level in the atmosphere at which the air contains a perceptible quantity of cloud particles for a given cloud or cloud layer. { 'klaȯd ˌtäp }

cloudy [METEOROL] The character of a day's weather when the average cloudiness, as determined from frequent observations, is more than 0.7 for the 24-hour period. { 'klaüd·ē }

clough [GEOGR] A cleft in a hill; a ravine or narrow valley. { kləf }

clubhead fungus *See* Cordyceps ophioglossoides. { ¦kləb¦hed 'fəŋ·gəs }

coagulation *See* agglomeration. { kō‚ag·yə'lā·shən }

coalescence [METEOROL] In cloud physics, merging of two or more water drops into a single larger drop. { ‚kō·ə'les·əns }

coal gas [PETR MIN] **1.** Flammable gas derived from coal either naturally in place, or by induced methods of industrial plants and underground gasification. **2.** Specifically, fuel gas obtained from carbonization of coal. { 'kōl ‚gas }

coalification [GEOL] Formation of coal from plant material by the processes of diagenesis and metamorphism. Also known as bituminization; carbonification; incarbonization; incoalation. { ‚kōl·ə·fə'kā·shən }

coal tar [PETR MIN] A tar obtained from carbonization of coal, usually in coke ovens or retorts, containing several hundred organic chemicals. { 'kōl ‚tär }

coal-tar dye [CHEM] Dye made from a coal-tar hydrocarbon or a derivative such as benzene, toluene, xylene, naphthalene, or aniline. { 'kōl ‚tär ‚dī }

coal-tar pitch [PETR MIN] Dark-brown to black amorphous residue from the redistillation of coal tar; melts at 150°F (66°C); used as a thermoplastic. { 'kōl ¦tär ‚pich }

coast [GEOGR] The general region of indefinite width that extends from the sea inland to the first major change in terrain features. { kōst }

coastal berm *See* berm. { 'kōs·təl 'bərm }

coastal current [OCEANOGR] An offshore current flowing generally parallel to the shoreline with a relatively uniform velocity. { 'kōs·təl 'kər·ənt }

coastal dune [GEOL] A mobile mound of windblown material found along many sea and lake shores. { 'kōs·təl 'dün }

coastal ice *See* fast ice. { 'kōs·təl 'īs }

coastal landform [GEOGR] The characteristic features and patterns of land in a coastal zone subject to marine and subaerial processes of erosion and deposition. { 'kōs·təl 'land‚fórm }

coastal plain [GEOL] An extensive, low-relief area that is bounded by the sea on one side and by a high-relief province on the landward side. Its geologic province actually extends beyond the shoreline across the continental shelf; it is linked to the stable part of a continent on the trailing edge of a tectonic plate. Typically, it has strata that dip gently and uniformly toward the sea. { 'kōs·təl 'plān }

coastal sediment [GEOL] The mineral and organic deposits of deltas, lagoons, and bays, barrier islands and beaches, and the surf zone. { 'kōs·təl 'sed·ə·mənt }

coast ice *See* fast ice. { 'kōst ‚īs }

coastline [GEOGR] **1.** The line that forms the boundary between the shore and the coast. **2.** The line that forms the boundary between the water and the land. { 'kōst‚līn }

coast shelf *See* submerged coastal plain. { 'kōst ‚shelf }

cobalt-60 [PHYS] A radioisotope of cobalt, symbol ^{60}Co, having a mass number of 60; emits gamma rays and has many medical and industrial uses; the most commonly used isotope for encapsulated radiation sources. { 'kō‚bòlt 'siks·tē }

Cobb's disease [PL PATH] A bacterial disease of sugarcane caused by *Xanthomonas vascularum* and characterized by a slime in the vascular bundles, dwarfing, streaking of leaves, and decay. Also known as sugarcane gummosis. { 'käbz di‚zēz }

cocarcinogen [MED] A noncarcinogenic agent which augments the carcinogenic process. { ¦kō·kär'sin·ə·jən }

Coccidioides immitis [MED] A mold primarily found in desert soil that converts into spherules containing endospores when growing within the body and that causes coccidioidomycosis or San Joaquin valley fever. { ‚käk·sid·ē¦òi‚dēz i'mīd·əs *or* i'mēd·əs }

coccidioidomycosis [MED] An infectious fungus disease of humans and animals of either a pulmonary or a cutaneous nature; caused by *Coccidioides immitis*. Also known as San Joaquin Valley fever. { käk¦sid·ē¦òid·ō·mī'kō·səs }

coccidiosis [MED] The state of or the conditions associated with being infected by coccidia. { käk¦sid·ē¦ō·səs }

cocculin *See* picrotoxin. { 'käk·yə·lən }

cockpit karst *See* cone karst. { 'käk‚pit 'karst }

cocoon [ZOO] **1.** A protective case formed by the larvae of many insects, in which they pass the pupa stage. **2.** Any of the various protective egg cases formed by invertebrates. { kə'kün }

cocurrent line [OCEANOGR] A line through places having the same tidal current hour. { kō'kər·ənt 'līn }

coffin [ENG] A box of heavy shielding material, usually lead, used for transporting radioactive objects and having walls thick enough to attenuate radiation from the contents to an allowable level. Also known as cask; casket. { 'kò·fən }

cohesionless [GEOL] Referring to a soil having low shear strength when dry, and low cohesion when wet. Also known as frictional; noncohesive. { kō'hē·zhən·ləs }

cohesiveness [GEOL] Property of unconsolidated fine-grained sediments by which the particles stick together by surface forces. { kō'hē·siv·nəs }

cohesive soil [GEOL] A sticky soil, such as clay or silt; its shear strength equals about half its unconfined compressive strength. { kō'hē·siv 'sòil }

coke [PETR MIN] A coherent, cellular, solid residue remaining from the dry (destructive) distillation of a coking coal or of pitch, petroleum, petroleum residues, or other carbonaceous materials; contains carbon as its principal constituent, together with mineral matter and volatile matter. { kōk }

coke-oven gas [PETR MIN] A gas produced during carbonization of coal to form coke. { 'kōk ‚òv·ən 'gas }

col [METEOROL] The point of intersection of a trough and a ridge in the pressure pattern of a weather map; it is the point of relatively lowest pressure between two highs and the point of relatively highest pressure between two lows. Also known as neutral point; saddle point. { käl }

cold-air drop *See* cold pool. { 'kōld ¦er ‚dräp }

cold-air outbreak *See* polar outbreak. { 'kōld ¦er 'aùt‚brāk }

cold anticyclone *See* cold high. { 'kōld ‚an·tē'sī‚klōn }

cold-blooded [BIOL] Having body temperature approximating that of the environment and not internally regulated. { 'kōld ¦bləd·əd }

cold-core cyclone *See* cold low. { ¦kōld ¦kòr 'sī‚klōn }

cold-core high *See* cold high. { ¦kōld ¦kòr 'hī }

cold-core low See cold low. { ¦kōld ¦kȯr 'lō }

cold desert See tundra. { ¦kōld 'dez·ərt }

cold dome [METEOROL] A cold air mass, considered as a three-dimensional entity. { 'kōld ˌdōm }

cold drop See cold pool. { 'kōld ˌdräp }

cold front [METEOROL] Any nonoccluded front, or portion thereof, that moves so that the colder air replaces the warmer air; the leading edge of a relatively cold air mass. { 'kōld ˌfrənt }

cold-front-like sea breeze [METEOROL] Sea breeze that forms over the ocean, moves slowly toward the land, and then moves inland quite suddenly. Also known as sea breeze of the second kind. { 'kōld ˌfrənt ˌlīk 'sē ˌbrēz }

cold-front thunderstorm [METEOROL] A thunderstorm attending a cold front. { 'kōld ˌfrənt 'thən·dərˌstȯrm }

cold glacier [GEOL] A glacier whose base is at a temperature much below 32°F (0°C) and frozen to the bedrock, resulting in insignificant movement and almost no erosion. { ¦kōld 'glā·shər }

cold high [METEOROL] At a given level in the atmosphere, any high that is generally characterized by colder air near its center than around its periphery. Also known as cold anticyclone; cold-core high. { 'kōld 'hī }

cold low [METEOROL] At a given level in the atmosphere, any low that is generally characterized by colder air near its center than around its periphery. Also known as cold-core cyclone; cold-core low. { 'kōld 'lō }

cold pole [CLIMATOL] The location which has the lowest mean annual temperature in its hemisphere. { 'kōld ˌpōl }

cold pool [METEOROL] A region of relatively cold air surrounded by warmer air; the term is usually applied to cold air of appreciable vertical extent that has been isolated in lower latitudes as part of the formation of a cutoff low. Also known as cold-air drop; cold drop. { 'kōld ˌpül }

cold tongue [METEOROL] In synoptic meteorology, a pronounced equatorward extension or protrusion of cold air. { 'kōld ˌtəŋ }

cold wall [OCEANOGR] The line or surface along which two water masses of significantly different temperature are in contact. { 'kōld ˌwȯl }

cold-water desert [GEOGR] An arid, often foggy region characterized by sparse precipitation because incoming airstreams are cooled over an offshore coastal current and deposit rain over the sea. { ¦kōld ˌwȯd·ər 'dez·ərt }

cold-water sphere [OCEANOGR] Those portions of the ocean water having a temperature below 8°C. Also known as oceanic stratosphere. { 'kōld ˌwȯd·ər ˌsfir }

cold wave [METEOROL] A rapid fall in temperature within 24 hours to a level requiring substantially increased protection to agriculture, industry, commerce, and social activities. { 'kōld ˌwāv }

Coleoptera [ZOO] The beetles, holometabolous insects making up the largest order of the animal kingdom; general features of the Insecta are found in this group. { ˌkō·lē'äp·tə·rə }

Coleosporaceae [MYCOL] A family of parasitic fungi in the order Uredinales. { ˌkō·lē·ō·spə'rās·ē‚ē }

coliphage [MICROBIO] Any bacteriophage able to infect *Escherichia coli*. { 'kä·ləˌfāj }

colloidal instability [METEOROL] A property attributed to clouds, by which the particles of the cloud tend to aggregate into masses large enough to precipitate. { kə'lȯid·əl in·stə‚bil·əd·ē }

colloider [CIV ENG] A device that removes colloids from sewage. { kə'lȯid·ər }

colon bacillus See Escherichia coli. { 'kō·lən bə'sil·əs }

colonization [ECOL] The establishment of an immigrant species in a peripherally unsuitable ecological area; occasional gene exchange with the parental population occurs, but generally the colony evolves in relative isolation and in time may form a distinct unit. { ‚käl·ə·nə'zā·shən }

colony [BIOL] A localized population of individuals of the same species which are living either attached or separately. [MICROBIO] A cluster of microorganisms growing on the surface of or within a solid medium; usually cultured from a single cell. { 'käl·ə·nē }

color lake See lake. { 'kəl·ər ‚lāk }

columnar stem [BOT] An unbranched, cylindrical stem bearing a set of large leaves at its summit, as in palms, or no leaves, as in cacti. { kə'ləm·nər ‚stem }

comber [OCEANOGR] A deep-water wave of long, curling character with a high, breaking crest pushed forward by a strong wind. { 'kōm·ər }

combined carbon [CHEM] Carbon that is chemically combined within a compound, as contrasted with free or uncombined elemental carbon. { kəm'bīnd 'kär·bən }

combined sewers [CIV ENG] A drainage system that receives both surface runoff and sewage. { kəm'bīnd 'sü·ərz }

combined water [GEOCHEM] Water attached to soil minerals by means of chemical bonds. { kəm'bīnd 'wȯd·ər }

combustible gas [PETR MIN] A gas that burns, including the fuel gases, hydrogen, hydrocarbon, carbon monoxide, or a mixture of these. { kəm'bəs·tə·bəl ‚gas }

combustion nucleus [METEOROL] A condensation nucleus formed as a result of industrial or natural combustion processes. { kəm'bəs·chən ‚nü·klē·əs }

cometabolism [ECOL] A process in which compounds not utilized for growth or energy are transformed to other products by microorganisms. { ‚kō·mə'tab·ə‚liz·əm }

comfort index See temperature-humidity index. { 'kəm·fərt ‚in‚deks }

comfort temperature [ENG] Any one of the indexes in which air temperatures have been adjusted to represent human comfort or discomfort under prevailing conditions of temperature, humidity, radiation, and wind. { 'kəm·fərt ‚tem·prə·chər }

commensal [ECOL] An organism living in a state of commensalism. { kə'men·səl }

commensalism [ECOL] An interspecific, symbiotic relationship in which two different species are associated, wherein one is benefited and the other neither benefited nor harmed. { kə'men·sə‚liz·əm }

communicable disease [MED] An infectious disease that can be transmitted from one individual to another either directly by contact or indirectly by fomites and vectors. { kə'myü·nə·kə·bəl di‚zēz }

community [ECOL] Aggregation of organisms characterized by a distinctive combination of two or more ecologically related species; an example is a deciduous forest. Also known as ecological community. { kə'myü·nə·dē }

community classification [ECOL] Arrangement of communities into classes with respect to their complexity and extent, their stage of ecological succession, or their primary production. { kə'myü·nə·dē ‚klas·ə·fə'kā·shən }

compaction [ENG] Increasing the dry density of a granular material, particularly soil, by means such as impact or by rolling the surface layers. [GEOL] Process by which soil and sediment mass loses pore space in response to the increasing weight of overlying material. { kəm'pak·shən }

compactor [ENG] **1.** Machine designed to consolidate earth and paving materials by kneading, weight, vibration, or impact, to sustain loads greater than those sustained in an uncompacted state. **2.** A machine that compresses solid waste material for convenience in disposal. { kəm'pak·tər }

compensation depth [OCEANOGR] In bodies of water, the depth at which the light intensity is just sufficient to bring about a balance between the oxygen produced and that consumed by algae. { ˌkäm·pən'sā·shən ˌdepth }

compensation point [BOT] The light intensity at which the amount of carbon dioxide released in respiration equals the amount used in photosynthesis, and the amount of oxygen released in photosynthesis equals the amount used in respiration. { ˌkäm·pən'sā·shən ˌpȯint }

competence [HYD] The ability of a stream, flowing at a given velocity, to move the largest particles. { 'käm·pəd·əns }

competition [ECOL] The inter- or intraspecific interaction resulting when several individuals share an environmental necessity that is in limited supply. { ˌkäm·pə'tish·ən }

competitive displacement [ECOL] The inability of a species to successfully live in an area because a second species dominates local resources. { kəmˌped·əd·iv di 'splās·mənt }

competitive exclusion [ECOL] The result of a competition in which one species is forced out of part of the available habitat by a more efficient species. { kəm'ped·əd·iv iks'klüzh·ən }

competitive-exclusion principle See Gause's principle. { kəm'ped·əd·iv iks'klüzh·ən ˌprin·sə·pəl }

complete flower [BOT] A flower having all four floral parts, that is, having sepals, petals, stamens, and carpels. { kəm'plēt 'flau̇·ər }

complete leaf [BOT] A dicotyledon leaf consisting of three parts: blade, petiole, and a pair of stipules. { kəm'plēt 'lēf }

complex climatology [CLIMATOL] Analysis of the climate of a single space, or comparison of the climates of two or more places, by the relative frequencies of various weather types or groups of such types; a type is defined by the simultaneous occurrence within specified narrow limits of each of several weather elements. { 'käm ˌpleks ˌklī·mə'täl·ə·jē }

complex low [METEOROL] An area of low atmospheric pressure within which more than one low-pressure center is found. { 'käm,pleks 'lō }

Compositae [BOT] The single family of the order Asterales; perhaps the largest family of flowering plants, it contains about 19,000 species. { kəm'päz·ə,tē }

compound leaf [BOT] A type of leaf with the blade divided into two or more separate parts called leaflets. { 'käm,pau̇nd 'lēf }

compound pistil [BOT] A pistil composed of two or more united carpels. { 'käm,pau̇nd 'pis·təl }

compound valley glacier [HYD] A glacier composed of several ice streams emanating from different tributary valleys. { 'käm,pau̇nd 'val·ē ,glā·shər }

compression process [CHEM ENG] The recovery of natural gasoline from gas containing a high proportion of hydrocarbons. { kəm'presh·ən ,prä·səs }

compression wood [BOT] Dense wood found at the base of some tree trunks and on the undersides of branches. { kəm'presh·ən ˌwud }

concentration [HYD] The ratio of the area of the sea covered by ice to the total area of sea surface. { ˌkän·sən'trā·shən }

concentration ratio [AGR] A measure of a plant's ability to take up a contaminant from soil; it is expressed as the concentration of the element of interest in the dried plant material divided by its concentration in the dried soil. { ˌkän·sən'trā·shən ˌrā·shō }

concentration time [HYD] The time required for water to travel from the most remote portion of a river basin to the basin outlet; it varies with the quantity of flow and channel conditions. { ˌkän·sən'trā·shən ˌtīm }

concordant coastline [GEOL] A coastline parallel to the land structures which form the margin of an ocean basin. { kən'kord·ənt 'kōst,līn }

concordant drainage *See* accordant drainage. { kən'kord·ənt 'drān·ij }

condensation [METEOROL] The process by which water vapor becomes a liquid such as dew, fog, or cloud or a solid like snow; condensation in the atmosphere is brought about by either of two processes: cooling of air to its dew point, or addition of enough water vapor to bring the mixture to the point of saturation (that is, the relative humidity is raised to 100). { ˌkän·dən'sā·shən }

condensation nucleus [METEOROL] A particle, either liquid or solid, upon which condensation of water vapor begins in the atmosphere. { ˌkän·dən'sā·shən 'nü·klē·əs }

condensation pressure [METEOROL] The pressure at which a parcel of moist air expanded dry adiabatically reaches saturation. Also called adiabatic condensation pressure; adiabatic saturation pressure. { ˌkän·dən'sā·shən ˌpresh·ər }

condensation temperature [METEOROL] The temperature at which a parcel of moist air expanded dry adiabatically reaches saturation. Also known as adiabatic condensation temperature; adiabatic saturation temperature. { ˌkän·dən'sā·shən 'tem·prə·chər }

condensation trail [METEOROL] A visible trail of condensed water vapor or ice particles left behind an aircraft, an airfoil, or such, in motion through the air. Also known as contrail; vapor trail. { ˌkän·dən'sā·shən ˌtrāl }

conditional instability [METEOROL] The state of a column of air in the atmosphere when its lapse rate of temperature is less than the dry adiabatic lapse rate but greater than the saturation adiabatic lapse rate. { kən'dish·ən·əl ˌin·stə'bil·əd·ē }

conductive equilibrium *See* isothermal equilibrium. { kən'dək·tiv ˌē·kwə'lib·rē·əm }

conductivity *See* permeability. { ˌkän,dək'tiv·əd·ē }

cone [BOT] The ovulate or staminate strobilus of a gymnosperm. [GEOL] A mountain, hill, or other landform having relatively steep slopes and a pointed top. { kōn }

cone karst [GEOL] A type of karst, typical of tropical regions, characterized by a pattern of steep, convex sides and slightly concave floors. Also known as cockpit karst; Kegel karst. { 'kōn ˌkärst }

cone of depression [HYD] The depression in the water table around a well defining the area of influence of the well. Also known as cone of influence. { 'kōn əv di'presh·ən }

cone of influence *See* cone of depression. { 'kōn əv 'in·flü·əns }

confined aquifer *See* artesian aquifer. { kən'fīnd 'ak·wə·fər }

confining bed [GEOL] An impermeable bed adjacent to an aquifer. { kən'fīn·iŋ ˌbed }

confluence [HYD] **1.** A stream formed from the flowing together of two or more streams. **2.** The place where such streams join. { 'kän,flü·əns }

confused sea |OCEANOGR| A highly disturbed water surface without a single, well-defined direction of wave travel. { kən'fyüzd 'sē }

congelifraction |GEOL| The splitting or disintegration of rocks as the result of the freezing of the water contained. Also known as frost bursting; frost riving; frost shattering; frost splitting; frost weathering; frost wedging; gelifraction; gelivation. { kən¦jel·ə¦frak·shən }

congeliturbate |GEOL| Soil or unconsolidated earth which has been moved or disturbed by frost action. { kən‚jel·ə'tər·bət }

congeliturbation |GEOL| The churning and stirring of soil as a result of repeated cycles of freezing and thawing; includes frost heaving and surface subsidence during thaws. Also known as cryoturbation; frost churning; frost stirring; geliturbation. { kən‚jel·ə·tər'bā·shən }

congeneric |SYST| Referring to the species of a given genus. { ¦kän·jə'ner·ik }

congestin |BIOL| A toxin produced by certain sea anemones. { kən'jes·tən }

conidiophore |MYCOL| A specialized aerial hypha that produces conidia in certain ascomycetes and imperfect fungi. { kə'nid·ē·ə‚för }

conidiospore See conidium. { kə'nid·ē·ə‚spòr }

conidium |MYCOL| Unicellular, asexual reproductive spore produced externally upon a conidiophore. Also known as conidiospore. { kə'nid·ē·əm }

conifer |BOT| The common name for cone-bearing plants of the order Pinales, such as pines, firs, spruces, and hemlocks. { 'kän·ə·fər }

Coniferales |BOT| The equivalent name for Pinales. { kə‚nif·ə'rā·lēz }

Coniferophyta |BOT| The equivalent name for Pinicae. { kə‚nif·ə'räf·əd·ə }

coniferous forest |ECOL| An area of wooded land predominated by conifers. { kə'nif·ə·rəs 'fär·əst }

connate water |HYD| Water entrapped in the interstices of igneous rocks when the rocks were formed; usually highly mineralized. { kə'nāt 'wòd·ər }

conservation |ECOL| The maintenance of environmental quality, resources, and biodiversity in an area. { ‚kän·sər'vā·shən }

conservative concentrations |OCEANOGR| Concentrations such as heat content or salinity occurring in bodies of water that are altered locally, except at the boundaries, by processes of diffusion and advection only. { kən'sər·və·tiv ‚kän·sən'trā·shənz }

consociation |ECOL| A climax community of plants which is dominated by a single species. { kən‚sō·sē'ā·shən }

consolidated ice |OCEANOGR| Ice which has been compacted into a solid mass by wind and ocean currents and covers an area of the ocean. { kən'säl·ə‚dād·əd 'īs }

consortism See symbiosis. { 'kän‚sòrd‚iz·əm }

conspecific |SYST| Referring to individuals or populations of a single species. { ¦kän·spə'sif·ik }

constancy See persistence. { 'kän·stən·sē }

constant |SCI TECH| A value that does not change during a particular process. { 'kän·stənt }

constant-height chart |METEOROL| A synoptic chart for any surface of constant geometric altitude above mean sea level (a constant-height surface), usually containing plotted data and analyses of the distribution of such variables as pressure, wind, temperature, and humidity at that altitude. Also known as constant-level chart; fixed-level chart; isohypsic chart. { ¦kän·stənt ¦hīt ‚chärt }

constant-level chart   *See* constant-height chart.   { ¦kän·stənt ¦lev·əl 'chärt }

constant-pressure chart   [METEOROL] The synoptic chart for any constant-pressure surface, usually containing plotted data and analyses of the distribution of height of the surface, wind temperature, humidity, and so on. Also known as isobaric chart; isobaric contour chart.   { ¦kän·stənt ¦presh·ər ,chärt }

constant-pressure surface   *See* isobaric surface.   { ¦kän·stənt ¦presh·ər ,sər·fəs }

constituent   [SCI TECH] An essential part or component of a system or group: examples are an ingredient of a chemical system, or a component of an alloy.   { kən'stich·ə·wənt }

consumer   [ECOL] A nutritional grouping in the food chain of an ecosystem, composed of heterotrophic organisms, chiefly animals, which ingest other organisms or particulate organic matter.   { kən'süm·ər }

consumption   *See* tuberculosis.   { kən'səm·shən }

consumptive use   [HYD] The total annual land water loss in an area, due to evaporation and plant use.   { kən'səm·div 'yüs }

contact adsorption   [CHEM ENG] Process for removal of minor constituents from fluids by stirring in direct contact with powdered or granulated adsorbents, or by passing the fluid through fixed-position adsorbent beds (activated carbon or ion-exchange resin); used to decolorize petroleum lubricating oils and to remove solvent vapors from air.   { 'kän,takt ad'sòrp·shən }

contact aerator   [CIV ENG] A tank in which sewage that is settled on a bed of stone, cement-asbestos, or other surfaces is treated by aeration with compressed air.   { 'kän ,takt 'er,ād·ər }

contact bed   [CIV ENG] A bed of coarse material such as coke, used to purify sewage.   { 'kän,takt ,bed }

contagious abortion   [VET MED] Brucellosis in cattle caused by *Brucella abortus* and inducing abortion. Also known as Bang's disease; infectious abortion.   { kən'tā·jəs ə'bòr·shən }

contagious disease   [MED] An infectious disease communicable by contact with a person suffering from it, with the bodily discharge, or with an object touched by the person.   { kən'tā·jəs di'zēz }

containment   [ENG] An enclosed space or facility to contain and prevent the escape of hazardous material.   [ENG] **1.** Provision of a gastight enclosure around the highly radioactive components of a nuclear power plant, to contain the radioactivity released by a possible major accident.   **2.** The use of remote-control devices (slave apparatus) to remove spent cores from nuclear power plants or, in shielded laboratory hoods, to perform chemical studies of dangerous radioactive materials.   { kən'tān·mənt }

contaminate   [SCI TECH] To render unfit or to soil by the introduction of foreign or unwanted material.   { kən'tam·ə,nāt }

contamination   [MICROBIO] The process or act of soiling with bacteria.   [SCI TECH] Something that contaminates.   { kən,tam·ə'nā·shən }

contamination monitor   [ENG] A radiation counter used to detect radioactive contamination of surface areas or of the atmosphere.   { kən,tam·ə'nā·shən ,män·əd·ər }

contemporary carbon   [CHEM] The isotopic carbon content of living matter, based on the assumption of a natural proportion of carbon-14.   { kən'tem·pə,rer·ē 'kär·bən }

continent   [GEOGR] A protuberance of the earth's crustal shell, with an area of several million square miles and sufficient elevation so that much of it is above sea level.   { 'känt·ən·ənt }

continental air [METEOROL] A type of air whose characteristics are developed over a large land area and which therefore has relatively low moisture content. { ¦känt·ən ¦ent·əl 'er }

continental anticyclone See continental high. { ¦känt·ən¦ent·əl ˌan·tē'sī͝klōn }

continental climate [CLIMATOL] Climate characteristic of the interior of a landmass of continental size, marked by large annual, daily, and day-to-day temperature ranges, low relative humidity, and a moderate or small irregular rainfall; annual extremes of temperature occur soon after the solstices. { ¦känt·ən¦ent·əl 'klī·mət }

continental displacement See continental drift. { ¦känt·ən¦ent·əl di'splās·mənt }

continental divide [GEOL] A drainage divide of a continent, separating streams that flow in opposite directions; for example, the divide in North America that separates watersheds of the Pacific Ocean from those of the Atlantic Ocean. { ¦känt·ən¦ent·əl di'vīd }

continental drift [GEOL] The concept of continent formation by the fragmentation and movement of land masses on the surface of the earth. Also known as continental displacement. { ¦känt·ən¦ent·əl 'drift }

continental glacier [HYD] A sheet of ice covering a large tract of land, such as the ice caps of Greenland and the Antarctic. { ¦känt·ən¦ent·əl 'glā·shər }

continental high [METEOROL] A general area of high atmospheric pressure which on mean charts of sea-level pressure is seen to overlie a continent during the winter. Also known as continental anticyclone. { ¦känt·ən¦ent·əl 'hī }

continentality [CLIMATOL] The degree to which a point on the earth's surface is in all respects subject to the influence of a land mass. { ˌkänt·ən·en'tal·əd·ē }

continental margin [GEOL] Those provinces between the shoreline and the deep-sea bottom; generally consists of the continental borderland, shelf, slope, and rise. { ¦känt·ən¦ent·əl 'mär·jən }

continental mass [GEOGR] The continental land rising more or less abruptly from the ocean floor and also the shallow submerged areas surrounding this land. { ¦känt·ən ¦ent·əl 'mas }

continental plateau See tableland. { ¦känt·ən¦ent·əl plə'tō }

continental platform See continental shelf. { ¦känt·ən¦ent·əl 'plat͵fórm }

continental polar air [METEOROL] Polar air having low surface temperature, low moisture content, and (especially in its source regions) great stability in the lower layers. { ¦känt·ən¦ent·əl ¦pō·lər 'er }

continental rise [GEOL] A transitional part of the continental margin; a gentle slope with a generally smooth surface, built up by the shedding of sediments from the continental block, and located between the continental slope and the abyssal plain. { ¦känt·ən¦ent·əl 'rīz }

continental shelf [GEOL] The zone around a continent, that part of the continental margin extending from the shoreline and the continental slope; composes with the continental slope the continental terrace. Also known as continental platform; shelf. { ¦känt·ən¦ent·əl 'shelf }

continental slope [GEOL] The part of the continental margin consisting of the declivity from the edge of the continental shelf extending down to the continental rise. { ¦känt·ən¦ent·əl 'slōp }

continental terrace [GEOL] The continental shelf and slope together. { ¦känt·ən¦ent·əl 'ter·əs }

continental tropical air [METEOROL] A type of tropical air produced over subtropical arid regions; it is hot and very dry. { ¦känt·ən¦ent·əl ¦träp·ə·kəl 'er }

continuity chart [METEOROL] A chart maintained for weather analysis and forecasting upon which are entered the positions of significant features (pressure centers, fronts, instability lines, through lines, ridge lines) of the regular synoptic charts at regular intervals in the past. { ˌkänt·ən'ü·əd·ē ˌchärt }

continuous permafrost zone [GEOL] Regional zone predominantly underlain by permanently frozen subsoil that is not interrupted by pockets of unfrozen ground. { kənˈtin·yə·wəs 'pər·məˌfròst ˌzōn }

contourite [OCEANOGR] A marine sediment deposited by swift ocean-bottom currents that generally flow along contours. { 'känˌtùˌrīt }

contour line [METEOROL] A line on a weather map connecting points of equal atmospheric pressure, temperature, or such. { 'känˌtùr ˌlīn }

contour microclimate [CLIMATOL] That portion of the microclimate which is directly attributable to the small-scale variations of ground level. { 'känˌtùr ¦mī·krō¦klī·mət }

contour plowing [AGR] Cultivation of land along lines connecting points of equal elevation, to prevent water erosion. Also known as terracing. { 'känˌtùr ˌplaù·iŋ }

contrail See condensation trail. { 'känˌtrāl }

contra solem [METEOROL] Characterizing air motion that is counterclockwise in the Northern Hemisphere and clockwise in the Southern Hemisphere; literally, against the sun. { 'kän·trə 'sō¸lem }

contributory See tributary. { kənˈtrib·yəˌtòr·ē }

controlled atmosphere [SCI TECH] A specified gas or mixture of gases at a predetermined temperature, and sometimes humidity, in which selected processes take place. { kənˈtrōld 'at·məˌsfir }

control-tower visibility [METEOROL] The visibility that is observed from an airport control tower. { kənˈtrōl ¦taù·ər ˌviz·ə'bil·əd·ē }

convection [METEOROL] Atmospheric motions that are predominantly vertical, resulting in vertical transport and mixing of atmospheric properties. [OCEANOGR] Movement and mixing of ocean water masses. { kən'vek·shən }

convectional stability See static stability. { kən'vek·shən·əl stə'bil·əd·ē }

convection cell [METEOROL] An atmospheric unit in which organized convective fluid motion occurs. { kən'vek·shən ˌsel }

convection current [METEOROL] Any current of air involved in convection; usually, the upward-moving portion of a convection circulation, such as a thermal or the updraft in cumulus clouds. Also known as convective current. { kən'vek·shən ˌkər·ənt }

convection stability See static stability. { kən'vek·shən stə'bil·əd·ē }

convection theory of cyclones [METEOROL] A theory of cyclone development proposing that the upward convection of air (particularly of moist air) due to surface heating can be of sufficient magnitude and duration that the surface inflow of air will attain appreciable cyclonic rotation. { kən'vek·shən ˌthē·ə·rē əv 'sī¸klōnz }

convective activity [METEOROL] Generally, manifestations of convection in the atmosphere, alluding particularly to the development of convective clouds and resulting weather phenomena, such as showers, thunderstorms, squalls, hail, and tornadoes. { kən'vek·div ak'tiv·əd·ē }

convective cloud [METEOROL] A cloud which owes its vertical development, and possibly its origin, to convection. { kən'vek·div 'klaùd }

convective current See convection current. { kən'vek·div ˌkər·ənt }

convective equilibrium See adiabatic equilibrium. { kən'vek·div ˌē·kwə'lib·rē·əm }

convective overturn   See overturn.   { kən'vek·div 'ō·vər,tərn }

convective precipitation   [METEOROL] Precipitation from convective clouds, generally considered to be synonymous with showers.   { kən'vek·div prə,sip·ə'tā·shən }

convective region   [METEOROL] An area particularly favorable for the formation of convection in the lower atmosphere, or one characterized by convective activity at a given time.   { kən'vek·div ,rē·jən }

convergence   [HYD] The line of demarcation between turbid river water and clear lake water.   [METEOROL] The increase in wind setup observed beyond that which would take place in an equivalent rectangular basin of uniform depth, caused by changes in platform or depth.   [OCEANOGR] A condition in the ocean in which currents or water masses having different densities, temperatures, or salinities meet; results in the sinking of the colder or more saline water.   { kən'vər·jəns }

convergent precipitation   [METEOROL] A synoptic type of precipitation caused by local updrafts of moist air.   { kən'vər·jənt prə,sip·ə'tā·shən }

convivium   [ECOL] A population exhibiting differentiation within the species and isolated geographically, generally a subspecies or ecotype.   { kən'viv·ē·əm }

Cooke unit   [BIOL] A unit for the standardization of pollen antigenicity.   { 'kúk ,yü·nət }

cooking snow   See water snow.   { 'kúk·iŋ ,snō }

cooling table   See hotbed.   { 'kül·iŋ ,tā·bəl }

cooperative observer   [METEOROL] An unpaid observer who maintains a meteorological station for the U.S. National Weather Service.   { kō'äp·rəd·iv əb'zər·vər }

Copenhagen water   See normal water.   { ¦kō·pən¦häg·ən ,wòd·ər }

copper arsenate   [CHEM] $Cu_3(AsO_4)_2 \cdot 4H_2O$ or $Cu_5H_2(AsO_4)_4 \cdot 2H_2O$ Bluish powder, soluble in ammonium hydroxide and dilute acids, insoluble in water and alcohol; used as a fungicide and insecticide.   { 'käp·ər 'ärs·ən,āt }

copper arsenite   [CHEM] $CuHAsO_3$ A toxic, light green powder which is soluble in acids and decomposes at the melting point; used as a pigment and insecticide. Also known as copper orthoarsenite; cupric arsenite; Scheele's green.   { 'käp·ər 'ärs·ən,īt }

copperas   See ferrous sulfate.   { 'käp·ə·rəs }

copper blight   [PL PATH] A leaf spot disease of tea caused by the fungus *Guignardia camelliae.*   { 'käp·ər ,blīt }

copper carbonate   [CHEM] $Cu_2(OH)_2CO_3$ A toxic, green powder; decomposes at 200°C and is soluble in acids; used in pigments and pyrotechnics and as a fungicide and feed additive. Also known as artificial malachite; cupric carbonate; mineral green.   { 'käp·ər 'kär·bə,nāt }

copper-8-quinolinolate   [CHEM] $C_{18}H_{14}N_2O_2Cu$ A khaki-colored, water-insoluble solid used as a fungicide in fruit-handling equipment.   { 'käp·ər ¦āt ¦kwīn·ə¦lin·ə,lāt }

copper oleate   [CHEM] $Cu[OOC(CH_2)_7CH=CH-(CH_2)_7CH_3]_2$ A green-blue liquid, used as a fungicide for fruits and vegetables.   { 'käp·ər 'ō·lē,āt }

copper orthoarsenite   See copper arsenite.   { 'käp·ər ¦òr·thō'ärs·ən,īt }

copper oxide   See cuprous oxide.   { 'käp·ər 'äk,sīd }

copper-64   [PHYS] Radioactive isotope of copper with mass number of 64; derived from pile-irradiation of metallic copper; used as a research aid to study diffusion, corrosion, and friction wear in metals and alloys.   { 'käp·ər ,sik·stē'fòr }

coppice   [ECOL] A growth of small trees that are repeatedly cut down at short intervals; the new shoots are produced by the old stumps.   { 'käp·əs }

coprophilous   [ECOL] Living in dung.   { kə'präf·ə·ləs }

coral |ZOO| The skeleton of certain solitary and colonial anthozoan cnidarians; composed chiefly of calcium carbonate. { 'kä·rəl }

coral head |GEOL| A small reef patch of coralline material. Also known as coral knoll. { 'kä·rəl ¦hed }

coral knoll See coral head. { 'kä·rəl ¦nōl }

coral mud |GEOL| Fine-grade deposits of coral fragments formed around coral islands and coasts bordered by coral reefs. { 'kär·əl ‚məd }

coral pinnacle |GEOL| A sharply upward-projecting growth of coral rising from the floor of an atoll lagoon. { 'kär·əl 'pin·ə·kəl }

coral reef |GEOL| A ridge or mass of limestone built up of detrital material deposited around a framework of skeletal remains of mollusks, colonial coral, and massive calcareous algae. { 'kär·əl ‚rēf }

coral-reef lagoon |GEOGR| The central, shallow body of water of an atoll or the water separating a barrier reef from the shore. { 'kär·əl ‚rēf lə'gün }

coral-reef shoreline |GEOL| A shoreline formed by reefs composed of coral polyps. Also known as coral shoreline. { 'kär·əl ‚rēf 'shȯr‚līn }

coral rock See reef limestone. { 'kär·əl ‚räk }

coral sand |GEOL| Coarse-grade deposits of coral fragments formed around coral islands and coasts bordered by coral reefs. { 'kär·əl ‚sand }

coral shoreline See coral-reef shoreline. { 'kär·əl 'shȯr‚līn }

cordillera |GEOGR| A mountain range or group of ranges, including valleys, plains, rivers, lakes, and so on, forming the main mountain axis of a continent. { ‚kȯrd·əl'er·ə }

Cordyceps ophioglossoides |MYCOL| A mushroom that is a parasite on the fruiting bodies of the truffle found in the soil of bamboo, oak, and pine woods that has antitumor properties and is an immune booster. Also known as clubhead fungus. { ¦kȯrd·ə‚seps ‚ō·fē·ə·glə'sȯi‚dēz }

Cordyceps sinensis |MYCOL| A type of mushroom found on the cold mountain tops and snowy grass marshlands of China that infects insect larvae with spores that germinate before the cocoons are formed; it has been successfully used in clinical trials to treat liver diseases, high cholesterol, and loss of sexual drive. Also known as caterpillar fungus. { ¦kȯrd·ə‚seps sī'nen·sis }

core |OCEANOGR| That area within a layer of ocean water where parameters such as temperature, salinity, or velocity reach extreme values. { kȯr }

core sample |GEOL| A sample of rock, soil, snow, or ice obtained by driving a hollow tube into the undisturbed medium and withdrawing it with its contained sample or core. { 'kȯr ‚sam·pəl }

Coriolis deflection See Coriolis effect. { kȯr·ē'ō·ləs di'flek·shən }

Coriolis effect |GEOPHYS| **1.** The deflection relative to the earth's surface of any object moving above the earth, caused by the Coriolis force; an object moving horizontally is deflected to the right in the Northern Hemisphere, to the left in the Southern. **2.** The effect of the Coriolis force in any rotating system. Also known as Coriolis deflection. { kȯr·ē'ō·ləs i'fekt }

Coriolus versicolor See Trametes versicolor. { kȯr·ē'ō·ləs 'vər·sə·kə·lər }

cork |BOT| A protective layer of cells that replaces the epidermis in older plant stems. { kȯrk }

corm |BOT| A short, erect, fleshy underground stem, usually broader than high and covered with membranous scales. { kòrm }

corn |BOT| *Zea mays*. A grain crop of the grass order Cyperales grown for its edible seeds (technically fruits). { kòrn }

corn smut |PL PATH| A fungus disease of corn caused by *Ustilago maydis*. { 'kòrn ˌsmət }

corn snow *See* spring snow. { 'kòrn ˌsnō }

corolla |BOT| Collectively, the petals of a flower. { kə'räl·ə }

corollate |BOT| Having a corolla. { kə'rä,lāt }

corona |METEOROL| A set of one or more prismatically colored rings of small radii, concentrically surrounding the disk of the sun, moon, or other luminary when veiled by a thin cloud; due to diffraction by numerous waterdrops. { kə'rō·nə }

coronavirus |MICROBIO| A major group of animal viruses including avian infectious bronchitis virus and mouse hepatitis virus. { kəˌrō·nəˌvī·rəs }

corrasion |GEOL| Mechanical wearing away of rock and soil by the action of solid materials moved along by wind, waves, running water, glaciers, or gravity. Also known as mechanical erosion. { kə'rā·zhən }

corrected establishment *See* mean high-water lunitidal interval. { kə'rek·təd i'stab·lish·mənt }

corridor |ECOL| A piece of habitat or land bridge that connects two otherwise separated larger areas. { 'kär·ə·dər }

corrie *See* cirque. { kòr·ē }

corrosion |GEOCHEM| Chemical erosion by motionless or moving agents. { kə'rō·zhən }

corrosive sublimate *See* mercuric chloride. { kə'rō·siv 'səb·lə,māt }

cortex |BOT| A primary tissue in roots and stems of vascular plants that extends inward from the epidermis to the phloem. { 'kòr,teks }

Corynebacterium diphtheriae |MICROBIO| A facultatively aerobic, nonmotile species of bacteria that causes diphtheria in humans. Also known as Klebs-Loeffler bacillus. { ˌkòr·əˌnē·bak'tir·ē·əm dif'thir·ē,ī }

cosmopolitan |ECOL| Having a worldwide distribution wherever the habitat is suitable, with reference to the geographical distribution of a taxon. { ˌkäz·məˌpäl·ət·ən }

cotinine |CHEM| The major metabolic product of nicotine which is excreted in the urine; used as a marker for environmental tobacco smoke. { 'kōt·ən,ēn }

cotton |AGR| The most economical natural fiber, obtained from plants of the genus *Gossypium*, used in making fabrics, cordage, and padding and for producing artificial fibers and cellulose. |BOT| Any plant of the genus *Gossypium* in the order Malvales; cultivated for the fibers obtained from its encapsulated fruits or bolls. { 'kät·ən }

cotton anthracnose |PL PATH| A fungus disease of cotton caused by *Glomerella gossypii* and characterized by reddish-brown to light-colored or necrotic spots. { 'kät·ən an'thrak,nōs }

cotton-belt climate |CLIMATOL| A type of warm climate characterized by dry winters and rainy summers; that is, a monsoon climate, in contrast to a Mediterranean climate. { 'kät·ən ,belt ,klī·mət }

cotton root rot |PL PATH| A fungus disease of cotton caused by *Phymatotrichum omnivorum* and marked by bronzing of the foliage followed by sudden wilting and death of the plant. { 'kät·ən 'rüt ,rät }

cotton rust [PL PATH] A fungus disease of cotton caused by *Puccinia stakmanii* producing low, greenish-yellow or orange elevations on the undersurface of leaves. { 'kät·ən ,rəst }

cotton wilt [PL PATH] **1.** A fungus disease of cotton caused by *Fusarium vasinfectum* growing in the water-conducting vessels and characterized by wilt, yellowing, blighting, and death. **2.** A fungus blight of cotton caused by *Verticillium albo-atrum* and characterized by yellow mottling of the foliage. { 'kät·ən ,wilt }

cottony rot [PL PATH] A fungus disease of many plants, especially citrus trees, marked by fluffy white growth caused by *Sclerotinia sclerotiorum*, in which there is stem wilt and rot. { 'kät·ən·ē ,rät }

cotyledon [BOT] The first leaf of the embryo of seed plants. { ,käd·əl'ēd·ən }

counter *See* radiation counter. { 'kaûnt·ər }

courtship [ECOL] A sequence of behavioral patterns that eventually may lead to completed mating. { 'kòrt,ship }

cove [GEOGR] **1.** A small, narrow, sheltered bay, inlet, or creek on a coast. **2.** A deep recess or hollow occurring in a cliff or steep mountainside. { kōv }

cover crops [AGR] Crops, especially grasses, grown for the express purpose of preventing and protecting a bare soil surface. { 'kəv·ər ,kräps }

covered smut [PL PATH] A seed-borne smut of certain grain crops caused by *Ustilago hordei* in barley and *U. avenae* in oats. { 'kəv·ərd 'smət }

covert [ECOL] A refuge or shelter, such as a coppice, for game animals. { 'kō·vərt }

cowpox *See* vaccinia. { 'kaù,päks }

cowpox virus [MICROBIO] The causative agent of cowpox in cattle. { 'kaù,päks ,vī·rəs }

coxsackievirus [MICROBIO] A large subgroup of the enteroviruses in the picornavirus group including various human pathogens. { kúk'säk·ē,vī·rəs }

C₃ plant [BOT] A plant that produces the 3-carbon compound phosphoglyceric acid as the first stage of photosynthesis. { ¦sē'thrē ,plant }

C₄ plant [BOT] A plant that produces the 4-carbon compound oxalocethanoic (ox-aloacetic) acid as the first stage of photosynthesis. { ¦sē'fòr ,plant }

Crassulaceae [BOT] A family of dicotyledonous plants in the order Rosales notable for their succulent leaves and resistance to desiccation. { ,kras·ə'lās·ē,ē }

crater lake [HYD] A fresh-water lake formed by the accumulation of rain and ground-water in a caldera or crater. { 'krād·ər ,lāk }

cream ice *See* sludge. { 'krēm ,īs }

creek [HYD] A natural stream of water, smaller than a river but larger than a brook. { krēk }

creosol [CHEM] $CH_3O(CH_3)C_6H_3OH$ A combination of isomers, derived from coal tar or petroleum; a yellowish liquid with a phenolic odor; used as a disinfectant, in the manufacture of resins, and in flotation of ore. Also known as hydroxymethylbenzene; methyl phenol. { 'krē·ə,sòl }

creosote oil [AGR] A coal tar fraction, boiling between 240 and 270°F (116–132°C); used for producing materials such as creosote and tar acids and used directly as a germicide, insecticide, or pesticide. { 'krē·ə,sōt ,òil }

crescentic lake *See* oxbow lake. { krə'sen·tik 'lāk }

cresol [CHEM] $CH_3C_6H_4OH$ One of three poisonous, colorless isomeric methyl phenols: *o*-cresol, *m*-cresol, *p*-cresol; used in the production of phenolic resins, tricresyl phosphate, disinfectants, and solvents. { 'krē,sòl }

crest cloud [METEOROL] A type of standing cloud which forms along a mountain ridge, either on the ridge, or slightly above and leeward of it, and remains in the same position relative to the ridge. Also known as cloud crest. { 'krest ,klaůd }

crest length [OCEANOGR] The length of a wave measured along its crest. Also known as crest width. { 'krest ,leŋkth }

crest stage [HYD] The highest stage reached at a point along a stream culminating a rise by waters of that stream. { 'krest ,stāj }

crest width *See* crest length. { 'krest ,width }

crevasse hoar [HYD] Ice crystals which form and grow in glacial crevasses and in other cavities where a large cooled space is formed and in which water vapor can accumulate under calm, still conditions. { krə'vas ,hȯr }

crevice [SCI TECH] A deep, narrow opening. { 'krev·əs }

Criconematoidea [ZOO] A superfamily of plant parasitic nematodes of the order Diplogasterida distinguished by their ectoparasitic habit and males that have atrophied mouthparts and do not feed. { ,krī·kō,nem·ə'tȯid·ē·ə }

critical temperature [AGR] The temperature below which a plant cannot grow. { 'krid·ə·kəl 'tem·prə·chər }

Cromwell Current [OCEANOGR] An eastward-setting subsurface current that extends about 1½° north and south of the equator, and from about 150°E to 92°W. { 'kräm ,wel ,kər·ənt }

crop [AGR] A plant or animal grown for its commercial value. { kräp }

crop dusting [AGR] Applying fungicides or insecticides in powder form to a crop; usually done from a low-flying aircraft. { 'kräp ,dəst·iŋ }

crop-flow sensor [AGR] An instrument used in precision agriculture to measure either the volume or the mass of the harvested portion of a crop using a variety of engineering principles, including light interception, radiation absorption, measurement of impact force, and directly weighing the crop. { 'kräp,flō,sen·sər }

crop micrometeorology [AGR] The branch of meteorology that deals with the interaction of crops and their immediate physical environment. { ¦kräp ,mī·krō ,mēt·ē·ə'räl·ə·jē }

crop rotation [AGR] A method of protecting the soil and replenishing its nutrition by planting a succession of different crops on the same land. { 'kräp rō'tā·shən }

cross *See* spider. { krȯs }

crossbreed [BIOL] To propagate new individuals by breeding two distinctive varieties of a species. Also known as outbreed. { 'krȯs,brēd }

cross-fertilization [BOT] Fertilization between two separate plants. { 'krȯs ,fərd·əl·ə 'zā·shən }

cross-pollination [BOT] Transfer of pollen from the anthers of one plant to the stigmata of another plant. { ¦krȯs ,pä·lə,nā·shən }

cross sea [OCEANOGR] A series of waves or swell crossing another wave system at an angle. { ¦krȯs ¦sē }

cross section [GEOL] **1.** A diagram or drawing that shows the downward projection of surficial geology along a vertical plane, for example, a portion of a stream bed drawn at right angles to the mean direction of the flow of the stream. **2.** An actual exposure or cut which reveals geological features. { 'krȯs ,sek·shən }

crosswind [METEOROL] A wind which has a component directed perpendicularly to the course (or heading) of an exposed, moving object. { 'krȯs,wind }

croup-associated virus [MICROBIO] A virus belonging to subgroup 2 of the parainfluenza viruses and found in children with croup. Also known as CA virus; laryngotracheobronchitis virus. { ¦krüp ə¦sō·sē̄,ād·əd 'vī·rəs }

crown [BOT] **1.** The topmost part of a plant or plant part. **2.** *See* corona. { kraün }

crown fire [FOR] A forest fire burning primarily in the tops of trees and shrubs. { 'kraün ,fīr }

crown gall [PL PATH] A bacterial disease of many plants induced by *Bacterium tumefaciens* and marked by abnormal enlargement of the stem near the root crown. { 'kraün ,gól }

crown rot [PL PATH] Any plant disease or disorder marked by deterioration of the stem at or near ground level. { 'kraün ,rät }

crown rust [PL PATH] A rust disease of oats and certain other grasses caused by varieties of *Puccinia coronata* and marked by light-orange masses of fungi on the leaves. { 'kraün ,rəst }

crude material *See* raw material. { 'krüd me,tir·ē·əl }

crude oil [GEOL] A comparatively volatile liquid bitumen composed principally of hydrocarbon, with traces of sulfur, nitrogen, or oxygen compounds; can be removed from the earth in a liquid state. { ¦krüd 'óil }

crumb structure [GEOL] A soil condition in which the particles are crumblike aggregates; suitable for agriculture. { 'krəm ,strək·chər }

crust [HYD] A hard layer of snow lying on top of a soft layer. { krəst }

crustal abundance *See* clarke. { ¦krəst·əl ə'bən·dəns }

crustose [BOT] Of a lichen, forming a thin crustlike thallus which adheres closely to the substratum of rock, bark, or soil. { 'krəs,tōs }

crust vegetation [ECOL] Zonal growths of algae, mosses, lichens, or liverworts having variable coverage and a thickness of only a few centimeters. { 'krəst ,vej·ə'tā·shən }

cry-, cryo- [SCI TECH] Combining form meaning cold, freezing. { krī, 'krī·ō }

cryobiosis [BIOL] A type of cryptobiosis induced by low temperatures. { ¦krī·ō·bī'ō·səs }

cryology [HYD] The study of ice and snow. { krī'äl·ə·jē }

cryophilic *See* cryophilous. { ,krī·ə'fil·ik }

cryophilous [ECOL] Having a preference for low temperatures. Also known as cryophilic. { krī'äf·ə·ləs }

cryophilous crop [AGR] A crop that will fully flower and seed only after it has experienced low temperatures early in its growth cycle. { krī¦äf·ə·ləs 'kräp }

cryophyte [ECOL] A plant that forms winter buds below the soil surface. { 'krī·ə,fīt }

cryoplanation [GEOL] Land erosion at high latitudes or elevations due to processes of intensive frost action. { ¦krī·ō·plə'nā·shən }

cryosphere [GEOL] That region of the earth in which the surface is perennially frozen. { 'krī·ə,sfir }

cryostat [ENG] An apparatus used to provide low-temperature environments in which operations may be carried out under controlled conditions. { 'krī·ə,stat }

cryostatic pressure [GEOL] Hydrostatic pressure exerted on soil and rocks when soil water freezes. { 'krī·ə,stad·ik 'presh·ər }

cryoturbation *See* congeliturbation. { ¦krī·ō·tər'bā·shən }

cryptobiosis [BIOL] A state in which metabolic rate of the organism is reduced to an imperceptible level. { ¦krip·tō·bī'ō·səs }

cryptobiotic [ECOL] Living in concealed or secluded situations. { ¦krip·tō·bī'äd·ik }

cryptoclimate [ENG] The climate of a confined space, such as inside a house, barn, or greenhouse, or in an artificial or natural cave; a form of microclimate. Also spelled kryptoclimate. { ¦krip·tō'klī·mət }

cryptoclimatology [CLIMATOL] The science of climates of confined spaces (crypto-climates); basically, a form of microclimatology. Also spelled kryptoclimatology. { ¦krip·tō₁klī·mə'täl·ə·jē }

cryptococcosis [MED] A yeast infection of humans, primarily of the central nervous system, caused by *Cryptococcus neoformans*. Also known as torulosis. { ₁krip·tə·kä 'kō·səs }

Cryptococcus [MYCOL] A genus of encapsulated pathogenic yeasts in the order Moniliales. { ₁krip·tə'käk·əs }

cryptophyte [BOT] A plant that produces buds either underwater or underground on corms, bulbs, or rhizomes. { 'krip·tə₁fīt }

crystalline frost [HYD] Hoarfrost that exhibits a relatively simple macroscopic crystalline structure. { 'kris·tə·lən 'fróst }

crystal violet *See* methyl violet. { ¦krist·əl 'vī·lət }

crystosphene [HYD] A buried sheet or mass of ice, as in the tundra of northern America, formed by the freezing of rising and spreading springwater beneath alluvial deposits. { ¦kris·tə¦sfēn }

Cs *See* cirrostratus cloud.

cucumber mildew [PL PATH] **1.** A downy mildew of cucumbers and melons caused by *Peronoplasmopara cubensis.* **2.** A powdery mildew of cucumbers and melons caused by *Erysiphe cichoracearum.* { 'kyü·kəm·bər 'mil₁dü }

cucumber mosaic [PL PATH] A virus disease of cucumbers and related fruits, producing mottling of terminal leaves and fruits and dwarfing of vines. { 'kyü·kəm·bər mō'zā·ik }

cucurbit wilt [PL PATH] A bacterial disease of cucumbers and related plants caused by *Erwinia tracheiphila*, characterized by sudden wilting of the plant. { kə'kər·bət ₁wilt }

cuesta [GEOGR] A gently sloping plain which terminates in a steep slope on one side. { 'kwes·tə }

culm [BOT] **1.** A jointed and usually hollow grass stem. **2.** The solid stem of certain monocotyledons, such as the sedges. { kəlm }

cultigen [BIOL] A cultivated variety or species of organism for which there is no known wild ancestor. Also known as cultivar. { 'kəl·tə·jən }

cultivar *See* cultigen. { 'kəl·tə₁vär }

cultivate [AGR] To prepare soil for the raising of crops. { 'kəl·tə₁vāt }

cultivator [AGR] A farm implement pulled behind a powered machine that is used to break up soil, kill weeds, and create a surface mulch for moisture. { 'kəl·tə₁vād·ər }

cultural ecology [ECOL] The branch of ecology that involves the study of the interaction of human societies with one another and with the natural environment. { ¦kəl·chər·əl ē'käl·ə·jē }

culture [BIOL] A growth of living cells or microorganisms in a controlled artificial environment. { 'kəl·chər }

culture community [ECOL] A plant community which is established or modified through human intervention; for example, a fencerow, hedgerow, or windbreak. { 'kəl·chər kə'myü·nəd·ē }

cumatophyte [ECOL] A plant that grows under surf conditions. { kyü'mad·ə,fīt }

cumulative dose [MED] The total dose resulting from repeated exposures to radiation. { 'kyü·myə·ləd·iv 'dōs }

cumuliform cloud [METEOROL] A fundamental cloud type, showing vertical development in the form of rising mounds, domes, or towers. { 'kyü·myə·lə,fórm ,klaúd }

cumulonimbus calvus cloud [METEOROL] A species of cumulonimbus cloud evolving from cumulus congestus: the protuberances of the upper portion have begun to lose the cumuliform outline; they loom and usually flatten, then transform into a whitish mass with a more or less diffuse outline and vertical striation; cirriform cloud is not present, but the transformation into ice crystals often proceeds with great rapidity. { ¦kyü·myə·lō'nim·bəs 'kal·vəs ,klaúd }

cumulonimbus capillatus cloud [METEOROL] A species of cumulonimbus cloud characterized by the presence of distinct cirriform parts, frequently in the form of an anvil, a plume, or a vast and more or less disorderly mass of hair, and usually accompanied by a thunderstorm. { ¦kyü·myə·lō'nim·bəs kap·ə'lad·əs ,klaúd }

cumulonimbus cloud [METEOROL] A principal cloud type, exceptionally dense and vertically developed, occurring either as isolated clouds or as a line or wall of clouds with separated upper portions. { ¦kyü·myə·lō'nim·bəs ,klaúd }

cumulus cloud [METEOROL] A principal type of cloud in the form of individual, detached elements which are generally dense and possess sharp nonfibrous outlines; these elements develop vertically, appearing as rising mounds, domes, or towers, the upper parts of which often resemble a cauliflower. { 'kyü·myə·ləs ,klaúd }

cumulus congestus cloud [METEOROL] A strongly sprouting cumulus species with generally sharp outline and sometimes a great vertical development, and with cauliflower or tower aspect. { 'kyü·myə·ləs kən'jes·təs ,klaúd }

cumulus humilis cloud [METEOROL] A species of cumulus cloud characterized by small vertical development and a generally flattened appearance, vertical growth is usually restricted by the existence of a temperature inversion in the atmosphere, which in turn explains the unusually uniform height of the cloud. Also known as fair-weather cumulus. { 'kyü·myə·ləs 'hyü·mə·ləs ,klaúd }

cumulus mediocris cloud [METEOROL] A cloud species unique to the species cumulus, of moderate vertical development, the upper protuberances or sproutings being not very marked; there may be a small cauliflower aspect; while this species does not give any precipitation, it frequently develops into cumulus congestus and cumulonimbus. { 'kyü·myə·ləs mē·dē'ō·krəs ,klaúd }

cup crystal [HYD] A crystal of ice in the form of a hollow hexagonal cup; a common form of depth hoar. { 'kəp ,krist·əl }

cupric arsenite *See* copper arsenite. { 'kyü·prik ärs·ən,īt }

cupric carbonate *See* copper carbonate. { 'kyü·prik 'kär·bə,nāt }

cuprous oxide [CHEM] Cu_2O An oxide of copper found in nature as cuprite and formed on copper by heat; used chiefly as a pigment and as a fungicide. Also known as copper oxide. { 'kyü·prəs 'äk,sīd }

curare [CHEM] Poisonous extract from the plant Strychnos toxifera containing a mixture of alkaloids that produce paralysis of the voluntary muscles by acting on synaptic junctions; used as an adjunct to anesthesia in surgery. { kyü'rä·rē }

currant leaf spot [PL PATH] **1.** An angular leaf spot of currants caused by the fungus *Cercospora angulata*. **2.** An anthracnose of currants caused by *Pseudopeziza ribis* and characterized by brown or black spots. { 'kər·ənt ˌlēf ˌspät }

current constants [OCEANOGR] Tidal current relations that remain practically constant for any particular locality. { 'kər·ənt 'kän·stənts }

current curve [OCEANOGR] In marine operations, a graphic representation of the flow of a current, consisting of a rectangular-coordinate graph on which speed is represented by the ordinates and time by the abscissas. { 'kər·ənt ˌkərv }

current cycle [OCEANOGR] A complete set of tidal current conditions, as those occurring during a tidal day, lunar month, or Metonic cycle. { 'kər·ənt ˌsī·kəl }

current drift [HYD] A broad, shallow, slow-moving ocean or lake current. { 'kər·ənt ˌdrift }

current hour [OCEANOGR] The average time interval between the moon's transit over the meridian of Greenwich and the time of the following strength of flood current modified by the times of slack water and strength of ebb. { 'kər·ənt ˌau̇·ər }

current rips [OCEANOGR] Small waves formed on the surface of water by the meeting of opposing ocean currents; vertical oscillation, rather than progressive waves, is characteristic of current rips. { 'kər·ənt ˌrips }

current tables [OCEANOGR] Tables listing predictions of the time and speeds of tidal currents at various places. { 'kər·ənt ˌtā·bəlz }

cutaneous anthrax [MED] The commonest form of anthrax, resulting from contamination of the skin; characterized by a pus-filled lesion surrounded by an area of edema and vesicles containing yellow fluid. { kyü'tā·nē·əs 'an̩ˌthraks }

cutaneous mycosis [MED] Any of a group of infections (collectively known as dermatophytoses, ringworms, or tineas) that are caused by keratinophilic fungi (dermatophytes). In general, the infections are limited to the nonliving keratinized layers of skin, hair, and nails, but a variety of pathologic changes can occur depending on the etiologic agent, site of infection, and immune status of the host. { kyü͝'tān·ē·əs mī'kō·səs }

cutoff high [METEOROL] A warm high which has become displaced out of the basic westerly current, and lies to the north of this current. { 'kətˌȯf ˌhī }

cutoff lake *See* oxbow lake. { 'kətˌȯf ˌlāk }

cutoff low [METEOROL] A cold low which has become displaced out of the basic westerly current, and lies to the south of this current. { 'kətˌȯf ˌlō }

cutting [BOT] A piece of plant stem with one or more nodes, which, when placed under suitable conditions, will produce roots and shoots resulting in a complete plant. { 'kəd·iŋ }

cutting-off process [METEOROL] A sequence of events by which a warm high or cold low, originally within the westerlies, becomes displaced either poleward (cutoff high) or equatorward (cutoff low) out of the westerly current; this process is evident at very high levels in the atmosphere, and it frequently produces, or is part of the production of, a blocking situation. { 'kəd·iŋ ˌȯf ˌpräs·əs }

cwm *See* cirque. { küm }

cyanazine [CHEM] $C_9H_{13}N_6Cl$ A white solid with a melting point of 166.5–167°C; used as a pre- and postemergence herbicide for corn, sorghum, soybeans, alfalfa, cotton, and wheat. { sī'an·əˌzēn }

cyanobacteria [MICROBIO] A group of one-celled to many-celled aquatic organisms. Also known as blue-green algae. { ˌsī·ə·noˌbak'tir·ē·ə }

cyanogen |CHEM| C_2N_2 A colorless, highly toxic gas with a pungent odor; a starting material for the production of complex thiocyanates used as insecticides. Also known as dicyanogen. { sī'an·ə·jən }

cyanogen chloride [CHEM] ClCN A poisonous, colorless gas or liquid, soluble in water; used in organic synthesis. { sī'an·ə·jən 'klȯr,īd }

cyanogen fluoride |CHEM| CNF A toxic, colorless gas, used as a tear gas. { sī'an·ə·jən 'flur,īd }

cyanometer [PHYS] An instrument designed to measure or estimate the degree of blueness of light, as of the sky. { sī·ə'näm·əd·ər }

cyanophage [MICROBIO] A virus that replicates in blue-green algae. Also known as algophage; blue-green algal virus. { sī'an·ə,fāj }

cycle of erosion See geomorphic cycle. { 'sī·kəl əv i'rō·zhən }

cycle of sedimentation [GEOL] **1.** A series of related processes and conditions appearing repeatedly in the same sequence in a sedimentary deposit. **2.** The sediments deposited from the beginning of one cycle to the beginning of a second cycle of the spread of the sea over a land area, consisting of the original land sediments, followed by those deposited by shallow water, then deep water, and then the reverse process of the receding water. Also known as sedimentary cycle. { 'sī·kəl əv ,sed·ə·mən'tā·shən }

cyclethrin [CHEM] $C_{21}H_{28}O_3$ A viscous, brown liquid, soluble in organic solvents; used as an insecticide. { sī'klē·thrən }

cyclic [SCI TECH] **1.** Pertaining to some cycle. **2.** Repeating itself in some manner in space or time. { 'sīk·lik }

cyclic salt [OCEANOGR] Salt removed from the sea as spray, blown inland, and returned to its source by land drainage. { 'sīk·lik 'sȯlt }

cycloate [CHEM] $C_{11}H_{21}NOS$ A yellow liquid with limited solubility in water; boiling point is 145–146°C; used as an herbicide to control weeds in sugarbeets, spinach, and table beets. { 'sī·klə,wāt }

cyclogenesis [METEOROL] Any development or strengthening of cyclonic circulation in the atmosphere. { ¦sī·klō'jen·ə·səs }

cyclohexanol [CHEM] $C_6H_{11}OH$ An oily, colorless, hygroscopic liquid with a camphor-like odor and a boiling point of 160.9°C; used in soapmaking, insecticides, dry cleaning, plasticizers, and germicides. Also known as hexahydrophenol. { ¦sī·klō'hek·sə,nȯl }

cyclolysis [METEOROL] The weakening or decay of cyclonic circulation in the atmosphere. { sī'kläl·ə·səs }

cyclomorphosis [ECOL] Cyclic recurrent polymorphism in certain planktonic fauna in response to seasonal temperature or salinity changes. { ,sī·klō'mȯr·fə·səs }

cyclone [METEOROL] A low-pressure region of the earth's atmosphere with roundish to elongated-oval ground plan, in-moving air currents, centrally upward air movement, and generally outward movement at various higher elevations in the troposphere. { 'sī,klōn }

cyclone family [METEOROL] A series of wave cyclones occurring in the interval between two successive major outbreaks of polar air, and traveling along the polar front, usually eastward and poleward. { 'sī,klōn ,fam·lē }

cyclone wave [METEOROL] **1.** A disturbance in the lower troposphere, of wavelength 1000–2500 kilometers; cyclone waves are recognized on synoptic charts as migratory high- and low-pressure systems. **2.** A frontal wave at the crest of which there is a center of cyclonic circulation, that is, the frontal wave of a wave cyclone. { 'sī,klōn ,wāv }

cyclonic scale |METEOROL| The scale of the migratory high-and low-pressure systems (or cyclone waves) of the lower troposphere, with wavelengths of 1000–2500 kilometers. Also known as synoptic scale. { sī'klän·ik 'skāl }

cyhexatin |CHEM| $C_{18}H_{34}OSn$ A whitish solid, insoluble in water; used as a miticide to control plant-feeding mites. { sī'hek·sə·tən }

Cyperaceae [BOT] The sedges, a family of monocotyledonous plants in the order Cyperales characterized by spirally arranged flowers on a spike or spikelet; a usually solid, often triangular stem; and three carpels. { ‚sip·ə'rās·ē‚ē }

cyrtosis [PL PATH] A virus disease of cotton characterized by stunting, distortion, and abnormal branching and coloration. { sər'tō·səs }

cytomegalovirus [MICROBIO] An animal virus belonging to subgroup B of the herpesvirus group; causes cytomegalic inclusion disease and pneumonia. { ¦sīd·ō ¦meg·ə·lō'vī·rəs }

cytotoxic |BIOL| Pertaining to an agent, such as a drug or virus, that exerts a toxic effect on cells. { ¦sīd·ə'täk·sik }

cytotoxic T cell [MED] A type of T cell which protects against pathogens that invade host cell cytoplasm, where they cannot be bound by antibodies, by recognizing and killing the host cell before the pathogens can proliferate and escape. { ¦sīd·ə‚täk·sik 'tē ‚sel }

D

2,4-D   *See* 2,4-dichlorophenoxyacetic acid.

daily forecast |METEOROL| A forecast for periods of from 12 to 48 hours in advance. { ¦dā·lē 'for,kast }

daily mean |METEOROL| The average value of a meteorological element over a period of 24 hours.   { ¦dā·lē ,mēn }

daily retardation |OCEANOGR| The amount of time by which corresponding tidal phases grow later day by day; averages approximately 50 minutes.   { ¦dā·lē ,re,tär'dā·shən }

dalapon |CHEM| Generic name for 2,2-dichloropropionic acid; a liquid with a boiling point of 185–190°C at 760 mmHg; soluble in water, alcohol, and ether; used as a herbicide.   { 'dal·ə,pän }

Dallis grass |BOT| The common name for the tall perennial forage grasses composing the genus *Paspalum* in the order Cyperales.   { 'da·ləs ,gras }

damp |PETR MIN| A poisonous gas in a coal mine.   { damp }

damp air |METEOROL| Air that has a high relative humidity.   { ¦damp ¦er }

damp haze |METEOROL| Small water droplets or very hygroscopic particles in the air, reducing the horizontal visibility somewhat, but to not less than 1¼ miles (2 kilometers); similar to a very thin fog, but the droplets or particles are more scattered than in light fog and presumably smaller.   { ¦damp 'hāz }

damping-off |PL PATH| A fungus disease of seedlings and cuttings in which the parasites invade the plant tissues near the ground level, causing wilting and rotting. { ¦dam·piŋ 'of }

dancing dervish   *See* dust whirl.   { ¦dan·siŋ 'dər·vish }

dancing devil   *See* dust whirl.   { ¦dan·siŋ ,dev·əl }

dangerous semicircle |METEOROL| The half of the circular area of a tropical cyclone having the strongest winds and heaviest seas, where a ship tends to be drawn into the path of the storm.   { 'dān·jə·rəs 'sem·i,sər·kəl }

Darwinism |BIOL| The theory of the origin and perpetuation of new species based on natural selection of those offspring best adapted to their environment because of genetic variation and consequent vigor. Also known as Darwin's theory.   { 'där·wə ,niz·əm }

Darwin's theory   *See* Darwinism.   { ¦där·winz 'thē·ə·rē }

data |SCI TECH| Numerical or qualitative values derived from scientific experiments. { 'dad·ə, 'dād·ə, *or* 'däd·ə }

dating |SCI TECH| The use of methods and techniques to fix dates, assign periods of time, and determine age in archeology, biology, and geology.   { 'dād·iŋ }

datum level   *See* datum plane.   { 'dad·əm ,lev·əl }

datum plane [ENG] A permanently established horizontal plane, surface, or level to which soundings, ground elevations, water surface elevations, and tidal data are referred. Also known as chart datum; datum level; reference level; reference plane. { 'dad·əm ˌplān }

Davidson Current [OCEANOGR] A coastal countercurrent of the Pacific Ocean running north, inshore of the California Current, along the western coast of the United States (from northern California to Washington to at least latitude 48°N) during the winter months. { 'dā·vəd·sən ˌkər·ənt }

day neutral [BOT] Reaching maturity regardless of relative length of light and dark periods. { 'dā¡nü·trəl }

day-neutral response [BIOL] A photoperiodic response that is independent or nearly independent of day length. { ¦dā¡nü·trəl ri'späns }

dazomet [CHEM] $C_5H_{10}N_2S_2$ A white, crystalline compound that decomposes at 100°C; used as a herbicide and nematicide for soil fungi and nematodes, weeds, and soil insects. Also known as tetrahydro-3,5-dimethyl-2H-1,3,5-thiadiazine-6-thione. { 'dā· zə·mət }

DBCP See dibromochloropropane.

DCNA See 2,6-dichloro-4-nitroaniline.

DCPA See dimethyl-2,3,5,6-tetrachloroterephthalate.

DDA value See depth-duration-area value. { ¦dē¡dē'ā¡val·yü }

DDD See 2,2-bis(*para*-chlorophenyl)-1,1-dichloroethane.

DDT [CHEM] Common name for an insecticide; melting point 108.5°C, insoluble in water, very soluble in ethanol and acetone, colorless, and odorless; especially useful against agricultural pests, flies, lice, and mosquitoes. It is very persistent in the environment and undergoes biomagnification in food chains. Toxic effects on top predators such as birds and the contamination of human food supplies led to an EPA ban on registration and interstate sale of DDT in the United States in 1972. Also known as dichlorodiphenyltrichloroethane.

DDVP See dichlorvos.

dead cave [GEOL] A cave where there is no moisture or no growth of mineral deposits associated with moisture. { ¦ded 'kāv }

dead sea [HYD] A body of water that has undergone precipitation of its rock salt, gypsum, or other evaporites. { ¦ded 'sē }

Dead Sea [GEOGR] A salt lake between Jordan and Israel. { ¦ded 'sē }

death rate See mortality rate. { 'deth ˌrāt }

debris [GEOL] Large fragments arising from disintegration of rocks and strata. { də'brē }

debris glacier [HYD] A glacier formed from ice fragments that have fallen from a larger and taller glacier. { də'brē ˌglā·shər }

debromoaplysiatoxin [BIOL] A bislactone toxin related to aplysiatoxin and produced by the blue-green alga *Lyngbya majuscula*. { dē,brō·mō·ə'plizh·ə,täk·sən }

deca- [SCI TECH] A prefix denoting 10. { 'dek·ə }

decay [OCEANOGR] In ocean-wave studies, the loss of energy from wind-generated ocean waves after they have ceased to be acted on by the wind; this process is accompanied by an increase in length and a decrease in height of the wave. { di'kā }

decay area [OCEANOGR] The area into which ocean waves travel (as swell) after leaving the generating area. { di'kā,er·ē·ə }

decay rate [PHYS] The time rate of disintegration of radioactive material, generally accompanied by emission of particles or gamma radiation. { di'kā,rāt }

deci- [SCI TECH] A prefix indicating 10^{-1}, 0.1, or a tenth. { 'des·ē }

deciduous [BIOL] Falling off or being shed at the end of the growing period or season. [BOT] Of plants, regularly losing their leaves at the end of each growing season. { di'sij·ə·wəs }

declining population [ECOL] A population in which old individuals outnumber young individuals. { də'klin·iŋ ,päp·yə'lā·shən }

decommissioning [ENG] The process of shutting down a nuclear facility such as a nuclear reactor or reprocessing plant so as to provide adequate protection from radiation exposure and to isolate radioactive contamination from the human environment. { dē·kə'mish·ən·iŋ }

decomposer [ECOL] A heterotrophic organism (including bacteria and fungi) which breaks down the complex compounds of dead protoplasm, absorbs some decomposition products, and releases substances usable by consumers. Also known as microcomposer; microconsumer; reducer. { de·kəm'pō·zər }

decomposition *See* chemical weathering. { dē,käm·pə'zish·ən }

decontamination [ENG] The removing of chemical, biological, or radiological contamination from, or the neutralizing of it on, a person, object, or area. { dē·kən,tam·ə'nā·shən }

decrement *See* groundwater discharge. { 'dek·rə·mənt }

deep [OCEANOGR] An area of great depth in the ocean, representing a depression in the ocean floor. { dēp }

deep-casting [OCEANOGR] Sampling ocean water at great depths by lowering a number of self-sealing bottles, usually made of brass or bronze, on a cable. { 'dēp ,kast·iŋ }

deep easterlies *See* equatorial easterlies. { ¦dēp 'ē·stər·lēz }

deep hibernation [BIOL] Profound decrease in metabolic rate and physiological function during winter, with a body temperature near 0°C, in certain warm-blooded vertebrates. Also known as hibernation. { 'dēp ,hī·bər'nā·shən }

deep inland sea [GEOGR] A sea adjacent to but in restricted communication with the sea; depth exceeds 660 feet (200 meters). { 'dēp 'in·lənd 'sē }

deep-marine sediments [GEOL] Sedimentary environments occurring in water deeper than 200 meters (660 feet), seaward of the continental shelf break, on the continental slope and the basin. { ,dēp mə¦rën 'sed·ə·mins }

deep-sea basin [GEOL] A depression of the sea floor more or less equidimensional in form and of variable extent. { ¦dēp ¦sē 'bās·ən }

deep-sea channel [GEOL] A trough-shaped valley of low relief beyond the continental rise on the deep-sea floor. Also known as mid-ocean canyon. { ¦dēp ¦sē 'chan·əl }

deep-sea plain [GEOL] A broad, almost level area forming the predominant portion of the ocean floor. { ¦dēp ¦sē 'plān }

deep-seated *See* plutonic. { ¦dēp 'sēd·əd }

deep-sea trench [GEOL] A long, narrow depression of the deep-sea floor having steep sides and containing the greatest ocean depths; formed by depression, to several kilometers' depth, of the high-velocity crustal layer and the mantle. { ¦dēp ¦sē 'trench }

deep trades *See* equatorial easterlies. { ¦dēp 'trādz }

deep water [OCEANOGR] An ocean area where depth of the water layer is greater than one-half the wave length. { 'dēp 'wȯd·ər }

deep-water wave [OCEANOGR] A surface wave whose length is less than twice the depth of the water. Also known as short wave. { 'dēp ,wȯd·ər ,wāv }

DEET   *See* diethyltoluamide.

definitive host |BIOL| The host in which a parasite reproduces sexually. Also known as primary host.   { də'fin·əd·iv 'hōst }

deflation |GEOL| The sweeping erosive action of the wind over the ground.   { di'flā·shən }

deflation lake |HYD| A lake in a basin that was formed primarily by wind erosion, especially in arid or semiarid regions.   { di'flā·shən ,lāk }

defoliate |BOT| To remove leaves or cause leaves to fall, especially prematurely.   { dē 'fō·lē,āt }

deforestation |FOR| The act or process of removing trees from or clearing a forest.   { dē ,fär·ə'stā·shən }

deglaciation |HYD| Exposure of an area from beneath a glacier or ice sheet as a result of shrinkage of the ice by melting.   { dē,glās·ē'ā·shən }

degradation |HYD| **1.** Lowering of a stream bed.   **2.** Shrinkage or disappearance of permafrost.   { ,deg·rə'dā·shən }

degrading stream |HYD| A stream actively deepening its channel or valley and capable of transporting more load than is presently provided.   { də'grād·iŋ ,strēm }

dehiscence |BOT| Spontaneous bursting open of a mature plant structure, such as fruit, anther, or sporangium, to discharge its contents.   { də'his·əns }

dehydration |CHEM| Removal of water from any substance.   { ,dē·hī'drā·shən }

dehydroacetic acid |CHEM| $C_8H_8O_4$ Crystals that melt at 108.5°C and are insoluble in water, soluble in acetone; used as a fungicide and bactericide. Abbreviated DHA.   { dē ¦hī·drō·ə¦sēd·ik 'as·əd }

deinking |CHEM ENG| The process of removing ink from recycled paper so that the fibers can be used again.   { dē'iŋk·iŋ }

dell |GEOGR| A small, secluded valley or vale.   { del }

delta |GEOL| An alluvial deposit, usually triangular in shape, at the mouth of a river, stream, or tidal inlet.   { 'del·tə }

deme |ECOL| A local population in which the individuals freely interbreed among themselves but not with those of other demes.   { dēm }

demersal |BIOL| Living at or near the bottom of the sea.   { də'mər·səl }

demeton-*S*-methyl |CHEM| $C_6H_{15}O_3PS_2$ An oily liquid with a 0.3% solubility in water; used as an insecticide and miticide to control aphids.   { 'dem·ə,tän ¦es 'meth·əl }

demeton-*S*-methyl sulfoxide |CHEM| $C_6H_{15}O_4PS_2$ A clear, amber liquid; limited solubility in water; used as an insecticide and miticide for pests of vegetable, fruit, and field crops, ornamental flowers, shrubs, and trees.   { 'dem·ə,tän ¦es 'meth·əl səl'fäk ,sīd }

demographic genetics |BIOL| A branch of population genetics and ecology concerned with genetic differences related to age, population size, genetic alteration in competitive ability, and viability.   { ¦dem·ə,graf·ik jə¦ned·iks }

demography |ECOL| The statistical study of populations with reference to natality (birth rate), mortality, migratory movements, age, and sex, among other social, ethnic, and economic factors.   { də'mäg·rə·fē }

demorphism   *See* weathering.   { dē'mòr·fiz· əm }

dendritic |SCI TECH| Having a branching, treelike structure or pattern.   { den'drid·ik }

dendritic drainage [HYD] Irregular stream branching, with tributaries joining the main stream at all angles. { den'drid·ik 'drān·ij }

dendrochemistry [CHEM] The analysis of the chemical composition of tree rings for naturally occurring or human-manufactured chemicals, especially the mineral elements, to understand the impact of pollution in the air, or surface-water or groundwater supply in ecosystems, or to detect environmental changes over time. { ,den·drō 'kem·i·strē }

dendrochronology [GEOL] The science of measuring time intervals and dating events and environmental changes by reading and dating growth layers of trees as demarcated by the annual rings. { ¦den·drō·krə'näl·ə·jē }

dendroclimatology [METEOROL] The study of the tree-ring record to reconstruct climate history, based on the fact that temperature, precipitation, and other climatic variables affect tree growth. { ,den·drō,klī·mə'täl·ə·jē }

dendroecology [ECOL] The use of tree rings to study changes in ecological processes over time such as defoliation by insect outbreaks; the effects of air, water, and soil pollution on tree growth and forest health; the age, maturity, and successional status of forest stands; and the effects of human disturbances and management on forest vitality. { ,den·drō·ē'käl·ə·jē }

dendrohydrology [HYD] The science of determining hydrologic occurrences by the comparison of tree ring thickness with streamflow or precipitation. Also known as tree-ring hydrology. { ,den·dro·hī'dräl·ə·jē }

dendrology [FOR] The division of forestry concerned with the classification, identification, and distribution of trees and other woody plants. { den'dräl·ə·jē }

dendrometer [FOR] A device used to measure a tree's height and diameter using principles based on the relation of the sides of similar triangles. { den'dräm·əd·ər }

dendrophagous [ZOO] Feeding on trees, referring to insects. { den'dräf·ə·gəs }

Dengue fever [MED] An infection borne by the *Aedes* female mosquito, and caused by one of four closely related but antigenically distinct Dengue virus serotypes (DEN-1, DEN-2, DEN-3, and DEN-4). It starts abruptly after an incubation period of 2–7 days with high fever, severe headache, myalgia, and rash. It is found throughout the tropical and subtropical zones. Also known as break-bone fever. { 'deŋ·gē ,fēv·ər }

denitrification [MICROBIO] The reduction of nitrate or nitrite to gaseous products such as nitrogen, nitrous oxide, and nitric oxide; brought about by denitrifying bacteria. { dē,nī·trə·fə'kā·shən }

denitrifying bacteria [MICROBIO] Bacteria that reduce nitrates to nitrites or nitrogen gas; most are found in soil. { dē'nī·trə,fī·iŋ bak'tir·ē·ə }

density current [METEOROL] Intrusion of a dense air mass beneath a lighter air mass; the usage applies to cold fronts. [OCEANOGR] *See* turbidity current. { 'den·səd·ē ,kər·ənt }

density-dependent factor [ECOL] A factor that affects the birth rate or mortality rate of a population in ways varying with the population density. { ¦den·səd·ē ,di¦pen·dənt ,fak·tər }

density-independent factor [ECOL] A factor that affects the birth rate or mortality rate of a population in ways that are independent of the population density. { ¦den·səd·ē ,in·də¦pen·dənt ,fak·tər }

density ratio [METEOROL] The ratio of the density of the air at a given altitude to the air density at the same altitude in a standard atmosphere. { 'den·səd·ē ,rā·shō }

denudation [GEOL] General wearing away of the land; laying bare of subjacent lands. { ,dē·nü'dā·shən }

deoxygenation |CHEM| Removal of oxygen from a substance, such as blood or polluted water. { dē,äk·sə·jə'nā·shən }

deoxyribonucleic acid [BIOL] Carrier of genetic material present in chromosomes, chromosomal material of cell organelles such as mitochondria and chloroplasts, and in some viruses, it is a linear polymer made up of deoxyribonucleotide repeating units (composed of the sugar 2-deoxyribose, phosphate, and a purine or pyrimidine base) linked by the phosphate group joining the 3' position of one sugar to the 5' position of the next; most molecules are double-stranded and antiparallel, resulting in a right-handed helix structure kept together by hydrogen bonds between a purine on one chain and a pyrimidine on another. Abbreviated DNA. { dē¦äk·sē¦rī·bō·nü ¦klē·ik 'as·əd }

2,4-DEP See tris[2-(2,4-dichlorophenoxy)ethyl]phosphite.

departure |METEOROL| The amount by which the value of a meteorological element differs from the normal value. { di'pär·chər }

depergelation |HYD| The act or process of thawing permafrost. { dē,pər·jə'lā·shən }

depletion |ECOL| Using a resource, such as water or timber, faster than it is replenished. { də'plē·shən }

depocenter |GEOL| A site of maximum deposition. { 'dep·ə,sen·tər }

deposit |GEOL| Consolidated or unconsolidated material that has accumulated by a natural process or agent. |SCI TECH| Any solid matter which is gradually laid down on a surface by a natural process. { də'päz·ət }

deposit dose |MED| The residual radioactivity deposited on the surface after a nuclear explosion, as by water falling as rain from the base surge of an underwater atomic explosion. { də'päz·ət ,dōs }

deposit gage |ENG| The general name for instruments used in air pollution studies for determining the amount of material deposited on a given area during a given time. { də'päz·ət ,gāj }

deposition |GEOL| The laying, placing, or throwing down of any material; specifically, the constructive process of accumulation into beds, veins, or irregular masses of any kind of loose, solid rock material by any kind of natural agent. { ,dep·ə'zish·ən }

depression |METEOROL| An area of low pressure; usually applied to a certain stage in the development of a tropical cyclone, to migratory lows and troughs, and to upper-level lows and troughs that are only weakly developed. Also known as low. { di'presh·ən }

depression spring |HYD| A type of gravity spring that flows onto the land surface because the surface slopes down to the water table. { di'presh·ən 'spriŋ }

depression storage |HYD| Water retained in puddles, ditches, and other depressions in the surface of the ground. { di'presh·ən ,stȯr·ij }

depth |OCEANOGR| The vertical distance from a specified sea level to the sea floor. { depth }

depth contour See isobath. { 'depth ,kän,túr }

depth curve See isobath. { 'depth ,kərv }

depth-duration-area value |METEOROL| The average depth of precipitation that has occurred within a specified time interval over an area of given size. Abbreviated DDA value. { ¦depth də¦rā·shən 'er·ē·ə ,val·yü }

depth hoar |HYD| A layer of ice crystals formed between the ground and snow cover by sublimation. Also known as sugar snow. { 'depth ,hȯr }

depth of compensation [HYD] The depth in a body of water at which illuminance has diminished to the extent that oxygen production through photosynthesis and oxygen consumption through respiration by plants are equal; it is the lower boundary of the euphotic zone. { 'depth əv ˌkäm·pən'sā·shən }

depth zone [OCEANOGR] Any one of four oceanic environments: the littoral, neritic, bathyal, and abyssal zones. { 'depth ˌzōn }

derecho See plow wind. { dā'rā·chō }

derelict land [ECOL] Land that, because of mining, drilling, or other industrial processes, or by serious neglect, is unsightly and cannot be beneficially utilized without treatment. { 'der·əˌlikt ˌland }

derivative rock See sedimentary rock. { də'riv·əd·iv 'räk }

dermatophyte [MYCOL] A fungus parasitic on skin or its derivatives. { dər'mad·əˌfīt }

Dermestidae [ZOO] The skin beetles, a family of coleopteran insects in the super-family Dermestoidea, including serious pests of stored agricultural grain products. { dər'mes·tə·dē }

desalination [CHEM ENG] Removal of salt, as from water or soil. Also known as desalting. { dēˌsal·ə'nā·shən }

desalinization See desalination. { dēˌsal·ə·nə'zā·shən }

desalting See desalination. { dē'sȯl·tiŋ }

descriptive botany [BOT] The branch of botany that deals with diagnostic characters or systematic description of plants. { di'skrip·tiv 'bät·ən·ē }

descriptive climatology [CLIMATOL] Climatology as presented by graphic and verbal description, without going into causes and theory. { di'skrip·tiv klī·mə'täl·ə·jē }

descriptive meteorology [METEOROL] A branch of meteorology which deals with the description of the atmosphere as a whole and its various phenomena, without going into theory. Also known as aerography. { di'skrip·tiv mēd·ē·ə'räl·ə·jē }

desensitization [MED] Loss or reduction of sensitivity to infection or an allergen accomplished by means of frequent, small doses of the antigen. Also known as hyposensitization. { dēˌsen·sə·tə'zā·shən }

desert [GEOGR] **1.** A wide, open, comparatively barren tract of land with few forms of life and little rainfall. **2.** Any waste, uninhabited tract, such as the vast expanse of ice in Greenland. { 'dez·ərt }

desert climate [CLIMATOL] A climate type which is characterized by insufficient moisture to support appreciable plant life; that is, a climate of extreme aridity. { ¦dez·ərt ¦klī·mət }

desert crust See desert pavement. { ¦dez·ərt ¦krəst }

desert devil See dust whirl. { 'dez·ərt ˌdev·əl }

desertification [ECOL] The creation of desiccated, barren, desertlike conditions due to natural changes in climate or possibly through mismanagement of the semiarid zone. { dəˌzərd·ə·fə'kā·shən }

desert pavement [GEOL] A mosaic of pebbles and large stones which accumulate as the finer dust and sand particles are blown away by the wind. Also known as desert crust. { ¦dez·ərt 'pāv·mənt }

desert soil [GEOL] In early United States classification systems, a group of zonal soils that have a light-colored surface soil underlain by calcareous material and a hardpan. { ¦dez·ərt 'sȯil }

desert varnish See rock varnish. { ¦dez·ərt 'vär·nish }

desert wind [METEOROL] A wind blowing off the desert, which is very dry and usually dusty, hot in summer but cold in winter, and with a large diurnal range of temperature. { ¦dez·ərt 'wind }

desiccation [HYD] The permanent decrease or disappearance of water from a region, caused by a decrease of rainfall, a failure to maintain irrigation, or deforestation or overcropping. { ˌdes·ə'kā·shən }

design climatology [CLIMATOL] The scientific analysis of climatic data for the purpose of improving the design of equipment and structures intended to operate in or withstand extremes of climate. { di'zīn klī·mə'täl·ə·jē }

design feature [ECOL] An organismal trait that can influence rates of death and reproduction, and hence Darwinian fitness. { di'zīn ˌfē·chər }

design for environment [ENG] A methodology for the design of products and systems that promotes pollution prevention and resource conservation by including within the design process the systematic consideration of the environmental implications of engineering designs. Abbreviated DFE. { di¦zīn fər in'vī·ərn·mənt }

design water depth [OCEANOGR] **1.** A value based on the sum of the vertical distance from the nominal water level to the ocean bottom and the height of the tides, both astronomical and storm. **2.** The greatest water depth in which an offshore drilling well is able to maintain its operations. { di'zīn 'wòd·ər ˌdepth }

desilication [GEOCHEM] Removal of silica, as from rock or a magma. { dē,sil·ə'kā·shən }

desmetryn [CHEM] $C_9H_{17}N_5S$ A white, crystalline compound with a melting point of 84–86°C; used as a postemergence herbicide for broadleaf and grassy weeds. { dez'me·trən }

desmochore [ECOL] A plant having sticky or barbed disseminules. { 'dez·mə,kòr }

DET *See* diethyltoluamide.

detector [SCI TECH] Apparatus or system used to detect the presence of an object, radiation, chemical compound, or such. { di'tek·tər }

detoxification [BIOL] The act or process of removing a poison or the toxic properties of a substance in the body. { dē,täk·sə·fə'kā·shən }

detrainment [METEOROL] The transfer of air from an organized air current to the surrounding atmosphere. { dē'trān·mənt }

detrital sediment [GEOL] Accumulations of the organic and inorganic fragmental products of the weathering and erosion of land transported to the place of deposition. { də'trīd·əl 'sed·ə·mənt }

detritivore [ECOL] An organism that consumes dead organic matter. { di'trid·ə,vòr }

detritus [ECOL] Dead plants and corpses or cast-off parts of various organisms. { də'trīd·əs }

detritus food web [ECOL] A trophic web that is based on the consumption of dead organic material. { di,trīd·əs 'füd ,web }

detritus tank [CIV ENG] A tank in which heavy suspended matter is removed in sewage treatment. { də'trīd·əs ,taŋk }

detrivorous [BIOL] Referring to an organism that feeds on dead animals or partially decomposed organic matter. { də'triv·ə·rəs }

development [METEOROL] The process of intensification of an atmospheric disturbance, most commonly applied to cyclones and anticyclones. { də'vel·əp·mənt }

developmental instability [GEN] Variation of development within a genotype due to local fluctuations in internal or external environmental conditions. { di‚vel·əp ¦men·təl ‚in·stə'bil·əd·ē }

developmental toxicity [MED] Adverse effects on the developing child which result from exposure to toxic chemicals or other toxic substances, can include birth defects, low birth weight, and functional or behavioral weaknesses that show up as the child develops. { di‚vel·əp¦ment·əl tak'sis·ə·dē }

devernalization [BOT] Annulment of the vernalization effect. { dē‚vərn·əl·ə'zā·shən }

De Vries effect [GEOCHEM] A relatively short-term oscillation, on the order of 100 years, in the radiocarbon content of the atmosphere, and the resulting variation in the apparent radiocarbon age of samples. { də'vrēz i'fekt }

devrinol [CHEM] $C_{17}H_{21}O_2N$ A brown solid with a melting point of 68.5–70.5°C; slight solubility in water; used as a herbicide for crops. Also known as 2-(α-naphthoxy)-N,N-diethylpropionamide. { 'dev·rə‚nȯl }

dew [HYD] Water condensed onto grass and other objects near the ground, the temperatures of which have fallen below the dew point of the surface air because of radiational cooling during the night but are still above freezing. { dü }

dew point [METEOROL] The temperature at which air becomes saturated when cooled without addition of moisture or change of pressure; any further cooling causes condensation. Also known as dew-point temperature. { 'dü ‚pȯint }

dew-point temperature *See* dew point. { 'dü ‚pȯint 'tem·prə·chər }

dew retting [MICROBIO] A type of retting process in which the stems of fiber plants are spread out in moist meadows, and the pectin decomposition is accomplished by molds and aerobic bacteria with the formation of CO_2 and H_2. { 'dü ‚red·iŋ }

DFE *See* design for environment.

DHA *See* dehydroacetic acid.

D horizon [GEOL] A soil horizon sometimes occurring below a B or C horizon, consisting of unweathered rock. { 'dē hə'rīz·ən }

di- [SCI TECH] Prefix meaning two. { dī }

DI *See* temperature-humidity index.

diadromous [ZOO] Of fish, migrating between salt and fresh waters. { dī'ad·rə·məs }

diagenesis [GEOL] Chemical and physical changes occurring in sediments during and after their deposition but before consolidation. { ‚dī·ə'jen·ə·səs }

diageotropism [BIOL] Growth orientation of a sessile organism or structure perpendicular to the line of gravity. { ¦dī·ə·jē'ä·trə‚piz·əm }

diaheliotropism [BOT] Movement of plant leaves which follow the sun such that they remain perpendicular to the sun's rays throughout the day. { ‚dī·ə‚hē·lē·ə'trä‚piz·əm }

dialifor [CHEM] $C_{14}H_{17}ClNO_4S_2P$ A white, crystalline compound with a melting point of 67–69°C; insoluble in water; used to control pests in citrus fruits, grapes, and pecans. { dī'al·ə‚fȯr }

diatom [ZOO] The common name for algae composing the class Bacillariophyceae; noted for the symmetry and sculpturing of the siliceous cell walls. { 'dī·ə‚täm }

diatropism [BOT] Growth orientation of certain plant organs that is transverse to the line of action of a stimulus. { dī'a·trə‚piz·əm }

diazinon [CHEM] $C_{12}H_{21}N_2O_3PS$ A light amber to dark brown liquid with a boiling point of 83–84°C; used as an insecticide for soil and household pests, and as an insecticide and nematicide for fruits and vegetables. { dī'a·zə₁nōn }

diazotroph [MICROBIO] An organism that carries out nitrogen fixation; examples are *Clostridium* and *Azotobacter.* { dī'az·ə₁träf }

dibromochloropropane [CHEM] $C_3H_5Br_2Cl$ A light yellow liquid with a boiling point of 195°C; used as a nematicide for crops. Abbreviated DBCP. { dī¦brō·mō₁klòr·ə'prō₁pān }

Dice's life zones [ECOL] Biomes proposed by L.R. Dice based on the concept of the biotic province. { 'dīs·əz 'līf ₁zōnz }

dichlamydeous [BOT] Having both calyx and corolla. { dī·klə'mid·ē·əs }

dichlobenil [CHEM] $C_7H_3Cl_2N$ A colorless, crystalline compound with a melting point of 139–145°C; used as a herbicide to control weeds in orchards and nurseries. { dī'klō·bə·nəl }

dichlofenthion [CHEM] $C_{10}H_{13}Cl_2O_3PS$ A white, liquid compound, insoluble in water; used as an insecticide and nematicide for ornamentals, flowers, and lawns. { dī₁klō·fən'thī₁än }

dichlofluanid [CHEM] $C_9H_{11}Cl_2FN_2O_2S$ A white powder with a melting point of 105–105.6°C; insoluble in water; used as a fungicide for fruits, garden crops, and ornamental flowers. { ₁dī·klō·flü'an·əd }

dichlone [CHEM] $C_{10}H_4O_2Cl_2$ A yellow, crystalline compound, used as a fungicide for foliage and as an algicide. { 'dī₁klōn }

dichlorobenzene [CHEM] $C_6H_4Cl_2$ Any of a group of substitution products of benzene and two atoms of chlorine; the three forms are *meta*-dichlorobenzene, colorless liquid boiling at 172°C, soluble in alcohol and ether, insoluble in water, or *ortho*-, colorless liquid boiling at 179°C, used as a solvent and chemical intermediate, or *para*-, volatile white crystals, insoluble in water, soluble in organic solvents, used as a germicide, insecticide, and chemical intermediate. { dī¦klòr·ō'ben₁zēn }

dichlorodiethylsulfide *See* mustard gas. { dī¦klòr·ō·dī¦eth·əl'səl₁fīd }

dichlorodifluoromethane [CHEM] CCl_2F_2 A nontoxic, nonflammable, colorless gas made from carbon tetrachloride; boiling point −30°C; used as a refrigerant and as a propellant in aerosols. { dī¦klòr·ō·dī¦flür·ō'me₁thān }

dichlorodiphenyltrichloroethane *See* DDT. { dī¦klòr·ō·dī¦fen·əl·trī¦klòr·ō'e₁thān }

dichlorofluoromethane [CHEM] $CHCl_2F$ A colorless, heavy gas with a boiling point of 8.9°C and a freezing point of −135°C; soluble in alcohol and ether; used in fire extinguishers and as a solvent, refrigerant, and aerosol propellant. Also known as fluorocarbon-21; fluorodichloromethane. { dī¦klòr·ō¦flür·ō'me₁thān }

2,6-dichloro-4-nitroaniline [CHEM] $C_6H_4Cl_2N_2O_2$ A yellow, crystalline compound that melts at 192–194°C; used as a fungicide for fruits, vegetables, and ornamental flowers. Abbreviated DCNA. { ¦tü ¦siks dī'klòr·ō ¦fòr ₁nī·trō'an·ə₁lēn }

dichloropentane [CHEM] $C_5H_{10}Cl_2$ Mixed dichloro derivatives of normal pentane and isopentane; clear, light-yellow liquid used as solvent, paint and varnish remover, insecticide, and soil fumigant. { dī¦klòr·ō'pen₁tān }

dichlorophen [CHEM] $C_{13}H_{10}Cl_2O_2$ A white, crystalline compound with a melting point of 177–178°C; used as an agricultural fungicide, germicide in soaps, and antihelminthic drug in humans. { dī'klòr·ə·fən }

2,4-dichlorophenoxyacetic acid [CHEM] $Cl_2C_6H_3OCH_2COOH$ Yellow crystals, melting at 142°C; used as a herbicide and pesticide. Abbreviated 2,4-D. { ¦tü ¦fòr dī¦klòr·ō·fə ¦näk·sē·ə'sēd·ik 'as·əd }

dichlorprop |CHEM| $C_9H_8Cl_2O_3$ A colorless, crystalline solid with a melting point of 117–118°C; used as a herbicide and fumigant for brush control on rangeland and rights-of-way. Abbreviated 2,4-DP. { dī'klȯr,präp }

dichlorvos |CHEM| $C_4H_7O_4Cl_2P$ An amber liquid, used as an insecticide and miticide on public health pests, stored products, and flies on cattle. Abbreviated DDVP. { dī'klȯr ,väs }

dichogamous |BOT| Referring to a type of flower in which the pistils and stamens reach maturity at different times. { dī'käg·ə·məs }

diclinous |BOT| Having stamens and pistils on different flowers. { dī'klī·nəs }

dicotyledon |BOT| Any plant of the class Magnoliopsida, all having two cotyledons. { ,dī,käd·əl'ēd·ən }

dicrotophos |CHEM| $C_8H_{16}O_2P$ The dimethyl phosphate of 3-hydroxy-N,N-dimethyl-*cis*-crotonamide; a brown liquid with a boiling point of 400°C; miscible with water; used as an insecticide and miticide for cotton, soybeans, seeds, and ornamental flowers. { dī'kräd·ə,fäs }

dicyanogen *See* cyanogen. { ¦dī¦sī'an·ə·jən }

dicyclohexylamine |CHEM| $(C_6H_{11})_2NH$ A clear, colorless liquid with a boiling point of 256°C; used for insecticides, corrosion inhibitors, antioxidants, and detergents, and as a plasticizer and catalyst. { dī¦sī·klō,hek'sil·ə,mēn }

dieback |ECOL| A large area of exposed, unprotected swamp or marsh deposits resulting from the salinity of a coastal lagoon. { 'dī,bak }

die down |BOT| Normal seasonal death of aboveground parts of herbaceous perennials. { 'dī ,daún }

diel |SCI TECH| Occurring on a 24-hour cycle, as opposed to diurnal (day) or nocturnal (night) occurrences. { 'dī,el }

dieldrin |CHEM| $C_{12}H_8Cl_6O$ A white, crystalline contact insecticide obtained by oxidation of aldrin; used in mothproofing carpets and other furnishings. { 'dēl·drən }

diet |BIOL| The food or drink regularly consumed. { 'dī·ət }

diethyl *para*-nitrophenyl phosphate *See para*-oxon. { dī,eth·əl ¦par·ə ,nī·trō'fen·əl 'fäs ,fāt }

diethyl phthalate |CHEM| $C_6H_4(CO_2C_2H_5)_2$ Clear, colorless, odorless liquid with bitter taste, boiling at 298°C; soluble in alcohols, ketones, esters, and aromatic hydrocarbons, partly soluble in aliphatic solvents; used as a cellulosic solvent, wetting agent, alcohol denaturant, mosquito repellent, and in perfumes. { dī'eth·əl 'tha,lāt }

diethyltoluamide |CHEM| $C_{12}H_{17}ON$ A liquid whose color ranges from off-white to light yellow; used as an insect repellent for people and clothing. Also known as DEET; DET; N,N-diethyl-*meta*-toluamide. { ,dī¦eth·əl,täl·yü'a,mīd }

N,N-diethyl-*meta*-toluamide *See* diethyltoluamide. { ¦en ¦en dī,eth·əl ¦med·ə 'tə·lü·ə ,mīd }

differential chart |METEOROL| A chart showing the amount and direction of change of a meteorological quantity in time or space. { ,dif·ə'ren·chəl 'chärt }

differential erosion |GEOL| Rapid erosion of one area of the earth's surface relative to another. { ,dif·ə'ren·chəl i'rō·zhən }

diffusion |PHYS| **1.** The spontaneous movement and scattering of particles (atoms and molecules), of liquids, gases, and solids. **2.** In particular, the macroscopic motion of the components of a system of fluids that is driven by differences in concentration. { də'fyü·zhən }

diffusion respiration [BIOL] Exchange of gases through the cell membrane, between the cells of unicellular or other simple organisms and the environment. { də'fyü·zhən res·pə'rā·shən }

Digenea [ZOO] A group of parasitic flatworms or flukes constituting a subclass or order of the class Trematoda and having two types of generations in the life cycle. { dī'jē·nē·ə }

digested sludge [CIV ENG] Sludge or thickened mixture of sewage solids with water that has been decomposed by anaerobic bacteria. { də'jes·təd 'sləj }

digestion [BIOL] The process of converting food to an absorbable form by breaking it down to simpler chemical compounds. [CHEM ENG] **1.** Liquefaction of organic waste materials by action of microbes. **2.** Removing lignin from wood in manufacture of chemical cellulose paper pulp. [CIV ENG] The process of sewage treatment by the anaerobic decomposition of organic matter. { də'jes·chən }

digestive efficiency [ECOL] A measure of the amount of ingested chemical energy actually absorbed by an animal. { dī¦jes·tiv i'fish·ən·sē }

diggings [SCI TECH] **1.** Excavated materials. **2.** A place of excavating. { ˌdig·iŋz }

digitalis [MED] The dried leaf of the purple foxglove plant (*Digitalis purpurea*), containing digitoxin and gitoxin; constitutes a powerful cardiac stimulant and diuretic. { dij·ə'tal·əs }

digitoxin [CHEM] $C_{41}H_{64}O_{13}$ A poisonous steroid glycoside found as the most active principle of digitalis, from the foxglove leaf. { ˌdij·ə'täk·sən }

5,6-dihydro-2-methyl-1,4-oxathiin-3-carboxanilide-4,4-dioxide See oxycarboxin. { ¦fīv ¦siks dī¦hī·drō ¦tü meth·əl ¦wən ¦fór äk·sə·'thī·ən ¦thrē ˌkär'bäks'an·əl·īd ¦fór ¦fór ¦dī'äk ˌsīd }

dihydroxyacetone [CHEM] $(HOCH_2)_2CO$ A colorless, crystalline solid with a melting point of 80°C; soluble in water and alcohol; used in medicine, fungicides, plasticizers, and cosmetics. Abbreviated DHA. { ¦dī,hī¦dräk·sē'as·ə,tōn }

diisopropanolamine [CHEM] $(CH_3CHOHCH_2)_2NH$ A white, crystalline solid with a boiling point of 248.7°C; used as an emulsifying agent for polishes, insecticides, and water paints. Abbreviated DIPA. { dī¦ī,sō,prō·pə'näl·ə,mēn }

dimetan [CHEM] The generic name for 5,5-dimethyldehydroresorcinol dimethylcarbamate, a synthetic carbamate insecticide. { 'dī·mə,tan }

dimethachlon [CHEM] $C_{10}H_7Cl_2NO_2$ A yellowish, crystalline solid with a melting point of 136.5–138°C; insoluble in water; used as a fungicide. { dī·mə'tha,klän }

dimethoate [CHEM] $C_5H_{12}NO_3PS_2$ A crystalline compound, soluble in most organic solvents; used as an insecticide. { dī'meth·ə,wāt }

dimethrin [CHEM] $C_{19}H_{28}O_2$ An amber liquid with a boiling point of 175°C; soluble in petroleum hydrocarbons, alcohols, and methylene chloride; used as an insecticide for mosquitoes, body lice, stable flies, and cattle flies. { dī'me·thrən }

dimethylbenzene See xylene. { ˌdī'meth·əl'ben,zēn }

dimethyl carbate [CHEM] $C_{11}H_{14}O_4$ A colorless liquid with a boiling point of 114–115°C; used as an insect repellent. { ˌdī'meth·əl 'kär,bāt }

dimethyl phthalate [CHEM] $C_6H_4(COOCH_3)_2$ Odorless, colorless liquid, boiling at 282°C; soluble in organic solvents, slightly soluble in water; used as a plasticizer, in resins, lacquers, and perfumes, and as an insect repellent. { ˌdī'meth·əl 'tha,lāt }

dimethyl-2,3,5,6-tetrachloroterephthalate [CHEM] $C_{10}H_6Cl_4O_4$ A colorless, crystalline compound with a melting point of 156°C; used as an herbicide for turf, ornamental flowers, and certain vegetables and berries. Abbreviated DCPA. { ˌdī'meth·əl ¦tü ¦thrē ¦fīv ¦siks ˌte·trə·klór·ō,ter·ə'tha,lāt }

dimictic lake [HYD] A lake which circulates twice a year. { dī'mik·tik 'lāk }

dimorphism [SCI TECH] Existing in two distinct forms, with reference to two members expected to be identical. { dī'mȯr,fiz·əm }

dinitramine [CHEM] $C_{11}H_{13}N_3O_4F_3$ A yellow solid with a melting point of 98–99°C; used as a preemergence herbicide for annual grass and broadleaf weeds in cotton and soybeans. { dī'nī·trə,mēn }

dinitrogen fixation *See* nitrogen fixation. { dī'nī·trə·jən fik'sā·shən }

dinoflagellate [ZOO] Unicellular, photosynthetic organism possessing two flagella; although primarily marine, some occur in fresh water. { ,dī·nō·'fla·jə·lət }

dinoseb [CHEM] $C_{10}H_{12}O_5N_2$ A reddish-brown liquid with a melting point of 32°C; used as an insecticide and herbicide for numerous crops and in fruit and nut orchards. { 'dī·nə,seb }

dinoterb acetate [CHEM] $C_{12}H_{14}N_2O_6$ A yellow, crystalline compound with a melting point of 133–134°C; used as a preemergence herbicide for sugarbeets, legumes, and cereals, and as a postemergence herbicide for maize, sorghum, and alfalfa. { 'dī·nə ,tərb 'as·ə,tāt }

dioctyl phthalate [CHEM] $(C_8H_{17}OOC)_2C_6H_4$ Pale, viscous liquid, boiling at 384°C; insoluble in water; used as a plasticizer for acrylate, vinyl, and cellulosic resins, and as a miticide in orchards. Abbreviated DOP. { dī,äkt·əl 'tha,lāt }

dioecious [BIOL] Having the male and female reproductive organs on different individuals. Also known as dioic. { dī'ē·shəs }

dioic *See* dioecious. { dī'ō·ik }

dioxin [CHEM] A member of a family of highly toxic chlorinated aromatic hydrocarbons; found in a number of chemical products as lipophilic contaminants. Also known as polychlorinated dibenzo-*para*-dioxin. { dī'äk·sən }

DIPA *See* diisopropanolamine.

diphenamid [CHEM] $C_{16}H_{17}ON$ An off-white, crystalline compound with a melting point of 134–135°C; used as a preemergence herbicide for food crops, fruits, and ornamentals. { dī'fen·ə·məd }

diphenatrile [CHEM] $C_{14}H_{11}N$ A yellow, crystalline compound with a melting point of 73–73.5°C; used as a preemergence herbicide for turf. { dī'fen·ə·trəl }

diphenylene oxide [CHEM] $C_{12}H_8O$ A crystalline solid derived from coal tar; melting point is 87°C; used as an insecticide. { dī,fen·əl,ēn 'äk,sīd }

Diphyllobothrium latum [ZOO] A large tapeworm that infects humans, dogs, and cats; causes anemia and disorders of the nervous and digestive systems in humans. { dī ,fil·ō'bäth·rē·əm 'läd·əm }

diploid [GEN] Having two complete chromosome pairs in a nucleus (2N). { 'di,plȯid }

dip oil [AGR] Oil containing about 25% tar acids; used as dip for animals to kill insect parasites. { 'dip ,ȯil }

dipping acid *See* sulfuric acid. { 'dip·iŋ ,as·əd }

dip stream [HYD] A consequent stream that flows in the direction of the dip of the strata it traverses. { 'dip ,strēm }

diquat [CHEM] $C_{12}H_{12}N_2Br_2$ A yellow water-soluble solid used as a herbicide. { 'dī ,kwät }

dirt bed [GEOL] A buried soil containing partially decayed organic material; sometimes occurs in glacial drift. { 'dərt ,bed }

117

Discellaceae [MYCOL] A family of fungi of the order Sphaeropsidales, including saprophytes and some plant pathogens. { ˌdis·ə'lās·ē,ē }

disclimax [ECOL] A climax community that includes foreign species following a disturbance of the natural climax by humans or domestic animals. Also known as disturbance climax. { dis'klī·maks }

discomfort index See temperature-humidity index. { dis'kəm·fərt ˌin,deks }

discontinuous construction [ENG] A building in which there is no solid connection between the rooms and the building structure or between different sections of the building; the design aims to reduce the transmission of noise. { ˌdis·kən'tin·yə·wəs kən'strək·shən }

discrete-film zone See belt of soil water. { di'skrēt ˌfilm ˌzōn }

disdrometer [ENG] Equipment designed to measure and record the size distribution of raindrops as they occur in the atmosphere. { diz'dräm·əd·ər }

disease [MED] An alteration of the dynamic interaction between an individual and his or her environment which is sufficient to be deleterious to the well-being of the individual and produces signs and symptoms. { di'zēz }

dishpan experiment [METEOROL] A model experiment carried out by differential heating of fluid in a flat, rotating pan; it establishes similarity with the atmosphere and is used to reproduce many important features of the general circulation and, on a smaller scale, atmospheric motion. { 'dish,pan ik,sper·ə·mənt }

disk cultivator [AGR] A cultivator consisting of pairs of oppositely inclined disks. { 'disk 'kəl·tə,vād·ər }

disk furrower [AGR] A furrower in which concave disks, at an angle to the direction of motion, are used to cut the soil. { 'disk ˌfər·ə·wər }

disk harrow [AGR] A harrow which has two or more opposed gangs of 3–12 disks for cutting clods and trash, destroying weeds, cutting in cover crops, and smoothing and preparing the surface for various farming operations. { 'disk ˌha·rō }

disodium methylarsonate [CHEM] $CH_3AsO(ONa)_2$ A colorless, hygroscopic, crystalline solid; soluble in water and methanol; used in pharmaceuticals and as a herbicide. Abbreviated DMA. { dī'sōd·ē·əm ¦meth·əl'ärs·ən,āt }

dispersal barrier [ECOL] A physical structure that prevents organisms from crossing into new space. { də'spər·səl ˌbar·ē·ər }

dispersal pattern [GEOCHEM] Distribution pattern of metals in soil, rock, water, or vegetation. { də'spər·səl ˌpad·ərn }

dispersed elements [GEOCHEM] Elements which form few or no independent minerals but are present as minor ingredients in minerals of abundant elements. { də'spərst 'el·ə·mənts }

dispersion [CHEM] A distribution of finely divided particles in a medium. { də'spər·zhən }

dissected topography [GEOGR] Physical features marked by erosive cutting. { də'sek·təd tə'päg·rə·fē }

disseminule [BIOL] An individual organism or part of an organism adapted for the dispersal of a population of organisms, such as seeds and spores. { də'sem·ə,nyül }

dissolved load [HYD] Material carried in solution by a stream or river. { dī¦zälvd 'lōd }

distemper [VET MED] Any of several contagious virus diseases of mammals, especially the form occurring in dogs, marked by fever, respiratory inflammation, and destruction of myelinated nerve tissue. { dis'tem·pər }

distributary [HYD] An irregular branch flowing out from a main stream and not returning to it, as in a delta. Also known as distributary channel. { də'strib·yə,ter·ē }

distributary channel See distributary. { də'strib·yə,ter·ē ,chan·əl }

distribution graph [HYD] A statistically derived hydrograph for a storm of specified duration, graphically representing the percent of total direct runoff passing a point on a stream, as a function of time; usually presented as a histogram or table of percent runoff within each of successive short time intervals. { ,dis·trə'byü·shən ,graf }

district forecast [METEOROL] In U.S. Weather Bureau usage, a general weather forecast for conditions over an established geographical "forecast district." { 'di·strikt 'fór ,kast }

disturbance [METEOROL] **1.** Any low or cyclone, but usually one that is relatively small in size and effect. **2.** An area where weather, wind, pressure, and so on show signs of the development of cyclonic circulation. **3.** Any deviation in flow or pressure that is associated with a disturbed state of the weather, such as cloudiness and precipitation. **4.** Any individual circulatory system within the primary circulation of the atmosphere. { də'stər·bəns }

disturbance climax See disclimax. { də'stər·bəns 'klī,maks }

ditch [CIV ENG] **1.** A small artificial channel cut through earth or rock to carry water for irrigation or drainage. **2.** A long narrow cut made in the earth to bury pipeline, cable, or similar installations. { dich }

ditching [ENG] The digging of ditches, as around storage tanks or process areas to hold liquids in the event of a spill or along the sides of a roadway for drainage. { 'dich·iŋ }

diurnal [BIOL] Active during daylight hours. [SCI TECH] Occuring during the daytime. { dī'ərn·əl }

diurnal inequality [OCEANOGR] The difference between the heights of the two high waters or the two low waters of a lunar day. { dī'ərn·əl ,in·ə'kwäl·əd·ē }

diurnal migration [BIOL] The daily rhythmic movements of organisms in the sea from deeper water to the surface at the approach of darkness and their return to deeper water before dawn. { dī'ərn·əl mī'grā·shən }

diurnal tide [OCEANOGR] A tide in which there is only one high water and one low water each lunar day. { dī'ərn·əl 'tīd }

divagation [HYD] Lateral shifting of the course of a stream caused by extensive deposition of alluvium in its bed and frequently accompanied by the development of meanders. { ,div·ə'gā·shən }

divergence [METEOROL] Horizontal net outflow of air caused by winds. [OCEANOGR] A horizontal flow of water, in different directions, from a common center or zone. { də'vər·jəns }

divergent adaptation [GEN] Adaptation to different kinds of environment that results in divergence from a common ancestral form. Also known as branching adaptation; cladogenic adaptation. { də'vər·jənt ,ad,ap'tā·shən }

divide [GEOGR] A ridge or section of high ground between drainage systems. { də'vīd }

divinyl ether See vinyl ether. { dī'vīn·əl 'ē·thər }

divinyl oxide See vinyl ether. { dī'vīn·əl 'äk,sīd }

DMA See disodium methylarsonate.

DMDT See methoxychlor.

DNA See deoxyribonucleic acid.

Dobson spectrophotometer [CHEM] A photoelectric spectrophotometer used in the determination of the ozone content of the atmosphere; compares the solar energy at two wavelengths in the absorption band of ozone by permitting the radiation of each to fall alternately upon a photocell. { 'däb·sən ˌspek·trō·fə'täm·əd·ər }

Dobson unit [METEOROL] The unit of measure for atmospheric ozone; one Dobson unit is equal to 2.7×10^{16} ozone molecules per square centimeter, which would be equivalent to a layer of ozone 0.001 centimeter thick, at 1 atmosphere and 0°C. { 'däb·sən ˌyü·nət }

doldrums [METEOROL] A nautical term for the equatorial trough, with special reference to the light and variable nature of the winds. Also known as equatorial calms. { 'dōl ˌdrəmz }

domestication [BIOL] The adaptation of an animal or plant through breeding in captivity to a life intimately associated with and advantageous to humans. { də ˌmes·tə'kā·shən }

dominance [ECOL] The influence that a controlling organism has on numerical composition or internal energy dynamics in a community. { 'däm·ə·nəns }

dominant species [ECOL] A species of plant or animal that is particularly abundant or controls a major portion of the energy flow in a community. { 'däm·ə·nənt 'spē,shēz }

DOP See dioctyl phthalate.

dormancy [BOT] A state of quiescence during the development of many plants characterized by their inability to grow, though continuing their morphological and physiological activities. { 'dȯr·mən·sē }

Dorngeholz See thornbush. { 'dȯrn·gəˌhōlts }

Dorngestrauch See thornbush. { 'dȯrn·gəˌstrau̲k̲ }

dornveld See thornbush. { 'dȯrnˌfelt }

dose rate [MED] The rate at which nuclear radiation is delivered. { 'dōs ˌrāt }

dose-rate meter [ENG] An instrument that measures radiation dose rate. { 'dōs ˌrāt ˌmēd·ər }

dosimetry [ENG] Measurement of the power, energy, irradiance, or radiant exposure of high-energy, ionizing radiation. Also known as radiation dosimetry. { dō'sim·ə·trē }

dosing tank [CIV ENG] A holding tank that discharges sewage at a rate required by treatment processes. { 'dōs·iŋ ˌtaŋk }

double ebb [OCEANOGR] An ebb current comprising two maxima of velocity that are separated by a smaller ebb velocity. { ¦dəb·əl 'eb }

double fertilization [BOT] In most seed plants, fertilization involving fusion between the egg nucleus and one sperm nucleus, and fusion between the other sperm nucleus and the polar nuclei. { ¦dəb·əl ˌfərd·əl·ə'zā·shən }

double tide [OCEANOGR] A high tide comprising two maxima of nearly identical height separated by a relatively small depression, or low tide comprising two minima separated by a relatively small elevation. { ¦dəb·əl 'tīd }

doubling dose [GEN] The radiation dose that would double the rate of spontaneous mutation. { ¦dəb·liŋ ¦dōs }

douse [PETR MIN] To locate and delineate subsurface resources such as water, oil, gas, or minerals. { daús }

downcomer [ENG] In an air-pollution control system, a pipe that conducts gases downward to a device that removes undesirable substances. { 'daúnˌkəm·ər }

downrush [METEOROL] A term sometimes applied to the strong downward-flowing air current that marks the dissipating stages of a thunderstorm. { 'daủn,rəsh }

downstream [HYD] In the direction of flow, as a current or waterway. { 'daủn,strēm }

downwelling *See* sinking. { 'daủn,wel·iŋ }

downy mildew [PL PATH] A fungus disease of higher plants caused by members of the family Peronosporaceae and characterized by a white, downy growth on the diseased plant parts. { ¦daủn·ē 'mil·dü }

2,4-DP *See* dichlorprop.

Dracunculoidea [ZOO] An order or superfamily of parasitic nematodes characterized by their habitat in host tissues and by the way larvae leave the host through a skin lesion. { drə,kəŋ·kyə'lóid·ē·ə }

Draeger escape apparatus [PETR MIN] A portable, self-contained oxygen-breathing apparatus that is carried on the back of the user; protects against poisonous gases or oxygen shortages for 1 hour. { 'drāg·ər ə'skāp ,ap·ə,rad·əs }

drain [CIV ENG] **1.** A channel which carries off surface water. **2.** A pipe which carries off liquid sewage. { drān }

drainage [HYD] The pattern followed by the waters of an area as they pass or flow off in surface or subsurface streams. { 'drān·ij }

drainage area *See* drainage basin. { 'drān·ij ,er·ē·ə }

drainage basin [HYD] An area in which surface runoff collects and from which it is carried by a drainage system, as a river and its tributaries. Also known as catchment area; drainage area; feeding ground; gathering ground; hydrographic basin. { 'drān·ij ,bā·sən }

drainage canal [CIV ENG] An artificial canal built to drain water from an area having no natural outlet for precipitation accumulation. { 'drān·ij kə,nal }

drainage density [HYD] Ratio of the total length of all channels in a drainage basin to the basin area. { 'drān·ij ,den·səd·ē }

drainage lake [HYD] An open lake which loses water via a surface outlet or whose level is essentially controlled by effluent discharge. { 'drān·ij ,lāk }

drainage pattern [HYD] The configuration of a natural or artificial drainage system; stream patterns reflect the topography and rock patterns of the area. { 'drān·ij ,pad·ərn }

drainage ratio [HYD] The ratio expressing runoff compared with precipitation in a specific area for a given time period. { 'drān·ij ,rā·shō }

drainage system [HYD] A surface stream or a body of impounded surface water, together with all other such streams and bodies that are tributary, by which a geographical area is drained. { 'drān·ij ,sis·təm }

drainage wind *See* gravity wind. { 'drān·ij ,wind }

drawdown [HYD] The magnitude of the change in water surface level in a well, reservoir, or natural body of water resulting from the withdrawal of water. { 'dró,daủn }

dredge [ENG] A cylindrical or rectangular device for collecting samples of bottom sediment and benthic fauna. { drej }

dressing [AGR] Manure or compost used as a fertilizer. { 'dres·iŋ }

drift *See* drift current. { drift }

drift bottle [OCEANOGR] A bottle which is released into the sea for studying currents; contains a card, identifying the date and place of release, to be returned by the finder with date and place of recovery. Also known as floater. { 'drift ,bäd·əl }

drift current |OCEANOGR| A wide, slow-moving ocean current principally caused by winds. Also known as drift; wind drift; wind-driven current. { 'drift ,kə·rənt }

drift glacier See snowdrift glacier. { 'drift ,glā·shər }

drift ice |OCEANOGR| Sea ice that has drifted from its place of formation. { 'drift ,īs }

drift ice foot See ramp. { 'drift ,īs ,fút }

drifting snow [METEOROL] Wind-driven snow raised from the surface of the earth to a height of less than 6 feet (1.8 meters). { 'drif·tiŋ 'snō }

drift station |OCEANOGR| **1.** A scientific station established on the ice of the Arctic Ocean, generally based on an ice flow. **2.** A set of observations made over a period of time from a drifting vessel. { 'drift ,stā·shən }

drip [HYD] Condensed or otherwise collected moisture falling from leaves, twigs, and so forth. { drip }

drip irrigation [AGR] A method of providing water to plants, almost continuously, through small-diameter tubes and emitters. { 'drip ,ir·i,gā·shən }

driven snow [METEOROL] Snow which has been moved by wind and collected into snowdrifts. { ¦driv·ən 'snō }

drizzle [METEOROL] Very small, numerous, and uniformly dispersed water drops that may appear to float while following air currents; unlike fog droplets, drizzle falls to the ground; it usually falls from low stratus clouds and is frequently accompanied by low visibility and fog. { 'driz·əl }

drizzle drop [METEOROL] A drop of water of diameter 0.2 to 0.5 millimeter falling through the atmosphere; however, all water drops of diameter greater than 0.2 millimeter are frequently termed raindrops, as opposed to cloud drops. { 'driz·əl ,dräp }

drop |PL PATH| A fungus disease of various vegetables caused by *Sclerotinia sclerotiorum* and characterized by wilt and stem rot. { dräp }

droplet [METEOROL] A water droplet in the atmosphere; there is no defined size limit separating droplets from drops of water, but sometimes a maximum diameter of 0.2 millimeter is the limit for droplets. { 'dräp·lət }

droplet infection [MED] Infection by contact with airborne droplets of sputum carrying infectious agents. { 'dräp·let in,fek·shən }

drop-size distribution [METEOROL] The frequency distribution of drop sizes (diameters, volumes) that is characteristic of a given cloud or rainfall. { 'dräp ,sīz ,dis·trə'byü·shən }

drop theory See barrier theory of cyclones. { 'dräp ,thē·ə·rē }

drought [CLIMATOL] A period of abnormally dry weather sufficiently prolonged so that the lack of water causes a serious hydrologic imbalance (such as crop damage, water supply shortage, and so on) in the affected area; in general, the term should be reserved for relatively extensive time periods and areas. { draút }

drowned coast [GEOL] A shoreline transformed from a hilly land surface to an archipelago of small islands by inundation by the sea. { ¦draúnd 'kōst }

drowned river mouth See estuary. { ¦draúnd 'riv·ər ,maúth }

drowned stream [HYD] A stream that has been flooded over by the ocean. Also known as flooded stream. { ¦draúnd 'strēm }

drowned valley [GEOL] A valley whose lower part has been inundated by the sea due to submergence of the land margin. { ¦draúnd 'val·ē }

droxtal |HYD| An ice particle measuring 10–20 micrometers in diameter, formed by direct freezing of supercooled water droplets at temperatures below −30°C. { 'dräk·stəl }

drug resistance |MICROBIO| A decreased reactivity of living organisms to the injurious actions of certain drugs and chemicals. { ¦drəg ri'zis·təns }

drupe [BOT] A fleshy fruit, such as cherry, having a single seed within a stony endocarp (or pit). Also known as stone fruit. { drüp }

dry |SCI TECH| Free from or deficient in moisture. { drī }

dry adiabat [METEOROL] A line of constant potential temperature on a thermodynamic diagram. { ¦drī 'ad·ē·ə¦bat }

dry adiabatic lapse rate [METEOROL] A special process lapse rate of temperature, defined as the rate of decrease of temperature with height of a parcel of dry air lifted adiabatically through an atmosphere in hydrostatic equilibrium. Also known as adiabatic lapse rate; adiabatic rate. { ¦drī ˌad·ē·ə¦bad·ik 'laps ˌrāt }

dry adiabatic process [METEOROL] An adiabatic process in a system of dry air. { ¦drī ˌad·ē·ə¦bad·ik 'präs·əs }

dry air [METEOROL] Air that contains no water vapor. { ¦drī 'er }

dry-bulb temperature [PHYS] The actual air temperature as measured by a dry-bulb thermometer. { ¦drī ˌbəlb 'tem·prə·chər }

dry-chemical fire extinguisher [CHEM ENG] A dry powder, consisting principally of sodium bicarbonate, which is used for extinguishing small fires, especially electrical fires. { ¦drī ˌkem·i·kəl 'fīr ik¸stiŋ·gwə·shər }

dry-cleaning fluid [CHEM ENG] An organic solvent such as chlorinated hydrocarbons or petroleum naphtha with narrow, carefully selected boiling points; used in dry cleaning. { 'drī ¦klēn·iŋ ˌflü·əd }

dry climate [CLIMATOL] **1.** In W. Köppen's climatic classification, the major category which includes steppe climate and desert climate, defined strictly by the amount of annual precipitation as a function of seasonal distribution and of annual temperature. **2.** In C. W. Thornwaite's climatic classification, any climate type in which the seasonal water surplus does not counteract seasonal water deficiency, and having a moisture index of less than zero; included are the dry subhumid, semiarid, and arid climates. { ¦drī 'klī·mət }

dry-dock iceberg *See* valley iceberg. { 'drī ˌdäk 'īs¸bərg }

dry farming [AGR] Production of crops in regions having sparse rainfall without the use of irrigation by employing cultivation techniques that conserve soil moisture. { ¦drī 'färm·iŋ }

dry firn *See* polar firn. { ¦drī 'fərn }

dry fog [METEOROL] A fog that does not moisten exposed surfaces. { ¦drī 'fäg }

dry forest [FOR] A type of forest characterized by relatively sparse distributions of pine, juniper, oak, olive, acacia, mesquite, and other drought-resistant species growing in scrub woodland, savanna, or chaparral settings, occurs in the southwestern United States, Mediterranean region, sub-Saharan Africa, and semiarid regions of Mexico, India, and Central and South America. { ¦drī 'fär·əst }

dry freeze [HYD] The freezing of the soil and terrestrial objects caused by a reduction of temperature when the adjacent air does not contain sufficient moisture for the formation of hoarfrost on exposed surfaces. { ¦drī 'frēz }

dry haze [METEOROL] Fine dust or salt particles in the air, too small to be individually apparent but in sufficient number to reduce horizontal visibility, and to give the atmosphere a characteristic hazy appearance. { ¦drī 'hāz }

dry limestone process [CHEM ENG] An air-pollution control method in which sulfur oxides are exposed to limestone to convert them to disposable residues. { 'drī 'līm ‚stōn ‚präs·əs }

dryline [METEOROL] The boundary separating warm dry air from warm moist air along which thunderstorms and tornadoes may develop. { 'drī‚līn }

dry rot [MICROBIO] A rapid decay of seasoned timber caused by certain fungi which cause the wood to be reduced to a dry, friable texture. [PL PATH] Any of various rot diseases of plants characterized by drying of affected tissues. { 'drī ‚rät }

dry sand [GEOL] **1.** A formation, underlying the production sand, into which oil has leaked due to careless drilling practices. **2.** A nonproductive oil sand. { ¦drī ¦sand }

dry season [CLIMATOL] In certain types of climate, an annually recurring period of one or more months during which precipitation is at a minimum for the region. { 'drī ‚sēz·ən }

dry spell [CLIMATOL] A period of abnormally dry weather, generally reserved for a less extensive, and therefore less severe, condition than a drought; in the United States, describes a period lasting not less than 2 weeks, during which no measurable precipitation was recorded. { 'drī ‚spel }

dry-steam energy system [ENG] **1.** A geothermal energy source that produces superheated steam. **2.** A hydrothermal convective system driven by vapor with a temperature in excess of 300°F (150°C). { 'drī ¦stēm 'en·ər·jē ‚sis·təm }

dry tongue [METEOROL] In synoptic meteorology, a pronounced protrusion of relatively dry air into a region of higher moisture content. { ¦drī ¦təŋ }

dry valley [GEOL] A valley, usually in a chalk or karst type of topography, that has no permanent water course along the valley floor. { ¦drī 'val·ē }

dry well [CIV ENG] **1.** A well that has been completely drained. **2.** An excavated well filled with broken stone and used to receive drainage when the water percolates into the soil. **3.** Compartment of a pumping station in which the pumps are housed. [ENG] The first containment tank surrounding a water-cooled nuclear reactor that uses the pressure-suppressing containment system. { 'drī ‚wel }

dune [GEOL] A mound or ridge of unconsolidated granular material, usually of sand size and of durable composition (such as quartz), capable of movement by transfer of individual grains entrained by a moving fluid. { dün }

duplicatus [METEOROL] A cloud variety composed of superposed layers, sheets, or patches, at slightly different levels and sometimes partly merged. { ‚dü·plə'käd·əs }

duration [OCEANOGR] The interval of time of the rising or falling tide, or the length of time of flood or ebb tidal currents. { də'rā·shən }

duricrust [GEOL] The case-hardened soil crust formed in semiarid climates by precipitation of salts; contains aluminous, ferruginous, siliceous, and calcareous material. { 'dúr·ə‚krəst }

dust and fume monitor [PETR MIN] An instrument designed to measure and record concentrations of dust, fume, and gas in mine environments over an extended period of time. { ¦dəst ən 'fyüm ‚män·əd·ər }

dust bowl [CLIMATOL] A name given, early in 1935, to the region in the south-central United States afflicted by drought and dust storms, including parts of Colorado, Kansas, New Mexico, Texas, and Oklahoma, and resulting from a long period of deficient rainfall combined with loosening of the soil by destruction of the natural vegetation; dust bowl describes similar regions in other parts of the world. { 'dəst ‚bōl }

dust control system [ENG] System to capture, settle, or inert dusts produced during handling, drying, or other process operations; considered important for safety and health. { 'dəst kən‚trōl ‚sis·təm }

dust counter [ENG] A photoelectric apparatus which measures the size and number of dust particles per unit volume of air. Also known as Kern counter. { 'dəst ‚kau̇nt·ər }

dust devil [METEOROL] A small but vigorous whirlwind, usually of short duration, rendered visible by dust, sand, and debris picked up from the ground; diameters range from about 10 to 100 feet (3 to 30 meters), and average height is about 600 feet (180 meters). { 'dəst ‚dev·əl }

dust horizon [METEOROL] The top of a dust layer which is confined by a low-level temperature inversion and has the appearance of the horizon when viewed from above, against the sky; the true horizon is usually obscured by the dust layer. { 'dəst hə‚rīz·ən }

dusting clay [AGR] Finely pulverized clay used as an extender or carrier in insecticide dust formulations. { 'dəst·iŋ ‚klā }

dust storm [METEOROL] A strong, turbulent wind carrying large clouds of dust. { 'dəst ‚stȯrm }

dust well [HYD] A pit in an ice surface produced when small, dark particles on the ice are heated by sunshine and sink down into the ice. { 'dəst ‚wel }

dust whirl [METEOROL] A rapidly rotating column of air over a dry and dusty or sandy area, carrying dust, leaves, and other light material picked up from the ground; when well developed, it is known as a dust devil. Also known as dancing dervish; dancing devil; desert devil; sand auger; sand devil. { 'dəst ‚wərl }

duty of water [HYD] The total volume of irrigation water required to mature a particular type of crop, including consumptive use, evaporation and seepage from ditches and canals, and the water eventually returned to streams by percolation and surface runoff. { 'düd·ē əv 'wȯd·ər }

dwarf disease [PL PATH] A virus disease marked by the inhibition of fruit production; common in plum trees. { 'dwȯrf di‚zēz }

dynamic climatology [CLIMATOL] The climatology of atmospheric dynamics and thermodynamics, that is, a climatological approach to the study and explanation of atmospheric circulation. { dī‚nam·ik ‚klī·mə'täl·ə·jē }

dynamic forecasting See numerical forecasting. { dī‚nam·ik 'fȯr‚kast·iŋ }

dynamic meteorology [METEOROL] The study of atmospheric motions as solutions of the fundamental equations of hydrodynamics or other systems of equations appropriate to special situations, as in the statistical theory of turbulence. { dī‚nam·ik mēd·ē·ə'räl·ə·jē }

dynamic roughness [OCEANOGR] A quantity, designated z_0, dependent on the shape and distribution of the roughness elements of the sea surface, and used in calculations of wind at the surface. Also known as roughness length. { dī‚nam·ik 'rəf·nəs }

dynamic thickness [OCEANOGR] The vertical separation between two isobaric surfaces in the ocean. { dī‚nam·ik 'thik·nəs }

dynamic trough [METEOROL] A pressure trough formed on the lee side of a mountain range across which the wind is blowing almost at right angles. Also known as lee trough. { dī‚nam·ik 'trȯf }

dystrophic [BIOL] Pertaining to an environment that does not supply adequate nutrition. { di'stäf·ik }

E

E   *See* exa-.

eager   *See* bore.   { 'ē·gər }

earlywood   [BOT] The portion of the annual ring that is formed during the early part of a tree's growing season.   { 'ər·lē‚wůd }

earplug   [ENG] A device made of a pliable substance which fits into the ear opening; used to protect the ear from excessive noise or from water.   { 'ir‚pləg }

ear protector   [ENG] A device, such as a plug or ear muff, used to protect the human ear from loud noise that may be injurious to hearing, such as that of jet engines.   { 'ir prə‚tek·tər }

ear rot   [PL PATH] Any of several fungus diseases of corn, occurring both in the field and in storage and marked by decay and molding of the ears.   { 'ir ‚rät }

earth   [GEOL] **1.** Solid component of the globe, distinct from air and water. **2.** Soil; loose material composed of disintegrated solid matter.   { ərth }

earth coal   *See* lignite.   { 'ərth ‚kōl }

earth crust   *See* crust.   { 'ərth ‚krəst }

earth hummock   [GEOL] A small, dome-shaped uplift of soil caused by the pressure of groundwater. Also known as earth mound.   { 'ərth ‚həm·ək }

earth mound   *See* earth hummock.   { 'ərth ‚maůnd }

earthquake   [GEOPHYS] A sudden movement of the earth caused by the abrupt release of accumulated strain along a fault in the interior. The released energy passes through the earth as seismic waves (low-frequency sound waves), which cause the shaking.   { 'ərth‚kwāk }

earthquake zone   [GEOL] An area of the earth's crust in which movements, sometimes with associated volcanism, occur. Also known as seismic area.   { 'ərth‚kwāk ‚zōn }

Earth Radiation Budget Experiment   [METEOROL] A satellite observational program to study the earth's radiation budget. Abbreviated ERBE.   { ‚ərth ‚rād·ē¦ā·shən ¦bəj·ət ik‚sper·ə·mənt }

earth resources technology satellite   [ENG] One of a series of satellites designed primarily to measure the natural resources of the earth; functions include mapping, cataloging water resources, surveying crops and forests, tracing sources of water and air pollution, identifying soil and rock formations, and acquiring oceanographic data. Abbreviated ERTS.   { ¦ərth ri¦sòr·səz tek¦näl·ə·je 'sad·əl‚īt }

earth science   [SCI TECH] The science that deals with the earth or any part thereof; includes the disciplines of geology, geography, oceanography, and meteorology, among others.   { 'ərth ‚sī·əns }

earth shadow   [METEOROL] Any shadow projecting into a hazy atmosphere from mountain peaks at times of sunrise or sunset.   { 'ərth ‚shad·ō }

earth system |GEOPHYS| The atmosphere, oceans, biosphere, cryosphere, and geosphere, together. { 'ərth ₁ sis·təm }

East Africa Coast Current |OCEANOGR| A current that is influenced by the monsoon drifts of the Indian Ocean, flowing southwestward along the Somalia coast in the Northern Hemisphere winter and northeastward in the Northern Hemisphere summer. Also known as Somali Current. { ¦ēst ¦af·rə·kə ¦kōst ₁kə·rənt }

East Australia Current |OCEANOGR| The current which is formed by part of the South Equatorial Current and flows southward along the eastern coast of Australia. { ¦ēst ȯ'strāl·yə ₁kə·rənt }

easterly wave |METEOROL| A long, weak migratory low-pressure trough occurring in the tropics. { 'ēs·tər·lē 'wāv }

Eastern Hemisphere |GEOGR| The half of the earth lying mostly to the east of the Atlantic Ocean, including Europe, Africa, and Asia. { ¦ē·stərn 'hem·ə₁sfir }

East Greenland Current |OCEANOGR| A current setting south along the eastern coast of Greenland and carrying water of low salinity and low temperature. { ¦ēst 'grēn·lənd ₁kə·rənt }

ebb current |OCEANOGR| The tidal current associated with the decrease in the height of a tide. { 'eb ₁kə·rənt }

ebb tide |OCEANOGR| The portion of the tide cycle between high water and the following low water. Also known as falling tide. { 'eb ₁tīd }

ecesis |ECOL| Successful naturalization of a plant or animal population in a new environment. { ə'sē·səs }

echolocation |BIOL| An animal's use of sound reflections to localize objects and to orient in the environment. { 'ek·ō·lō₁kā·shən }

echo sounder *See* sonic depth finder. { 'ek·ō₁saúnd·ər }

eclosion |ZOO| The process of an insect hatching from its egg. { ē'klō·zhən }

ecocline |ECOL| A genetic gradient of adaptability to an environmental gradient; formed by the merger of ecotypes. { 'ek·ō₁klīn }

ecofallow |AGR| A system for destroying weeds and conserving soil moisture in crop rotation with minimum disturbance of crop residue and soil. { ¦ek·ō¦fa·lō }

ecological association |ECOL| A complex of communities, such as an elm-hackberry association, which develops in accord with variations in physiography, soil, and successional history within the major subdivision of a biotic realm. { ek·ə'läj·ə·kəl ə₁sō·shē'ā·shən }

ecological climatology |BIOL| A branch of bioclimatology, including the physiological adaptation of plants and animals to their climate, and the geographical distribution of plants and animals in relation to climate. { ek·ə'läj·ə·kəl klī·mə'täl·ə·jē }

ecological community *See* community. { ek·ə'läj·ə·kəl kə'myün·əd·ē }

ecological energetics |ECOL| The study of the flow of energy within an ecological system from the time the energy enters the living system until it is ultimately degraded to heat and irretrievably lost from the system. Also known as production ecology. { ₁ek·ə₁läj·ə·kəl ₁en·ər'jed·iks }

ecological interaction |ECOL| The relation between species that live together in a community; specifically, the effect an individual of one species may exert on an individual of another species. { ek·ə'läj·ə·kəl in·tər'ak·shən }

ecological modeling |ECOL| Representing the interaction and dynamics of ecological systems using mathematics, computer simulations, or conceptual flowcharts. { ₁ek·ə₁läj·ə·kəl 'mäd·əl·iŋ }

ecological physiology [BIOL] The science of the interrelationships between the physiology of organisms and their environment. { ,ē·kə¦läj·ə·kəl fiz·ē'äl·ə·jē }

ecological pyramid [ECOL] A pyramid-shaped diagram representing quantitatively the numbers of organisms, energy relationships, and biomass of an ecosystem; numbers are high for the lowest trophic levels (plants) and low for the highest trophic level (carnivores). { ek·ə'läj·ə·kəl 'pir·ə·mid }

ecological succession [ECOL] An orderly sequential change in community composition, such that the original plant and animal species are gradually replaced with new plant and animal species. Also known as succession. { ek·ə'läj·ə·kəl sək'sesh·ən }

ecological system See ecosystem. { ek·ə'läj·ə·kəl 'sis·təm }

ecological zoogeography [ECOL] The study of animal distributions in terms of their environments. { ,ek·ə,läj·ə·kəl ,zō·ō·jē'äg·rə·fē }

ecology [BIOL] A study of the interrelationships which exist between organisms and their environment. Also known as bionomics; environmental biology. { ē'käl·ə·jē }

economic entomology [ECOL] The study of insects that have a direct influence on humanity, with an emphasis on pest management. { ,ek·ə'näm·ik ,en·tə'mäl·ə·jē }

economic geography [GEOGR] A branch of geography concerned with the relations of physical environment and economic conditions to the manufacture and distribution of commodities. { ,ek·ə'näm·ik jē'äg·rə·fē }

ecophene [GEN] The range of phenotypic modifications produced by one genotype within the limits of the habitat under which the genotype is found in nature. { 'ē·kə ,fēn }

ecophenotype [ECOL] A nongenetic phenotypic modification in response to environmental conditions. { ,ē·kō'phēn·ə,tīp }

ecospecies [ECOL] A group of ecotypes capable of interbreeding without loss of fertility or vigor in the offspring. { 'ē·kō,spē·shēz }

ecosystem [ECOL] A functional system which includes the organisms of a natural community together with their environment. Derived from ecological system. { 'ek· o,sis·təm or 'ē·kō,sis·təm }

ecosystem mapping [ECOL] The drawing of maps that locate different ecosystems in a geographic area. { 'ek·o,sis·təm ,map·iŋ }

ecotone [ECOL] A zone of intergradation between ecological communities. { 'ek· ə,tōn }

ecotrine [ECOL] A metabolite produced by one kind of organism and utilized by another. { 'ek·ə,trēn }

ecotype [ECOL] A subunit, race, or variety of a plant ecospecies that is restricted to one habitat; equivalent to a taxonomic subspecies. { 'ek·ə,tīp }

ectocommensal [ECOL] An organism living on the outer surface of the body of another organism, without affecting its host. { ¦ek·tō·kə'men·səl }

ectohumus [GEOL] An accumulation of organic matter on the soil surface with little or no mixing with mineral material. Also known as mor; raw humus. { ¦ek·tō'hyü·məs }

ectomycorrhizae [ECOL] A type of mycorrhizae composed of a fungus sheath around the outside of root tips, with individual hyphae penetrating between the cortical cells of the root to absorb photosynthates. { ,ek·tō·mī'kór·ə,zī }

ectoparasite [ECOL] A parasite that lives on the exterior of its host. { ¦ek·tō'par·ə,sīt }

ectophagous [ZOO] The larval stage of a parasitic insect which is in the process of development externally on a host. { ek'täf·ə·gəs }

ectophyte |ECOL| A plant which lives externally on another organism. { 'ek·tə,fīt }

ectosymbiont |ECOL| A symbiont that lives on the surface of or is physically separated from its host. { ¦ek·tō'sim·bē,änt }

ectotherm |BIOL| An animal that obtains most of its heat from the environment and therefore has a body temperature very close to that of its environment. { 'ek·tə,thərm }

ectotrophic |BIOL| Obtaining nourishment from outside; applied to certain parasitic fungi that live on and surround the roots of the host plant. { ¦ek·tə'träf·ik }

ectozoa |ECOL| Animals which live externally on other organisms. { ,ek·tə'zō·ə }

ED₅₀ See effective dose 50.

edaphic community |ECOL| A plant community that results from or is influenced by soil factors such as salinity and drainage. { ē'daf·ik kə'myün·əd·ē }

edaphon |BIOL| Flora and fauna in soils. { 'ed·ə,fän }

eddy correlation |METEOROL| A method of studying the effects of sea surface on the air above it by measuring simultaneous fluctuations of the horizontal and vertical components of the airflow from the mean. { 'ed·ē 'kä·rə,lā·shən }

edge effect |ECOL| The influence of adjacent plant communities on the number of animal species present in the direct vicinity. { 'ej i,fekt }

edge water |GEOL| In reservoir structures, the subsurface water that surrounds the gas or oil. { 'ej ,wòd·ər }

edge wave |OCEANOGR| An ocean wave moving parallel to the coast, with crests normal to the coastline; maximum amplitude is at shore, with amplitude falling off exponentially farther from shore. { 'ej ,wāv }

EDTC See S-ethyl-N,N-dipropylthiocarbamate.

eel grass See tape grass. { ēl ,gras }

effective dose 50 |MED| The amount of a drug required to produce a response in 50% of the subjects to whom the drug is given. Abbreviated ED₅₀. Also known as median effective dose. { ə¦fek·tiv ¦dōs 'fif·tē }

effective precipitation |HYD| **1.** The part of precipitation that reaches stream channels as runoff. Also known as effective rainfall. **2.** In irrigation, the portion of the precipitation which remains in the soil and is available for consumptive use. { ə¦fek·tiv prə,sip·ə'tā·shən }

effective rainfall See effective precipitation. { ə¦fek·tiv 'rān,fòl }

effective snowmelt |HYD| The part of snowmelt that reaches stream channels as runoff. { ə¦fek·tiv 'snō,melt }

effective temperature |METEOROL| The temperature at which motionless, saturated air would induce, in a sedentary worker wearing ordinary indoor clothing, the same sensation of comfort as that induced by the actual conditions of temperature, humidity, and air movement. { ə¦fek·tiv 'tem·prə·chər }

efflorescence |BOT| The period or process of flowering. { ,ef·lə'res·əns }

effluent |HYD| **1.** Flowing outward or away from. **2.** Liquid which flows away from a containing space or a main waterway. { ə'flü·ənt }

effluent stream |HYD| A stream that is fed by seeping groundwater. { ə'flü·ənt 'strēm }

effluvium |ENG| By-products of food and chemical processes, in the form of wastes. { ə'flü·vē·əm }

effusion |SCI TECH| **1.** The act or process of leaking or pouring out. **2.** Any material that is effused. { e'fyü·zhən }

egest |BIOL| **1.** To discharge indigestible matter from the digestive tract. **2.** To rid the body of waste. { ē'jest }

EGT *See* ethylene glycol bis(trichloroacetate).

ejecta |GEOL| Material which is discharged by a volcano. |SCI TECH| Material which is cast out. { ē'jek·tə }

Ekman spiral [METEOROL] A theoretical representation that a wind blowing steadily over an ocean of unlimited depth and extent and uniform viscosity would cause, in the Northern Hemisphere, the immediate surface water to drift at an angle of 45° to the right of the wind direction, and the water beneath to drift further to the right, and with slower and slower speeds, as one goes to greater depths. { 'ek·mən ˌspī·rəl }

Ekman transport [OCEANOGR] The movement of ocean water caused by wind blowing steadily over the surface; occurs at right angles to the wind direction. { 'ek·mən ˌtrans ˌpȯrt }

Ekman water bottle [ENG] A cylindrical tube fitted with plates at both ends and used for deep-water samplings; when hit by a messenger it turns 180°, closing the plates and capturing the water sample. { 'ek·mən 'wōd·ər ˌbäd·əl }

Elaphomycetaceae [MYCOL] A family of underground, saprophytic or mycorrhiza-forming fungi in the order Eurotiales characterized by ascocarps with thick, usually woody walls. { ˌel·ə·fō͵mī·sə'tās·ē͵ē }

Elapidae |ZOO| A family of poisonous reptiles, including cobras, kraits, mambas, and coral snakes; all have a pteroglyph fang arrangement. { ə'lap·ə͵dē }

elater [BOT] A spiral, filamentous structure that functions in the dispersion of spores in certain plants, such as liverworts and slime molds. { 'el·ə·tər }

elbow [GEOGR] A sharp change in direction of a coast line, channel, bank, or so on. { 'el͵bō }

ELDORA *See* Electra Doppler Radar. { el'dȯr·ə }

Electra Doppler Radar [METEOROL] An airborne Doppler radar used for detecting and measuring weather phenomena, as well as meteorological research. Abbreviated ELDORA. { iˈlek·trə ˌdäp·lər 'rā͵där }

electrical noise [ENG] Noise generated by electrical devices, for example, motors, engine ignition, power lines, and so on, and propagated to the receiving antenna direct from the noise source. { i'lek·trə·kəl 'nȯiz }

electrical storm [METEOROL] A popular term for a thunderstorm. { i'lek·trə·kəl 'stȯrm }

electric power plant *See* power plant. { iˈlek·trik 'paů·ər ˌplant }

electric vehicle [ENG] A ground vehicle propelled by a motor powered by electrical energy from rechargeable batteries or other source onboard the vehicle, or from an external source in, on, or above the roadway; examples include the electrically powered automobile and trolley bus. { iˈlek·trik 'vē·ə·kəl }

electrogram [METEOROL] A record, usually automatically produced, which shows the time variations of the atmospheric electric field at a given point. { i'lek·trə͵gram }

electrostatic coalescence [METEOROL] **1.** The coalescence of cloud drops induced by electrostatic attractions between drops of opposite charges. **2.** The coalescence of two cloud or rain drops induced by polarization effects resulting from an external electric field. { i͵lek·trə'stad·ik kō·ə'les·əns }

electrotaxis [BIOL] Movement of an organism in response to stimulation by electric charges. { iˈlek·trō'tak·səs }

electrotropism [BIOL] Orientation response of a sessile organism to stimulation by electric charges. { i͵lek'trä·trə͵piz·əm }

element [CHEM] A substance made up of atoms with the same atomic number; common examples are hydrogen, gold, and iron. Also known as chemical element. { 'el·ə·mənt }

elevation of ivory point *See* barometer elevation. { ,el·ə'vā·shən əv 'īv·rē ,point }

elfinwood *See* krummholz. { 'el·fən,wu̇d }

El Niño [METEOROL] A warming of the tropical Pacific Ocean that occurs roughly every 4–7 years. { el 'nēn·yō }

El Niño Southern Oscillation [OCEANOGR] **1.** The irregular cyclic swing in atmospheric pressure in the tropical Pacific. **2.** The irregular cyclic swing of warm and cold phases in the tropical Pacific. Abbreviated ENSO. { el ¦nēn·yō ¦səth·ərn ,äs·ə'lā·shən }

ELR scale *See* equal listener response scale. { ¦ē¦el¦är ,skāl }

Elsasser's radiation chart [METEOROL] A radiation chart developed by W. M. Elsasser for the graphical solution of the radiative transfer problems of importance in meteorology: given a radiosonde record of the vertical variation of temperature and water vapor content, one can find with this chart such quantities as the effective terrestrial radiation, net flux of infrared radiation at a cloud base or a cloud top, and radiative cooling rates. { 'el·zə·sərz rād·ē'ā·shən ,chärt }

eluviation [HYD] The process of transporting dissolved or suspended materials in the soil by lateral or downward water flow when rainfall exceeds evaporation. { ē,lü·ve 'ā·shən }

elve [METEOROL] A transient luminous event that occurs over a thunderstorm, constituting a broad disk of illumination typically at an altitude of 85–90 kilometers (51–54 miles) with a thickness of about 6 kilometers (4 miles). { elv }

embacle [HYD] The piling up of ice in a stream after a refreeze, and the pile so formed. { em'bak·əl }

embayed [GEOGR] Formed into a bay. { em'bād }

embayment [GEOGR] Indentation in a shoreline forming a bay. [GEOL] Act or process of forming a bay. { em'bā·mənt }

embryo [BOT] **1.** The young sporophyte of a seed plant. **2.** embryology **3.** An early stage of development in multicellular organisms. { 'em·brē·ō }

Embryobionta [BOT] The land plants, a subkingdom of the Plantae characterized by having specialized conducting tissue in the sporophyte (except bryophytes), having multicellular sex organs, and producing an embryo. { ¦em·brē·ō·bī'än·tə }

Embryophyta [BOT] The equivalent name for Embryobionta. { ,em·brē'äf·əd·ə }

embryo sac [BOT] The female gametophyte of a seed plant, containing the egg, synergids, and polar and antipodal nuclei; fusion of the antipodals and a pollen generative nucleus forms the endosperm. { 'em·brē·ō,sak }

emerged bog [ECOL] A bog which grows vertically above the water table by drawing water up through the mass of plants. { ə¦mərjd 'bäg }

emerged shoreline *See* shoreline of emergence. { ə¦mərjd 'shȯr,līn }

emergence *See* resurgence. { ə'mər·jəns }

emigration [ECOL] The movement of individuals or their disseminules out of a population or population area. { ,em·ə'grā·shən }

emissary sky [METEOROL] A sky of cirrus clouds which are either isolated or in small, separated groups; so called because this formation often is one of the first indications of the approach of a cyclonic storm. { 'em·ə,ser·ē ,skī }

emission [METEOROL] A natural or anthropogenic discharge of particulate, gaseous, or soluble waste material or pollution into the air. { i'mish·ən }

emission control [METEOROL] A strategy for reducing or preventing atmospheric pollution, such as a catalytic converter used for pollutant removal from automotive exhaust. { i'mish·ən kən,trōl }

emission inventory [ECOL] A quantitative detailed compilation of pollutants emitted into the atmosphere of a given community. { i'mish·ən 'in·vən,tòr·ē }

emission standard [ENG] The maximum legal quantity of pollutant permitted to be discharged from a single source. { i'mish·ən ,stan·dərd }

empirical [SCI TECH] Based on actual measurement, observation, or experience, rather than on theory. { em'pir·ə·kəl }

empirical rule [SCI TECH] A rule which is derived from measurements or observations, and is not based on any theory. { em'pir·ə·kəl 'rül }

emulsion [CHEM] A stable dispersion of one liquid in a second immiscible liquid, such as oil dispersed in water. { ə'məl·shən }

Endamoeba [ZOO] The type genus of the Endamoebidae comprising insect parasites and, in some systems of classification, certain vertebrate parasites. { ¦end·ə'mē·bə }

endemic rural plague *See* sylvatic plague. { en'dem·ik ¦rür·əl 'plāg }

endo- [SCI TECH] Prefix denoting within or inside. { 'en·dō }

endobiotic [ECOL] Referring to an organism living in the cells or tissues of a host. { ¦en·dō·bī'äd·ik }

endocarp [BOT] The inner layer of the wall of a fruit or pericarp. { 'en·dō,kärp }

endocommensal [ECOL] A commensal that lives within the body of its host. { ¦en·dō·kə'men·səl }

endocytobiosis [ECOL] Symbiosis in which the symbionts live within host cells. { ,en·dō,sī·tō·bī'ō·səs }

endodermis [BOT] The innermost tissue of the cortex of most plant roots and certain stems consisting of a single layer of at least partly suberized or cutinized cells; functions to control the movement of water and other substances into and out of the stele. { ¦en·dō¦dər·məs }

endolithic [ECOL] Living within rocks, as certain algae and coral. { ¦en·də¦lith·ik }

endoparasite [ECOL] A parasite that lives inside its host. { ¦en·dō'par·ə,sīt }

endophagous [ZOO] Of an insect larva, living within and feeding upon the host tissues. { en'däf·ə·gəs }

endophyte [ECOL] A plant that lives within, but is not necessarily parasitic on, another plant. { 'en·də,fīt }

endoreism *See* endorheism. { ,en·dō'rē,iz·əm }

endorheism [HYD] A drainage pattern of a basin or region in which little or none of the surface drainage reaches the ocean. Also spelled endoreism. { ,en·dō'rē,iz·əm }

endosperm [BOT] **1.** The nutritive protein material within the embryo sac of seed plants. **2.** Storage tissue in the seeds of gymnosperms. { 'en·də,spərm }

endosymbiosis [ECOL] A mutually beneficial relationship in which one organism lives inside the other. { ,en·dō,sim·bē'ō·səs }

endotherm [BIOL] An animal that produces enough heat from its own metabolism and employs devices to retard heat loss so that it is able to keep its body temperature higher than that of its environment. { 'en·də,thərm }

endotoxin |MICROBIO| A biologically active substance produced by gram-negative bacteria and consisting of lipopolysaccharide, a complex macromolecule containing a polysaccharide covalently linked to a unique lipid structure, termed lipid A. { ˌen·dō 'täk·sən }

endrin |CHEM| $C_{12}H_8OCl_6$ Poisonous, white crystals that are insoluble in water; it is used as a pesticide and is a stereoisomer of dieldrin, another pesticide. { 'en·drən }

energy budget |CLIMATOL| The energy pools, the directions of energy flow, and the rates of energy transformations quantified within a physical or ecological system. { 'en·ər·jē ˌbəj·ət }

energy pyramid |ECOL| An ecological pyramid illustrating the energy flow within an ecosystem. { 'en·ər·jē ˌpir·ə·mid }

englacial |HYD| Of or pertaining to the inside of a glacier. { en'glā·shəl }

enphytotic |PL PATH| 1. A disease that occurs regularly among plants of a specific region. 2. An outbreak of such a disease. { ˌen·fī'täd·ik }

ensiling |AGR| The anaerobic fermentation process used to preserve immature green corn, legumes, grasses, and grain plants; the crop is chopped and packed while at about 70–80% moisture and put into silos or other containers to exclude air. { en'sīl·iŋ }

ENSO *See* El Niño Southern Oscillation. { 'enˌsō }

ensonification field |OCEANOGR| The area of the sea floor that is acoustically imaged in the course of a sonar survey. { enˌsän·ə·fə'kā·shən ˌfēld }

enteric bacilli |MICROBIO| Microorganisms, especially the gram-negative rods, found in the intestinal tract of humans and animals. { en'ter·ik bə'sil·ī }

Enterobacter |MICROBIO| A genus of bacteria in the family Enterobacteriaceae; motile rods found in the intestine of humans and other animals; some strains are encapsulated. { ˌent·ə·rō'bak·tər }

enterotoxin |MICROBIO| A toxin produced by *Micrococcus pyogenes* var. *aureus* (*Staphylococcus aureus*) which gives rise to symptoms of food poisoning in humans and monkeys. { ˌent·ə·rō'täk·sən }

enterovirus |MICROBIO| One of the two subgroups of human picornaviruses; includes the polioviruses, the coxsackieviruses, and the echoviruses. { ˌent·ə·rō'vī·rəs }

Entisol |GEOL| An order of soil having few or faint horizons. { 'entˌəˌsòl }

entombment |ENG| A method of decommissioning a nuclear facility in which radioactive contamination is made inaccessible by demolition techniques and then the residue is covered with reinforced concrete. { en'tüm·mənt }

entomogenous |BIOL| Growing on or in an insect body, as certain fungi. { ˌent·ə'mäj·ə·nəs }

entomology |ZOO| A branch of the biological sciences that deals with the study of insects. { ˌent·ə'mäl·ə·jē }

entomophagous |ZOO| Feeding on insects. { ˌent·ə'mäf·ə·gəs }

entomophilic fungi |MYCOL| Fungi that parasitize insects. { ˌen·tə·məˌfil·ik 'fən·jī }

entomophilous |ECOL| Pollinated by insects. { ˌent·ə'mäf·ə·ləs }

entrainment |HYD| The pickup and movement of sediment as bed load or in suspension by current flow. |METEOROL| The mixing of environmental air into a preexisting organized air current so that the environmental air becomes part of the current. |OCEANOGR| The transfer of fluid by friction from one water mass to another, usually occurring between currents moving in respect to each other. { en'trān·mənt }

entrenched meander [HYD] A deepened meander of a river which is carried downward further below the valley surface in which the meander originally formed. Also known as inherited meander. { en'trencht mē'an·dər }

entrenched stream [HYD] A stream that flows in a valley or narrow trench cut into a plain or relatively level upland. Also spelled intrenched stream. { en'trencht 'strēm }

envelope orography [METEOROL] A method for developing a numerical model for weather forecasting in which it is assumed that mountain passes and valleys are filled mostly with stagnant air, thus increasing the average height of the model mountains and enhancing the blocking effect. { 'en·və,lōp ȯ'räg·rə·fē }

environment [ECOL] The sum of all external conditions and influences affecting the development and life of organisms. { in'vī·ərn·mənt *or* in'vī·rən·ment }

environmental biology *See* ecology. { in¦vī·ərn¦ment·əl bī'äl·ə·jē }

environmental control [ENG] Modification and control of soil, water, and air environments of humans and other living organisms. { in¦vī·ərn¦mənt·əl kən'trōl }

environmental engineering [ENG] The technology concerned with the reduction of pollution, contamination, and deterioration of the surroundings in which humans live. { in¦vī·ərn¦mənt·əl en·jə'nir·iŋ }

environmental fluid mechanics [PHYS] The study of the flows of air and water, of the species carried by them (especially pollution), and of their interactions with geological, biological, social, and engineering systems in the vicinity of a planet's surface. { in,vī·ərn,ment·əl ,flü·əd mi'kan·iks }

environmental impact analysis [ECOL] Predetermination of the extent of pollution or environmental degradation which will be involved in a mining or processing project. { in¦vī·ərn¦mənt·əl 'im,pakt ə,nal·ə·səs }

environmental impact statement [ENG] A report of the potential effect of plans for land use in terms of the environmental, engineering, esthetic, and economic aspects of the proposed objective. { in¦vī·ərn¦mənt·əl 'im,pakt ,stāt·mənt }

environmental lapse rate [METEOROL] The rate of decrease of temperature with elevation in the atmosphere. Also known as atmospheric lapse rate. { in¦vī·ərn ¦mənt·əl 'laps ,rāt }

environmental pathology [MED] A branch of pathology concerned with nonliving environmental agents that adversely influence human health. { in,vī·ərn¦ment· əl pa'thäl·ə·jē }

environmental protection [ENG] The protection of humans and equipment against stresses of climate and other elements of the environment. { in¦vī·ərn¦ment· əl prə'tek·shən }

Environmental Protection Agency [ENG] The governmental agency responsible for the development and enforcement of regulations that protect environmental quality. Abbreviated EPA. { in¦vī·ərn¦mənt·əl prə·'tek·shən 'ā·jən·sē }

environmental resistance [ECOL] The effect of physical and biological factors in preventing a species from reproducing at its maximum rate. { in,vī·ərn¦men· təl ri'zis·təns }

environmental toxicology [MED] A broad field of study encompassing the production, fate, and effects of natural and synthetic pollutants in the environment. { in,vī·ərn ,ment·əl ,täk·sə'käl·ə·jē }

environmental variance [GEN] That portion of the phenotypic variance caused by differences in the environments to which the individuals in a population have been exposed. { in¦vī·ərn,ment·əl 'ver·ē·əns }

environment of sedimentation [GEOL] A more or less destructive geomorphologic setting in which sediments are deposited as beach environment. { in¦vī·ərn¦mənt əv ,sed·ə·men'tā·shən }

environment simulator [ENG] Any machine or artificial device that simulates all or some of the attributes of an environment, such as the solar simulators with artificial suns used in testing spacecraft. { in'vī·ərn'mənt 'sim·yə,lād·ər }

enzootic [VET MED] **1.** A disease affecting animals in a limited geographic region. **2.** Pertaining to such a disease. { |en·zō|äd·ik }

eolation [GEOL] Any action of wind on the land. { ,ē·ə'lā·shən }

eolian [METEOROL] Pertaining to the action or the effect of the wind, as in eolian sounds or eolian deposits (of dust). Also spelled aeolian. { ē'ōl·yən }

eolian erosion [GEOL] Erosion due to the action of wind. { ē'ōl·yən ə'rō·zhən }

eolian sand [GEOL] Deposits of sand arranged by the wind. { ē'ōl·yən 'sand }

eolian soil [GEOL] A type of soil ranging from sand dunes to loess deposits whose particles are predominantly of silt size. { ē'ōl·yən 'sȯil }

EPA *See* Environmental Protection Agency.

epeiric sea *See* epicontinental sea. { ə'pīr·ik 'sē }

ephemeral gully [GEOL] A channel that forms in a cultivated field when precipitation exceeds the rate of soil infiltration. { ə|fem·ə·rəl |gəl·ē }

ephemeral plant [BOT] An annual plant that completes its life cycle in one short moist season; desert plants are examples. { ə'fem·ə·rəl 'plant }

ephemeral stream [HYD] A stream channel which carries water only during and immediately after periods of rainfall or snowmelt. { ə'fem·ə·rəl 'strēm }

epi- [SCI TECH] Prefix denoting upon, beside, near to, over, outer, anterior, prior to, or after. { 'ep·ē }

epibiosis [ECOL] The arrangement in which organisms live on top of each other. { ,ep·ə·bī'ō·səs }

epibiotic [ECOL] Living, usually parasitically, on the surface of plants or animals; used especially of fungi. { ,ep·ə·bī'äd·ik }

epicontinental [GEOL] Located upon a continental plateau or platform. { |ep·ə,kant·ən'ent·əl }

epicontinental sea [OCEANOGR] That portion of the sea lying upon the continental shelf, and the portions which extend into the interior of the continent with similar shallow depths. Also known as epeiric sea; inland sea. { |ep·ə,kant·ən'ent·əl 'sē }

epidemic hepatitis *See* infectious hepatitis. { |ep·ə|dem·ik ,hep·ə'tīd·əs }

epidemic jaundice *See* infectious hepatitis. { |ep·ə|dem·ik 'jȯn·dəs }

epidemic roseola *See* rubella. { |ep·ə|dem·ik ,rō·zē'ō·lə }

epidemiological study [MED] A population study designed to examine associations (commonly, hypothesized causal relations) between personal characteristics and environmental exposures that increase the risk of disease. { ,ep·ə,dē·mē·ə|läj·ə·kəl 'stəd·ē }

epidermis [BOT] The outermost layer (sometimes several layers) of cells on the primary plant body. { ,ep·ə'dər·məs }

epigean [BOT] Pertaining to a plant or plant part that grows above the ground surface. { |ep·ə|jē·ən }

epilimnion [HYD] A fresh-water zone of relatively warm water in which mixing occurs as a result of wind action and convection currents. { ,ep·ə'lim·nē,än }

epipelagic [OCEANOGR] Of or pertaining to the portion of oceanic zone into which enough light penetrates to allow photosynthesis. { |ep·ə·pə'laj·ik }

epipelagic zone [OCEANOGR] The region of an ocean extending from the surface to a depth of about 600 feet (200 meters); light penetrates this zone, allowing photosynthesis. { ¦ep·ə·pə'laj·ik 'zōn }

epiphyll [ECOL] A plant that grows on the surface of leaves. { 'ep·ə‚fil }

epiphyte [ECOL] A plant which grows nonparasitically on another plant or on some nonliving structure, such as a building or telephone pole, deriving moisture and nutrients from the air. Also known as aerophyte. { 'ep·ə‚fīt }

epiphytotic [PL PATH] 1. Any infectious plant disease that occurs sporadically in epidemic proportions. 2. Of or pertaining to an epidemic plant disease. { ¦ep·ə‚fī¦täd·ik }

epiplankton [BIOL] Plankton occurring in the sea from the surface to a depth of about 100 fathoms (180 meters). { ¦ep·ə'plaŋk·tən }

episperm See testa. { 'ep·ə‚spərm }

epixylous [ECOL] Growing on wood; used especially of fungi. { ¦ep·ə¦zī·ləs }

epizoic [BIOL] Living on the body of an animal. { ¦ep·ə¦zō·ik }

epizootic [VET MED] 1. Affecting many animals of one kind in one region simultaneously; widely diffuse and rapidly spreading. 2. An extensive outbreak of an epizootic disease. { ¦ep·ə·zō¦äd·ik }

EPN See O-ethyl-O-*para*-nitrophenyl phenylphosphonothioate.

equal listener response scale [PHYS] An arbitrary scale of noisiness which measures the average response of a listener to a noise when allowance is made for the apparent increase of intensity of a noise as its frequency increases. Abbreviated ELR scale. { ¦ē·kwəl ¦lis·nər ri'späns ‚skāl }

equatorial air [METEOROL] The air of the doldrums or the equatorial trough; distinguished somewhat vaguely from the tropical air of the trade-wind zones. { ‚e·kwə'tȯr·ē·əl 'er }

equatorial calms See doldrums. { ‚e·kwə'tȯr·ē·əl 'kämz }

equatorial convergence zone See intertropical convergence zone. { ‚e·kwə'tȯr·ē·əl kən'vər·jəns ‚zōn }

Equatorial Countercurrent [OCEANOGR] An ocean current flowing eastward (counter to and between the westward-flowing North Equatorial Current and South Equatorial Current) through all the oceans. { ‚e·kwə'tȯr·ē·əl 'kaůnt·ər‚kər·ənt }

Equatorial Current See North Equatorial Current ; South Equatorial Current. { ‚e·kwə 'tȯr·ē·əl 'kə·rənt }

equatorial dry zone [CLIMATOL] An arid region existing in the equatorial trough; the most famous dry zone is situated a little south of the equator in the central Pacific. Also known as arid zone. { ‚e·kwə'tȯr·ē·əl 'drī ‚zōn }

equatorial easterlies [METEOROL] The trade winds in the summer hemisphere when they are very deep, extending at least 5 to 6 miles (8 to 10 kilometers) in altitude, and generally not topped by upper westerlies; if upper westerlies are present, they are too weak and shallow to influence the weather. Also known as deep easterlies; deep trades. { ‚e·kwə'tȯr·ē·əl 'ēs·tər·lēz }

equatorial front See intertropical front. { ‚e·kwə'tȯr·ē·əl 'frənt }

equatorial tide [OCEANOGR] 1. A lunar fortnightly tide. 2. A tidal component with a period of 328 hours. { ‚e·kwə'tȯr·ē·əl 'tīd }

equatorial trough [METEOROL] The quasicontinuous belt of low pressure lying between the subtropical high-pressure belts of the Northern and Southern hemispheres. Also known as meteorological equator. { ‚e·kwə'tȯr·ē·əl 'trȯf }

Equatorial Undercurrent [OCEANOGR] **1.** A subsurface current flowing from west to east in the Indian Ocean near the 450-foot (150-meter) depth at the equator during the time of the Northeast Monsoon. **2.** A permanent subsurface current in the equatorial region of the Atlantic and Pacific oceans. { ‚e·kwə'tȯr·ē·əl 'ən·dər‚kə·rənt }

equatorial vortex [METEOROL] A closed cyclonic circulation within the equatorial trough. { ‚e·kwə'tȯr·ē·əl 'vȯr‚teks }

equatorial wave [METEOROL] A wavelike disturbance of the equatorial easterlies that extends across the equatorial trough. { ‚e·kwə'tȯr·ē·əl 'wāv }

equatorial westerlies [METEOROL] The westerly winds occasionally found in the equatorial trough and separated from the mid-latitude westerlies by the broad belt of easterly trade winds. { ‚e·kwə'tȯr·ē·əl 'wes·tər·lēz }

equilibrium line [HYD] The level on a glacier where the net balance equals zero and accumulation equals ablation. { ‚ē·kwə'lib·rē·əm ‚līn }

equilibrium theory [OCEANOGR] An ocean water model which assumes instantaneous response of water bodies to the tide-producing forces of the moon and sun to form an equilibrium surface, and disregards the effects due to friction, inertia, and irregular distribution of land masses. { ‚ē·kwə'lib·rē·əm ‚thē·ə·rē }

equilibrium tide [OCEANOGR] The hypothetical tide due to the tide-producing forces of celestial bodies, particularly the sun and moon. { ‚ē·kwə'lib·rē·əm ‚tīd }

equinoctial rains [METEOROL] Rainy seasons which occur regularly at or shortly after the equinoxes in many places within a few degrees of the equator. { ‚ē·kwə'näk·shəl 'rānz }

equinoctial tide [OCEANOGR] A tide occurring near an equinox. { ‚ē·kwə'näk·shəl 'tīd }

equipotent [SCI TECH] Equal in capacity or effect. { ¦e·kwə¦pōt·ənt }

equivalent temperature [METEOROL] **1.** The temperature that an air parcel would have if all water vapor were condensed out at constant pressure, the latent heat released being used to heat the air. Also known as isobaric equivalent temperature. **2.** The temperature that an air parcel would have after undergoing the following theoretical process: dry-adiabatic expansion until saturated, pseudoadiabatic expansion until all moisture is precipitated out, and dry adiabatic compression to the initial pressure; this is the equivalent temperature as read from a thermodynamic chart and is always greater than the isobaric equivalent temperature. Also known as adiabatic equivalent temperature; pseudoequivalent temperature. { i'kwiv·ə·lənt 'tem·prə·chər }

ERBE *See* Earth Radiation Budget Experiment. { 'ər‚bē }

erbon [CHEM] $C_{11}H_9Cl_5O_3$ A white solid with a melting point of 49–50°C; insoluble in water; used as a herbicide for perennial broadleaf weeds. { 'ər‚bän }

erg [GEOGR] A large expanse of the earth's surface that is covered with sand, generally blown by wind into dune formations. { ərg }

ergosterin *See* ergosterol. { ər'gäs·tə·rən }

ergosterol [BIOL] $C_{28}H_{44}O$ A crystalline, water-insoluble, unsaturated sterol found in ergot, yeast, and other fungi, and which may be converted to vitamin D_2 on irradiation with ultraviolet light or activation with electrons. Also known as ergosterin. { ər'gäs·tə‚rȯl }

ergot [MYCOL] The dark purple or black sclerotium of the fungus *Claviceps purpurea*. { 'ər·gət }

ergotamine [CHEM] $C_{33}H_{35}N_5O_5$ An alkaloid found in the fungal parasite ergot; causes smooth muscles in peripheral blood vessels to constrict, limiting blood flow; used to treat migraine headaches. { ər'gäd·ə‚mēn }

ergotism [MED] Acute or chronic intoxication resulting from ingestion of grain infected with ergot fungus, or from chronic use of drugs containing ergot. { 'ər·gə₁tiz·əm }

ericophyte [ECOL] A plant that grows on a heath or moor. { 'er·ək·ə₁fīt }

eroding velocity [GEOL] The minimum average velocity required for eroding homogeneous material of a given particle size. { ə'rōd·iŋ və'läs·əd·ē }

erosion [GEOL] **1.** The loosening and transportation of rock debris at the earth's surface. **2.** The wearing away of the land, chiefly by rain and running water. { ə'rō·zhən }

erosion cycle [GEOL] A postulated sequence of conditions through which a new landmass proceeds as it wears down, classically the concept of youth, maturity, and old age, as stated by W. M. Davis; an original landmass is uplifted above base level, cut by canyons, gradually converted into steep hills and wide valleys, and is finally reduced to a flat lowland at or near base level. { ə'rō·zhən ₁sī·kəl }

erosion pavement [GEOL] A layer of pebbles and small rocks that prevents the soil underneath from eroding. { ə'rō·zhən ₁pāv·mənt }

erosion ridge [HYD] One of a group of ridges on the surface of snow; formed by the corrosive action of wind-blown snow. { ə'rō·zhən ₁rij }

erosion surface [GEOL] A land surface shaped by agents of erosion. { ə'rō·zhən ₁sər·fəs }

ertor [METEOROL] The effective (radiational) temperature of the ozone layer (region). { 'ər₁tór }

ERTS *See* earth resources technology satellite.

erysipelas [MED] An acute, infectious bacterial disease caused by *Streptococcus pyogenes* and characterized by inflammation of the skin and subcutaneous tissues. { ₁er·ə'sip·ə·ləs }

Erysiphales [MYCOL] An order of ascomycetous fungi which are obligate parasites of seed plants, causing powdery mildew and sooty mold. { ₁er·ə·sə'fā·lēz }

erythromycin [MICROBIO] A crystalline antibiotic produced by *Streptomyces erythreus* and used in the treatment of gram-positive bacterial infections. { ə₁rith·rə'mīs·ən }

escarpment [GEOL] A cliff or steep slope of some extent, generally separating two level or gently sloping areas, and produced by erosion or by faulting. Also known as scarp. { ə'skärp·mənt }

Escherichia [MICROBIO] A genus of bacteria in the family Enterobacteriaceae; straight rods occurring singly or in pairs. { ₁esh·ə'rik·ē·ə }

Escherichia coli [MICROBIO] The type species of the genus, occurring as part of the normal intestinal flora in vertebrates. Also known as colon bacillus. { ₁esh·ə'rik·ē·ə 'kō₁lī }

Escherichia coli O157:H7 [MICROBIO] An unusually virulent food-borne pathogen that is found primarily in cattle and causes severe, sometimes life-threatening illness; symptoms include hemorrhagic colitis, hemolytic uremic syndrome, and thrombotic thrombocytopenic purpura. { ₁es·kə₁rēk·ē·ə 'kō₁lī ₁ō'wən₁fīv₁sev·ən 'āch₁sev·ən }

espalier drainage *See* trellis drainage. { e'spal·yər ₁drān·ij }

establishment [OCEANOGR] The interval of time between the transit (upper or lower) of the moon and the next high water at a place. { i'stab·lish·mənt }

estivation [BIOL] A period of dormancy or torpidity experienced by some organisms under very hot or dry weather conditions. Also spelled aestivation. { ₁es·tə'vā·shən }

estuarine circulation [OCEANOGR] In an estuary, the outflow (seaward) of low-salinity surface water over a deeper inflowing layer of dense, high-salinity water. { 'es·chə·wə ₁rēn ₁sər·kyə₁lā·shən }

estuarine environment [OCEANOGR] The physical conditions and influences of an estuary. { 'es·chə·wə,rēn en'vī·rən·mənt }

estuarine oceanography [OCEANOGR] The study of the chemical, physical, biological, and geological properties of estuaries. { 'es·chə·wə,rēn ,ō·shə'näg·rə·fē }

estuary [GEOGR] A semienclosed coastal body of water which has a free connection with the open sea and within which sea water is measurably diluted with fresh water. Also known as branching bay; drowned river mouth; firth. { 'es·chə ,wer·ē }

ESU See evolutionarily significant unit.

etesian climate See Mediterranean climate. { ə'tē·zhən 'klī·mət }

ethane [CHEM] CH_3CH_3 A colorless, odorless gas belonging to the alkane series of hydrocarbons, with freezing point of $-183.3°C$ and boiling point of $-88.6°C$; used as a fuel and refrigerant and for organic synthesis. { 'eth,ān }

ethanol [CHEM] C_2H_5OH A colorless liquid, miscible with water, boiling point 78.32°C; used as a reagent and solvent. Also known as ethyl alcohol; grain alcohol. { 'eth·ə,nȯl }

ethanolamine [CHEM] $NH_2(CH_2)_2OH$ A colorless liquid, miscible in water; used in scrubbing hydrogen sulfide (H_2S) and carbon dioxide (CO_2) from petroleum gas streams, for dry cleaning, in paints, and in pharmaceuticals. { ,eth·ə'näl·ə,mēn }

ethiolate [CHEM] $C_7H_{15}ONS$ A yellow liquid with a boiling point of 206°C; used as a preemergence herbicide for corn. { ə'thī·ə,lāt }

Ethiopian zoogeographic region [ECOL] A geographic unit of faunal homogeneity including all of Africa south of the Sahara. { ,ē·thē'ō·pē·ən ¦zō·ō,jē·ə·¦graf·ik 'rē·jən }

ethogram [ECOL] An extensive list, inventory, or description of the behavior of an organism. { 'ē·thə,gram }

ethohexadiol [CHEM] $C_8H_{18}O_2$ A slightly oily liquid, used as an insect repellent. { ¦eth·ō,hek·sə'dī·ȯl }

ethological isolation See behavioral isolation. { ,ē·thə'läj·ə·kəl ī·sə'lā·shən }

ethoprop [CHEM] $C_8H_{19}O_2PS_2$ A pale yellow liquid compound, insoluble in water; used as an insecticide for soil insects and as a nematicide for plant parasitic nematodes. { 'ē·thō,präp }

ethyl alcohol See ethanol. { 'eth·əl 'al·kə,hȯl }

ethyl carbamate See urethane. { ¦eth·əl 'kär·bə,māt }

S-ethyl-N,N-dipropylthiocarbamate [CHEM] $C_9H_{19}NOS$ An amber liquid soluble in water at 370 parts per million; used as a pre- and postemergence herbicide on vegetable crops. Abbreviated EDTC. { ¦es ¦eth·əl ¦en ¦en dī¦prō·pəl,thī·ō'kär·bə,māt }

ethylene bromide See ethylene dibromide. { 'eth·ə,lēn 'bro,mīd }

ethylene dibromide [CHEM] $BrCH_2CH_2Br$ A colorless, poisonous liquid, boiling at 131°C; insoluble in water; used in medicine, as a solvent in organic synthesis, and in antiknock gasoline. Also known as ethylene bromide. { 'eth·ə·lēn dī'brō,mīd }

ethylene glycol bis(trichloroacetate) [CHEM] $C_4H_4\text{-}Cl_6O_4$ A white solid with a melting point of 40.3°C; used as a herbicide for cotton and soybeans. Abbreviated EGT. { 'eth·ə·lēn 'glī,kȯl ,bis·trī,klȯr·ō'as·ə,tāt }

ethylethanolamine [CHEM] $C_2H_5NHCH_2CH_2OH$ Water-white liquid with amine odor; soluble in alcohol, ether, and water; used in dyes, insecticides, fungicides, and surface-active agents. { ¦eth·əl,eth·ə'näl·ə,mēn }

ethyl formate [CHEM] $HCOOC_2H_5$ A colorless liquid, boiling at 54.4°C; used as a solvent, fumigant and larvicide and in flavors, resins, and medicines. { ¦eth·əl 'fȯr,māt }

O-ethyl-O-para-nitrophenyl phenylphosphonothioate [CHEM] $C_2H_5O_4NPS$ A yellow, crystalline compound with a melting point of 36°C; used as an insecticide and miticide on fruit crops. Abbreviated EPN. { ¦ō ¦eth·əl ¦ō ¦par·ə ,nī·trō'fen·əl ,fen·əl·fäs¦fä·nō'thī·ə,wāt }

ortho-ethyl(O-2,4,5-trichlorophenyl)ethylphosphonothioate [CHEM] $C_{10}H_{12}OPSCl_2$ An amber liquid with a boiling point of 108°C at 0.01 mmHg; solubility in water is 50 parts per million; used as an insecticide for vegetable crops and soil pests on meadows. Also known as trichloronate. { ¦òr·thō ¦eth·əl ¦ō ¦tü ¦fòr ¦fiv ¦trī,klòr·ō¦fen·əl¦eth·əl,fäs¦fan·ō'thī·ə,wāt }

ethyl urethane See urethane. { ¦eth·əl 'yùr·ə,thān }

etiolation [BOT] The yellowing or whitening of green plant parts grown in darkness. { ,ed·ē·ə'lā·shən }

Eucaryota [BIOL] Primitive, unicellular organisms having a well-defined nuclear membrane, chromosomes, and mitotic cell division. { yü·kar·ē'ōd·ə }

eucaryote See eukaryote. { yü'kar·ē,ōt }

Euglenophyta [BOT] A division of the plant kingdom including one- celled, chiefly aquatic flagellate organisms having a spindle-shaped or flattened body, naked or with a pellicle. { ,yü·glə'näf·əd·ə }

Eukarya [BIOL] The domain that contains all the eukaryotic kingdoms (such as plants, animals, fungi, and protists). { yü'kar·ē·ə }

eukaryote [BIOL] A cell with a definitive nucleus (one which has a nuclear membrane). Also spelled eucaryote. { yü'kar·ē,ōt }

Eulerian nutation See Chandler wobble. { òi'ler·ē·ən nyü'tā·shən }

eulittoral [OCEANOGR] A subdivision of the benthic division of the littoral zone of the marine environment, extending from high-tide level to about 200 feet (60 meters), the lower limit for abundant growth of attached plants. { yü'lid·ə·rəl }

eupelagic See pelagic. { yü·pə'laj·ik }

euphotic [OCEANOGR] Of or constituting the upper levels of the marine environment down to the limits of effective light penetration for photosynthesis. { yü'fäd·ik }

Europe [GEOGR] A great western peninsula of the Eurasian landmass, usually called a continent; its eastern limits are arbitrary and are conventionally drawn along the water divide of the Ural Mountains, the Ural River, the Caspian Sea, and the Caucasus watershed to the Black Sea. { 'yùr·əp }

European boreal faunal region [ECOL] A zoogeographic region describing marine littoral faunal regions of the northern Atlantic Ocean between Greenland and the northwestern coast of Europe. { ¦yùr·ə¦pē·ən ¦bòr·ē·əl 'fòn·əl 'rē·jən }

European canker [PL PATH] **1.** A fungus disease of apple, pear, and other fruit and shade trees caused by Nectria galligena and characterized by cankers with concentric rings of callus on the trunk and branches. **2.** A fungus disease of poplars caused by Dothichiza populea. { ¦yùr·ə¦pē·ən 'kaŋ·kər }

eurybathic [ECOL] Living at the bottom of a body of water. { ¦yùr·ə¦bath·ik }

euryhaline [ECOL] Pertaining to the ability of marine organisms to tolerate a wide range of saline conditions, and therefore a wide variation of osmotic pressure, in the environment. { ¦yùr·ə¦ha,līn }

euryplastic [BIOL] Referring to an organism with a marked ability to change and adapt to a wide spectrum of environmental conditions. { ¦yùr·ə¦plas·tik }

eurytherm [BIOL] An organism that is tolerant of a wide range of temperatures. { 'yùr·ə,thərm }

eurytopic [ECOL] Referring to organisms which are widely distributed. { ‚yŭr·ə'täp·ik }

eusocial [ZOO] Pertaining to animal societies, such as those of certain insects, in which sterile individuals work on behalf of reproductive individuals. { ‚yü'sō·shəl }

eustacy [OCEANOGR] Worldwide fluctuations of sea level due to changing capacity of the ocean basins or the volume of ocean water. { 'yü·stə·sē }

euthenics [BIOL] The science that deals with the improvement of the future of humanity by changing the environment. { yü'then·iks }

eutrophic [HYD] Pertaining to a lake containing a high concentration of dissolved nutrients; often shallow, with periods of oxygen deficiency. { yü'träf·ik }

eutrophication [ECOL] The process by which a body of water becomes, either by natural means or by pollution, excessively rich in dissolved nutrients, resulting in increased primary productivity that often leads to a seasonal deficiency in dissolved oxygen. { yü·trə·fə'kā·shən }

euxinic [HYD] Of or pertaining to an environment of restricted circulation and stagnant or anaerobic conditions. { yük'sin·ik }

evacuate [SCI TECH] To remove something, especially gases and vapors, from an enclosure, such as from the envelope of an electron tube, or from a well. Also known as exhaust. { i'vak·yə‚wāt }

evaporation capacity See evaporative power. { i‚vap·ə'rā·shən kə‚pas·əd·ē }

evaporation current [OCEANOGR] An ocean current resulting from the accumulation of water through precipitation and river runoff at one point, and loss by evaporation at another point. { i‚vap·ə'rā·shən ‚kə·rənt }

evaporation gage See atmometer. { i‚vap·ə'rā·shən ‚gāj }

evaporation pan [ENG] A type of atmometer consisting of a pan, used in the measurement of the evaporation of water into the atmosphere. { i‚vap·ə'rā·shən ‚pan }

evaporation power See evaporative power. { i‚vap·ə'rā·shən ‚paŭ·ər }

evaporative capacity See evaporative power. { i'vap·ə‚rād·iv kə'pas·əd·ē }

evaporative power [METEOROL] A measure of the degree to which the weather or climate of a region is favorable to the process of evaporation; it is usually considered to be the rate of evaporation, under existing atmospheric conditions, from a surface of water which is chemically pure and has the temperature of the lowest layer of the atmosphere. Also known as evaporation capacity; evaporation power; evaporative capacity; evaporativity; potential evaporation. { i'vap·ə‚rād·iv 'paŭ·ər }

evaporativity See evaporative power. { i‚vap·ə·rə'tiv·əd·ē }

evaporimeter See atmometer. { i‚vap·ə'rim·əd·ər }

evaporite [GEOL] Deposits of mineral salts from sea water or salt lakes due to evaporation of the water. { i'vap·ə‚rīt }

evapotranspiration [HYD] Discharge of water from the earth's surface to the atmosphere by evaporation from lakes, streams, and soil surfaces and by transpiration from plants. Also known as fly-off; total evaporation; water loss. { i‚vap·ō‚tranz·pə'rā·shən }

everglade [ECOL] A type of wetland in southern Florida usually containing sedges (Cyperaceae) and at least seasonally covered by slowly moving water. { 'ev·ər‚glād }

evergreen [BOT] Pertaining to a perennially green plant. Also known as aiophyllous. { 'ev·ər‚grēn }

evolution   [BIOL] The processes of biological and organic change in organisms by which descendants come to differ from their ancestors.   { ‚ev·ə'lü·shən }

evolutionarily significant unit   [ECOL] A distinct local population within a species that has very different behavioral and phenological traits and thus harbors enough genetic uniqueness to warrant its own management and conservation agenda. Abbreviated ESU.   { ‚ev·ə‚lü·shə¦ner·ə·lē sig¦nif·i·kənt 'yü·nət }

evolutionary progress   [GEN] The acquisition of new macromolecular and metabolic processes by which competitive superiority is achieved.   { ‚ev·ə¦lü·shə‚ner·ē 'prä‚gres }

exa-   [SCI TECH] A prefix indicating 10^{18}. Abbreviated E.   { 'ek·sə }

excessive precipitation   [METEOROL] Precipitation (generally in the form of rain) of an unusually high rate of fall; although often used qualitatively, several meteorological services have adopted quantitative limits.   { ek'ses·iv prə‚sip·ə'tā·shən }

exchange capacity   [GEOL] The ability of a soil material to participate in ion exchange as measured by the quantity of exchangeable ions in a given unit of the material.   { iks'chānj kə‚pas·əd·ē }

excision enzyme   [BIOL] A bacterial enzyme that removes damaged dimers from the deoxyribonucleic acid molecule of a bacterial cell following light or ultraviolet radiation or nitrogen mustard damage.   { ek'sizh·ən 'en‚zīm }

exclusion area   [ENG] The area around a nuclear operation (reactor, bomb test, and so on) where human habitation is restricted.   { ik'sklü·zhən ‚er·ē·ə }

exclusion principle   [ECOL] The principle according to which two species cannot coexist in the same locality if they have identical ecological requirements.   { ik'sklü·zhən ‚prin·sə·pəl }

exclusive species   [ECOL] A species which is completely or nearly limited to one community.   { ik'sklü·siv 'spē·shēz }

exfiltration   [SCI TECH] A gradual escape of fluid, for example, through a membrane or a wall.   { ‚eks‚fil'trā·shən }

exfoliation   [SCI TECH] Flaking away or peeling off in scales.   { eks‚fō·lē'ā·shən }

exhaust   [ENG] **1.** The working substance discharged from an engine cylinder or turbine after performing work on the moving parts of the machine. **2.** The phase of the engine cycle concerned with this discharge. **3.** A duct for the escape of gases, fumes, and odors from an enclosure, sometimes equipped with an arrangement of fans. [SCI TECH] *See* evacuate.   { ig'zȯst }

exhaust head   [ENG] A device placed on the end of an exhaust pipe to remove oil and water and to reduce noise.   { ig'zȯst ‚hed }

exhaust trail   [METEOROL] A visible condensation trail (contrail) that forms when the water vapor of an aircraft exhaust is mixed with and saturates (or slightly supersaturates) the air in the wake of the aircraft.   { ig'zȯst ‚trāl }

exhumation   [GEOL] The uncovering or exposure through erosion of a former surface, landscape, or feature that had been buried by subsequent deposition.   { ‚eks·yü'mā·shən }

exo-   [SCI TECH] A prefix denoting outside or outer.   { 'ek·sō }

exodermis   *See* hypodermis.   { ‚ek·sō'dər·məs }

exogenous   [BIOL] **1.** Due to an external cause; not arising within the organism. **2.** Growing by addition to the outer surfaces.   { ‚ek'säj·ə·nəs }

exopathogen   [PL PATH] An external, nonparasitic plant pathogen.   { ¦ek·sō'path·ə·jən }

exopathogenesis   [PL PATH] The external incitement of disease by a nonparasitic pathogen.   { ¦ek·sō‚path·ə'jen·ə·səs }

exosphere [METEOROL] An outermost region of the atmosphere, estimated at 300–600 miles (500–1000 kilometers), where the density is so low that the mean free path of particles depends upon their direction with respect to the local vertical, being greatest for upward-traveling particles. Also known as region of escape. { 'ek·sō¦sfir }

exotic [ECOL] Not native to an area. { ig'zäd·ik }

exotic stream [HYD] A stream that crosses a desert as it flows to the sea, or any stream which derives most of its water from the drainage system of another region. { ig'zäd·ik 'strēm }

exotic viral disease [MED] A viral disease that occurs only rarely in human populations of developed countries. { ig¦zäd·ik ¦vī·rəl diz¦ēz }

exotoxin [MICROBIO] A toxin that is excreted by a microorganism. { ¦ek·sə¦täk·sən }

expanded foot [HYD] A broad, bulblike or fan-shaped ice mass formed where a valley glacier flows beyond its confining walls and extends onto an adjacent lowland at the bottom of a mountain slope. { ik'spand·əd 'fút }

expanding population [ECOL] A population containing a large proportion of young individuals. { ik'spand·iŋ ¦päp·yə'lā·shən }

experiment [SCI TECH] The test of a hypothesis under controlled conditions. { ik'sper·ə·mənt }

experimental ecology [ECOL] The manipulation of organisms or their environments to discover the underlying mechanisms governing distribution and abundance. { ik¦sper·ə¦ment·əl ē'käl·ə·jē }

explosive fuel [ENG] Any substance which combines with oxygen and other explosive ingredients to produce explosion energy, including aluminum, silicon, carbon, sulfur, glycerol, glucol, paraffin wax, diesel oil, and guar gum. { ik'splō·siv 'fyül }

exponential atmosphere See isothermal atmosphere. { ¦ek·spə'nen·chəl 'at·mə¦sfir }

exponential growth [MICROBIO] The period of bacterial growth during which cells divide at a constant rate. Also known as logarithmic growth. [SCI TECH] The increase of a quantity x with time t according to the equation $x = Ka^t$, where K and a are constants, a is greater than 1, and K is greater than 0. { ¦ek·spə'nen·chəl 'grōth }

exposure [METEOROL] The general surroundings of a site, with special reference to its openness to winds and sunshine. { ik'spō·zhər }

exposure dose [MED] A measure of the radiation in a certain place based upon its ability to produce ionization in air. { ik'spō·zhər ¦dōs }

exposure limit [MED] The maximum radiation dose equivalent permitted under specified conditions. { ik'spō·zhər ¦lim·ət }

exposure time [CIV ENG] The time period of interest for seismic hazard calculations such as the design lifetime of a building or the time over which the numbers of casualties should be estimated. { ik'spō·zhər ¦tīm }

exsiccate [SCI TECH] To dry by driving off, or draining of, moisture. { 'ek·sə¦kāt }

exsurgence See resurgence. { ek'sər·jəns }

extended forecast [METEOROL] In general, a forecast of weather conditions for a period extending beyond 2 days from the day of issue. Also known as long-range forecast. { ik'stend·əd 'fór¦kast }

extended-range forecast See medium-range forecast. { ik¦stend·əd ¦rānj 'fór¦kast }

extended stream [HYD] A stream lengthened by the extension of its downstream course; the course is through a newly emerged land such as a coastal plain. { ik¦stend·əd 'strēm }

extending flow |HYD| A glacial flow pattern in which velocity increases as the distance downstream becomes greater. { ik'stend·iŋ ,flō }

external forcing |CLIMATOL| The influence on the earth system by solar radiation. { ik ,stərn·əl 'fòrs·iŋ }

external respiration |BIOL| The processes by which oxygen is carried into living cells from the outside environment and by which carbon dioxide is carried in the reverse direction. { ek'stərn·əl ,res·pə'rā·shən }

extinction |GEN| The worldwide death and disappearance of a specific organism or group of organisms. |HYD| The drying up of a lake by either water loss or destruction of the lake basin. { ek'stiŋk·shən }

extirpate |BIOL| To uproot, destroy, make extinct, or exterminate. { 'ek·stər,pāt }

extracellular |BIOL| Outside the cell. { ¦ek·strə'sel·yə·lər }

extratropical cyclone |METEOROL| Any cyclone-scale storm that is not a tropical cyclone. Also known as extratropical low; extratropical storm. { ¦ek·strə¦träp·i·kəl 'sī ,klōn }

extratropical low See extratropical cyclone. { ¦ek·strə¦träp·i·kəl 'lō }

extratropical storm See extratropical cyclone. { ¦ek·strə¦träp·i·kəl 'stòrm }

extreme |CLIMATOL| The highest, and in some cases the lowest, value of a climatic element observed during a given period or during a given month or season of that period; if this is the whole period for which observations are available, it is the absolute extreme. { ek'strēm }

extremophiles |BIOL| Microorganisms belonging to the domains Bacteria and Archaea that can live and thrive in environments with extreme conditions such as high or low temperatures and pH levels, high salt concentrations, and high pressure. { ek'trem· ə,fīlz }

eye of the storm |METEOROL| The center of a tropical cyclone, marked by relatively light winds, confused seas, rising temperature, lowered relative humidity, and often by clear skies. { 'ī əv thə 'stòrm }

eye of the wind |METEOROL| The point or direction from which the wind is blowing. { 'ī əv thə 'wind }

eyespot |BOT| **1.** A small photosensitive pigment body in certain unicellular algae. **2.** A dark area around the hilum of certain seeds, as some beans. |PL PATH| A fungus disease of sugarcane and certain other grasses which is caused by *Helminthosporium sacchari* and characterized by yellowish oval lesions on the stems and leaves. { 'ī,spät }

eye wall |METEOROL| A zone at the periphery of the eye of the storm where winds reach their highest speed. { 'ī ,wòl }

F

facet |GEOGR| Any part of an intersecting surface that constitutes a unit of geographic study, for example, a flat or a slope. { 'fas·ət }

faciation |ECOL| A part of a climax association that lacks some of the dominant species of the normal association due to slight differences in the environment. { 'fā·shē·ā·shən }

facies |ECOL| The makeup or appearance of a community or species population. { 'fā·shēz }

facultative aerobe |MICROBIO| An anaerobic microorganism which can grow under aerobic conditions. { 'fa·kəl,tād·iv 'er,ōb }

facultative anaerobe |MICROBIO| A microorganism that grows equally well under aerobic and anaerobic conditions. { 'fak·əl,tād·iv 'an·ə,rōb }

facultative parasite |ECOL| An organism that can exist independently but may be parasitic on certain occasions, such as the flea. { 'fak·əl,tād·iv 'par·ə,sīt }

facultative photoheterotroph |MICROBIO| Any bacterium that usually grows anaerobically in light but can also grow aerobically in the dark. { 'fak·əl,tād·iv ¦fōd·ō¦hed·ə·rə ,träf }

Fahrenholz's rule |ECOL| The rule that in groups of permanent parasites the classification of the parasites usually corresponds directly to the natural relationships of the hosts. { 'fär·ən,hōlt·səz ,rül }

fair |METEOROL| Generally descriptive of pleasant weather conditions, with regard for location and time of year; it is subject to popular misinterpretation, for it is a purely subjective description; when this term is used in forecasts of the U.S. Weather Bureau, it is meant to imply no precipitation, less than 0.4 sky cover of low clouds, and no other extreme conditions of cloudiness or windiness. { fer }

fair-weather cumulus See cumulus humilis cloud. { 'fer ,we͜th·ər 'kyü·myə·ləs }

Falkland Current |OCEANOGR| An ocean current flowing northward along the Argentine coast. { 'fōk·lənd 'kə·rənt }

falling tide See ebb tide. { 'fōl·iŋ 'tīd }

fallout |PHYS| The material that descends to the earth or water well beyond the site of a surface or subsurface nuclear explosion. Also known as atomic fallout; radioactive fallout. { 'fōl,aut }

fallout shelter |CIV ENG| A structure that affords some protection against fallout radiation and other effects of nuclear explosion; maximum protection is in reinforced concrete shelters below the ground. Also known as radiation shelter. { 'fōl,aut ,shel·tər }

fallout winds |METEOROL| Tropospheric winds that carry radioactive fallout materials, observed by standard winds-aloft observation techniques. { 'fōl,aut ,winz }

fallow [AGR] Pertaining to land normally used for crop production but left unsown for one or more growing seasons. { 'fal·ō }

fall-streak hole [METEOROL] A hole occurring in a cloud layer of supercooled water droplets; produced by the local freezing of some of the droplets and their conversion into fallout, frequently in a streak form. { 'fól ,strēk ,hōl }

fall streaks See virga. { 'fól ,strēks }

Fallstreifen See virga. { 'fäl,strīf·ən }

fall wind [METEOROL] A strong, cold, downslope wind, differing from a foehn in that the initially cold air remains relatively cold despite adiabatic warming upon descent, and from the gravity wind in that it is a larger-scale phenomenon prerequiring an accumulation of cold air at high elevations. { 'fól ,wind }

false cirrus cloud [METEOROL] Cirrus composed of the debris of the upper frozen parts of a cumulonimbus cloud. { ¦fóls 'sir·əs ,klaůd }

false ice foot [OCEANOGR] Ice that forms along a beach terrace and attaches to it just above the high-water mark; derived from water coming from melting snow above the terrace. { ¦fóls 'īs ,fůt }

false smut [PL PATH] A fungus disease of palm caused by *Graphiola phoenicis* and characterized by small cylindrical protruding pustules, often surrounded by yellowish leaf tissue. See green smut. { ¦fóls 'smət }

family [SYST] A taxonomic category based on the grouping of related genera. { 'fam·lē }

famphur [CHEM] $C_{10}H_{16}NO_5PS_2$ A crystalline compound with a melting point of 55°C; slightly soluble in water; used as an insecticide for lice and grubs of reindeer and cattle. { 'fam·fər }

fan [AGR] A mechanical device used for winnowing grain. [BIOL] Any structure, such as a leaf or the tail of a bird, resembling an open fan. [GEOL] A gently sloping, fan-shaped feature usually found near the lower termination of a canyon. { fan }

farcy See glanders. { 'fär·sē }

farm [AGR] A tract of land used for cultivating crops or raising animals. { färm }

farming [AGR] The skills and practices of agriculture. { 'fär·miŋ }

fascicle [BOT] A small bundle, as of fibers or leaves. { fas·i·kəl }

fastest mile [METEOROL] Over a specified period (usually the 24-hour observational day), the fastest speed, in miles per hour, of any mile of wind, with its accompanying direction. { ¦fas·təst 'mīl }

fast ice [OCEANOGR] Sea ice generally remaining in the position where originally formed and sometimes attaining a considerable thickness; it is attached to the shore or over shoals where it may be held in position by islands, grounded icebergs, or polar ice. Also known as coastal ice; coast ice. { ¦fast 'īs }

fathom [OCEANOGR] The common unit of depth in the ocean, equal to 6 feet (1.8288 meters). { 'fath·əm }

fathom curve See isobath. { 'fath·əm ,kərv }

fault [GEOL] A fracture in rock which the adjacent rock surfaces are differentially displaced. Earthquakes are caused by sudden movement along a fault. { fólt }

fault line See fault. { 'fólt,līn }

fault trace See fault line. { 'fólt ,trās }

fauna [ZOO] **1.** Animals. **2.** The animal life characteristic of a particular region or environment. { 'fón·ə }

faunal region [ECOL] A division of the zoosphere, defined by geographic and environmental barriers, to which certain animal communities are bound. { 'fȯn·əl ˌrē·jən }

febrile disease [MED] Any disease associated with or characterized by fever. { 'feb·rəl di₁zēz }

fecundity [BIOL] The innate potential reproductive capacity of the individual organism, as denoted by number of offspring per female in a given time. { fə'kən·dəd·ē }

feeder *See* tributary. { 'fēd·ər }

feeder current [OCEANOGR] A current which flows parallel to the shore before converging with other such currents and forming the neck of a rip current. { 'fēd·ər ₁kə·rənt }

feeding ground *See* drainage basin. { 'fēd·iŋ ₁grau̇nd }

fell-field [ECOL] A culture community of dwarfed, scattered plants or grasses above the timberline. { 'fel ₁fēld }

female [BOT] A flower lacking stamens. [ZOO] An individual that bears young or produces eggs. { 'fē₁māl }

fen [GEOGR] Peat land covered by water, especially in the upper regions of old estuaries and around lakes, that can be drained only artificially. { fen }

fenaminosulf [CHEM] $C_8H_{10}N_3SO_3Na$ A yellow-brown powder, decomposing at 200°C; used as a fungicide for seeds and seedlings in crops. { ₁fen'am·ə·nō₁səlf }

fenazaflor [CHEM] $C_{15}H_7Cl_2F_3N_2O_2$ A greenish-yellow, crystalline compound with a melting point of 103°C; used as an insecticide and miticide for spider mites and eggs. { fə'naz·ə₁flȯr }

fenitrothion [CHEM] $C_9H_{12}NO_5PS$ A yellow-brown liquid, insoluble in water; used as a miticide and insecticide for rice, orchards, vegetables, cereals, and cotton, and for fly and mosquito control. { ₁fen·ə·trō'thī₁än }

fensulfothion [CHEM] $C_{11}H_{17}S_2O_2P$ A brown liquid with a boiling point of 138–141°C; used as an insecticide and nematicide in soils. { ₁fen₁səl·fō'thī₁än }

fentinacetate [CHEM] $C_{20}H_{18}O_2Sn$ A yellow to brown, crystalline solid that melts at 124–125°C; used as a fungicide, molluscicide, and algicide for early and late blight on potatoes, sugarbeets, peanuts, and coffee. Also known as triphenyltinacetate. { ₁fent·ən'as·ə₁tāt }

fenuron [CHEM] $C_9H_{12}N_2O$ A white, crystalline compound with a melting point of 133–134°C; soluble in water; used as a herbicide to kill weeds and bushes. { ₁fen'yu̇₁rän }

fenuron-TCA [CHEM] $C_{11}H_{13}Cl_3N_2O_3$ A white, crystalline compound with a melting point of 65–68°C; moderately soluble in water; used as a herbicide for noncrop areas. { ₁fen'yu̇₁rän ¦tē¦sē¦ā }

ferbam [CHEM] $C_9H_{18}FeN_3S_6$ [iron(III) dimethyldithiocarbamate] A fungicide for protecting fruits, vegetables, melons, and ornamental plants. { 'fər·bəm }

fermentation [MICROBIO] An enzymatic transformation of organic substrates, especially carbohydrates, generally accompanied by the evolution of gas; a physiological counterpart of oxidation, permitting certain organisms to live and grow in the absence of air; used in various industrial processes for the manufacture of products such as alcohols, acids, and cheese by the action of yeasts, molds, and bacteria; alcoholic fermentation is the best-known example. Also known as zymosis. { ₁fər·mən'tā·shən }

fern [BOT] Any of a large number of vascular plants composing the division Polypodiophyta. { fərn }

ferric arsenate |CHEM| $FeAsO_4 \cdot 2H_2O$ A green or brown powder, insoluble in water, soluble in dilute mineral acids; used as an insecticide. { 'fer·ik 'ärs·ən,āt }

ferric chloride [CHEM] $FeCl_3$ Brown crystals, melting at 300°C, that are soluble in water, alcohol, and glycerol; used as a coagulant for sewage and industrial wastes, as an oxidizing and chlorinating agent, as a disinfectant, in copper etching, and as a mordant. Also known as anhydrous ferric chloride; ferric trichloride; flores martis; iron chloride. { 'fer·ik 'klòr,īd }

ferric hydrate See ferric hydroxide **2.** { 'fer·ik 'hī,drāt }

ferric hydroxide [CHEM] $Fe(OH)_3$ A brown powder, insoluble in water; used as arsenic poisoning antidote, in pigments, and in pharmaceutical preparations. Also known as ferric hydrate; iron hydroxide. { 'fer·ik hī'dräk,sīd }

ferric sulfate |CHEM| $Fe_2(SO_4)_3 \cdot 9H_2O$ Yellow, water-soluble, rhombohedral crystals, decomposing when heated; used as a chemical intermediate, disinfectant, soil conditioner, pigment, and analytical reagent, and in medicine. Also known as iron sulfate. { 'fer·ik 'səl,fāt }

ferric trichloride See ferric chloride. { 'fer·ik trī'klòr,īd }

ferriferous [GEOL] Of a sedimentary rock, iron-rich. { fə'rif·ə·rəs }

ferrisulphas See ferrous sulfate. { ¦fe·ri'səl·fəs }

ferrous arsenate [CHEM] $Fe_3(AsO_4)_2 \cdot 6H_2O$ Water-insoluble, toxic green amorphous powder, soluble in acids; used in medicine and as an insecticide. Also known as iron arsenate. { 'fer·əs 'ärs·ən,āt }

ferrous chloride [CHEM] $FeCl_2 \cdot 4H_2O$ Green, monoclinic crystals, soluble in water; used as a mordant in dyeing, for sewage treatment, in metallurgy, and in pharmaceutical preparations. Also known as iron chloride; iron dichloride. { 'fer·əs 'klòr,īd }

ferrous sulfate [CHEM] $FeSO_4 \cdot 7H_2O$ Blue-green, water-soluble, monoclinic crystals; used as a mordant in dyeing wool, in the manufacture of ink, and as a disinfectant. Also known as copperas; ferrisulphas; green copperas; green vitriol; iron sulfate. { 'fer·əs 'səl,fāt }

ferruginous [SCI TECH] **1.** Pertaining to or containing iron. **2.** Having the appearance or color of iron rust (ferric oxide). { fə'rü·jə·nəs }

fertigation |AGR| The practice of fertilizing plants via a drip irrigation system. { ,fər·tə'gā·shən }

fertility [BIOL] The state of or capacity for abundant productivity. { fər'til·əd·ē }

fescue [BOT] A group of grasses of the genus *Festuca*, used for both hay and pasture. { 'fes,kyü }

fetch |OCEANOGR| **1.** The distance traversed by waves without obstruction. **2.** An area of the sea surface over which seas are generated by a wind having a constant speed and direction. **3.** The length of the fetch area, measured in the direction of the wind in which the seas are generated. Also known as generating area. { fech }

fiard See fjard. { fē'ärd }

fiber [BOT] **1.** An elongate, thick-walled, tapering plant cell that lacks protoplasm and has a small lumen. **2.** A very slender root. { 'fī·bər }

fiber crops [AGR] Plants, such as flax, hemp, jute, and sisal, cultivated for their content or yield of fibrous material. { 'fī·bər ,kräps }

fiber flax [BOT] The flax plant grown in fertile, well-drained, well-prepared soil and cool, humid climate; planted in the early spring and harvested when half the seed pods turn yellow; used in the manufacture of linen. { 'fī·bər ,flaks }

fibratus [METEOROL] A cloud species characterized by a fine hairlike or striated composition, the filaments of which are usually distinctly separated from each other; the extremities of these filaments are always thin and never terminated by tufts or hooks. Also known as filosus. { fi'bräd·əs }

fibrous ice *See* acicular ice. { 'fī·brəs 'īs }

FIDO [METEOROL] A system for artificially dissipating fog, in which gasoline or other fuel is burned at intervals along an airstrip to be cleared. Derived from fog investigation dispersal operations. { 'fī·dō }

fiducial temperature [METEOROL] That temperature at which, in a specified latitude, the reading of a particular barometer does not require temperature or latitude correction. { fə'dü·shəl 'tem·prə·chər }

field [GEOL] A region or area with a particular mineral resource, for example, a gold field. { fēld }

field capacity [HYD] The maximum amount of water that a soil can retain after gravitational water has drained away. { 'fēld kə,pas·əd·ē }

field changes [METEOROL] With regard to thunderstorm electricity, the rapid variations in the vertical component of the electric field strength at the earth's surface. { 'fēld ,chānj·əz }

field investigation [SCI TECH] An investigation carried out in the field; usually applied to an investigation made by someone not domiciled at the site. { 'fēld ,in·ves·tə,gā·shən }

field laboratory [SCI TECH] Usually a temporary or portable laboratory facility set up at the site of an operation to conduct chemical or physical evaluations. { 'fēld ,lab·rə ,tòr·ē }

field moisture [HYD] Water in the ground above the water table. { 'fēld ,mòis·chər }

field test [SCI TECH] A nonformal experiment, that is, one with fewer controls than a laboratory experiment, conducted under field conditions. { 'fēld ,test }

fieldwork [SCI TECH] Work done, such as surveying or making geological observations, in the field. { 'fēld,wərk }

Fiji disease [PL PATH] A virus disease of sugarcane; elongated swellings on the underside of leaves precede death of the plant. { 'fē,jē di,zēz }

Fijivirus [MICROBIO] A genus in the viral family Reoviridae that is the causative agent of Fiji disease in plants and insects. { 'fē·jē,vī·rəs }

film badge [ENG] A device worn for the purpose of indicating the absorbed dose of radiation received by the wearer; usually made of metal, plastic, or paper and loaded with one or more pieces of x-ray film. Also known as badge meter. { 'film ,baj }

film cohesion *See* apparent cohesion. { 'film kō'hē·zhən }

filosus *See* fibratus. { fī'lō·səs }

filtrate [SCI TECH] The discharge liquor in filtration. Also known as mother liquor; strong liquor. { 'fil,trāt }

filtration [SCI TECH] A process of separating particulate matter from a fluid, such as air or a liquid, by passing the fluid carrier through a medium that will not pass the particulates. { fil'trā·shən }

finger lake [HYD] A long, comparatively narrow lake, generally glacial in origin; may occupy a rock basin in the floor of a glacial trough or be confined by a morainal dam across the lower end of the valley. { 'fiŋ·gər ,lāk }

fingerprint |CHEM| Evidence for the presence or the identity of a substance that is obtained by techniques such as spectroscopy, chromatography, or electrophoresist. { 'fiŋ·gər‚print }

finite closed aquifer |HYD| The part of a subterranean reservoir containing water (aquifer) in which the aquifer is limited (finite), with no water flow across the exterior reservoir boundary. { ¦fī‚nīt ¦klōzd 'ak·wə·fər }

fiord *See* fjord. { fyȯrd }

fire blight |PL PATH| A bacterial disease of apple, pear, and related pomaceous fruit trees caused by *Erwinia amylovora*; leaves are blackened, cankers form on the trunk, and flowers and fruits become discolored. { 'fīr ‚blīt }

firebreak |FOR| A cleared area of land intended to check the spread of forest or prairie fire. { 'fīr‚brāk }

firedamp |PETR MIN| **1.** A gas formed in mines by decomposition of coal or other carbonaceous matter; consists chiefly of methane and is combustible. **2.** An airtight stopping to isolate an underground fire and to prevent the inflow of fresh air and the outflow of foul air. Also known as fire wall. { 'fīr‚damp }

fire disclimax |ECOL| A community that is perpetually maintained at an early stage of succession through recurrent destruction by fire followed by regeneration. { 'fīr dis'klī‚maks }

fire extinguisher |ENG| Any of various portable devices used to extinguish a fire by the ejection of a fire-inhibiting substance, such as water, carbon dioxide, gas, or chemical foam. { 'fīr ik‚stiŋ·gwish·ər }

fire foam |ENG| A colloidal solution of small gas bubbles produced by chemical reaction or mechanical agitation and used to extinguish hydrocarbon fire. { 'fīr ‚fōm }

fire tower |FOR| A tower used to watch for fires, especially forest fires. { 'fīr ‚taů·ər }

fire wall |CIV ENG| A fire-resisting wall surrounding an oil storage tank to retain oil that may escape and to confine fire. |PETR MIN| *See* firedamp. { 'fīr ‚wȯl }

fire weather |METEOROL| The state of the weather with respect to its effect upon the kindling and spreading of forest fires. { 'fīr ‚we<u>th</u>·ər }

firn |HYD| Material transitional between snow and glacier ice; it is formed from snow after existing through one summer melt season and becomes glacier ice when its permeability to liquid water drops to zero. Also known as firn snow. { fərn }

firn basin *See* firn field. { 'fərn ‚bās·ən }

firn field |HYD| The accumulation area or upper region of a glacier where snow accumulates and firn is secreted. Also known as firn basin. { 'fərn ‚fēld }

firn ice *See* iced firn. { 'fərn ‚īs }

firnification |HYD| The process of firn formation from snow and of transformation of firn into glacier ice. { ‚fər·nə·fə'kā·shən }

firn snow *See* firn ; old snow. { 'fərn ‚snō }

first bottom |GEOL| The floodplain of a river, below the first terrace. { ¦fərst 'bäd·əm }

first gust |METEOROL| The sharp increase in wind speed often associated with the early mature stage of a thunderstorm cell; it occurs with the passage of the discontinuity zone which is the boundary of the cold-air downdraft. { ¦fərst 'gəst }

first-order climatological station |METEOROL| A meteorological station at which autographic records or hourly readings of atmospheric pressure, temperature, humidity, wind, sunshine, and precipitation are made, together with observations at fixed hours of the amount and form of clouds and notes on the weather. { ¦fərst ‚ȯrd·ər klī·mə·tə¦läj·ə·kəl 'stā·shən }

first-order relief   [GEOGR] Relief features on the largest scale, consisting of continental platforms and ocean basins.   { 'fərst ¦ȯr·dər 'ri·lēf }

first-order station   [METEOROL] After U.S. National Weather Service practice, any meteorological station that is staffed in whole or in part by National Weather Service (Civil Service) personnel, regardless of the type or extent of work required of that station.   { ¦fərst ˌȯrd·ər 'stā·shən }

firth   *See* estuary.   { fərth }

fisheries conservation   [ECOL] Those measures concerned with the protection and preservation of fish and other aquatic life, particularly in sea waters.   { 'fish·ə·rēz ˌkän·sər'vā·shən }

fishery   [ECOL] A place for harvesting fish or other aquatic life, particularly in sea waters.   { 'fish·ə·rē }

fish stock   [ECOL] Any natural population of fish which is an isolated and self-perpetuating group of the same species.   { 'fish ˌstäk }

fission   [PHYS] The division of an atomic nucleus into parts of comparable mass; usually restricted to heavier nuclei such as isotopes of uranium, plutonium, and thorium. Also known as atomic fission; nuclear fission.   { 'fish·ən }

fissure spring   *See* artesian spring.   { 'fish·ər ˌspriŋ }

fitness   [GEN] A measure of reproductive success for a genotype, based on the average number of surviving progeny of this genotype as compared to the average number of other, competing genotypes.   { 'fit·nəs }

five-day forecast   [METEOROL] A forecast of the average weather conditions and large-scale synoptic features in a 5-day period; a type of extended forecast.   { 'fīv ˌdā'fȯr ˌkast }

fixed-level chart   *See* constant-height chart.   { ¦fikst ˌlev·əl 'chärt }

fjard   [GEOGR] A small, narrow, and irregular inlet of the sea with low banks on either side. Also spelled fiard.   { fē'ärd }

fjord   [GEOGR] A narrow, deep inlet of the sea between high cliffs or steep slopes. Also spelled fiord.   { fyȯrd }

fjord valley   [GEOGR] A deep, narrow channel occupied by the sea and extending inland about 50–100 miles (80–160 kilometers).   { 'fyȯrd ˌval·ē }

flag smut   [PL PATH] A smut affecting the leaves and stems of cereals and other grasses, characterized by formation of sori within the tissues, which rupture releasing black spore masses and causing fraying of the infected area.   { 'flag ˌsmət }

flame cultivator   [AGR] A flamethrower for destroying weeds between rows of crops.   { 'flām ˌkəl·tə,vād·ər }

flamethrower   [AGR] A device used to project ignited fuel from a nozzle to destroy material such as weeds or insects.   { 'flām,thrō·ər }

flash coloration   [ECOL] A type of protective coloration in which the prey is cryptic when at rest but reveals brilliantly colored parts while escaping.   { ˌflash kəl·ə'rā·shən }

flash flood   [HYD] A sudden local flood of short duration and great volume; usually caused by heavy rainfall in the immediate vicinity.   { ¦flash ¦fləd }

flat   [GEOGR] A level tract of land.   [GEOL] *See* mud flat.   { flat }

flatwood   [ECOL] An almost-level zone containing mostly imperfectly drained, acid soils and vegetation consisting of wiregrass and saw palmetto at ground level, shrubs such as gallberry and waxmyrtle, and trees such as longleaf and slash pines.   { 'flat,wu̇d }

flavescence |PL PATH| Yellowing or blanching of green plant parts due to diminution of chlorophyll accompanying certain virus disease. { flə'ves·əns }

flaw |METEOROL| An English nautical term for a sudden gust or squall of wind. |OCEANOGR| **1.** The seaward edge of fast ice. **2.** A shore lead lust outside fast ice. { flò }

flax |BOT| *Linum usitatissimum*. An erect annual plant with linear leaves and blue flowers; cultivated as a source of flaxseed and fiber. { flaks }

flea |ZOO| Any of the wingless insects composing the order Siphonaptera; most are ectoparasites of mammals and birds. { flē }

Flehmen response |ECOL| A courtship behavior displayed by the males of some mammalian species in which the upper lip is curled and the neck is extended, facilitating the reception of olfactory cues. { 'flā·mən ri,späns }

fleshy fruit |BOT| A fruit having a fleshy pericarp that is usually soft and juicy, but sometimes hard and tough. { ¦flesh·ē ¦früt }

float |AGR| A device consisting of one or more blades used to level a seedbed. |BIOL| An air-filled sac in many pelagic flora and fauna that serves to buoy up the body of the organism. |GEOL| An isolated, displaced rock or ore fragment. { flōt }

floater *See* drift bottle. { 'flōd·ər }

floating ice |OCEANOGR| Any form of ice floating in water, including grounded ice and drifting land ice. { ¦flōd·iŋ 'īs }

floccus |METEOROL| A cloud species in which each element is a small tuft with a rounded top and a ragged bottom. { 'fläk·əs }

floe |OCEANOGR| A piece of floating sea ice other than fast ice or glacier ice; may consist of a single fragment or of many consolidated fragments, but is larger than an ice cake and smaller than an ice field. Also known as ice floe. { flō }

floeberg |OCEANOGR| A mass of hummocked ice formed by the piling up of many ice floes by lateral pressure; an extreme form of pressure ice; may be more than 50 feet (15 meters) high and resemble an iceberg. { 'flō,bərg }

flood |HYD| The condition that occurs when water overflows the natural or artificial confines of a stream or other body or water, or accumulates by drainage over low-lying areas. |OCEANOGR| The highest point of a tide. { fləd }

flood basin |GEOL| **1.** The tract of land actually submerged during the highest known flood in a specific region. **2.** The flat, wide area lying between a low, sloping plain and the natural levee of a river. { 'fləd ,bās·ən }

flood current |OCEANOGR| The tidal current associated with the increase in the height of a tide. { 'fləd ,kə·rənt }

flooded stream *See* drowned stream. { ¦fləd·əd 'strēm }

flood flow |HYD| Stream discharge during a flood. { 'fləd ,flō }

flood fringe *See* pondage land. { 'fləd ,frinj }

floodgate |CIV ENG| **1.** A gate used to restrain a flow or, when opened, to allow a flood flow to pass. **2.** The lower gate of a lock. { 'fləd,gāt }

flood icing *See* icing. { 'fləd ,īs·iŋ }

flooding |AGR| Filling of ditches or covering of land with water during the raising of crops; rice, for example, must have occasional flooding to grow properly. { 'fləd·iŋ }

flooding ice *See* icing. { 'fləd·iŋ ,īs }

floodplain |GEOL| The relatively smooth valley floors adjacent to and formed by alluviating rivers which are subject to overflow. { 'fləd,plān }

flood plane [HYD] The position of a stream's water surface during a particular flood. { 'fləd ,plān }

flood routing [HYD] The process of computing the progressive time and shape of a flood wave at successive points along a river. Also known as storage routing; streamflow routing. { 'fləd ,rüd·iŋ }

flood stage [HYD] The stage, on a fixed river gage, at which overflow of the natural banks of the stream begins to cause damage in any portion of the reach for which the gage is used as an index. { 'fləd ,stāj }

flood tide [OCEANOGR] **1.** That period of tide between low water and the next high water. **2.** A tide at its highest point. { 'fləd ,tīd }

flora [BOT] **1.** Plants. **2.** The plant life characterizing a specific geographic region or environment. { 'flȯr·ə }

flores martis *See* ferric chloride. { 'flȯr·ēz 'märd·əs }

floret [BOT] A small individual flower that is part of a compact group of flowers, such as the head of a composite plant or inflorescence. { 'flȯr·ət }

floriculture [AGR] A segment of horticulture concerned with commercial production, marketing, and retail sale of cut flowers and potted plants, as well as home gardening and flower arrangement. { 'flȯr·ə,kəl·chər }

Florida Current [OCEANOGR] A fast current that sets through the Straits of Florida to a point north of Grand Bahama Island, where it joins the Antilles Current to form the Gulf Stream. { 'flär·ə·də ,kə·rənt }

floriferous [BOT] Blooming freely, used principally of ornamental plants. { flȯ 'rif·ə·rəs }

florula [ECOL] Plants which grow in a small, confined habitat, for example, a pond. { 'flȯr·yə·lə }

flowage [HYD] Flooding of water onto adjacent land. { 'flō·ij }

flowage line [GEOL] A contour line at the edge of a body of water, such as a reservoir, representing a given water level. { 'flō·ij ,līn }

flow bog [ECOL] A peat bog with a surface level that fluctuates in accordance with rain and tides. { 'flō ,bäg }

flower [BOT] The characteristic reproductive structure of a seed plant, particularly if some or all of the parts are brightly colored. { 'flaů·ər }

flow line [HYD] A contour of the water level around a body of water. { 'flō ,līn }

flow regime [HYD] A range of streamflows having similar bed forms, flow resistance, and means of transporting sediment. { 'flō rə,zhēm }

flowstone [GEOL] Deposits of calcium carbonate that accumulated against the walls of a cave where water flowed on the rock. { 'flō,stōn }

flow velocity [GEOL] In soil, a vector point function used to indicate rate and direction of movement of water through soil per unit of time, perpendicular to the direction of flow. { 'flō və'läs·əd·ē }

fluctuation [SCI TECH] **1.** Variation, especially back and forth between successive values in a series of observations. **2.** Variation of data points about a smooth curve passing among them. { ,flək·chə'wā·shən }

fluoridation [ENG] The addition of the fluorine ion (F^-) to municipal water supplies in a final concentration of 0.8–1.6 parts per million to help prevent dental caries in children. [GEOCHEM] Formation in rocks of fluorine-containing minerals such as fluorite or topaz. { flůr·ə'dā·shən }

fluorine [CHEM] A gaseous or liquid chemical element, symbol F, atomic number 9, atomic weight 18.998403; a member of the halide family, it is the most electronegative element and the most chemically energetic of the nonmetallic elements; highly toxic, corrosive, and flammable; used in rocket fuels and as a chemical intermediate. { 'flùr ‚ēn }

fluoroacetic acid [CHEM] CH_2FCOOH A poisonous, crystalline compound obtained from plants, such as those of the Dichapetalaceae family, South Africa, soluble in water and alcohol, and burns with a green flame; the sodium salt is used as a water-soluble rodent poison. Also known as gifblaar poison. { ¦flùr·ō·ə'sēd·ik 'as·əd }

fluorocarbon-11 See trichlorofluoromethane. { ¦flùr·ō'kär·bən ə'lev·ən }

fluorocarbon-21 See dichlorofluoromethane. { ¦flùr·ō'kär·bən ‚twen·tē'wən }

fluorochlorocarbon See chlorofluorocarbon. { ¦flùr·ō¦klòr·ō'kär·bən }

fluorodichloromethane See dichlorofluoromethane. { ¦flùr·ō·dī¦klòr·ō'meth‚ān }

fluorodifen [CHEM] $C_{13}H_7F_3N_2O_4$ A yellow, crystalline compound with a melting point of 93°C; used as a pre- and postemergence herbicide for food crops. { flù'räd·ə·fen }

fluorotrichloromethane See trichlorofluoromethane. { ¦flùr·ō·trī¦klòr·ō'meth‚ān }

fluosilicic acid [CHEM] H_2SiF_6 A colorless acid, soluble in water, which attacks glass and stoneware; highly corrosive and toxic; used in water fluoridation and electroplating. Also known as hydrofluorosilicic acid; hydrofluosilicic acid. { ¦flü·ə·sə'lis·ik 'as·əd }

flurenol [CHEM] $C_{18}H_{18}O_3$ A solid, crystalline compound with a melting point of 70–71°C; used as an herbicide for vegetables, cereals, and ornamental flowers. { 'flùr·ə ‚nòl }

flurry [METEOROL] A brief shower of snow accompanied by a gust of wind, or a sudden, brief wind squall. { 'flər·ē }

flush [ECOL] An evergreen herbaceous or nonflowering vegetation growing in habitats where seepage water causes the surface to be constantly wet but rarely flooded. { fləsh }

flushing period [HYD] The interval of time required for a quantity of water equal to the volume of a lake to pass through the lake outlet; computed by dividing lake volume by mean flow rate of the outlet. { 'fləsh·iŋ ‚pir·ē·əd }

flush tank [CIV ENG] **1.** A tank in which water or sewage is retained for periodic release through a sewer. **2.** A small water-filled tank for flushing a water closet. { 'fləsh ‚taŋk }

fluvial [HYD] **1.** Pertaining to or produced by the action of a stream or river. **2.** Existing, growing, or living in or near a river or stream. { 'flü·vē·əl }

fluvial cycle of erosion See normal cycle. { 'flü·vē·əl 'sī·kəl əv ə'rō·zhən }

fluvial deposit [GEOL] A sedimentary deposit of material transported by or suspended in a river. { ¦flü·vē·əl di'päz·ət }

fluvial sand [GEOL] Sand laid down by a river or stream. { ¦flü·vē·əl 'sand }

fluvial soil [GEOL] Soil laid down by a river or stream. { ¦flü·vē·əl 'sòil }

fluviatile [GEOL] Resulting from river action. { 'flü·vē·ə‚tīl }

fluviology [HYD] The science of rivers. { flü·vē'äl·ə·jē }

fluviomorphology See river morphology. { ¦flü·vē·ō·mòr'fäl·ə·jē }

fly [ZOO] The common name for a number of species of the insect order Diptera characterized by a single pair of wings, antennae, compound eyes, and hindwings modified to form knoblike balancing organs, the halters. { flī }

fly-off See evapotranspiration. { 'flī,óf }

foam crust [HYD] A snow surface feature that looks like small overlapping waves, like sea foam on a beach, occurring during the ablation of the snow surface and may further develop into a more pronounced wedge-shaped form, known as plowshares. { 'fōm ,krəst }

foam line [OCEANOGR] The front of a wave as it moves toward the shore, after the wave has broken. { 'fōm ,līn }

foehn [METEOROL] A warm, dry wind on the lee side of a mountain range, the warmth and dryness being due to adiabatic compression as the air descends the mountain slopes. Also spelled föhn. { fān }

foehn air [METEOROL] The warm, dry air associated with foehn winds. { 'fān ,er }

foehn cloud [METEOROL] Any cloud form associated with a foehn, but usually signifying only those clouds of the lenticularis species formed in the lee wave parallel to the mountain ridge. { 'fān ,klaud }

foehn cyclone [METEOROL] A cyclone formed (or at least enhanced) as a result of the foehn process on the lee side of a mountain range. { 'fān 'sī,klōn }

foehn island [METEOROL] An isolated area where the foehn has reached the ground, in contrast to the surrounding area where foehn air has not replaced colder surface air. { 'fān 'ī·lənd }

foehn pause [METEOROL] **1.** A temporary cessation of the foehn at the ground, due to the formation or intrusion of a cold air layer which lifts the foehn above the valley floor. **2.** The boundary between foehn air and its surroundings. { 'fān ,póz }

foehn period [METEOROL] The duration of continuous foehn conditions at a given location. { 'fān ,pir·ē·əd }

foehn phase [METEOROL] One of three stages to describe the development of the foehn in the Alps: the preliminary phase, when cold air at the surface is separated from warm dry air aloft by a subsidence inversion; the anticyclonic phase, when the warm air reaches a station as the result of the cold air flowing out from the plain; and the stationary phase or cyclonic phase, when the foehn wall forms and the downslope wind becomes appreciable. { 'fān ,fāz }

foehn storm [METEOROL] A type of destructive storm which frequently occurs in October in the Bavarian Alps. { 'fān ,stórm }

foehn trough [METEOROL] The dynamic trough formed in connection with the foehn. { 'fān ,tróf }

foehn wall [METEOROL] The steep leeward boundary of flat, cumuliform clouds formed on the peaks and upper windward sides of mountains during foehn conditions. { 'fān ,wól }

fog [METEOROL] Water droplets or, rarely, ice crystals suspended in the air in sufficient concentration to reduce visibility appreciably. { fäg }

fogbank [METEOROL] A fairly well-defined mass of fog observed in the distance, most commonly at sea. { 'fäg,baŋk }

fog climax [ECOL] A community that deviates from a climatic climax because of the persistent occurrence of a controlling fog blanket. { ¦fäg 'klī,maks }

fog deposit [HYD] The deposit of an ice coating on exposed surfaces by a freezing fog. { 'fäg dī,päz·ət }

fog dispersal [METEOROL] Artificial dissipation of a fog by means such as seeding or heating. { 'fäg di,spərs·əl }

fog drip [HYD] Water dripping to the ground from trees or other objects which have collected the moisture from drifting fog; the dripping can be as heavy as light rain, as sometimes occurs among the redwood trees along the coast of northern California. { 'fäg ‚drip }

fog drop [METEOROL] An elementary particle of fog, physically the same as a cloud drop. Also known as fog droplet. { 'fäg ‚dräp }

fog droplet See fog drop. { 'fäg ‚dräp·lət }

fog forest [ECOL] The dense, rich forest growth which is found at high or medium-high altitudes on tropical mountains; occurs when the tropical rain forest penetrates altitudes of cloud formation, and the climate is excessively moist and not too cold to prevent plant growth. { 'fäg ‚fär·əst }

fog horizon [METEOROL] The top of a fog layer which is confined by a low-level temperature inversion so as to give the appearance of the horizon when viewed from above against the sky; the true horizon is usually obscured by the fog in such instances. { 'fäg hə‚rīz·ən }

fog scale [METEOROL] A classification of fog intensity based on its effectiveness in decreasing horizontal visibility; such practice is not current in United States weather observing procedures. { 'fäg ‚skāl }

föhn See foehn. { fān }

foliage [BOT] The leaves of a plant. { 'fō·lē·ij }

foliated ice [HYD] Large masses of ice which grow in thermal contraction cracks in permafrost. Also known as ice wedge. { 'fō·lē‚ād·əd 'īs }

foliation [BOT] **1.** The process of developing into a leaf. **2.** The state of being in leaf. { ‚fō·lē'ā·shən }

foliferous [BOT] Producing leaves. { fə'lif·ə·rəs }

foliicolous [BIOL] Growing or parasitic upon leaves, as certain fungi. { 'fä·lē·ə'kə·ləs }

folimat [CHEM] $C_5H_{12}NO_4PS$ An oily liquid that decomposes at 135°C; soluble in water; used as an insecticide and miticide on fruit and vegetable crops and on ornamental flowers. Also known as omethioate. { 'fä·lə‚mat }

follicle [BOT] A type of dehiscent fruit composed of one carpel opening along a single suture. { 'fäl·ə·kəl }

following wind [METEOROL] **1.** A wind blowing in the direction of ocean-wave advance. **2.** See tailwind. { ¦fäl·ə·win 'wind }

folpet [CHEM] $C_9H_4Cl_3NO_2S$ A buff or white, crystalline compound with a melting point of 177–178°C; insoluble in water; used as a fungicide on fruits, vegetables, and ornamental flowers. { 'fäl·pet }

fomite [MED] An inanimate object contaminated with an infectious organism (for example, a dish, clothing, towel, needle, or dust). { 'fō‚mīt }

food [BIOL] A material that can be ingested and utilized by the organism as a source of nutrition and energy. { füd }

food chain [ECOL] The transfer of energy through a series of organisms in different trophic levels. { 'füd ‚chān }

food infection [MED] A type of bacterial food poisoning in which the host is infected by organisms carried by food. { 'füd in‚fek·shən }

food irradiation [ENG] The treatment of fresh or processed foods with ionizing radiation that inactivates biological contaminants (insects, molds, parasites, or bacteria), rendering foods safe to consume and extending their storage lifetime. { 'füd i‚rā·dē¦ā·shən }

food microbiology [MICROBIO] The science that deals with the microorganisms involved in the spoilage, contamination, and preservation of food. { 'füd ˌmī·krō· bī'äl·ə·jē }

food poisoning [MED] Poisoning due to intake of food contaminated by bacteria or poisonous substances produced by bacteria. { 'füd ˌpȯiz·ən·iŋ }

food pyramid [ECOL] An ecological pyramid representing the food relationship among the animals in a community. { 'füd ˌpir·əˌmid }

foodstuff [AGR] Any substance that can be used to feed animals. { 'füdˌstəf }

food web [ECOL] A modified food chain that expresses feeding relationships at various, changing trophic levels. { 'füd ˌweb }

foot-and-mouth disease [VET MED] A highly contagious virus disease of cattle, pigs, sheep, and goats that is transmissible to humans; characterized by fever, salivation, and formation of vesicles in the mouth and pharynx and on the feet. Also known as hoof-and-mouth disease. { ˈfu̇t ən 'mau̇th diˌzēz }

foothills [GEOGR] A region of relatively low, rounded hills at the base of, or on the periphery of, a mountain range. { 'fu̇tˌhilz }

foot rot [PL PATH] Any disease that involves rotting of the stem or trunk of a plant. { 'fu̇t ˌrät }

forage [AGR] A vegetable food for domestic animals. { fär·ij }

forb [BOT] A weed or broadleaf herb. { fȯrb }

ford [HYD] A shallow and usually narrow part of a stream, estuary, or other body of water that may be crossed; for example, by wading or by a wheeled land vehicle. { fȯrd }

forecast [METEOROL] A statement of expected future meteorological occurrences. { 'fȯrˌkast }

forecasting [METEOROL] Procedures for extrapolation of the future characteristics of weather on the basis of present and past conditions. { 'fȯrˌkast·iŋ }

forecast period [METEOROL] The time interval for which a forecast is made. { 'fȯrˌkast ˌpir·ē·əd }

forecast-reversal test [METEOROL] A test used to evaluate the adequacy of a given method of forecast verification; the same verification method is applied, simultaneously, to a given forecast and to a fabricated forecast of opposite conditions; comparison of the verification scores gives an indication of the value of the verification system. { ˈfȯrˌkast ri'vər·səl ˌtest }

forecast verification [METEOROL] Any process for determining the accuracy of a weather forecast by comparing the predicted weather with the observed weather of the forecast period; used to test forecasting skills and methods. { ˈfȯrˌkast ˌver·ə·fə 'kā·shən }

foredune [GEOL] A coastal dune or ridge that is parallel to the shoreline of a large lake or ocean and is stabilized by vegetation. { 'fȯrˌdün }

foreland [GEOGR] An extensive area of land jutting out into the sea. [GEOL] **1.** A lowland area onto which piedmont glaciers have moved from adjacent mountains. **2.** A stable part of a continent bordering an orogenic or mobile belt. { 'fȯr·lənd }

Forel scale [OCEANOGR] A scale of yellows, greens, and blues for recording the color of sea water as seen against the white background of a Secchi disk. { fȯ'rel ˌskāl }

forerunner [OCEANOGR] Low, long-period ocean swell which commonly precedes the main swell from a distant storm, especially a tropical cyclone. { 'fȯrˌrən·ər }

forest [ECOL] An ecosystem consisting of plants and animals and their environment, with trees as the dominant form of vegetation. { 'fär·əst }

forest climate   *See* humid climate.   { 'fär·əst ‖klī·mət }

forest conservation   [ECOL] Those measures concerned with the protection and preservation of forest lands and resources.   { 'fär·əst ‚kän·sər'vā·shən }

forest ecology   [ECOL] The science that deals with the relationship of forest trees to their environment, to one another, and to other plants and to animals in the forest.   { 'fär·əst i‚käl·ə·jē }

forest ecosystem   [ECOL] The entire assemblage of forest organisms (trees, shrubs, herbs, bacteria, fungi, and animals, including people) together with their environmental substrate (the surrounding air, soil, water, organic debris, and rocks), interacting inside a defined boundary.   { ‖fär·əst 'ek·ō‚sis·təm }

forest engineering   [ENG] A branch of engineering concerned with the solution of forestry problems with regard to long-range environmental and economic effects.   { 'fär·əst ‚en·jə‚nir·iŋ }

forest fire   [FOR] Uncontrolled combustion of forest fuels, such as dead leaves, grasses, pine needles, and branches on the ground..   { 'fär·əst ‚fīr }

forest genetics   [FOR] The study of variation and inheritance in forest trees; it provides the knowledge necessary to breed trees through traditional methods of selection and hybridization, and also through the newer biotechnologies.   { ‚fär·əst jə'ned·iks }

forest management   [FOR] Measures concerned with the effective organization of a forest to ensure continued production of its goods and services.   { 'fär·əst ‚man·ij·mənt }

forest mapping   [FOR] The branch of forestry dealing with the preparation of maps showing the distribution and conformation of individual forest stands.   { 'fär·əst ‚map·iŋ }

forest measurement   [FOR] The branch of forestry concerned with the measurement of standing trees, cut roundwood, and lumber products.   { ‖fär·əst 'mezh·ər·mənt }

forest product   [FOR] Any material afforded by a forest for commercial use, such as tree products and forage.   { ‖fär·əst ‖präd·əkt }

forest resources   [FOR] Forest land and the trees on it.   { ‖fär·əst ri'sȯrs·əs }

forestry   [ECOL] The management of forest lands for wood, forages, water, wildlife, and recreation.   { 'fär·ə·strē }

forest soil   [FOR] The natural medium for growth of tree roots and associated forest vegetation.   { 'fär·əst ‚sȯil }

forest stand   [FOR] The basic unit of forest mapping; a group of trees that are more or less homogeneous with regard to species composition, density, size, and sometimes habitat.   { 'fär·əst ‚stand }

forest-tundra   [ECOL] A temperate and cold savanna which occurs at high altitudes and consists of scattered or clumped trees and a shrub layer of varying coverage.   { ‖fär·əst ‖tən·drə }

forest wind   [METEOROL] A light breeze which blows from forests toward open country on calm clear nights.   { ‖fär·əst 'wind }

formaldehyde   [CHEM] HCHO The simplest aldehyde; a gas at room temperature, and a poisonous, clear, colorless liquid solution with pungent odor; used to make synthetic resins by reaction with phenols, urea, and melamine, as a chemical intermediate, as an embalming fluid, and as a disinfectant. Also known as formol; methanal; methylene oxide.   { fȯr'mal·də‚hīd }

formation water   [HYD] Water present with petroleum or gas in reservoirs. Also known as oil-reservoir water.   { fȯr'mā·shən ‚wȯd·ər }

formic acid [CHEM] HCOOH A colorless, pungent, toxic, corrosive liquid melting at 8.4°C; soluble in water, ether, and alcohol; used as a chemical intermediate and solvent, in dyeing and electroplating processes, and in fumigants. Also known as methanoic acid. { ¦fȯr·mik 'as·əd }

formol See formaldehyde. { 'fȯr‚mȯl }

formonitrile See hydrocyanic acid. { ¦fȯr·mō¦nī·trəl }

Forrel cell [METEOROL] A type of atmospheric circulation in which air moves away from the thermal equator at low latitude levels and in the opposite direction in higher latitudes. { fə'rel ‚sel }

fortnightly tide [OCEANOGR] A tide occurring at intervals of one-half the period of oscillation of the moon, approximately 2 weeks. { ‚fȯrt¦nīt·lē 'tīd }

fossil fuel [GEOL] Any hydrocarbon deposit that may be used for fuel; examples are petroleum, coal, and natural gas. { ¦fäs·əl 'fyül }

fossil ice [HYD] **1.** Relatively old ground ice found in regions of permafrost. **2.** Underground ice in regions where present-day temperatures are not low enough to have formed it. { ¦fäs·əl 'īs }

fossil permafrost See passive permafrost. { ¦fäs·əl 'pər·mə‚frȯst }

fossil turbulence [METEOROL] Inhomogeneities of temperature and humidity remaining in the air after the motion which produced them has subsided and the density has become uniform; causes scattering of radio waves, and lumpy clouds when air is rising. { ¦fäs·əl 'tər·byə·ləns }

fouling organism [ECOL] Any aquatic organism with a sessile adult stage that attaches to and fouls underwater structures of ships. { 'faůl·iŋ ‚ȯr·gə‚niz·əm }

fouling plates [ENG] Metal plates submerged in water to allow attachment of fouling organisms, which are then analyzed to determine species, growth rate, and growth pattern, as influenced by environmental conditions and time. { 'faůl·iŋ ‚plāts }

founder [GEOL] To sink under water either by depression of the land or by rise of sea level, especially in reference to large crustal masses, islands, or significant portions of continents. { 'faůn·dər }

4-D chart [METEOROL] A chart showing the field of D values (deviations of the actual altitudes along a constant-pressure surface from the standard atmosphere altitude of that surface) in terms of the three dimensions of space and one of time; it is a form of a four-dimensional display of pressure altitude; the space dimensions are represented by D-value contours, and the time dimension is provided by tau-value lines. { ¦fȯr ¦dē ‚chärt }

fourré See temperate and cold scrub ; tropical scrub. { fů'rā }

fowl [AGR] A domestic cock or hen, especially an adult hen, such as among chickens or several other gallinaceous birds. { faůl }

fowl pox [VET MED] A disease of birds caused by a virus and characterized by wartlike nodules on the skin, particularly on the head. { 'faůl ‚päks }

fractus [METEOROL] A cloud species in which the cloud elements are irregular but generally small in size, and which presents a ragged, shredded appearance, as if torn; these characteristics change ceaselessly and often rapidly. { 'frak·təs }

fragmentation nucleus [METEOROL] A tiny ice particle broken from a large ice crystal, serving as an ice nucleus; that is, a growth center for a new ice crystal. { ‚frag·mən'tā·shən ‚nü·klē·əs }

Francisella [MICROBIO] A genus of gram-negative, aerobic bacteria of uncertain affiliation; cells are small, coccoid to ellipsoidal, pleomorphic rods and can be parasitic on mammals, birds, and arthropods. { ‚fran·si'sel·ə }

frazil [HYD] Ice crystals which form in supercooled water that is too turbulent to permit coagulation of the crystals into sheet ice. { 'fra·zəl }

frazil ice [HYD] A spongy or slushy accumulation of frazil in a body of water. Also known as needle ice. { 'fra·zəl ˌīs }

free-air temperature [METEOROL] Temperature of the atmosphere, obtained by a thermometer located so as to avoid as completely as practicable the effects of extraneous heating. { 'frē ˌer 'tem·prə·chər }

free convection *See* thermal convection. { ¦frē kən'vek·shən }

free foehn *See* high foehn. { ¦frē ¦fān }

free meander [HYD] A stream meander that displaces itself very easily by lateral corrasion (erosion by scraping). { ¦frē mē'an·dər }

free-water content *See* water content. { 'frē ˌwōd·ər ˌkän·tent }

free-water elevation *See* water table. { 'frē ˌwōd·ər el·ə‚vā·shən }

free-water surface *See* water table. { 'frē ˌwōd·ər ¦sər·fəs }

freeze-out lake [HYD] A shallow lake which may be deeply frozen over for long periods of time. { 'frēz ˌaút ˌlāk }

freeze-up [HYD] The formation of a continuous ice cover on a body of water. { 'frēz ˌəp }

freezing drizzle [METEOROL] Drizzle that falls in liquid form but freezes upon impact with the ground to form a coating of glaze. { ¦frēz·iŋ ¦driz·əl }

freezing level [METEOROL] The lowest altitude in the atmosphere over a given location, at which the air temperature is 0°C; the height of the 0°C constant-temperature surface. { ¦frēz·iŋ ¦lev·əl }

freezing-level chart [METEOROL] A synoptic chart showing the height of the 0°C constant-temperature surface by means of contour lines. { 'frēz·iŋ ˌlev·əl ˌchärt }

freezing nucleus [METEOROL] Any particle which, when present within a mass of supercooled water, will initiate growth of an ice crystal about itself. { 'frēz·iŋ ˌnü·klē·əs }

freezing precipitation [METEOROL] Any form of liquid precipitation that freezes upon impact with the ground or exposed objects; that is, freezing rain or freezing drizzle. { ¦frēz·iŋ prə‚sip·ə'tā·shən }

freezing rain [METEOROL] Rain that falls in liquid form but freezes upon impact to form a coating of glaze upon the ground and on exposed objects. { ¦frēz·iŋ 'rān }

French measles *See* rubella. { ¦french 'mē·zəlz }

frequency [STAT] The number of times an event or item falls into or is expected to fall into a certain class or category. { 'frē·kwən·sē }

fresh breeze [METEOROL] In the Beaufort wind scale, a wind whose speed is 17 to 21 knots (19 to 24 miles per hour, or 31 to 39 kilometers per hour). { ¦fresh 'brēz }

freshet [HYD] **1.** The annual spring rise of streams in cold climates as a result of melting snow. **2.** A flood resulting from either rain or melting snow; usually applied only to small streams and to floods of minor severity. **3.** A small fresh-water stream. { 'fresh·ət }

fresh gale [METEOROL] In the Beaufort wind scale, a wind whose speed is from 34 to 40 knots (39 to 46 miles per hour, or 63 to 74 kilometers per hour). { ¦fresh 'gāl }

fresh ice *See* newly formed ice. { ¦fresh 'īs }

fresh water |HYD| Water containing no significant amounts of salts, such as in rivers and lakes. { ¦fresh 'wȯd·ər }

fresh-water ecosystem [ECOL] The living organisms and nonliving materials of an inland aquatic environment. { 'fresh ˌwōd·ər 'ek·ō¸sis·təm }

friable |GEOL| Referring to the property of a soil that is easily crumbled or pulverized. { 'frī·ə·bəl }

frictional *See* cohesionless. { 'frik·shən·əl }

friction layer *See* surface boundary layer. { 'frik·shən ˌlā·ər }

Friedlander's bacillus *See* Klebsiella pneumoniae. { 'frēd¸lan·dərz bə¸sil·əs }

fringing reef |GEOL| A coral reef attached directly to or bordering the shore of an island or continental landmass. { ¦frin·jiŋ 'rēf }

frog |ZOO| The common name for a number of tailless amphibians in the order Anura; most have hindlegs adapted for jumping, scaleless skin, and large eyes. { fräg }

frog storm [METEOROL] The first bad weather in spring after a warm period. Also known as whippoorwill storm. { 'fräg ˌstȯrm }

frond |BOT| **1.** The leaf of a palm or fern. **2.** A foliaceous thallus or thalloid shoot. { fränd }

front |METEOROL| A sloping surface of discontinuity in the troposphere, separating air masses of different density or temperature. { frənt }

frontal contour |METEOROL| The line of intersection of a front (frontal surface) with a specified surface in the atmosphere, usually a constant-pressure surface; with respect to only one surface, this line is usually called the front. { ¦frənt·əl 'kän·tür }

frontal cyclone |METEOROL| Any cyclone associated with a front; often used synonymously with wave cyclone or with extratropical cyclone (as opposed to tropical cyclones, which are nonfrontal). { ¦frənt·əl 'sī¸klōn }

frontal fog [METEOROL] Fog associated with frontal zones and frontal passages. { ¦frənt·əl 'fäg }

frontal inversion |METEOROL| A temperature inversion in the atmosphere, encountered upon vertical ascent through a sloping front (or frontal zone). { ¦frənt·əl in'vər·zhən }

frontal lifting [METEOROL] The forced ascent of the warmer, less-dense air at and near a front, occurring whenever the relative velocities of the two air masses are such that they converge at the front. { ¦frənt·əl 'lift·iŋ }

frontal occlusion *See* occluded front. { ¦frənt·əl ə'klü·zhən }

frontal passage [METEOROL] The passage of a front over a point on the earth's surface. { ¦frənt·əl 'pas·ij }

frontal precipitation [METEOROL] Any precipitation attributable to the action of a front; used mainly to distinguish this type from air-mass precipitation and orographic precipitation. { ¦frənt·əl prə¸sip·ə'tā·shən }

frontal profile [METEOROL] The outline of a front as seen on a vertical cross section oriented normal to the frontal surface. { ¦frənt·əl 'prō¸fīl }

frontal strip [METEOROL] The presentation of a front, on a synoptic chart, as a frontal zone; that is, two lines, rather than a single line, are drawn to represent the boundaries of the zone; a rare usage. { ¦frənt·əl ˌstrip }

frontal system [METEOROL] A system of fronts as they appear on a synoptic chart. { 'frənt·əl ˌsis·təm }

frontal thunderstorm [METEOROL] A thunderstorm associated with a front; limited to thunderstorms resulting from the convection induced by frontal lifting. { 'frənt·əl 'thən·dər₁stórm }

frontal wave [METEOROL] A horizontal, wavelike deformation of a front in the lower levels, commonly associated with a maximum of cyclonic circulations in the adjacent flow; it may develop into a wave cyclone. { 'frənt·əl ₁wāv }

frontal zone [METEOROL] The three-dimensional zone or layer of large horizontal density gradient, bounded by frontal surfaces and surface front. { 'frənt·əl ₁zōn }

frontogenesis [METEOROL] **1.** The initial formation of a frontal zone or front. **2.** The increase in the horizontal gradient of an air mass property, mainly density, and the formation of the accompanying features of the wind field that typify a front. { ¦frən·tō ¦jen·ə·səs }

frontolysis [METEOROL] **1.** The dissipation of a front or frontal zone. **2.** In general, a decrease in the horizontal gradient of an air mass property, principally density, and the dissipation of the accompanying features of the wind field. { ₁frən'täl·ə·səs }

frost [HYD] A covering of ice in one of its several forms, produced by the sublimation of water vapor on objects colder than 32°F (0°C). { fróst }

frost action [GEOL] **1.** The weathering process caused by cycles of freezing and thawing of water in surface pores, cracks, and other openings. **2.** Alternate or repeated cycles of freezing and thawing of water contained in materials; the term is especially applied to disruptive effects of this action. { 'fróst ₁ak·shən }

frostbite [MED] Injury to skin and subcutaneous tissues, and in severe cases to deeper tissues also, from exposure to extreme cold. { 'fróst₁bīt }

frost boil [GEOL] **1.** An accumulation of water and mud released from ground ice by accelerated spring thawing. **2.** A low mound formed by local differential frost heaving at a location most favorable for the formation of segregated ice and accompanied by the absence of an insulating cover of vegetation. { 'fróst ₁bóil }

frost bursting See congelifraction. { 'fróst ₁bərst·iŋ }

frost churning See congeliturbation. { 'fróst ₁chərn·iŋ }

frost climate [CLIMATOL] The coldest temperature province in C. W. Thornthwaite's climatic classification: the climate of the ice cap regions of the earth, that is, those regions perennially covered with snow and ice. { 'fróst ₁klī·mət }

frost day [METEOROL] An observational day on which frost occurs. { 'fróst ₁dā }

frost feathers See ice feathers. { 'fróst ₁fe<u>th</u>·ərz }

frost flakes See ice fog. { 'fróst ₁flāks }

frost flowers See ice flowers. { 'fróst ₁flaú·ərz }

frost fog See ice fog. { 'fróst ₁fäg }

frost hazard [METEOROL] The risk of damage by frost, expressed as the probability or frequency of killing frost on different dates during the growing season, or as the distribution of dates of the last killing frost of spring or the first of autumn. { 'fróst ₁haz·ərd }

frost heaving [GEOL] The lifting and distortion of a surface due to internal action of frost resulting from subsurface ice formation; affects soil, rock, pavement, and other structures. { 'fróst ₁hēv·iŋ }

frost hollow [METEOROL] A small, low-lying zone which experiences frequent and severe frosts owing to the accumulation of cold night air; often severe where hills block the afternoon sunshine. { 'fróst ₁häl·ō }

frostless zone |METEOROL| The warmest part of a slope above a valley floor, lying between the layer of cold air which forms over the valley floor on calm clear nights and the cold hill tops or plateaus; the air flowing down the slopes is warmed by mixing with the air above ground level, and to some extent also by adiabatic compression. Also known as green belt; verdant zone. { 'fròst·ləs ‚zōn }

frost line |GEOL| 1. The maximum depth of frozen ground during the winter. 2. The lower limit of the permafrost. { 'fròst ‚līn }

frost mound |GEOL| A hill and knoll associated with frozen ground in a permafrost region, containing a core of ice. Also known as soffosian knob; soil blister. { 'fròst ‚maúnd }

frost pocket |METEOROL| A parcel of cold air in a hollow or at a valley floor, occurring when nighttime terrestrial radiation is greatest on valley slopes. { 'fròst ‚päk·ət }

frost ring |BOT| A false annual growth ring in the trunk of a tree due to out-of-season defoliation by frost and subsequent regrowth of foliage. { 'fròst ‚riŋ }

frost riving See congelifraction. { 'fròst ‚rīv·iŋ }

frost shattering See congelifraction. { 'fròst ‚shad·ə·riŋ }

frost smoke |METEOROL| 1. A rare type of fog formed in the same manner as a steam fog, but at colder temperatures so that it is composed of ice particles instead of water droplets. 2. See steam fog. { 'fròst ‚smōk }

frost splitting See congelifraction. { 'fròst ‚splid·iŋ }

frost stirring See congelifraction. { 'fròst ‚stər·iŋ }

frost weathering See congelifraction. { 'fròst ‚weth·ə·riŋ }

frost wedging See congelifraction. { 'fròst ‚wej·iŋ }

frost zone See seasonally frozen ground. { 'fròst ‚zōn }

frozen fog See ice fog. { ¦frōz·ən 'fäg }

frozen ground |GEOL| Soil having a temperature below freezing, generally containing water in the form of ice. Also known as gelisol; merzlota; taele; tjaele. { ¦frōz·ən 'graúnd }

frozen precipitation |METEOROL| Any form of precipitation that reaches the ground in frozen form; that is, snow, snow pellets, snow grains, ice crystals, ice pellets, and hail. { ¦frōz·ən prə‚sip·ə'tā·shən }

fructescence |BOT| The period of fruit maturation. { ‚frək'tes·əns }

fructification |BOT| 1. The process of producing fruit. 2. A fruit and its appendages. |MYCOL| A sporogenous structure. { 'frək·tə·fə'kā·shən }

fruit |BOT| A fully matured plant ovary with or without other floral or shoot parts united with it at maturity. { früt }

fruit bud |BOT| A fertilized flower bud that matures into a fruit. { 'früt ‚bəd }

fruiting body |BOT| A specialized, spore-producing organ. { 'früd·iŋ ‚bäd·ē }

frutescent |BIOL| See fruticose. |BOT| Shrublike in habit. { frü'tes·ənt }

fruticose |BIOL| Resembling a shrub; applied especially to lichens. Also known as frutescent. { 'früd·ə‚kōs }

fucoxanthin |BIOL| $C_{40}H_{60}O_6$ A carotenoid pigment; a partial xanthophyll ester found in diatoms and brown algae. { ¦fyü·kō¦zan·thən }

fuel cell |CHEM| An electrochemical device in which the reaction between a fuel, such as hydrogen, and an oxidant, such as oxygen or air, converts the chemical energy of the fuel directly into electrical energy without combustion. { 'fyül ‚sel }

fugacious [BOT] Lasting a short time; used principally to describe plant parts that fall soon after being formed. { fyü'gā·shəs }

fully arisen sea See fully developed sea. { 'fùl·ē ə,riz·ən 'sē }

fully developed sea [OCEANOGR] The maximum ocean waves or sea state that can be produced by a given wind force blowing over sufficient fetch, regardless of duration. Also known as fully arisen sea. { 'fùl·ē di,vel·əpt 'sē }

fumigant [CHEM] A chemical compound which acts in the gaseous state to destroy insects and their larvae and other pests; examples are dichlorethyl ether, *p*-dichlorobenzene, and ethylene oxide. { 'fyü·mə·gənt }

fumigating [ENG] The use of a chemical compound in a gaseous state to kill insects, nematodes, arachnids, rodents, weeds, and fungi in confined or inaccessible locations; also used to control weeds, nematodes, and insects in the field. { 'fyü·mə,gād·iŋ }

fumulus [METEOROL] A very thin cloud veil at any level, so delicate that it may be almost invisible. { 'fyü·myə·ləs }

fungal ecology [ECOL] The subdiscipline in mycology and ecology that examines fungal community composition and structure; responses, activities, and interactions of single fungus species; and the functions of fungi in ecosystems. { ,fəŋ·gəl i'käl·ə·jē }

fungal sheath [MYCOL] A compact layer of fungal hyphae that surrounds the young root surface of the host plant and prevents direct contact between the root and the soil. { 'fəŋ·gəl ,shēth }

fungi [MYCOL] Nucleated, usually filamentous, sporebearing organisms devoid of chlorophyll. { 'fən,jī }

fungus gall [PL PATH] A plant gall resulting from an attack of a parasitic fungus. { 'fəŋ·gəs ,gòl }

funnel cloud [METEOROL] The popular term for the tornado cloud, often shaped like a funnel with the small end nearest the ground. { 'fən·əl ,klaùd }

2-furaldehyde See furfural. { ¦tü fə'ral·də,hīd }

furancarboxylic acid See furoic acid. { ¦fyùr·ən,kar,bäk'sil·ik 'as·əd }

furfural [CHEM] C_4H_3OCHO When pure, a colorless liquid, soluble in organic solvents, slightly soluble in water; used as a lube oil-refining solvent, in cellulosic formulations, in making resins, as a weed killer, as a fungicide, and as a chemical intermediate. Also known as 2-furaldehyde; furfuraldehyde; furfurol; furol. { 'fər·fə,ral }

furfuraldehyde See furfural. { 'fər·fə'ral·də,hīd }

furfurol See furfural. { 'fər·fə,ròl }

furoic acid [CHEM] $C_5H_4O_3$ Long monoclinic prisms crystallized from the water solution, soluble in ether and alcohol; used as a preservative and bactericide. Also known as furancarboxylic acid; pyromucic acid. { fyü'rō·ik 'as·əd }

furrow irrigation [AGR] Irrigation via furrows between rows of crops. { 'fər·ō,ir·ə'gā·shən }

Fusarium [MYCOL] A genus of fungi in the family Tuberculariaceae having sickle-shaped, multicelled conidia; includes many important plant pathogens. { fyü'za·rē·əm }

Fusarium oxysporum [MYCOL] A pathogenic fungus causing a variety of plant diseases, including cabbage yellows and wilt of tomato, flax, cotton, peas, and muskmelon. { fyü'za·rē·əm ,äk·sə'spór·əm }

Fusarium solani [MYCOL] A pathogenic fungus implicated in root rot and wilt diseases of several plants, including sisal and squash. { fyü'za·rē·əm sō'lan·ē }

fusicoccin [PL PATH] A nonselective pathotoxin with growth-regulator properties that is produced by *Fusicoccum amygdale* and causes a wilt disease of peach and almond trees. { ‚fyüz·i'käk·sən }

Fusicoccum amygdali [MYCOL] A fungal pathogen that produces fusicoccin, the cause of wilt disease in peach and almond trees. { ‚fyüz·i‚käk·əm ə'mig·də‚lē }

G

G *See* giga-.

gale |METEOROL| **1.** An unusually strong wind. **2.** In storm-warning terminology, a wind of 28–47 knots (52–87 kilometers per hour). **3.** In the Beaufort wind scale, a wind whose speed is 28–55 knots (52–102 kilometers per hour). { 'gāl }

gale warning |METEOROL| A storm warning for marine interests of impending winds from 28 to 47 knots (52–87 kilometers per hour), signaled by two triangular red pennants by day, and a white lantern over a red lantern by night. { 'gāl ,wȯrn·iŋ }

gall |PL PATH| A large swelling on plant tissues caused by the invasion of parasites, such as fungi or bacteria, following puncture by an insect; insect oviposit and larvae of insects are found in galls. { gȯl }

galleria forest |ECOL| A modified tropical deciduous forest occurring along stream banks. { ‚gal·ə'rē·ə ‚fär·əst }

gallery forest |FOR| A forest occurring on both banks of a river in a region that is otherwise treeless. { 'gal·rē ‚fär·əst }

gallivorous |ZOO| Feeding on the tissues of galls, especially certain insect larvae. { gȯ'liv·ə·rəs }

galvanotaxis |BIOL| Movement of a free-living organism in response to an electrical stimulus. { ¦gal·və·nō¦tak·səs }

galvanotropism |BIOL| Response of an organism to electrical stimulation. { ‚gal·və'nä·trə‚piz·əm }

gamete |BIOL| A cell which participates in fertilization and development of a new organism. Also known as germ cell; sex cell. { 'ga‚mēt }

gametophyte |BOT| **1.** The haploid generation producing gametes in plants exhibiting metagenesis. **2.** An individual plant of this generation. { gə'mēd·ə‚fīt }

gamodeme |ECOL| An isolated breeding community. { 'gam·ə‚dēm }

Ganoderma lucidum |MYCOL| A mushroom found throughout the United States, Europe, South America, and Asia that appears to have antiallergic, anti-inflammatory, antibacterial, antioxidant, antitumor, and immunostimulating activity. Also known as ling-zhi; reishi mushroom. { ¦gen·ə‚dər·mə 'lüs·ə·dəm }

gap |GEOGR| Any sharp, deep notch in a mountain ridge or between hills. { gap }

garbin |METEOROL| A sea breeze; in southwest France it refers to a southwesterly sea breeze which sets in about 9 a.m., reaches its maximum toward 2 p.m., and ceases about 5 p.m. { gär'ba }

garigue |ECOL| A low, open scrubland restricted to limestone sites in the Mediterranean area; characterized by small evergreen shrubs and low trees. { gə'rēg }

garúa |METEOROL| A dense fog or drizzle from low stratus clouds on the west coast of South America, creating a raw, cold atmosphere that may last for weeks in winter, and supplying a limited amount of moisture to the area. Also known as camanchaca. { gä′rü·ə }

gas black [CHEM] Fine particles of carbon formed by partial combustion or thermal decomposition of natural gas; used to reinforce rubber products such as tires. Also known as carbon black; channel black. { ′gas ‚blak }

gas cleaning [ENG] Removing ingredients, pollutants, or contaminants from domestic and industrial gases. { ′gas ‚klēn·iŋ }

gas mask [ENG] A device to protect the eyes and respiratory tract from noxious gases, vapors, and aerosols, by removing contamination with a filter and a bed of adsorbent material. { ′gas ‚mask }

gasoline [PETR MIN] A fuel for internal combustion engines consisting essentially of volatile flammable liquid hydrocarbons; derived from crude petroleum by processes such as distillation reforming, polymerization, catalytic cracking, and alkylation; the common name is gas. Also known as petrol. { ′gas·ə‚lēn }

gas reservoir [GEOL] An accumulation of natural gas found with or near accumulations of crude oil in the earth's crust. { ‚gas ¦rez·əv‚wär }

gas trap [CIV ENG] A bend or chamber in a drain or sewer pipe that prevents sewer gas from escaping. { ′gas ‚trap }

gathering ground See drainage basin. { ′ga<u>th</u>·ə·riŋ ‚graủnd }

Gause's principle [ECOL] A statement that two species cannot occupy the same niche simultaneously. Also known as competitive-exclusion principle. { ′gaủz·əz ‚prin·sə·pəl }

Geiger counter See Geiger-Müller counter. { ′gī·gər ‚kaủnt·ər }

Geiger counter tube See Geiger-Müller tube. { ′gī·gər ‚kaủnt·ər ‚tüb }

Geiger-Müller counter [ENG] **1.** A radiation counter that uses a Geiger-Müller tube in appropriate circuits to detect and count ionizing particles; each particle crossing the tube produces ionization of gas in the tube which is roughly independent of the particle's nature and energy, resulting in a uniform discharge across the tube. Abbreviated GM counter. Also known as Geiger counter. **2.** See Geiger-Müller tube. { ¦gī·gər ′myül·ər ‚kaủnt·ər }

Geiger-Müller counter tube See Geiger-Müller tube. { ¦gī·gər ′myül·ər ‚kaủnt·ər ‚tüb }

Geiger-Müller tube [ENG] A radiation-counter tube operated in the Geiger region; it usually consists of a gas-filled cylindrical metal chamber containing a fine-wire anode at its axis. Also known as Geiger counter tube; Geiger-Müller counter; Geiger-Müller counter tube. { ¦gī·gər ′myül·ər ‚tüb }

geitonogamy [BOT] Pollination and fertilization of one flower by another on the same plant. { ‚gīt·ən′äg·ə·mē }

gelifraction See congeliturbation. { ¦jel·ə¦frak·shən }

gelisol See frozen ground. { ′jel·ə‚sȯl }

geliturbation See congeliturbation. { ‚jel·ə‚ter′bāsh·ən }

gelivation See congelifraction. { ¦jel·ə¦vā·shən }

gemmation See budding. { je′mā·shən }

gene [GEN] The basic unit of inheritance; composed of a deoxyribonucleic acid (DNA) sequence that contains the elements required for transcription of a complementary ribonucleic acid (RNA) which is sometimes the functional gene product but more often is converted into messenger RNAs that specify the amino acid sequence of a protein product. { jēn }

genecology [BIOL] The study of species and their genetic subdivisions, their place in nature, and the genetic and ecological factors controlling speciation. { ¦jēn·ə'käl·ə·jē }

gene flow [GEN] The passage and establishment of alleles characteristic of one breeding population into the gene pool of another population through hybridization and backcrossing. { 'jēn ,flō }

gene pool [GEN] The totality of the genes of a specific population at a given time. { 'jēn ,pül }

general circulation [METEOROL] The complete statistical description of atmospheric motions over the earth. Also known as planetary circulation. { ¦jen·rəl ,sər·kyə'lā·shən }

general paresis [MED] An inflammatory and degenerative disease of the brain caused by infection with *Treponema pallidum*. Also known as syphilitic meningoencephalitis. { ¦jen·rəl pə'rē·səs }

generating area *See* fetch. { 'jen·ə,rād·iŋ ,er·ē·ə }

generating plant *See* generating station. { 'jen·ə,rād·iŋ ,plant }

generating station [ENG] A stationary plant containing apparatus for large-scale conversion of some form of energy (such as hydraulic, steam, chemical, or nuclear energy) into electrical energy. Also known as generating plant; power station. { 'jen·ə,rād·iŋ ,stā·shən }

generation [BIOL] A group of organisms having a common parent or parents and comprising a single level in line of descent. { ,jen·ə'rā·shən }

generative nucleus [BOT] A haploid nucleus in a pollen grain that produces two sperm nuclei by mitosis. { 'jen·rəd·iv 'nü·klē·əs }

genetic drift [GEN] The random fluctuation of gene frequencies from generation to generation that occurs predominantly in small populations. { jə¦ned·ik 'drift }

genetic isolation [GEN] The absence of genetic exchange between populations or species as a result of geographic separation or of mechanisms that prevent reproduction. { jə¦ned·ik īs·əl'ā·shən }

genetic material [GEN] The nuclear (chromosomal) and cytoplasmic (mitochondrial and chloroplast) material that plays a fundamental role in determining the nature of all cell substances, cell structures, and cell effects; the genes have properties of self-propagation and variation. { jə¦ned·ik mə'tir·ē·əl }

genetics [BIOL] The science that is concerned with the study of biological inheritance. { jə'ned·iks }

genicide [CHEM] $C_{13}H_8O_2$ A compound with needlelike crystals and a melting point of 174°C; insoluble in water; used as an insecticide, miticide, and ovicide. Also known as oxoxanthone; 9-xanthenone; xanthone. { 'jen·ə,sīd }

genome [GEN] **1.** The genetic endowment of a species. **2.** The haploid set of chromosomes. { 'jē,nōm }

genomic stress [GEN] Any influence that may disrupt the stability of the genome by fostering chromosome damage or mutation, such as environmental factors or altered genetic background. { jə¦nōm·ik 'stres }

genotoxant [BIOL] An agent that induces toxic, lethal, or heritable effects to nuclear and extranuclear genetic material in cells. { ,jēn·ə'täk·sənt }

genotype [GEN] The type species of a genus. { 'jē·nə,tīp }

genotype frequency [GEN] The proportion or frequency of any particular genotype among the individuals of a population. { 'jēn·ə,tīp ,frē·kwən·sē }

gentamicin |MICROBIO| A broad-spectrum antibiotic produced by a species of *Micromonospora*. { ¦jent·ə¦mīs·ən }

gentian violet *See* methyl violet. { 'jen·chən 'vī·lət }

gentle breeze |METEOROL| In the Beaufort wind scale, a wind whose speed is from 7 to 10 knots (13–19 kilometers per hour). { ¦jent·əl 'brēz }

genus |SYST| A taxonomic category that includes groups of closely related species; the principal subdivision of a family. { 'jē·nəs }

geo |GEOGR| A narrow coastal inlet bordered by steep cliffs. Also spelled gio. { 'gyō }

geobotanical prospecting |GEOL| The use of the distribution, appearance, and growth anomalies of plants in locating ore deposits. { ¦jē·ō·bə¦tan·ə·kəl 'präs·pek·tiŋ }

geobotany |BOT| The study of plants as related to their geologic environment. { ¦jē·ō 'bät·ən·ē }

geochemical anomaly |GEOCHEM| Above-average concentration of a chemical element in a sample of rock, soil, vegetation, stream, or sediment; indicative of nearby mineral deposit. { ¦jē·ō¦kem·ə·kəl ə'näm·ə·lē }

geochemical balance |GEOCHEM| The proportional distribution, and the migration rate, in the global fractionation of elements, minerals, or compounds; for example, the distribution of quartz in igneous rocks, its liberation by weathering, and its redistribution into sediments and, in solution, into lakes, rivers, and oceans. { ¦jē·ō ¦kem·ə·kəl 'bal·əns }

geochemical cycle |GEOCHEM| During geologic changes, the sequence of stages in the migration of elements between the lithosphere, hydrosphere, and atmosphere. { ¦jē·ō¦kem·ə·kəl 'sī·kəl }

geochemical evolution |GEOCHEM| **1.** A change in any constituent of a rock beyond that amount present in the parent rock. **2.** A change in chemical composition of a major segment of the earth during geologic time, as the oceans. { ¦jē·ō¦kem·ə·kəl ˌev·ə'lü·shən }

geochemical prospecting |ENG| The use of geochemical and biogeochemical principles and data in the search for economic deposits of minerals, petroleum, and natural gases. { ¦jē·ō¦kem·ə·kəl 'prä·spek·tiŋ }

geochemistry |GEOL| The study of the chemical composition of the various phases of the earth and the physical and chemical processes which have produced the observed distribution of the elements and nuclides in these phases. { ¦jē·ō¦kem·ə·strē }

geodesy |GEOPHYS| A subdivision of geophysics which includes determination of the size and shape of the earth, the earth's gravitational field, and the location of points fixed to the earth's crust in an earth-referred coordinate system. { jē'äd·ə·sē }

geodetic coordinates |GEOGR| The quantities latitude, longitude, and elevation which define the position of a point on the surface of the earth with respect to the reference spheroid. { ¦jē·ə¦ded·ik kō'ȯrd·ən,ats }

geoeconomy |GEOGR| The study of economic conditions that are influenced by geographic factors. { ¦jē·ō·i¦kän·ə·mē }

geoengineering |SCI TECH| Artificial modification of earth systems to counteract anthropogenic effects, such as increasing carbon dioxide uptake by fertilizing ocean surface waters or screening out sunlight with orbiting mirrors. { ˌjē·ō,en·jə'nir·iŋ }

geographical botany *See* plant geography. { ¦jē·ə¦graf·ə·kəl 'bät·ən·ē }

geographical coordinates |GEOGR| Spherical coordinates, designating both astronomical and geodetic coordinates, defining a point on the surface of the earth, usually latitude and longitude. Also known as terrestrial coordinates. { ¦jē·ə¦graf·ə·kəl kō 'ȯrd·ən·ats }

geographical cycle   *See* geomorphic cycle.   { ¦¦jē·ə¦graf·ə·kəl 'sī·kəl }

geographical position   [GEOGR] Any position on the earth defined by means of its geographical coordinates, either astronomical or geodetic.   { ¦¦jē·ə¦graf·ə·kəl pə'zish·ən }

geographic position   [GEOGR] The position of a point on the surface of the earth expressed in terms of geographical coordinates either geodetic or astronomical.   { ¦¦jē·ə¦graf·ik pə'zish·ən }

geography   [SCI TECH] The study of all aspects of the earth's surface, comprising its natural and political divisions, the differentiation of areas, and, sometimes people in relationship to the environment.   { ˌjē'äg·rə·fē }

geohydrology   [HYD] The science dealing with underground water, often referred to as hydrogeology.   { ¦jē·ō˛hī'dräl·ə·jē }

geological oceanography   [GEOL] The study of the floors and margins of the oceans, including descriptions of topography, composition of bottom materials, interaction of sediments and rocks with air and sea water, the effects of movements in the mantle on the sea floor, and action of wave energy in the submarine crust of the earth. Also known as marine geology; submarine geology.   { ¦jē·ə¦läj·ə·kəl ˌō·shə'näg·rə·fē }

geological survey   [GEOL] **1.** An organization making geological surveys and studies. **2.** A systematic geologic mapping of a terrain.   { ¦jē·ə¦läj·ə·kəl 'sər˛vā }

geological transportation   [GEOL] Shifting of material by the action of moving water, ice, or air.   { ¦jē·ə¦läj·ə·kəl ˌtranz·pər'tā·shən }

geologic erosion   *See* normal erosion.   { ¦jē·ə¦läj·ik ə'rō·zhən }

geologic thermometer   *See* geothermometer.   { ¦jē·ə¦läj·ik thər'mäm·əd·ər }

geologic thermometry   *See* geothermometry.   { ¦jē·ə¦läj·ik thər'mäm·ə·trē }

geology   [SCI TECH] The study or science of the earth, its history, and its life as recorded in the rocks; includes the study of geologic features of an area, such as the geometry of rock formations, weathering and erosion, and sedimentation.   { jē'äl·ə·jē }

geomorphic cycle   [GEOL] The cycle of change in the surface configuration of the earth. Also known as cycle of erosion; geographical cycle.   { ¦jē·ō¦mȯr·fik 'sī·kəl }

geophilous   [ECOL] Living or growing in or on the ground.   { jē'äf·ə·ləs }

geophyte   [ECOL] A perennial plant that is deeply embedded in the soil substrata.   { 'jē·ə˛fīt }

geoscience   *See* earth science.   { 'jē·ō˛sī·əns }

geosensing   [BOT] The sensing or detecting of gravity by a plant relative to its longitudinal axis.   { 'jē·ō˛sens·iŋ }

geosere   [GEOL] A series of ecological climax communities following each other in geologic time and changing in response to changing climate and physical conditions.   { 'jē·ō˛sir }

geosophy   [GEOGR] The study of the nature and expression of geographical knowledge, both past and present.   { jē'äs·ə·fē }

geostrophic   [GEOPHYS] Pertaining to deflecting force resulting from the earth's rotation.   { ¦jē·ō¦sträf·ik }

geostrophic flux   [METEOROL] The transport of an atmospheric property by means of the geostrophic wind.   { ¦jē·ō¦sträf·ik 'fləks }

geostrophic wind   [METEOROL] That horizontal wind velocity for which the Coriolis acceleration exactly balances the horizontal pressure force.   { ¦jē·ō¦sträf·ik 'wind }

geostrophic-wind level [METEOROL] The lowest level at which the wind becomes geostrophic in the theory of the Ekman spiral. Also known as gradient-wind level. { ¦jē·ō¦sträf·ik 'wind ,lev·əl }

geostrophic-wind scale [METEOROL] A graphical device used for the determination of the speed of the geostrophic wind from the isobar or contour-line spacing on a synoptic chart. { ¦jē·ō¦sträf·ik 'wind ,skāl }

geotechnology [ENG] Application of the methods of engineering and science to exploitation of natural resources. { ¦jē·ō·tek'näl·ə·jē }

geothermal energy [GEOPHYS] Thermal energy contained in the earth; can be used directly to supply heat or can be converted to mechanical or electrical energy. { ¦jē·ō ,thərm·əl 'en·ər·jē }

geothermal system [GEOL] Any regionally localized geological setting where naturally occurring portions of the earth's internal heat flow are transported close enough to the earth's surface by circulating steam or hot water to be readily harnessed for use; examples are the Geysers Region of northern California and the hot brine fields in the Imperial Valley of southern California. { ¦jē·ō¦thər·məl 'sis·təm }

geothermometer [ENG] A thermometer constructed to measure temperatures in boreholes or deep-sea deposits. { ¦jē·ō·thər'mäm·əd·ər }

geothermometry [GEOL] Measurement of the temperatures at which geologic processes occur or occurred. Also known as geologic thermometry. { ¦jē·ō·thər'mäm·ə·trē }

geotropism [BOT] Response of a plant to the force of gravity. Also known as gravitropism. { jē'ä·trə,piz·əm }

germanium [CHEM] A brittle, water-insoluble, silvery-gray metallic element in the carbon family, symbol Ge, atomic number 32, atomic weight 72.59, melting at 959°C. { jər'mān·ē·əm }

German measles See rubella. { ¦jər·mən 'mē·zəlz }

germ cell See gamete. { 'jərm ,sel }

germicide [AGR] An agent that destroys germs. { 'jər·mə,sīd }

germination [BOT] The beginning or the process of development of a spore or seed. { ,jer·mə'nā·shən }

gestation period [BIOL] The period in mammals from fertilization to birth. { jə'stā·shən ,pir·ē·əd }

geyser [HYD] A natural spring or fountain which discharges a column of water or steam into the air at more or less regular intervals. { 'gī·zər }

ghost spot [PL PATH] A disease of tomato characterized by small white rings on the fruit. { 'gōst ,spät }

Gibberella fujikuroi [MYCOL] A fungal pathogen that causes bakanae disease, a seed-borne disease of rice that is characterized by the growth of excessively long internodes, through its production of plant growth hormones called gibberellins. { ,jib·ə,rel·ə ,fü·jē'kü,roi }

gid [VET MED] A chronic brain disease of sheep, less frequently of cattle, characterized by forced movements of circling or rolling, caused by the larval form of the tapeworm *Multiceps multiceps*. { gid }

gifblaar poison See fluoroacetic acid. { 'gif,blär ,pȯiz·ən }

giga- [SCI TECH] A prefix representing 10^9, which is 1,000,000,000, or a billion. Abbreviated G. Also known as kilomega- (deprecated usage). { 'gig·ə }

gio See geo. { gyō }

glacial   [HYD] Pertaining to ice, especially in great masses such as sheets of land ice or glaciers.   { 'glā·shəl }

glacial anticyclone   [METEOROL] A type of semipermanent anticyclone which overlies the ice caps of Greenland and Antarctica. Also known as glacial high.   { ¦glā·shəl ‚an·ti'sī‚klōn }

glacial erosion   [GEOL] Movement of soil or rock from one point to another by the action of the moving ice of a glacier. Also known as ice erosion.   { ¦glā·shəl ə'rō·zhən }

glacial flow   See glacier flow.   { ¦glā·shəl 'flō }

glacial high   See glacial anticyclone.   { ¦glā·shəl 'hī }

glacial ice   [HYD] Ice that is flowing or that exhibits evidence of having flowed.   { ¦glā·shəl 'īs }

glacial lobe   [HYD] A tonguelike projection from a continental glacier's main mass.   { ¦glā·shəl 'lōb }

glacial mill   See moulin.   { ¦glā·shəl 'mil }

glacial till   See till.   { ¦glā·shəl 'til }

glaciation   [METEOROL] The transformation of cloud particles from waterdrops to ice crystals, as in the upper portion of a cumulonimbus cloud.   { ‚glā·shē'ā·shən }

glacier   [HYD] A mass of land ice flowing slowly (at present or in the past) from an accumulation area to an area of ablation.   { 'glā·shər }

glacieret   See snowdrift ice.   { ¦glā·shə‚ret }

glacier flow   [HYD] The motion that exists within a glacier's body. Also known as glacial flow.   { 'glā·shər ‚flō }

glacier front   [HYD] The leading edge of a glacier.   { 'glā·shər ‚frənt }

glacier ice   [HYD] Any ice that is or was once a part of a glacier, consolidated from firn by further melting and refreezing and by static pressure; for example, an iceberg.   { 'glā·shər ‚īs }

glacier mill   See moulin.   { 'glā·shər ‚mil }

glacier pothole   See moulin.   { 'glā·shər 'pät‚hōl }

glacier well   See moulin.   { 'glā·shər ‚wel }

glacier wind   [METEOROL] A shallow gravity wind along the icy surface of a glacier, caused by the temperature difference between the air in contact with the glacier and free air at the same altitude.   { 'glā·shər ‚wind }

glacioeustasy   [GEOL] Changes in sea level due to storage or release of water from glacier ice.   { ¦glās·ē·ō'yü·stə‚sē }

glaciofluvial   [GEOL] Pertaining to streams fed by melting glaciers, or to the deposits and landforms produced by such streams.   { ¦glā·shē·ō¦flü·vē·əl }

glaciolacustrine   [GEOL] Pertaining to lakes fed by melting glaciers, or to the deposits forming therein.   { ¦glā·shē·ō·lə'kəs·trən }

glaciology   [GEOL] A broad field encompassing all aspects of the study of ice: glaciers, the largest ice masses on earth; ice that forms on rivers, lakes, and the sea; ice in the ground, including both permafrost and seasonal ice such as that which disrupts roads; ice that crystallizes directly from the air on structures such as airplanes and antennas, and all forms of snow research, including hydrological and avalanche forecasting.   { ‚glā·shē'äl·ə·jē }

glaçon   [OCEANOGR] A piece of sea ice which is smaller than a medium-sized floe.   { gla'sōn }

glanders [VET MED] A bacterial disease of equines caused by *Actinobacillus mallei*; involves the respiratory system, skin, and lymphatics. Also known as farcy. { 'glan·dərz }

glare ice [HYD] Ice with a smooth, shiny surface. { 'gler ‚īs }

glareous [ECOL] Growing in gravelly soil; refers specifically to plants. { 'gla·rē·əs }

Glasser's disease [VET MED] A generalized bacterial infection of swine caused by *Mycoplasma hyorhinis*. { 'glas·ərz di‚zēz }

glaze [HYD] A coating of ice, generally clear and smooth but usually containing some air pockets, formed on exposed objects by the freezing of a film of supercooled water deposited by rain, drizzle, or fog, or possibly condensed from supercooled water vapor. Also known as glaze ice; glazed frost; verglas. { glāz }

glazed frost *See* glaze. { ¦glāzd 'fròst }

glaze ice *See* glaze. { 'glāz ‚īs }

glime [HYD] An ice coating with a consistency intermediate between glaze and rime. { glīm }

glimmer ice [HYD] Ice newly formed within the cracks or holes of old ice, or on the puddles on old ice. { 'glim·ər ‚īs }

global climate change [CLIMATOL] The periodic fluctuations in global temperatures and precipitation, such as the glacial (cold) and interglacial (warm) cycles. { ¦glō·bəl 'klī·mət ‚chānj }

global radiation [GEOPHYS] The total of direct solar radiation and diffuse sky radiation received by a horizontal surface of unit area. { 'glō·bəl ‚rād·ē'ā·shən }

global sea [OCEANOGR] All the seawater of the earth considered as a single ocean constantly intermixing. { 'glō·bəl 'sē }

global warming [CLIMATOL] The gradual rise in the surface temperature of the earth believed to be caused by increasing concentrations of greenhouse gases in the atmosphere. { ¦glō·bəl 'wòrm·iŋ }

global warming potential [METEOROL] The ratio of global warming or radiative forcing from 1 kilogram of a greenhouse gas to 1 kilogram of carbon dioxide over 100 years, expressed per mole or per kilogram; it provides a way to calculate the contribution of each greenhouse gas to the annual increase in radiative forcing. { ¦glō·bəl 'wòrm·iŋ pə‚ten·chəl }

gloom [METEOROL] The condition existing when daylight is very much reduced by dense cloud or smoke accumulation above the surface, the surface visibility not being materially reduced. { glüm }

gloup [GEOL] An opening in the roof of a sea cave. { glüp }

glove box [ENG] A sealed box with gloves attached and passing through openings into the box, so that workers can handle materials in the box; used to handle certain radioactive and biologically dangerous materials and to prevent contamination of materials and objects such as germfree rats or lunar rocks. { 'gləv ‚bäks }

glucochloral *See* chloralose. { ¦glü·kō¦klòr·əl }

glucochloralose *See* chloralose. { ¦glü·kō¦klòr·ə‚lōs }

glume [BOT] One of two bracts at the base of a spikelet of grass. { glüm }

glutinous [BOT] Having a sticky surface. { 'glüt·ən·əs }

glycerinated vaccine virus *See* smallpox vaccine. { 'glis·ə·rə‚nād·əd 'vak‚sēn ‚vī·rəs }

glycophyte [BOT] A plant requiring more than 0.5% sodium chloride solution in the substratum. { 'gli·kə‚fīt }

glyphosate [CHEM] $C_3H_8NO_5P$ A white solid with a melting point of 200°C; slight solubility in water; used as a herbicide in postharvest treatment of crops. { 'glif·ə,sāt }

GM counter See Geiger-Müller counter. { ¦jē'em ,kaúnt·ər }

gnat [ZOO] The common name for a large variety of biting insects in the order Diptera. { nat }

gnotobiote [MICROBIO] **1.** An individual (host) living in intimate association with another known species (microorganism). **2.** The known microorganism living on a host. { ¦nō·dō'bī,ōt }

gold [CHEM] A chemical element, symbol Au, atomic number 79, atomic weight 196.96765; soluble in aqua regia; melts at 1065°C. { gōld }

golden algae [BOT] The common name for members of the class Chrysophyceae. { 'gol·dən 'al·jē }

golden-brown algae [BOT] The common name for members of the division Chrysophyta. { 'gol·dən ,braún 'al·jē }

gonococcus See Neisseria gonorrhoeae. { ,gän·ō'käk·əs }

gonorrhea [MED] A bacterial infection of humans caused by the gonococcus (Neisseria gonorrhoeae) which invades the mucous membrane of the urogenital tract. { ,gän·ə'rē·ə }

gonyautoxin [BIOL] One of a group of saxitoxin-related compounds that are produced by the dinoflagellates Gonyaulax catenella and G. tamarensis. { ¦gō·nē·ō¦täk·sən }

gorge [OCEANOGR] A collection of solid matter obstructing a channel or a river, as an ice gorge. { górj }

gorge wind See canyon wind. { 'górj ,wind }

graded stream [HYD] A stream in which, over a period of years, slope is adjusted to yield the velocity required for transportation of the load supplied from the drainage basin. { ¦grād·əd 'strēm }

graded topocline [ECOL] A topocline having a wide range, or ranging into different kinds of environment, thus subjecting its members to differential selection so that divergence between local races may become sufficient to warrant creation of varietal, or even specific, names. { ¦grād·əd 'täp·ə,klīn }

gradient-wind level See geostrophic-wind level. { 'grād·ē·ənt ,wind ,lev·əl }

graft [BOT] To unite a scion to a stock in such manner that the two grow together and continue development as a single plant without change in scion or stock. { graft }

grain [HYD] The particles which make up settled snow, firn, and glacier ice. { grān }

grain alcohol See alcohol. { 'grān 'al·kə,hól }

graminicolous [ECOL] Living upon grass. { ¦gram·ə¦nik·ə·ləs }

graminoid [BOT] Of or resembling the grasses. { 'gram·ə,nóid }

gram-negative [MICROBIO] Of bacteria, decolorizing and staining with the counterstain when treated with Gram's stain. { 'gram ¦neg·əd·iv }

gram-positive [MICROBIO] Of bacteria, holding the color of the primary stain when treated with Gram's stain. { 'gram ¦päs·əd·iv }

granular [SCI TECH] Having a grainy texture. { 'gran·yə·lər }

granular ice [HYD] Ice composed of many tiny, opaque, white or milky pellets or grains frozen together and presenting a rough surface; this is the type of ice deposited as rime and compacted as névé. { 'gran·yə·lər 'īs }

granular snow   *See* snow grains.   { 'gran·yə·lər 'snō }

granulation |PL PATH| Dry, tasteless condition of citrus fruit due to hardening of the juice sacs when fruit is left on trees too late in the season. |SCI TECH| The state or process of reducing a material to grains or small particles.   { ˌgran·yə'lā·shən }

granulosis |ZOO| A virus disease of lepidopteran larvae characterized by the accumulation of small granular inclusion bodies (capsules) in the infected cells.   { ˌgran·yə'lō·səs }

Granville wilt |PL PATH| A bacterial wilt of tobacco caused by *Pseudomonas solanacearum*.   { 'gran·vəl 'wilt }

grapevine drainage   *See* trellis drainage.   { 'grāp,vīn ˌdrān·ij }

Graphiolaceae |MYCOL| A family of parasitic fungi in the order Ustilaginales in which teleutospores are produced in a cuplike fruiting body.   { ˌgraf·ē·ō'lās·ē,ē }

grass |BOT| The common name for all members of the family Gramineae; moncotyledonous plants having leaves that consist of a sheath which fits around the stem like a split tube, and a long, narrow blade.   { gras }

grassland |ECOL| Any area of herbaceous terrestrial vegetation dominated by grasses and graminoid species.   { 'gras,land }

grassland climate   *See* subhumid climate.   { 'gras,land ˌklī·mət }

grass sickness |VET MED| A disease of horses occurring mainly in Scotland; thought to be caused by a virus similar to the one that causes poliomyelitis in humans.   { 'gras ˌsik·nəs }

grass temperature |METEOROL| The temperature registered by a thermometer with its bulb at the level of the tops of the blades of grass in short turf.   { 'gras ˌtem·prə·chər }

graupel   *See* snow pellets.   { 'grau̇·pəl }

gravel |GEOL| A loose or unconsolidated deposit of rounded pebbles, cobbles, or boulders.   { 'grav·əl }

gravitational convection   *See* thermal convection.   { ˌgrav·ə'tā·shən·əl kən'vek·shən }

gravitational tide |OCEANOGR| An atmospheric tide due to gravitational attraction of the sun and moon.   { ˌgrav·ə'tā·shən·əl 'tīd }

gravitational water |HYD| Soil water of a temporary character that results from prolonged infiltration from above and which moves downward to the groundwater zone in response to gravity.   { ˌgrav·ə'tā·shən·əl 'wȯd·ər }

gravitropism   *See* geotropism.   { grə'vi·trə,piz·əm }

gravity drainage |HYD| Withdrawal of water from soil or rock strata as a result of gravitational forces.   { 'grav·əd·ē 'drān·ij }

gravity flow |HYD| A form of glacier movement in which the flow of the ice results from the downslope gravitational component in an ice mass resting on a sloping floor.   { 'grav·əd·ē ˌflō }

gravity-flow gathering system |PETR MIN| The use of gravity (downhill flow) through pipelines to transport and collect liquid at a central location; used for gathering of waste water from waterflooding operations for treatment prior to reuse or disposal.   { 'grav·əd·ē ˌflō'gath·ə·riŋ ˌsis·təm }

gravity spring |HYD| A spring that issues under the influence of gravity, not internal pressure.   { 'grav·əd·ē ˌspriŋ }

gravity wind |METEOROL| A wind (or component thereof) directed down the slope of an incline and caused by greater air density near the slope than at the same levels some distance horizontally from the slope. Also known as drainage wind; katabatic wind. { 'grav·əd·ē ,wind }

gray blight [PL PATH] A fungus disease of tea caused by *Pestalotia* (*Pestalozzia*) *theae*, which invades the tissues and causes the formation of black dots on the leaves. { ¦grā 'blīt }

gray leaf spot [PL PATH] A fungus disease of tomatoes caused by *Stemphylium solani* and characterized by water-soaked brown spots on the leaves that become gray with age. { 'grā 'lēf 'spät }

gray mold [PL PATH] Any fungus disease characterized by a gray surface appearance of the affected part. { 'grā ,mōld }

grazing food web [ECOL] A trophic web that is based on the consumption of the tissues of living organisms. { 'graz·iŋ ,füd ,web }

grease ice [HYD] A kind of slush with a greasy appearance, formed from the congelation of ice crystals in the early stages of freezing. Also known as ice fat; lard ice. { 'grēs ,īs }

grease trap [CIV ENG] A trap in a drain or waste pipe to stop grease from entering a sewer system. { 'grēs ,trap }

Great Basin high [METEOROL] A high-pressure system centered over the Great Basin of the western United States; it is a frequent feature of the surface chart in the winter season. { ¦grāt ¦bas·ən 'hī }

greater ebb [OCEANOGR] The stronger of two ebb currents occurring during a tidal day. { 'grād·ər ¦eb }

greater flood [OCEANOGR] The stronger of two flood currents occurring during a tidal day. { 'grād·ər ¦fləd }

great soil group [GEOL] A group of soils having common internal soil characteristics; a subdivision of a soil order. { ¦grāt 'soil ,grüp }

green algae [BOT] The common name for members of the plant division Chlorophyta. { 'grēn ¦al·jē }

green belt *See* frostless zone. { 'grēn ,belt }

green copperas *See* ferrous sulfate. { ¦grēn 'käp·ə·rəs }

green design *See* industrial ecology. { ¦grēn di'zīn }

greenhouse [BOT] Glass-enclosed, climate-controlled structure in which young or out-of-season plants are cultivated and protected. { 'grēn,haús }

greenhouse effect [METEOROL] The effect created by the earth's atmosphere in trapping heat from the sun; the atmosphere acts like a greenhouse. { 'grēn,haús i,fekt }

greenhouse gases [METEOROL] Gases whose concentration is small and varies, mostly due to anthropogenic factors; they absorb heat from incoming solar radiation but do not allow long-wave radiation to reflect back into space. They include carbon dioxide, methane, and nitrous oxide, as well as, water vapor, carbon monoxide, nitrogen oxide, nitrogen dioxide, and ozone. { 'grēn,haús ,gas·əz }

Greenland anticyclone [METEOROL] The glacial anticyclone which is supposed to overlie Greenland; analogous to the Antarctic anticyclone. { 'grēn·lənd ,ant·i'sī ,klōn }

green manure [AGR] Herbaceous plant material plowed into the soil while still green. { ¦grēn mə'nü·ər }

green mud [GEOL] **1.** A fine-grained, greenish terrigenous mud or oceanic ooze found near the edge of a continental shelf at depths of 300–7500 feet (90–2300 meters).

2. A deep-sea terrigenous deposit characterized by the presence of a considerable proportion of glauconite and calcium carbonate. { ¦grēn 'məd }

green muscardine   [ZOO] A disease of the European corn borer, the wheat cockchafer, and other insects caused by the fungus Metarhizium anisopliae. { ˌgrēn 'məs·kər,dēn }

green rosette   [PL PATH] A virus disease of the peanut characterized by bunching and yellowing of the leaves with severe stunting of the plant. { ¦grēn rō'zet }

greensand   [GEOL] A greenish sand consisting principally of grains of glauconite and found between the low-water mark and the inner mud line. { 'grēn,sand }

green sky   [METEOROL] A greenish tinge to part of the sky, supposed by seamen to herald wind or rain, or in some cases, a tropical cyclone. { 'grēn ¦skī }

green smut   [PL PATH] A fungus disease of rice characterized by enlarged grains covered with a green powder consisting of conidia, and caused by Ustilaginoidea virens. Also known as false smut. { 'grēn 'smət }

green snow   [HYD] A snow surface that has attained a greenish tint as a result of the growth within it of certain microscopic algae. { 'grēn ¦snō }

green sulfur bacteria   [MICROBIO] A physiologic group of green photosynthetic bacteria of the Chloraceae that are capable of using hydrogen sulfide (H_2S) and other inorganic electron donors. { 'grēn ¦səl·fər bak'tir·ē·ə }

green vitriol   See ferrous sulfate. { 'grēn ¦vi,trē,ȯl }

Grifola frondosa   [MYCOL] A type of mushroom found in parts of the eastern United States, Europe, and Asia, growing in masses at the base of stumps and on roots; has an anticancer effect in patients with lung and stomach cancers or leukemia. Also known as maitake mushroom. { grə,fō·lə frän'dōs·ə }

grit chamber   [CIV ENG] A chamber designed to remove sand, gravel, or other heavy solids that have subsiding velocities or specific gravities substantially greater than those of the organic solids in waste water. { 'grit ,chām·bər }

gross primary production   [ECOL] The total accumulation of organic material by autotrophs, including the proportion used for respiration. { ¦grōs 'prī,mer·ē prə'dək·shən }

gross production rate   [ECOL] The speed of assimilation of organisms belonging to a specific trophic level. { ¦grōs prə'dək·shən ,rāt }

ground cover   [BOT] Prostrate or low plants that cover the ground instead of grass. [FOR] All forest plants except trees. { 'graůnd ,kəv·ər }

grounded ice   See stranded ice. { 'graůnd·əd ¦īs }

ground fog   [METEOROL] A fog that hides less than 0.6 of the sky and does not extend to the base of any clouds that may lie above it. { 'graůnd ,fäg }

ground frost   [METEOROL] In British usage, a freezing condition injurious to vegetation, which is considered to have occurred when a minimum thermometer exposed to the sky at a point just above a grass surface records a temperature (grass temperature) of 30.4°F (−0.9°C) or below. { 'graůnd ,frȯst }

ground ice   [HYD] **1.** A body of clear ice in frozen ground, most commonly found in more or less permanently frozen ground (permafrost), and may be of sufficient age to be termed fossil ice. Also known as stone ice; subsoil ice; subterranean ice; underground ice. **2.** See anchor ice. { 'graůnd ,īs }

ground inversion   See surface inversion. { 'graůnd in,vər·zhən }

ground layer   See surface boundary layer. { 'graůnd ,lā·ər }

ground streamer [METEOROL] An upward advancing column of high-ion density which rises from a point on the surface of the earth toward which a stepped leader descends at the start of a lightning discharge. { 'graúnd ˌstrē·mər }

ground swell [OCEANOGR] A swell passing through shallow water, characterized by a marked increase in height in water shallower than one-tenth wavelength. { 'graúnd ˌswel }

ground visibility [METEOROL] In aviation terminology, the horizontal visibility observed at the ground, that is, surface visibility or control-tower visibility. { 'graúnd ˌviz·ə'bil·əd·ē }

groundwater [HYD] All subsurface water, especially that part that is in the zone of saturation. { 'graúnd,wȯd·ər }

groundwater decrement *See* groundwater discharge. { 'graúnd,wȯd·ər ¦dek·rə·mənt }

groundwater depletion curve [HYD] A recession curve of streamflow, so adjusted that the slope of the curve represents the runoff (depletion rate) of the groundwater; it is formed by the observed hydrograph during prolonged periods of no precipitation. Also known as groundwater recession. { 'graúnd,wȯd·ər di'plē·shən ˌkərv }

groundwater discharge [HYD] **1.** Water released from the zone of saturation. **2.** Release of such water. Also known as decrement; groundwater decrement; phreatic-water discharge. { 'graúnd,wȯd·ər 'dis,chärj }

groundwater flow [HYD] That portion of the precipitation that has been absorbed by the ground and has become part of the groundwater. { 'graúnd,wȯd·ər ˌflō }

groundwater hydrology [HYD] The study of the occurrence, circulation, distribution, and properties of any liquid water residing beneath the surface of the earth. { ˌgraúnd ˌwȯd·ər hī'dräl·ə·jē }

groundwater increment *See* recharge. { 'graúnd,wȯd·ər ¦iŋ·krə·mənt }

groundwater level [HYD] The level below which the rocks and subsoil are full of water. { 'graúnd,wȯd·ər ˌlev·əl }

groundwater recession *See* groundwater depletion curve. { 'graúnd,wȯd·ər ri'sesh·ən }

groundwater recharge *See* recharge. { 'graúnd,wȯd·ər 'rē,chärj }

groundwater replenishment *See* recharge. { 'graúnd,wȯdər ri'plen·ish·mənt }

groundwater surface *See* water table. { 'graúnd,wȯd·ər ¦sər·fəs }

groundwater table *See* water table. { 'graúnd,wȯd·ər ¦tā·bəl }

group selection [GEN] Selection in which changes in gene frequency are brought about by the differential extinction and proliferation of populations. { ¦grüp si¦lek·shən }

growing season [AGR] The period of the year when climatic conditions are favorable for plant growth, common to a place or an area. { 'grō·iŋ ˌsēz·ən }

growler [OCEANOGR] A small piece of floating sea ice, usually a fragment of an iceberg or floeberg; it floats low in the water, and its surface often is heavily pitted; it often appears greenish in color. Also known as bergy-bit. { 'graúl·ər }

growth form [ECOL] The habit of a plant determined by its appearance of branching and periodicity. { 'grōth ˌfȯrm }

growth lattice [GEOL] The rigid, reef-building, inplace framework of an organic reef, consisting of skeletons of sessile organisms and excluding reef-flank and other associated fragmental deposits. Also known as organic lattice. { 'grōth ˌlad·əs }

guard cell [BOT] Either of two specialized cells surrounding each stoma in the epidermis of plants; functions in regulating stoma size. { 'gärd ˌsel }

guest element   See trace element.   { 'gest ¦el·ə·mənt }

Guiana Current   [OCEANOGR] A current flowing northwestward along the northeastern coast of South America.   { gī'an·ə ¦kə·rənt }

guild   [ECOL] A group of species that utilize the same kinds of resources, such as food, nesting sites, or places to live, in a similar manner.   { gild }

Guinea Current   [OCEANOGR] A current flowing eastward along the southern coast of northwestern Africa into the Gulf of Guinea.   { 'gin·ē ¦kə·rənt }

guinea worm   [ZOO] *Dracunculus medinensis*. A parasitic nematode that infects the subcutaneous tissues of humans and other mammals.   { 'gin·ē ˌwərm }

gulch   [GEOGR] A gulley, sometimes occupied by a torrential stream.   { gəlch }

gulf   [GEOGR] **1.** An abyss or chasm.   **2.** A large extension of the sea partially enclosed by land.   { gəlf }

Gulf Stream   [OCEANOGR] A relatively warm, well-defined, swift, relatively narrow, northward-flowing ocean current which originates north of Grand Bahama Island where the Florida Current and the Antilles Current meet, and which eventually becomes the eastward-flowing North Atlantic Current.   { 'gəlf ¦strēm }

Gulf Stream Countercurrent   [OCEANOGR] **1.** A surface current opposite to the Gulf Stream, one current component on the Sargasso Sea side and the other component much weaker, on the inshore side.   **2.** A predicted, but as yet unobserved, large current deep under the Gulf Stream but opposite to it.   { 'gəlf ¦strēm 'kaúnt·ər ˌkə·rənt }

Gulf Stream system   [OCEANOGR] The Florida Current, Gulf Stream, and North Atlantic Current, collectively.   { 'gəlf ¦strēm 'sis·təm }

gully   [GEOGR] A narrow ravine.   { 'gəl·ē }

gully erosion   [GEOL] Erosion of soil by running water.   { 'gəl·ē i¦rō·zhən }

gumbo   [GEOL] A soil that forms a sticky mud when wet.   { 'gəm·bō }

gumbotil   [GEOL] Deoxidized, leached clay that contains siliceous stones.   { 'gəm·bō ˌtil }

gunbarrel   [CHEM ENG] An atmospheric vessel used for treatment of waterflood waste water.   { 'gən ˌbar·əl }

gust   [METEOROL] A sudden, brief increase in the speed of the wind; it is of a more transient character than a squall and is followed by a lull or slackening in the wind speed.   { gəst }

gustiness factor   [METEOROL] A measure of the intensity of wind gusts; it is the ratio of the total range of wind speeds between gusts and the intermediate periods of lighter wind to the mean wind speed, averaged over both gusts and lulls.   { 'gəs·tē·nəs ˌfak·tər }

guttation   [BOT] The discharge of water from a plant surface, especially from a hydathode.   { ˌgə'tā·shən }

Gymnoascaceae   [MYCOL] A family of ascomycetous fungi in the order Eurotiales including dermatophytes and forms that grow on dung, soil, and feathers.   { ¦jim·nō· ə'skās·ē ˌē }

gymnosperm   [BOT] The common name for members of the division Pinophyta; seed plants having naked ovules at the time of pollination.   { 'jim·nə ˌspərm }

Gymnospermae   [BOT] The equivalent name for Pinophyta.   { ˌjim·nə'spər·mē }

Gymnosporangium juniperi-virginianae   [MYCOL] A heteroecious fungal pathogen that is the cause of apple-cedar rust.   { ˌjim·nō·spəˌran·jē·əm ˌjü·niˌper·ē ˌvərˌjin·ē'a ˌnī }

gynaecandrous [BOT] Having staminate and pistillate flowers on the same spike. { ¦gī·nə¦kan·drəs }

gynoecious [BOT] Pertaining to plants that have only female flowers. { gī'nē·shəs }

gynoecium [BOT] The aggregate of carpels in a flower. { gī'nē·sē·əm }

gynomonoecious [BOT] Having complete and pistillate flowers on the same plant. { ¦gīn·ō·mä'nē·shəs }

gypsophilous |ECOL] Flourishing on a gypsum-rich substratum. { ,jip'säf·ə·ləs }

gypsy moth [ZOO] *Porthetria dispar.* A large lepidopteran insect of the family Lymantri-idae that was accidentally imported into New England from Europe in the late 19th century; larvae are economically important as pests of deciduous trees. { 'jip·sē ¦moth }

gyre [OCEANOGR] A closed circulatory system that is larger than a whirlpool or eddy. { jīr }

gyttja [GEOL] A fresh-water anaerobic mud containing an abundance of organic matter; capable of supporting aerobic life. { 'yi,chä }

H

habitat [ECOL] The part of the physical environment in which a plant or animal lives. { 'hab·ə,tat }

habitus [BIOL] General appearance or constitution of an organism. { 'hab·ə·təs }

hadal [OCEANOGR] Pertaining to the environment of the ocean trenches, over 4 miles (6.5 kilometers) in depth. { 'hād·əl }

Hadley cell [METEOROL] A direct, thermally driven, and zonally symmetric circulation first proposed by George Hadley as an explanation for the trade winds; it consists of the equatorward movement of the trade winds between about latitude 30° and the equator in each hemisphere, with rising wind components near the equator, poleward flow aloft, and finally descending components at about latitude 30° again. { 'had·lē ,sel }

hadromycosis [PL PATH] Any plant disease resulting from infestation of the xylem by a fungus. { ¦had·rō,mī'kō·səs }

Haemophilus (para) gallinarum [MICROBIO] A bacterial pathogen that causes infectious coryza in chickens and some birds. { hē,mäf·ə·ləs ,par·ə ,gal·ə'när·əm }

Haemophilus [MICROBIO] A genus of gram-negative coccobacilli or rod-shaped bacteria of uncertain affiliation; cells may form threads and filaments and are aerobic or facultatively anaerobic; strictly blood parasites. { hē'mä·fə·ləs }

Haemophilus aegyptius [MICROBIO] A pathogenic bacterium associated with acute contagious forms of conjunctivitis and Brazilian purpuric fever. { hē,mäf·ə·ləs ə'jip·tē·əs }

Haemophilus ducreyi [MICROBIO] A bacterial pathogen that causes the sexually transmitted disease soft chancre, or chancroid. { hē¦mäf·ə·ləs dü'krā,ī }

Haemophilus parasuis [MICROBIO] A bacterial pathogen that frequently inhabits the normal upper respiratory tract, can cause secondary pneumonias and, in young or otherwise susceptible animals, generalized illness with arthritis, meningitis, pleuritis, and peritonitis. { hē,mäf·ə·ləs ,par·ə'sü·is }

haff [GEOGR] A freshwater lagoon separated from the sea by a sandbar. { haf }

hail [METEOROL] Precipitation composed of lumps of ice formed in strong updrafts in cumulonimbus clouds, having a diameter of at least 0.2 inch (5 millimeters), most hailstones are spherical or oblong, some are conical, and some are bumpy and irregular. { hāl }

hailstone [METEOROL] A single unit of hail, ranging in size from that of a pea to that of a grapefruit, or from less than $\frac{1}{4}$ inch (6 millimeters) to more than 5 inches (13 centimeters) diameter; may be spheroidal, conical, or generally irregular in shape. { 'hāl,stōn }

hair ball *See* lake ball. { 'her ,bȯl }

halazone [CHEM] $COOHC_6H_4SO_2NCl_2$ White crystals, with strong chlorine aroma; slightly soluble in water and chloroform; used as water disinfectant. { 'hal·ə‚zōn }

halcyon days [METEOROL] A period of fine weather. { 'hal·sē·ən ¦dāz }

half-arc angle [METEOROL] The elevation angle of that point which a given observer regards as the bisector of the arc from his zenith to his horizon; a measure of the apparent degree of flattening of the dome of the sky. { 'haf ‚ärk 'aŋ·gəl }

half-hardy plant [BOT] A plant that can withstand relatively low temperatures but cannot survive severe freezing in cold climates unless carefully protected. { ¦haf ‚här·dē 'plant }

half tide [OCEANOGR] The condition when the tide is at the level between any given high tide and the following or preceding low tide. Also known as mean tide. { 'haf ‚tīd }

half-tide level [OCEANOGR] The level midway between mean high water and mean low water. { 'haf ‚tīd ‚lev·əl }

halmyrolysis [GEOCHEM] Postdepositional chemical changes that occur while sediment is on the sea floor. { ‚hal·mə'räl·ə·səs }

halo [METEOROL] Any one of a large class of atmospheric optical phenomena which appear as colored or whitish rings and arcs about the sun or moon when seen through an ice crystal cloud or in a sky filled with falling ice crystals. { 'hā·lō }

halo blight [PL PATH] A bacterial blight of beans and sometimes other legumes caused by *Pseudomonas phaseolicola* and characterized by water-soaked lesions surrounded by a yellow ring on the leaves, stems, and pods. { 'hā·lō‚blīt }

halocline [OCEANOGR] A well-defined vertical gradient of salinity in the oceans and seas. { 'hal·ə‚klīn }

halomorphic [GEOCHEM] Referring to an intrazonal soil whose features have been strongly affected by either neutral or alkali salts, or both. { ¦hal·ə¦mȯr·fik }

halophile [BIOL] An organism that requires high salt concentrations for growth and maintenance. { 'hal·ə‚fīl }

halophilism [BIOL] The phenomenon of demand for high salt concentrations for growth and maintenance. { ¦hal·ə¦fil·iz·əm }

halophyte [ECOL] A plant or microorganism that grows well in soils having a high salt content. { 'hal·ə‚fīt }

halosere [ECOL] The series of communities succeeding one another, from the pioneer stage to the climax, and commencing in salt water or on saline soil. { 'hal·ə‚sir }

hammock See hummock. { 'ham·ək }

hand-foot-and-mouth disease [MED] An infectious disease of humans caused by a coxsackie virus and characterized by maculopapular and vesicular eruptions in the mouth and on the hands and feet. { ¦hand ¦fut ən 'mauth di‚zēz }

hanging glacier [HYD] A glacier lying above a cliff or steep mountainside; as the glacier advances, calving can cause ice avalanches. { 'haŋ·iŋ ¦glā·shər }

HAP See hazardous air pollutants. { hap or ¦āch¦ā'pē }

haplobiont [BOT] A plant that produces only sexual haploid individuals. { ¦ha·plō¦bī ‚änt }

haploid [GEN] Having half of the diploid or full complement of chromosomes, that is, one complete set, as in mature gametes. { 'ha‚plȯid }

haptotropism [BIOL] Movement of sessile organisms in response to contact, especially in plants. { hap'tä·trə‚piz·əm }

hard data |SCI TECH| Data in the form of numbers or graphs, as opposed to qualitative information.   { 'härd ¦dad·ə }

hard detergent |CHEM| A nonbiodegradable detergent.   { 'härd di'tər·jənt }

hardening |BOT| Treatment of plants designed to increase their resistance to extremes in temperature or drought.   { 'hard·ən·iŋ }

hard fiber |BOT| A heavily lignified leaf fiber used in making cordage, twine, and textiles.   { 'härd ‚fī·bər }

hard freeze |HYD| A freeze in which seasonal vegetation is destroyed, the ground surface is frozen solid underfoot, and heavy ice is formed on small water surfaces such as puddles and water containers.   { 'härd ¦frēz }

hard frost   See black frost.   { 'härd ¦fròst }

hardness |CHEM| The amount of calcium carbonate dissolved in water, usually expressed as parts of calcium carbonate per million parts of water.   { 'härd·nəs }

hard rime |HYD| Opaque, granular masses of rime deposited chiefly on vertical surfaces by a dense super-cooled fog; it is more compact and amorphous than soft rime, and may build out into the wind as glazed cones or feathers.   { 'härd ¦rīm }

hard rot |PL PATH| **1.** Any plant disease characterized by lesions with hard surfaces. **2.** A fungus disease of gladiolus caused by *Septoria gladioli* which produces hard-surfaced lesions on the leaves and corms.   { 'härd ‚rät }

hardwood |BOT| Dense, close-grained wood of an angiospermous tree, such as oak, walnut, cherry, and maple.   { 'härd‚wúd }

hardwood forest |ECOL| **1.** An ecosystem having deciduous trees as the dominant form of vegetation.   **2.** An ecosystem consisting principally of trees that yield hardwood.   { 'härd‚wúd ¦fär·əst }

hardy plant |BOT| A plant able to withstand low temperatures without artificial protection.   { ¦här·dē ¦plant }

Hardy-Weinberg law |GEN| The concept that frequencies of both genes and genotypes will remain constant from generation to generation in an idealized population where mating is random and evolutionary forces (such as mutation, migration, selection, or genetic drift) are absent.   { ¦här·dē 'wīn‚bərg ‚lò }

harmonic prediction |OCEANOGR| A method used in predicting the tides and tidal currents by combining the harmonic constituents into a single tide curve.   { här'män·ik prə'dik·shən }

harmonic tide plane   See Indian spring low water.   { här'män·ik 'tīd ‚plan }

harrow |AGR| An implement that is pulled over plowed soil to break clods, level the surface, and destroy weeds.   { 'här·ō }

harrowing |AGR| Cultivation of the soil with a harrow.   { 'här·ə·wiŋ }

Hartig net |MYCOL| A complex network of fungal hyphae that is the site of nutrient exchange between the fungus and the host plant.   { 'här·tig ‚net }

haustoria |MYCOL| Specialized branches of hyphae that penetrate host cells and absorb nutrients from them.   { haú'stòr·ē·ə }

haustorial |MYCOL| Pertaining to fungi that have food-absorbing cells in the host.   { haú'stòr·ē·əl }

haustorium |BOT| **1.** An outgrowth of certain parasitic plants which serves to absorb food from the host.   **2.** Food-absorbing cell of the embryo sac in nonparasitic plants.   { hò'stòr·ē·əm }

Haverhill fever [MED] An acute bacterial infection caused by *Streptobacillus moniliformis*, usually acquired by rat bite, and characterized by acute onset, intermittent fever, erythematous rash, and polyarthritis. Also known as streptobacillary fever. { 'hāv·ə·rəl ,fē·vər }

haycock [HYD] An isolated ice cone created on land ice or shelf ice because of pressure or ice movement. { 'hā,käk }

hay fever [MED] An allergic disorder of the nasal membranes and related structures due to sensitization by certain plant pollens. Also known as allergic rhinitis; pollinosis. { 'hā,fē·vər }

hazardous air pollutants [ENG] Chemicals that are known or suspected to cause cancer or other serious health effects, such as reproductive effects or birth defects, or adverse environmental effects. Listed hazardous air pollutants include benzene, found in gasoline; perchlorethlyene, emitted from some dry cleaning facilities; and methylene chloride, used as a solvent and paint stripper in industry; as well as dioxin, asbestos, toluene, and metals such as cadmium, mercury, chromium, and lead compounds. Also known as air toxics. Abbreviated HAP. { ,haz·ər·dəs 'er pə,lüt·əns }

hazardous material [ENG] A poison, corrosive agent, flammable substance, explosive, radioactive chemical, or any other material which can endanger human health or well-being if handled improperly. { 'haz·ərd·əs mə'tir·ē·əl }

haze [METEOROL] Fine dust or salt particles dispersed through a portion of the atmosphere; the particles are so small that they cannot be felt, or individually seen with the naked eye, but they diminish horizontal visibility and give the atmosphere a characteristic opalescent appearance that subdues all colors. { hāz }

haze horizon [METEOROL] The top of a haze layer which is confined by a low-level temperature inversion and has the appearance of the horizon when viewed from above against the sky. { 'hāz hə,riz·ən }

haze layer [METEOROL] A layer of haze in the atmosphere, usually bounded at the top by a temperature inversion and frequently extending downward to the ground. { 'hāz lā·ər }

haze level See haze line. { 'hāz ,lev·əl }

haze line [METEOROL] The boundary surface in the atmosphere between a haze layer and the relatively clean, transparent air above the top of a haze layer. Also known as haze level. { 'hāz ,līn }

HCB See hexachlorobenzene.

head [BOT] A dense cluster of nearly sessile flowers on a very short stem. [GEOGR] See headland. { hed }

headland [GEOGR] **1.** A high, steep-faced promontory extending into the sea. Also known as head; mull. **2.** High ground surrounding a body of water. { 'hed·lənd }

head smut [PL PATH] A fungus disease of corn and sorghum caused by *Sphacelotheca reiliana* which destroys the head of the plant. { 'hed ,smət }

headwaters [HYD] The source and upstream waters of a stream. { 'hed,wȯd·ərz }

health [MED] A state of dynamic equilibrium between an organism and its environment in which all functions of mind and body are normal. { helth }

health physics [MED] The study of the protection of personnel from harmful effects of ionizing radiation by such means as routine radiation surveys, area and personnel monitoring, and protective equipment and procedures. { 'helth ,fiz·iks }

heartwater disease [VET MED] A septicemic infectious disease of cattle, sheep, and goats in Africa caused by the rickettsial microorganism *Cowdria ruminantium*. { 'härt ,wȯd·ər di,zēz }

heartwood [BOT] Xylem of an angiosperm. { 'härt,wu̇d }

heat equator [METEOROL] **1.** The line which circumscribes the earth and connects all points of highest mean annual temperature for their longitudes. **2.** The parallel of latitude of 10°N, which has the highest mean temperature of any latitude. Also known as thermal equator. { 'hēt i¦kwād·ər }

heath *See* temperate and cold scrub. { hēth }

heather [BOT] *Calluna vulgaris.* An evergreen heath of northern and alpine regions distinguished by racemes of small purple-pink flowers. { 'he<u>th</u>·ər }

heating degree-day [METEOROL] A form of degree-day used as an indication of fuel consumption; in United States usage, one heating degree-day is given for each degree that the daily mean temperature departs below the base of 65°F (where the Celsius scale is used, the base is usually 19°C). { 'hēd·iŋ di'grē ,dā }

heat island effect [METEOROL] In urban areas with tall buildings, an atmospheric condition in which heat and pollutants create a haze dome that prevents warm air from rising and being cooled at a normal rate, especially in the absence of strong winds. { 'hēt ,ī·lənd i,fekt }

heat low *See* thermal low. { 'hēt ¦lō }

heat radiation [PHYS] The energy radiated by solids, liquids, and gases in the form of electromagnetic waves as a result of their temperature. Also known as thermal radiation. { 'hēt ,rād·ē'ā·shən }

heat shock response [BIOL] A cellular reaction to a stimulus such as elevated temperatures or abrupt environmental changes, in which there is cessation or slowdown of normal protein synthesis and activation of previously inactive genes, resulting in the production of heat shock proteins. { 'hēt ,shäk ri,späns }

heat storage [OCEANOGR] The tendency of the ocean to act as a heat reservoir; results in smaller daily and annual variations in temperature over the sea. { 'hēt ,stȯr·ij }

heat thunderstorm [METEOROL] In popular terminology, a thunderstorm of the air mass type which develops near the end of a hot, humid summer day. { 'hēt ¦thən·dər ,stȯrm }

heat transfer [PHYS] The movement of heat from one body to another (gas, liquid, solid, or combinations thereof) by means of radiation, convection, or conduction. { 'hēt ¦tranz·fər }

heat wave [METEOROL] A period of abnormally and uncomfortably hot and usually humid weather; the condition must prevail at least 1 day to be a heat wave, but conventionally the term is reserved for periods of several days to several weeks. Also known as hot wave; warm wave. { 'hēt ,wāv }

heave [OCEANOGR] The motion imparted to a floating body by wave action. { hēv }

heavy crude [PETR MIN] Crude oil having a high proportion of viscous, high-molecular-weight hydrocarbons, and often having a high sulfur content. { 'hev·ē 'krüd }

heavy floe [OCEANOGR] A mass of sea ice that is more than 10 feet (3 meters) thick. Also known as heavy ice. { 'hev·ē 'flō }

heavy ice *See* heavy floe. { 'hev·ē 'īs }

heavy-mineral prospecting [PETR MIN] Locating the source of an economic mineral by determining the relative amounts of the mineral in stream sediments and tracing the drainage upstream. { 'hev·ē ¦min·rəl 'präs,pek·tiŋ }

hecto- [SCI TECH] A prefix representing 10^2 or 100. { 'hek·tō }

height-change chart [METEOROL] A chart indicating the change in height of a constant-pressure surface over a specified previous time interval; comparable to a pressure-change chart. { 'hīt ,chānj ,chärt }

height of tide [OCEANOGR] Vertical distance from the chart datum to the level of the water at any time; it is positive if the water level is higher than the chart datum. { 'hīt əv 'tīd }

height pattern [METEOROL] The general geometric characteristics of the distribution of height of a constant-pressure surface as shown by contour lines on a constant-pressure chart. Also known as baric topography; isobaric topography; pressure topography. { 'hīt ‚pad·ərn }

Heine-Medin disease *See* poliomyelitis. { 'hī·nə 'med·ən di‚zēz }

hekistotherm [ECOL] Plant adapted for conditions of minimal heat; can withstand long dark periods. { he'kis·tō‚thərm }

heliophilous [ECOL] Attracted by and adapted for a high intensity of sunlight. { ¦hē·lē ¦äf·ə·ləs }

heliophyte [ECOL] A plant that thrives in full sunlight. { 'hē·lē·ə‚fīt }

heliotaxis [BIOL] Orientation movement of an organism in response to the stimulus of sunlight. { ¦hē·lē·ō¦tak·səs }

heliotrope [BOT] A plant whose flower or stem turns toward the sun. [ENG] An instrument that reflects the sun's rays over long distances; used in geodetic surveys. { 'hē·lē·ə‚trōp }

heliotropic wind [METEOROL] A subtle, diurnal component of the wind velocity leading to a diurnal shift of the wind or turning of the wind with the sun, produced by the east-to-west progression of daytime surface heating. { ‚hē·lē·ə'träp·ik 'wind }

heliotropism [BIOL] Growth or orientation movement of a sessile organism or part, such as a plant, in response to the stimulus of sunlight. { ‚hē·lē'ä·trə‚piz·əm }

Helminthosporium [MYCOL] A genus of parasitic fungi of the family Dematiaceae having conidiophores which are more or less irregular or bent and bear conidia successively on new growing tips. { hel‚min·thə'spȯr·ē·əm }

Helminthosporium victoriae [MYCOL] A fungal pathogen that produces victorin, the cause of Victoria blight of oats. { hel‚min·thə‚spȯr·ē·əm vik'tȯr·ē‚ī }

helophyte [ECOL] A marsh plant; buds overwinter underwater. { 'he·lə‚fīt }

helophytia [ECOL] Differences in ecological control by fluctuations in water level such as in marshes. { ‚he·lə'fī·shə }

helotism [ECOL] Symbiosis in which one organism is a slave to the other, as between certain species of ants. { 'hel·ə‚tiz·əm }

hematochrome [BIOL] A red pigment occurring in green algae, especially when plants are exposed to intense light on subaerial habitats. { hi'mad·ə‚krōm }

hemicryptophyte [ECOL] A plant having buds at the soil surface and protected by scales, snow, or litter. { ¦he·mē'krip·tə‚fīt }

Hemileia vastatrix [MYCOL] A fungus of the order Uredinales which is the causative agent of orange coffee rust. { ‚hem·ə‚lē·yə 'vas·tə‚triks }

hemimetabolous metamorphosis [ZOO] An incomplete metamorphosis (lacking a pupal stage) in insects; gills are present in aquatic larvae, or naiads. { ¦he·mē·me'tab·ə·ləs ‚med·ə'mȯr·fə·səs }

hemiparasite [ECOL] A parasite capable of a saprophytic existence, especially certain parasitic plants containing some chlorophyll. Also known as semiparasite. { ¦he·mē'par·ə‚sīt }

hemipelagic [ECOL] Of the biogeographic environment of the hemipelagic region with both neritic and pelagic qualities. { ¦he·mē·pə'laj·ik }

hemipelagic region [OCEANOGR] The region of the ocean extending from the edge of a shelf to the pelagic environment; roughly corresponds to the bathyal zone, in which the bottom is 660 to 3300 feet (200 to 1000 meters) below the surface. { ¦he·mē·pə'laj·ik 'rē·jən }

hemipelagic sediment [GEOL] Deposits containing terrestrial material and the remains of pelagic organisms, found in the ocean depths. { ¦he·mē·pə'laj·ik 'sed·ə·mənt }

hemipelagite [OCEANOGR] Deep-sea mud deposits in which more than 25% of the fraction of particles coarser than 5 micrometers is of terrigenous, volcanogenic, or neritic origin. { ‚hem·ē'pel·ə‚jīt }

hemisphere [GEOGR] A half of the earth divided into north and south sections by the equator, or into an east section containing Europe, Asia, and Africa, and a west section containing the Americas. { 'he·mē‚sfir }

hemolysin [MED] A substance that lyses erythrocytes. { ‚hē·mə'līs·ən }

hemorrhagic fever virus [MICROBIO] Any of several arboviruses causing acute infectious human diseases characterized by fever, prostration, vomiting, and hemorrhage. { ‚hem·ə'raj·ik ¦fē·vər ‚vī·rəs }

hemorrhagic septicemia [VET MED] An infectious bacterial disease of fowl, rabbit, buffalo, and other animals caused by *Pasteurella mulfocida*. Also known as pasteurellosis. { ‚hem·ə'raj·ik ‚sep·tə'sē·mē·ə }

hepatitis [MED] Inflammation of the liver; commonly of viral origin but also occurring in association with syphilis, typhoid fever, malaria, toxemias, and parasitic infestations. { ‚hep·ə'tīd·əs }

hepatitis virus [MICROBIO] Any of several viruses causing hepatitis in humans and lower mammals. { 'hep·ə'tīd·əs 'vī·rəs }

Hepsogastridae [ZOO] A family of parasitic insects in the order Mallophaga. { ‚hep·sə'gas·trə‚dē }

heptachlor [CHEM] $C_{10}H_7Cl_7$ An insecticide; a white to tan, waxy solid; insoluble in water, soluble in alcohol and xylene; melts at 95–96°C. { 'hep·tə‚klór }

herb [BOT] **1.** A seed plant that lacks a persistent, woody stem aboveground and dies at the end of the season. **2.** An aromatic plant or plant part used medicinally or for food flavoring. { hərb }

herbaceous [BOT] **1.** Resembling or pertaining to a herb. **2.** Pertaining to a stem with little or no woody tissue. { hər'bā·shəs }

herbarium [BOT] **1.** A collection of plant specimens, pressed and mounted on paper or placed in liquid preservatives, and systematically arranged. **2.** A building where a herbarium is housed. { hər'ber·ē·əm }

herbicolous [ECOL] Living on herbs. { hər'bik·əl·əs }

herbivore [ZOO] An animal that eats only vegetation. { 'hər·bə‚vór }

herbivory [ECOL] The consumption of plants, algae, or other primary producers by heterotrophs. { hər'biv·ə·rē }

herd immunity [MED] Immunity of a sufficient number of individuals in a population such that infection of one individual will not result in an epidemic. { 'hərd i‚myü·nəd·ē }

heritability [GEN] A measure of the degree to which a particular trait is the result of the genotype and can be modified by selection. { ‚her·əd·ə'bil·əd·ē }

hermatype *See* hermatypic coral. { 'hər·mə‚tīp }

hermatypic coral [ZOO] Reef-building coral characterized by the presence of symbiotic algae within their endodermal tissue. Also known as hermatype. { ‚hər·mə'tip·ik 'kär·əl }

herpes simplex virus   [MICROBIO] Either of two types of subgroup A herpesviruses that are specific for humans; given the binomial designation *Herpesvirus hominis*.   { ¦hər¦pēz 'sim¦pleks 'vī·rəs }

herpesvirus   [MICROBIO] A major group of deoxyribonucleic acid-containing animal viruses, distinguished by a cubic capsid, enveloped virion, and affinity for the host nucleus as a site of maturation.   { 'hər¦pēz¦vī·rəs }

herpes zoster   [MED] A systemic virus infection affecting spinal nerve roots, character-ized by vesicular eruptions distributed along the course of a cutaneous nerve. Also known as shingles; zoster.   { ¦hər¦pēz 'zäs·tər }

herpetology   [ZOO] The study of amphibians and reptiles.   { ˌhər·pə·'tä·lə·jē }

heterauxesis   *See* allometry.   { ¦hed·ər¦óg'zē·səs }

heterocarpous   [BOT] Producing two distinct types of fruit.   { ˌhed·ə·rō'kär·pəs }

heterochthonous   [SCI TECH] Not indigenous to the area of present occurrence.   { ˌhed·ə'räk·thə·nəs }

heteroecious   *See* heteroxenous.   { ˌhed·ə'rē·shəs }

heterogeneity   [BIOL] The condition or state of being different in kind or nature. [SCI TECH] The condition of a sample of matter that is composed of particles or aggregates of different substances of dissimilar composition.   { ¦hed·ə·rə·jə'nē·əd·ē }

heterogeneous   [SCI TECH] Composed of dissimilar or nonuniform constituents.   { ˌhed·ə'räj·ə·nəs }

heterogony   [BIOL] **1.** Alteration of generations in a complete life cycle, especially of a dioecious and hermaphroditic generation. **2.** *See* allometry.[BOT] Having heteromorphic perfect flowers with respect to lengths of the stamens or styles.   { ˌhed·ə'räg·ə·nē }

heterophyte   [BOT] A plant that depends upon living or dead plants or their products for food materials.   { 'hed·ə·rə¦fīt }

heterosphere   [METEOROL] The upper portion of a two-part division of the atmosphere (the lower portion is the homosphere) according to the general homogeneity of atmospheric composition; characterized by variation in composition, and in mean molecular weight of constituent gases; starts at 50–62 miles (80–100 kilometers) above the earth and therefore closely coincides with the ionosphere and the thermosphere.   { 'hed·ə·rə¦sfir }

heterospory   [BOT] Development of more than one type of spores, especially relating to the microspores and megaspores in ferns and seed plants.   { ˌhed·ə'räs·pə·rē }

heterotherm   [ECOL] An animal that is endothermic part of the time but can reduce metabolic heat production and lower body temperature when conservation of food energy supplies is necessary.   { 'hed·ə·rə¦thərm }

heterotopia   [ECOL] An abnormal habitat.   { ˌhed·ə·rō'tō·pē·ə }

heterotroph   [BIOL] An organism that obtains nourishment from the ingestion and breakdown of organic matter.   { 'hed·ə·rō¦träf }

heterotrophic ecosystem   [ECOL] An ecosystem that depends upon preformed organic matter that is imported from autotrophic ecosystems elsewhere.   { ¦hed·ə·rə¦träf·ik 'ek·ō¦sis·təm }

heterotrophic succession   [ECOL] A type of ecological succession that involves decomposer organisms.   { ¦hed·ə·rə¦träf·ik sək'sesh·ən }

heteroxenous   [BIOL] Requiring more than one host to complete a life cycle. Also known as heteroecious.   { ¦hed·ə¦räk·sə·nəs }

heterozygote   [GEN] An individual that has different alleles at one or more loci and therefore produces gametes of two or more different kinds with respect to their loci.   { ¦hed·ə·rō'zī‚gōt }

hexachlorobenzene   [CHEM] C_6Cl_6 Colorless, needlelike crystals with a melting point of 231°C; used in organic synthesis and as a fungicide. Abbreviated HCB.   { ¦hek·sə ¦klór·ō'ben‚zēn }

1,2,3,4,5,6-hexachlorocyclohexane   [CHEM] $C_6H_6Cl_6$ A white or yellow powder or flakes with a musty odor; a systemic insecticide toxic to flies, cockroaches, aphids, and boll weevils. Abbreviated TBH.   { ¦wən ¦tü ¦thrē ¦fór ¦fīv ¦siks ¦hek·sə¦klór·ō‚sī·klō'hek‚sān }

hexachlorophene   [CHEM] $(C_6HCl_3OH)_2CH_2$ A white powder melting at 161°C; soluble in alcohol, ether, acetone, and chloroform, insoluble in water; bacteriostat used in antiseptic soaps, cosmetics, and dermatologicals.   { ‚hek·sə'klór·ə‚fēn }

hexagonal column   [METEOROL] One of the many forms in which ice crystals are found in the atmosphere; this crystal habit is characterized by hexagonal cross-section in a plane perpendicular to the long direction (principal axis, optic axis, or c axis) of the columns; it differs from that found in hexagonal platelets only in that environmental conditions have favored growth along the principal axis rather than perpendicular to that axis.   { hek'sag·ə·nəl 'käl·əm }

hexagonal platelet   [METEOROL] A small ice crystal of the hexagonal tabular form; the distance across the crystal from one side of the hexagon to the opposite side may be as large as about 1 millimeter, and the thickness perpendicular to this dimension is of the order of one-tenth as great; this crystal form is usually formed at temperatures of −10 to −20°C by sublimation; at higher temperatures the apices of the hexagon grow out and develop dendritic forms.   { hek'sag·ə·nəl 'plāt·lət }

hexahydrophenol   See cyclohexanol.   { ¦hek·sə¦hī·drō'fē‚nól }

n-hexaldehyde   [CHEM] $CH_3(CH_2)_4CHO$ Colorless liquid with sharp aroma, boiling at 128.6°C; used as an intermediate for plasticizers, dyes, insecticides, resins, and rubber chemicals.   { ¦en ‚heks'al·də‚hīd }

hexametapol   [CHEM] $C_6H_{18}N_3OP$ A liquid used as a solvent in organic synthesis, as a deicing additive for jet engine fuel, and as an insect pest chemosterilant and chemical mutagen.   { ‚hek·sə'med·ə‚pól }

1-hexene   [CHEM] $CH_3(CH_2)_3HC{:}CH_2$ Colorless, olefinic hydrocarbon boiling at 64°C; soluble in alcohol, acetone, ether, and hydrocarbons, insoluble in water; used as a chemical intermediate and for resins, drugs, and insecticides. Also known as hexylene.   { ¦wən 'hek‚sēn }

hexylene   See 1-hexene.   { 'hek·sə‚lēn }

hibernaculum   [BIOL] A winter shelter for plants or dormant animals.   [BOT] A winter bud or other winter plant part.   { ‚hī·bər'nak·yə·ləm }

hibernal   [METEOROL] Of or pertaining to winter.   { hī'bərn·əl }

hibernation   [BIOL] **1.** Condition of dormancy and torpor found in cold-blooded vertebrates and invertebrates.   **2.** See deep hibernation.   { ‚hī·bər'nā·shən }

hiemal climate   [CLIMATOL] Climate pertaining to winter.   { 'hī·ə·məl ¦klī·mət }

high   [METEOROL] An area of high pressure, referring to a maximum of atmospheric pressure in two dimensions (closed isobars) in the synoptic surface chart, or a maximum of height (closed contours) in the constant-pressure chart; since a high is, on the synoptic chart, always associated with anticyclonic circulation, the term is used interchangeably with anticyclone.   { hī }

high aloft   See upper-level anticyclone.   { ¦hī ə'lóft }

high-altitude station   [METEOROL] A weather observing station at a sufficiently high elevation to be nonrepresentative of conditions near sea level; 6500 feet (about 2000 meters) has been given as a reasonable lower limit.   { 'hī ¦al·tə‚tüd 'stā·shən }

high clouds [METEOROL] Types of clouds whose mean lower level is above 20,000 feet (6100 meters); principal clouds in this group are cirrus, cirrocumulus, and cirrostratus. { 'hī ¦klaùdz }

high-energy environment [GEOL] An aqueous sedimentary environment which features a high energy level and turbulent motion, created by waves, currents, or surf, which prevents the settling and piling up of fine-grained sediment. { 'hī ‚en·ər· jē in'vī·ərn·mənt }

higher high water [OCEANOGR] The higher of two high tides occurring during a tidal day. { 'hī·ər ¦hī ‚wòd·ər }

higher low water [OCEANOGR] The higher of two low tides occurring during a tidal day. { 'hi·ər ¦lō‚wòd·ər }

high foehn [METEOROL] The occurrence of warm, dry air above the level of the general surface, accompanied by clear skies, resembling foehn conditions; it is due to subsiding air in an anticyclone, above a cold surface layer; in such circumstances the mountain peaks may be warmer than the lowlands. Also known as free foehn. { ¦hī 'fān }

high index [METEOROL] A relatively high value of the zonal index which, in middle latitudes, indicates a relatively strong westerly component of wind flow and the characteristic weather features attending such motion; a synoptic circulation pattern of this type is commonly called a high-index situation. { 'hī ¦in‚deks }

highland [GEOGR] **1.** Any relatively large area of elevated or mountainous land standing prominently above adjacent low areas. **2.** The higher land of a region. [GEOL] **1.** A lofty headland, cliff, or other high platform. **2.** A dissected mountain region composed of old folded rocks. { 'hī·lənd }

highland climate See mountain climate. { 'hī·lənd ¦klī·mət }

highland glacier [HYD] A semicontinuous ice cap or glacier that covers the highest or central portion of a mountainous area and partly reflects irregularities of the land surface lying beneath it. Also known as highland ice. { 'hī·lənd ¦glā·shər }

highland ice See highland glacier. { 'hī·lənd ¦īs }

high-level anticyclone See upper-level anticyclone. { 'hī ‚lev·əl an·tē'sī‚klōn }

high-level cyclone See upper-level cyclone. { 'hī ‚lev·əl 'sī‚klōn }

high-level ridge See upper-level ridge. { 'hī ‚lev·əl 'rij }

high-level thunderstorm [METEOROL] Generally, a thunderstorm based at a comparatively high altitude in the atmosphere, roughly 8000 feet (2400 meters) or higher. { 'hī ‚lev·əl 'thən·dər‚stòrm }

high-level trough See upper-level trough. { 'hī ‚lev·əl 'tròf }

highmoor bog [ECOL] A bog whose surface is covered by sphagnum mosses and is not dependent upon the water table. { 'hī‚mür 'bäg }

high plain [GEOGR] A large area of level land situated above sea level. { 'hī ¦plān }

high-pressure area See anticyclone. { 'hī ¦presh·ər 'er·ē·ə }

high tide [OCEANOGR] The maximum height reached by a rising tide. Also known as high water. { 'hī ¦tīd }

high water See high tide. { 'hī ¦wòd·ər }

high-water inequality [OCEANOGR] The difference between the heights of the two high tides during a tidal day. { 'hī ¦wòd·ər ‚in·ə'kwäl·əd·ē }

high-water line [OCEANOGR] The intersection of the plane of mean high water with the shore. { 'hī ¦wòd·ər ‚līn }

high-water quadrature [OCEANOGR] The average high-water interval when the moon is at quadrature. { 'hī ¦wȯd·ər'kwäd·rə·chər }

high-water springs *See* mean high-water springs. { 'hī ¦wȯd·ər 'spriŋz }

high-water stand [OCEANOGR] The condition at high tide when there is no change in the height of the water. { 'hī ¦wȯd·ər 'stand }

hill [GEOGR] A land surface feature characterized by strong relief; it is a prominence smaller than a mountain. { hil }

histiocyte *See* macrophage. { 'his·tē·ə,sīt }

histomycosis [MED] Infection of deep tissues by a fungus. { ,his·tə,mī'kō·səs }

Histoplasma [MYCOL] A genus of parasitic fungi. { ,his·tə'plaz·mə }

Histoplasma capsulatum [MYCOL] The parasitic fungus that causes histoplasmosis in humans. { ,his·tə'plaz·mə ,kap·sə'läd·əm }

histoplasmosis [MED] An infectious fungus disease of the lungs of humans caused by *Histoplasma capsulatum*. { ,his·tə,plaz'mō·səs }

historical biogeography [ECOL] The study of how species' distributions have changed over time in relationship to the history of landforms, ocean basins, and climate, as well as how those changes have contributed to the evolution of biotas. { his,tär·i·kəl ,bī·ō·jē'ag·rə·fē }

historical climate [CLIMATOL] A climate of the historical period (the past 7000 years). { hi'stär·ə·kəl 'klī·mət }

historical zoogeography [ECOL] The study of animal distributions in terms of evolutionary history. { his,tär·i·kəl ,zō·ō·jē'äg·rə·fē }

Histosol [GEOL] An order of wet soils consisting mostly of organic matter, popularly called peats and mucks. { 'his·tə,sȯl }

HIV *See* human immunodeficiency virus.

hoar crystal [HYD] An individual ice crystal in a deposit of hoarfrost; always grows by sublimation. { 'hȯr ,krist·əl }

hoarfrost [HYD] A deposit of interlocking ice crystals formed by direct sublimation on objects. Also known as white frost. { 'hȯr,frȯst }

hog cholera [VET MED] A fatal infectious virus disease of swine characterized by fever, diarrhea, and inflammation and ulceration of the intestine; secondary infection by *Salmonella cholerae suis* is common. Also known as African swine fever. { 'häg ,käl·ə·rə }

hogging [ENG] Mechanical chipping of wood waste for fuel. { 'häg·iŋ }

holarctic zoogeographic region [ECOL] A major unit of the earth's surface extending from the North Pole to 30-45°N latitude and characterized by faunal homogeneity. { hō'lärd·ik ,zō·ō,jē·ə¦graf·ik 'rē·jən }

Holland formula [ENG] A formula used to calculate the height of a plume formed by pollutants emitted from a stack in terms of the diameter of the stack exit, the exit velocity and heat emission rate of the stack, and the mean wind speed. { 'häl·ənd ,fȯr·myə·lə }

holocoenosis [ECOL] The nature of the action of the environment on living organisms. { ¦häl·ō·sə¦nō·səs }

holometabolous metamorphosis [ZOO] Complete metamorphosis in insects, during which there are four stages; the egg, larva, pupa, and imago or adult. { ¦häl·ō·mə 'tab·ə·ləs ,med·ə'mȯr·fə·səs }

holomictic lake [HYD] A lake whose water circulates completely from top to bottom. { ¦häl·ō¦mik·tik 'lāk }

holoplankton |ZOO| Organisms that live their complete life cycle in the floating state. { ,häl·ō'plaŋk·tən }

homeohydric |BIOL| Pertaining to the ability to restrict cellular water loss regardless of environmental conditions. { ,hō·mē·o 'hī·drik }

homeostasis |BIOL| In higher animals, the maintenance of an internal constancy and an independence of the environment. { ,hō·mē·ō'stā·səs }

home range |ECOL| The physical area of an organism's normal activity. { 'hōm ¦rānj }

homobront *See* isobront. { ¦häm·ə¦bränt }

homoecious |BIOL| Of parasites, having one host for all stages of the life cycle. { hō 'mē·shəs }

homoeomerous |BOT| Having algae distributed uniformly throughout the thallus of a lichen. { ¦hō·mē¦äm·ə·rəs }

homogamous |BIOL| Of or pertaining to homogamy. { hə'mäg·ə·məs }

homogamy |BIOL| Inbreeding due to isolation. |BOT| Condition of having all flowers alike. { hə'mäg·ə·mē }

homogeneous |SCI TECH| Uniform in structure or composition. { ,hä·mə'jē·nē·əs }

homogeneous atmosphere |METEOROL| A hypothetical atmosphere in which the density is constant with height. { ,hä·mə'jē·nē·əs 'at·mə,sfir }

homoiothermal |BIOL| Referring to an organism which maintains a constant internal temperature which is often higher than that of the environment; common among birds and mammals. Also known as warm-blooded. { hō¦mȯi·ō¦thər·məl }

homosphere |METEOROL| The lower portion of a two-part division of the atmosphere (the upper portion is the heterosphere) according to the general homogeneity of atmospheric composition; the region in which there is no gross change in atmospheric composition, that is, all of the atmosphere from the earth's surface to about 50 to 62 miles (80–100 kilometers). { 'hä·mə,sfir }

homospory |BOT| Production of only one kind of asexual spore. { hə'mäs·pə·rē }

homozygote |GEN| An individual who has identical alleles at one or more loci and therefore produces identical gametes with respect to these loci. { ¦hō·mə'zī,gōt }

honeydew |ZOO| The viscous secretion deposited on leaves by many aphids and scale insects; an attractant for ants. { 'hən·ē,dü }

hoof-and-mouth disease *See* foot-and-mouth disease. { ¦hůf ən 'maůth di,zēz }

hook |GEOGR| The end of a spit of land that is turned toward shore. Also known as recurved spit. { hůk }

hopperburn |PL PATH| A disease of potato and peanut plants caused by a leafhopper which secretes a toxic substance on the leaves, causing browning and shriveling. { 'häp·ər ,bərn }

horizon |GEOL| 1. The surface separating two beds. 2. One of the layers, each of which is a few inches to a foot thick, that make up a soil. Starting from the soil surface, the major horizons include the O (organic) horizon, A (topsoil) horizon, B (subsoil) horizon, C (mineral) horizon, and R (bedrock) horizon. { hə'rīz·ən }

hormesis |BIOL| Providing stimulus by nontoxic amounts of a toxic agent. { 'hȯr· mə·səs }

horsehair blight |PL PATH| A fungus disease of tea and certain other tropical plants caused by *Marasmius equicrinis* and characterized by black festoons of mycelia hanging from the branches. { 'hȯrs,her ,blīt }

horse latitudes [METEOROL] The belt of latitudes over the oceans at approximately 30–35°N and S where winds are predominantly calm or very light and weather is hot and dry. { 'hȯrs ¦lad·ə‚tüdz }

horseshoe bend See oxbow. { 'hȯr‚shü ‚bend }

horseshoe lake See oxbow lake. { 'hȯr‚shü ‚lāk }

horticultural crop [AGR] Any food-producing plant. { ¦hȯrd·ə¦kəlch·ə·rəl 'kräp }

horticulture [BOT] The art and science of growing plants. { 'hȯrd·ə‚kəl·chər }

host [BIOL] **1.** An organism on or in which a parasite lives. **2.** The dominant partner of a symbiotic or commensal pair. { hōst }

host structure See host. { ¦hōst 'strək·chər }

hotbed [AGR] A bed of soil enclosed by a low frame with glass panels and heated by fermented manure or electric cables; used for forcing tender plants to grow out of season or to protect tender exotic plants. { 'hät‚bed }

hot belt [CLIMATOL] The belt around the earth within which the annual mean temperature exceeds 20°C. { 'hät ‚belt }

hothouse [ENG] A greenhouse heated to grow plants out of season. { 'hät‚haüs }

hot spot [FOR] A forest region where fires occur at frequent intervals. [GEOL] An area of localized volcanic activity, such as Iceland and the Hawaiian Islands, believed to be caused by plumes of magna rising from the earth's mantle to its surface. [PHYS] **1.** A surface area of higher than average radioactivity. **2.** A part of a reactor fuel surface element that has been overheated. { 'hät ‚spät }

hot spring [HYD] A thermal spring whose water temperature is above 98°F (37°C). { 'hät ¦spriŋ }

hot wave See heat wave. { 'hät ‚wāv }

hot wind [METEOROL] General term for winds characterized by intense heat and low relative humidity, such as summertime desert winds or an extreme foehn. { 'hät ¦wind }

hourly observation See record observation. { 'aü·ər·lē ‚äb·sər'vā·shən }

house drain [CIV ENG] Horizontal drain in a basement receiving waste from stacks. { 'haüs ‚drān }

housefly [ZOO] *Musca domestica*. A dipteran insect with lapping mouthparts commonly found near human habitations; a vector in the transmission of many disease pathogens. { 'haüs‚flī }

house sewer [CIV ENG] Connection between house drain and public sewer. { 'haüs ‚sü·ər }

Hubbard Glacier [GEOGR] A valley glacier which reaches tidewater from a source area of Mount St. Elias of Alaska and the Yukon. { 'həb·ərd 'glā·shər }

Hudsonian life zone [ECOL] A zone comprising the climate and biotic communities of the northern portions of North American coniferous forests and the peaks of high mountains. { ¦həd¦sō·nē·ən 'līf ‚zōn }

hull [BOT] The outer, usually hard, covering of a fruit or seed. { həl }

human biogeography [ECOL] The science concerned with the distribution of human populations on the earth. { 'hyü·mən ¦bī·ō·jē'äg·rə·fē }

human community [ECOL] That portion of a human ecosystem composed of human beings and associated plant and animal species. { 'hyü·mən kə'myün·əd·ē }

human ecology [ECOL] The branch of ecology that considers the relations of individual persons and of human communities with their particular environment. { 'hyü·mən ē'käl·ə·jē }

human geography [GEOGR] The study of the characteristics and phenomena of the earth's surface that relate directly to or are due to human activities. Also known as anthropogeography. { ¦hyü·mən jē¦äg·rə·fē }

human immunodeficiency virus [MICROBIO] The retrovirus that causes acquired immune deficiency syndrome. Abbreviated HIV. { 'hyü·mən ¦im·yə·nō·di'fish·ən·sē ˌvī·rəs }

Humboldt Current See Peru Current. { 'həm,bōlt ¦kə·rənt }

Humboldt Glacier [HYD] The largest Arctic iceberg, at latitude 79°, with a seaward front extending 65 miles (105 kilometers). { 'həm,bōlt ¦glā·shər }

humic [GEOL] Pertaining to or derived from humus. { 'hyü·mik }

humicolous [ECOL] Of or pertaining to plant species inhabiting medium-dry ground. { hyü'mik·ə·ləs }

humid climate [CLIMATOL] A climate whose typical vegetation is forest. Also known as forest climate. { 'hyü·məd ¦klī·mət }

humidity [METEOROL] Atmospheric water vapor content, expressed in any of several measures, such as relative humidity. { hyü'mid·əd·ē }

humidity coefficient [METEOROL] A measure of the precipitation effectiveness of a region; it recognizes the exponential relationship of temperature versus plant growth and is expressed as humidity coefficient = $P/(1.07)^t$, where P is the precipitation in centimeters, and t is the mean temperature in degrees Celsius for the period in question; the denominator approximately doubles with each 10°C rise in temperature. { hyü'mid·əd·ē ˌkō·i'fish·ənt }

humidity index [CLIMATOL] An index of the degree of water surplus over water need at any given station; it is calculated as humidity index = $100s/n$, where s (the water surplus) is the sum of the monthly differences between precipitation and potential evapotranspiration for those months when the normal precipitation exceeds the latter, and where n (the water need) is the sum of monthly potential evapotranspiration for those months of surplus. { hyü'mid·əd·ē ,in,deks }

humidity indicator [CHEM] Cobalt salt (for example, cobaltous chloride) that changes color as the surrounding humidity changes; changes from pink when hydrated, to greenish-blue when anhydrous. { hyü'mid·əd·ē ,in·də,kād·ər }

humidity mixing ratio [METEOROL] The amount of water vapor mixed with one unit mass of dry air, usually expressed as grams of water vapor per kilogram of air. { hyü'mid·əd·ē 'mik·siŋ ,rā·shō }

humidity province [CLIMATOL] A region in which the precipitation effectiveness of its climate produces a definite type of biological consequence, in particular the climatic climax formations of vegetation (rain forest, tundra, and the like). { hyü'mid·əd·ē ,präv·əns }

humid transition life zone [ECOL] A zone comprising the climate and biotic communities of the northwest moist coniferous forest of the north-central United States. { 'hyü·məd tran¦zish·ən 'līf ,zōn }

humifuse [BIOL] Spread over the ground surface. { 'hyü·mə,fyüs }

humivore [ECOL] An organism that feeds on humus. { 'hyü·mə,vȯr }

hummock [ECOL] A rounded or conical knoll frequently formed of earth and covered with vegetation. [HYD] A mound, hillock, or pile of broken floating ice, either fresh or weathered, that has been forced upward by pressure, as in an ice field or ice floe. { 'həm·ək }

hummocked ice [OCEANOGR] Pressure ice, characterized by haphazardly arranged mounds or hillocks; it has less definite form, and show the effects of greater pressure, than either rafted ice or tented ice, but in fact may develop from either of those. { 'həm·əkt 'īs }

humus [GEOL] The amorphous, ordinarily dark-colored, colloidal matter in soil; a complex of the fractions of organic matter of plant, animal, and microbial origin that are most resistant to decomposition. { 'hyü·məs }

hundred-year flood [HYD] A flood that has a 1% chance of occurring in a given year. { ¦hən·drəd ¦yir 'fləd }

hurricane [METEOROL] A tropical cyclone of great intensity; any wind reaching a speed of more than 73 miles per hour (117 kilometers per hour) is said to have hurricane force. { 'hər·ə,kān }

hurricane band See spiral band. { 'hər·ə,kān ,band }

hurricane-force wind [METEOROL] In the Beaufort wind scale, a wind whose speed is 64 knots (117 kilometers per hour) or higher. { 'hər·ə,kān ,fȯrs ,wind }

hurricane monitoring buoy [METEOROL] A free-floating automatic weather station designed as an expendable instrument in connection with hurricane and typhoon monitoring and forecasting services. { 'hər·ə,kān 'män·ə·triŋ ,bȯi }

hurricane radar band See spiral band. { 'hər·ə,kān 'rā,där ,band }

hurricane surge See hurricane wave. { 'hər·ə,kān ¦sərj }

hurricane tide See hurricane wave. { 'hər·ə,kān ¦tīd }

hurricane tracking [ENG] Recording of the movement of individual hurricanes by means of airplane sightings and satellite photography. { 'hər·ə,kān ,trak·iŋ }

hurricane warning [METEOROL] A warning of impending winds of hurricane force; for maritime interests, the storm warning signals for this condition are two square red flags with black centers by day, and a white lantern between two red lanterns by night. { 'hər·ə,kān ,wȯrn·iŋ }

hurricane watch [METEOROL] An announcement for a specific area that hurricane conditions pose a threat; residents are cautioned to take stock of their preparedness needs but, otherwise, are advised to continue normal activities. { 'hər·ə,kān ,wäch }

hurricane wave [OCEANOGR] As experienced on islands and along a shore, a sudden rise in the level of the sea associated with a hurricane. Also known as hurricane surge; hurricane tide. { 'hər·ə,kān ¦wāv }

hurricane wind [METEOROL] In general, the severe wind of an intense tropical cyclone (hurricane or typhoon); the term has no further technical connotation, but is easily confused with the strictly defined hurricane-force wind. Also known as typhoon wind. { 'hər·ə,kān ¦wind }

husk [BOT] The outer coat of certain seeds, particularly if it is a dry, membranous structure. { həsk }

hybrid [SCI TECH] Having two or more different characteristics or types of structure. { 'hī·brəd }

hybridization [BIOL] The production of viable hybrid somatic cells following experimentally induced cell fusion. [GEN] **1.** Production of a hybrid by pairing complementary ribonucleic acid and deoxyribonucleic acid (DNA) strands. **2.** Production of a hybrid by pairing complementary DNA single strands. { ,hī·brəd·ə'zā·shən }

hybrid zone [ECOL] A geographic zone in which two populations hybridize after the breakdown of the geographic barrier that separated them. { ¦hī·brəd ¦zōn }

hydathode [BOT] An opening of the epidermis of higher plants specialized for exudation of water. { 'hīd·ə,thōd }

hydrargyrism *See* mercurialism. { hī'drär·jə,riz·əm }

hydraulic current [OCEANOGR] A current in a channel, due to a difference in the water level at the two ends. { hī'dro·lik 'kə·rənt }

hydraulic discharge [HYD] The direct discharge of groundwater from the zone of saturation upon the land or into a body of surface water. { hī'dro·lik 'dis,chärj }

hydraulic engineering [CIV ENG] A branch of civil engineering concerned with the design, erection, and construction of sewage disposal plants, waterworks, dams, water-operated power plants, and such. { hī'dro·lik ,en·jə'nir·iŋ }

hydraulic grade line [PHYS] **1.** In a closed channel, a line joining the elevations that water would reach under atmospheric pressure. **2.** The free water surface in an open channel. { hī'dro·lik 'grād ,līn }

hydraulic gradient [HYD] The slope of the hydraulic grade line of a stream. { hī'dro·lik 'grād·ē·ənt }

hydraulic sprayer [ENG] A machine that sprays large quantities of insecticide or fungicide on crops. { hi'dro·lik 'sprā·ər }

hydraulic turbine [ENG] A machine which converts the energy of an elevated water supply into mechanical energy of a rotating shaft. { hī'dro·lik 'tər·bən }

hydric [ECOL] Characterized by or thriving in abundance of moisture. { 'hī·drik }

hydrocarbon [CHEM] One of a very large group of chemical compounds composed only of carbon and hydrogen; the largest source of hydrocarbons is from petroleum crude oil. { ¦hī·drə'kär·bən }

hydrocast [OCEANOGR] A series of water samplers on a single hydrographic wire (electric-powered hoist) which obtain samples simultaneously. { 'hī·drə,kast }

hydrochory [BIOL] Dispersal of disseminules by water. { 'hī·drə,kòr·ē }

hydrocyanic acid [CHEM] HCN A highly toxic liquid that has the odor of bitter almonds and boils at 25.6°C; used to manufacture cyanide salts, acrylonitrile, and dyes, and as a fumigant in agriculture. Also known as formonitrile; hydrogen cyanide; prussic acid. { ¦hī·drō·sī'an·ik 'as·əd }

hydrodynamics [PHYS] The study of the motion of a fluid and of the interactions of the fluid with its boundaries, especially in the incompressible inviscid case. { ,hī·drō·dī'nam·iks }

hydroelectric generator [ENG] An electric rotating machine that transforms mechanical power from a hydraulic turbine or water wheel into electric power. { ¦hī·drō·i'lek·trik 'jen·ə,rād·ər }

hydroelectricity [ENG] Electric power produced by hydroelectric generators. Also known as hydropower. { ¦hī·drō·i,lek'tris·əd·ē }

hydroelectric plant [ENG] A facility at which electric energy is produced by hydroelectric generators. Also known as hydroelectric power station. { ¦hī·drō·i'lek·trik 'plant }

hydroelectric power [ENG] The generation of electricity by flowing water; potential energy from the weight of water falling through a vertical distance is converted to electrical energy. { ¦hī·drō·i'lek·trik 'pau̇·ər }

hydroelectric power station *See* hydroelectric plant. { ¦hī·drō·i'lek·trik 'pau̇·ər ,stā·shən }

hydrofluorosilicic acid *See* fluosilicic acid. { ¦hī·drō¦flu̇r·ō·sə'lis·ik 'as·əd }

hydrofluosilicic acid *See* fluosilicic acid. { ¦hī·drō¦flü·ə·sə'lis·ik 'as·əd }

hydrogen bacteria [MICROBIO] Bacteria capable of obtaining energy from the oxidation of molecular hydrogen. { 'hī·drə·jən bak'tir·ē·ə }

hydrogen cyanide See hydrocyanic acid. { 'hī·drə·jən 'sī·ə‚nīd }

hydrogeochemistry [GEOCHEM] The study of the chemical characteristics of ground and surface waters as related to areal and regional geology. { ‚hī·drō‚jē·ō'kem·ə·strē }

hydrogeology [HYD] The science dealing with the occurrence of surface and ground water, its utilization, and its functions in modifying the earth, primarily by erosion and deposition. { ‚hī·drō·jē'äl·ə·jē }

hydrograph [HYD] A graphical representation of stage, flow, velocity, or other characteristics of water at a given point as a function of time. { 'hī·drə‚graf }

hydrographic basin See drainage basin. { 'hī·drə'graf·ik 'bās·ən }

hydrographic cruise [OCEANOGR] Exploration of a body of water for hydrographic surveys. { 'hī·drə'graf·ik 'krüz }

hydrographic survey [OCEANOGR] Survey of a water area with particular reference to tidal currents, submarine relief, and any adjacent land. { 'hī·drə'graf·ik 'sər‚vā }

hydrographic table [OCEANOGR] Tabular arrangement of data relating sea-water density to salinity, temperature, and pressure. { 'hī·drə'graf·ik 'tā·bəl }

hydrography [GEOGR] Science which deals with the measurement and description of the physical features of the oceans, lakes, rivers, and their adjoining coastal areas, with particular reference to their control and utilization. { hī'dräg·rə·fē }

hydrologic accounting [HYD] A systematic summary of the terms (inflow, outflow, and storage) of the storage equation as applied to the computation of soil-moisture changes, groundwater changes, and so forth; an evaluation of the hydrologic balance of an area. Also known as basin accounting; water budget. { ‚hī·drə‚läj·ik ə'kaúnt·iŋ }

hydrologic cycle [HYD] The complete cycle through which water passes, from the oceans, through the atmosphere, to the land, and back to the ocean. Also known as water cycle. { ‚hī·drə‚läj·ik 'sī·kəl }

hydrologist [HYD] An individual who specializes in hydrology. { hī'dräl·ə·jəst }

hydrometeor [HYD] **1.** Any product of condensation or sublimation of atmospheric water vapor, whether formed in the free atmosphere or at the earth's surface. **2.** Any water particles blown by the wind from the earth's surface. { ‚hī·drō'mēd·ē·ər }

hydrometeorology [METEOROL] That part of meteorology of direct concern to hydrologic problems, particularly to flood control, hydroelectric power, irrigation, and similar fields of engineering and water resources. { ‚hī·drō‚mēd·ē·ə'räl·ə·jē }

hydrophilous [ECOL] Inhabiting moist places. { hī'dräf·ə·ləs }

hydrophobia See rabies. { 'hī·drə'fō·bē·ə }

hydrophyte [BOT] **1.** A plant that grows in a moist habitat. **2.** A plant requiring large amounts of water for growth. Also known as hygrophyte. { 'hī·drə‚fīt }

hydroponics [BOT] Growing of plants in a nutrient solution with the mechanical support of an inert medium such as sand. { ‚hī·drə'pän·iks }

hydropower See hydroelectricity. { 'hī·drə‚paú·ər }

hydrosere [ECOL] Community in which pioneer plants invade open water, eventually forming some kind of soil such as peat or muck. { 'hī·drə‚sir }

hydrosphere [HYD] The water portion of the earth as distinguished from the solid part (lithosphere) and from the gaseous outer envelope (atmosphere). { 'hī·drə‚sfir }

hydrostatic stability See static stability. { ‚hī·drə'stad·ik stə'bil·əd·ē }

hydrothermal |GEOL| Of or pertaining to heated water, to its action, or to the products of such action. { ,hī·drə'thər·məl }

hydrothermal vent |OCEANOGR| A hot spring on the ocean floor, found mostly along mid-oceanic ridges, where heated fluids exit from cracks in the earth's crust. Iron, sulfur, and other materials precipitate from these waters to form dark clouds. Also known as black smoker. { ,hī·drə|thərm·əl 'vent }

hydrotropism |BIOL| Orientation involving growth or movement of a sessile organism or part, especially plant roots, in response to the presence of water. { hī'drä·trə ,piz·əm }

hydroxisoxazole |CHEM| $C_4H_5NO_2$ A colorless, crystalline compound with a melting point of 86–87°C; used as a fungicide in soil and as a growth regulator for seeds. Also known as 3-hydroxy-5-methylisoxazole; hymexazol. { hī|dräk·sə|säk·sə,zōl }

hydroxycholine See muscarine. { hī|dräk·sē'kō,lēn }

2-(hydroxydiphenyl)methane |CHEM| $C_6H_5CH_2C_6H_4OH$ A crystalline substance with a melting point of 20.2–20.9°C, or a liquid; used as a germicide, preservative, and antiseptic. { |tü hī|dräk·sē·dī'fen·əl'meth,ān }

hydroxymethylbenzene See creosol. { |hī|dräk·sē|meth·əl'ben,zēn }

3-hydroxy-5-methylisoxazole See hydroxisoxazole. { |thre ,hī'dräks·ē |fīv 'meth·əl ə'säks·ə·,zōl }

hyetal coefficient See pluviometric coefficient. { 'hī·əd·əl ,kō·i'fish·ənt }

hyetal equator |CLIMATOL| A line (or transition zone) which encircles the earth (north of the geographical equator) and lies between two belts that typify the annual time distribution of rainfall in the lower latitudes of each hemisphere; a form of meteorological equator. { 'hī·əd·əl i'kwād·ər }

hyetal region |CLIMATOL| A region in which the amount and seasonal variation of rainfall are of a given type. { 'hī·əd·əl ,rē·jən }

hyetograph |CLIMATOL| A map or chart displaying temporal or areal distribution of precipitation. { hī'ed·ə,graf }

hyetography |CLIMATOL| The study of the annual variation and geographic distribution of precipitation. { ,hī·ə'täg·rə·fē }

hyetology |METEOROL| The science which treats of the origin, structure, and various other features of all the forms of precipitation. { ,hī·ə'täl·ə·jē }

hygiene |MED| The science that deals with the principles and practices of good health. { 'hī,jēn }

hygrokinematics |METEOROL| The descriptive study of the motion of water substances in the atmosphere. { |hī·grə,kin·ə'mad·iks }

hygrology |METEOROL| The study which deals with the water vapor content (humidity) of the atmosphere. { hī'gräl·ə·jē }

hygrophyte See hydrophyte. { 'hī·grə,fīt }

hygroscopic |BOT| Being sensitive to moisture, such as certain tissues. { |hī·grə |skäp·ik }

hygroscopic coefficient |HYD| The percentage of water that a soil will absorb and hold in equilibrium in a saturated atmosphere. { |hī·grə|skäp·ik ,kō·i'fish·ənt }

hygroscopic water |HYD| The component of soil water that is held adsorbed on the surface of soil particles and is not available to vegetation. { |hī·grə|skäp·ik 'wȯd·ər }

hylaea See tropical rainforest. { hī·lē·ə }

hylotomous |ZOO| Cutting wood, as wood-boring insects. { hī|läd·ə·məs }

hymenolepiasis [MED] Intestinal infection by tapeworms of the genus *Hymenolepis*. { ¦hī·mə·nō·lə'pī·ə·səs }

hymexazol *See* hydroxisoxazole. { hī'mek·sə,zȯl }

hyperparasite [ECOL] An organism that is parasitic on other parasites. { ¦hī·pər'par·ə ,sīt }

hyperpycnal inflow [HYD] A denser inflow that occurs when a sediment-laden fluid flows down the side of a basin and along the bottom as a turbidity current. { ¦hī·pər ¦pik·nəl 'in,flō }

hyperthermophile [MICROBIO] An extremophile that thrives in high-temperature (above 60°C or 140°F) environments. { ,hī·pər'thər·mə,fīl }

hypha [MYCOL] One of the filaments composing the mycelium of a fungus. { 'hī·fə }

hypodermis [BOT] The outermost cell layer of the cortex of plants. Also known as exodermis. { ,hī·pə'dər·mis }

hypogeal *See* hypogeous. { ¦hī·pə¦jē·əl }

hypogeous [BIOL] Living or maturing below the surface of the ground. Also known as hypogeal. { ,hī·pə'jē·əs }

hypolimnion [HYD] The lower level of water in a stratified lake, characterized by a uniform temperature that is generally cooler than that of other strata in the lake. { ¦hī·pō'lim·nē,än }

hypoplankton [BIOL] Forms of marine life whose swimming ability lies somewhere between that of the plankton and the nekton; includes some mysids, amphipods, and cumacids. { ¦hī·pō¦plaŋk·tən }

hypopycnal inflow [HYD] Flowing water of lower density than the body of water into which it flows. { ,hī·pō'pik·nəl 'in,flō }

hyporheic zone [ECOL] The saturated sediment environment below a stream that exchanges water, nutrients, and fauna with surface flowing waters. { ,hī·pə'rē·ik ,zōn }

hyposensitization *See* desensitization. { ¦hī·pō,sen·səd·ə'zā·shən }

hypothermal [GEOL] Referring to the high-temperature (300–500°C) environment of hypothermal deposits. { ¦hī·pō'thər·məl }

hypothermal deposit [GEOL] Mineral deposit formed at great depths and high (300–500°C) temperatures. { ¦hī·pō'thər·məl di'päz·ət }

hypoxemia *See* hypoxia. { ,hī,päk'sē·mē·ə }

hypoxia [ECOL] A condition characterized by a low level of dissolved oxygen in an aquatic environment. { hī'päk·sē·ə }

hypsography [GEOGR] The science of measuring or describing elevations of the earth's surface with reference to a given datum, usually sea level. { hip'säg·rə·fē }

IAC   *See* international analysis code.

ice accretion   [HYD] The process by which a layer of ice builds up on solid objects which are exposed to freezing precipitation or to supercooled fog or cloud droplets.   { 'īs ə̩krē·shən }

ice apron   [HYD] **1.** The snow and ice attached to the walls of a cirque.   **2.** The ice that is flowing from an ice sheet over the edge of a plateau.   **3.** A piedmont glacier's lobe. **4.** Ice that adheres to a wall of a valley below a hanging glacier.   { 'īs ̩ā·prən }

ice band   [HYD] A layer of ice in firn or snow.   { 'īs ̩band }

ice barrier   [HYD] The periphery of the Antarctic ice sheet; or used generally for any ice dam.   { 'īs ̩bar·ē·ər }

ice bay   [OCEANOGR] A baylike recess in the edge of a large ice floe or ice shelf. Also known as ice bight.   { 'īs ̩bā }

ice belt   [OCEANOGR] A band of fragments of sea ice in otherwise open water. Also known as ice strip.   { 'īs ̩belt }

iceberg   [OCEANOGR] A large mass of glacial ice broken off and drifted from parent glaciers or ice shelves along polar seas; it is distinguished from polar pack ice, which is sea ice, and from frozen seawater, whose rafted or hummocked fragments may resemble small icebergs.   { 'īs̩bərg }

ice bight   *See* ice bay.   { 'īs ̩bīt }

ice blink   [METEOROL] A relatively bright, usually yellowish-white glare on the underside of a cloud layer, produced by light reflected from an ice-covered surface such as pack ice; used in polar regions with reference to the sky map; ice blink is not as bright as snow blink, but much brighter than water sky or land sky.   { 'īs̩bliŋk }

ice boundary   [HYD] At any given time, the boundary between fast ice and pack ice or between areas of different concentrations of pack ice.   { 'īs ̩baùn·drē }

ice bridge   [OCEANOGR] Surface river ice of sufficient thickness to impede or prevent navigation.   { 'īs ̩brij }

ice cake   [HYD] A single, usually relatively flat piece of ice of any size in a body of water. { 'īs ̩kāk }

ice canopy   *See* pack ice.   { 'īs ̩kam·ə·pē }

ice cap   [HYD] **1.** A perennial cover of ice and snow in the shape of a dome or plate on the summit area of a mountain through which the mountain peaks emerge.   **2.** A perennial cover of ice and snow on a flat land mass such as an Arctic island.   { 'īs ̩kap }

ice-cap climate   *See* perpetual frost climate.   { 'īs ̩kap ̩klīm·ət }

ice cascade   *See* icefall.   { 'īs ka'skād }

ice cave [HYD] A cave in ice such as a glacier formed by a stream of melted water. { 'īs ‚kāv }

ice crust [HYD] A type of snow crust; a layer of ice, thicker than a film crust, upon a snow surface, formed by the freezing of meltwater or rainwater which has flowed onto it. { 'īs ‚krəst }

ice-crystal cloud [METEOROL] A cloud consisting entirely of ice crystals, such as cirrus (in this sense distinguished from water clouds and mixed clouds), and having a diffuse and fibrous appearance quite different from that typical of water droplet clouds. { 'īs ‚krist·əl ‚klaůd }

ice-crystal fog See ice fog. { 'īs ‚krist·əl ‚fäg }

ice-crystal haze [METEOROL] A type of very light ice fog composed only of ice crystals and at times observable to altitudes as great as 20,000 feet (6100 meters), and usually associated with precipitation of ice crystals. { 'īs ‚krist·əl ‚hāz }

ice-crystal theory See Bergeron-Findeisen theory. { 'īs ‚krist·əl ‚thē·ə·rē }

ice day [CLIMATOL] A day on which the maximum air temperature in a thermometer shelter does not rise above 32°F (0°C), and ice on the surface of water does not thaw. { 'īs ‚dā }

ice desert [CLIMATOL] Any polar area permanently covered by ice and snow, with no vegetation other than occasional red snow or green snow. { 'īs ‚dez·ərt }

iced firn [HYD] A mixture of glacier ice and firn; firn permeated with meltwater and then refrozen. Also known as firn ice. { īst ¦fərn }

ice erosion [GEOL] **1.** Erosion due to freezing of water in rock fractures. **2.** See glacial erosion. { 'īs i'rō·zhən }

icefall [HYD] That portion of a glacier where a sudden steepening of descent causes a chaotic breaking up of the ice. Also known as ice cascade. { 'īs ‚föl }

ice fat See grease ice. { 'īs ¦fat }

ice feathers [HYD] A type of hoarfrost formed on the windward side of terrestrial objects and on aircraft flying from cold to warm air layers. Also known as frost feathers. { 'īs ‚feth·ərz }

ice field [HYD] A mass of land ice resting on a moutain region and covering all but the highest peaks. [OCEANOGR] A flat sheet of sea ice that is more than 5 miles (8 kilometers) across. { 'īs ‚fēld }

ice floe See floe. { 'īs ‚flō }

ice flowers [HYD] **1.** Formations of ice crystals on the surface of a quiet, slowly freezing body of water. **2.** Delicate tufts of hoarfrost that occasionally form in great abundance on an ice or snow surface. Also known as frost flowers. **3.** Frost crystals resembling a flower, formed on salt nuclei on the surface of sea ice as a result of rapid freezing of sea water. Also known as salt flowers. { 'īs ‚flaů·ərz }

ice fog [METEOROL] A type of fog composed of suspended particles of ice, partly ice crystals 20–100 micrometers in diameter but chiefly, especially when dense, droxtals 12–20 micrometers in diameter; occurs at very low temperatures and usually in clear, calm weather in high latitudes. Also known as frost flakes; frost fog; frozen fog; ice-crystal fog; pogonip; rime fog. { 'īs ‚fäg }

ice foot [OCEANOGR] Sea ice firmly frozen to a polar coast at the high-tide line and unaffected by tide; this fast ice is formed by the freezing of seawater during ebb tide, and of spray, and it is separated from the floating sea ice by a tide crack. { 'īs ¦fůt }

ice-free [HYD] **1.** Referring to a harbor, river, estuary, and so on, when there is not sufficient ice present to interfere with navigation. **2.** Descriptive of a water surface completely free of ice. { 'īs ¦frē }

ice fringe [HYD] An ice deposit on plant surfaces, not of hoarfrost from atmospheric water vapor, but of moisture exuded from the stems of plants and appearing as frosted fringes or ribbons. Also known as ice ribbon. [OCEANOGR] A belt of sea ice extending a short distance from the shore. { 'īs ˌfrinj }

ice front [HYD] The floating vertical cliff forming the seaward face or edge of an ice shelf or other glacier that enters water. { 'īs ˌfrənt }

ice gland [HYD] A column of ice in the granular snow at the top of a glacier. { 'īs ˌgland }

ice gruel [HYD] A type of slush formed by the irregular freezing together of ice crystals. { 'īs ˌgrül }

ice island [OCEANOGR] A large tabular fragment of shelf ice found in the Arctic Ocean and having an irregular surface, thickness of 15–50 meters (50–165 feet), and an area between a few thousand square meters and 500 square kilometers (200 square miles) or more. { 'īs ˌī·lənd }

ice-island iceberg [OCEANOGR] An iceberg having a conical or dome-shaped summit, often mistaken by mariners for ice-covered islands. { 'īs ˌī·lənd 'īsˌbərg }

ice jam [HYD] **1.** An accumulation of broken river ice caught in a narrow channel, frequently producing local floods during a spring breakup. **2.** Fields of lake or sea ice thawed loose from the shores in early spring, and blown against the shore, sometimes exerting great pressure. { 'īs ˌjam }

ice-laid drift See till. { 'īs ˌlād ˌdrift }

Icelandic low [METEOROL] **1.** The low-pressure center located near Iceland (mainly between Iceland and southern Greenland) on mean charts of sea-level pressure. **2.** On a synoptic chart, any low centered near Iceland. { 'īs'land·ik 'lō }

ice layer [HYD] An ice crust covered with new snow; when exposed at a glacier front or in crevasses, the ice layers viewed in cross section are termed ice bands. { 'īs ˌlā·ər }

ice mantle See ice sheet. { 'īs ˌmant·əl }

ice nucleus [METEOROL] Any particle which may act as a nucleus in formation of ice crystals in the atmosphere. { 'īs ˌnü·klē·əs }

ice pack See pack ice. { 'īs ˌpak }

ice pellets [METEOROL] A type of precipitation consisting of transparent or translucent pellets of ice 0.2 inch (5 millimeters) or less in diameter; may be spherical, irregular, or (rarely) conical in shape. { 'īs ˌpel·əts }

ice period [CLIMATOL] The interval between the first appearance and the final dissipation of ice during any year in a given locale. { 'īs ˌpir·ē·əd }

ice pillar [HYD] A column of glacial ice covered with stones or debris which tend to protect the ice from melting. { 'īs ˌpil·ər }

ice pole [GEOGR] The approximate center of the most consolidated portion of the arctic pack ice, near 83 or 84°N and 160°W. Also known as pole of inaccessibility. { 'īs ˌpōl }

icequake [HYD] The crash or concussion that accompanies the breakup of ice masses, frequently owing to contraction from the extreme cold. { 'īsˌkwāk }

ice ribbon See ice fringe. { 'īs ˌrib·ən }

ice rind [HYD] A thin but hard layer of sea ice, river ice, or lake ice, which is either a new encrustation upon old ice or a single layer of ice usually found in bays and fiords, where fresh water freezes on top of slightly colder sea water. { 'īs ˌrīnd }

ice run [HYD] The initial stage in the spring or summer breakup of river ice, being an exceedingly rapid process, seldom taking more than 1 day. { 'īs ˌrən }

ice sheet [HYD] A thick glacier, more than 19,300 square miles (50,000 square kilometers) in area, forming a cover of ice and snow that is continuous over a land surface and moving outward in all directions. Also known as ice mantle. { 'īs ,shēt }

ice shelf [OCEANOGR] A thick sheet of ice with a fairly level or undulating surface, formed along a polar coast and in shallow bays and inlets, fastened to the shore along one side but mostly afloat and nourished by annual accumulation of snow and by the seaward extension of land glaciers. { 'īs ,shelf }

ice storm [METEOROL] A storm characterized by a fall of freezing precipitation, forming a glaze on terrestrial objects that creates many hazards. Also known as silver storm. { 'īs ,storm }

ice stream [HYD] A current of ice flowing in an ice sheet or ice cap; usually moves toward an ocean or to an ice shelf. { 'īs ,strēm }

ice strip See ice belt. { 'īs ,strip }

ice tongue [HYD] Any narrow extension of a glacier or ice shelf, such as a projection floating in the sea or an outlet glacier of an ice cap. { 'īs ,təŋ }

ice wall [HYD] A cliff of ice forming the seaward margin of a glacier that is not afloat. { 'īs ,wȯl }

ice wedge See foliated ice. { 'īs ,wej }

ichthyology [ZOO] A branch of vertebrate zoology that deals with the study of fishes. { ,ik·thē'äl·ə·jē }

ichthyosarcotoxism [MED] Poisoning caused by eating the flesh of fish containing toxic substances. { ,ik·thē·ō,sär·kə'täk,siz·əm }

icicle [HYD] Ice shaped like a narrow cone, hanging point downward from a roof, fence, or other sheltered or heated source from which water flows and freezes in below-freezing air. { ī,sik·əl }

icing [HYD] **1.** Any deposit or coating of ice on an object, caused by the impingement and freezing of liquid (usually supercooled) hydrometeors. **2.** A mass or sheet of ice formed on the ground surface during the winter by successive freezing of sheets of water that may seep from the ground, from a river, or from a spring. Also known as flood icing; flooding icing. { 'ī·siŋ }

ICL See lifting condensation level.

icterohematuria [VET MED] A disease of sheep caused by the protozoan *Babesia ovis* and characterized by hemolysis of erythrocytes accompanied by jaundice. { ¦ik·tə·rō ,hē·mə'tur·ē·ə }

ID$_{50}$ See infective dose 50.

igneous [GEOL] Pertaining to rocks which have congealed from a molten mass. { 'ig·nē·əs }

illumination climate [METEOROL] **1.** The worldwide distribution of natural light from the sun and sky (direct solar radiation plus diffuse sky radiation) as received on a horizontal surface. **2.** The character of total illumination at any given place. Also known as light climate. { ə,lü·mə,nā·shən ,klī·mət }

illuvial [GEOL] Pertaining to a region or material characterized by the accumulation of soil by the illuviation of another zone or material. { i'lü·vē·əl }

illuvial horizon See B horizon. { i'lü·vē·əl hə'rīz·ən }

illuviation [GEOL] The deposition of colloids, soluble salts, and small mineral particles in an underlying layer of soil. { i,lü·vē'ā·shən }

illuvium [GEOL] Material leached by chemical or other processes from one soil horizon and deposited in another. { i'lü·vē·əm }

imaging radar [ENG] Radar carried on aircraft which forms images of the terrain. { 'im·i·jiŋ 'rā,där }

Imhoff cone [CIV ENG] A graduated glass vessel for measuring settled solids in testing the composition of sewage. { 'im,hóf ,kōn }

Imhoff tank [CIV ENG] A sewage treatment tank in which digestion and settlement take place in separate compartments, one below the other. { 'im,hóf ,taŋk }

immature soil See azonal soil. { |im·ə'chúr 'sóil }

immersion [SCI TECH] Placement into or within a fluid, usually water. { ə'mər·zhən }

immigrant [ECOL] An organism that settles in a zone where it was previously unknown. { 'im·ə·grənt }

immigration [ECOL] The one-way inward movement of individuals or their disseminules into a population or population area. { ,im·ə'grā·shən }

immune response [MED] The physiological responses stemming from activation of the immune system by antigens, consisting of a primary response in which the antigen is recognized as foreign and eliminated, and a secondary response to subsequent contact with the same antigen. { i'myün ri,späns }

immunity [MED] The condition of a living organism whereby it resists and overcomes an infection or a disease. { i'myü·nəd·ē }

immunological deficiency [MED] A state wherein the immune mechanisms are inadequate in their ability to perform their normal function, that is, the elimination of foreign materials (usually infectious agents such as bacteria, viruses, and fungi). { ,im·yə·nə,läj·ə·kəl di'fish·ən·sē }

immunotoxicity [MED] Adverse effects on the normal functioning of the immune system, caused by exposure to a toxic chemical. The result can be higher rates of infectious diseases or cancer, more severe cases of such autoimmune disease, or allergic reactions. { ,im·yə·nō·täk'sis·əd·ē }

immunotoxin [MED] Conjugate of antibody and toxic protein such that the specificity of the antibody molecule is combined with the cytotoxic property of the toxin. { ,im·yə·nō'täk·sən }

imperfect flower [BOT] A flower lacking either stamens or carpels. { im'pər·fikt 'flaü·ər }

impermeable [SCI TECH] Not permitting water or other fluid to pass through. Also known as impervious. { im'pər·mē·ə·bəl }

impervious See impermeable. { im'pər·vē·əs }

impetigo [MED] An acute, contagious, inflammatory skin disease caused by streptococcal or staphylococcal infections and characterized by vesicular or pustular lesions. { ,im·pə'tī,gō }

impregnated timber [FOR] Timber which has been made flame-resistant, fungi-resistant, or insect-proof by forcing into it under vacuum or pressure a flame retardant or a fungal or insect poison. { im'preg,nād·əd 'tim·bər }

impurity [SCI TECH] An undesirable foreign material in a pure substance. { im'pyúr·əd·ē }

inactive front [METEOROL] A front, or portion thereof, that produces very little cloudiness and no precipitation, as opposed to an active front. Also known as passive front. { in'ak·tiv 'frənt }

inbreeding [GEN] Reproduction behavior between closely related individuals; self-fertilization, as in some plants, is the most extreme form. { 'in,brēd·iŋ }

incarbonization See coalification. { in,kär·bə·nə'zā·shən }

Inceptisol |GEOL| A soil order characterized by soils that are usually moist, with pedogenic horizons of alteration of parent materials but not of illuviation. { in'sep·tə ,sȯl }

inclination |GEOL| The angle at which a geological body or surface deviates from the horizontal or vertical; often used synonymously with dip. { ,iŋ·klə'nā·shən }

incline |SCI TECH| An upward-or downward-sloping surface. { 'in,klīn }

incoalation *See* coalification. { ,in·kō'lā·shən }

incomplete flower |BOT| A flower lacking one or more modified leaves, such as petals, sepals, pistils, or stamens. { ,in·kəm'plēt 'flaü·ər }

increment *See* recharge. { 'iŋ·krə·mənt }

increment borer |FOR| An augerlike instrument with a hollow bit, used to extract thin radial cylinders of wood from trees to determine age and growth rate. { 'iŋ·krə·mənt ,bȯr·ər }

incubation period |MED| The period of time required for the development of symptoms of a disease after infection, or of altered reactivity after exposure to an allergen. { ,iŋ·kyə'bā·shən ,pir·ē·əd }

incubatory carrier |MED| A person infected with a certain microorganism but in such an early stage of disease that clinical manifestations are not apparent. { 'iŋ·kyə·bə ,tȯr·ē ¦kar·ē·ər }

incumbent |ECOL| Referring to the occupation and utilization of resources to the exclusion of other species. { in'kəm·bənt }

incus |METEOROL| A supplementary cloud feature peculiar to cumulonimbus capillatus; the spreading of the upper portion of cumulonimbus when this part takes the form of an anvil with a fibrous or smooth aspect. Also known as anvil; thunderhead. { 'iŋ·kəs }

indefinite ceiling |METEOROL| After United States weather observing practice, the ceiling classification applied when the reported ceiling value represents the vertical visibility upward into surface-based, atmospheric phenomena (except precipitation), such as fog, blowing snow, and all of the lithometeors. Formerly known as ragged ceiling. { in'def·ə·nət 'sēl·iŋ }

indehiscent |BOT| **1.** Remaining closed at maturity, as certain fruits. **2.** Not splitting along regular lines. { ¦in·də'his·ənt }

indeterminate growth |BOT| Growth of a plant in which the axis is not limited by development of a reproductive structure, and therefore growth continues indefinitely. { ,in·də'tərm·ə·nət 'grōth }

index cycle |METEOROL| A roughly cyclic variation in the zonal index. { 'in,deks ,sī·kəl }

index forest |FOR| A forest reaching the highest average in a given locality for density, volume, and increment. { 'in,deks ,fär·əst }

index of aridity |CLIMATOL| A measure of the precipitation effectiveness or aridity of a region, given by the following relationship: index of aridity $= P/(T + 10)$, where P is the annual precipitation in centimeters, and T the annual mean temperature in degrees Celsius. { 'in,deks əv ə'rid·əd·ē }

Indian Ocean |GEOGR| The smallest and geologically the most youthful of the three oceans, whose surface area is 29,300,000 square miles (75,900,000 square kilometers); it is bounded on the north by India, Pakistan, and Iran; on the east by the Malay Peninsula; on the south by Antarctica; and on the west by the Arabian peninsula and Africa. { 'in·dē·ən 'ō·shən }

Indian spring low water |OCEANOGR| An arbitrary tidal datum approximating the level of the mean of the lower low waters at spring time, first used in waters surrounding

India. Also known as harmonic tide plane; Indian tide plane. { 'in·dē·ən ¦spriŋ ¦lō 'wód·ər }

Indian summer [CLIMATOL] A period, in mid-or late autumn, of abnormally warm weather, generally clear skies, sunny but hazy days, and cool nights; the term is most often heard in the northeastern United States, but its usage extends throughout English-speaking countries. { 'in·dē·ən 'səm·ər }

Indian tide plane *See* Indian spring low water. { 'in·dē·ən 'tīd ˌplān }

indicated air temperature [METEOROL] The uncorrected reading from a free air temperature gage. Also known as outside air temperature. { 'in·dəˌkād·əd 'er ˌtem·prə·chər }

indicator plant [BOT] A plant used in geobotanical prospecting as an indicator of a certain geological phenomenon. { 'in·dəˌkād·ər ˌplant }

indicator species [ECOL] A species whose presence is indicative of a particular environmental condition or association of plants and animals. { 'in·dəˌkād·ər ˌspē·shēz }

indifferent equilibrium *See* neutral stability. { in'dif·ərnt ˌē·kwə'lib·rē·əm }

indifferent stability *See* neutral stability. { in'dif·ərnt stə'bil·əd·ē }

indigenous [SCI TECH] Existing and having originated naturally in a particular region or environment. { in'dij·ə·nəs }

Indo-Pacific faunal region [ECOL] A marine littoral faunal region extending eastward from the east coast of Africa, passing north of Australia and south of Japan, and ending in the east Pacific south of Alaska. { ¦in·dō·pə'sif·ik 'fón·əl ˌrē·jən }

induction inoculation [PL PATH] Repeated inoculation of plants to induce a maximum level of systemic resistance to disease. { in'dək·shən inˌäk·yəˌlā·shən }

induction silencer [ENG] A device for reducing engine induction noise, which consists essentially of a low-pass acoustic filter with the inertance of the air-entrance tube and the acoustic compliance of the annular and central volumes providing acoustic filtering elements. { in'dək·shən ¦sī·lən·sər }

industrial climatology [CLIMATOL] A type of applied climatology which studies the effect of climate and weather on industry's operations; the goal is to provide industry with a sound statistical basis for all administrative and operational decisions which involve a weather factor. { in'dəs·trē·əl ˌklī·mə'täl·ə·jē }

industrial ecology [ECOL] The development and use of industrial processes that result in products based on simultaneous consideration of product functionality and competitiveness, natural-resource conservation, and environmental preservation. Also known as design for environment; green design. { in¦dəs·trē·əl ē'käl·ə·jē }

industrial geography [GEOGR] A branch of geography that deals with location, raw materials, products, and distribution, as influenced by geography. { in'dəs·trē·əl jē'äg·rə·fē }

industrial hygiene [MED] The science that deals with the anticipation and control of unhealthy conditions in workplaces in order to prevent illness among employees. { in'dəs·trē·əl 'hī·jēn }

industrial meteorology [METEOROL] The application of meteorological information and techniques to industrial problems. { in'dəs·trē·əl ˌmē·dē·ə'räl·ə·jē }

industrial microbiology [MICROBIO] The study, utilization, and manipulation of those microorganisms capable of economically producing desirable substances or changes in substances, and the control of undesirable microorganisms. { in'dəs·trē·əl ˌmī·krō·bī'äl·ə·jē }

industrial yeast [MICROBIO] Any yeast used for the production of fermented foods and beverages, for baking, or for the production of vitamins, proteins, alcohol, glycerol, and enzymes. { in'dəs·trē·əl ¦yēst }

inertia currents [OCEANOGR] Currents resulting after the cessation of wind in a generating area or after the water movement has left the generating area; circular currents with a period of one-half pendulum day. { i'nər·shə ¸kə·rəns }

inertial theory [OCEANOGR] The theory associated with the motion of an ocean current under the influences of inertia and the Coriolis force, which cause it to take a circular path. { i'nər·shəl 'thē·ə·rē }

infant botulism [MED] Botulism that involves ingestion of *Clostridium botulinum* spores with subsequent germination and toxin production in the gastrointestinal tract, found mostly in children aged 6 months or younger. { ¸in·fənt 'bäch·ə¸liz·əm }

infantile diarrhea [MED] An acute gastrointestinal disease in infants resulting from damage of the intestinal mucosa by an infectious organism. { 'in·fən¸tīl ¸dī·ə'rē·ə }

infantile eczema [MED] An allergic inflammation of the skin in young children, usually due to common antigens such as food or inhalants. { 'in·fən¸tīl 'ek·sə·mə }

infantile paralysis *See* poliomyelitis. { 'in·fən¸tīl pə'ral·ə·səs }

infect [MED] To cause an infection, as by contamination with or invasion by a pathogen. { in'fekt }

infection [MED] **1.** Invasion of the body by a pathogenic organism, with or without disease manifestation. **2.** Pathologic condition resulting from invasion of a pathogen. { in'fek·shən }

infectious [MED] Caused by infection. { in'fek·shəs }

infectious abortion *See* contagious abortion. { in'fek·shəs ə'bòr·shən }

infectious chlorosis [PL PATH] A virus disease of plants characterized by yellowing of the green parts. { in'fek·shəs klə'rō·səs }

infectious disease [MED] Any disease caused by invasion by a pathogen which subsequently grows and multiplies in the body. { in'fek·shəs di'zēz }

infectious drug resistance [MICROBIO] A type of drug resistance that is transmissible from one bacterium to another by infectivelike agents referred to as resistance factors. { in'fek·shəs 'drəg ri¸zis·təns }

infectious endocarditis [MED] Inflammation of the endocardium due to an infectious microorganism. { in'fek·shəs ¸en·dō¸kär'dīd·əs }

infectious hepatitis [MED] Type A viral hepatitis, an acute infectious virus disease of the liver associated with hepatic inflammation and characterized by fever, liver enlargement, and jaundice. Also known as catarrhal jaundice; epidemic hepatitis; epidemic jaundice; virus hepatitis. { in'fek·shəs ¸hep·ə'tīd·əs }

infectious myocarditis [MED] Inflammation of the myocardium due to an infectious microorganism. { in'fek·shəs ¸mī·ō¸kär'dīd·əs }

infectious rhinitis [MED] Inflammation of the nasal mucous membrane due to an infectious microorganism. { in'fek·shəs rī'nīd·əs }

infective dose 50 [MICROBIO] The dose of microorganisms required to cause infection in 50% of the experimental animals; a special case of the median effective dose. Abbreviated ID$_{50}$. Also known as median infective dose. { in'fek·tiv ¦dōs 'fif·tē }

infiltration [HYD] Movement of water through the soil surface into the ground. { ¸in·fil'trā·shən }

infiltration capacity [HYD] The maximum rate at which water enters the soil or other porous material in a given condition. { ¸in·fil'trā·shən kə'pas·əd·ē }

infinite aquifer [HYD] The portion of a formation that contains water, and for which the exterior boundary is at an effectively infinite distance from the oil reservoir. { 'in·fə·nət 'ak·wə·fər }

inflorescence [BOT] **1.** A flower cluster. **2.** The arrangement of flowers on a plant. { ‚in·flə'res·əns }

influent [SCI TECH] An input stream of a fluid, as water into a reservoir, or liquid into a process vessel. { 'in‚flü·ənt }

influent stream [HYD] A stream that contributes water to the zone of saturation of groundwater and develops bank storage. Also known as losing stream. { 'in‚flü·ənt 'strēm }

influenza [MED] An acute virus disease of the respiratory system characterized by headache, muscle pain, fever, and prostration. { ‚in·flü'en·zə }

influenzal pneumonia [MED] Pneumonia resulting from infection by *Hemophilus influenzae*. { ‚in·flü'enz·əl nə'mōn·yə }

influenza vaccine [MED] A vaccine prepared from formaldehyde-attenuated mixtures of strains of influenza virus. { ‚in·flü'en·zə vak'sēn }

influenza virus [MICROBIO] Any of three immunological types, designated A, B, and C, belonging to the myxovirus group which cause influenza. { ‚in·flü'en·zə ‚vī·rəs }

influx *See* mouth. { 'in‚fləks }

infrared [PHYS] Pertaining to infrared radiation. { ¦in·frə¦red }

infrared radiation [PHYS] Electromagnetic radiation whose wavelengths lie in the range from 0.75 or 0.8 micrometer (the long-wavelength limit of visible red light) to 1000 micrometers (the shortest microwaves). { ¦in·frə¦red ‚rād·ē'ā·shən }

infructescence [BOT] An inflorescence's fruiting stage. { ‚in‚frək'tes·əns }

ingesta [BIOL] Food and other substances taken into an animal body. { in'jes·tə }

ingestion [BIOL] The act or process of taking food and other substances into the animal body. { in'jes·chən }

ingress [SCI TECH] The act of entering, as of air into the lungs or a liquid into an orifice. { 'in‚gres }

inhabited building distance [ENG] The minimum distance permitted between an ammunition or explosive location and any building used for habitation or where people are accustomed to assemble, except operating buildings or magazines. { in'hab·əd·əd ¦bil·diŋ ‚dis·təns }

inherited meander *See* entrenched meander. { in'her·əd·əd mē'an·dər }

initial detention *See* surface storage. { i'nish·əl di'ten·chən }

injection well *See* recharge well. { in'jek·shən ‚wel }

inland [GEOGR] Interior land, not bordered by the sea. { 'in·lənd }

inland ice [HYD] Ice composing the inner portion of a continental glacier or large ice sheet; applied particularly to Greenland ice. { 'in·lənd 'īs }

inland sea *See* epicontinental sea. { 'in·lənd 'sē }

inland water [GEOGR] **1.** A lake, river, or other body of water wholly within the boundaries of a state. **2.** An interior body of water not bordered by the sea. { 'in·lənd 'wȯd·ər }

inlet [GEOGR] **1.** A short, narrow waterway connecting a bay or lagoon with the sea. **2.** A recess or bay in the shore of a body of water. **3.** A waterway flowing into a larger body of water. { 'in‚let }

inner mantle   *See* lower mantle.   { ¦in·ər 'mant·əl }

inorganic |CHEM| Pertaining to or composed of chemical compounds that do not contain carbon as the principal element (excepting carbonates, cyanides, and cyanates), that is, matter other than plant or animal.   { ¦in·ȯr¦gan·ik }

insect |ZOO| **1.** A member of the Insecta.   **2.** An invertebrate that resembles an insect, such as a spider, mite, or centipede.   { 'in,sekt }

insect attractant |CHEM ENG| A chemical agent, usually associated with an insect's sexual drive, which may be used to attract pests to poisoned bait or for insect surveys.   { 'in,sekt ə,trak·tənt }

insect control |ECOL| Regulation of insect populations by biological or chemical means.   { 'in,sekt kən,trōl }

insecticide |AGR| A chemical agent that destroys insects.   { in'sek·tə,sīd }

insectistasis |ECOL| The use of pheromones to trap, confuse, or inhibit insects in order to hold populations below a level where they can cause significant economic damage.   { in¦sek·tə¦stā·səs }

insectivorous |BIOL| Feeding on a diet of insects.   { in,sek'tiv·ə·rəs }

insectivorous plant |BOT| A plant that captures and digests insects as a source of nutrients by using specialized leaves. Also known as carnivorous plant.   { in ,sek'tiv·ə·rəs 'plant }

insemination |BIOL| Internal fertilization.   { in,sem·ə'nā·shən }

insequent stream |HYD| A stream that has developed on the present surface, but not consequent upon it, and seemingly not controlled or adjusted by the rock structure and surface features.   { in'sē·kwənt 'strēm }

inshore |GEOGR| **1.** Located near the shore.   **2.** Indicating a shoreward position.   { 'in'shȯr }

inshore current |OCEANOGR| The horizontal movement of water inside the surf zone, including longshore and rip currents.   { 'in,shȯr 'kə·rənt }

inshore zone |GEOL| The zone of variable width extending from the shoreline at low tide through the breaker zone.   { 'in,shȯr 'zōn }

in situ |SCI TECH| In the original location.   { in 'si·chü }

instability line |METEOROL| Any nonfrontal line or band of convective activity in the atmosphere; this is the general term and includes the developing, mature, and dissipating stages; however, when the mature stage consists of a line of active thunderstorms, it is properly termed a squall line; therefore, in practice, instability line often refers only to the less active phases.   { ,in·stə'bil·əd·ē ,līn }

instrumented buoy |OCEANOGR| An uncrewed floating structure for the mounting, operation, data collection, and transmission of meteorological and oceanographic parameter-measuring systems.   { ¦in·strə¦men·təd 'bȯi }

intake   *See* recharge.   { 'in,tāk }

intake area   *See* recharge area.   { 'in,tāk ,er·ē·ə }

integrated drainage |HYD| Drainage resulting after folding and faulting of a surface under arid conditions; the streams by working headward have joined basins across intervening mountains or ridges.   { 'in·tə,grād·əd 'drān·ij }

interception ı |HYD| **1.** The process by which precipitation is caught and retained on vegetation or structures and subsequently evaporated without reaching the ground.   **2.** That part of the precipitation intercepted by vegetation. |METEOROL| **1.** The loss of sunshine, a part of which may be intercepted by hills, trees, or tall buildings.

2. The depletion of part of the solar spectrum by atmospheric gases and suspensoids; this commonly refers to the absorption of ultraviolet radiation by ozone and dust. { ‚in·tər'sep·shən }

interceptometer [ENG] A rain gage which is placed under trees or in foliage to determine the rainfall in that location; by comparing this catch with that from a rain gage set in the open, the amount of rainfall which has been intercepted by foliage is found. { ‚in·tər‚sep'täm·əd·ər }

intercontinental sea [GEOGR] A large body of salt water extending between two continents. { in·tər‚kant·ən'ent·əl 'sē }

intercropping [AGR] A form of multiple cropping in which two or more crops simultaneously occupy the same field. { ¦in·tər'kräp·iŋ }

interferon [BIOL] A protein produced by intact animal cells when infected with viruses; acts to inhibit viral reproduction and to induce resistance in host cells. { ‚in·tər'fir ‚än }

interflow [HYD] The water, derived from precipitation, that infiltrates the soil surface and then moves laterally through the upper layers of soil above the water table until it reaches a stream channel or returns to the surface at some point downslope from its point of infiltration. { 'in·tər‚flō }

interfluve [GEOL] The area of land between two rivers, usually an upland or ridge between two adjacent valleys that contain streams flowing in approximately the same direction. { 'in·tər‚flüv }

intermediate host [BIOL] The host in which a parasite multiplies asexually. { ‚in·tər'mēd·ē·ət 'hōst }

intermittent current [OCEANOGR] A unidirectional current interrupted at intervals. { ¦in·tər¦mit·ənt 'kə·rənt }

intermittent spring [HYD] A spring that ceases flow after a long dry spell but flows again after heavy rains. { ¦in·tər¦mit·ənt 'spriŋ }

intermittent stream [HYD] A stream which carries water a considerable portion of the time, but which ceases to flow occasionally or seasonally because water loss due to bed seepage and evapotranspiration exceed the available water supply. { ¦in·tər ¦mit·ənt 'strēm }

intermontane [GEOL] Located between or surrounded by mountains. { ¦in·tər¦män ‚tān }

internal drift current [OCEANOGR] Motion in an underlying layer of water caused by shearing stresses and friction created by current in a top layer that has different density. { in'tərn·əl 'drift ‚kə·rənt }

international analysis code [METEOROL] An internationally recognized code for communicating details of synoptic chart analyses. Abbreviated IAC. { ¦in·tər¦nash·ən·əl ə'nal·ə·səs ‚kōd }

International Ice Patrol [OCEANOGR] An organization established in 1914 to protect shipping by providing iceberg warnings. { ¦in·tər¦nash·ən·əl 'īs pə‚trōl }

international index numbers [METEOROL] A system of designating meteorological observing stations by number, established and administered by the World Meteorological Organization; under this scheme, specified areas of the world are divided into blocks, each bearing a two-number designator; stations within each block have an additional unique three-number designator, the numbers generally increasing from east to west and from south to north. { ¦in·tər¦nash·ən·əl 'in‚deks ‚nəm·bərz }

International Polar Year [METEOROL] The years 1882 and 1932, during which participating nations undertook increased observations of geophysical phenomena in polar (mostly arctic) regions; the observations were largely meteorological, but included such as auroral and magnetic studies. { ¦in·tər¦nash·ən·əl 'pō·lər ˌyir }

international synoptic code [METEOROL] A synoptic code approved by the World Meteorological Organization in which the observable meteorological elements are encoded and transmitted in words of five numerical digits length. { ¦in·tər¦nash·ən·əl sə¦näp·tik ˌkōd }

interspersion [ECOL] **1.** An intermingling of different organisms within a community. **2.** The level or degree of intermingling of one kind of organism with others in the community. { ˌin·tər'spər·zhən }

interstitial plasma-cell pneumonia See Pneumocystis carinii pneumonia. { ¦in·tər ¦stish·əl 'plaz·mə ˌsel nə'mō·nē·ə }

interstitial water [HYD] Subsurface water contained in pore spaces between the grains of rock and sediments. { ¦in·tər¦stish·əl 'wȯd·ər }

interstitial water saturation [HYD] The water content of a subterranean reservoir formation. { ¦in·ter¦stish·əl ¦wȯd·ər 'sach·ə'rā·shən }

intertidal zone [OCEANOGR] The part of the littoral zone above low-tide mark. { ¦in·tər'tīd·əl ˌzōn }

intertropical convergence zone [METEOROL] The axis, or a portion thereof, of the broad trade-wind current of the tropics; this axis is the dividing line between the southeast trades and the northeast trades (of the Southern and Northern hemispheres, respectively). Also known as equatorial convergence zone; meteorological equator. { ¦in·tər'träp·ə·kəl kən'vər·jəns ˌzōn }

intertropical front [METEOROL] The interface or transition zone occurring within the equatorial trough between the Northern and Southern hemispheres. Also known as equatorial front; tropical front. { ¦in·tər'träp·ə·kəl 'frənt }

interurban [GEOGR] Connecting or extending between urban areas. { ¦in·tər'ər·bən }

intraspecific [BIOL] Being within or occurring among the members of the same species. { ¦in·trə·spə·'sif·ik }

intrazonal soil [GEOL] A group of soils with well-developed characteristics that reflect the dominant influence of some local factor of relief, parent material, or age over the usual effect of vegetation and climate. { ˌin·trə'zōn·əl 'sȯil }

introgressive hybridization [GEN] The spreading of genes of a species into the gene complex of another due to hybridization between numerically dissimilar populations and the extensive backcrossing. { ¦in·trə¦gres·iv ˌhī·brəd·ə'zā·shən }

inundation [HYD] Flooding, by the rise and spread of water, of a land surface that is not normally submerged. { ˌi·nən'dā·shən }

inundative control [AGR] The mass production and periodic release of large numbers of biocontrol agents to achieve controlling densities. { ˌi·nən¦dād·iv kən'trōl }

invasion [MED] **1.** The phase of an infectious disease during which the pathogen multiplies and is distributed; precedes signs and symptoms. **2.** The process by which microorganisms enter the body. { in'vā·zhən }

inversion [METEOROL] A departure from the usual decrease or increase with altitude of the value of an atmospheric property, most commonly temperature. { in'vər·zhən }

inversion layer [METEOROL] The atmosphere layer through which an inversion occurs. { in'vər·zhən ˌlā·ər }

inverted tide See reversed tide. { in'vərd·əd 'tīd }

in vitro [BIOL] Pertaining to a biological reaction taking place in an artificial apparatus. { in 'vē·trō }

in vivo [BIOL] Pertaining to a biological reaction taking place in a living cell or organism. { in 'vē·vō }

involucre [BOT] Bracts forming one or more whorls at the base of an inflorescence or fruit in certain plants. { 'in·və‚lü·kər }

iodine [CHEM] A nonmetallic halogen element, symbol I, atomic number 53, atomic weight 126.9045; melts at 114°C, boils at 184°C; the poisonous, corrosive, dark plates or granules are readily sublimed; insoluble in water, soluble in common solvents; used as germicide and antiseptic, in dyes, tinctures, and pharmaceuticals, in engraving lithography, and as a catalyst and analytical reagent. { 'ī·ə‚dīn }

iodine-131 [PHYS] A radioactive, artificial isotope of iodine, mass number 131; its half-life is 8 days with beta and gamma radiation; used in medical and industrial radioactive tracer work; moderately radiotoxic. { 'ī·ə‚dīn ‚wən¦thərd·ē'wən }

iodoethylene See tetraiodoethylene. { ī‚ō·dō'eth·ə‚lēn }

ionic ratio [OCEANOGR] The ratio by weight of a major constituent of seawater to the chloride ion content; for example, $SO_4/Cl = 0.1396$, $Ca/Cl = 0.02150$, $Mg/Cl = 0.06694$. { ī'än·ik 'rā·shō }

ionization radiation See ionizing radiation. { ‚ī·ə·nə'zā·shən ‚rād·ē'ā·shən }

ionizing radiation [PHYS] **1.** Particles or photons that have sufficient energy to produce ionization directly in their passage through a substance. Also known as ionization radiation. **2.** Particles that are capable of nuclear interactions in which sufficient energy is released to produce ionization. { 'ī·ə‚nīz·iŋ ‚rād·ē'ā·shən }

ioxynil octanoate [CHEM] $C_{15}H_{17}I_2NO_2$ A waxy solid with a melting point of 59–60°C; insoluble in water; used as an insecticide for cereals and sugarcane. { ī'äk·sə‚nil ‚äk·tə 'nō·ət }

IPC See propham.

iridescent cloud [METEOROL] An ice-crystal cloud which exhibits brilliant spots or borders of colors, usually red and green, observed up to about 30° from the sun. { ‚ir·ə'des·ənt 'klaůd }

irisation [METEOROL] The coloration exhibited by iridescent clouds and at times along the borders of lenticular clouds. { ī·rə'sā·shən }

Irish moss See carrageen. { 'ī‚rish 'mȯs }

Irish Sea [GEOGR] A marginal sea of the Atlantic Ocean between Ireland and England, approximately 53°N latitude and 5°W longitude. { 'ī·rish 'sē }

Irminger Current [OCEANOGR] An ocean current that is one of the terminal branches of the Gulf Stream system, flowing west off the southern coast of Iceland. { 'ər·miŋ·ər 'kə·rənt }

iron arsenate See ferrous arsenate. { 'ī·ərn 'ärs·ən‚āt }

iron bacteria [MICROBIO] The common name for bacteria capable of oxidizing ferrous iron to the ferric state. { 'ī·ərn bak'tir·ē·ə }

iron chloride See ferric chloride; ferrous chloride. { 'ī·ərn 'klȯr‚īd }

iron dichloride See ferrous chloride. { 'ī·ərn dī'klȯr‚īd }

iron hydroxide See ferric hydroxide. { 'ī·ərn hī'dräk‚sīd }

iron sulfate See ferric sulfate; ferrous sulfate. { 'ī·ərn 'səl‚fāt }

iron winds [METEOROL] Northeasterly winds of Central America, prevalent during February and March, and blowing steadily for several days at a time. { 'ī·ərn 'winz }

irradiation [ENG] The exposure of a material, object, or patient to x-rays, gamma rays, ultraviolet rays, or other ionizing radiation. { i‚rād·ē'ā·shən }

irregular [BOT] Lacking symmetry, as of a flower having petals unlike in size or shape. { i'reg·yə·lər }

irregular crystal [METEOROL] A snow particle, sometimes covered by a coating of rime, composed of small crystals randomly grown together; generally, component crystals are so small that the crystalline form of the particle can be seen only through a magnifying glass or microscope. { i'reg·yə·lər 'krist·əl }

irregular iceberg See pinnacled iceberg. { i'reg·yə·lər 'īs‚bərg }

irrespirable atmosphere [PETR MIN] Atmosphere in a coal mine requiring workers to wear breathing apparatus because of poisonous gas or insufficient oxygen as a result of an explosion from firedamp or coal dust, or mine fires. { i'res·pə·rə·bəl 'at·mə‚sfir }

isanthous [BOT] Having regular flowers. { ī'san·thəs }

isarithm See isopleth. { 'ī·sə‚rith·əm }

isentropic chart [METEOROL] A constant-entropy chart; a synoptic chart presenting the distribution of meteorological elements in the atmosphere on a surface of constant potential temperature (equivalent to an isentropic surface); it usually contains the plotted data and analysis of such elements as pressure (or height), wind, temperature, and moisture at that surface. { ‚īs·ən‚träp·ik 'chärt }

isentropic condensation level See lifting condensation level. { ‚īs·ən‚träp·ik ‚känd·ən'sā·shən ‚lev·əl }

isentropic mixing [METEOROL] Any atmospheric mixing process which occurs within an isentropic surface. { ‚īs·ən'träp·ik 'mik·siŋ }

isentropic surface [METEOROL] A surface in space in which potential temperature is everywhere equal. { ‚īs·ən'träp·ik 'sər·fəs }

island [GEOGR] A tract of land smaller than a continent and surrounded by water; normally in an ocean, sea, lake, or stream. { 'ī·lənd }

island arc [GEOGR] A group of islands usually with a curving archlike pattern, generally convex toward the open ocean, having a deep trench or trough on the convex side and usually enclosing a deep basin on the concave side. { 'ī·lənd ‚ärk }

isoactyl thioglycolate [CHEM] HSCH₂COOCH₂ C₇H₁₅ A colorless liquid with a slight fruity odor and a boiling point of 125°C; used in antioxidants, insecticides, oil additives, and plasticizers. { ‚ī·sō'akt·əl ‚thī·ə'glī·kə‚lāt }

isobar [METEOROL] A line drawn through all points of equal atmospheric pressure along a given reference surface, such as a constant-height surface (notably mean sea level on surface charts), an isentropic surface, or the vertical plan of a synoptic cross section. { 'ī·sə‚bär }

isobaric chart See constant-pressure chart. { ‚ī·sə‚bär·ik 'chärt }

isobaric contour chart See constant-pressure chart. { ‚ī·sə‚bär·ik 'kän·túr ‚chärt }

isobaric equivalent temperature See equivalent temperature. { ‚ī·sə‚bär·ik i‚kwiv·ə·lənt 'tem·prə·chər }

isobaric map [METEOROL] A map depicting points in the atmosphere of equal barometric pressure. { ‚ī·sə‚bär·ik 'map }

isobaric surface [METEOROL] A surface on which the pressure is uniform. Also known as constant-pressure surface. { ‚ī·sə‚bär·ik 'sər·fəs }

isobaric topography See height pattern. { ‚ī·sə‚bär·ik tə'päg·rə·fē }

isobath [OCEANOGR] A contour line connecting points of equal water depths on a chart. Also known as depth contour; depth curve; fathom curve. { 'ī·sə‚bath }

isobathytherm [OCEANOGR] A line or surface showing the depth in oceans or lakes at which points have the same temperatures. { ‚ī·sə'bath·ə‚thərm }

isobront [METEOROL] A line drawn through geographical points at which a given phase of thunderstorm activity occurred simultaneously. Also known as homobront. { 'ī·sə ‚bränt }

isobutyric acid [CHEM] $(CH_3)_2CHCOOH$ Colorless liquid boiling at 154°C; soluble in water, alcohol, and ether; used as a chemical intermediate and disinfectant, in flavor and perfume bases, and for leather treating. { ¦ī·sō·byü'tir·ik 'as·əd }

isoceraunic [METEOROL] Indicating or having equal frequency or intensity of thunderstorm activity. Also spelled isokeraunic. { ¦ī·sō·sə'rȯn·ik }

isoceraunic line [METEOROL] A line drawn through geographical points at which some phenomenon connected with thunderstorms has the same frequency or intensity; used for lines of equal frequency of lightning discharges. { ¦ī·sō·sə'rȯn·ik 'līn }

isodecyl chloride [CHEM] $C_{10}H_{21}Cl$ A colorless liquid with a boiling point of 210.6°C; used as a solvent and in extractants, cleaning compounds, pharmaceuticals, insecticides, and plasticizers. { ‚ī·sə'des·əl 'klȯr‚īd }

isofronts-preiso code [METEOROL] A code in which data on isobars and fronts at sea level (or earth's surface) are encoded and transmitted; a modified form of the international analysis code. { ¦ī·sə‚frəns ¦prē¦ī·sō kōd }

isogenic [GEN] Having the same genotype, as all organisms of an inbred strain. { ‚ī·sə·jə'nē·ik }

isogradient [METEOROL] A line connecting points having the same horizontal gradient of atmospheric pressure, temperature, and so on. { ¦ī·sə'grād·ē·ənt }

isogram *See* isopleth. { 'ī·sə‚gram }

isohaline [OCEANOGR] **1.** Of equal or constant salinity. **2.** A line on a chart connecting all points of equal salinity. { ‚ī·sō'hā‚lēn }

isohel [METEOROL] A line drawn through geographical points having the same duration of sunshine (or other function of solar radiation) during any specified time period. { 'ī·sō‚hel }

isohume [METEOROL] A line drawn through points of equal humidity on a given surface; an isopleth of humidity; the humidity measures used may be the relative humidity or the actual moisture content (specific humidity or mixing ratio). { 'ī·sə‚hyüm }

isohyet [METEOROL] A line drawn through geographic points recording equal amounts of precipitation for a specified period or for a particular storm. { ¦ī·sə¦hī·ət }

isohypsic chart *See* constant-height chart. { ‚ī·sə'hip·sik 'chärt }

isokeraunic *See* isoceraunic. { ¦ī·sō·kə'rȯn·ik }

isokinetic [BIOL] Pertaining to the force of a human muscle that is applied during constant velocity of motion. { ‚i·sə·ki'ned·ik }

isolate [GEN] A population so cut off from others that mating occurs only within the group. { 'ī·sə‚lāt }

isolating mechanism [GEN] A geographic barrier or biological difference that prevents mating or genetic exchange between individuals of different populations or species. { 'ī·sə‚lād·iŋ mek·ə‚niz·əm }

isolation [GEN] The restriction or limitation of gene flow between distinct populations due to barriers to interbreeding. [MED] Separation of an individual with a communicable disease from other, healthy individuals. [MICROBIO] Separation of a pure chemical substance from a compound or mixture; as in distillation, precipitation, or absorption. { ‚ī·sə'lā·shən }

isolator [ENG] Any device that absorbs vibration or noise, or prevents its transmission. { 'ī·sə‚lād·ər }

isomorphism [SCI TECH] The quality or state of being identical or similar in form, shape, or structure, such as between organisms resulting from evolutionary convergence, or crystalline forms of similar composition. { ‚ī·sə‚mȯr‚fiz·əm }

isoneph [METEOROL] A line drawn through all points on a map having the same amount of cloudiness. { 'ī·sə‚nef }

isopectic [CLIMATOL] A line on a map connecting points at which ice begins to form at the same time of winter. { ‚i·sə'pek·tik }

isopleth [METEOROL] **1.** A line of equal or constant value of a given quantity with respect to either space or time. Also known as isogram. **2.** More specifically, a line drawn through points on a graph at which a given quantity has the same numerical value (or occurs with the same frequency) as a function of the two coordinate variables. Also known as isarithm. { 'ī·sə‚pleth }

isopluvial [METEOROL] A line on a map drawn through geographical points having the same amount of precipitation. { ‚ī·sō'plü·vē·əl }

isopotential level *See* potentiometric surface. { ‚ī·sō·pə'ten·chəl 'lev·əl }

isopropaline [CHEM] $C_{15}H_{23}N_3O_4$ An orange liquid with limited solubility in water; used as a preemergence herbicide for control of grass and broadleaf weeds on tobacco. { ‚ī·sə'prō·pə‚lēn }

isopropylamine [CHEM] $(CH_3)_2CHNH_2$ A volatile, colorless liquid with a boiling point of 32.4°C; used as a solvent and in the manufacture of pharmaceuticals, dyes, insecticides, and bactericides. Also known as 2-aminopropane. { ‚ī·sə·prō'pil·ə‚mēn }

isopropyl 4,4'-dibromobenzilate [CHEM] $C_{17}H_{16}O_3Br_2$ A brownish solid with a melting point of 77°C; solubility in water is less than 0.5 part per million at 20°C; used as a miticide for deciduous fruit and citrus. { ‚ī·sə‚prō·pəl ‚fȯr ‚fȯr‚prīm dī‚brō·mō'ben·zə ‚lāt }

isopropyl 4,4'-dichlorobenzilate [CHEM] $C_{17}H_{16}O_3Cl_2$ A white powder with a melting point of 70–72°C; solubility in water is less than 10 parts per million at 20°C; used as a miticide for spider mites on apple and pear trees. { ‚ī·sə‚prō·pəl ‚fȯr ‚fȯr‚prīm dī ‚klȯr·ō'ben·zə‚lāt }

2-isopropoxyphenyl *N*-methylcarbamate [CHEM] $C_{11}H_{15}O_3N$ A colorless solid with a melting point of 91°C; used as an insecticide for cockroaches, flies, mosquitoes, and lawn insects. { ‚tü ‚ī·sō·prə‚päk·sē'fen·əl ‚en ‚meth·əl'kär·bə‚māt }

***N*-4-isopropylphenyl-*N*',*N*'-dimethylurea** [CHEM] $(CH_3)_2CHC_6H_4NHCON(CH_3)_2$ A crystalline solid with a melting point of 151–153°C; solubility in water is 170 parts per million; used as an herbicide for wheat, barley, and rye. { ‚en ‚fȯr ‚ī·sə‚prō·pəl ‚fen·əl ‚en‚prīm ‚en‚prīm dī‚meth·əl·yü'rē·ə }

***ortho*-isopropylphenyl-methylcarbamate** [CHEM] $C_{11}H_{15}O_2N$ A white, crystalline compound with a melting point of 88–89°C; used as an insecticide for rice and cacao crops. Also known as MIPC. { ‚ȯr·thō‚ī·sə‚prō·pəl‚fen·əl ‚meth·əl'kär·bə‚māt }

isopycnic [METEOROL] A line on a chart connecting all points of equal or constant density. { ‚ī·sō‚pik·nik }

isopycnic level [METEOROL] Specifically, a level surface in the atmosphere, at about 5 miles (8 kilometers) altitude, where the air density is approximately constant in space and time. { ‚ī·sō‚pik·nik ‚lev·əl }

isoquinoline [CHEM] $C_6H_4CHNCHCH$ Colorless liquid boiling at 243°C; soluble in most organic solvents and dilute mineral acids, insoluble in water; derived from coal tar or made synthetically; used to make dyes, insecticides, pharmaceuticals, and rubber accelerators, and as a chemical intermediate. { ‚ī·sə'kwin·ə‚lēn }

isotach [METEOROL] A line in a given surface connecting points with equal wind speed. Also known as isokinetic; isovel. { 'ī·sə,tak }

isotach chart [METEOROL] A synoptic chart showing the distribution of wind by means of isotachs. { 'ī·sə,tak ,chärt }

isothere [CLIMATOL] A line on a map connecting points having the same mean summer temperature. { 'ī·sə,thir }

isotherm [GEOPHYS] A line on a chart connecting all points of equal or constant temperature. { 'ī·sə,thərm }

isothermal atmosphere [METEOROL] An atmosphere in hydrostatic equilibrium, in which the temperature is constant with height and the pressure decreases exponentially upward. Also known as exponential atmosphere. { ¦ī·sə¦thər·məl 'at·mə ,sfir }

isothermal equilibrium [METEOROL] The state of an atmosphere at rest, uninfluenced by any external agency, in which the conduction of heat from one part to another has produced, after a sufficient length of time, a uniform temperature throughout its entire mass. Also known as conductive equilibrium. { ¦ī·sə¦thər·məl ,ē·kwə'lib·rē·əm }

isothermal layer [METEOROL] The approximately isothermal region of the atmosphere immediately above the tropopause. { ¦ī·sə¦thər·məl 'lā·ər }

isothermobath [OCEANOGR] A line connecting points having the same temperature in a diagram of a vertical section of the ocean. { ,ī·sə'thər·mə,bath }

isotope farm [BOT] A carbon-14 (^{14}C) growth chamber, or greenhouse, arranged as a closed system in which plants can be grown in an atmosphere of carbon dioxide (CO_2) containing ^{14}C and thus become labeled with ^{14}C; isotope farms also can be used with other materials, such as heavy water (D_2O), phosphorus-35 (^{35}P), and so forth, to produce biochemically labeled compounds. { 'ī·sə,tōp ,färm }

isotopic age determination *See* radiometric dating. { ¦ī·sə¦täp·ik 'aj di,tər·mə,nā·shən }

isovel *See* isotach. { 'ī·sə,vel }

isthmus [BIOL] A passage or constricted part connecting two parts of an organ. [GEOGR] A narrow strip of land having water on both sides and connecting two large land masses. { 'is·məs }

Itonididae [ZOO] The gall midges, a family of orthorrhaphous dipteran insects in the series Nematocera; most are plant pests. { ,id·ə'nid·ə,dē }

J

Jacobshavn Glacier [HYD] A glacier on the west coast of Greenland at latitude 68°N; it is the most productive glacier in the Northern Hemisphere, calving about 1400 icebergs yearly. { 'yä·kəps,häf·ən 'glā·shər }

Japanese encephalitis [MED] A human viral infection epidemic in Japan, transmitted by the common house mosquito (*Culex pipiens*) and characterized by severe inflammation of the brain. { ¦jap·ə¦nēz in,sef·ə'līd·əs }

Java citronella oil *See* citronella oil. { 'jäv·ə ,si·trə'nel·ə 'öil }

Jennerian vaccine *See* smallpox vaccine. { jə'nir·ē·ən vak'sēn }

jet-effect wind [METEOROL] A wind which is increased in speed through the channeling of air by some mountainous configuration, such as a narrow mountain pass or canyon. { ¦jet i¦fekt ,wind }

jet stream [METEOROL] A relatively narrow, fast-moving wind current flanked by more slowly moving currents; observed principally in the zone of prevailing westerlies above the lower troposphere, and in most cases reaching maximum intensity with regard to speed and concentration near the troposphere. { 'jet ,strēm }

Jevons effect [METEOROL] The effect upon the measurement of rainfall caused by the presence of the rain gage; in 1861 W.S. Jevons pointed out that the rain gage causes a disturbance in airflow past it, and this carries part of the rain past the gage which would normally be captured. { 'jev·ənz i,fekt }

jodfenphos [CHEM] $C_8H_8O_3Cl_2IPS$ A crystalline compound with a melting point of 76°C; slight solubility in water; used as an insecticide in homes, farm buildings, and industrial sites. { 'yöd·fən,fäs }

jordanon *See* microspecies. { 'jörd·ən,än }

juglone [CHEM] $C_{10}H_6O_3$ A naphthoquinone derivative that occurs naturally in black walnuts and is toxic to plants. { 'jəg,lōn }

jump fire [FOR] A fire carried ahead of a forest fire by wind-borne burning material. { 'jəmp ,fīr }

jungle [ECOL] An impenetrable thicket of second-growth vegetation replacing tropical rain forest that has been disturbed; lower growth layers are dense. { 'jəŋ·gəl }

juvenile water *See* magmatic water. { 'jü·vən·əl 'wöd·ər }

K

k *See* kilo-.

kainite |GEOL| $MgSO_4 \cdot KCl \cdot 3H_2O$ A white, gray, pink, or black monoclinic mineral, occurring in irregular granular masses; used as a fertilizer and as a source of potassium and magnesium compounds. { 'kī,nīt }

Kansasii disease |MED| A mycobacterial tuberculosislike infection caused by *Mycobacterium kansasii*, an orange-yellow acid-fast bacterium. { kan'zas·ē,ī di,zēz }

karbutilate |CHEM| $C_{14}H_{21}N_3O_3$ An off-white solid with a melting point of 176–177°C; used as a herbicide on noncroplands, railroad rights-of-way, and plant sites. { kär'byüd·əl,āt }

karoo *See* karroo. { kə'rü }

karroo |GEOGR| A dry, broad, level, elevated area found especially in southern Africa, often rising to considerable elevations in terrace formations; does not support vegetation in the dry season but supports grass during the wet season.Also spelled karoo. { kə'rü }

karst |GEOL| A topography formed over limestone, dolomite, or gypsum and characterized by sinkholes, caves, and underground drainage. { kärst }

katabaric *See* katallobaric. { ¦kad·ə¦bar·ik }

katabatic wind *See* gravity wind. { ¦kad·ə¦bad·ik 'wind }

katafront |METEOROL| A front (usually a cold front) at which warm air descends the frontal surface (except, presumably, in the lowest layers). { 'kad·ə,frənt }

katallobaric |METEOROL| Of or pertaining to a decrease in atmospheric pressure. Also known as katabaric. { kə,tal·ə'bar·ik }

kay *See* key. { kā }

Kegel karst *See* cone karst. { 'kā·gəl ,kärst }

kelp |BOT| The common name for brown seaweed belonging to the Laminariales and Fucales. { kelp }

Kelvin wave |OCEANOGR| **1.** An eastward-propagating internal gravity wave that crosses the Pacific Ocean along the equator and has no north-south velocity component. **2.** A type of wave progression in relatively confined water bodies where, because of Coriolis force, the wave is higher to the right of direction of advance (in the Northern Hemisphere). { 'kel·vən ,wāv }

Kerguelen faunal region |ECOL| A marine littoral faunal region comprising a large area surrounding Kerguelen Island in the southern Indian Ocean. { 'kər·gə·lən 'fon·əl ,rē·jən }

Kern counter *See* dust counter. { 'kərn ¦kaun·tər }

kernel |BOT| **1.** The inner portion of a seed. **2.** A whole grain or seed of a cereal plant, such as corn or barley. { 'kərn·əl }

kernel blight [PL PATH] Any of several fungus diseases of barley caused chiefly by *Gibberella zeae*, *Helminthosporium sativum*, and *Alternaria* species shriveling and discoloring the grain. { 'kərn·əl ‚blīt }

kerosene *See* kerosine. { 'ker·ə‚sēn }

kerosine [PETR MIN] Also spelled kerosene. A refined petroleum fraction used as a fuel for heating and cooking, jet engines, lamps, and weed burning and as a base for insecticides; specific gravity is about 0.8; components are mostly paraffinic and naphthenic hydrocarbons in the C_{10} to C_{14} range. Also known as lamp oil. { 'ker·ə ‚sēn }

key [GEOL] A cay, especially one of the islets off the south of Florida. Also spelled kay. [SYST] An arrangement of the distinguishing features of a taxonomic group to serve as a guide for establishing relationships and names of unidentified members of the group. { kē }

kilo- [SCI TECH] A prefix representing 10^3 or 1000. Abbreviated k. { 'ki·lō,'kē·lō }

kilomega- *See* giga-. { 'kil·ə'meg·ə }

Kimberley reefs [GEOL] Gold-bearing reefs in southern Africa that lie above the Main reef and Bird reef groups. Also known as battery reefs. { 'kim·bər·lē ‚rēfs }

kinetic [SCI TECH] Pertaining to or producing motion. { kə'ned·ik }

kingdom [SYST] One of the primary divisions that include all living organisms: most authorities recognize two, the animal kingdom and the plant kingdom, while others recognize three or more, such as Protista, Plantae, Animalia, and Mycota. { 'kiŋ·dəm }

kite observation [METEOROL] An atmospheric sounding by means of instruments carried aloft by a kite. { 'kīt ‚äb·sər'vā·shən }

Klebsiella pneumoniae [MICROBIO] An encapsulated pathogenic bacterium that causes severe pneumonitis in humans. Formerly known as Friedlander's bacillus; pneumobacillus. { ‚kleb·zē'el·ə nə'mō·nē‚ī }

Klebsiella rhinoscleromatis [MICROBIO] A gram-negative, nonmotile, pathogenic species of bacteria that causes the upper respiratory disease rhinoscleroma. { ‚kleb·zē‚el·ə ‚rī·nō‚skler·ə'mäd·əs }

Klebs-Loeffler bacillus *See* Corynebacterium diphtheriae. { 'kläps 'lef·lər bə‚sil·əs }

klendusity [BOT] The tendency of a plant to resist disease due to a protective covering, such as a thick cuticle, that prevents inoculation. { klen'dü·səd·ē }

klinotaxis [BIOL] Positive orientation movement of a motile organism induced by a stimulus. { ¦klī·nə'tak·səs }

klint [GEOL] An exhumed coral reef or bioherm that is more resistant to the processes of erosion than the rocks that enclose it so that the core remains in relief as hills and ridges. { klint }

knife harrow [AGR] A type of harrow that consists of a frame holding a number of knives which scrape and partly invert the soil surface to smooth it and destroy small weeds. { 'nīf ‚har·ō }

knoll [GEOL] A mound rising less than 3300 feet (1000 meters) from the sea floor. Also known as sea knoll. { ‚nōl }

Knudsen's tables [OCEANOGR] Hydrographical tables published by Martin Knudsen in 1901 to facilitate the computation of results of seawater chlorinity titrations and hydrometer temperature readings, and their conversion to salinity and density. { kə'nüd·sən ‚tā·bəlz }

Kojic acid [CHEM] $C_6H_6O_4$ A crystalline antibiotic with a melting point of 152–154°C; soluble in water, acetone, and alcohol; used in insecticides and as an antifungal and antimicrobial agent. { 'kō·jik ‚as·əd }

kona [METEOROL] A stormy, rain-bringing wind from the southwest or south-southwest in Hawaii; it blows about five times a year on the southwest slopes, which are in the lee of the prevailing northeast trade winds. { 'kō·nə }

kona cyclone [METEOROL] A slow-moving extensive cyclone which forms in subtropical latitudes during the winter season. Also known as kona storm. { 'kō·nə 'sī,klōn }

kona storm See kona cyclone. { 'kō·nə 'stōrm }

Köppen climate classification system [CLIMATOL] The most widely used method for classifying the world's climates. the system has five major climate categories based on annual and monthly average temperature and precipitation: A tropical rainy; B, dry; C, mild midlatitude; D, severe midlatitude; and E, polar. { ku·pən 'klīm·ət ,klas·ə·fə'kā·shən ,sis·təm }

Köppen-Supan line [METEOROL] The isotherm connecting places which have a mean temperature of 10°C (50°F) for the warmest month of the year. { 'kep·ən sü'pän ,līn }

Krakatao winds [METEOROL] A layer of easterly winds over the tropics at an altitude of about 11 to 14.5 miles (18 to 24 kilometers), which tops the mid-tropospheric westerlies (the antitrades), is at least 3.5 miles (6 kilometers) deep, and is based at about 1.2 miles (2 kilometers) above the tropopause. { 'krak·ə,taü ,winz }

kremastic water See vadose water. { krə'mas·tik ¦wòd·ər }

krill [ZOO] A name applied to planktonic crustaceans that constitute the diet of many whales. { kril }

krummholz [ECOL] Stunted alpine forest vegetation. Also known as elfinwood. { 'krúm ,hōlts }

kryptoclimate See cryptoclimate. { ¦krip·tō'klī·mət }

kryptoclimatology See cryptoclimatology. { ¦krip·tō,kli·mə'täl·ə·jē }

K selection [ECOL] Selection favoring species that reproduce slowly where a resource is constant but available in limited quantities; population is maintained at or near the carrying capacity (K) of the habitat. { 'kāsi,lek·shən }

kudzu [BOT] Any of various perennial vine legumes of the genus *Pueraria* in the order Rosales cultivated principally as a forage crop. { 'kúd,zü }

Kuroshio [OCEANOGR] A fast ocean current originating off the southeast coast of Luzon, Philippines, and flowing northeastward off the coasts of China and Japan into the upper waters of the north Pacific Ocean. It carries large quantities of warm water from the tropics into the midlatitude regions, and is an important agent in redistributing global heat. { ,kù·rə'shē·ō }

Kuroshio Countercurrent [OCEANOGR] A component of the Kuroshio system flowing south and southwest between latitudes 155° and 160°E about 44 miles (70 kilometers) from the coast of Japan on the right-hand side of the Kuroshio Current. { ,kù·rə'shē·ō 'kaúnt·ər,kə·rənt }

Kuroshio extension [OCEANOGR] A general term for the warm, eastward-transitional flow that connects the Kuroshio and the North Pacific currents. { ,kù·rə'shē·ō ik'sten·shən }

Kuroshio system [OCEANOGR] A system of ocean currents which includes part of the North Equatorial Current, the Tsushima Current, the Kuroshio Current, and the Kuroshio extension. { ,kù·rə'shē·ō,sis·təm }

Kyasanur Forest virus [MICROBIO] A group B arbovirus recognized as an agent that causes hemorrhagic fever. { kī'az·ə·núr ¦fär·əst 'vī·rəs }

kyrohydratic point [OCEANOGR] The temperature at which a particular salt crystallizes in brine which is trapped by frozen seawater. { ¦kī·rō·hī'drad·ik ,póint }

L

Labarraque's solution [CHEM ENG] Aqueous solution of 4–6% sodium hypochlorite and 4–6% sodium chloride with sodium hydroxide or sodium carbonate stabilizer; used as disinfectant. { ¦la·ba¦raks sə¸lü·shən }

labile [SCI TECH] **1.** Readily changed, as by heat, oxidation, or other processes. **2.** Moving from place to place. Also known as metastable. { 'lā¸bīl }

Laboulbeniales [MYCOL] An order of ascomycetous fungi made up of species that live primarily on the external surfaces of insects. { lə¸bül·ben·ē'ā·lēz }

Labrador Current [OCEANOGR] A current that flows southward from Baffin Bay, through the Davis Strait, and southwestward along the Labrador and Newfoundland coasts. { 'lab·rə¸dȯr ¸kə·rənt }

lacunaris See lacunosus. { ¸lak·yə'nar·əs }

lacunosus [METEOROL] A cloud variety characterized more by the appearance of the spaces between the cloud elements than by the elements themselves, the gaps being generally rounded, often with fringed edges, and the overall appearance being that of a honeycomb or net; it is the negative of clouds composed of separate rounded elements.Formerly known as lacunaris. { ¸lak·yə'nō·səs }

lacustrine [GEOL] Belonging to or produced by lakes. { lə'kəs·trən }

lagoon [GEOGR] **1.** A shallow sound, pond, or lake generally near but separated from or communicating with the open sea. **2.** A shallow fresh-water pond or lake generally near or communicating with a larger body of fresh water. { lə'gün }

Lagrangian current measurement [OCEANOGR] Observation of the speed direction of an ocean current by means of a device, such as a parachute drogue, which follows the water movement. { lə'grän·jē·ən 'kə·rənt ¸mezh·ər·mənt }

lake [HYD] An inland body of water, small to moderately large, with its surface water exposed to the atmosphere. { lāk }

lake ball [ECOL] A spherical mass of tangled, waterlogged fibers and other filamentous material of living or dead vegetation, produced mechanically along a lake bottom by wave action, and usually impregnated with sand and fine-grained mineral fragments. Also known as burr ball; hair ball. { 'lāk ¸bȯl }

lake breeze [METEOROL] A wind, similar in origin to the sea breeze but generally weaker, blowing from the surface of a large lake onto the shores during the afternoon; it is caused by the difference in surface temperature of land and water, as in the land and sea breeze system. { 'lāk ¸brēz }

lake effect [METEOROL] Generally, the effect of any lake in modifying the weather about its shore and for some distance downwind; in the United States, this term is applied specifically to the region about the Great Lakes. { 'lāk i¸fekt }

lake effect storm [METEOROL] A severe snowstorm over a lake caused by the interaction between the warmer water and unstable air above it. { 'lāk i¸fekt ¸stȯrm }

laminar |SCI TECH| **1.** Arranged in thin layers. **2.** Pertaining to viscous streamline flow without turbulence. { 'lam·ə·nər }

lamination |SCI TECH| Arrangement in layers. { 'lam·ə,nā·shən }

lamp oil *See* kerosine. { 'lamp ,óil }

land |GEOGR| The portion of the earth's surface that stands above sea level. { land }

land accretion |CIV ENG| Gaining land in a wet area, such as a marsh or by the sea, by planting maritime plants to encourage silt deposition or by dumping dredged materials in the area. Also known as land reclamation. { 'land ə,krē·shən }

land and sea breeze |METEOROL| The complete cycle of diurnal local winds occurring on seacoasts due to differences in surface temperature of land and sea; the land breeze component of the system blows from land to sea, and the sea breeze blows from sea to land. { ¦land ən 'sē ,brēz }

land blink |METEOROL| A yellowish glow observed over snow-covered land in the polar regions. { 'land ,bliŋk }

land breeze |METEOROL| A coastal breeze blowing from land to sea, caused by the temperature difference when the sea surface is warmer than the adjacent land; therefore, the land breeze usually blows by night and alternates with a sea breeze which blows in the opposite direction by day. { 'land ,brēz }

land bridge |GEOGR| A strip of land linking two landmasses, often subject to temporary submergence, but permitting intermittent migration of organisms. { 'land ,brij }

landfast ice *See* fast ice. { 'lan,fast 'īs }

landfill |CIV ENG| Disposal of solid waste by burying in layers of earth in low ground. { 'lan,fil }

landform |GEOGR| All the physical, recognizable, naturally formed features of land, having a characteristic shape; includes major forms such as a plain, mountain, or plateau, and minor forms such as a hill, valley, or alluvial fan. { 'lan,fórm }

landform map *See* physiographic diagram. { 'lan,fórm ,map }

land hemisphere |GEOGR| The half of the globe, with its pole located at 47.25°N 2.5°W, in which most of the earth's land area is concentrated. { 'land ¦hem·ə,sfir }

land ice |HYD| Any part of the earth's seasonal or perennial ice cover which has formed over land as the result, principally, of the freezing of precipitation. { 'land ,īs }

landlocked |GEOGR| Pertaining to a harbor which is surrounded or almost completely surrounded by land. { 'land,läkt }

land plaster |GEOCHEM| Finely ground gypsum, used as a fertilizer and as a corrective for soil with excess sodium and potassium carbonates. { 'land ,plas·tər }

land reclamation *See* land accretion.|PETR MIN| The process by which seriously disturbed land surfaces are stabilized against the hazards of water and wind erosion. { 'land ,rek·lə'mā·shən }

landscape |GEOGR| The distinct association of landforms that can be seen in a single view. { 'lan,skāp }

landscape ecology |ECOL| Landscapes, including the ecology of their biological inhabitants. { ¦lan,skāp ē'käl·ə·jē }

land sky |METEOROL| The relatively dark appearance of the underside of a cloud layer when it is over land that is not snow-covered, used largely in polar regions with reference to the sky map; it is brighter than water sky, but much darker than ice blink or snow blink. { 'land ,skī }

landslide [GEOL] The perceptible downward sliding or falling of a relatively dry mass of earth, rock, or combination of the two under the influence of gravity. Also known as landslip. { 'lan,slīd }

landslip *See* landslide. { 'lan,slip }

land-use map [GEOGR] A map showing land-use classes as well as other earth surface features such as roads, manufacturing plants, and harbors. { 'land ,yüs ,map }

Langmuir circulation [OCEANOGR] A form of motion found in the near-surface layer of lakes and oceans under windy conditions, and observed as streaks of bubbles, seaweed, or debris forming into lines running roughly parallel to the wind, called windrows. { 'laŋ·myür ,sər·kyə,lā·shən }

lansan [METEOROL] A strong southeast trade wind of the New Hebrides and East Indies. { ¦län¦sän }

lapidicolous [ECOL] Living under a stone. { ¦lap·ə¦dik·ə·ləs }

lapse line [METEOROL] A curve showing the variation of temperature with height in the free air. { 'laps ,līn }

lapse rate [METEOROL] **1.** The rate of decrease of temperature in the atmosphere with height. **2.** Sometimes, the rate of change of any meteorological element with height. { 'laps ,rāt }

lard ice *See* grease ice. { 'lärd ,īs }

large nuclei [OCEANOGR] Particles of concentrated seawater or crystalline salt in the marine atmosphere having radii larger than 10^{-5} centimeter. { 'lärj 'nü·klē,ī }

large scale [METEOROL] A scale such that the curvature of the earth may not be considered negligible; this scale is applicable to the high tropospheric long-wave patterns, with four or five waves around the hemisphere in the middle latitudes. { 'lärj 'skāl }

large-scale convection [METEOROL] Organized vertical motion on a larger scale than atmospheric free convection associated with cumulus clouds; the patterns of vertical motion in hurricanes or in migratory cyclones are examples of such convection. { 'lärj ¦skāl kən'vek·shən }

larva [ZOO] An independent, immature, often vermiform stage that develops from the fertilized egg and must usually undergo a series of form and size changes before assuming characteristic features of the parent. { 'lär·va }

larvicide [AGR] A pesticide used to kill larvae. { 'lär·və,sīd }

larviporous [ZOO] Feeding on larva, referring especially to insects. { lär'vip·ə·rəs }

laryngotracheobronchitis virus *See* croup-associated virus. { lə¦riŋ·gō¦trā·kē·ō ,braŋ'kīd·əs 'vī·rəs }

Lassa fever [MED] An acute, highly communicable exotic infection that is endemic in western Africa. Caused by an arenavirus (the Lassa virus), it is characterized by high fever, weakness, headaches, mouth ulcers, hemorrhages under the skin, heart and kidney failure, and a high mortality rate. { 'läs·ə ,fē·vər }

late blight [PL PATH] A fungus blight disease in which symptoms do not appear until late in the growing season and vary for different species. { 'lāt ¦blīt }

latency [MED] The stage of an infectious disease, other than the incubation period, in which there are neither clinical signs nor symptoms. { 'lat·ən·sē }

latent instability [METEOROL] The state of that portion of a conditionally unstable air column lying above the level of free convection; latent instability is released only if an initial impulse on an air parcel gives it sufficient kinetic energy to carry it through the layer below the level of free convection, within which the environment is warmer than the parcel. { 'lāt·ənt ˌin·stə'bil·əd·ē }

latent period [MED] Any stage of an infectious disease in which there are no clinical signs of symptoms of the infection. { 'lāt·ənt ¦pir·ē·əd }

latent virus [MICROBIO] A virus that remains dormant within body cells but can be reactivated by conditions such as reduced host defenses, toxins, or irradiation, to cause disease. { ¦lāt·ənt 'vī·rəs }

latent-virus infection [MED] A chronic, inapparent virus infection in which a virus-host equilibrium is established. { 'lāt·ənt ¦vī·rəs in,fek·shən }

lateral meristem [BOT] Strips or cylinders of dividing cells located parallel to the long axis of the organ in which they occur; the lateral meristem functions to increase the diameter of the organ. { 'lad·ə·rəl 'mer·ə,stem }

lateral root [BOT] A root branch arising from the main axis. { 'lad·ə·rəl 'rüt }

laterite [GEOL] Weathered material composed principally of the oxides of iron, aluminum, titanium, and manganese; laterite ranges from soft, earthy, porous soil to hard, dense rock. { 'lad·ə,rīt }

laterization [GEOL] Those conditions of weathering that lead to removal of silica and alkalies, resulting in a soil or rock with high concentrations of iron and aluminum oxides (laterite). { ˌlad·ə·rə'zā·shən }

latewood [BOT] The portion of the annual ring that is formed after formation of earlywood has ceased. { 'lāt,wúd }

laurel forest See temperate rainforest. { 'lȯr·əl 'fär·əst }

Laurentide ice sheet [HYD] A major recurring glacier that at its maximum completely covered North America east of the Rockies from the Arctic Ocean to a line passing through the vicinity of New York, Cincinnati, St. Louis, Kansas City, and the Dakotas. { 'lȯr·ən,tīd 'īs ,shēt }

lauric acid [CHEM] $CH_3(CH_2)_{10}COOH$ A fatty acid melting at 44°C, boiling at 225°C (100 mmHg; 13,332 pascals); colorless needles soluble in alcohol and ether, insoluble in water; found as the glyceride in vegetable fats, such as coconut and laurel oils; used for wetting agents, in cosmetics, soaps, resins, and insecticides, and as a chemical intermediate. { 'lȯr·ik 'as·əd }

laurisilva See temperate rainforest. { ¦lȯr·ə¦sil·və }

lauryl mercaptan [CHEM] $C_{12}H_{25}SH$ Pale-yellow or water-white liquid with mild odor; insoluble in water, soluble in organic solvents; used to manufacture plastics, pharmaceuticals, insecticides, fungicides, and elastomers. { 'lȯr·əl mər'kap,tan }

lava [GEOL] **1.** Molten extrusive material that reaches the earth's surface through volcanic vents and fissures. **2.** The rock mass formed by consolidation of molten rock issuing from volcanic vents and fissures, consisting chiefly of magnesium silicate; used for insulators. { 'lä·və }

law of minimum [BIOL] The law that those essential elements for which the ratio of supply to demand (A/N) reaches a minimum will be the first to be removed from the environment by life processes; it was proposed by J. von Liebig, who recognized phosphorus, nitrogen, and potassium as minimum in the soil; in the ocean the corresponding elements are phosphorus, nitrogen, and silicon. Also known as Liebig's law of minimum. { 'lȯ əv 'min·ə·məm }

law of rational intercepts See Miller law. { 'lȯ əv ¦rash·ən·əl 'int·ər,seps }

law of storms [METEOROL] Historically, the general statement of the manner in which the winds of a cyclone rotate about the cyclone's center, and the way that the entire disturbance moves over the earth's surface. { 'lȯ əv 'stȯrmz }

layer [GEOL] A tabular body of rock, ice, sediment, or soil lying parallel to the supporting surface and distinctly limited above and below. { 'lā·ər }

layer depth [OCEANOGR] **1.** The thickness of the mixed layer in an ocean. **2.** The depth to the top of the thermocline. { 'lā·ər ,depth }

layering [ECOL] A stratum of plant forms in a community, such as mosses, shrubs, or trees in a bog area. { 'lā·ə·riŋ }

layer of no motion [OCEANOGR] A layer, assumed to be at rest, at some depth in the ocean. { 'lā·ər əv ¦nō'mō·shən }

LCL *See* lifting condensation level.

LD₅₀ *See* lethal dose 50.

leachate [GEOCHEM] A liquid that has percolated through soil and dissolved some soil materials in the process. { 'lē,chāt }

leaching [GEOCHEM] The separation or dissolving out of soluble constituents from a rock or ore body by percolation of water. { 'lēch·iŋ }

lead [CHEM] A chemical element, symbol Pb, atomic number 82, atomic weight 207.19. [ENG] A soft, heavy metal with a silvery-bluish color; when freshly cut it is malleable and ductile; occurs naturally, mostly in combination; used principally in alloys in pipes, cable sheaths, type metal, and shields against radioactivity. [GEOL] A small, narrow passage in a cave. { led }

lead arsenate [CHEM] $Pb_3(AsO_4)_2$ Poisonous, water-insoluble white crystals; soluble in nitric acid; used as an insecticide. { led 'ärs·ən,āt }

lead encephalopathy [MED] Degeneration of the neurons of the brain accompanied by cerebral edema, due to lead poisoning. { led ,en·sef·ə'läp·ə·thē }

lead palsy *See* lead polyneuropathy. { led ¦pȯl·zē }

lead poisoning [MED] Poisoning due to ingestion or absorption of lead over a prolonged period of time; characterized by colic, brain disease, anemia, and inflammation of peripheral nerves. { led 'pȯiz·ən·iŋ }

lead polyneuropathy [MED] Nerve disorder, affecting mainly the neurons of the wrist and hand, seen principally in adults with chronic lead poisoning; characterized by weakness, paresthesias (abnormal sensations), pain, and glove-and-stocking anesthesia. Also known as lead palsy. { led ¦päl·ē·nü'räp·ə·thē }

leaf [BOT] A modified aerial appendage which develops from a plant stem at a node, usually contains chlorophyll, and is the principal organ in which photosynthesis and transpiration occur. { lēf }

leaf blight [PL PATH] Any of various blight diseases which cause browning, death, and falling of the leaves. { 'lēf ,blīt }

leaf blotch [PL PATH] A plant disease characterized by discolored areas in the leaves with indistinct or diffuse margins. { 'lēf ,bläch }

leaf bud [BOT] A bud that produces a leafy shoot. { 'lēf ,bəd }

leaf curl [PL PATH] A fungus or viral disease of plants marked by the curling of leaves. { 'lēf ,kərl }

leaf drop [PL PATH] Premature falling of leaves, associated with disease. { 'lēf ,dräp }

leaf fiber [BOT] A long, multiple-celled fiber extracted from the leaves of many plants that is used for cordage, such as sisal for binder, and abaca for manila hemp. { 'lēf ,fī·bər }

leafhopper [ZOO] The common name for members of the homopteran family Cicadellidae. { 'lēf,häp·ər }

leaflet [BOT] **1.** A division of a compound leaf. **2.** A small or young foliage leaf. { 'lēf·lət }

leaf miner [ZOO] Any of the larvae of various insects which burrow into and eat the parenchyma of leaves. { 'lēf ,mīn·ər }

leaf mold [GEOL] A soil layer or compost consisting principally of decayed vegetable matter. { 'lēf ,mōld }

leaf rot [PL PATH] Any plant disease characterized by breakdown of leaf tissues; for example, caused by *Pellicularia koleroga* in coffee. { 'lēf ,rät }

leaf rust [PL PATH] Any rust disease that primarily affects leaves; common in coffee, alfalfa, and wheat, barley, and other cereals. { 'lēf ,rəst }

leaf scald [PL PATH] A bacterial disease of sugarcane caused by *Bacterium albilineans* which invades the vascular tissues, causing creamy or grayish streaking and withering of the leaves. { 'lēf ,skȯld }

leaf scar [BOT] A mark on a stem, formed by secretion of suberin and a gumlike substance, showing where a leaf has abscised. { 'lēf ,skär }

leaf scorch [BOT] Any of several disorders and fungus diseases marked by a burned appearance of the leaves; for example, caused by the fungus *Diplocarpon earliana* in strawberry. { 'lēf ,skȯrch }

leaf spot [PL PATH] Any of various diseases or disorders characterized by the appearance of well-defined discolored spots on the leaves. { 'lēf ,spät }

leak [PL PATH] A watery rot of fruits and vegetables caused by various fungi, such as *Rhizopus nigricans* in strawberry. { lēk }

leakage [ENG] Undesired and gradual escape or entry of a quantity, such as loss of neutrons by diffusion from the core of a nuclear reactor, escape of electromagnetic radiation through joints in shielding, flow of electricity over or through an insulating material, and flow of magnetic lines of force beyond the working region. { 'lēk·ij }

leather rot [PL PATH] A hard rot of strawberry caused by the fungus *Phytophthora cactorum*. { 'leth·ər ,rät }

lectin [BIOL] Any of various proteins that agglutinate erythrocytes and other types of cells and also have other properties, including mitogenesis, agglutination of tumor cells, and toxicity toward animals; found widely in plants, predominantly in legumes, and also occurring in bacteria, fish, and invertebrates. { 'lek·tən }

lee [SCI TECH] The side of an object, such as an island or a ship, away from the direction in which the wind is coming, and sheltered from wind or waves. { lē }

lee tide *See* leeward tidal current. { 'lē ,tīd }

lee trough *See* dynamic trough. { 'lē ,trȯf }

leeward [SCI TECH] **1.** Situated away from the wind. **2.** On the lee side. { 'lü·ərd, 'lē·wərd }

leeward tidal current [OCEANOGR] A tidal current setting in the same direction as that in which the wind is blowing. Also known as lee tide; leeward tide. { 'lē·wərd ¦tīd·əl 'kə·rənt }

leeward tide *See* leeward tidal current. { 'lē·wərd ,tīd }

left bank [GEOGR] The bank of a stream or river on the left of an observer when he is facing in the direction of flow, or downstream. { 'left ¦baŋk }

Legionella pneumonia *See* Legionnaire's disease. { ,lē·jə,nel·ə nə'mō·nyə }

Legionnaire's disease [MED] A type of pneumonia usually caused by infection with the bacterium *Legionella pneumophila* that was first observed at an American Legion convention in Philadelphia, Pennsylvania, in 1976. Symptoms include headache, fever reaching 102–105°F (32–41°C), muscle aches, a generalized feeling of discomfort, cough, shortness of breath, chest pains, and sometimes abdominal pain and diarrhea. Also known as Legionella pneumonia. { ˌlē·jə'nerz di‚zēz }

legume [BOT] A dry, dehiscent fruit derived from a single simple pistil; common examples are alfalfa, beans, peanuts, and vetch. { lə'gyüm }

Leiodidae [ZOO] The round carrion beetles, a cosmopolitan family of coleopteran insects in the superfamily Staphylinoidea; commonly found under decaying bark. { lī'äd·ə‚dē }

Leishman-Donovan bodies [MED] Small, oval protozoans lacking flagella and undulating membranes, found within macrophages of the skin, liver, and spleen in leishmanial infections such as kala-azar and mucocutaneous leishmaniasis. { 'lēsh·mən 'dän·ə·vən ‚bäd·ēz }

Leishmania [ZOO] A genus of flagellated protozoan parasites that are the etiologic agents of several diseases of humans, such as leishmaniasis. { lēsh'man·ē·ə }

leishmaniasis [MED] Any of several infections caused by *Leishmania* species. { ˌlēsh·mə'nī·ə·səs }

length of record [CLIMATOL] The period during which observations have been maintained at a meteorological station, and which serves as the frame of reference for climatic data at that station. { 'leŋkth əv 'rek·ərd }

lenitic *See* lentic. { lə'nid·ik }

lentic [ECOL] Of or pertaining to still waters such as lakes, reservoirs, ponds, and bogs. Also spelled lenitic. { 'len·tik }

lenticular cloud *See* lenticularis. { len'tik·yə·lər 'klaúd }

lenticularis [METEOROL] A cloud species, the elements of which have the form of more or less isolated, generally smooth lenses; the outlines are sharp. Also known as lenticular cloud. { len‚tik·yə'lar·əs }

Lentinula edodes [MYCOL] The second most widely cultivated mushroom in the world, it is native to Asia and is touted for its medicinal properties (cholesterol reduction and antitumor and immunostimulating activities) and flavorful addition to foods. Also known as the shiitake mushroom. { lə¦tin·yə·lə ē'dō‚dēz }

leptophos [CHEM] $C_{13}H_{10}BrCl_2O_2PS$ A white solid with a melting point of 70.2–70.6°C; slight solubility in water; used as an insecticide on vegetables, fruit, turf, and ornamentals. Also known as O-(4-bromo-2,5-dichlorophenyl) O-methyl phenylphosphorothioate. { 'lep·tə‚fäs }

leptophyll [ECOL] A growth-form class of plants having a leaf surface area of 0.04 square inch (25 square millimeters) or less; common in alpine and desert habitats. { 'lep·tə ‚fil }

lesser ebb [OCEANOGR] The weaker of two ebb currents occurring during a tidal day. { 'les·ər ‚eb }

lesser flood [OCEANOGR] The weaker of two flood currents occurring during a tidal day. { 'les·ər ‚fləd }

lethal dose 50 [MED] The dose of a substance which is fatal to 50% of test animals. Abbreviated LD_{50}. Also known as median lethal dose. { 'lē·thəl ¦dōs 'fif·tē }

Letinula edodes [MYCOL] The second most widely cultivated mushroom in the world; native to Asia, it is touted for its medicinal properties (cholesterol reduction and antitumor and immunostimulating activities) and flavorful addition to foods. Also known as the shiitake mushroom. { lə¦tin·yə·lə ē'dō‚dēz }

leukocidin |BIOL| A toxic substance released by certain bacteria which destroys leukocytes. { ˌlü·kə'sīd·ən }

leurocristine See vincristine. { ˌlü·rə'kris,tēn }

Levantine Basin |OCEANOGR| A basin in the Mediterranean Ocean between Asia Minor and Egypt. { 'le·vən,tēn 'bās·ən }

level of free convection |METEOROL| The level at which a parcel of air lifted dry and adiabatically until saturated, and lifted saturated and adiabatically thereafter, would first become warmer than its surroundings in a conditionally unstable atmosphere. Abbreviated LFC. { 'lev·əl əv ,frē kən'vek·shən }

level of saturation See water table. { 'lev·əl əv ,sach·ə'rā·shən }

LFC See level of free convection.

liana |BOT| A woody or herbaceous climbing plant with roots in the ground. { lē'än·ə }

Libby effect |GEOCHEM| The increase, since about 1950, in the carbon- 14 content of the atmosphere, produced by the detonation of thermonuclear devices. { 'lib·ē i,fekt }

lichen |BOT| The common name for members of the Lichenes. { 'lī·kən }

Lichenes |BOT| A group of organisms consisting of fungi and algae growing together symbiotically. { lī'kē·nēz }

lichenology |BOT| The study of lichens. { ˌlī·kə'näl·ə·jē }

Liebig's law of the minimum See law of minimum. { 'lē·bigz ¦lò əv thə 'min·ə·məm }

life cycle |BIOL| The major functional and morphological stages through which an organism passes over its lifetime (for example, egg-larva-pupa-adult in some insects). { 'līf ,sī·kəl }

life expectancy |BIOL| The expected number of years that an organism will live based on statistical probability. { 'līf ik'spek·tən·sē }

life form |ECOL| The form characteristically taken by a plant at maturity. { 'līf ,fórm }

life zone |ECOL| A portion of the earth's land area having a generally uniform climate and soil, and a biota showing a high degree of uniformity in species composition and adaptation. { 'līf ,zōn }

lifting condensation level |METEOROL| The level at which a parcel of moist air lifted dry adiabatically would become saturated. Abbreviated LCL. Also known as isentropic condensation level (ICL). { 'lift·iŋ ,kän,den'sā·shən ,lev·əl }

light |PHYS| **1.** Electromagnetic radiation with wavelengths capable of causing the sensation of vision, ranging approximately from 400 (extreme violet) to 770 nanometers (extreme red). Also known as light radiation; visible radiation. **2.** More generally, electromagnetic radiation of any wavelength; thus, the term is sometimes applied to infrared and ultraviolet radiation. { 'līt }

light climate See illumination climate. { 'līt ,klī·mət }

light freeze |METEOROL| The condition when the surface temperature of the air drops to below the freezing point of water for a short time period, so that only the tenderest plants and vines are adversely affected. { 'līt ¦frēz }

light frost |HYD| A thin and more or less patchy deposit of hoarfrost on surface objects and vegetation. { 'līt ¦fròst }

lightning recorder See sferics receiver. { 'līt·niŋ ri,kórd·ər }

light pillar See sun pillar. { 'līt ,pil·ər }

light radiation See light. { 'līt ,rād·ē,ā·shən }

lignify [BOT] To convert cell wall constituents into wood or woody tissue by chemical and physical changes. { 'lig·nə,fī }

lignite [GEOL] Coal of relatively recent origin, intermediate between peat and bituminous coal; often contains patterns from the wood from which it formed. Also known as brown coal; earth coal. { 'lig,nīt }

lignosa [BOT] Woody vegetation. { lig'nō·sə }

lignosulfonate [CHEM] Any of several substances manufactured from waste liquor of the sulfate pulping process of soft wood; used in the petroleum industry to reduce the viscosity of oil well muds and slurries, and as extenders in glues, synthetic resins, and cements. { ¦lig·nō'səl·fə,nāt }

lily-pad ice See pancake ice. { 'lil·ē ,pad ,īs }

limb [BOT] A large primary tree branch. { limb }

limestone [GEOL] **1.** A sedimentary rock composed dominantly (more than 95) of calcium carbonate, principally in the form of calcite; examples include chalk and travertine. **2.** Any rock containing 80% or more of calcium carbonate or magnesium carbonate. { 'līm,stōn }

limicolous [ECOL] Living in mud. { lī'mik·ə·ləs }

liming [AGR] Treating soil with lime (calcium-containing compounds) to reduce its acidity. { 'līm·iŋ }

limnetic [ECOL] Of, pertaining to, or inhabiting the pelagic region of a body of fresh water. { lim'ned·ik }

limnology [ECOL] The science of the life and conditions for life in lakes, ponds, and streams. { lim'näl·ə·jē }

limnoplankton [BIOL] Plankton found in fresh water, especially in lakes. { ¦lim·nō 'plaŋk·tən }

lindane [CHEM] The gamma isomer of 1,2,3,4,5,6-hexachlorocyclohexane, constituting a persistent, bioaccumulative pesticide and a neurotoxin. { 'lin,dān }

line of strike See strike. { 'līn əv 'strīk }

line squall [METEOROL] A squall that occurs along a squall line. { 'līn ,skwȯl }

ling-zhi See Ganoderma lucidum. { ¦liŋ¦tsē }

Linke scale [METEOROL] A type of cyanometer; used to measure the blueness of the sky; it is simply a set of eight cards of different standardized shades of blue, numbered (evenly) 2 to 16; the odd numbers are used by the observer if the sky color lies between any of the given shades. Also known as blue-sky scale. { 'liŋk ,skāl }

lipper [OCEANOGR] **1.** Slight ruffling or roughness appearing on a water surface. **2.** Light spray originating from small waves. { 'lip·ər }

liquid extraction See solvent extraction. { 'lik·wəd ik'strak·shən }

liquid-water content See water content. { 'lik·wəd ¦wȯd·ər ,kän,tent }

Listeria [MICROBIO] A genus of small, gram-positive, motile coccoid rods of uncertain affiliation; found in animal and human feces. { li'stir·ē·ə }

listeriosis [MED] A bacterial disease of humans and some animals caused by *Listeria monocytogenes*; occurs primarily as meningitis or granulomatosis infantiseptica in humans, and takes many forms, such as meningoencephalitis, distemperlike disease, or generalized infection, in animals. { li,stir·ē'ō·səs }

lithogeochemical survey [GEOCHEM] A geochemical survey that involves the sampling of rocks. { ,lith·ō,jē·ə'kem·ə·kəl 'sər,vā }

lithology |GEOL| The description of the physical character of a rock as determined by eye or with a low-power magnifier, and based on color, structures, mineralogic components, and grain size. { lə'thäl·ə·jē }

lithometeor |METEOROL| The general term for dry atmospheric suspensoids, including dust, haze, smoke, and sand. { ,lith·ə'mēd·ē·ər }

lithophyte |ECOL| A plant that grows on rock. { 'lith·ə,fīt }

lithosere |ECOL| A succession of plant communities that originate on rock. { 'lith·ə ,sir }

lithosphere |GEOGR| The solid, rocky portion of the earth as distinguished from the water portion (hydrosphere) and gaseous outer envelope (atmosphere). |GEOL| The rigid outer layer of solid rock, encompassing the earth's crust and upper mantle, that overlies the asthenosphere and forms tectonic plates. { 'lith·ə,sfir }

little brother |METEOROL| A subsidiary tropical cyclone that sometimes follows a more severe disturbance. { 'lid·əl 'brə<u>th</u>·ər }

littoral current |OCEANOGR| A current, caused by wave action, that sets parallel to the shore; usually in the nearshore region within the breaker zone. Also known as alongshore current; longshore current. { 'lit·ə·rəl 'kə·rənt }

littoral drift |GEOL| Materials moved by waves and currents of the littoral zone. Also known as longshore drift. { 'lit·ə·rəl 'drift }

littoral transport |GEOL| The movement of littoral drift. { 'lit·ə·rəl 'tranz,pȯrt }

littoral zone |ECOL| Of or pertaining to the biogeographic zone between the high- and low-water marks. { 'lit·ə·rəl ,zōn }

liverwort |BOT| The common name for members of the Marchantiatae. { 'liv·ər,wȯrt }

loaded concrete |ENG| Concrete to which elements of high atomic number or capture cross section have been added to increase its effectiveness as a radiation shield in nuclear reactors. { 'lōd·əd kän'krēt }

loam |GEOL| Soil mixture of sand, silt, clay, and humus. { lōm }

loaming |GEOCHEM| In geochemical prospecting, a method in which samples of material from the surface are tested for traces of a sought-after metal; its presence on the surface presumably indicates a near-surface ore body. { ,lōm·iŋ }

lobe |HYD| A curved projection on the margin of a continental ice sheet. { lōb }

local change |OCEANOGR| The time rate of change of a scalar quantity (such as temperature, salinity, pressure, or oxygen content) in a fixed locality. { 'lō·kəl 'chānj }

local forecast |METEOROL| Generally, any weather forecast of conditions over a relatively limited area, such as a city or airport. { 'lō·kəl 'fȯr,kast }

local inflow |HYD| The water that enters a stream between two stream-gaging stations. { 'lō·kəl 'in,flō }

local peat |GEOL| Peat formed by groundwater. Also known as basin peat. { 'lō·kəl ,pēt }

local storm |METEOROL| A storm of mesometeorological scale; thus, thunderstorms, squalls, and tornadoes are often put in this category. { 'lō·kəl 'stȯrm }

local winds |METEOROL| Winds which, over a small area, differ from those which would be appropriate to the general pressure distribution, or which possess some other peculiarity. { 'lō·kəl 'winz }

lockjaw *See* tetanus. { 'läk,jȯ }

loco disease |VET MED| Poisoning in livestock resulting from ingestion of selenium-containing plants (loco weed); characterized by atrophy, delirium, convulsions, and stupor, often terminating in death. { 'lō·kō di,zēz }

loco weed [BOT] Any species of *Astragalus* containing selenium taken up from the soil. { 'lō·kō ,wēd }

locust |BOT| Either of two species of commercially important trees, black locust (*Robinia pseudoacacia*) and honey locust (*Gladitsia triacanthos*), in the family Leguminosae. { 'lō·kəst }

logarithmic growth See exponential growth. { 'läg·ə,rith·mik 'grōth }

logging [FOR] The cutting and removal of the woody stem portions of forest trees. { 'läg·iŋ }

logistic growth [BIOL] Population growth in which the growth rate decreases with increasing number of individuals until it becomes zero when the population reaches a maximum. { lə'jis·tik 'grōth }

log volume [FOR] The cubic volume of a log computed inside the bark as determined by any of several formulas; parameters are the cross-sectional areas of log midpoint, large end and small end of log, and log length. { läg ¦väl·yəm }

lolly ice [OCEANOGR] Saltwater frazil, a heavy concentration of which is called sludge. { 'läl·ē ,īs }

loment [BOT] A dry, indehiscent single-celled fruit that is formed from a single superior ovary; splits transversely in numerous segments at maturity. { 'lō,ment }

Lomonosov ridge |GEOGR| An undersea ridge which subdivides the Arctic Basin, extending from Ellesmere Land to the New Siberian Islands. { lō·mō'nȯ,sȯf ,rij }

long-day plant [BOT] A plant that flowers in response to a long photoperiod. { 'lȯŋ ¦dā,plant }

long-day response [BIOL] A photoperiodic response that is evoked by increasing day lengths and decreasing night lengths. { ¦lȯŋ,dā ri'späns }

longitudinal [SCI TECH] Pertaining to the lengthwise dimension. { ,län·jə'tüd·ən·əl }

longitudinal stream See subsequent stream. { ,län·jə'tüd·ən·əl 'strēm }

long-period tide [OCEANOGR] A tide or tidal current constituent with a period which is independent of the rotation of the earth but which depends upon the orbital movement of the moon or of the earth. { 'lȯŋ ,pir·ē·əd 'tīd }

long-range forecast [METEOROL] A weather forecast covering periods from 48 hours to a week in advance (medium-range forecast), and ranging to even longer forecasts over periods of a month, a season, and so on. { 'lȯŋ ,rānj 'fȯr,kast }

longshore bar |GEOL| A ridge of sand, gravel, or mud built on the seashore by waves and currents, generally parallel to the shore and submerged by high tides. Also known as offshore bar. { 'lȯŋ,shȯr ,bär }

longshore current See littoral current. { 'lȯŋ,shȯr ,kə·rənt }

longshore drift See littoral drift. { 'lȯŋ,shȯr ,drift }

long wave [METEOROL] With regard to atmospheric circulation, a wave in the major belt of westerlies which is characterized by large length (thousands of kilometers) and significant amplitude; the wavelength is typically longer than that of the rapidly moving individual cyclonic and anticyclonic disturbances of the lower troposphere. Also known as major wave; planetary wave. { 'lȯŋ ¦wāv }

loom [METEOROL] The glow of light below the horizon produced by greater-than-normal refraction in the lower atmosphere; it occurs when the air density decreases more rapidly with height than in the normal atmosphere. { lüm }

loop lake   See oxbow lake.   { 'lüp ,lāk }

loop rating   [HYD] A rating curve that has higher values of discharge for a certain stage when the river is rising than it does when the river is falling; thus, the curve (stage versus discharge) describes a loop with each rise and fall of the river.   { 'lüp ,rād·iŋ }

losing stream   See influent stream.   { 'lüs·iŋ ,strēm }

lost stream   [HYD] **1.** A stream that disappears from the surface into an underground channel without reappearing in the same or even a neighboring drainage basin. **2.** An evaporated stream in a desertlike region.   { 'lȯst 'strēm }

lotic   [ECOL] Of or pertaining to swiftly moving waters.   { 'lōd·ik }

loudness   [PHYS] The magnitude of the physiological sensation produced by a sound, which varies directly with the physical intensity of sound but also depends on frequency of sound and waveform.   { 'lau̇d·nəs }

loudness level   [PHYS] The level of a sound, in phons, equal to the sound pressure level in decibels, relative to 0.0002 microbar, of a pure 1000-hertz tone that is judged to be equally loud by listeners.   { 'lau̇d·nəs ,lev·əl }

louping ill   [VET MED] A virus disease of sheep, similar to encephalomyelitis, transmitted by the tick *Ixodes racinus*. Also known as ovine encephalomyelitis; trembling ill.   { 'lüp·iŋ ,il }

louping-ill virus   [MICROBIO] A group B arbovirus that is infectious in sheep, monkeys, mice, horses, and cattle.   { 'lüp·iŋ ,il ,vī·rəs }

low   See depression.   { lō }

low aloft   See upper-level cyclone.   { 'lō ə'lȯft }

low clouds   [METEOROL] Types of clouds, the mean level of which is between the surface and 6500 feet (1980 meters); the principal clouds in this group are stratocumulus, stratus, and nimbostratus.   { 'lō 'klau̇dz }

low-energy environment   [GEOL] An aqueous sedimentary environment in which there is standing water with a general lack of wave or current action, permitting accumulation of very fine-grained sediments.   { 'lō, en·ər·jē in'vī·ərn·mənt }

lower atmosphere   [METEOROL] That part of the atmosphere in which most weather phenomena occur (that is, the troposphere and lower stratosphere); in other contexts, the term implies the lower troposphere.   { 'lō·ər 'at·mə,sfir }

lower high water   [OCEANOGR] The lower of two high tides occurring during a tidal day.   { 'lō·ər 'hī ,wȯd·ər }

lower low water   [OCEANOGR] The lower of two low tides occurring during a tidal day.   { 'lō·ər 'lō,wȯd·ər }

lower mantle   [GEOL] The portion of the mantle below a depth of about 600 miles (1000 kilometers). Also known as inner mantle; mesosphere; pallasite shell.   { 'lō·ər mant·əl }

Lower Sonoran life zone   [ECOL] A life zone characterized by an arid to semiarid climate, mild winters and hot summers, low elevations, scant rainfall, and dessert vegetation, such as cactus, agave, creosote, bush, and mesquite.   { 'lō·ər sə'nȯr·ən 'līf ,zōn }

low index   [METEOROL] A relatively low value of the zonal index which, in middle latitudes, indicates a relatively weak westerly component of wind flow (usually implying stronger north-south motion), and the characteristic weather attending such motion; a circulation pattern of this type is commonly called a low-index situation.   { 'lō ¦in,deks }

low-population zone [ENG] An area of low population density sometimes required around a nuclear installation; the number and density of residents is of concern in providing, with reasonable probability, that effective protection measures can be taken if a serious accident should occur. { 'lō ˌpäp·ə'lā·shən ˌzōn }

low-temperature coke [PETR MIN] Coke produced at temperatures of 500–750°C, used chiefly for house heating, particularly in England. Also known as char. { 'lō ˌtem·prə·chər 'kōk }

low tide See low water. { 'lō 'tīd }

low water [OCEANOGR] The lowest limit of the surface water level reached by the lowering tide. Also known as low tide. { 'lō 'wȯd·ər }

low-water inequality [OCEANOGR] The difference between the heights of two successive low tides. { 'lō ˌwȯd·ər ˌin·i'kwäl·əd·ē }

low-water neaps See mean low-water neaps. { 'lō ˌwȯd·ər 'nēps }

low-water springs See mean low-water springs. { 'lō ˌwȯd·ər 'spriŋz }

lucerne See alfalfa. { lü'sərn }

luminous meteor [METEOROL] According to United States weather observing practice, any one of a number of atmospheric phenomena which appear as luminous patterns in the sky, including halos, coronas, rainbows, aurorae, and their many variations, but excluding lightning (an igneous meteor or electrometeor). { 'lü·mə·nəs 'mēd·ē·ər }

lumpy jaw See actinomycosis. { 'ləm·pē ˌjȯ }

lunar atmospheric tide [METEOROL] An atmospheric tide due to the gravitational attraction of the moon; the only detectable components are the 12-lunar-hour or semidiurnal component, as in the oceanic tides, and two others of very nearly the same period; the amplitude of this atmospheric tide is so small that it is detected only by careful statistical analysis of a long record. { 'lü·nər ˌat·mə,sfir·ik 'tīd }

lunar tide [OCEANOGR] The portion of a tide produced by forces of the moon. { 'lü·nər 'tīd }

lunisolar tides [OCEANOGR] Harmonic tidal constituents attributable partly to the development of both the lunar tide and the solar tide and partly to the lunisolar synodic fortnightly constituent. { ¦lü·nə'sō·lər 'tīdz }

lunitidal interval [OCEANOGR] The period between the moon's upper or lower transit over a specified meridian and a specified phase of the tidal current following the transit. { ¦lü·nə'tīd·əl 'in·tər·vəl }

Lyme borreliosis See Lyme disease. { ˌlīm bə,rel·ē'ō·səs }

Lyme disease [MED] A complex multisystem human illness caused by the tick-borne spirochete Borrelia burgdorferi. Also known as Lyme borreliosis. { 'līm di,zēz }

lyngbyatoxin A [BIOL] An indole alkaloid toxin produced by Lyngbya majuscula. { 'liŋ·bē·ə,täk·sən 'ā }

lysocline [OCEANOGR] The level or ocean depth at which the rate of solution of calcium carbonate increases significantly. { 'lī·sə,klīn }

Lyssavirus [MICROBIO] A genus of the viral family Rhabdoviridae that is characterized by a bullet-shaped enveloped virion covered with projections that contains one molecule of linear, negative-sense, single-stranded ribonucleic acid, the causative agent of rabies. { 'līs·ə,vī·rəs }

M   *See* mega-.

mackerel sky [METEOROL] A sky with considerable cirrocumulus or small-element altocumulus clouds, resembling the scales on a mackerel. { 'mak·rəl 'skī }

Macquer's salt *See* potassium arsenate. { mə'kerz ,sȯlt }

macrandrous [BOT] Having both antheridia and oogonia on the same plant; used especially for certain green algae. { ma'kran·drəs }

macro- [SCI TECH] Prefix meaning large. { 'mak·rō *or* ,mak·rə }

macroclimate [CLIMATOL] The climate of a large geographic region. { ¦mak·rō'klī·mət }

macroconsumer [ECOL] A large consumer which ingests other organisms or particulate organic matter. Also known as biophage. { ¦mak·rō·kən'sü·mər }

macrocyclic [MYCOL] Of a rust fungus, having binuclear spores as well as teliospores and sporidia, or having a life cycle that is long or complex. { ¦mak·rō'sī·klik }

macrofauna [ZOO] Animals visible to the naked eye. { ¦mak·rō'fȯn·ə }

macroflora [ECOL] Large plant mataerial such as tree roots. { ¦mak·rō'flȯr·ə }

macrohabitat [ECOL] An extensive habitat presenting considerable variation of the environment, containing a variety of ecological niches, and supporting a large number and variety of complex flora and fauna. { ¦mak·rō'hab·ə,tat }

macrometeorology [METEOROL] The study of the largest-scale aspects of the atmosphere, such as the general circulation, and weather types. { ¦mak·rō,mēd·ē·ə'räl·ə·jē }

macrophage [MED] A large phagocyte of the reticuloendothelial system. Also known as a histiocyte. { 'mak·rə,fāj }

macrophagy [BIOL] Feeding on large particulate matter. { 'mak·rə,fā·jē }

macrophyllous [BOT] Having large or long leaves. { ¦mak·rō'fil·əs }

macrophyte [ECOL] A macroscopic plant, especially one in an aquatic habitat. { 'mak·rə,fīt }

macropore [GEOL] A pore in soil of a large enough size so that water is not held in it by capillary attraction. { 'mak·rə,pȯr }

macroscopic [SCI TECH] Large enough to be observed by the naked eye. { ¦mak·rə ¦skäp·ik }

macrosporangium [BOT] A spore case in which macrospores are produced. Also known as megasporangium. { ¦mak·rə·spə'ran·jē·əm }

macrospore [BOT] The larger of two spore types produced by heterosporous plants; the female gamete. Also known as megaspore. { 'mak·rə,spȯr }

macrosporogenesis [BOT] In angiosperms, the formation of macrospores and the production of the embryo sac from one or occasionally several cells of the subepidermal cell layer within the ovule of a closed ovary. Also known as megasporogenesis. { ˌmak·rō¦spór·ō'jen·ə·səs }

macrothermophyte See megathermophyte. { ˌmak·rə'thər·mə¸fīt }

maelstrom [OCEANOGR] A powerful and often destructive water current caused by the combined effects of high, wind-generated waves and a strong, opposing tidal current. { 'māl·strəm }

magma [GEOL] The molten rock material from which igneous rocks are formed. { 'mag·mə }

magmatic water [HYD] Water derived from or existing in molten igneous rock or magma. Also known as juvenile water. { mag'mad·ik 'wód·ər }

magnesium [CHEM] A metallic element, symbol Mg, atomic number 12, atomic weight 24.305. { mag'nē·zē·əm }

magnesium arsenate [CHEM] $Mg_3(AsO_4)_2 \cdot xH_2O$ A white, poisonous, water-insoluble powder used as an insecticide. { mag'nē·zē·əm 'ärs·ən¸āt }

magnesium borate [CHEM] $3MgO \cdot B_2O_3$ Crystals that are white or colorless and transparent; soluble in alcohol and acids, slightly soluble in water; used as a fungicide, antiseptic, and preservative. { mag'nē·zē·əm 'bór¸āt }

magnesium chloride [CHEM] $MgCl_2 \cdot 6H_2O$ Deliquescent white crystals; soluble in water and alcohol; used in disinfectants and fire extinguishers, and in ceramics, textiles, and paper manufacture. { mag'nē·zē·əm 'klór¸īd }

magnetic wind direction [METEOROL] The direction, with respect to magnetic north, from which the wind is blowing; distinguished from true wind direction. { mag'ned·ik 'wind də¸rek·shən }

magnetohydrodynamic generator [ENG] A system for generating electric power in which the kinetic energy of a flowing conducting fluid is converted to electric energy by a magnetohydrodynamic interaction. Abbreviated MHD generator. { mag¦nēd·ō ¸hī·drə·dī'nām·ik 'jen·ə¸rād·ər }

Magnoliophyta [BOT] The angiosperms, a division of vascular seed plants having the ovules enclosed in an ovary and well-developed vessels in the xylem. { mag¸nō·lē'äf·əd·ə }

mainland [GEOGR] A continuous body of land that constitutes the main part of a country or continent. { 'mān·lənd }

main stream [HYD] The principal or largest stream of a given area or drainage system. Also known as master stream; trunk stream. { 'mān 'strēm }

main thermocline [OCEANOGR] A thermocline that is deep enough in the ocean to be unaffected by seasonal temperature changes in the atmosphere. Also known as permanent thermocline. { 'mān 'thər·mə¸klīn }

maitake mushroom See Grifola frondosa. { ˌmä·i¸tä·ke 'məsh¸rüm }

maize [BOT] Zea mays. Indian corn, a tall cereal grass characterized by large ears. { māz }

major trough [METEOROL] A long-wave trough in the large-scale pressure pattern of the upper troposphere. { 'mā·jər 'tróf }

major wave See long wave. { 'mā·jər 'wāv }

malaria [MED] A group of human febrile diseases with a chronic relapsing course caused by hemosporidian blood parasites of the genus Plasmodium, transmitted by the bite of the Anopheles mosquito. { mə'ler·ē·ə }

malathion [CHEM] $C_{10}H_{19}O_6PS_2$ A yellow liquid, slightly soluble in water; malathion is the generic name for S-1,2-bis(ethoxycarbonyl)ethyl O,O-dimethylphosphorodithioate; used as an insecticide. { ,mal·ə'thī,än }

male [BOT] A flower lacking pistils. { māl }

malenclave [HYD] A body of contaminated groundwater surrounded by uncontaminated water. { ,mal'än,klāv }

malezal swamp [ECOL] A swamp resulting from drainage of water over an extensive plain with a slight, almost imperceptible slope. { mə'lēz·əl ,swämp }

malignant [MED] **1.** Endangering the life or health of an individual. **2.** Pertaining to the growth and proliferation of cancer cells which terminate in death if not checked by treatment. { mə'līg·nənt }

malignant catarrh [VET MED] A catarrhal fever of cattle caused by a virus and characterized by acute inflammation and edema of the respiratory and digestive systems. { mə'līg·nənt kə'tär }

mallee See tropical scrub. { 'mä·lē }

malm See marl. { mäm }

Malta fever See brucellosis. { 'mȯl·tə 'fē·vər }

Malthusianism [BIOL] The theory that population increases more rapidly than the food supply unless held in check by epidemics, wars, or similar phenomena. { mal'thü·zhə ,niz·əm }

maneb [CHEM] Mn[SSCH(CH$_2$)$_2$NHCSS] A generic term for manganese ethylene-1,2-bisdithiocarbamate; irritating to eyes, nose, skin, and throat; used as a fungicide. { 'ma,neb }

manganese [CHEM] A metallic element, symbol Mn, atomic weight 54.938, atomic number 25; a transition element whose properties fall between those of chromium and iron. [ENG] A hard, brittle, grayish-white metal used chiefly in making steel. { 'maŋ·gə,nēs }

manganese sulfate See manganous sulfate. { 'maŋ·gə,nēs 'səl,fāt }

manganous sulfate [CHEM] MnSO$_4$·4H$_2$O Water-soluble, translucent, efflorescent rose-red prisms; melts at 30°C; used in medicine, textile printing, and ceramics, as a fungicide and fertilizer, and in paint manufacture. Also known as manganese sulfate. { 'maŋ·gə·nəs 'səl,fāt }

mange [VET MED] Infestation of the skin of mammals by certain mites (Sarcoptoidea) which burrow into the epidermis; characterized by multiple lesions accompanied by severe itching. { mānj }

mangrove [BOT] A tropical tree or shrub of the genus *Rhizophora* characterized by an extensive, impenetrable system of prop roots which contribute to land building. { 'maŋ,grōv }

mangrove swamp [ECOL] A tropical or subtropical marine swamp distinguished by the abundance of low to tall trees, especially mangrove trees. { 'maŋ,grōv ,swämp }

mansonelliasis [MED] A parasitic infection of humans by the filarioid nematode *Mansonella ozzardi*. { ,man·sən·ə'lī·ə·səs }

mantle [GEOL] The intermediate shell zone of the earth below the crust and above the core (to a depth of 2160 miles or 3480 kilometers). { 'mant·əl }

mantle rock See regolith. { 'mant·əl ,räk }

map plotting [METEOROL] The process of transcribing weather information onto maps, diagrams, and so on; it usually refers specifically to decoding synoptic reports and entering those data in conventional station-model form on synoptic charts. Also known as map spotting. { 'map ˌpläd·iŋ }

map spotting See map plotting. { 'map ˌspäd·iŋ }

maquis [ECOL] A type of vegetation composed of shrubs, or scrub, usually not exceeding 10 feet (3 meters) in height, the majority having small, hard, leathery, often spiny or needlelike drought-resistant leaves and occurring in areas with a Mediterranean climate. { mä'kē }

Marburg virus [MICROBIO] A large virus transmitted to humans by the grivet monkey (*Cercopithecus aethiops*). { 'mär,bûrg ,vī·rəs }

marcescent [BOT] Withering without falling off. { mär'ses·ənt }

march [METEOROL] The variation of any meteorological element throughout a specific unit of time, such as a day, month, or year; as the daily march of temperature, the complete cycle of temperature during 24 hours. { märch }

Marchantiales [BOT] The thallose liverworts, an order of the class Marchantiopsida having a flat body composed of several distinct tissue layers, smooth-walled and tuberculate-walled rhizoids, and male and female sex organs borne on stalks on separate plants. { mär,shan·tē'ā·lēz }

Marchantiopsida [BOT] The liverworts, a class of lower green plants; the plant body is usually a thin, prostrate thallus with rhizoids on the lower surface. { mär ,shan·tē'äp·sə·də }

margin [GEOGR] The boundary around a body of water. [SCI TECH] An outside limit. { 'mär·jən }

marginal blight [PL PATH] A bacterial disease of lettuce caused by *Pseudomonas marginalis*, characterized by brownish marginal discoloration of the foliage. { 'mär·jən·əl 'blīt }

marginal chlorosis [PL PATH] A virus disease characterized by yellowing or blanching of leaf margins; common disease of peanut plants. { 'mär·jən·əl klə'rō·səs }

marginal sea [GEOGR] A semiclosed sea adjacent to a continent and connected with the ocean at the water surface. { 'mär·jən·əl 'sē }

Margules equation See Witte-Margules equation. { mär'gü·ləs i,kwā·zhən }

mariculture [AGR] The cultivation of marine organisms, plant and animal, for purposes of human consumption. { 'mar·ə,kəl·chər }

marigram [OCEANOGR] A graphic record of the rising and falling movements of the tide expressed as a curve. { 'mar·ə,gram }

marine [OCEANOGR] Pertaining to the sea. { mə'rēn }

marine abrasion [GEOL] Erosion of the ocean floor by sediment moved by ocean waves. Also known as wave erosion. { mə'rēn ə'brā·zhən }

marine biocycle [ECOL] A major division of the biosphere composed of all biochores of the sea. { mə'rēn 'bī·ō,sī·kəl }

marine biology [BIOL] A branch of biology that deals with those living organisms which inhabit the sea. { mə'rēn bī'äl·ə·jē }

marine climate [CLIMATOL] A regional climate which is under the predominant influence of the sea, that is, a climate characterized by oceanity; the antithesis of a continental climate. Also known as maritime climate; oceanic climate. { mə'rēn 'klī·mət }

marine ecology [ECOL] An integrative science that studies the basic structural and functional relationships within and among living populations and their physical-chemical environments in marine ecosystems. { mə,rēn ē'käl·ə·jē }

marine forecast [METEOROL] A forecast, for a specified oceanic or coastal area, of weather elements of particular interest to maritime transportation, including wind, visibility, the general state of the weather, and storm warnings. { mə'rēn 'fȯr₁kast }

marine geology *See* geological oceanography. { mə'rēn jē'äl·ə·jē }

marine littoral faunal region [ECOL] A geographically determined division of that portion of the zoosphere composed of marine animals. { mə'rēn 'lit·ə·rəl 'fȯn·əl ₁rē·jən }

marine marsh [ECOL] A flat, savannalike land expanse at the edge of a sea; usually covered by water during high tide. { mə'rēn 'märsh }

marine meteorology [METEOROL] That part of meteorology which deals mainly with the study of oceanic areas, including island and coastal regions; in particular, it serves the practical needs of surface and air navigation over the oceans. { mə'rēn ₁mēd·ē·ə'räl·ə·jē }

marine salina [GEOGR] A body of salt water found along an arid coast and separated from the sea by a sand or gravel barrier. { mə'rēn səl'lēn·ə }

marine snow [OCEANOGR] A concentration of living and dead organic material and inorganic debris of the sea suspended at density boundaries such as the thermocline. { mə'rēn 'snō }

marine swamp [GEOGR] An area of low, salty, or brackish water found along the shore and characterized by abundant grasses, mangrove trees, and similar vegetation.Also known as paralic swamp. { mə'rēn 'swämp }

marine weather observation [METEOROL] The weather as observed from a ship at sea, usually taken in accordance with procedures specified by the World Meteorological Organization. { mə'rēn 'weth·ər ₁äb·zər₁vā·shən }

maritime air [METEOROL] A type of air whose characteristics are developed over an extensive water surface and which, therefore, has the basic maritime quality of high moisture content in at least its lower levels. { 'mar·ə₁tīm 'er }

maritime climate *See* marine climate. { 'mar·ə₁tīm 'klī·mət }

maritime polar air [METEOROL] Polar air initially possessing similar properties to those of continental polar air, but in passing over warmer water it becomes unstable with a higher moisture content. { 'mar·ə₁tīm 'pō·lər ¦er }

maritime tropical air [METEOROL] The principal type of tropical air, produced over the tropical and subtropical seas; it is very warm and humid, and is frequently carried poleward on the western flanks of the subtropical highs. { 'mar·ə₁tīm 'träp·ə·kəl ¦er }

marl [GEOL] A deposit of crumbling earthy material composed principally of clay with magnesium and calcium carbonate; used as a ertilizer for lime-deficient soils. Also known as malm. { märl }

Marsden chart [METEOROL] A system for showing the distribution of meteorological data on a chart, especially over the oceans; using a Mercator map projection, the world between 80°N and 70°S latitudes is divided into Marsden "squares," each of 10° latitude by 10° longitude and systematically numbered to indicate position; each square may be divided into quarter squares, or into 100 one-degree subsquares numbered from 00 to 99 to give the position to the nearest degree. { 'märz·dən ₁chärt }

marsh [ECOL] A transitional land-water area, covered at least part of the time by estuarine or coastal waters, and characterized by aquatic and grasslike vegetation, especially without peatlike accumulation. { märsh }

marsh gas [GEOCHEM] Combustible gas, consisting chiefly of methane, produced as a result of decay of vegetation in stagnant water. { 'märsh ₁gas }

mascaret *See* bore. { ¦mas·kə¦ret }

masculinize [BIOL] To cause a female or a sexually immature animal to take on male secondary sex characteristics. { 'mas·kyə·lə‚nīz }

mass bleaching [ZOO] A disease affecting coral reefs in which a reduction in the number of zooxanthellae (symbiotic plants) causes corals to lose their characteristic brown color over a period of several weeks and take on a brilliant white appearance. { ¦mas 'blēch·iŋ }

master stream *See* main stream. { 'mas·tər ‚strēm }

mate [BIOL] **1.** To pair for breeding. **2.** To copulate. { māt }

mathematical biology [BIOL] A discipline that encompasses all applications of mathematics, computer technology, and quantitative theorizing to biological systems, and the underlying processes within the systems. { ¦math·ə¦mad·ə·kəl bī'äl·ə·jē }

mathematical climate [CLIMATOL] An elementary generalization of the earth's climatic pattern, based entirely on the annual cycle of the sun's inclination; this early climatic classification recognized three basic latitudinal zones (the summerless, intermediate, and winterless), which are now known as the Frigid, Temperate, and Torrid Zones, and which are bounded by the Arctic and Antarctic Circles and the Tropics of Cancer and Capricorn. { ¦math·ə¦mad·ə·kəl 'klī·mət }

mathematical ecology [ECOL] The application of mathematical theory and technique to ecology. { ¦math·ə¦mad·ə·kəl ē'käl·ə·jē }

mathematical forecasting *See* numerical forecasting. { ¦math·ə¦mad·ə·kəl 'fòr‚kast·iŋ }

mathematical geography [GEOGR] The branch of geography that deals with the features and processes of the earth, and their representations on maps and charts. { ¦math·ə¦mad·ə·kəl jē'äg·rə·fē }

matinal [METEOROL] The morning winds, that is, an east wind. { 'mat·ən·əl }

mating [BIOL] The meeting of individuals for sexual reproduction. { 'mād·iŋ }

mature [BIOL] **1.** Being fully grown and developed. **2.** Ripe. [GEOL] **1.** Pertaining to a topography or region, and to its landforms, having undergone maximum development and accentuation of form. **2.** Pertaining to the third stage of textural maturity of a clastic sediment. { mə'chùr }

mature soil *See* zonal soil. { mə'chùr 'sòil }

maximum ebb [OCEANOGR] The greatest speed of an ebb current. { 'mak·sə·məm 'eb }

maximum flood [OCEANOGR] The greatest speed of a flood current. { 'mak·sə·məm 'fləd }

maximum permissible concentration [MED] The maximum quantity/unit volume of radioactive material in air, water, and foodstuffs that is not considered an undue risk to human health. { 'mak·sə·məm pər'mis·ə·bəl ‚kän·sən'trā·shən }

maximum permissible dose [MED] The dose of ionizing radiation that a person may receive in his lifetime without appreciable bodily injury. { 'mak·sə·məm pər'mis·ə·bəl 'dōs }

maximum sustainable yield [OCEANOGR] **1.** In fishery management, the highest average fishing level over time that does not reduce a stock's abundance in balance with the stock's reproductive and growth capacities under a given set of environmental conditions. **2.** A level of fishing that, if approached, should signal caution rather than increased fishing. { ¦mak·sə·məm sə¦stān·ə·bul 'yēld }

maximum-wind and shear chart [METEOROL] A synoptic chart on which are plotted the altitudes of the maximum wind speed, the maximum wind velocity (wind direction optional), plus the velocity of the wind at mandatory levels both above and below the level of maximum wind. Also known as max-wind and shear chart. { 'mak·sə·məm 'wind ən 'shir ‚chärt }

maximum-wind level [METEOROL] The height at which the maximum wind speed occurs, determined in a winds-aloft observation. Also known as max-wind level. { 'mak·sə·məm 'wind ˌlev·əl }

maximum-wind topography [METEOROL] The topography of the surface of maximum wind speed. Also known as max-wind topography. { 'mak·sə·məm ˌwind tə'päg·rə·fē }

maximum zonal westerlies [METEOROL] The average west-to-east component of wind over the continuous 20° belt of latitude in which this average is a maximum; it is usually found, in the winter season, in the vicinity of 40–60° north latitude. { 'mak·sə·məm ˌzōn·əl 'wes·tər,lēz }

max-wind and shear chart See maximum-wind and shear chart. { 'maks 'wind ən 'shir ˌchärt }

max-wind level See maximum-wind level. { 'maks 'wind ˌlev·əl }

max-wind topography See maximum-wind topography. { 'maks 'wind tə'päg·rə·fē }

mazaedium [BOT] The fruiting body of certain lichens, with the spores lying in a powdery mass in the capitulum. [MYCOL] A slimy layer on the hymenial surface of some ascomycetous fungi. { mə'zē·dē·əm }

meadow [ECOL] A vegetation zone which is a low grassland, dense and continuous, variously interspersed with forbs but few if any shrubs. Also known as Wiesen. { 'med·ō }

mealybug [ZOO] Any of various scale insects of the family Pseudococcidae which have a powdery substance covering the dorsal surface; all are serious plant pests. { 'mē·lē ˌbəg }

mean chart [METEOROL] Any chart on which isopleths of the mean value of a given meteorological element are drawn. Also known as mean map. { 'mēn ˌchärt }

mean depth [HYD] Average water depth in a stream channel or conduit computed by dividing the cross-sectional area by the surface width. { 'mēn 'depth }

meander [HYD] A sharp, sinuous loop or curve in a stream, usually part of a series. [OCEANOGR] A deviation of the flow pattern of a current. { mē'an·dər }

meandering stream [HYD] A stream having a pattern of successive meanders. Also known as snaking stream. { mē'an·də·riŋ 'strēm }

mean high water [OCEANOGR] The average height of all high waters recorded at a given place over a 19-year period. { 'mēn 'hī 'wȯd·ər }

mean high-water lunitidal interval [OCEANOGR] The average interval of time between the transit (upper or lower) of the moon and the next high water at a place. Also known as corrected establishment. { 'mēn ˌhī ˌwȯd·ər ˌlü·nəˌtīd·əl 'in·tər·vəl }

mean high-water neaps [OCEANOGR] The average height of the high waters of neap tides. Also known as neap high water. { 'mēn ˌhī ˌwȯd·ər 'nēps }

mean high-water springs [OCEANOGR] The average height of the high waters of spring tides. Also known as high-water springs; spring high water. { 'mēn ˌhī ˌwȯd·ər 'spriŋz }

mean low water [OCEANOGR] The average height of all low waters recorded at a given place over a 19-year period. { 'mēn 'lō 'wȯd·ər }

mean low-water lunitidal interval [OCEANOGR] The average interval of time between the transit (upper or lower) of the moon and the next low water at a place. { 'mēn ˌlō ˌwȯd·ər ˌlü·nəˌtīd·əl 'in·tər·vəl }

mean low-water neaps [OCEANOGR] The average height of the low water at neap tides. Also known as low-water neaps; neap low water. { 'mēn ˌlō ˌwȯd·ər 'nēps }

mean low-water springs [OCEANOGR] The average height of the low waters of spring tides; this level is used as a tidal datum in some areas. Also known as low-water springs; spring low water. { 'mēn ¦lō ¦wȯd·ər 'spriŋz }

mean map See mean chart. { 'mēn ˌmap }

mean range [OCEANOGR] The difference in the height between mean high water and mean low water. { 'mēn 'rānj }

mean rise interval [OCEANOGR] The average interval of time between the transit (upper or lower) of the moon and the middle of the period of rise of the tide at a place. { 'mēn 'rīz ˌin·tər·vəl }

mean river level [HYD] The average height of the surface of a river at any point for all stages of the tide over a 19-year period. { 'mēn 'riv·ər ˌlev·əl }

mean sea level [OCEANOGR] The average sea surface level for all stages of the tide over a 19-year period, usually determined from hourly height readings from a fixed reference level. { 'mēn 'sē ˌlev·əl }

mean temperature [METEOROL] The average temperature of the air as indicated by a properly exposed thermometer during a given time period, usually a day, month, or year. { 'mēn 'tem·prə·chər }

mean tide See half tide. { 'mēn 'tīd }

mean tide level [OCEANOGR] The tide level halfway between mean high water and mean low water. { 'mēn 'tīd ˌlev·əl }

mean water level [OCEANOGR] The average surface level of a body of water. { 'mēn 'wȯd·ər ˌlev·əl }

measles [MED] An acute, highly infectious viral disease with cough, fever, and maculopapular rash (having both flat and raise areas). Also known as rubeola. { 'mē·zəlz }

mechanical erosion See corrasion. { mi'kan·ə·kəl i'rō·zhən }

mechanical instability See absolute instability. { mi'kan·ə·kəl ˌin·stə'bil·əd·ē }

mechanical turbulence [METEOROL] Irregular air movement in the lower atmosphere resulting from obstructions, for example, tall buildings. { mi'kan·ə·kəl 'tər·byə·ləns }

mechanical weathering [GEOL] The process of weathering by which physical forces break down or reduce a rock to smaller and smaller fragments, involving no chemical change. Also known as physical weathering. { mi'kan·ə·kəl 'weth·ə·riŋ }

mechanoreceptor [BIOL] A receptor that provides the organism with information about such mechanical changes in the environment as movement, tension, and pressure. { ¦mek·ə·nō·ri'sep·tər }

medial [SCI TECH] Located in the middle. { 'mē·dē·əl }

median effective dose See effective dose 50. { 'mē·dē·ən i'fek·tiv 'dōs }

median infective dose See infective dose 50. { 'mē·dē·ən in'fek·tiv 'dōs }

median lethal dose See lethal dose 50. { 'mē·dē·ən 'lēth·əl 'dōs }

median lethal time [MICROBIO] The period of time required for 50% of a large group of organisms to die following a specific dose of an injurious agent, such as a drug or radiation. { 'mē·dē·ən 'lēth·əl ˌtīm }

medical bacteriology [MED] A branch of medical microbiology that deals with the study of bacteria which affect human health, especially those which produce disease. { 'med·ə·kəl bak·tir·ē'äl·ə·jē }

medical entomology [MED] The study of insects that are vectors for diseases and parasitic infestations in humans and domestic animals. { 'med·ə·kəl ˌen·tə'mäl·ə·jē }

medical microbiology |MED| The study of microorganisms which affect human health. { 'med·ə·kəl ¦mī·krō·bī'äl·ə·jē }

medical mycology |MED| A branch of medical microbiology that deals with fungi that are pathogenic to humans. { 'med·ə·kəl mī'käl·ə·jē }

Mediterranean climate |CLIMATOL| A type of climate characterized by hot, dry, sunny summers and a winter rainy season; basically, this is the opposite of a monsoon climate. Also known as etesian climate. { ¸med·ə·tə'rā·nē·ən 'klī·mət }

Mediterranean faunal region |ECOL| A marine littoral faunal region including that offshore portion of the Atlantic Ocean from northern France to near the Equator. { ¸med·ə·tə'rā·nē·ən 'fȯn·əl ¸rē·jən }

Mediterranean fever *See* brucellosis. { ¸med·ə·tə'rā·nē·ən 'fē·vər }

mediterranean sea |GEOGR| A deep epicontinental sea that is connected with the ocean by a narrow channel. { ¸med·ə·tə'rā·nē·ən 'sē }

Mediterranean Sea |GEOGR| A sea that lies between Europe, Asia Minor, and Africa and is completely landlocked except for the Strait of Gibraltar, the Bosporus, and the Suez Canal; total water area is 965,000 square miles (2,501,000 square kilometers). { ¸med·ə·tə'rā·nē·ən 'sē }

medium-range forecast |METEOROL| A forecast of weather conditions for a period of 48 hours to a week in advance. Also known as extended-range forecast. { 'mē·dē·əm ¦rānj 'fȯr¸kast }

mega- |SCI TECH| A prefix representing 10^6, or one million. Abbreviated M. { 'meg·ə }

Megachilidae |ZOO| The leaf-cutting bees, a family of hymenopteran insects in the superfamily Apoidea. { ¸meg·ə'kil·ə¸dē }

megagametophyte |BOT| The female gametophyte in plants having two types of spores. { ¦meg·ə·gə'mēd·ə¸fīt }

megaphyllous |BOT| Having large leaves or leaflike extensions. { ¦meg·ə'fil·əs }

megasporangium *See* macrosporangium. { ¦meg·ə·spə'ran·jē·əm }

megaspore *See* macrospore. { 'meg·ə¸spȯr }

megaspore mother cell *See* megasporocyte. { 'meg·ə¸spȯr 'math·ər ¸sel }

megasporocyte |BOT| A diploid cell from which four megaspores are produced by meiosis. Also known as megaspore mother cell. { ¦meg·ə'spȯr·ə¸sīt }

megasporogenesis *See* macrosporogenesis. { ¸meg·ə¸spȯr·ə'jen·ə·səs }

megasporophyll |BOT| A leaf bearing megasporangia. { ¸meg·ə'spȯr·ə¸fil }

megathermophyte |ECOL| A plant that requires great heat and abundant moisture for normal growth. Also known as macrothermophyte. { ¸meg·ə'thər·mə¸fīt }

meiofauna |ECOL| Small benthic animals ranging in size between macrofauna and microfauna; includes interstitial animals. { ¦mī·ə'fȯn·ə }

meioflora |ECOL| Small benthic plants ranging in size between macroflora and microflora; includes interstitial plants. { ¦mī·ə'flȯr·ə }

Melampsoraceae |MYCOL| A family of parasitic fungi in the order Uredinales in which the teleutospores are laterally united to form crusts or columns. { ¸mel·əm·sə 'rās·ē¸ē }

Melanconiales |MYCOL| An order of the class Fungi Imperfecti including many plant pathogens commonly causing anthracnose; characterized by densely aggregated cnidophores on an acervulus. { ¸mel·ən¸kō·nē'ā·lēz }

Melanesia |GEOGR| A group of islands in the Pacific Ocean northeast of Australia. { mel·ə'nē·zhə }

melanin |BIOL| Any of a group of brown or black pigments occurring in plants and animals. { 'mel·ə·nən }

melioidosis |VET MED| An endemic bacterial disease, primarily of rodents but occasionally communicable to humans, caused by *Pseudomonas pseudomallei* and characterized by infectious granulomas. { ˌmel·ē,ȯi'dō·səs }

mel-, melo- |SCI TECH| A combining form denoting dark or black; denoting or pertaining to melanin. { mel, 'mel·ō }

melting level |METEOROL| The altitude at which ice crystals and snowflakes melt as they descend through the atmosphere. { 'melt·iŋ ˌlev·əl }

meltwater |HYD| Water derived from melting ice or snow, especially glacier ice. { 'melt ˌwȯd·ər }

MEMC *See* methoxyethylmercury chloride.

meningitis |MED| Inflammation of the meninges of the brain and spinal cord, caused by viral, bacterial, and protozoan agents. { ˌmen·ən'jīd·əs }

meningococcus *See* Neisseria meningitidis. { məˌniŋ·gəˌkäk·əs }

Mercator projection |GEOGR| A conformal cylindrical map projection in which the surface of a sphere or spheroid, such as the earth, is conceived as developed on a cylinder tangent along the Equator; meridians appear as equally spaced vertical lines, and parallels as horizontal lines drawn farther apart as the latitude increases, such that the correct relationship between latitude and longitude scales at any point is maintained. { mər'kād·ər prəˌjek·shən }

merchantable tree height |FOR| The usable portion of the tree stem, for single-stemmed trees this is the length from an assumed stump height to an arbitrary upper-stem diameter. { ˈmər·chənt·ə·bəl 'trē ˌhīt }

mercurialism |MED| Chronic type of mercury poisoning. Also known as hydrargyrism. { mər'kyúr·ē·əˌliz·əm }

mercurial nephrosis |MED| Nephrosis (degenerative lesions of the kidney) caused by poisoning with mercury bichloride. { mər'kyúr·ē·əl ne'frō·səs }

mercurial tremor |MED| A fine muscular tremor observed in persons with mercurialism or poisoning by other heavy metals. { mər'kyür·ē·əl 'trem·ər }

mercuric chloride |CHEM| $HgCl_2$ An extremely toxic compound that forms white, rhombic crystals which sublime at 300°C and are soluble in alcohol or benzene; used for the manufacture of other mercuric compounds, as a fungicide, and in medicine and photography. Also known as bichloride of mercury; corrosive sublimate. { mər'kyür·ik 'klȯrˌīd }

mercuric cyanide |CHEM| $Hg(CN)_2$ Poisonous, colorless, transparent crystals that darken in light, decompose when heated; soluble in water and alcohol; used in photography, medicine, and germicidal soaps. Also known as mercury cyanide. { mər'kyür·ik 'sī·əˌnīd }

mercuric oleate |CHEM| $Hg(C_{18}H_{33}O_2)_2$ A poisonous yellowish-to-red liquid or solid mass; insoluble in water; used in medicine and antifouling paints, and as an antiseptic. Also known as mercury oleate. { mər'kyür·ik 'ōl·ēˌāt }

mercuric stearate |CHEM| $Hg(C_{17}H_{35}CO_2)_2$ Poisonous yellow powder; soluble in fatty acids, slightly soluble in alcohol; used as a germicide and in medicine. Also known as mercury stearate. { mər'kyür·ik 'stirˌāt }

mercury cyanide *See* mercuric cyanide. { 'mər·kyə·rē 'sī·əˌnīd }

mercury oleate *See* mercuric oleate. { 'mər·kyə·rē 'ōl·ē,āt }

mercury stearate *See* mercuric stearate. { 'mər·kyə·rē 'stir,āt }

mere [HYD] A large pond or a shallow lake. { mīr }

Merian's formula [OCEANOGR] A formula for the period of a seiche, $T = (1/n) (2L/\sqrt{gd})$, where n is the number of nodes, L is the horizontal dimension of the basin measured in the direction of wave motion, g is the acceleration of gravity, and d is the depth of the water. { 'mer·ē·ənz ,fȯr·myə·lə }

meridional circulation [METEOROL] An atmospheric circulation in a vertical plane oriented along a meridian; it consists, therefore, of the vertical and the meridional (north or south) components of motion only. [OCEANOGR] The exchange of water masses between northern and southern oceanic regions. { mə'rid·ē·ən·əl ,sər·kyə'lā·shən }

meridional flow [METEOROL] A type of atmospheric circulation pattern in which the meridional (north and south) component of motion is unusually pronounced; the accompanying zonal component is usually weaker than normal. [OCEANOGR] Current moving along a meridian. { mə'rid·ē·ən·əl 'flō }

meridional front [METEOROL] A front in the South Pacific separating successive migratory subtropical anticyclones; such fronts are essentially in the form of great arcs with meridians of longitudes as chords; they have the character of cold fronts. { mə'rid·ē·ən·əl 'frənt }

meridional index [METEOROL] A measure of the component of air motion along meridians, averaged, without regard to sign, around a given latitude circle. { mə'rid·ē·ən·əl 'in,deks }

meridional wind [METEOROL] The wind or wind component along the local meridian, as distinguished from the zonal wind. { mə'rid·ē·ən·əl 'wind }

meristem [BOT] Formative plant tissue composed of undifferentiated cells capable of dividing and giving rise to other meristematic cells as well as to specialized cell types; found in growth areas. { 'mer·ə,stem }

meromictic [HYD] Of or pertaining to a lake whose water is permanently stratified and therefore does not circulate completely throughout the basin at any time during the year. { ¦mer·ə¦mik·tik }

meroplankton [BIOL] Plankton composed of floating developmental stages (that is, eggs and larvae) of the benthos and nekton organisms. Also known as temporary plankton. { ¦mer·ə'plaŋk·tən }

merzlota *See* frozen ground. { ,merz'lō·tə }

mesa [GEOGR] A broad, isolated, flat-topped hill bounded by a steep cliff or slope on at least one side; represents an erosion remnant. { 'mā·sə }

mesa-butte [GEOGR] A butte formed as the result of erosion and reduction of a mesa. { 'mā·sə ,byüt }

mesa plain [GEOGR] A flat-topped summit of a hilly mountain. { 'mā·sə ,plān }

mesic [ECOL] **1.** Of or pertaining to a habitat characterized by a moderate amount of water. **2.** Of or pertaining to a mesophyte. { 'me·zik }

mesobenthos [OCEANOGR] The sea bottom at depths of 100–500 fathoms (180–900 meters). { ¦me·zō¦ben,thäs }

mesocarp [BOT] The middle layer of the pericarp. { 'mez·ə,kärp }

mesoclimate [CLIMATOL] **1.** The climate of small areas of the earth's surface which may not be representative of the general climate of the district. **2.** A climate characterized by moderate temperatures, that is, in the range 20–30°C. Also known as mesothermal climate. { ¦me·zō¦klī·mət }

mesoclimatology [CLIMATOL] The study of mesoclimates. { ˌme·zō‚klī·mə'täl·ə·jē }

mesohaline [OCEANOGR] 1. Referring to estuarine water with salinity ranging 5–18 parts per thousand. 2. Referring to moderately brackish water. { ˌme·sō'ha·lēn }

mesometeorology [METEOROL] That portion of the science of meteorology concerned with the study of atmospheric phenomena on a scale larger than that of micrometeorology, but smaller than the cyclonic scale. { ˌme·zō‚mē·dē·ə'räl·ə·jē }

mesopause [METEOROL] The top of the mesosphere; corresponds to the level of minimum temperature at 50 to 60 miles (80 to 95 kilometers). { 'mez·ə‚pöz }

mesopeak [METEOROL] The temperature maximum at about 30 miles (50 kilometers) in the mesosphere. { 'me·zō‚pēk }

mesophile [BIOL] An organism, as certain bacteria, that grows at moderate temperature. { 'mez·ə‚fīl }

mesophily [ECOL] Physiological response of organisms living in environments with moderate temperatures and a fairly high, constant amount of moisture. { 'mez·ə ‚fil·ē }

mesophyll [BOT] Parenchymatous tissue between the upper and lower epidermal layers in foliage leaves. { 'mez·ə‚fil }

mesophyte [ECOL] A plant requiring moderate amounts of moisture for optimum growth. { 'mez·ə‚fīt }

mesosphere [METEOROL] The atmospheric shell between about 28–35 and 50–60 miles (45–55 and 80–95 kilometers), extending from the top of the stratosphere to the mesopause; characterized by a temperature that generally decreases with altitude. { 'mez·ə‚sfir }

mesotherm [ECOL] A plant that grows successfully at moderate temperatures. { 'mez·ə ‚thərm }

mesothermal [GEOL] Of a hydrothermal mineral deposit, formed at great depth at temperatures of 200–300°C. { ¦mez·ə¦thər·məl }

mesothermal climate *See* mesoclimate. { ¦mez·ə¦thər·məl 'klī·mət }

metabiosis [ECOL] An ecological association in which one organism precedes and prepares a suitable environment for a second organism. { ˌmed·ə·bī'ō·səs }

metagenesis [BIOL] The phenomenon in which one generation of certain plants and animals reproduces asexually, followed by a sexually reproducing generation. Also known as alternation of generations. { ¦med·ə'jen·ə·səs }

metaldehyde [CHEM] $(CH_3CHO)_n$ White acetaldehyde-polymer prisms; soluble in organic solvents, insoluble in water; used as a pesticide or fuel. { me'tal·də‚hīd }

metalimnion *See* thermocline. { ˌmed·ə'lim·nē‚än }

metallic soap [CHEM] A salt of stearic, oleic, palmitic, lauric, or erucic acid with a heavy metal such as cobalt or copper; used as a drier in paints and inks, in fungicides, decolorizing varnish, and waterproofing. { mə'tal·ik 'sōp }

metallothionein [BIOL] A group of vertebrate and invertebrate proteins that bind heavy metals; it may be involved in zinc homeostasis and resistance to heavy-metal toxicity. { mə‚tal·ō'thī·ə‚nēn }

metamorphic rock [GEOL] A rock formed from preexisting solid rocks by mineralogical, structural, and chemical changes, in response to extreme changes in temperature, pressure, and shearing stress. { ¦med·ə¦mȯr·fik 'räk }

metamorphism [GEOL] The mineralogical and structural changes of solid rock in response to environmental conditions at depth in the earth's crust. { ¦med·ə¦mȯr‚fiz·əm }

metamorphosis [BIOL] **1.** A structural transformation. **2.** A marked structural change in an animal during postembryonic development. { ‚med·ə'mȯr·fə·səs }

Metaphyta [BIOL] A kingdom set up to include mosses, ferns, and other plants in some systems of classification. { mə'taf·əd·ə }

metastable *See* labile. { ¦med·ə'stā·bəl }

***meta*-toluidine** [CHEM] $CH_3C_6H_4NH_2$ A combustible, colorless, toxic liquid soluble in alcohol and ether, slightly soluble in water, boils at 203°C; used for dyes and as a chemical intermediate. { ¦med·ə tə'lü·ə‚dēn }

***meta*-xylene** [CHEM] $1,3\text{-}C_6H_4(CH_3)_2$ A flammable, toxic liquid; insoluble in water, soluble in alcohol and ether; boils at 139°C; used as an intermediate for dyes, a chemical intermediate, and a solvent, and in insecticides and aviation fuel. { ¦med·ə 'zī‚lēn }

meteoric water [HYD] Groundwater which originates in the atmosphere and reaches the zone of saturation by infiltration and percolation. { ‚mēd·ē·ȯr·ik 'wȯd·ər }

meteorogram [METEOROL] A chart in which meteorological variables are plotted against time. { ‚med·ē·ȯr·ə‚gram }

meteorological [METEOROL] Of or pertaining to meteorology or weather. { ‚med·ē·ə·rə'läj·ə·kəl }

meteorological chart [METEOROL] A weather map showing the spatial distribution, at an instant of time, of atmospheric highs and lows, rain clouds, and other phenomena. { ‚med·ē·ə·rə'läj·ə·kəl 'chärt }

meteorological data [METEOROL] Facts pertaining to the atmosphere, especially wind, temperature, and air density. { ‚med·ē·ə·rə'läj·ə·kəl 'dad·ə }

meteorological equator [METEOROL] **1.** The parallel of latitude 5° north, so called because it is the annual mean latitude of the equatorial trough. **2.** *See* equatorial trough ; intertropical convergence zone. { ‚med·ē·ə·rə'läj·ə·kəl i'kwäd·ər }

meteorological radar [METEOROL] A remote sensing device that transmits and receives microwave radiation for the purpose of detecting and measuring weather phenomena; includes Doppler radar, which is used to determine air motions (to detect tornadoes), and multiparameter radar, which provides information on the phase (ice or liquid), shapes, and sizes of hydrometeors. { ‚mēd·ē·ə·rə·'läj·ə·kəl 'rā‚där }

meteorological tide [OCEANOGR] A change in water level caused by local meteorological conditions, in contrast to an astronomical tide, caused by the attractions of the sun and moon. { ‚med·ē·ə·rə'läj·ə·kəl 'tīd }

meteorology [SCI TECH] The science concerned with the atmosphere and its phenomena; the meteorologist observes the atmosphere's temperature, density, winds, clouds, precipitation, and other characteristics and aims to account for its observed structure and evolution (weather, in part) in terms of external influence and the basic laws of physics. { ‚med·ē·ə'räl·ə·jē }

methanal *See* formaldehyde. { 'meth·ə‚nal }

methane [CHEM] CH_4 A colorless, odorless, and tasteless gas, lighter than air and reacting violently with chlorine and bromine in sunlight, a chief component of natural gas; used as a source of methanol, acetylene, and carbon monoxide. Also known as methyl hydride. { 'meth‚ān }

methane-oxidizing bacteria [MICROBIO] Bacteria that derive energy from oxidation of methane. { 'meth‚ān ¦äk·sə‚dīz·iŋ bak'tir·ē·ə }

methanogen [BIOL] A single-celled organism belonging to domain Archaea that produces methane gas as a product of anaerobic metabolism. { mə'than·ə·jən }

methanogenesis [BIOL] The biosynthesis of the hydrocarbon methane; common in certain bacteria. Also known as bacterial methanogenesis. { ¦meth·ə·nō'jen·ə·səs }

methanoic acid *See* formic acid. { ‚meth·ə'nō·ik 'as·əd }

methanotroph [MICROBIO] A bacterial organism that can use methane as its only source of carbon and energy. { mə'than·ə‚träf }

methidathion [CHEM] $C_4H_{11}O_4N_2PS_3$ A colorless, crystalline compound with a melting point of 39–40°C; used as an insecticide and miticide for pests on alfalfa, citrus, and cotton. { mə‚thid·ə'thī‚än }

methoxy- [CHEM] OCH_3 –A combining form indicating the oxygen-containing methane radical, found in many organic solvents, insecticides, and plasticizer intermediates. { mə'thäk·sē }

methoxychlor [CHEM] $Cl_3CCH(C_6H_4OCH_3)_2$ White, water-insoluble crystals melting at 89°C; used as an insecticide. Also known as DMDT; methoxy DDT. { me'thäk·si‚klȯr }

methoxy DDT *See* methoxychlor. { me'thäk·sē ¦de¦dē'tē }

methoxyethylmercury chloride [CHEM] $CH_3OCH_2CH_2HgCl$ A white, crystalline compound with a melting point of 65°C; used as a fungicide in diseases of sugarcane, pineapples, seed potatoes, and flower bulbs, and as seed dressings for cereals, legumes, and root crops. Abbreviated MEMC. { ma¦thäk·sē¦eth·əl¦mər·kyə·rē 'klȯr‚īd }

methyl allyl chloride [CHEM] $CH_2{:}C(CH_3)CH_2Cl$ Volatile, flammable, colorless liquid boiling at 72°C; has disagreeable odor; used as an insecticide and fumigant, and for chemical synthesis. { 'meth·əl 'al·əl 'klȯr‚īd }

2-methyl anthraquinone *See* tectoquinone. { tü ¦meth·əl ‚an·thrə‚kwē'nōn }

methyl bromide [CHEM] CH_3Br A toxic, colorless gas that forms a crystalline hydrate with cold water; used in synthesis of organic compounds, and as a fumigant. { 'meth·əl 'brō‚mīd }

methyl chloride *See* chloromethane. { 'meth·əl 'klȯr‚īd }

methyl chloroform *See* trichloroethane. { 'meth·əl 'klȯr·ə‚fȯrm }

methyl-*N*-(3,4-dichlorophenyl)carbamate *See* swep. { 'meth·əl ¦en ¦thrē ¦fȯr dī‚klȯr·ō'fen·əl'kär·bə‚māt }

methylene oxide *See* formaldehyde. { 'meth·ə‚lēn 'äk‚sīd }

methyl hydride *See* methane. { 'meth·əl 'hī‚drīd }

methyl isothiocyanate [CHEM] C_2H_3NS A crystalline compound, with a melting point of 35–36°C; soluble in alcohol and ether; used as a pesticide and in amino acid sequence analysis. { ¦meth·əl ¦ī·sō‚thī·ə'sī·ə‚nāt }

methylmercury compound [CHEM] Any member of a class of toxic compounds containing the methyl-mercury group, CH_3Hg. { ¦meth·əl'mər·kyə·rē ‚käm‚paúnd }

methylmercury cyanide *See* methylmercury nitrile. { ¦meth·əl'mər·kyə·rē 'sī·ə‚nīd }

methylmercury nitrile [CHEM] CH_3HgCN A crystalline solid with a melting point of 95°C; soluble in water; used as a fungicide to treat seeds of cereals, flax, and cotton. Also known as methylmercury cyanide. { ¦meth·əl'mər·kyə·rē 'nī·trəl }

methylnaphthalene [CHEM] $C_{10}H_7CH_3$ A solid melting at 34°C; used in insecticides and organic synthesis. { ‚meth·əl'naf·thə‚lēn }

methylotrophic bacteria [MICROBIO] Bacteria that are capable of growing on methane derivatives as their sole source of carbon and metabolic energy. { ¦meth·ə·lə¦trä·fik bak'tir·ē·ə }

methyl styrene   See vinyltoluene.   { 'meth·əl 'stī,rēn }

methyl violet   [CHEM] A derivative of pararosaniline, used as an antiallergen and bactericide, acid-base indicator, biological stain, and textile dye. Also known as crystal violet; gentian violet.   { 'meth·əl 'vī·lət }

MHD generator   See magnetohydrodynamic generator.   { ¦em¦āch'dē 'jen·ə·rād·ər }

micro-   [SCI TECH] **1.** A prefix indicating smallness, as in microwave. **2.** A prefix indicating extreme sensitivity, as in microradiometer and microphone.   { 'mī·krō }

microaerophilic   [MICROBIO] Pertaining to those microorganisms requiring free oxygen but in very low concentration for optimum growth.   { ¦mī·krō¦er·ə¦fil·ik }

microbe   [MICROBIO] A microorganism, especially a bacterium of a pathogenic nature.   { ¦mī,krōb }

microbial ecology   [ECOL] The study of interrelationships between microorganisms and their living and nonliving environments.   { mī,krōb·ē·əl ē'käl·ə·jē }

microbial insecticide   [MICROBIO] Species-specific bacteria which are pathogenic for and used against injurious insects.   { mī'krō·bē·əl in'sek·tə,sīd }

microburst   [METEOROL] A downdraft with horizontal extent of about 2.5 miles (4 kilometers) or less, associated with atmospheric convection, often a thundershower.   { 'mī·krō,bərst }

microclimate   [CLIMATOL] The local, rather uniform climate of a specific place or habitat, compared with the climate of the entire area of which it is a part.   { ¦mī·krō'klī·mət }

microclimatology   [CLIMATOL] The study of a microclimate, including the study of profiles of temperature, moisture and wind in the lowest stratum of air, the effect of the vegetation and of shelterbelts, and the modifying effect of towns and buildings.   { ¦mī·krō,klī·mə'täl·ə·jē }

microconsumer   See decomposer.   { ¦mī·krō·kən'sü·mər }

microcyclic   [MYCOL] Referring to a rust fungus with a short life cycle.   { ¦mī·krə¦sī·klik }

microenvironment   [ECOL] The specific environmental factors in a microhabitat.   { ¦mī·krō·in'vī·ərn·mənt }

microfauna   [ECOL] Microscopic animals such as protozoa and nematodes.   { 'mī·krə ,fȯn·ə }

microflora   [ECOL] The flora of a microhabitat.   { ¦mī·krō'flȯr·ə }

microgametophyte   [BOT] The male gametophyte in plants having two types of spores.   { ¦mī·krō·gə'mēd·ə,fīt }

microgeography   [GEOGR] The detailed empirical geographical study on a small scale of a specific locale.   { ,mī·krə·jē'äg·rə·fē }

microhabitat   [ECOL] A small, specialized, and effectively isolated location.   { ¦mī·krō 'hab·ə,tat }

microlayer   [OCEANOGR] The thin zone beneath the surface of the ocean or any free water surface within which physical processes are modified by proximity to the air-water boundary.   { 'mī·krō,lā·ər }

micrometeorology   [METEOROL] That portion of the science of meteorology that deals with the observation and explanation of the smallest-scale physical and dynamic occurrences within the atmosphere; studies are confined to the surface boundary layer of the atmosphere, that is, from the earth's surface to an altitude where the effects of the immediate underlying surface upon air motion and composition become negligible.   { ¦mī·krō,mē·dē·ə'räl·ə·jē }

micronekton   [ECOL] Active pelagic crustaceans and other forms intermediate between thrusting nekton and feebler-swimming plankton.   { ,mī·krə'nek,tän }

micronutrient [BIOL] An element required by animals or plants in small amounts. { ¦mī·krō'nü·trē·ənt }

microorganism [MICROBIO] A microscopic organism, including bacteria, protozoans, yeast, viruses, and algae. { ¦mī·krō'òr·gə,niz·əm }

microphagy [BIOL] Feeding on minute organisms or particles. { mī'kräf·ə·jē }

microphyllous [BOT] **1.** Having small leaves. **2.** Having leaves with a single, unbranched vein. { ¦mī·krō¦fil·əs }

microphyte [ECOL] **1.** A microscopic plant. **2.** A plant that is dwarted due to unfavorable environmental conditions. { 'mī·krə,fīt }

microplankton [ECOL] Zooplankton between 20 and 200 micrometers in size. { 'mī·krə ,plaŋk·tən }

micropore [GEOL] A soil pore small enough to hold water against the pull of gravity and to retard water flow. { 'mī·krə,pòr }

microrelief [GEOGR] Irregularities of the land surface causing variations in elevation amounting to no more than a few feet. { 'mī·krō·ri,lēf }

microspecies [ECOL] A small, localized species population that is clearly differentiated from related forms. Also known as jordanon. { ¦mī·krō'spē·shēz }

microsporangium [BOT] A sporangium bearing microspores. { ¦mī·krō·spə 'ran·jē·əm }

microspore [BOT] The smaller spore of heterosporous plants; gives rise to the male gametophyte. { 'mī·krə,spòr }

microspore mother cell See microsporocyte. { 'mī·krə,spòr 'məth·ər ,sel }

microsporocyte [BOT] A diploid cell from which four microspores are produced by meiosis. Also known as microspore mother cell. { ¦mī·krō'spòr·ə,sīt }

microsporogenesis [BOT] In angiosperms, formation of microspores and production of the male gametophyte. { ,mī·krə,spòr·ə'jen·ə·səs }

microsporophyll [BOT] A sporophyll bearing microsporangia. { ¦mī·krō'spòr·ə,fil }

microtherm [ECOL] A plant requiring a mean annual temperature range of 0–14°C for optimum growth. { 'mī·krə,thərm }

microthermal climate [CLIMATOL] A temperature province in both of C.W. Thornthwaite's climatic classifications, generally described as a "cool" or "cold winter" climate. { ¦mī·krə'thər·məl 'klī·mət }

middle clouds [METEOROL] Types of clouds the mean level of which is between 6500 and 20,000 feet (1980 and 6100 meters); the principal clouds in this group are altocumulus and altostratus. { 'mid·əl 'klaùdz }

middle latitude [GEOGR] **1.** A point of latitude that is midway on a north-and-south line between two parallels. Also known as mid-latitude. { 'mid·əl 'lad·ə,tüd }

middle-latitude westerlies See westerlies. { 'mid·əl ¦lad·ə,tüd 'wes·tər,lēz }

mid-extreme tide [OCEANOGR] A level midway between the extreme high water and extreme low water occurring at a place. { 'mid ik¦strēm 'tīd }

midge [ZOO] Any of various dipteran insects, principally of the families Ceratopogonidae, Cecidomyiidae, and Chironomidae; many are biting forms and are vectors of parasites of man and other vertebrates. { mij }

mid-latitude See middle latitude. { 'mid ¦lad·ə,tüd }

mid-latitude westerlies See westerlies. { 'mid ¦lad·ə,tüd 'wes·tər,lēz }

mid-ocean canyon See deep-sea channel. { 'mid ¦ō·shən 'kan·yən }

mid-oceanic ridge [GEOL] A continuous, median, seismic mountain range on the floor of the ocean, extending through the North and South Atlantic oceans, the Indian Ocean, and the South Pacific Ocean; the topography is rugged, elevation is 0.6–1.8 miles (1–3 kilometers), width is about 900 miles (1500 kilometers), and length is over 52,000 miles (84,000 kilometers). Also known as mid-ocean ridge; mid-ocean rise; oceanic ridge. { 'mid‚ō·shē¦an·ik 'rij }

mid-ocean ridge *See* mid-oceanic ridge. { 'mid¦ō·shən 'rij }

mid-ocean rise *See* mid-oceanic ridge. { 'mid¦ō·shən 'rīs }

migrant [BIOL] An organism that moves from one habitat to another. { 'mī·grənt }

migration [HYD] Slow, downstream movement of a system of stream meanders. { mī'grā·shən }

Milankovitch cycles [GEOPHYS] Periodic variations in the earth's position relative to the sun as the earth orbits, affecting the distribution of the solar radiation reaching the earth and causing climatic changes that have profound impacts on the abundance and distribution of organisms, best seen in the fossil record of the Quaternary Period (the last 1.6 million years). { mē·lən'kō·vich ‚sīk·əlz }

mildew [MYCOL] **1.** A whitish growth on plants, organic matter, and other materials caused by a parasitic fungus. **2.** Any fungus producing such growth. { 'mil‚dü }

milky weather *See* whiteout. { 'mil·kē 'weth·ər }

Miller law [AGR] A law administered by the Food and Drug Administration that regulates the production and use of agricultural fungicides in the United States, and will not allow materials to leave poisonous residues on edible crops. { 'mil·ər ‚lò }

millet [BOT] A common name applied to at least five related members of the grass family grown for their edible seeds. { 'mil·ət }

milli-micro- *See* nano-. { ¦mil·ə¦mī·krō }

Mima mound [GEOGR] A circular or oval domelike structure composed of loose silt and soil that is believed to be generated by a combination of geomorphic processes and burrowing by animals; found in northwest North America, Africa, and southern South America. { 'mē·mə ‚maúnd }

mimetic camouflage [ECOL] Protective coloration of a prey such that it resembles some other object, which is recognized by the predator but not associated in its mind with feeding. { mə‚med·ik 'kam·ə‚fläzh }

mine inspector [PETR MIN] Generally, the state mine inspector, as contrasted to the Federal mine inspector; inspects mines to find fire and dust hazards and inspects the safety of working areas, electric circuits, and mine equipment. { 'mīn in'spek·tər }

mineral [GEOL] A naturally occurring substance with a characteristic chemical composition expressed by a chemical formula; may occur as individual crystals or may be disseminated in some other mineral or rock; most mineralogists include the requirements of inorganic origin and internal crystalline structure. { 'min·rəl }

mineral deposit [GEOL] A mass of naturally occurring mineral material, usually of economic value. { 'min·rəl di‚päz·ət }

mineral fuel [PETR MIN] A carbonaceous fuel mined or stripped from the earth, such as petroleum, coal, peat, shale oil, or tar sands. { 'min·rəl ¦fyül }

mineral green *See* copper carbonate. { 'min·rəl ¦grēn }

mineralization [GEOL] **1.** The process of fossilization whereby inorganic materials replace the organic constituents of an organism. **2.** The introduction of minerals into a rock, resulting in a mineral deposit. { ‚min·rə·lə'zā·shən }

mineralize [GEOL] To convert to, or impregnate with, mineral material; applied to processes of ore vein deposition and of fossilization. { 'min·rə,līz }

mineral resources [GEOL] Valuable mineral deposits of an area that are presently recoverable and may be so in the future; includes known ore bodies and potential ore. { 'min·rəl ri'sȯrs·əz }

mineral soil [GEOL] Soil composed of mineral or rock derivatives with little organic matter. { 'min·rəl ,sȯil }

mineral spring [HYD] A spring whose water has a definite taste due to a high mineral content. { 'min·rəl ¦spriŋ }

mineral water [HYD] Water containing naturally or artificially supplied minerals or gases. { 'min·rəl ,wȯd·ər }

miner's self-rescuer [PETR MIN] A pocket gas mask effective against carbon monoxide; air passes through a cannister containing fused calcium chloride before entering the mouth. { 'mīn·ərz ¦self 'res·kyü·ər }

Minimata disease [MED] A disorder resulting from methyl mercury poisoning, which occurred in epidemic proportions in 1956 in Minimata Bay, a Japanese coastal town, where the inhabitants ate fish contaminated by industrial pollution; the most obvious symptoms are tremors and involuntary movements. { ,min·ē'mäd·ə di,zēz }

minimum ebb [OCEANOGR] The least speed of a current that runs continuously ebb. { 'min·ə·məm 'eb }

minimum flood [OCEANOGR] The least speed of a current that runs continuously flood. { 'min·ə·məm 'fləd }

minimum thermometer [ENG] A thermometer that automatically registers the lowest temperature attained during an interval of time. { 'min·ə·məm thər'mäm·əd·ər }

MIPC *See ortho*-isopropylphenyl-methylcarbamate.

mire [GEOL] Wet spongy earth, as of a marsh, swamp, or bog. { mīr }

misfit stream [HYD] A stream whose meanders are either too large or too small to have eroded the valley in which it flows. { 'mis,fit ¦strēm }

mist [METEOROL] A hydrometeor consisting of an aggregate of microscopic and more or less hygroscopic (condensation-producing) water droplets suspended in the atmosphere; it produces, generally, a thin, grayish veil over the landscape; it reduces visibility to a lesser extent than fog; the relative humidity with mist is often less than 95. { mist }

mist droplet [METEOROL] A particle of mist, intermediate between a haze droplet and a fog drop. { 'mist ,dräp·lət }

miticide [AGR] An agent that kills mites. Also known as acaricide. { 'mīd·ə,sīd }

mixed cloud [METEOROL] A cloud containing both water drops and ice crystals, hence a cloud whose composition is intermediate between that of a water cloud and that of an ice-crystal cloud. { 'mikst 'klaüd }

mixed current [OCEANOGR] A type of tidal current characterized by a conspicuous difference in speed between the two flood currents or two ebb currents usually occurring each tidal day. { 'mikst 'kə·rənt }

mixed forest [FOR] A forest consisting of two or more types of trees, with no more than 80% of the most common tree. { 'mikst 'fär·əst }

mixed layer [OCEANOGR] The layer of water which is mixed through wave action or thermohaline convection. Also known as surface water. { 'mikst 'lā·ər }

mixed tide   [OCEANOGR] A tide in which the presence of a diurnal wave is conspicuous by a large inequality in the heights of either the two high tides or the two low tides usually occurring each tidal day.   { 'mikst 'tīd }

mixing ratio   [METEOROL] In a system of moist air, the dimensionless ratio of the mass of water vapor to the mass of dry air; for many purposes, the mixing ratio may be approximated by the specific humidity.   { 'mik·siŋ ˌrā·shō }

mixolimnion   [HYD] The upper layer of a meromictic lake, characterized by low density and free circulation; this layer is mixed by the wind.   { ¦mik·sō'lim·nē,än }

mixotrophic   [BIOL] Referring to an organism that uses both organic and inorganic compounds as sources of carbon or energy.   { ¦mik·sə¦träf·ik }

moat   [HYD] **1.** A glacial channel in the form of a deep, wide trench. **2.** See oxbow lake.   { mōt }

mobile belt   [GEOL] A long, relatively narrow crustal region of tectonic acitivity.   { 'mō·bəl ¦belt }

mock fog   [METEOROL] A simulation of true fog by atmospheric refraction.   { 'mäk ˌfäg }

model atmosphere   [METEOROL] Any theoretical representation of the atmosphere, particularly of vertical temperature distribution.   { 'mäd·əl 'at·mə,sfir }

moderate breeze   [METEOROL] In the Beaufort wind scale, a wind whose speed is from 11 to 16 knots (13 to 18 miles per hour or 20 to 30 kilometers per hour).   { 'mäd·ə·rət 'brēz }

moderate gale   [METEOROL] In the Beaufort wind scale, a wind whose speed is from 28 to 33 knots (32 to 38 miles per hour or 52 to 61 kilometers per hour).   { 'mäd·ə·rət 'gāl }

moist air   [METEOROL] **1.** In atmospheric thermodynamics, air that is a mixture of dry air and any amount of water vapor. **2.** Generally, air with a high relative humidity   { 'mȯist 'er }

moist climate   [CLIMATOL] In C.W. Thornthwaite's climatic classification, any type of climate in which the seasonal water surplus counteracts seasonal water deficiency; thus it has a moisture index greater than zero.   { 'mȯist 'klīm·ət }

moisture   [METEOROL] The water vapor content of the atmosphere, or the total water substance (gaseous, liquid, and solid) present in a given volume of air.   { 'mȯis·chər }

moisture content   [ENG] The quantity of water in a mass of soil, sewage, sludge, or screenings; expressed in percentage by weight of water in the mass.   { 'mȯis·chər ˌkän·tent }

moisture inversion   [METEOROL] An increase with height of the moisture content of the air; specifically, the layer through which this increase occurs, or the altitude at which the increase begins.   { 'mȯis·chər in,vər·zhən }

mold   [GEOL] Soft, crumbling friable earth. [MYCOL] Any of various woolly fungus growths.   { mōld }

molecular fossils   See biomarkers.   { mə¦lek·yə·lər 'fäs·əlz }

molinate   [CHEM] $C_9H_{17}NOS$ A light yellow liquid with limited solubility in water; used as a herbicide to control watergrass in rice.   { 'mäl·ə,nāt }

Mollisol   [GEOL] An order of soils having dark or very dark, friable, thick A horizons high in humus and bases such as calcium and magnesium; most have lighter-colored or browner B horizons that are less friable and about as thick as the A horizons; all but a few have paler C horizons, many of which are calcareous.   { 'mal·ə,säl }

Monera   [BIOL] A kingdom in the old five-kingdom classification scheme, which is now renamed the domain Bacteria.   { mə'nir·ə }

Moniliales [MYCOL] An order of fungi of the Fungi Imperfecti containing many plant pathogens; asexual spores are always formed free on the surface of the material on which the organism is living. { mə‚nil·ē'ā·lēz }

moniliasis See candidiasis. { mə‚nil·ē'ī·ə·səs }

Monilinia fructicola [MYCOL] A fungal pathogen in the class Discomycetes that causes brown rot of stone fruits. { ‚mō·nə‚lin·ē·ə ‚frük·ti'kō·lə }

monimolimnion [HYD] The dense bottom stratum of a meromictic lake; it is stagnant and does not mix with the water above. { ¦man·ə·mō'lim·nē‚än }

monkeypox [VET MED] An animal virus that causes a smallpox-like eruption but only rarely infects humans and has little potential for interhuman spread. { 'məŋ·kē‚päks }

monocarpic [BOT] Bearing fruit once and then dying. { ¦män·ō¦kär·pik }

monoclimax [ECOL] A climax community controlled primarily by one factor, as climate. { ¦män·ō¦klī‚maks }

monoclinic [BOT] Having both stamens and pistils in the same flower. { ¦män·ə'klin·ik }

monocotyledon [BOT] Any plant of the class Liliopsida; all have a single cotyledon. { ¦män·ə‚käd·əl'ēd·ən }

monoecious [BOT] **1.** Having both staminate and pistillate flowers on the same plant. **2.** Having archegonia and antheridia on different branches. { mə'nē·shəs }

monophagous [ZOO] Subsisting on a single kind of food. Also known as monotrophic. { mə'näf·ə·gəs }

monosodium acid methanearsonate [CHEM] CH_4AsNaO_3 A white, crystalline solid; melting point is 132–139°C; soluble in water; used as an herbicide for grassy weeds on rights-of-way, storage areas, and noncrop areas, and as preplant treatment for cotton, citrus trees, and turf. Abbreviated MSMA. { ¦män·ə'sōd·ē·əm ¦as·əd ¦meth‚än'ärs·ən ‚āt }

monotrophic See monophagous. { ¦män·ə¦träf·ik }

monoxide [CHEM] A compound that contains a single oxygen atom, such as carbon monoxide, CO. { mə'näk‚sīd }

monsoon [METEOROL] A large-scale wind system which predominates or strongly influences the climate of large regions, and in which the direction of the wind flow reverses from winter to summer; an example is the wind system over the Asian continent. { män'sün }

monsoon climate [CLIMATOL] The type of climate which is found in regions subject to monsoons. { män'sün ‚klī·mət }

monsoon current [OCEANOGR] A seasonal wind-driven current occurring in the northern part of the Indian Ocean. { män'sün ‚kə·rənt }

monsoon fog [METEOROL] An advection type of fog occurring along a coast where monsoon winds are blowing, when the air has a high specific humidity and there is a large difference in the temperature of adjacent land and sea. { män'sün ‚fäg }

monsoon forest [ECOL] A tropical forest occurring in regions where a marked dry season is followed by torrential rain; characterized by vegetation adapted to withstand drought. { ¦män‚sün 'fär·əst }

monsoon low [METEOROL] A seasonal low found over a continent in the summer and over the adjacent sea in the winter. { män'sün ‚lō }

montane [ECOL] Of, pertaining to, or being the biogeographic zone composed of moist, cool slopes below the timberline and having evergreen trees as the dominant life-form. { män'tān }

moon pillar |METEOROL| A halo consisting of a vertical shaft of light through the moon. { 'mün ,pil·ər }

moor   *See* bog.   { mur }

mor   *See* ectohumus.   { mor }

morainal lake |HYD| A glacial lake filling a depression resulting from irregular deposition of drift in a terminal or ground moraine of a continental glacier.   { mə'rān· əl 'lāk }

Moraxella |MICROBIO| A genus of bacteria that are parasites of mucous membranes. { mə'rak·sə·lə }

morbidity |MED| **1.** The quantity or state of being diseased.   **2.** The conditions inducing disease.   **3.** The ratio of the number of sick individuals to the total population of a community.   { mor'bid·əd·ē }

morel |MYCOL| Any fungus belonging to the genus *Morchella*, distinguished by a large, pitted, spongelike cap; it is a highly prized food, but may be poisonous when taken with alcohol.   { mə'rel }

mores |ECOL| Groups of organisms preferring the same physical environment and having the same reproductive season.   { 'mor,āz }

morphogenetic region |GEOL| A region in which, under certain climatic conditions, the predominant geomorphic processes will contribute regional characteristics to the landscape that contrast with those of other regions formed under different climatic conditions.   { ,mor·fə·jə¦ned·ik 'rē·jən }

mortality rate |MED| For a given period of time, the ratio of the number of deaths occurring per 1000 population. Also known as death rate.   { mor'tal·əd·ē ,rāt }

mortlake   *See* oxbow lake.   { 'mort,lāk }

mosquito |ZOO| Any member of the dipterous subfamily Culicinae; a slender fragile insect, with long legs, a long slender abdomen, and narrow wings.   { mə'skēd·ō }

moss |BOT| Any plant of the class Bryatae, occurring in nearly all damp habitats except the ocean.   { mos }

moss forest   *See* temperate rainforest.   { 'mos 'fär·əst }

moss land |ECOL| An area which contains abundant moss but is not wet enough to be a bog.   { 'mos ,land }

mother liquor   *See* filtrate.   { 'məth·ər ,lik·ər }

mother-of-pearl clouds   *See* nacreous clouds.   { 'məth·ər əv 'pərl 'klaudz }

motile |BIOL| Being capable of spontaneous movement.   { mōd·əl }

motility symbiosis |ECOL| A symbiotic relationship in which motility is conferred upon an organism by its symbiont.   { mō¦til·əd·ē ,sim·bē'ō·səs }

motu |GEOGR| One of a series of closely spaced coral islets separated by narrow channels; the group of islets forms a ring-shaped atoll.   { 'mō·tü }

moulin |HYD| A shaft or hole in the ice of a glacier which is roughly cylindrical and nearly vertical, formed by swirling meltwater pouring down from the surface. Also known as glacial mill; glacier mill; glacier pothole; glacier well; pothole.   { mü'lan }

mountain |GEOGR| A feature of the earth's surface that rises high above the base and has generally steep slopes and a relatively small summit area.   { 'maunt·ən }

mountain and valley winds |METEOROL| A system of diurnal winds along the axis of a valley, blowing uphill and upvalley by day, and downhill and downvalley by night; they prevail mostly in calm, clear weather.   { 'maunt·ən ən 'val·ē 'winz }

mountain breeze [METEOROL] A breeze that blows down a mountain slope due to the gravitational flow of cooled air. Also known as mountain wind. { 'maunt·ən 'brēz }

mountain chain See mountain system. { 'maunt·ən ˌchān }

mountain climate [CLIMATOL] Very generally, the climate of relatively high elevations; mountain climates are distinguished by the departure of their characteristics from those of surrounding lowlands, because great variety is introduced by differences in latitude, elevation, and exposure to the sun, there exists no single, clearly defined, mountain climate. Also known as highland climate. { 'maunt·ən ˈklī·mət }

mountain-gap wind [METEOROL] A local wind blowing through a gap between mountains. { 'maunt·ən ˈgap ˌwind }

mountain glacier See alpine glacier. { 'maunt·ən 'glā·shər }

mountain meteorology [METEOROL] The branch of meteorology that studies the effects of mountains on the atmosphere, ranging over all scales of motion. { 'maunt·ən ˌmēd·ē·ə'räl·ə·jē }

mountain range [GEOGR] A succession of mountains or narrowly spaced mountain ridges closely related in position, direction, and geologic features. { 'maunt·ən ˌrānj }

mountain system [GEOGR] A group of mountain ranges tied together by common geological features. Also known as mountain chain. { 'maunt·ən ˌsis·təm }

mountain wave [METEOROL] An undulating flow of wind on the downwind, or lee, side of a mountain ridge caused by wind blowing strongly over the ridge. { 'maunt·ən ˌwāv }

mountain wind See mountain breeze. { 'maunt·ən ˌwind }

mountain wood [GEOL] A compact, fibrous, gray to brown type of asbestos which has an appearance similar to dry wood. Also known as rock wood. { 'maunt·ən ˌwud }

mouth [GEOGR] **1.** The place where one body of water discharges into another. Also known as influx. **2.** The entrance or exit of a geomorphic feature, such as of a cave or valley. { mauth }

Mozambique Current [OCEANOGR] The portion of the South Equatorial Current that turns and flows along the coast of Africa in the Mozambique Channel, forming one of the western boundary currents in the Indian Ocean. { ˌmō·zəm'bēk 'kə·rənt }

MSMA See monosodium acid methanearsonate.

mucigel [BIOL] A complex polysaccharide material that is composed of root mucilage and bacterial slime and acts to control aggregation of soil particles in the rhizosphere in the vicinity of older portions of plant roots. { 'myü·sə,jel }

mucilage [BOT] A sticky, gelatinous substance produced by some plants; in carnivorous plants, may be secreted on leaves to capture prey. { 'myü·sə·lij }

muck [GEOL] Dark, finely divided, well-decomposed, organic matter intermixed with a high percentage of mineral matter, usually silt, forming a surface deposit in some poorly drained areas. { mək }

mucormycosis [MED] An acute, usually fulminating fungus infection of humans caused by several genera of Mucorales, including A*bsidia*, *Rhizopus*, and M*ucor*. { ˈmyü·kòr,mī 'kō·səs }

mud [GEOL] An unindurated mixture of clay and silt with water; it is slimy with a consistency varying from that of a semifluid to that of a soft and plastic sediment. { məd }

mud flat [GEOL] A relatively level, sandy or muddy coastal strip along a shore or around an island; may be alternately covered and uncovered by the tide or may be covered by shallow water. Also known as flat. { 'məd ˌflat }

mudslide [GEOL] A slow-moving mudflow in which movement is mainly by sliding upon a discrete boundary shear surface. { 'məd,slīd }

muggy [METEOROL] Referring to warm and especially humid weather. { 'məg·ē }

mull [GEOGR] *See* headland. [GEOL] Granular forest humus that is incorporated with mineral matter. { məl }

multicycle [GEOL] Pertaining to a landscape or landform produced by more than one cycle of erosion. { 'məl·tə,sī·kəl }

multideck clarifiers [ENG] Extraction units which remove pollutants from recycled plant waste water. { 'məl·tə,dek 'klar·ə,fī·ərz }

multiple-current hypothesis [OCEANOGR] The hypothesis that the Gulf Stream, instead of being composed of a single tortuous current, actually consists of many quasipermanent currents, countercurrents, and eddies. { 'məl·tə·pəl ¦kə·rənt hī ,päth·ə·səs }

multiple fruit [BOT] Any fruit derived from the ovaries and accessory structures of several flowers consolidated into one mass, such as a pineapple and mulberry. { 'məl·tə·pəl 'früt }

murine plague [VET MED] Infection of the rat by the bacterium *Pasteurella pestis*; transmitted from rat to rat and from rat to human by a flea. { 'myu̇,rīn ¦plāg }

muscardine diseases [ZOO] A group of insect diseases caused by the muscardine fungi, in which the fungal pathogen emerges from the body of the insect and covers the animal with a characteristic fungus mat. { 'məs·kər,dēn di,zēz·əs }

muscarine [CHEM] $C_8H_{19}NO_3$ A quaternary ammonium compound, the toxic ingredient of certain mushrooms, as *Amanita muscaria*. Also known as hydroxycholine. { 'məs·kə ,rēn }

Musci *See* Bryopsida. { 'mə,sī }

musculoskeletal toxicity [MED] Adverse effects to the structure and/or function of the muscles, bones, and joints caused by exposure to a toxic chemical, such as coal dust or cadmium. Also, the bone disorders arthritis, fluorosis, and osteomalacia can result. { ,məs·kyə·lō¦skel·ət·əl ,tak'sis·əd·ē }

muskeg [ECOL] A peat bog or tussock meadow, with variably woody vegetation. { 'mə ,skeg }

mustard gas [CHEM] $HS(CH_2ClCH_2)_2S$ An oil with density 1.28, boiling point 215°C; used in chemical warfare. Also known as dichlorodiethylsulfide. { 'məs·tərd ,gas }

mustard oil *See* allyl isothiocyanate. { 'məs·tərd ,oil }

mutagen [GEN] An agent that raises the frequency of mutation above the spontaneous or background rate. { 'myüd·ə·jən }

mutagen persistence [GEN] The stability of a mutagen in the environment or in the human body. { ¦myüd·ə·jən pər'sis·təns }

mutualism [ECOL] Mutual interactions between two species that are beneficial to both species. { 'myü·chə·wə,liz·əm }

mycelium [BIOL] A mass of filaments, or hyphae, composing the vegetative body of many fungi and some bacteria. { mī'sē·lē·əm }

mycetome [ZOO] One of the specialized structures in the body of certain insects for holding endosymbionts. { 'mī·sə,tōm }

mycobiont [BOT] The fungal component of a lichen, commonly an ascomycete. { ¦mī·kə'bī,änt }

mycology [BOT] The branch of botany that deals with the study of fungi. { mī 'käl·ə·jē }

mycorrhiza [BOT] A mutual association in which the mycelium of a fungus invades the roots of a seed plant. { ‚mīk·ə'rīz·ə }

mycorrhizal fungi [MYCOL] Fungi that form symbiotic relationships in and on the roots of host plants. { ‚mī·kə‚rīz·əl 'fən‚jī }

mycosis [MED] An infection with or a disease caused by a fungus. { mī'kō·səs }

mycotic stomatitis *See* thrush. { mī'käd·ik ‚stō·mə'tīd·əs }

mycotoxicosis [MED] Any of a group of diseases caused by accidental or recreational ingestion of toxic fungal metabolites, such as mushroom poisoning. { ‚mī·kō ‚täk·sə'kō·səs }

mycotoxin [MYCOL] A toxin produced by a fungus. { 'mī·kə‚täk·sən }

myrmecophile [ECOL] A species that relies on ants for food or protection. { mər'mek·ə ‚fīl }

myrmecophyte [ECOL] A plant that houses and benefits from the habitation of ants. { mər'mek·ə‚fīt }

myxomatosis [VET MED] A virus disease of rabbits producing fever, skin lesions resembling myxomas, and mucoid swelling of mucous membranes. { mik‚sō·mə'tō· səs }

myxovirus [MICROBIO] A group of ribonucleic-acid animal viruses characterized by hemagglutination and hemadsorption; includes influenza and fowl plague viruses and the paramyxoviruses. { 'mik·sə‚vī·rəs }

N

nacreous clouds [METEOROL] Clouds of unknown composition, whose form resembles that of cirrus or altocumulus lenticularis, and which show very strong irisation similar to that of mother-of-pearl, especially when the sun is several degrees below the horizon; they occur at heights of about 12 or 18 miles (20 or 30 kilometers). Also known as mother-of-pearl clouds. { 'nā·krē·əs 'klaùdz }

nadir [OCEANOGR] The point on the sea floor that lies directly below the sonar during a survey. { 'nā·dər }

nailhead spot [PL PATH] A fungus rot of tomato caused by *Alternaria tomato* and marked by small brown to black sunken spots on the fruit. { 'nāl,hed ,spät }

Nairobi sheep disease [VET MED] A tick-borne viral disease of sheep and goats that is caused by a ribonucleic acid-containing virus of the genus *Nairovirus*, characterized by hemorrhagic gastroenteritis and high mortality. { nī,rō·bē 'shēp di,zēz }

Nairovirus [MICROBIO] A genus of the viral family Bunyaviridae that causes Nairobi sheep disease. { 'nī·rə,vī·rəs }

naked bud [BOT] A bud covered only by rudimentary foliage leaves. { 'nā·kəd 'bəd }

nannoplankton [BIOL] Minute plankton; the smallest (usually from 2 to 20 nanometers) plankton, including algae, bacteria, and protozoans. Also spelled nanoplankton. { ¦nan·ō'plaŋk·tən }

nano- [BIOL] A prefix meaning dwarfed. { 'nan·ō }

nanophanerophyte [ECOL] A shrub not exceeding 6.6 feet (2 meters) in height. { ¦nan·ō'fan·ə·rə,fīt }

nanoplankton *See* nannoplankton. { ¦nan·ō'plaŋk·tən }

Nansen bottle [ENG] A bottlelike water-sampling device with valves at both ends that is lowered into the water by wire; at the desired depth it is activated by a messenger which strikes the reversing mechanism and inverts the bottle, closing the valves and trapping the water sample inside. Also known as Petterson-Nansen water bottle; reversing water bottle. { 'nan·sən ,bäd·əl }

naphthalene [CHEM] $C_{10}H_8$ White, volatile crystals with coal tar aroma; insoluble in water, soluble in organic solvents; structurally it is represented as two benzenoid rings fused together; boiling point 218°C, melting point 80.1°C; used for moth repellents, fungicides, lubricants, and resins, and as a solvent. Also known as naphthalin; tar camphor. { 'naf·thə,lēn }

naphthalin *See* naphthalene. { 'naf·thə·lən }

2-(α-naphthoxy)-N,N-diethylpropionamide *See* devrinol. { ¦tü ¦al·fə naf'thäk·sē ¦en ¦en dī,eth·əl,pro·pē'än·ə·məd }

N-1-naphthylphthalamic acid [CHEM] $C_{10}H_7NHCOC_6H_4COOH$ A crystalline solid with a melting point of 185°C; used as a preemergence herbicide. { ¦en ¦wən ¦naf·thil·thə'lam·ik 'as·əd }

1-(1-naphthyl)-2-thiourea [CHEM] $C_{10}H_7NHCSNH_2$ A crystalline compound with a melting point of 198°C; soluble in water, acetone, triethylene glycol, and hot alcohol; used as a poison to control the adult Norway rat. { ¦wən ¦wən 'naf·thil ¦tü ¸thī·ə·yü'rē·ə }

narrow [GEOGR] A constricted section of a mountain pass, valley, or cave, or a gap or narrow passage between mountains. { 'nar·ō }

narrows [GEOGR] A navigable narrow part of a bay, strait, or river. { 'nar·ōz }

narrow-spectrum antibiotic [MICROBIO] An antibiotic effective against a limited number of microorganisms. { 'nar·ō¦spek·trəm ¸ant·i¸bī'äd·ik }

nastic movement [BOT] Movement of a flat plant part, oriented relative to the plant body and produced by diffuse stimuli causing disproportionate growth or increased turgor pressure in the tissues of one surface. { 'nas·tik 'müv·mənt }

Nathansohn's theory [OCEANOGR] The theory that nutrient salts in the lighted surface layers of the ocean are consumed by plants, accumulate in the deep ocean through sinking of dead plant and animal bodies, and eventually return to the euphotic layer through diffusion and vertical circulation of the water. { 'nā·thən·sənz ¸thē·ə·rē }

native [BIOL] Grown, produced or originating in a specific region or country. { 'nād·iv }

native element [GEOL] Any of 20 elements, such as copper, gold, and silver, which occur naturally uncombined in a nongaseous state; there are three groups—metals, semimetals, and nonmetals. { 'nād·iv 'el·ə·mənt }

native metal [GEOCHEM] A metallic native element; includes silver, gold, copper, iron, mercury, iridium, lead, palladium, and platinum. { 'nād·iv 'med·əl }

native uranium [GEOCHEM] Uranium as found in nature; a mixture of the fertile uranium-238 isotope (99.3%), the fissionable uranium-235 isotope (0.7%), and a minute percentage of other uranium isotopes.Also known as natural uranium; normal uranium. { 'nād·iv yə'rā·nē·əm }

natron lake *See* soda lake. { 'nā·trən ¸lāk }

natural gas [PETR MIN] A combustible, gaseous mixture of low-molecular-weight paraffin hydrocarbons, generated below the surface of the earth; contains mostly methane and ethane with small amounts of propane, butane, and higher hydrocarbons, and sometimes nitrogen, carbon dioxide, hydrogen sulfide, and helium. { 'nach·rəl 'gas }

natural gasoline [PETR MIN] The liquid paraffin hydrocarbon contained in natural gas and recovered by compression, distillation, and absorption. { 'nach·rəl ¸gas·ə'lēn }

natural immunity [MED] Native immunity possessed by the individuals of a race, strain, or species. { 'nach·rəl i'myü·nəd·ē }

naturalized [ECOL] Of a species, having become permanently established after being introduced. { 'nach·rə¸līzd }

natural load [HYD] The quantity of sediment carried by a stable stream. { 'nach·rəl 'lōd }

natural radiation *See* background radiation. { 'nach·rəl ¸rād·ē'ā·shən }

natural selection [GEN] Darwin's theory of evolution, according to which organisms tend to produce progeny far above the means of subsistence; in the struggle for existence that ensues, only those progeny with favorable variations survive; the favorable variations accumulate through subsequent generations, and descendants diverge from their ancestors. { 'nach·rəl si'lek·shən }

natural uranium *See* native uranium. { 'nach·rəl yü'rā·nē·əm }

naval meteorology [METEOROL] The branch of meteorology which studies the interaction between the ocean and the overlying air mass, and which is concerned with atmospheric phenomena over the oceans, the effect of the ocean surface on these phenomena, and the influence of such phenomena on shallow and deep seawater. { 'nā·vəl ˌmē·dē·ə'räl·ə·jē }

Navy Oceanographic and Meteorological Automatic Device [OCEANOGR] A 6-meter-long, boat-shaped, moored instrumented buoy. Abbreviated NOMAD. { ¦nāv·ē ˌō·shə·nə¦graf·ik en ˌmēd·ē·ə·rə¦läj·ə·kəl ¦òd·ə‚mad·ik di'vīs }

neap high water *See* mean high-water neaps. { 'nēp 'hī ˌwòd·ər }

neap low water *See* mean low-water neaps. { 'nēp 'lō ˌwòd·ər }

neaps *See* neap tide. { nēps }

neap tidal currents [OCEANOGR] Tidal currents of decreased speed occurring at the time of neap tides. { 'nēp 'tīd·əl ˌkə·rəns }

neap tide [OCEANOGR] Tide of decreased range occurring about every 2 weeks when the moon is in quadrature, that is, during its first and last quarter. Also known as neaps. { 'nēp ˌtīd }

Nearctic fauna [ECOL] The indigenous animal communities of the Nearctic zoogeographic region. { nē'ärd·ik 'fòn·ə }

Nearctic zoogeographic region [ECOL] The zoogeographic region that includes all of North America to the edge of the Mexican Plateau. { nē'ärd·ik ¦zō·ō‚jē·ə'graf·ik ˌrē·jən }

nearshore [OCEANOGR] An indefinite zone which extends from the shoreline seaward to a point beyond the breaker zone. { 'nir‚shòr }

nearshore circulation [OCEANOGR] Ocean circulation consisting of both the nearshore currents and the coastal currents. { 'nir‚shòr ˌsər·kyə'lā·shən }

nearshore current system [OCEANOGR] A current system, caused mainly by wave action in and near the breaker zone, which contains four elements: the shoreward mass transport of water; longshore currents; seaward return flow, including rip currents; and the longshore movement of the expanded heads of rip currents. { 'nir‚shòr 'kə·rənt ˌsis·təm }

near wilt [PL PATH] A fungus disease of peas caused by *Fusarium oxysporum pisi*; affects scattered plants and develops more slowly than true wilt. { 'nir ˌwilt }

nebulosus [METEOROL] A cloud species with the appearance of a nebulous veil, showing no distinct details; found principally in the genera cirrostratus and stratus. { 'neb·yə'lō·səs }

neck [GEOGR] A narrow strip of land, especially one connecting two larger areas. { nek }

neck rot [PL PATH] A fungus disease of onions caused by species of *Botrytis* and characterized by rotting of the leaves just above the bulb. { 'nek ˌrät }

necrosis [MED] Death of a cell or group of cells as a result of injury, disease, or other pathologic state. { nə'krō·səs }

necrotic enteritis [VET MED] A bacterial infection of young swine caused by *Salmonella suipestifer* or *S. choleraesuis* and characterized by fever and necrotic and ulcerative inflammation of the intestine. { nə'kräd·ik ‚ent·ə'rīd·əs }

necrotic ring spot [PL PATH] A virus leaf spot of cherries marked by small, dark water-soaked rings which may drop out, giving the leaf a tattered appearance. { nə'kräd·ik 'riŋ ˌspät }

nectar [BOT] A sugar-containing liquid secretion of the nectaries of many flowers. { 'nek·tər }

needle [HYD] A long, slender snow crystal that is at least five times as long as it is broad. { 'nēd·əl }

needle ice *See* frazil ice; pipkrake. { 'nēd·əl ˌīs }

negative area   *See* negative element.|GEOGR| An area that is almost uncultivable or uninhabitable.   { 'neg·əd·iv 'er·ē·ə }

negative element |GEOL| A large structural feature or part of the earth's crust, characterized through a long geologic time period by frequent and conspicuous downward movement (subsidence) or by extensive erosion, or by an uplift that is considerably less rapid or less frequent than that of adjacent positive elements. Also known as negative area.   { 'neg·əd·iv 'el·ə·mənt }

negative feedback |SCI TECH| Feedback which tends to reduce the output in a system.   { 'neg·əd·iv 'fēd,bak }

negative landform |GEOL| **1.** A relatively depressed or low-lying topographic form, such as a valley, basin, or plain.   **2.** A volcanic feature formed by a lack of material (such as a caldera).   { 'neg·əd·iv 'land,fórm }

negative rain |METEOROL| Rain which exhibits a net negative electric charge.   { 'neg·əd·iv 'rān }

negative shoreline   *See* shoreline of emergence.   { 'neg·əd·iv 'shór,līn }

Neisseriaceae |MICROBIO| The single family of gram-negative aerobic cocci and coccobacilli; some species are human parasites and pathogens.   { 'nī·sər·ē'ās·ē,ē }

Neisseria gonorrhoeae |MICROBIO| A gram-negative coccus pathogen that causes the sexually transmitted disease gonorrhea. Also known as gonococcus.   { nī·sə,rē·ə ,gän·ə'rē,ī }

Neisseria meningitidis |MICROBIO| A gram-negative, nonmotile, coccal bacterium commonly known as meningococcus. Pathogenic to humans, it is the causative agent of meningococcal meningitis.   { nī·sə,rē·ə ,men·ən'jīd·əs }

nektobenthos |ECOL| Those forms of marine life that exist just above the ocean bottom and occasionally rest on it.   { ¦nek·tə'ben,thós }

nekton |ZOO| Free-swimming aquatic animals, essentially independent of water movements.   { 'nek·tən }

nematicide |AGR| A chemical used to kill plant-parasitic nematodes. Also spelled nematocide.   { nə'mad·ə,sīd }

nematocide   *See* nematicide.   { nə'mad·ə,sīd }

Nematoda |ZOO| A group of unsegmented worms which have been variously recognized as an order, class, and phylum.   { ,nem·ə'tō·də }

nematode |ZOO| **1.** Any member of the Nematoda.   **2.** Of or pertaining to the Nematoda.   { 'nem·ə,tōd }

Nematospora coryli |MICROBIO| A mycelial species with needle-shaped ascospores that causes yeast spot disease of various crops.   { nə,mad·ə,spór·ə 'kór·ə,lē }

neoformation   *See* neogenesis.   { ¦nē·ō·fór'mā·shən }

neogenesis |GEOL| The formation of new minerals, as by diagenesis or metamorphism. Also known as neoformation.   { ¦nē·ō'jen·ə·səs }

Neogregarinida |ZOO| An order of sporozoan protozoans in the subclass Gregarinia which are insect parasites.   { ¦nē·ō,greg·ə'rin·ə·də }

neomineralization |GEOCHEM| Chemical interchange within a rock whereby its mineral constituents are converted into entirely new mineral species.   { ¦nē·ō,min·rə·lə'zā·shən }

neo-, ne- |SCI TECH| Prefix meaning new, or different in form; indicating a compound related to an older one, or a precursor.   { 'nē·ō }

neoplasm [MED] An aberrant new growth of abnormal cells or tissues; a tumor. { 'nē·ə,plaz·əm }

Neotropical zoogeographic region [ECOL] A zoogeographic region that includes Mexico south of the Mexican Plateau, the West Indies, Central America, and South America. { ¦nē·ō'träp·ə·kəl ¦zō·ō,jē·ə'graf·ik 'rē·jən }

nephanalysis [METEOROL] The analysis of a synoptic chart in terms of the types and amount of clouds and precipitation; cloud systems are identified both as entities and in relation to the pressure pattern, fronts, and other aspects. { ,nef·ə'nal·ə·səs }

nepheloid zone [OCEANOGR] A layer of water near the bottom of the continental rise and slope of the North Atlantic Ocean that contains suspended sediment of the clay fraction and organic matter. { 'nef·ə,lȯid ,zōn }

nephology [METEOROL] The study of clouds. { ne'fäl·ə·jē }

nephsystem *See* cloud system. { 'nef,sis·təm }

neptunic rock [GEOL] A rock that is formed in the sea. { nep'tün·ik 'räk }

neritic [OCEANOGR] Of or pertaining to the region of shallow water adjoining the seacoast and extending from low-tide mark to a depth of about 660 feet (200 meters). { nə'rid·ik }

nervous system [BIOL] A coordinating and integrating system which functions in the adaptation of an organism to its environment; in vertebrates, the system consists of the brain, brainstem, spinal cord, cranial and peripheral nerves, and ganglia. { 'nər·vəs ,sis·təm }

ness *See* cape. { 'nes }

net aerial production [ECOL] The biomass or biocontent that is incorporated into the aerial parts, that is, the leaf, stem, seed, and associated organs, of a plant community. { 'net 'er·ē·əl prə'dək·shən }

net balance [HYD] The change in mass of a glacier from the time of minimum mass in one year to the time of minimum mass in the succeeding year. Also known as net budget. { 'net 'bal·əns }

net blotch [PL PATH] A fungus disease of barley caused by *Helminthosporium teres* and marked by spotting of the foliage. { 'net 'bläch }

net budget *See* net balance. { 'net 'bəj·ət }

net plankton [ECOL] Plankton that can be removed from sea water by the process of filtration through a fine net. { 'net 'plaŋk·tən }

net primary production [ECOL] The production of biomass by autotrophs, excluding the biomass used for respiration. { 'net 'prīm·ə·rē prə'dək·shən }

net production rate [ECOL] The assimilation rate (gross production rate) of an ecosystem minus the amount of matter lost through predation, respiration, and decomposition. { 'net prə'dək·shən ,rāt }

neurotoxicity [MED] Adverse effects on the structure or function of the central and/or peripheral nervous system caused by exposure to a toxic chemical; symptoms include muscle weakness, loss of sensation and motor control, tremors, cognitive alterations, and autonomic nervous system dysfunction. { ,nü·ro·täk'sis·əd·ē }

neurotoxin [BIOL] A substance that has an adverse effect on the structure or function of the nervous system. { ¦nür·ō'täk·sən }

neutral estuary [GEOGR] An estuary in which neither fresh-water inflow nor evaporation dominates. { 'nü·trəl 'es·chə,wer·ē }

neutralism [ECOL] A neutral interaction between two species, that is, one having no evident effect on either species. { 'nü·trə,liz·əm }

neutralize   [CHEM] To make a solution neutral (neither acidic nor basic, pH of 7) by adding a base to an acidic solution, or an acid to a basic solution.   { 'nü·trǝ,līz }

neutral point   See col.   { 'nü·trǝl ,póint }

neutral stability   [METEOROL] The state of an unsaturated or saturated column of air in the atmosphere when its environmental lapse rate of temperature is equal to the dry-adiabatic lapse rate or the saturation-adiabatic lapse rate respectively; under such conditions a parcel of air displaced vertically will experience no buoyant acceleration. Also known as indifferent equilibrium; indifferent stability.   { 'nü·trǝl stǝ'bil·ǝd·ē }

neutron shield   [ENG] A shield that protects personnel from neutron irradiation. { 'nü,trän ,shēld }

neutron soil-moisture meter   [ENG] An instrument for measuring the water content of soil and rocks as indicated by the scattering and absorption of neutrons emitted from a source, and resulting gamma radiation received by a detector, in a probe lowered into an access hole.   { 'nü,trän 'sóil ,móis·chǝr ,mēd·ǝr }

neutrophilous   [BIOL] Preferring an environment free of excess acid or base. { nü'träf·ǝ·lǝs }

neutrosphere   [METEOROL] The atmospheric shell from the earth's surface upward, in which the atmospheric constituents are for the most part un-ionized, that is, electrically neutral; the region of transition between the neutrosphere and the ionosphere is somewhere between 42 and 54 miles (70 and 90 kilometers), depending on latitude and season.   { 'nü·trǝ,sfir }

nevada   [METEOROL] A cold wind descending from a mountain glacier or snowfield, for example, in the higher valleys of Ecuador.   { nǝ'väd·ǝ }

névé   [HYD] An accumulation of compacted, granular snow in transition from soft snow to ice; it contains much air; the upper portions of most glaciers and ice shelves are usually composed of névé.   { nā'vā }

Newcastle disease   [VET MED] An acute viral disease of fowls, with respiratory, gastrointestinal, and central nervous system involvement; may be transmitted to human beings as a mild conjunctivitis. Also known as avian pneumoencephalitis; avian pseudoplague; Philippine fowl disease.   { 'nü,kas·ǝl di,zēz }

Newcastle virus   [MICROBIO] A ribonucleic acid hemagglutinating myxovirus responsible for Newcastle disease.   { 'nü,kas·ǝl ,vī·rǝs }

newly formed ice   [HYD] Ice in the first stage of formation and development. Also known as fresh ice.   { 'nü·lē ,fórmd 'īs }

new snow   [METEOROL] **1.** Fallen snow whose original crystalline structure has been retained and is therefore recognizable. **2.** Snow which has fallen in a single day. { 'nü 'snō }

NEXRAD   See next-generation radar.   { 'neks,rad }

next-generation radar   [METEOROL] A Doppler radar, called WSR-88D, that enables forecasters to detect and give early warning for potentially severe weather. Abbreviated NEXRAD.   { ,nekst ,jen·ǝ,rā·shǝn 'rā,där }

niche   [ECOL] The unique role or way of life of a plant or animal species.   { nich }

niche glacier   [HYD] A common type of small mountain glacier occupying a funnel-shaped hollow or irregular recess in a mountain slope.   { 'nich ,glā·shǝr }

nickel-63   [PHYS] Radioactive nickel with beta radiation and 92-year half-life; derived by pile-irradiation of nickel; used in radioactive composition studies and tracer studies. { 'nik·ǝl ¦sik·stē'thrē }

nicotine   [CHEM] $C_{10}H_{14}N_2$ A colorless liquid alkaloid derived from the tobacco plant, which is cultivated in many parts of the world for the preparation of cigarettes, cigars, and pipe tobacco. It is used as a contact insecticide fumigant in closed spaces. { 'nik·ə,tēn }

nidus [ZOO] A nest or breeding place. { 'nīd·əs }

night wind [METEOROL] Dry squalls which occur at night in southwest Africa and the Congo; the term is loosely applied to other diurnal local winds such as mountain wind, land breeze, and midnight wind. { 'nīt ,wind }

nilas [HYD] A thin elastic crust of gray-colored ice formed on a calm sea; characterized by a matte surface, and easily bent by waves and thrust into a pattern of interlocking fingers. { 'nī·ləs }

nimbostratus [METEOROL] A principal cloud type, or cloud genus, gray-colored and often dark, rendered diffuse by more or less continuously falling rain, snow, or sleet of the ordinary varieties, and not accompanied by lightning, thunder, or hail; in most cases the precipitation reaches the ground. { ¦nim·bō¦strad·əs }

nimbus [METEOROL] A characteristic rain cloud; the term is not used in the international cloud classification except as a combining term, as cumulonimbus. { 'nim·bəs }

nitrification [MICROBIO] Formation of nitrous and nitric acids or salts by oxidation of the nitrogen in ammonia; specifically, oxidation of ammonium salts to nitrites and oxidation of nitrites to nitrates by certain bacteria. { ,nī·trə·fə'kā·shən }

nitrifying bacteria [MICROBIO] Members of the family Nitrobacteraceae. { 'nī·trə,fī·iŋ bak'tir·ē·ə }

nitrilotriacetic acid [CHEM] $N(CH_2COOH)_3$ A white powder, melting point 240°C, with some decomposition; soluble in water; it is toxic, and birth abnormalities may result from ingestion; may be used as a chelating agent in the laboratory. Also known as NTA; TGA. { ¦nī·trə·lō,trī·ə'sēd·ik 'as·əd }

nitrobenzene [CHEM] $C_6H_5NO_2$ Greenish crystals or a yellowish liquid, melting point 5.70°C; a toxic material; used in aniline manufacture. Also known as oil of mirbane. { ¦nī·trō'ben,zēn }

nitrochloroform See chloropicrin. { ¦nī·trō'klór·ə,fórm }

nitrogen balance [GEOCHEM] The net loss or gain of nitrogen in a soil. { 'nī·trə·jən ,bal·əns }

nitrogen fixation [CHEM ENG] Conversion of atmospheric nitrogen into compounds such as ammonia, calcium cyanamide, or nitrogen oxides by chemical or electric-arc processes. { 'nī·trə·jən ,fik¦sā·shən }

nitrogen oxides [CHEM] NO^x Chemical compounds of nitrogen and oxygen; produced primarily from the combustion of fossil fuels, they contribute to the formation of ground-level ozone. { ¦nī·trə·jən 'äk,sīdz }

nitrophyte [BOT] A plant that requires nitrogen-rich soil for growth. { ¦nī·trə,fīt }

nival [ECOL] **1.** Characterized by or living in or under the snow. **2.** Of or pertaining to a snowy environment. { 'nī·vəl }

nivation glacier [HYD] A small, newly formed glacier; represents the initial stage of glaciation. Also known as snowbank glacier. { nī'vā·shən ,glā·shər }

noctilucent cloud [METEOROL] A cloud of unknown composition which occurs at great heights and high altitudes; photometric measurements have located such clouds between 45 and 54 miles (75 and 90 kilometers); they resemble thin cirrus, but usually with a bluish or silverish color, although sometimes orange to red, standing out against a dark night sky. { ¦näk·tə¦lü·sənt 'klaúd }

noct-, nocti-, nocto-, noctu- [SCI TECH] Combining form meaning night. { näkt, 'näk·tē, 'näk·tō, 'näk·tə }

Noctuidae [ZOO] A large family of dull-colored, medium-sized moths in the superfamily Noctuoidea; larva are mostly exposed foliage feeders, representing an important group of agricultural pests. { näk'tü·ə,dē }

nocturnal [BIOL] Active during the nighttime. { näk'tərn·əl }

node [BOT] A site on a plant stem at which leaves and axillary buds arise. { nōd }

nodule [BOT] A bulbous enlargement found on roots of legumes and certain other plants, whose formation is stimulated by symbiotic, nitrogen- fixing bacteria that colonize the roots. [GEOL] A small, hard mass or lump of a mineral or mineral aggregate characterized by a contrasting composition from and a greater hardness than the surrounding sediment or rock matrix in which it is embedded. { 'näj·ül }

noise [PHYS] Sound which is unwanted, either because of its effect on humans, its effect on fatigue or malfunction of physical equipment, or its interference with the perception or detection of other sounds. { nȯiz }

noise control [PHYS] The process of obtaining an acceptable noise environment for a particular observation point or receiver, involving control of the noise source, transmission path, or receiver, or all three. { 'nȯiz kən,trōl }

noise immission level [PHYS] A measure of the cumulative noise energy to which an individual is exposed over time; equal to the average noise level to which the person has been exposed, in decibels, plus 10 times the logarithm of the number of years for which the individual is exposed. { ¦nȯiz i¦mish·ən ,lev·əl }

noise measurement [PHYS] The process of quantitatively determining one or more properties of acoustic noise. { 'nȯiz ,mezh·ər·mənt }

noise pollution [PHYS] Excessive noise in the human environment. { 'nȯiz pə,lü·shən }

noise rating number [PHYS] The perceived noise level of the noise that can be tolerated under specified conditions; for example, the noise rating number of a bedroom is 25, that of a workshop is 65. { 'nȯiz 'rād·iŋ ,nəm·bər }

noise reduction coefficient [PHYS] The average over the logarithm of frequency, in the frequency range from 256 to 2048 hertz inclusive, of the sound absorption coefficient of a material. { 'nȯiz ri,dək·shən ,kō·i'fish·ənt }

noise reduction rating [PHYS] A common method for expressing values of noise reduction or attenuation provided by different types of hearing protectors; values range from 0 to approximately 30, with higher values indicating greater amounts of noise reduction. Abbreviated NRR. { 'nȯiz ri,dək·shən ,rād·iŋ }

NOMAD *See* Navy Oceanographic and Meteorological Automatic Device. { 'nō,mad }

noncohesive *See* cohesionless. { ,nän·kō'hē·siv }

noncontributing area [HYD] An area with closed drainage. { ¦nän·kən'trib·yəd·iŋ 'er·ē·ə }

nongraded [GEOL] Pertaining to a soil or an unconsolidated sediment consisting of particles of essentially the same size. { ¦nän'grād·əd }

nonrecording rain gage [ENG] A rain gage which indicates but does not record the amount of precipitation. { ¦nän·ri'kȯrd·iŋ 'rān ,gāj }

nontidal current [OCEANOGR] Any current due to causes other than tidal, as a permanent ocean current. { 'nän,tīd·əl 'kə·rənt }

no observed effect level [MED] The highest dose at which no effects can be observed; used as a measure of chronic toxicity. { ¦nō əb¦zərvd i'fekt ,lev·əl }

noosphere *See* anthroposphere. { 'nō·ə,sfir }

Nordenskjöld line [CLIMATOL] The line connecting all places at which the mean temperature of the warmest month is equal (in degrees Celsius) to $9 - 0.1k$, where *k* is the mean temperature of the coldest month (in degrees Fahrenheit it becomes $51.4 - 0.1k$). { 'nȯrd·ən,shēld ,līn }

nor'easter *See* northeaster. { nȯr'ē·stər }

normal [METEOROL] The average value of a meteorological element over any fixed period of years that is recognized as standard for the country and element concerned. { 'nȯr·məl }

normal aeration [GEOL] The complete renewal of soil air to a depth of 8 inches (20 centimeters) about once each hour. { 'nȯr·məl e'rā·shən }

normal chart [METEOROL] Any chart that shows the distribution of the official normal values of a meteorological element. Also known as normal map. { 'nȯr·məl 'chärt }

normal cycle [GEOL] A cycle of erosion whereby a region is reduced to base level by running water, especially by the action of rivers. Also known as fluvial cycle of erosion. { 'nȯr·məl 'sī·kəl }

normal erosion [GEOL] Erosion effected by prevailing agencies of the natural environment, including running water, rain, wind, waves, and organic weathering. Also known as geologic erosion. { 'nȯr·məl i'rō·zhən }

normal map *See* normal chart. { 'nȯr·məl 'map }

normal pressure *See* standard pressure. { 'nȯr·məl 'presh·ər }

normal soil [GEOL] A soil having a profile that is more or less in equilibrium with the environment. { 'nȯr·məl 'sȯil }

normal uranium *See* native uranium. { 'nȯr·məl yü'rā·nē·əm }

normal water [OCEANOGR] Water whose chlorinity lies between 19.30 and 19.50 parts per thousand and has been determined to within ±0.001 per thousand. Also known as Copenhagen water; standard seawater. { 'nȯr·məl 'wȯd·ər }

North America [GEOGR] The northern of the two continents of the New World or Western Hemisphere, extending from narrow parts in the tropics to progressively broadened portions in middle latitudes and Arctic polar margins. { 'nȯrth ə'mer·i·kə }

North American anticyclone *See* North American high. { 'nȯrth ə'mer·i·kən ,ant·i'sī ,klōn }

North American high [METEOROL] The relatively weak general area of high pressure which, as shown on mean charts of sea-level pressure, covers most of North America during winter. Also known as North American anticyclone. { 'nȯrth ə'mer·i·kən 'hī }

North Atlantic Current [OCEANOGR] A wide, slow-moving continuation of the Gulf Stream originating in the region east of the Grand Banks of Newfoundland. { 'nȯrth at'lan·tik 'kə·rənt }

North Cape Current [OCEANOGR] A warm current flowing northeastward and eastward around northern Norway, and curving into the Barents Sea. { 'nȯrth 'kāp ,kə·rənt }

Northeast Drift Current [OCEANOGR] A North Atlantic Ocean current flowing northeastward toward the Norwegian Sea, gradually widening and, south of Iceland, branching and continuing as the Irminger Current and the Norwegian Current; it is the northern branch of the North Atlantic Current. { 'nȯrth,ēst ¦drift 'kə·rənt }

northeaster [METEOROL] A northeast wind, particularly a strong wind or gale. Also spelled nor'easter. { nȯr'thē·stər *or* nȯ'rē·stər }

northeast storm [METEOROL] A cyclonic storm of the east coast of North America, so called because the winds over the coastal area are from the northeast; they may occur at any time of year but are most frequent and most violent between September and April. { 'nȯr'thēst 'stȯrm }

northeast trades [METEOROL] The trade winds of the Northern Hemisphere. { 'nòr'thēst 'trādz }

North Equatorial Current [OCEANOGR] Westward ocean currents driven by the northeast trade winds blowing over tropical oceans of the Northern Hemisphere. Also known as Equatorial Current. { 'nòrth ‚ē·kwə'tòr·ē·əl 'kə·rənt }

norther [METEOROL] A northerly wind. { 'nòr·ther }

Northern Hemisphere [GEOGR] The half of the earth north of the Equator. { 'nòr·thərn 'hem·i‚sfir }

Northern Hemisphere annular mode See Arctic Oscillation. { ¦nòr·thərn ¦hem·ə‚sfir ¦an·yə·lər 'mōd }

north frigid zone [GEOGR] That part of the earth north of the Arctic Circle. { 'nòrth 'frij·əd ‚zōn }

north geographic pole See North Pole. { 'nòrth ‚jē·ə'graf·ik 'pōl }

North Pacific Current [OCEANOGR] The warm branch of the Kuroshio Extension flowing eastward across the Pacific Ocean. { 'nòrth pə'sif·ik 'kə·rənt }

North Pole [GEOGR] The geographic pole located at latitude 90°N in the Northern Hemisphere of the earth; it is the northernmost point of the earth, and the northern extremity of the earth's axis of rotation. Also known as north geographic pole. { 'nòrth 'pōl }

north temperate zone [CLIMATOL] That part of the earth between the Tropic of Cancer and the Arctic Circle. { 'nòrth 'tem·prət ‚zōn }

northwester [METEOROL] A northwest wind. Also spelled nor'wester. { nòrth'wes·tər or nòr'wes·tər }

Norway Current [OCEANOGR] A continuation of the North Atlantic Current, which flows northward along the coast of Norway. Also known as Norwegian Current. { 'nòr‚wā ‚kə·rənt }

Norwegian Current See Norway Current. { nòr'wē·jən ‚kə·rənt }

nor'wester See northwester. { nòr'wes·tər }

nosocomial [MED] **1.** Pertaining to a hospital. **2.** Of disease, caused or aggravated by hospital life. { ¦näz·ə¦kō·mē·əl }

notch [GEOL] A deep, narrow cut near the high-water mark at the base of a sea cliff. [GEOGR] A narrow passage between mountains or through a ridge, hill, or mountain. { näch }

Notogaean [ECOL] Pertaining to or being a biogeographic region including Australia, New Zealand, and the southwestern Pacific islands. { ¦nōd·ə¦jē·ən }

nowcasting [METEOROL] **1.** The detailed description of the current weather along with forecasts obtained by extrapolation up to about 2 hours ahead. **2.** Any area-specific forecast for the period up to 12 hours ahead that is based on very detailed observational data. { 'naú‚kast·iŋ }

noy [PHYS] A unit of perceived noisiness equal to the perceived noisiness of random noise occupying the frequency band 910–1090 hertz at a sound pressure level of 40 decibels above 0.0002 microbar; a sound that is n times as noisy as this sound has a perceived noisiness of n noys, under the assumption that the perceived noisiness of a sound increases with physical intensity at the same rate as the loudness. { nòi }

N-P-K [CHEM ENG] The code identifying the components in a fertilizer mixture: nitrogen (N), phosphorus pentoxide (P), and potassium oxide (K). Fertilizers are graded in the order N-P-K, with the numbers indicating the percentage of the total weight of each component. For example, 5-10-10 represents a mixture containing by weight 5% nitrogen, 10% phosphorus pentoxide, and 10% potassium oxide.

NPV See nuclear polyhedrosis virus.

NRM wind scale [METEOROL] A wind scale adapted by the United States Forest Service for use in the forested areas of the Northern Rocky Mountains (NRM); it is an adaptation of the Beaufort wind scale; the difference between these two scales lies in the specification of the visual effects of the wind; the force numbers and the corresponding wind speeds are the same in both. { ¦en¦är'em 'wind ¸skāl }

NRR See noise reduction rating.

NTA See nitrilotriacetic acid.

nuclear electric power generation [ENG] Large-scale generation of electric power in which the source of energy is nuclear fission, generally in a nuclear reactor, or nuclear fusion. { 'nü·klē·ər i¦lek·trik 'paů·ər ¸jen·ə¸rā·shən }

nuclear fission See fission. { 'nü·klē·ər 'fish·ən }

nuclear polyhedrosis virus [MICROBIO] A Baculovirus subgroup characterized by the multiplication and formation of polyhedron-shaped inclusion bodies in the nuclei of infected host cells, used in the control of agriculture and forest insects. Abbreviated NPV. { ¦nü·klē·ər ¸päl·ē·hē'drōs·əs ¸vī·rəs }

nuclear power plant [ENG] A power plant in which nuclear energy is converted into heat for use in producing steam for turbines, which in turn drive generators that produce electric power. Also known as atomic power plant. { 'nü·klē·ər 'paů·ər ¸plant }

nuclear radiation [PHYS] A term used to denote alpha particles, neutrons, electrons, photons, and other particles which emanate from the atomic nucleus as a result of radioactive decay and nuclear reactions. { 'nü·klē·ər ¸rād·ē'ā·shən }

nuclear spontaneous reaction See radioactive decay. { 'nü·klē·ər spän'tā·nē·əs rē'ak·shən }

nuclear twin-probe gage See profiling snow gage. { 'nü·klē·ər ¦twin ¸prōb ¸gāj }

nuclear winter [METEOROL] Predicted global-scale changes resulting from a nuclear war, in which dust raised by nuclear bursts and smoke generated in fires would cause reductions in solar energy reaching the earth's surface and reductions in surface temperatures for periods of months. { 'nü·klē·ər 'win·tər }

nucleus [HYD] A particle of any nature upon which, or a locus at which, molecules of water or ice accumulate as a result of a phase change to a more condensed state. { 'nü·klē·əs }

nugget [GEOL] A small mass of metal found free in nature. { 'nəg·ət }

numerical forecasting [METEOROL] The forecasting of the behavior of atmospheric disturbances by the numerical solution of the governing fundamental equations of hydrodynamics, subject to observed initial conditions. Also known as dynamic forecasting; mathematical forecasting; numerical weather prediction; physical forecasting. { nü'mer·i·kəl 'fȯr¸kast·iŋ }

numerical weather prediction See numerical forecasting. { nü'mer·i·kəl 'we<u>th</u>·ər pri ¸dik·shən }

nut [BOT] **1.** A fruit which has at maturity a hard, dry shell enclosing a kernel consisting of an embryo and nutritive tissue. **2.** An indehiscent, one-celled, one-seeded, hard fruit derived from a single, simple, or compound ovary. { nət }

nutrient [BIOL] A chemical substance that an organism must obtain from its environment in order to maintain life and reproduce. { 'nü·trē·ənt }

nutrient biopurification [ECOL] A process taking place within a nutrient cycle that maintains the pools of nutrient substances at optimum concentrations, to the exclusion of nonnutrient substances. { 'nü·trē·ənt ¦bī·ō,pyür·ə·fə'kā·shən }

nutrient cycle [ECOL] The pattern of use, transformation, movement, and reuse of chemical elements and compounds among nonliving and living components of an ecosystem. { 'nü·trē·ənt ,sī·kəl }

nutrition [BIOL] The science of nourishment, including the study of nutrients that each organism must obtain from its environment in order to maintain life and reproduce. { nü'trish·ən }

nyctinasty [BOT] A nastic movement in higher plants associated with diurnal light and temperature changes. { 'nik·tə,nas·tē }

nymph [ZOO] An immature life stage of hemimetabolous insects. { nimf }

oasis [GEOGR] An isolated fertile area, usually limited in extent and surrounded by desert, and marked by vegetation and a water supply. { ō'ā·səs }

oat [BOT] Any plant of the genus *Avena* in the family Graminae, cultivated as an agricultural crop for its seed, a cereal grain, and for straw. { ōt }

oberwind [METEOROL] A night wind from mountains or the upper ends of lakes; a wind of Salzkammergut in Austria. { 'ō·bər,vint }

obligate [BIOL] Restricted to a specified condition of life, as an obligate parasite. { 'äb·lə·gət }

obligate aerobe [MICROBIO] A microorganism that uses oxygen for cellular respiration and requires some free molecular oxygen in its surroundings to support growth. { ¦äb·lə,gāt 'er·ōb }

obligate anaerobe [MICROBIO] A microorganism that cannot use oxygen and can grow only in the absence of free oxygen. { ¦äb·li,gāt 'an·ə,rōb }

obscuration [METEOROL] In United States weather observing practice, the designation for the sky cover when the sky is completely hidden by surface-based obscuring phenomena, such as fog. Also known as obscured sky cover. { ,äb·skyü'rā·shən }

obscured sky cover *See* obscuration. { əb'skyürd 'skī ,kəv·ər }

obscuring phenomenon [METEOROL] In United States weather observing practice, any atmospheric phenomenon (not including clouds) which restricts the vertical visibility or slant visibility, that is, which obscures a portion of the sky from the point of observation. { əb'skyür·iŋ fə,näm·ə,nän }

obsequent [GEOL] Of a stream, valley, or drainage system, being in a direction opposite to that of the original consequent drainage. { 'äb·sə·kwənt }

obstruction to vision [METEOROL] In United States weather observing practice, one of a class of atmospheric phenomena, other than the weather class of phenomena, which may reduce horizontal visibility at the earth's surface; examples are fog, smoke, and blowing snow. { əb'strək·shən tə 'vizh·ən }

occluded cyclone [METEOROL] Any cyclone (or low) within which there has developed an occluded front. { ə'klüd·əd 'sī,klōn }

occluded front [METEOROL] A composite of two fronts, formed as a cold front overtakes a warm front or quasi-stationary front. Also known as frontal occlusion; occlusion. { ə'klüd·əd 'frənt }

occlusion *See* occluded front. { ə'klü·zhən }

occupational ecology [ECOL] A discipline concerned with the interaction of workers with the environment, and with matching humans with the environment in the most ergonomically efficient way and with minimal disturbance of the environment. { ,ä·kyə'pā·shen·əl i'käl·ə·jē }

occupational medicine [MED] The branch of medicine which deals with the relationship of humans to their occupations, for the purpose of the prevention of disease and injury and the promotion of optimal health, productivity, and social adjustment. { ˌä·kyə'pā·shən·əl 'med·i·sən }

Occupational Safety and Health Administration [ECOL] A governmental agency within the Department of Labor that sets and enforces health and safety standards for the workplace. Abbreviated OSHA. { ˌä·kyə·'pā·shən·əl 'saf·tē and 'helth ad·ˌmi·nə·'strā·shən }

ocean [OCEANOGR] The interconnected body of salt water that occupies almost three-quarters of the earth's surface. { 'ō·shən }

ocean basin [GEOL] The great depression occupied by the ocean on the surface of the lithosphere. { 'ō·shən 'bā·sən }

ocean circulation [OCEANOGR] **1.** Water current flow in a closed circular pattern within an ocean. **2.** Large-scale horizontal water motion within an ocean. { 'ō·shən ˌsər·kyə'lā·shən }

ocean current [OCEANOGR] A net transport of ocean water along a definable path. { 'ō·shən 'kə·rənt }

ocean engineering [ENG] A subfield of engineering involved with the development of new equipment concepts and the methodical improvement of techniques which allow humans to operate successfully beneath the ocean surface in order to develop and utilize marine resources. { 'ō·shən ˌen·jə'nir·iŋ }

ocean floor [GEOL] The near-horizontal surface of the ocean basin. { 'ō·shən 'flȯr }

ocean-floor spreading See sea-floor spreading. { 'ō·shən ¦flȯr ˌspred·iŋ }

Oceanian [ECOL] Of or pertaining to the zoogeographic region that includes the archipelagos and islands of the central and south Pacific. { ˌō·shē'an·ē·ən }

oceanic anticyclone See subtropical high. { ˌō·shē'an·ik ˌant·i'sī,klōn }

oceanic basalt [GEOL] Rocks of the oceanic island volcanoes. { ˌō·shē'an·ik bə'sȯlt }

oceanic climate See marine climate. { ˌō·shē'an·ik 'klī·mət }

oceanic crust [GEOL] A thick mass of igneous rock which lies under the ocean floor. { ˌō·shē'an·ik 'krəst }

oceanic high See subtropical high. { ˌō·shē'an·ik 'hī }

oceanic island [GEOL] Any island which rises from the deep-sea floor rather than from shallow continental shelves. { ˌō·shē'an·ik 'ī·lənd }

oceanicity [CLIMATOL] The degree to which a point on the earth's surface is in all respects subject to the influence of the sea; it is the opposite of continentality; oceanicity usually refers to climate and its effects; one measure for this characteristic is the ratio of the frequencies of maritime to continental types of air mass. Also known as oceanity. { ˌō·shē·ə'nis·əd·ē }

oceanic province [OCEANOGR] The water of the ocean that lies seaward of the break in the continental shelf. { ˌō·shē'an·ik 'präv·əns }

oceanic ridge See mid-oceanic ridge. { ˌō·shē'an·ik 'rij }

oceanic rise [GEOL] A long, broad elevation of the bottom of the ocean. { ˌō·shē'an·ik 'rīz }

oceanic stratosphere See cold-water sphere. { ˌō·shē'an·ik 'strad·ə,sfir }

oceanic zone [OCEANOGR] The biogeographic area of the open sea. { ˌō·shē'an·ik 'zōn }

oceanity See oceanicity. { ˌo·shē'an·əd·ē }

oceanization |GEOL| Process by which continental crust is converted into oceanic crust. { ˌō·shə·nə'zā·shən }

oceanographic equator |OCEANOGR| **1.** The region of maximum temperature of the ocean surface. **2.** The region in which the temperature of the ocean surface is greater than 28°C. { ˌō·shə·nəˌgraf·ik i'kwād·ər }

oceanographic model |OCEANOGR| A theoretical representation of the marine environment which relates physical, chemical, geological, biological, and other oceanographic properties. { ˌō·shə·nəˌgraf·ik 'mäd·əl }

oceanographic station |OCEANOGR| A geographic location at which oceanographic observations are taken from a stationary ship. { ˌō·shə·nəˌgraf·ik 'stā·shən }

oceanographic survey |OCEANOGR| A study of oceanographic conditions with reference to physical, chemical, biological, geological, and other properties of the ocean. { ˌō·shə·nəˌgraf·ik 'sər,vā }

oceanography |OCEANOGR| The science of the sea, including physical oceanography (the study of the physical properties of seawater and its motion in waves, tides, and currents), marine chemistry, marine geology, and marine biology. Also known as oceanology. { ˌō·shə'näg·rə·fē }

oceanology *See* oceanography. { ˌō·shə'näl·ə·jē }

ocean thermal-energy conversion |ENG| The conversion of energy arising from the temperature difference between warm surface water of oceans and cold deep-ocean current into electrical energy or other useful forms of energy. Abbreviated OTEC. { 'ō·shən 'thər·məl 'en·ər·jē kən,vər·zhən }

ocean tomography |PHYS| A form of acoustic tomography in which an array of acoustic sources and receivers transmits and detects a pulse; the pulse travel times are used to determine temperature distributions in the ocean. { 'ō·shən tō'mäg·rə·fē }

ocean weather station |METEOROL| As defined by the World Meteorological Organization, a specific maritime location occupied by a ship equipped and staffed to observe weather and sea conditions and report the observations by international exchange. { 'ō·shən 'weth·ər ,stā·shən }

offshore |GEOL| The comparatively flat zone of variable width extending from the outer margin of the shoreface to the edge of the continental shelf. { 'ȯfˌshȯr }

offshore bar *See* longshore bar. { 'ȯfˌshȯr 'bär }

offshore beach *See* barrier beach. { 'ȯfˌshȯr 'bēch }

offshore current |OCEANOGR| **1.** A prevailing nontidal current usually setting parallel to the shore outside the surf zone. **2.** Any current flowing away from shore. { 'ȯfˌshȯr 'kə·rənt }

offshore water |OCEANOGR| Water adjacent to land in which the physical properties are slightly influenced by continental conditions. { 'ȯfˌshȯr 'wȯd·ər }

offshore wind |METEOROL| Wind blowing from the land toward the sea. { 'ȯfˌshȯr 'wind }

off-site facility |CHEM ENG| In a chemical process plant, any supporting facility that is not a direct part of the reaction train, such as utilities, steam, and waste-treatment facilities. { 'ȯfˌsīt fə'sil·əd·ē }

oil *See* petroleum. { ȯil }

oil accumulation *See* oil pool. { 'ȯil ə,kyü·myə,lā·shən }

oil of mirbane *See* nitrobenzene. { 'ȯil əv 'mər,bān }

oil of vitriol *See* sulfuric acid. { 'ȯil əv 'vit·rē,ōl }

oil pool |GEOL| An accumulation of petroleum locally confined by subsurface geologic features. Also known as oil accumulation; oil reservoir. { 'ȯil ˌpül }

oil reservoir See oil pool. { 'ȯil 'rez·əv,wär }

oil-reservoir water See formation water. { 'ȯil ¦rez·əv,wär ,wȯd·ər }

oil seep |GEOL| The emergence of liquid petroleum at the land surface as a result of slow migration from its buried source through minute pores or fissure networks. Also known as petroleum seep. { 'ȯil ,sēp }

oil zone |GEOL| The formation or horizon from which oil is produced, usually immediately under the gas zone and above the water zone if all three fluids are present and segregated. { 'ȯil ,zōn }

old ice |OCEANOGR| Floating sea ice that is more than 2 years old. { 'ōld 'īs }

old snow |HYD| Deposited snow in which the original crystalline forms are no longer recognizable, such as firn or spring snow. Also known as firn snow. { 'ōld ¦snō }

old wives' summer |METEOROL| A period of calm, clear weather, with cold nights and misty mornings but fine warm days, which sets in over central Europe toward the end of September; comparable to Indian summer. { 'ōld ,wīvz 'səm·ər }

oligodynamic action |MICROBIO| The inhibiting or killing of microorganisms by use of very small amounts of a chemical substance. { ¦äl·ə·gō·dī¦nam·ik 'ak·shən }

oligomictic |HYD| Pertaining to a lake that circulates only at rare, irregular intervals during abnormal cold spells. { ə,lig·ə'mik·tik }

oligotrophic |HYD| Of a lake, lacking plant nutrients and usually containing plentiful amounts of dissolved oxygen without marked stratification. { ¦äl·ə·gō¦träf·ik }

ombrometer See rain gage. { äm'bräm·əd·ər }

ombrophilous |ECOL| Able to thrive in areas of abundant rainfall. { äm'bräf·ə·ləs }

ombrophobous |ECOL| Unable to live in the presence of long, continuous rain. { äm'bräf·ə·bəs }

omethioate See folimat. { ,ō·mə'thī·ə,wāt }

omnivore |ZOO| An organism that eats both animal and vegetable matter. { 'äm·nə ,vȯr }

one-year ice |OCEANOGR| Sea ice formed the previous season, not yet I year old. { 'wən ,yir 'īs }

onion scab See onion smudge. { 'ən·yən ,skab }

onion smudge |PL PATH| A fungus disease of the onion caused by *Colletotrichum circinans* and characterized by black concentric integral rings or smutty spots on the bulb scales. Also known as onion scab. { 'ən·yən ,sməj }

onion smut |PL PATH| A fungus disease of onion, especially seedlings, caused by *Urocystis cepulae* and characterized by elongate black blisters on the scales and foliage. { 'ən·yən ,smət }

onshore |GEOGR| Pertaining to, in the direction toward, or located on the shore. Also known as shoreside. { 'ȯn,shȯr }

onshore wind |METEOROL| Wind blowing from the sea toward the land. { 'ȯn,shȯr ,wind }

ooze |GEOL| **1.** A soft, muddy piece of ground, such as a bog, usually resulting from the flow of a spring or brook. **2.** A marine pelagic sediment composed of at least 30% skeletal remains of pelagic organisms, the rest being clay minerals. **3.** Soft mud or slime, typically covering the bottom of a lake or river. { üz }

opacus [METEOROL] A variety of cloud (sheet, layer, or patch), the greater part of which is sufficiently dense to obscure the sun; found in the genera altocumulus, altostratus, stratocumulus, and stratus; cumulus and cumulonimbus clouds are inherently opaque. { ō′pā·kəs }

opalized wood *See* silicified wood. { ′ō·pə,līzd ′wu̇d }

opaque sky cover [METEOROL] In United States weather observing practice, the amount (in tenths) of sky cover that completely hides all that might be above it; opposed to transparent sky cover. { ō′pāk ′skī ,kəv·ər }

open bay [GEOGR] An indentation between two capes or headlands which is so broad and open that waves coming directly into it are nearly as high near its center as they are in adjacent parts of the open sea. { ′ō·pən ′bā }

open coast [GEOGR] A coast that is not sheltered from the sea. { ′ō·pən ′kōst }

open community [ECOL] A community which other organisms readily colonize because some niches are unoccupied. { ′ō·pən kə′myü·nəd·ē }

open harbor [GEOGR] An unsheltered harbor exposed to the sea. { ′ō·pən ′här·bər }

open ice [OCEANOGR] On navigable waters, ice that has broken apart sufficiently to permit passage of vessels. { ′ō·pən ′īs }

opening [OCEANOGR] Any break in sea ice which reveals the water. { ′ōp·ə·niŋ }

open lake [HYD] **1.** A lake that has a stream flowing out of it. **2.** A lake whose water is free of ice or emergent vegetation. { ′ō·pən ′lāk }

open pack ice [OCEANOGR] Floes of sea ice that are seldom in contact with each other, generally covering between four-tenths and six-tenths of the sea surface. { ′ō·pən ′pak ,īs }

open sea [GEOGR] **1.** That part of the ocean not enclosed by headlands, not within narrow straits, and so on. **2.** That part of the ocean outside the territorial jurisdiction of any country. { ′ō·pən ′sē }

open system [HYD] A condition of freezing of the ground in which additional groundwater is available either through free percolation or through capillary movement. { ′ō·pən ′sis·təm }

open water [ECOL] Lake water that is free from emergent vegetation, artificial obstructions, or tangled masses of underwater vegetation at very shallow depths. [HYD] Lake water that does not freeze during the winter. [OCEANOGR] Water that is less than one-tenth covered with floating ice. { ′ō·pən ′wȯd·ər }

operational weather limits [METEOROL] The limiting values of ceiling, visibility, and wind, or runway visual range, established as safety minima for aircraft landings and takeoffs. { ,äp·ə′rā·shən·əl ′weth·ər ,lim·əts }

opium [MED] A narcotic obtained from the unripe capsules of the opium poppy (*Papaver somniferum*); crude extract contains alkaloids such as morphine (5–15%), narcotine (2–8%), and codeine (0.1–2.5%). { ′ō·pē·əm }

opportunistic microorganism [MICROBIO] A normally harmless endogenous (usually found in healthy individuals) microorganism that produces disease due to fortuitous events that affect the host. { ¦äp·ər,tü¦nis·tik ¦mī·krō′ȯr·gə,niz·əm }

opportunistic species [ECOL] Species characterized by high reproduction rates, rapid development, early reproduction, small body size, and uncertain adult survival. { ¦äp·ər,tü¦nis·tik ′spē·shēz }

opposing wind [OCEANOGR] In wave forecasting, a wind blowing in opposition to the direction that the waves are traveling. { ə′pōz·iŋ ′wind }

opposite tide |OCEANOGR| A high tide at a corresponding place on the opposite side of the earth which accompanies a direct tide. { 'äp·ə·zət 'tīd }

optical oceanography |OCEANOGR| That aspect of physical oceanography which deals with the optical properties of sea water and natural light in sea water. { 'äp·tə·kəl ‚ō·shə'näg·rə·fē }

optical thickness |METEOROL| Subjectively, the degree to which a cloud prevents light from passing through it; depends upon the physical constitution (crystals, drops, droplets), the form, the concentration of particles, and the vertical extent of the cloud. { 'äp·tə·kəl 'thik·nəs }

orange coffee rust |PL PATH| A disease of the coffee plant caused by the rust fungus *Hemileia vastatrix*, characterized by the formation of small, powdery, pale yellow to orange spots on the lower leaf surface, followed by defoliation. { ¦är·ənj 'ȯ·fē ·əst }

orbit |OCEANOGR| The path of a water particle affected by wave motion; it is almost circular in deep-water waves and almost elliptical in shallow-water waves. { 'ȯr·bət }

orbital current |OCEANOGR| The flow of water which follows the orbital motion of water particles in a wave. { 'ȯr·bəd·əl 'kə·rənt }

Orbivirus |MICROBIO| A genus in the family Reoviridae that is the causative agent of bluetongue. { 'ȯrb·ə‚vī·rəs }

order |SYST| A taxonomic category ranked below the class and above the family, made up either of families, subfamilies, or suborders. { 'ȯrd·ər }

ordinary tides |OCEANOGR| Tides which have cycles of 12 to 24 hours. { 'ȯrd·ən‚er·ē 'tīdz }

ore |GEOL| **1.** The naturally occurring material from which economically valuable minerals can be extracted. **2.** Specifically, a natural mineral compound of the elements, of which one element at least is a metal. **3.** More loosely, all metalliferous rock, though it contains the metal in a free state. **4.** Occasionally, a compound of nonmetallic substances, as sulfur ore. { ȯr }

ore bed |GEOL| An economic aggregation of minerals occurring between or in rocks of sedimentary origin. { 'ȯr ‚bed }

ore chimney *See* pipe. { 'ȯr ‚chim·nē }

ore deposit |GEOL| Rocks containing minerals of economic value in such amount that they can be profitably exploited. { 'ȯr di‚päz·ət }

ore of sedimentation *See* placer. { 'ȯr əv ‚sed·ə·mən'tā·shən }

ore pipe *See* pipe. { 'ȯr ‚pīp }

organic |BIOL| Relating to or derived from living organisms. |CHEM| Of chemical compounds, based on carbon chains or rings and also containing hydrogen with or without oxygen, nitrogen, or other elements. { ȯr'gan·ik }

organic chelates |AGR| Chelates formed by reaction of organic compounds with the mineral end products of weathering; they enhance nutrient richness of soils by forming organomineral complexes that are easy for plants to absorb. { ȯr‚gan·ik 'kē‚lāts }

organic geochemistry |GEOCHEM| A branch of geochemistry which deals with naturally occurring carbonaceous and biologically derived substances which are of geological interest. { ȯr'gan·ik ‚jē·ō'kem·ə·strē }

organic lattice *See* growth lattice. { ȯr'gan·ik 'lad·əs }

organic mound *See* bioherm. { ȯr'gan·ik 'maúnd }

organic reef |GEOL| A sedimentary rock structure of significant dimensions erected by, and composed almost exclusively of the remains of, corals, algae, bryozoans, sponges, and other sedentary or colonial organisms. { ȯr'gan·ik 'rēf }

organic soil [GEOL] Any soil or soil horizon consisting chiefly of, or containing at least 30% of, organic matter; examples are peat soils and muck soils. { ȯr'gan·ik 'sȯil }

organic texture [GEOL] A sedimentary texture resulting from the activity of organisms such as the secretion of skeletal material. { ȯr'gan·ik 'teks·chər }

organic weathering [GEOL] Biological processes and changes that contribute to the breakdown of rocks. Also known as biological weathering. { ȯr'gan·ik 'weth·ə·riŋ }

organism [BIOL] An individual constituted to carry out all life functions. { 'ȯr·gə‚niz·əm }

organogenic [GEOL] Property of a rock or sediment derived from organic substances. { ȯr¦gan·ə¦jen·ik }

organolite [GEOL] Any rock consisting mainly of organic material. { ȯr'gan·ə‚līt }

organotropic [MICROBIO] Of microorganisms, localizing in or entering the body by way of the viscera or, occasionally, somatic tissue. { ȯr¦gan·ə¦träp·ik }

Oriental zoogeographic region [ECOL] A zoogeographic region which encompasses tropical Asia from the Iranian Peninsula eastward through the East Indies to, and including, Borneo and the Philippines. { ‚ȯr·ē'ent·əl ‚zō·ə‚jē·ə'graf·ik ‚rē·jən }

ornithology [ZOO] The study of birds. { ‚ȯr·nə'thäl·ə·jē }

orogenesis See orogeny. { ‚ȯr·ə'jen·ə·səs }

orogeny [GEOL] The process or processes of mountain formation, especially the intense deformation of rocks by folding and faulting which, in many mountainous regions, has been accompanied by metamorphism, invasion of molten rock, and volcanic eruption; in modern usage, orogeny produces the internal structure of mountains, and epeirogeny produces the mountainous topography. Also known as orogenesis; tectogenesis. { ȯ'räj·ə·nē }

orographic cloud [METEOROL] A cloud whose form and extent is determined by the disturbing effects of orography upon the passing flow of air; because these clouds are linked with the form of the terrestrial relief, they generally move very slowly, if at all, although the winds at the same level may be very strong. { ¦ȯr·ə¦graf·ik 'klaůd }

orographic lifting [METEOROL] The lifting of an air current caused by its passage up and over surface elevations. { ¦ȯr·ə¦graf·ik 'lift·iŋ }

orographic occlusion [METEOROL] An occluded front in which the occlusion process has been hastened by the retardation of the warm front along the windward slopes of a mountain range. { ¦ȯr·ə¦graf·ik ə'klü·zhən }

orographic precipitation [METEOROL] Precipitation which results from the lifting of moist air over an orographic barrier such as a mountain range. { ¦ȯr·ə¦graf·ik prə‚sip·ə'tā·shən }

orography [GEOGR] The branch of geography dealing with mountains. [GEOL] The relief features of mountains. { ȯ'räg·rə·fē }

orohydrography [HYD] A branch of hydrography dealing with the relations of mountains to drainage. { ¦ȯr·ō·hī'dräg·rə·fē }

orophyte [ECOL] Any plant that grows in the subalpine region. { 'ȯr·ə‚fīt }

orthoarsenic acid See arsenic acid. { ¦ȯr·thō·är¦sen·ik 'as·əd }

orthochem [GEOCHEM] A precipitate formed within a depositional basin or within the sediment itself by direct chemical action. { 'ȯr·thə‚kem }

orthokinesis [BIOL] Random movement of a motile cell or organism in response to a stimulus. { ¦ȯr·thə·ki'nē·səs }

orthophosphate [CHEM] One of the possible salts of orthophosphoric acid; the general formula is M_3PO_4, where M may be potassium as in potassium orthophosphate, K_3PO_4. { ¦ȯr·thə'fäs¸fāt }

orthotropism [BOT] The tendency of a plant to grow with the longer axis oriented vertically. { ȯr'thä·trə¸piz·əm }

OSHA *See* Occupational Safety and Health Administration.

osmophile [MICROBIO] A microorganism adapted to media with high osmotic pressure. { 'äz·mə¸fīl }

OTEC *See* ocean thermal-energy conversion. { 'ō¸tek }

outbreed *See* crossbreed. { 'aut¸brēd }

outcrop water [HYD] Rain and surface water which seeps downward through outcrops of porous and fissured rock, fault planes, old shafts, or surface drifts. { 'aut¸kräp ¸wȯd·ər }

outer atmosphere [METEOROL] Very generally, the atmosphere at a great distance from the earth's surface; possibly best usage of the term is as an approximate synonym for exosphere. { 'aud·ər 'at·mə¸sfir }

outer beach [GEOL] The part of a beach that is ordinarily dry and reached only by the waves generated by a violent storm. { 'aud·ər 'bēch }

outer core [GEOL] The outer or upper zone of the earth's core, extending to a depth of 3160 miles (5100 kilometers), and including the transition zone. { 'aud·ər 'kȯr }

outfall [HYD] The narrow part of a stream, lake, or other body of water where it drops away into a larger body. { 'aut¸fȯl }

outlet glacier [HYD] A stream of ice from an ice cap to the sea. { 'aut¸let ¸glā·shər }

outlet head [HYD] The place where water leaves a lake and enters an effluent. { 'aut ¸let ¸hed }

outside air temperature *See* indicated air temperature. { 'aut¸sīd 'er ¸tem·prə·chər }

ovary [BOT] The enlarged basal portion of a pistil that bears the ovules in angiosperms. { 'ōv·ə·rē }

overbank stage [HYD] The height of the surface of a river as the river floods over its banks. { 'ō·vər¸baŋk ¸stāj }

overburden [GEOL] **1.** Rock material overlying a mineral deposit or coal seam. Also known as baring; top. **2.** Material of any nature, consolidated or unconsolidated, that overlies a deposit of useful materials, ores, or coal, especially those deposits that are mined from the surface by open cuts. **3.** Loose soil, sand, or gravel that lies above the bedrock. { 'ō·vər¸bərd·ən }

overburdened stream *See* overloaded stream. { 'ō·vər¸bərd·ənd 'strēm }

overcast [METEOROL] **1.** Pertaining to a sky cover of 1.0 (95% or more) when at least a portion of this amount is attributable to clouds or obscuring phenomena aloft, that is, when the total sky cover is not due entirely to surface-based obscuring phenomena. **2.** Cloud layer that covers most or all of the sky; generally, a widespread layer of clouds such as that which is considered typical of a warm front. { 'ō·vər¸kast }

overfalls [OCEANOGR] Short, breaking waves occurring when a strong current passes over a shoal or other submarine obstruction or meets a contrary current or wind. { 'ō·vər¸fȯlz }

overflow [CIV ENG] Any device or structure that conducts excess water or sewage from a conduit or container. [SCI TECH] Excess liquid that overflows its given limits. { 'ō·vər¸flō }

overflow channel [CIV ENG] An artificial waterway for conducting water away from an overflowing structure such as a reservoir or canal. { 'ō·vər₁flō ₁chan·əl }

overflow ice [HYD] Ice formed during high spring tides by water rising through cracks in the surface ice and then freezing. { 'ō·vər₁flō ₁īs }

overflow stream [HYD] **1.** A stream containing water that has overflowed the banks of a river or another stream. Also known as spill stream. **2.** An effluent from a lake, carrying water to a stream, a sea, or another lake. { 'ō·vər₁flō ₁strēm }

overland flow [HYD] Water flowing over the ground surface toward a channel; upon reaching the channel, it is called surface runoff. Also known as surface flow. { 'ō·vər·lənd 'flō }

overloaded stream [HYD] A stream so heavily loaded with sediment that its velocity is lessened and it is forced to deposit part of its load. Also known as overburdened stream. { ¦ō·vər¦lōd·əd 'strēm }

overrunning [METEOROL] A condition existing when an air mass is in motion aloft above another air mass of greater density at the surface; this term usually is applied in the case of warm air ascending the surface of a warm front or quasi-stationary front. { 'ō·və₁rən·iŋ }

overseeding [METEOROL] Cloud seeding in which an excess of nucleating material is released; as the term is normally used, the excess is relative to that amount of nucleating material which would, theoretically, maximize the precipitation received at the ground. { ¦ō·vər¦sēd·iŋ }

overturn [HYD] Renewal of bottom water in lakes and ponds in regions where winter temperatures are cold; in the fall, cooled surface waters become denser and sink, until the whole body of water is at 4°C; in the spring, the surface is warmed back to 4°C, and the lake is homothermous. Also known as convective overturn. { 'ō·vər₁tərn }

overwash pool [OCEANOGR] A tidal pool between a berm and a beach scarp which water enters only at high tide. { 'ō·vər₁wäsh pül }

ovex [CHEM] ClC₆H₄OSO₂C₆H₄Cl A white, crystalline solid with a melting point of 86.5°C; soluble in acetone and aromatic solvents; used as an insecticide and acaricide. { 'ō₁veks }

ovine encephalomyelitis *See* louping ill. { 'ō₁vīn in¦sef·ə·lō₁mī·ə'līd·əs }

oviposit [ZOO] To lay or deposit eggs, especially by means of a specialized organ, as found in certain insects and fishes. { 'ō·və₁päz·ət }

ovule [BOT] A structure in the ovary of a seed plant that develops into a seed following fertilization. { 'äv₁yül }

oxadiazon [CHEM] C₁₃H₁₈Cl₂N₂O₃ A white solid with a melting point of 88–90°C; slight solubility in water; used as a pre- and postemergence herbicide to control weeds in rice, turf, soybeans, peanuts, and orchards. { ₁äk·sə'dī·ə₁zän }

oxamyl [CHEM] C₇H₁₃N₃O₃S A white, crystalline compound with a melting point of 100–102°C; used to control pests of tobacco, ornamentals, fruits, and crops. { 'äk·sə ₁mil }

oxbow [HYD] **1.** A closely looping, U-shaped stream meander whose curvature is so extreme that only a neck of land remains between the two parts of the stream. Also known as horseshoe bend. **2.** *See* oxbow lake. { 'äks₁bō }

oxbow lake [HYD] The crescent-shaped body of water located alongside a stream in an abandoned oxbow after a neck cutoff is formed and the ends of the original bends are silted up. Also known as crescentic lake; cutoff lake; horseshoe lake; loop lake; moat; mortlake; oxbow. { 'äks₁bō ¦lāk }

oxidation pond [CIV ENG] A shallow lagoon or basin in which wastewater is purified by sedimentation and aerobic and anaerobic treatment. { ˌäk·sə'dā-shən ˌpänd }

oxidized zone [GEOL] A region of mineral deposits which has been altered by oxidizing surface waters. { 'äk·səˌdīzd ˌzōn }

Oxisol [GEOL] A soil order characterized by residual accumulations of inactive clays, free oxides, kaolin, and quartz; mostly tropical. { 'äk·səˌsȯl }

oxoferrite [GEOL] A variety of naturally occurring iron with some ferrous oxide in solid solution. { ¦äk·sō'feˌrīt }

***para*-oxon** [CHEM] $(C_2H_5O)_2P(O)C_6H_4NO_2$ A reddish-yellow oil with a boiling point of 148–151°C; soluble in most organic solvents; used as an insecticide. Also known as diethyl *para*-nitrophenyl phosphate. { ¦par·ə 'äkˌsän }

oxoxanthone See genicide. { ¦äk·sō'zanˌthōn }

oxycarboxin [CHEM] $C_{12}H_{13}NO_4S$ An off-white, crystalline compound with a melting point of 127.5–130°C; used to control rust disease in greenhouse carnations. Also known as 5,6-dihydro-2-methyl-1,4-oxathiin-3-carboxanilide-4,4-dioxide. { ¦äk·sē·kär'bäk·sən }

oxygen deficit [GEOCHEM] The difference between the actual amount of dissolved oxygen in lake or sea water and the saturation concentration at the temperature of the water mass sampled. { 'äk·sə·jən ˌdef·ə·sət }

oxygen distribution [OCEANOGR] The concentration of dissolved oxygen in ocean water as a function of depth, ranging from as much as 5 milliliters of oxygen per liter at the surface to a fraction of that value at great depths. { 'äk·sə·jən ˌdis·trə'byü·shən }

oxygen isotope fractionation [GEOCHEM] The use of temperature- dependent variations of the oxygen-18/oxygen-16 ratio in the carbonate shells of marine organisms, to measure water temperature at the time of deposition. { 'äk·sə·jən 'īs·əˌtōp ˌfrak·shə'nā·shən }

oxygen minimum layer [HYD] A subsurface layer of water in which the content of dissolved oxygen is very low (or absent), lower than in the layers above and below. { 'äk·sə·jən 'min·ə·məm 'lā·ər }

oxygen ratio See acidity coefficient. { 'äk·sə·jən ˌrā·shō }

oxygen toxicity [BIOL] **1.** Harmful effects of breathing oxygen at pressures greater than atmospheric. **2.** A toxic effect in a living organism caused by a species of oxygen-containing reactive intermediate produced during the reduction of dioxygen. { 'äk·sə·jən täk'sis·əd·ē }

oxyphytia [ECOL] Discordant habitat control due to an excessively acidic substratum. { ˌäk·sə'fīd·ē·ə }

Oyashio [OCEANOGR] A cold current flowing from the Bering Sea southwest along the coast of Kamchatka, past the Kuril Islands, continuing close to the northeast coast of Japan, and reaching nearly 35°N. { ō'yä·shē·ō }

ozone [CHEM] O_3 Unstable blue gas with pungent odor; an allotropic form of oxygen; a powerful oxidant boiling at −112°C; used as an oxidant, bleach, and water purifier, and to treat industrial wastes. Ozone formed in the stratosphere protects life on Earth by absorbing most incoming ultraviolet solar radiation. { 'ōˌzōn }

ozone hole See Antarctic ozone hole. { 'ōˌzōn ˌhōl }

ozone layer See stratospheric ozone. { 'ōˌzōn ˌlā·ər }

ozonesonde [METEOROL] A balloon-borne instrument for measuring the ozone concentration at various altitudes and transmitting the data by radio. { 'ōˌzōnˌsänd }

ozonide   [CHEM] Any of the oily, thick, unstable compounds formed by reaction of ozone with unsaturated compounds; an example is oleic ozonide from the reaction of oleic acid and ozone.   { 'äz·ə,nīd }

ozonosphere   [METEOROL] The general stratum of the upper atmosphere in which there is an appreciable ozone concentration and in which ozone plays an important part in the radiative balance of the atmosphere; lies roughly between 6 and 30 miles (10 and 50 kilometers), with maximum ozone concentration at about 12 to 15 miles (20 to 25 kilometers). Also known as ozone layer.   { ō'zō·nə,sfir }

P

Pacific anticyclone   *See* Pacific high.   { pə'sif·ik ˌant·i'sī͵klōn }

Pacific Equatorial Countercurrent   [OCEANOGR] The Equatorial Countercurrent flowing east across the Pacific Ocean between 3° and 10°N.   { pə'sif·ik ˌek·wə'tȯr·ē·əl 'kaȯnt·ər͵kə·rənt }

Pacific faunal region   [ECOL] A marine littoral faunal region including offshore waters west of Central America, running from the coast of South America at about 5° south latitude to the southern tip of California.   { pə'sif·ik 'fȯn·əl ˌrē·jən }

Pacific high   [METEOROL] The nearly permanent subtropical high of the North Pacific Ocean, centered, in the mean, at 30–40°N and 140–150°W. Also known as Pacific anticyclone.   { pə'sif·ik 'hī }

Pacific North Equatorial Current   [OCEANOGR] The North Equatorial Current which flows westward between 10° and 20°N in the Pacific Ocean.   { pə'sif·ik 'nȯrth ˌek·wə'tȯr·ē·əl 'kə·rənt }

Pacific Ocean   [GEOGR] The largest division of the hydrosphere, having an area of 63,690 square miles (165,000,000 square kilometers) and covering 46% of the surface of the total extent of the oceans and seas; it is bounded by Asia and Australia on the west and North and South America on the east.   { pə'sif·ik 'ō·shən }

Pacific South Equatorial Current   [OCEANOGR] The South Equatorial Current flowing westward between 3°N and 10°S in the Pacific Ocean.   { pə'sif·ik 'saȯth ˌek·wə'tȯr·ē·əl 'kə·rənt }

Pacific temperate faunal region   [ECOL] A marine littoral faunal region including a narrow zone in the North Pacific Ocean, from Indochina to Alaska and along the west coast of the United States to about 40° north latitude.   { pə'sif·ik 'tem·prət 'fȯn·əl ˌrē·jən }

pack   *See* pack ice.   { pak }

pack ice   [OCEANOGR] Any area of sea ice, except fast ice, composed of a heterogeneous mixture of ice of varying ages and sizes, and formed by the packing together of pieces of floating ice. Also known as ice canopy; ice pack; pack.   { 'pak ˌīs }

palaeotropical   *See* paleotropical.   { ¦pāl·ē·ō'träp·ə·kəl }

Palearctic   [ECOL] Pertaining to a biogeographic region including Europe, northern Asia and Arabia, and Africa north of the Sahara.   { ¦pāl·ē'ärd·ik }

paleoceanography   [OCEANOGR] The study of the history of the circulation, chemistry, biogeography, fertility, and sedimentation of the oceans.   { ¦pāl·ē·ō·shə'näg·rə·fē }

paleocrystic ice   [HYD] Sea ice generally considered to be at least 10 years old, especially well-weathered polar ice.   { ¦pāl·ē·ō¦kris·tik 'īs }

paleosere   [ECOL] A series of ecologic communities that have led to a climax community.   { 'pāl·ē·ə͵sir }

paleotropical [ECOL] Of or pertaining to a biogeographic region that includes the Oriental and Ethiopian regions. Also spelled palaeotropical. { ¦pāl·ē·ō'träp·ə·kəl }

pallasite shell *See* lower mantle. { 'pal·ə₁sīt ₁shel }

palm [BIOL] The flexor or volar surface of the hand. [BOT] Any member of the monocotyledonous family Arecaceae; most are trees with a slender, unbranched trunk and a terminal crown of large leaves that are folded between the veins. { päm }

palmate [BOT] Having lobes, such as on leaves, that radiate from a common point. { 'pä₁māt }

Palouse [ECOL] A prairie in eastern Washington. { pə'lüz }

palouser [METEOROL] A dust storm of northwestern Labrador. { pə'lüz·ər }

paludal [ECOL] Relating to swamps or marshes and to material that is deposited in a swamp environment. { pə'lüd·əl }

paludification [ECOL] Bog expansion resulting from the gradual rising of the water table as accumulation of peat impedes water drainage. { pə₁lüd·ə·fə'kā·shən }

palustrine [ECOL] Being, living, or thriving in a marsh. { pə'ləs·trən }

palytoxin [BIOL] A water-soluble toxin produced by several species of *Palythoa*; considered to be one of the most poisonous substances known. { 'pal·ə₁täk·sən }

pamaquine naphthoate [MED] $C_{42}H_{45}N_3O_7$ A yellow to orange-yellow powder, soluble in alcohol and acetone; used as an antimalarial drug. { 'pam·ə₁kwēn 'naf·thə₁wāt }

pampa [ECOL] An extensive plain in South America, usually covered with grass. { 'päm·pə }

pan *See* pancake ice. { pan }

Panama disease [PL PATH] A fungus disease of banana caused by invasion of the vascular system by *Fusarium oxysporum cubense*, resulting in yellowing and wilting of the foliage and ultimate death of the shoots. { 'pan·ə₁mä di₁zēz }

pancake *See* pancake ice. { 'pan₁kāk }

pancake ice [OCEANOGR] One or more small, newly formed pieces of sea ice, generally circular with slightly raised edges and about 1 to 10 feet (0.3 to 3 meters) across. Also known as lily-pad ice; pan; pancake; pan ice; plate ice. { 'pan₁kāk ¦īs }

panclimax [ECOL] Two or more related climax communities or formations having similar climate, life forms, and genera or dominants. Also known as panformation. { pan'klī₁maks }

pan coefficient [METEOROL] The ratio of the amount of evaporation from a large body of water to that measured in an evaporation pan. { 'pan ₁kō·i₁fish·ənt }

panformation *See* panclimax. { 'pan·fər₁mā·shən }

pan ice *See* pancake ice. { 'pan ₁īs }

panmixis [BIOL] Random mating within a breeding population; in a closed population this results in a high degree of uniformity. { pan'mik·səs }

pannus [METEOROL] Numerous cloud shreds below the main cloud; may constitute a layer separated from the main part of the cloud or attached to it. { 'pan·əs }

panzootic [VET MED] Affecting many animals of different species. { ₁pan·zō'äd·ik }

papovavirus [MICROBIO] A deoxyribonucleic acid-containing group of animal viruses, including papilloma and vacuolating viruses. { ¦pap·ə·və'vī·rəs }

pappataci fever *See* phlebotomus fever. { ¦päp·ə¦tä·chē ₁fē·vər }

paracoccidioidomycosis *See* South American blastomycosis. { ¦par·ə·käk‚sid·ē¦ȯid·ō· mī‚kō·səs }

paraformaldehyde [CHEM] (HCHO)$_n$ Polymer of formaldehyde where *n* is greater than 6; white, alkali-soluble solid, insoluble in alcohol, ether, and water; used as a disinfectant, fumigant, and fungicide, and to make resins. { ¦par·ə·fȯr'mal·də‚hīd }

paralic swamp *See* marine swamp. { pə'ral·ik 'swämp }

paralimnion [HYD] The littoral part of a lake, extending from the margin to the deepest limit of rooted vegetation. { ¦par·ə'lim·nē‚än }

parallax inequality [OCEANOGR] The variation in the range of tide or in the speed of tidal currents due to the continual change in the distance of the moon from the earth. { 'par·ə‚laks ‚in·i'kwäl·əd·ē }

parallel drainage pattern [HYD] A drainage pattern characterized by regularly spaced streams flowing parallel to one another over a large area. { 'par·ə‚lel 'drān·ij ‚pad·ərn }

parametric hydrology [HYD] That branch of hydrology dealing with the development and analysis of relationships among the physical parameters involved in hydrologic events and the use of these relationships to generate, or synthesize, hydrologic events. { ¦par·ə¦me·trik hī'dräl·ə·jē }

paramo [ECOL] A biological community, essentially a grassland, covering extensive high areas in equatorial mountains of the Western Hemisphere. { 'pär·ə‚mō }

Paramyxoviridae [MICROBIO] A family of negative-strand ribonucleic acid (RNA) viruses characterized by an enveloped spherical virion containing a single-stranded, nonfragmented molecule of RNA, contains the genera *Paramyxovirus* (sendai, mumps), *Morbillivirus* (measles), and P*neumovirus* (respiratory syncytial virus). { ‚par·ə ‚mik·sə'vir·ə‚dī }

paramyxovirus [MICROBIO] A subgroup of myxoviruses, including the viruses of mumps, measles, parainfluenza, and Newcastle disease; all are ribonucleic acid-containing viruses and possess an ether-sensitive lipoprotein envelope. { ¦par·ə ‚mik·sō'vī·rəs }

parapatric [ECOL] Referring to populations or species that occupy nonoverlapping but adjacent geographical areas without interbreeding. { ¦par·ə¦pa·trik }

parapertussis [MED] An acute bacterial respiratory infection similar to mild pertussis and caused by *Bordetella pertussis*. { ¦par·ə·pər'təs·əs }

paraquat [CHEM] [CH$_3$(C$_5$H$_4$N)$_2$CH$_3$]·2CH$_3$SO$_4$ A yellow, water-soluble solid, used as a herbicide. { 'par·ə‚kwät }

parasite [BIOL] An organism that lives in or on another organism of different species from which it derives nutrients and shelter. { 'par·ə‚sīt }

parasitic castration [BIOL] Destruction of the reproductive organs by parasites. { ¦par· ə¦sid·ik ka'strā·shən }

parasitic stomatitis *See* thrush. { ¦par·ə¦sid·ik ‚stō·mə'tīd·əs }

parasitism [ECOL] A symbiotic relationship in which the host is harmed, but not killed immediately, and the species feeding on it is benefited. { 'par·ə·sə‚tiz·əm }

parasitoidism [BIOL] Systematic feeding by an insect larva on living host tissues so that the host will live until completion of larval development. { ¦par·ə·sə'tȯid‚iz·əm }

parasitology [BIOL] A branch of biology which deals with those organisms, plant or animal, which have become dependent on other living creatures. { ‚par·ə·sə'täl·ə·jē }

paratonic movement [BOT] The movement of the whole or parts of a plant due to the influence of an external stimulus, such as gravity, chemicals, heat, light, or electricity. { ¦par·ə¦tän·ik 'müv·mənt }

paraxial |SCI TECH| Lying near the axis. { par'ak·sē·əl }

parcel method |METEOROL| A method of testing for instability in which a displacement is made from a steady state under the assumption that only the parcel or parcels displaced are affected, the environment remaining unchanged. { 'pär·səl ˌmeth·əd }

parenchyma |BOT| A tissue of higher plants consisting of living cells with thin walls that are agents of photosynthesis and storage; abundant in leaves, roots, and the pulp of fruit, and found also in leaves and stems. { pə'reŋ·kə·mə }

parkland See temperate woodland ; tropical woodland. { 'pärkˌland }

parsimony |SCI TECH| The principle that the simplest scientific explanation is best. { 'pär·səˌmō·nē }

parthenocarpy |BOT| Production of fruit without fertilization. { 'pär·thə·nōˌkär·pē }

partial obscuration |METEOROL| In United States weather observing practice, the designation for sky cover when part (0.1 to 0.9) of the sky is completely hidden by surface-based obscuring phenomena. { 'pär·shəl äb·skyü'rā·shən }

partial potential temperature |METEOROL| The temperature that the dry-air component of an air parcel would attain if its actual partial pressure were changed to 1000 millibars (10^5 pascals). { 'pär·shəl pə'ten·chəl 'tem·prə·chər }

partial tide |OCEANOGR| One of the harmonic components composing the tide at any point. Also known as tidal component; tidal constituent. { 'pär·shəl 'tīd }

particle detector |ENG| A device used to indicate the presence of fast-moving charged atomic or nuclear particles by observation of the electrical disturbance created by a particle as it passes through the device. Also known as radiation detector. { 'pärd·ə·kəl diˌtek·tər }

particle velocity |OCEANOGR| In ocean wave studies, the instantaneous velocity of a water particle undergoing orbital motion. { 'pärd·ə·kəl vəˌläs·əd·ē }

partly cloudy |METEOROL| **1.** The character of a day's weather when the average cloudiness, as determined from frequent observations, has been from 0.1 to 0.5 for the 24-hour period. **2.** In popular usage, the state of the weather when clouds are conspicuously present, but do not completely dull the day or the sky at any moment. { 'pärt·lē 'klaůd·ē }

pass |GEOGR| **1.** A natural break, depression, or other low place providing a passage through high terrain, such as a mountain range. **2.** A navigable channel leading to a harbor or river. **3.** A narrow opening through a barrier reef, atoll, or sand bar. { pas }

passage |GEOGR| A navigable channel, especially one through reefs or islands. { 'pas·ij }

passive front See inactive front. { 'pas·iv 'frənt }

passive glacier |HYD| A glacier with sluggish movement, generally occurring in a continental environment at a high latitude, where both accumulation and ablation are minimal. { 'pas·iv 'glā·shər }

passive permafrost |GEOL| Permafrost that will not refreeze under present climatic conditions after being disturbed or destroyed. Also known as fossil permafrost. { 'pas·iv 'pər·məˌfròst }

passive solar system |ENG| A solar heating or cooling system that operates by using gravity, heat flows, or evaporation rather than mechanical devices to collect and transfer energy. { 'pas·iv 'sō·lər ˌsis·təm }

Pasteurella |MICROBIO| A genus of gram-negative, nonmotile, nonsporulating, facultatively anaerobic coccobacillary to rod-shaped bacteria which are parasitic and often pathogens in many species of mammals, birds, and reptiles, it was named to honor Louis Pasteur in 1887. { ˌpas·chə'rel·ə }

pasteurellosis   See hemorrhagic septicemia.   { ˌpa·stə·rə'lō·səs }

pasteurization   [SCI TECH] The application of heat to matter for a specified time to destroy harmful microorganisms or other undesirable species.   { ˌpas·chə·rə·'zā·shən }

patch reef   [GEOL] **1.** A small, irregular organic reef with a flat top forming a part of a reef complex. **2.** A small, thick, isolated lens of limestone or dolomite surrounded by rocks of different facies.See reef patch.   { 'pach ˌrēf }

patent period   [MED] The period of an infective disease during which the causative agent can be detected.   { 'pat·ənt ˌpir·ē·əd }

paternoster lake   [HYD] One of a linear chain or series of small circular lakes, usually at different levels, which occupy rock basins in a glacial valley and are separated by morainal dams or riegels, but connected by streams, rapids, or waterfalls to resemble a rosary or string of beads. Also known as beaded lake; rock-basin lake; step lake.   { 'päd·ər,näs·tər ,lāk }

pathogenic   [MED] **1.** Producing or capable of producing disease. **2.** Pertaining to pathogenesis.   { ¦path·ə¦jen·ik }

pathotoxin   [PL PATH] A chemical of biological origin, other than an enzyme, that plays an important causal role in a plant disease.   { ¦path·ə¦täk·sən }

pathovar   [MICROBIO] A pathological variant of a nonpathological bacterial species.   { 'path·ə,vär }

patulin   [MED] $C_7H_6O_4$ An antibiotic derived from several fungi (*Aspergillus, Penicillium* species); crystalline compound soluble in water and most organic solvents; melting point is 111°C; used as an antimicrobial agent; also appears to be a potent carcinogenic mycotoxin. Also known as penicidin.   { 'pach·ə·lən }

PCB   See polychlorinated biphenyl.

PCNB   See pentachloronitrobenzene.

peak   [METEOROL] The point of intersection of the cold and warm fronts of a mature extra-tropical cyclone.   { pēk }

peak gust   [METEOROL] After United States weather observing practice, the highest instantaneous wind speed recorded at a station during a specified period, usually the 24-hour observational day; therefore, a peak gust need not be a true gust of wind.   { 'pēk 'gəst }

pearl moss   See carrageen.   { 'pərl ,mòs }

pea-soup fog   [METEOROL] Any particularly dense fog.   { 'pē ,süp 'fäg }

peat   [GEOL] A dark-brown or black residuum produced by the partial decomposition and disintegration of mosses, sedges, trees, and other plants that grow in marshes and other wet places.   { pēt }

peat ball   [ECOL] A lake ball containing an abundance of peaty fragments.   { 'pēt ,bòl }

peat formation   [GEOCHEM] Decomposition of vegetation in stagnant water with small amounts of oxygen, under conditions intermediate between those of putrefaction and those of moldering.   { 'pēt fòr'mā·shən }

peat moss   [ECOL] Moss, especially sphagnum moss, from which peat has been produced.   { 'pēt ,mòs }

peat soil   [GEOL] Soil containing a large amount of peat; it is rich in humus and gives an acid reaction.   { 'pēt ,sòil }

pebble   [GEOL] A clast, larger than a granule and smaller than a cobble, having a diameter in the range of 0.16–2.6 inches (4–64 millimeters). Also known as pebblestone.   { 'peb·əl }

pebblestone   See pebble.   { 'peb·əl,stōn }

ped |GEOL| A naturally formed unit of soil structure. { ped }

pedicel |BOT| **1.** The stem of a fruiting or sporebearing organ. **2.** The stem of a single flower. { 'ped·ə‚sel }

pedogenesis See soil genesis. { ¦ped·ō'jen·ə·səs }

pedogeochemical survey |GEOCHEM| A geochemical prospecting survey in which the materials sampled are soil and till. { ¦ped·ō‚jē·ō'kem·ə·kəl 'sər‚vā }

pedologic age |GEOL| The relative maturity of a soil profile. { ¦ped·ō¦läj·ik 'āj }

pedology See soil science. { pe'däl·ə·jē }

peduncle |BOT| **1.** A flower-bearing stalk. **2.** A stalk supporting the fruiting body of certain thallophytes. { 'pē‚dəŋ·kəl }

pelagic |OCEANOGR| Pertaining to water of the open portion of an ocean, above the abyssal zone and beyond the outer limits of the littoral zone. { pə'laj·ik }

pellicularia disease |PL PATH| A fungus disease of coffee and other tropical plants caused by Pellicularia koleroga and characterized by leaf spots. { pə‚lik·yə'lar·ē·ə di‚zēz }

pellicular water [HYD] Films of groundwater adhering to particles or cavities above the water table. { pə'lik·yə·lər 'wȯd·ər }

pelogloea |GEOL| Marine detrital slime from settled plankton. { ¦pel·ə¦glē·ə }

pelphyte |GEOL| A lake-bottom deposit consisting mainly of fine, nonfibrous plant remains. { 'pel‚fīt }

pendant cloud See tuba. { 'pen·dənt ‚klaȯd }

pendular water [HYD] Capillary water ringing the contact points of adjacent rock or soil particles in the zone of aeration. { 'pen·jə·lər ‚wȯd·ər }

peneplain See base-leveled plain. { 'pēn·ə‚plān }

penesaline |ECOL| Referring to an environment intermediate between normal marine and saline, characterized by evaporitic carbonates often interbedded with gypsum or anhydrite, and by a salinity high enough to be toxic to normal marine organisms. { ‚pēn·ə'sā‚lēn }

penicidin See patulin. { ‚pen·ə'sīd·ən }

penicillin |MICROBIO| **1.** The collective name for salts of a series of antibiotic organic acids produced by a number of Penicillium and Aspergillus species; active against most gram-positive bacteria and some gram-negative cocci. **2.** See benzyl penicillin sodium. { ‚pen·ə'sil·ən }

peninsula |GEOGR| A body of land extending into water from the mainland, sometimes almost entirely separated from the mainland except for an isthmus. { pə'nin·sə·lə }

penitent ice [HYD] A jagged spike or pillar of compacted firn caused by differential melting and evaporation; necessary for this formation are air temperature near freezing, dew point much below freezing, and strong insolation. { 'pen·ə·tənt 'īs }

penitent snow [HYD] A jagged spike or pillar of compacted snow caused by differential melting and evaporation. { 'pen·ə·tənt 'snō }

pennant |METEOROL| A means of representing wind speed in the plotting of a synoptic chart; it is a triangular flag, drawn pointing toward lower pressure from a wind-direction shaft. { 'pen·ənt }

penstock |CIV ENG| A valve or sluice gate for regulating water or sewage flow. { 'pen ‚stäk }

pentachloronitrobenzene |CHEM| $C_6Cl_5NO_2$ crystals or cream color with a melting point of 142–145°C; slightly soluble in alcohols; used as a fungicide and herbicide.

Abbreviated PCNB. Also known as quintozene; terrachlor. { ¦pen·tə¦klȯr·ō,nī·trə'ben ,zēn }

pentachlorophenol [CHEM] C_6Cl_5OH A toxic white powder, decomposing at 310°C, melting at 190°C; soluble in alcohol, acetone, ether, and benzene; used as a fungicide, bactericide, algicide, herbicide, and chemical intermediate. { ¦pen·tə¦klȯr·ō·'fē,nȯl }

pentad [CLIMATOL] A period of 5 consecutive days, often preferred to the week for climatological purposes since it is an exact factor of the 365-day year. { 'pen,tad }

pentamerous [BOT] Having each whorl of the flower consisting of five members, or a multiple of five. { pen'tam·ə·rəs }

***para*-pentyloxyphenol** [CHEM] $C_{11}H_{16}O_2$ Compound melting at 49–50°C; used as a bactericide. { ¦par·ə ¦pent·əl¦äk·sē'fē,nȯl }

pepo [BOT] A fleshy indehiscent berry with many seeds and a hard rind; characteristic of the Cucurbitaceae (for example, cucumber). { 'pē,pō }

peracetic acid [CHEM] CH_3COOOH A toxic, colorless liquid with strong aroma; boils at 105°C; explodes at 110°C; miscible with water, alcohol, glycerin, and ether; used as an oxidizer, bleach, catalyst, bactericide, fungicide, epoxy-resin precursor, and chemical intermediate. Also known as peroxyacetic acid. { ¦par·ə¦sēd·ik 'as·əd }

P-E ratio See precipitation-evaporation ratio. { ¦pē'ē ,rā·shō }

perceived noise decibel [PHYS] A unit of perceived noise level. Abbreviated PNdB. { pər'sēvd ¦nȯiz 'des·ə,bel }

perceived noise level [PHYS] In perceived noise decibels, the noise level numerically equal to the sound pressure level, in decibels, of a band of random noise of width one-third to one octave centered on a frequency of 1000 hertz which is judged by listeners to be equally noisy. { pər'sēvd ¦nȯiz ,lev·əl }

perched aquifer [HYD] An aquifer that is separated from another water-bearing stratum by an impermeable layer. { 'pərcht 'ak·wə·fər }

perched groundwater See perched water. { 'pərcht 'grau̇nd,wȯd·ər }

perched lake [HYD] A perennial lake whose surface level lies at a considerably higher elevation than those of other bodies of water, including aquifers, directly or closely associated with the lake. { 'pərcht 'lāk }

perched spring [HYD] A spring that arises from a body of perched water. { 'pərcht 'spriŋ }

perched stream [HYD] A stream whose surface level is above that of the water table and that is separated from underlying groundwater by an impermeable bed in the zone of aeration. { 'pərcht 'strēm }

perched water [HYD] Groundwater that is unconfined and separated from an underlying main body of groundwater by an unsaturated zone. Also known as perched groundwater. { 'pərcht 'wȯd·ər }

perched water table [HYD] The water table or upper surface of a body of perched water. Also known as apparent water table. { 'pərcht 'wȯd·ər ,tā·bəl }

percolation [SCI TECH] Slow movement of a liquid through a porous material. { pər·kə'lā·shən }

percolation zone [HYD] The area on a glacier or ice sheet where a limited amount of surface melting occurs, but the meltwater refreezes in the same snow layer and the snow layer is not completely soaked or brought up to the melting temperature. { pər·kə'lā·shən ,zōn }

perennial [BOT] A plant that lives for an indefinite period, dying back seasonally and producing new growth from a perennating part. { pə'ren·ē·əl }

perennial lake [HYD] A lake that retains water in its basin throughout the year and is not usually subject to extreme water-level fluctuations. { pə'ren·ē·əl 'lāk }

perennial spring [HYD] A spring that flows continuously, as opposed to an intermittent spring or a periodic spring. { pə'ren·ē·əl 'spriŋ }

perennial stream [HYD] A stream which contains water at all times except during extreme drought. { pə'ren·ē·əl 'strēm }

perfect flower [BOT] A flower having both stamens and pistils. { 'pər·fikt 'flaù·ər }

perfect prognostic [METEOROL] The observed pressure pattern at the verifying time of a forecast of some element other than pressure; used in objective forecast studies in which a forecast of the element is based on a simultaneous relation between this element and the pressure pattern plus a forecast of the pressure pattern at some future time. { 'pər·fikt präg'näs·tik }

perforated crust [HYD] A type of snow crust containing pits and hollows produced by ablation. { 'pər·fə,rād·əd 'krəst }

pergelation [HYD] The act or process of forming permafrost. { ,pər·jə'lā·shən }

pergelic [GEOL] Referring to a soil temperature regime in which the mean annual temperature is less than 0°C and there is permafrost. { pər'jel·ik }

perhumid climate [CLIMATOL] As defined by C. W. Thornthwaite in his climatic classification, a type of climate which has humidity index values of +100 and above; this is his wettest type of climate (designated A), and compares closely to the "wet climate" which heads his 1931 grouping of humidity provinces. { ¦pər'hyü·məd 'klī·mət }

perianth [BOT] The calyx and corolla considered together. { 'per·ē,anth }

pericarp [BOT] The wall of a fruit, developed by ripening and modification of the ovarian wall. { 'per·ə,kärp }

pericycle [BOT] The outer boundary of the stele of plants; may not be present as a distinct layer of cells. { 'per·ə,sī·kəl }

periderm [BOT] A group of secondary tissues forming a protective layer which replaces the epidermis of many plant stems, roots, and other parts; composed of cork cambium, phelloderm, and cork. { 'per·ə,dərm }

perigean range [OCEANOGR] The average range of tide at the time of perigean tides, when the moon is near perigee; the perigean range is greater than the mean range. { ¦per·ə¦jē·ən 'rānj }

perigean tidal currents [OCEANOGR] Tidal currents of increased speed occurring at the time of perigean tides. { ¦per·ə¦jē·ən 'tīd·əl ,kə·rəns }

perigean tide [OCEANOGR] Tide of increased range occurring when the moon is near perigee (the point in the moon's orbit when it is nearest the earth). { ¦per·ə¦jē·ən 'tīd }

periglacial [GEOL] Of or pertaining to the outer perimeter of a glacier, particularly to the fringe areas immediately surrounding the great continental glaciers of the geologic ice ages, with respect to environment, topography, areas, processes, and conditions influenced by the low temperature of the ice. { ¦per·ə'glā·shəl }

periglacial climate [CLIMATOL] The climate which is characteristic of the regions immediately bordering the outer perimeter of an ice cap or continental glacier; the principal climatic feature is the high frequency of very cold and dry winds off the ice area; it is also thought that these regions offer ideal conditions for the maintenance of a belt of intense cyclonic activity. { ¦per·ə'glā·shəl 'klī·mət }

periodic current [OCEANOGR] Current produced by the tidal influence of moon and sun or by any other oscillatory forcing function. { ¦pir·ē¦äd·ik 'kə·rənt }

periodic discing [AGR] A type of soil tillage involving a series of disc-shaped plows. { ¦pir·ē̩äd·ik 'disk·iŋ }

periodic spring [HYD] A spring that ebbs and flows periodically, apparently due to natural siphon action. { ¦pir·ē̩äd·ik 'spriŋ }

peripheral stream [HYD] A stream that flows parallel to the edge of a glacier, usually just beyond the moraine. { pə'rif·ə·rəl 'strēm }

periphyton [ECOL] Sessile biotal components of a fresh-water ecosystem. { pə'rif·ə ̩tän }

perlucidus [METEOROL] A cloud variety, usually of the species stratiformis, in which distinct spaces between its elements permit the sun, moon, blue sky, or higher clouds to be seen. { pər'lü·səd·əs }

permafrost [GEOL] Perennially frozen ground, occurring wherever the temperature remains below 0°C for several years, whether the ground is actually consolidated by ice or not and regardless of the nature of the rock and soil particles of which the earth is composed. { 'pər·mə̩fròst }

permanent current [OCEANOGR] A current which continues with relatively little periodic or seasonal change. { 'pər·mə·nənt 'kə·rənt }

permanent ice foot [HYD] An ice foot that does not melt completely in summer. { 'pər·mə·nənt 'īs ̩fút }

permanent thermocline See main thermocline. { 'pər·mə·nənt 'thər·mə̩klīn }

permanent water [HYD] A source of water that remains constant throughout the year. { 'pər·mə·nənt 'wòd·ər }

permeability [GEOL] The capacity of a porous rock, soil, or sediment for transmitting a fluid without damage to the structure of the medium. Also known as conductivity; perviousness. { ̩pər·mē·ə'bil·əd·ē }

permissible dose [MED] The amount of radiation that may be safely received by an individual within a specified period. Formerly known as tolerance dose. { pər'mis·ə·bəl 'dōs }

peroxyacetic acid See peracetic acid. { pə¦räk·sē·ə¦sēd·ik 'as·əd }

peroxydol See sodium perborate. { pə'räk·sə̩dòl }

perpetual frost climate [CLIMATOL] The climate of the ice cap regions of the world; thus, it requires temperatures sufficiently cold so that the annual accumulation of snow and ice is never exceeded by ablation. Also known as ice-cap climate. { pər'pech·ə·wəl 'fròst ̩klī·mət }

persistence [METEOROL] With respect to the long-term nature of the wind at a given location, the ratio of the magnitude of the mean wind vector to the average speed of the wind without regard to direction. Also known as constancy; steadiness. { pər'sis·təns }

persistence forecast [METEOROL] A forecast that the future weather condition will be the same as the present condition; often used as a standard of comparison in measuring the degree of skill of forecasts prepared by other methods. { pər'sis·təns ̩fòr̩kast }

persistent [BOT] Of a leaf, withering but remaining attached to the plant during the winter. { pər'sis·tənt }

personnel monitoring [ENG] Determination of the degree of radioactive contamination on individuals, using standard survey meters, and determination of the dose received by means of dosimeters. { ̩pərs·ən'el ̩män·ə·triŋ }

pertussis [MED] An infectious inflammatory bacterial disease of the air passages, caused by *Hemophilus pertussis* and characterized by explosive coughing ending in a whooping inspiration. Also known as whooping cough. { pər'təs·əs }

Peru Current |OCEANOGR| The cold ocean current flowing north along the coasts of Chile and Peru. Also known as Humboldt Current. { pə'rü 'kə·rənt }

perviousness *See* permeability. { 'pər·vē·əs·nəs }

pesticide |AGR| A chemical agent that destroys pests. Also known as biocide. { 'pes·tə ‚sīd }

petal |BOT| One of the sterile, leaf-shaped flower parts that make up the corolla. { 'ped·əl }

petiole |BOT| The stem which supports the blade of a leaf. { 'ped·ē‚ōl }

petrifaction |GEOL| A fossilization process whereby inorganic matter dissolved in water replaces the original organic materials, converting them to a stony substance. { ‚pe·trə'fak·shən }

petrified wood *See* silicified wood. { 'pe·trə‚fīd 'wùd }

petrochemistry |GEOCHEM| An aspect of geochemistry that deals with the study of the chemical composition of rocks. { ¦pe·trō'kem·ə·strē }

petrol *See* gasoline. { 'pe·trəl }

petroleum |GEOL| A naturally occurring complex liquid hydrocarbon which after distillation yields combustible fuels, petrochemicals, and lubricants; can be gaseous (natural gas), liquid (crude oil, crude petroleum), solid (asphalt, tar, bitumen), or a combination of states. { pə'trō·lē·əm }

petroleum coke |PETR MIN| A carbonaceous solid material made by the destructive heating of high-molecular-weight petroleum-refining residues. { pə'trō·lē·əm ‚kōk }

petroleum engineering |ENG| The application of almost all types of engineering to the drilling for and production of oil, gas, and liquefiable hydrocarbons. { pə'trō·lē·əm ‚en·jə'nir·iŋ }

petroleum microbiology |MICROBIO| Those aspects of microbiological science and engineering of interest to the petroleum industry, including the role of microbes in petroleum formation, and the exploration, production, manufacturing, storage, and food synthesis from petroleum. { pə'trō·lē·əm ‚mī·krō·bī'äl·ə·jē }

petroleum seep *See* oil seep. { pə'trō·lē·əm ‚sēp }

petroleum tar |PETR MIN| A viscous, black or dark-brown product of petroleum refining; yields substantial quantity of solid residue when partly evaporated or fractionally distilled. { pə'trō·lē·əm ‚tär }

petroliferous |GEOL| Containing petroleum. { ‚pe·trə'lif·ə·rəs }

Petterson-Nansen water bottle *See* Nansen bottle. { 'ped·ər·sən 'nan·sən 'wȯd·ər ‚bäd·əl }

Phaeophyta |BOT| The brown algae, constituting a division of plants; the plant body is multicellular, varying from a simple filamentous form to a complex, sometimes branched body having a basal attachment. { fē'äf·əd·ə }

phage *See* bacteriophage. { fāj }

phagotrophic |ZOO| Referring to a form of feeding in which an organism engulfs large solid objects, such as bacteria, and then delivers them to special digesting vacuoles. { 'fag·ə'trōf·ik }

phallotoxin |BIOL| One of a group of toxic peptides produced by the mushroom *Amanita phalloides*. { ¦fal·ō¦täk·sən }

phanerophyte |ECOL| A perennial tree or shrub with dormant buds borne on aerial shoots. { 'fan·ə·rō‚fīt }

phantom [ENG] A volume of material approximating as closely as possible the density and effective atomic number of living tissue, used in biological experiments involving radiation. { 'fan·təm }

phantom bottom [OCEANOGR] A false ocean bottom indicated by an echo sounder, some distance above the actual bottom; such an indication, quite common in the deeper parts of the ocean, is due to large quantities of small organisms. { 'fan·təm 'bäd·əm }

pharmaceutical biotechnology [MED] A field that uses micro- and macroorganisms and hybridomas to create pharmaceuticals that are safer and more cost-effective than conventionally produced pharmaceuticals. { ,fär·mə,süd·i·kəl ,bī·ō·tek'näl·ə·jē }

pharmacognosy [MED] A subfield of pharmacology which studies the biological and chemical components of medically useful substances that occur naturally (primarily those synthesized by plants). { ,fär·mə'käg·nə·sē }

phase inequality [OCEANOGR] Variations in the tide or tidal currents associated with changes in the phase of the moon. { 'fāz ,in·i'kwäl·əd·ē }

phase lag [OCEANOGR] Angular retardation of the maximum of a constituent of the observed tide behind the corresponding maximum of the same constituent of the hypothetical equilibrium tide. Also known as tidal epoch. { 'fāz ,lag }

phellem [BOT] Cork; the outer tissue layer of the periderm. { 'fel·əm }

phenol [CHEM] **1.** C_6H_5OH White, poisonous, corrosive crystals with sharp, burning taste; melts at 43°C, boils at 182°C; soluble in alcohol, water, ether, carbon disulfide, and other solvents; used to make resins and weed killers, and as a solvent and chemical intermediate. Also known as carbolic acid; phenylic acid. **2.** A chemical compound based on the substitution product of phenol, for example, ethylphenol ($C_2H_4C_4H_5OH$), the ethyl substitute of phenol. { 'fē,nȯl }

phenol-coefficient method [CHEM] A method for evaluating water-miscible disinfectants in which a test organism is added to a series of dilutions of the disinfectant; the phenol coefficient is the number obtained by dividing the greatest dilution of the disinfectant killing the test organism by the greatest dilution of phenol showing the same result. { 'fē,nȯl ,kō·i,fish·ənt ,meth·əd }

phenological shift [ECOL] A change in the timing of growth and breeding events in the life of an individual organism. { ,fēn·ə,läj·i·kəl 'shift }

phenology [CLIMATOL] The science which treats of periodic biological phenomena with relation to climate, especially seasonal changes; from a climatologic viewpoint, these phenomena serve as bases for the interpretation of local seasons and the climatic zones, and are considered to integrate the effects of a number of bioclimatic factors. { fə'näl·ə·jē }

phenotype [GEN] The observable characters of an organism, dependent upon genotype and environment. { 'fē·nə,tīp }

phenotypic plasticity [GEN] The extent of genotype expression in different environments. { ¦fē·nə,tip·ik plas'tis·əd·ē }

phenotypic sex determination [BIOL] Control of the development of gonads by environmental stimuli, such as temperature. { ¦fē·nə,tip·ik 'seks di,tər·mə,nā·shən }

phenyl [CHEM] C_6H_5- A functional group consisting of a benzene ring from which a hydrogen has been removed. { 'fen·əl }

phenylglyoxylonitriloxime *O,O*-**diethyl phosphorothioate** [CHEM] $(H_5C_2O)_2PSONC-CNC_6H_5$ A yellow liquid with a boiling point of 102°C at 0.01 mmHg (1.333 pascals); solubility in water is 7 parts per million at 20°C; used as an insecticide for stored products. Also known as phoxim. { ¦fen·əl·glī¦äk·sē¦län·ə·trəl'äk,sēm ¦ō¦ō dī'eth·əl ,fäs·fə·rō'thī·ə,āt }

phenylic acid   See phenol.   { fe'nil·ik 'as·əd }

phenylmercuric acetate   [CHEM] $C_8H_8O_2Hg$ White to cream-colored prisms with a melting point of 148–150°C; soluble in alcohol, benzene, and glacial acetic acid; used as an antiseptic, fungicide, herbicide, and mildewcide.   { ¦fen·əl·mər'kyür·ik 'as·ə‚tāt }

phenylmercuric oleate   [CHEM] $C_{41}H_{21}O_2Hg$ A white, crystalline powder with a melting point of 45°C; soluble in organic solvents; used in paints as a mildew-proofing agent, and as a fungicide.   { ¦fen·əl·mər'kyür·ik 'ō·lē‚āt }

phenylphenol   [CHEM] $C_6H_5C_6H_4OH$ Almost white crystals, soluble in alcohol, insoluble in water; the ortho form, melting at 56–58°C, is used to manufacture dyes, as germicide and fungicide, and in the rubber industry, and is also known as 2-hydroxybiphenyl, *ortho*-xenol; the para form, melting at 164–165°C, is used to manufacture dyes, resins, and rubber chemicals, and as a fungicide.   { ¦fen·əl'fē‚nòl }

pheoplast   [BIOL] A plastid containing brown pigment and found in diatoms, dinoflagellates, and brown algae.   { 'fē·ə‚plast }

Philippine fowl disease   See Newcastle disease.   { 'fil·ə‚pēn 'faùl di‚zēz }

philopatry   [ECOL] A dispersal method in which their reproductive particles remain near their point of origin.   { ‚fī·lə'pa·trē }

phlebotomus fever   [MED] An acute viral infection, transmitted by the fly *Phlebotomus papatosii* and characterized by fever, pains in the head and eyes, inflammation of the conjunctiva, leukopenia, and general malaise. Also known as Chitral fever; pappataci fever; sandfly fever; three-day fever.   { flə'bäd·ə·məs ‚fēv·ər }

Phlebovirus   [MICROBIO] A genus of the family Bunyaviridae that causes sandfly fever.   { 'flē·bə‚vī·rəs }

phloem   [BOT] A complex, food-conducting vascular tissue in higher plants; principal conducting cells are sieve elements. Also known as bast; sieve tissue.   { 'flō·əm }

phorate   [CHEM] $C_7H_{17}O_2PS_2$ A clear liquid with slight solubility in water; used as an insecticide for a wide range of insects on a wide range of crops.   { 'fòr‚āt }

phoresy   [ECOL] A relationship between two different species of organisms in which the larger, or host, organism transports a smaller organism, the guest.   { 'fòr·ə·sē }

phosgene   [CHEM] $COCl_2$ A highly toxic, colorless gas that condenses at 0°C to a fuming liquid; used as a war gas and in manufacture of organic compounds.   { 'fäz‚jēn }

phosphate fertilizer   [AGR] Fertilizer compound or mixture containing available (soluble) phosphate; examples are phosphate rock (phosphorite), superphosphates or triple superphosphates, nitrophosphate, potassium phosphates, or N-P-K mixtures.   { 'fä‚sfāt 'fərd·əl‚īz·ər }

phosphatization   [GEOCHEM] Conversion to a phosphate or phosphates; for example, the diagenetic replacement of limestone, mudstone, or shale by phosphate-bearing solutions, producing phosphates of calcium, aluminum, or iron.   { ‚fäs·fəd·ə'zā·shən }

phospholan   [CHEM] $C_6H_{14}O_3PNS_2$ A colorless to yellow solid with a melting point of 37–45°C; used as an insecticide and miticide for cotton.   { 'fä·sfə‚lan }

phosphorization   [GEOCHEM] Impregnation or combination with phosphorus or a compound of phosphorus; for example, the diagenetic process of phosphatization.   { ‚fäs·fə·rə'zā·shən }

phosphorus   [CHEM] A nonmetallic element, symbol P, atomic number 15, atomic weight 30.97376; used to manufacture phosphoric acid, in phosphor bronzes, incendiaries, pyrotechnics, matches, and rat poisons; the white (or yellow) allotrope is a soft waxy solid melting at 44.5°C, is soluble in carbon disulfide, insoluble in water and alcohol, and is poisonous and self-igniting in air; the red allotrope is an amorphous powder subliming at 416°C, igniting at 260°C, is insoluble in all solvents, and is nonpoisonous; the black allotrope comprises lustrous crystals similar to graphite, and is insoluble in most solvents. { 'fäs·fə·rəs }

phosphorus-nitrogen ratio [OCEANOGR] The proportion, by weight, of phosphorus to nitrogen in seawater or in plankton; the ratio is approximately 7:1. { 'fäs·fə·rəs 'nī·trə·jən 'rā·shō }

photic zone [ECOL] The uppermost layer of a body of water (approximately the upper 330 feet or 100 meters) that receives enough sunlight to permit the occurrence of photosynthesis. { 'fōd·ik }

photoautotroph [ECOL] An autotroph that uses energy from light to produce organic molecules. { ‚fōd·ō'ód·ō‚träf }

photoautotrophic [BIOL] Pertaining to organisms which derive energy from light and manufacture their own food. { ¦fōd·ō‚ód·ō'träf·ik }

photobiont [ECOL] A photosynthetic partner of a symbiotic pair, such as the algal component of the fungal-algal association in lichens. { ¦fōd·ō'bī‚änt }

photochemical smog [METEOROL] Chemical pollutants in the atmosphere resulting from chemical reactions involving hydrocarbons and nitrogen oxides in the presence of sunlight. { ¦fōd·ō'kem·ə·kəl 'smäg }

photoecology [ENG] The application of air photography to ecology, integrated land resource studies, and forestry. { ¦fōd·ō·i'käl·ə·jē }

photoelectric smoke-density control [ENG] A photoelectric control system used to measure, indicate, and control the density of smoke in a flue or stack. { ¦fōd·ō·i'lek·trik 'smōk ‚den·sad·ē kən‚trōl }

photoinhibition [BOT] Damage to the light-harvesting reactions of the photosynthetic apparatus caused by excess light energy trapped by the chloroplast. { ¦fōd·ō ‚in·ə'bish·ən }

photomorphogenesis [BOT] The control exerted by light over growth, development, and differentiation of plants that is independent of photosynthesis. { ¦fōd·ō‚mòr·fō 'jen·ə·səs }

photooxidation [CHEM] **1.** The loss of one or more electrons from a photoexcited chemical species. **2.** The reaction of a substance with oxygen and light. When oxygen remains in the product, the reaction is also known as photooxygenation. { ‚fōd·ō ‚äk·sə'dā·shən }

photoperiodism [BIOL] The physiological responses of an organism to the length of night or day or both. { ¦fōd·ō'pir·ē·ə‚diz·əm }

photophilic [BIOL] Thriving in full light. { ¦fōd·ō¦fil·ik }

photophobic [BIOL] **1.** Avoiding light. **2.** Exhibiting negative phototropism. { ‚fōd· ə'fō·bik }

photophosphorylase [BIOL] An enzyme that is associated with the surface of a thylakoid membrane and is involved in the final stages of adenosine triphosphate production by photosynthetic phosphorylation. { ¦fōd·ō‚fä'sfòr·ə‚lās }

photophosphorylation [BIOL] Phosphorylation that is induced by light energy in photosynthesis. { ¦fōd·ō‚fä·sfə·rə'lā·shən }

photophygous [BIOL] Thriving in shade. { fə'täf·ə·gəs }

photopigment [BIOL] A pigment that is unstable in the presence of light of appropriate wavelengths, such as the chromophore pigment which combines with opsins to form rhodopsin in the rods and cones of the vertebrate eye. { ¦fōd·ō¦pig·mənt }

photoreactive chlorophyll |BIOL| Chlorophyll molecules which receive light quanta from antenna chlorophyll and constitute a photoreaction center where light energy conversion occurs. { ¦fōd·ō·rē'ak·tiv 'klȯr·ə,fil }

photoreception |BIOL| The process of absorption of light energy by plants and animals and its utilization for biological functions, such as photosynthesis and vision. { ¦fōd·ō·ri'sep·shən }

photoreceptor |BIOL| A highly specialized, light-sensitive cell or group of cells containing photopigments. { ¦fōd·ō·ri'sep·tər }

photosynthesis |BIOL| Synthesis of chemical compounds in light, especially the manufacture of organic compounds (primarily carbohydrates) from carbon dioxide and a hydrogen source (such as water), with simultaneous liberation of oxygen, by chlorophyll-containing plant cells. { ¦fōd·ō'sin·thə·səs }

photosystem I |BIOL| One of two reaction sequences of the light phase of photosynthesis in green plants that involves a pigment system which is excited by wavelengths shorter than 700 nanometers and which transfers this energy to energy carriers such as NADPH that are subsequently utilized in carbon dioxide fixation. { 'fōd·ō,sis·təm 'wən }

photosystem II |BIOL| One of two reaction sequences of the light phase of photosynthesis in green plants which involves a pigment system excited by wavelengths shorter than 685 nanometers and which is directly involved in the splitting or photolysis of water. { 'fōd·ō,sis·təm 'tü }

phototaxis |BIOL| Movement of a motile organism or free plant part in response to light stimulation. { ¦fōd·ə¦tak·səs }

phototroph |BIOL| An organism that utilizes light as a source of metabolic energy. { 'fōd·ə,träf }

phototrophic bacteria |MICROBIO| Primarily aquatic bacteria comprising two principal groups: purple bacteria and green sulfur bacteria; all contain bacteriochlorophylls. { ¦fōd·ə¦träf·ik bak'tir·ē·ə }

phototropism |BOT| A growth-mediated response of a plant to stimulation by visible light. { fō'tä·trə,piz·əm }

phoxim See phenylglyoxylonitriloxime O,O-diethyl phosphorothioate. { 'fäk,sim }

phreatic |GEOL| Of a volcanic explosion of material such as steam or mud, not being incandescent. { frē'ad·ik }

phreatic cycle |HYD| The period of time during which the water table rises and then falls. { frē'ad·ik 'sī·kəl }

phreatic surface See water table. { frē'ad·ik 'sər·fəs }

phreatic water |HYD| Groundwater in the zone of saturation. { frē'ad·ik 'wȯd·ər }

phreatic-water discharge See groundwater discharge. { frē'ad·ik ¦wȯd·ər 'dis,chärj }

phreatic zone See zone of saturation. { frē'ad·ik ,zōn }

phreatophyte |ECOL| A plant with a deep root system which obtains water from the groundwater or the capillary fringe above the water table. { frē'ad·ə,fīt }

phthalic anhydride |CHEM| $C_6H_4(CO)_2O$ White crystals, melting at 131°C; sublimes when heated; slightly soluble in ether and hot water, soluble in alcohol; used to make dyes, resins, plasticizers, and insect repellents. { 'thal·ik an'hī,drīd }

phycobilin |BIOL| Any of various protein-bound pigments which are open-chain tetrapyrroles and occur in some groups of algae. { ,fī·kō'bī·lən }

phycobiliprotein [BIOL] A water-soluble photosynthetic membrane protein that covalently binds with phycobilins (photosynthetic pigments) in some groups of algae. { ‚fī·kō‚bil·ē'prō‚tēn }

phycobilisome [BIOL] A light-harvesting structure containing aggregates of photosynthetic accessory pigments that is located on the surface of thylakoid membranes in all cyanobacteria and red algae. { ‚fī·kō'bil·ē‚sōm }

phycobiont [BOT] The algal component of a lichen, commonly the green unicell of the genus *Trebouxia*. { ‚fī·kō'bī‚änt }

phycology *See* algology. { fī'käl·ə·jē }

phycourobilin [BIOL] A blue-green light (495-nanometer) absorbing pigment found in some cyanobacteria and red algae. { ‚fī·kō·yú'rō·bi·lin }

phyletic evolution [GEN] The gradual evolution of population without separation into isolated parts. { fī'led·ik ‚ev·ə'lü·shən }

phylum [SYST] A major taxonomic category in classifying animals (and plants in some systems), composed of groups of related classes. { 'fī·ləm }

physical climate [CLIMATOL] The actual climate of a place, as distinguished from a hypothetical climate, such as the solar climate or mathematical climate. { 'fiz·ə·kəl 'klī·mət }

physical climatology [CLIMATOL] The major branch of climatology, which deals with the explanation of climate, rather than with presentation of it (climatography). { 'fiz·ə·kəl ‚klī·mə'täl·ə·jē }

physical forecasting *See* numerical forecasting. { 'fiz·ə·kəl 'fȯr‚kast·ing }

physical geography [GEOGR] The branch of geography which deals with the description, analysis, classification, and genetic interpretation of the natural features and phenomena of the earth's surface. { 'fiz·ə·kəl jē'äg·rə·fē }

physical meteorology [METEOROL] That branch of meteorology which deals with optical, electrical, acoustical, and thermodynamic phenomena of the atmosphere, its chemical composition, the laws of radiation, and the explanation of clouds and precipitation. { 'fiz·ə·kəl ‚mēd·ē·ə'räl·ə·jē }

physical oceanography [OCEANOGR] The study of the physical aspects of the ocean, the movements of the sea, and the variability of these factors in relationship to the atmosphere and the ocean bottom. { 'fiz·ə·kəl ‚ō·shə'näg·rə·fē }

physical weathering *See* mechanical weathering. { 'fiz·ə·kəl 'weth·ə·riŋ }

physiographic diagram [GEOL] A small-scale map showing landforms by the systematic application of a standardized set of simplified pictorial symbols that represent the appearance such forms would have if viewed obliquely from the air at an angle of about 45°. Also known as landform map; morphographic map. { ¦fiz·ē·ə¦graf·ik 'dī·ə ‚gram }

physiological acoustics [PHYS] The study of the responses to acoustic stimuli that take place in the ear or in the associated central neural auditory pathways of humans and animals. { ‚fiz·ē·ə‚läj·ə·kəl ə'kü·stiks }

physiological ecology [ECOL] The study of biophysical, biochemical, and physiological processes used by animals to cope with factors of their physical environment, or employed during ecological interactions with other organisms. { ‚fiz·ē·ə'läj·ə·kəl ē'käl·ə·jē }

phytal zone [ECOL] The part of a lake bottom covered by water shallow enough to permit the growth of rooted plants. { 'fīd·əl ‚zōn }

phyteral [GEOL] Morphologically recognizable forms of vegetal matter in coal. { 'fīd·ə·rəl }

phytoalexin [BIOL] A natural substance that is toxic to fungi and is synthesized by a plant as a response to fungal infection. { 'fīd·ō·ə'lek·sən }

phytochemistry [BOT] The study of the chemistry of plants, plant products, and processes taking place within plants. { 'fīd·ō͝,kem·ə·strē }

phytochorology See plant geography. { ͝fīd·ō·kȯ'räl·ə·jē }

phytochrome [BIOL] A protein plant pigment which serves to direct the course of plant growth and development in response variously to the presence or absence of light, to photoperiod, and to light quality. { 'fīd·ə͝,krōm }

phytoclimatology [CLIMATOL] The study of the microclimate in the air space occupied by plant communities, on the surfaces of the plants themselves and, in some cases, in the air spaces within the plants. { ͝fīd·ō͝,klī·mə'täl·ə·jē }

phytocoenosis [ECOL] The entire plant population of a particular habitat. { ͝fīd·ō·sē'nō·səs }

phytocollite [GEOL] A black, gelatinous, nitrogenous humic body occurring beneath or within peat deposits. { fī'täk·ə͝,līt }

phytogenic dam [ECOL] A natural dam consisting of plants and plant remains. { ͝fīd·ə ͝jen·ik 'dam }

phytogenic dune [ECOL] Any dune in which the growth of vegetation influences the form of the dune, for example, by arresting the drifting of sand. { ͝fīd·ə͝jen·ik 'dün }

phytogeography See geobotany ; plant geography. { ͝fīd·ō·jē'äg·rə·fē }

phytopathogen [ECOL] An organism that causes a disease in a plant. { ͝fīd·ō 'path·ə·jən }

phytophagous [ZOO] Feeding on plants. { fī'täf·ə·gəs }

Phytophthra citrophthora [MYCOL] A water mold that causes citrus gummosis. { fī ͝däf·thrə ͝si·trəf'thȯr·ə }

phytoplankton [ECOL] Planktonic plant life. { ͝fīd·ə'plaŋk·tən }

phytoremediation [AGR] The use of green plants to manage or reduce high levels of soil and groundwater contaminants. { ͝fīd·ō·ri͝mēd·ē'ā·shən }

phytosociology [ECOL] A broad study of plants that includes the study of all phenomena affecting their lives as social units. { ͝fīd·ō͝sō·sē'äl·ə·jē }

phytotoxin [BIOL] **1.** A substance toxic to plants. **2.** A toxin produced by plants. { ͝fīd·ə'täk·sən }

phytotron [BOT] A research tool used to study whole plants; contains a large number of individually controlled environments that provide the means of studying the effect of each environmental factor, such as temperature or light, at many levels simultaneously. { 'fīd·ə͝,trän }

pibal See pilot-balloon observation. { 'pī͝,bal }

Picornaviridae [MICROBIO] A viral family made up of the small (18–30 nanometers) ether-sensitive viruses that lack an envelope and have a Togaviridae genome; contains the genera Enterovirus (human polio), Cardiovirus (mengo), Rhinovirus (common cold), and Aphtovirus (foot-and-mouth disease). { pē͝,kȯr·nə'vir·ə͝,dī }

picornavirus [MICROBIO] A viral group made up of small (18–30 nanometers), ether-sensitive viruses that lack an envelope and have a ribonucleic acid genome; among subgroups included are enteroviruses and rhinoviruses, both of human origin. { pē'kȯr·nə͝,vī·rəs }

picrotoxin [BIOL] $C_{30}H_{34}O_{13}$ A poisonous, crystalline plant alkaloid found primarily in Cocculus indicus; used as a stimulant and convulsant drug. Also known as cocculin. { ͝pik·rə'täk·sən }

piedmont bulb [HYD] The lobe or fan of ice formed when a glacier spreads out on a plain at the lower end of a valley. { 'pēd,mänt ¦bəlb }

piedmont glacier [HYD] A thick, continuous ice sheet formed at the base of a mountain range by the spreading out and coalescing of valley glaciers from higher mountain elevations. { 'pēd,mänt ¦glā·shər }

piedmont ice [HYD] An ice sheet formed by the joining of two or more glaciers on a comparatively level plain at the base of the mountains down which the glaciers descended; it may be partly afloat. { 'pēd,mänt ¦īs }

piedmont lake [HYD] An oblong lake occupying a partly overdeepened basin excavated from rock by a piedmont glacier, or dammed by a glacial moraine. { 'pēd,mänt ¦lāk }

Pierce's disease [PL PATH] A virus disease of grapes in which there is mottling between the veins of leaves, early defoliation, and early ripening and withering of the fruit. { 'pirs·əz di,zēz }

piezometric surface See potentiometric surface. { pē¦ā·zō¦me·trik 'sər·fəs }

pileus [METEOROL] An accessory cloud of small horizontal extent, often cirriform, in the form of a cap, hood, or scarf, which occurs above or attached to the top of a cumuliform cloud that often pierces it; several pileus clouds fairly often are observed above each other. Also known as scarf cloud. { 'pil·ē·əs }

pilot balloon [ENG] A small balloon whose ascent is followed by a theodolite in order to obtain data for the computation of the speed and direction of winds in the upper air. { 'pī·lət bə,lün }

pilot-balloon observation [METEOROL] A method of winds-aloft observation, that is, the determination of wind speeds and directions in the atmosphere above a station; involves reading the elevation and azimuth angles of a theodolite while visually tracking a pilot balloon. Also known as pibal. { 'pī·lət bə,lün ,äb·zər'vā·shən }

Pinales [BOT] An order of gymnospermous woody trees and shrubs in the class Pinopsida, including pine, spruce, fir, cypress, yew, and redwood; the largest plants are the conifers. { pī'nā·lēz }

pine [BOT] Any of the cone-bearing trees composing the genus *Pinus*; characterized by evergreen leaves (needles), usually in tight clusters of two to five. { 'pīn }

pine oil [BOT] Any of a group of volatile essential oils with pinaceous aromas distilled from cones, needles, or stumps of various pine or other conifer species; used as solvents, emulsifying agents, wetting agents, deodorants, germicides, and sources of chemicals. { 'pīn ,öil }

pingo [HYD] A frost mound resembling a volcano, being a relatively large and conical mound of soil-covered ice, elevated by hydrostatic pressure of water within or below the permafrost of arctic regions. { 'piŋ·gō }

pingo ice [HYD] Clear or relatively clear ice that occurs in permafrost; originates from groundwater under pressure. { 'piŋ·gō,īs }

Pinicae [BOT] A large subdivision of the Pinophyta, comprising woody plants with a simple trunk and excurrent branches, simple, usually alternative, needlelike or scalelike leaves, and wood that lacks vessels and usually has resin canals. { 'pī·nə ,sē }

pink root [PL PATH] A fungus disease of onion and garlic caused by various organisms, especially species of *Phoma* and *Fusarium*; marked by red discoloration of the roots. { 'piŋk ,rüt }

pink rot [PL PATH] **1.** A fungus disease of potato tubers caused by *Phytophtora erythroseptica* and characterized by wet rot and pink color of the cut surfaces of the tuber upon exposure to air. **2.** A rot disease of apples caused by the fungus *Tricothecium roseum*. **3.** A watery soft rot of celery caused by the fungus *Sclerotinia sclerotiorum*. { 'piŋk ,rät }

pinnacled iceberg [OCEANOGR] An iceberg weathered in such manner as to produce spires or pinnacles. Also known as irregular iceberg; pyramidal iceberg. { 'pin·ə·kəld 'īs,bərg }

pinnate [BOT] Having parts arranged like a feather, branching from a central axis. { 'pi ,nāt }

pinnate drainage [HYD] A dendritic drainage pattern in which the main stream receives many closely spaced, subparallel tributaries that join it at acute angles; resembles a feather in plan view. { 'pi,nāt 'drā·nij }

Pinophyta [BOT] The gymnosperms, a division of seed plants characterized as vascular plants with roots, stems, and leaves, and with seeds that are not enclosed in an ovary but are borne on cone scales or exposed at the end of a stalk. { pə'näf·əd·ə }

pioneer [ECOL] An organism that is able to establish itself in a barren area and begin an ecological cycle. { ,pī·ə'nir }

pipe [GEOL] **1.** A vertical, cylindrical ore body. Also known as neck; ore chimney; ore pipe; stock. **2.** A tubular cavity of varying depth in calcareous rocks, often filled with sand and gravel. **3.** A vertical conduit through the crust of the earth below a volcano, through which magmatic materials have passed. { pīp }

piperazine dihydrochloride [CHEM] $C_4H_{10}N_2 \cdot 2HCl$ White, water-soluble needles; used for insecticides and pharmaceuticals. { pī'par·ə,zēn dī,hī·drə'klòr,īd }

piperazine hexahydrate [CHEM] $C_4H_{10}N_2 \cdot 6H_2O$ White crystals with a melting point of 44°C; soluble in alcohol and water; used for pharmaceuticals and insecticides. { pī'par·ə,zēn ¦hek·sə'hī,drāt }

piping [HYD] Erosive action of water passing through or under a dam, which may result in leakage or failure. { 'pīp·iŋ }

pipkrake [HYD] A small, thin needlelike crystal of ice formed just below ground level and growing perpendicular to the soil surface. Also known as needle ice. { 'pip,krāk }

piracy See capture. { 'pī·rə·sē }

pirimiphosethyl [CHEM] $C_{13}H_{24}N_3O_3PS$ A straw-colored liquid which decomposes at 130°C; used as an insecticide for the control of soil insects in vegetables and other crops. { ,pir·əm·fäs'eth·əl }

pistil [BOT] The ovule-bearing organ of angiosperms; consists of an ovary, a style, and a stigma. { 'pist·əl }

pistillate [BOT] **1.** Having a pistil. **2.** Having pistils but no stamens. { 'pist·əl,āt }

pit [BOT] **1.** A cavity in the secondary wall of a plant cell, formed where secondary deposition has failed to occur, and the primary wall remains uncovered; two main types are simple pits and bordered pits. **2.** The stone of a drupaceous fruit. { pit }

pith [BOT] A central zone of parenchymatous tissue that occurs in most vascular plants and is surrounded by vascular tissue. { pith }

placer [GEOL] A mineral deposit at or near the surface of the earth, formed by mechanical concentration of mineral particles from weathered debris. Also known as ore of sedimentation. { 'plās·ər }

placic horizon [GEOL] A black to dark red soil horizon that is usually cemented with iron and is not very permeable. { 'plā·sik hə'rīz·ən }

plagioclimax [ECOL] A plant community which is in equilibrium under present conditions, but which has not reached its natural climax, or has regressed from it, due to biotic factors such as human intervention. { ¦plā·jē·ō'klī,maks }

plagiogravitropism [BOT] A response of root and shoot branches to gravity where growth is at different angles from the vertical. { ¦plā·jē·ō,grav·ə'trō·piz·əm }

plagiosere [ECOL] A plant succession deflected from its normal course by biotic factors. { 'plā·jē·ə,sir }

plague [MED] **1.** An infectious bacterial disease of rodents and humans caused by *Pasteurella pestis*, transmitted to humans by the bite of an infected flea (*Xenopsylla cheopis*) or by inhalation. Also known as black death; bubonic plague. **2.** Any contagious, malignant, epidemic disease. { plāg }

plain [GEOGR] An extensive, broad tract of level or rolling, almost treeless land with a shrubby vegetation, usually at a low elevation. [GEOL] A flat, gently sloping region of the sea floor.Also known as submarine plain. { plān }

planation [GEOL] Erosion resulting in flat surfaces, caused by meandering streams, waves, ocean currents, wind, or glaciers. { plā'nā·shən }

planation stream piracy [HYD] Stream capture effected by the lateral planation of a stream invading and diverting the upper part of a smaller stream. { plā'nā·shən ¦strēm ,pī·rə·sē }

plane atmospheric wave [METEOROL] An atmospheric wave represented in two-dimensional rectangular cartesian coordinates, in contrast to a wave considered on the spherical earth. { 'plān ¦at·mə¦sfir·ik 'wāv }

plane jet [HYD] A stream flow pattern characteristic of hyperpycnal inflow, in which the inflowing water spreads as a parabola whose width is about three times the square root of the distance downstream from the mouth. { 'plān ,jet }

plane of saturation *See* water table. { 'plān əv ,sach·ə'rā·shən }

planetary boundary layer [METEOROL] That layer of the atmosphere from the earth's surface to the geostrophic wind level, including, therefore, the surface boundary layer and the Ekman layer; above this layer lies the free atmosphere. { 'plan·ə,ter·ē 'baún·drē ,lā·ər }

planetary circulation *See* general circulation. { 'plan·ə,ter·ē ,sər·kyə'lā·shən }

planetary wave *See* Rossby wave. { 'plan·ə,ter·ē 'wāv }

planetary wind [METEOROL] Any wind system of the earth's atmosphere which owes its existence and direction to solar radiation and to the rotation of the earth. { 'plan·ə ,ter·ē 'wind }

planform [GEOGR] A body of water's outline or morphology as defined by the still water line. { 'plan,fórm }

plankton [ECOL] Generally tiny, passively floating or weakly motile aquatic plants and animals. { 'plaŋk·tən }

planktonic [ECOL] Free-floating. { plaŋk'tän·ik }

plankton net [OCEANOGR] A net for collecting plankton. { 'plaŋk·tən ,net }

plant [ENG] The land, buildings, and equipment used in an industry. { plant }

Plantae [SYST] The plant kingdom. { 'plan,tē }

plant geography [BOT] A major division of botany, concerned with all aspects of the spatial distribution of plants.Also known as geographical botany; phytochorology; phytogeography. { 'plant jē,äg·rə·fē }

plant key [BOT] An analytical guide to the identification of plants, based on the use of contrasting characters to subdivide a group under study into branches. { 'plant ,kē }

plant kingdom [SYST] The worldwide array of plant life constituting a major division of living organisms. { 'plant ,kiŋ·dəm }

plant pathology [BOT] The branch of botany concerned with diseases of plants. { 'plant pə'thäl·ə·jē }

plant physiology |BOT| The branch of botany concerned with the processes which occur in plants. { 'plant ,fiz·ē'äl·ə·jē }

plant societies |ECOL| Assemblages of plants which constitute structural parts of plant communities. { 'plant sə,sī·əd·ēz }

plant virus |MICROBIO| A virus that replicates only within plant cells. { 'plant ,vī·rəs }

plash |HYD| A shallow, standing, usually short-lived pool or small pond resulting from a flood, heavy rain, or melting snow. { plash }

plastic limit |GEOL| The water content of a sediment, such as a soil, at the point of transition between the plastic and semisolid states. { 'plas·tik 'lim·ət }

plastid |BIOL| One of the specialized cell organelles containing pigments or protein materials, often serving as centers of special metabolic activities; examples are chloroplasts and leukoplasts. { 'plas·təd }

plastoquinone |BIOL| Any of a group of quinones that are involved in electron transport in chloroplasts during photosynthesis. { ,plas·tə·kwə'nōn }

plateau |GEOGR| An extensive, flat-surfaced upland region, usually more than 45–90 meters (150–300 feet) in elevation and considerably elevated above the adjacent country and limited by an abrupt descent on at least one side. |GEOL| A broad, comparatively flat and poorly defined elevation of the sea floor, commonly over 60 meters (200 feet) in elevation. { pla'tō }

plateau glacier |HYD| A highland glacier that overlies a generally flat mountain tract; usually overflows its edges in hanging glaciers. { pla'tō¦glā·shər }

plate crystal |HYD| An ice crystal exhibiting typical hexagonal (rarely triangular) symmetry and having comparatively little thickness parallel to its principal axis (*c* axis); as such crystals fall through the clouds in which they form, they may encounter conditions causing them to develop dendritic extensions, that is, to become plane-dendritic crystals. { 'plāt ¦krist·əl }

plate ice *See* pancake ice. { 'plāt ,īs }

platelet |HYD| A small ice crystal which, when united with other such crystals, forms a layer of floating ice, especially sea ice, and serves as seed crystals for further thickening of the ice cover. { 'plāt·lət }

plate tectonics |GEOL| Global tectonics based on a model of the earth characterized by a small number (10–25) of semirigid plates which float on some viscous underlayer in the mantle; each plate moves more or less independently and grinds against the others, concentrating most deformation, volcanism, and seismic activity along the periphery. Also known as raft tectonics. { 'plāt tek'tän·iks }

platform reef |GEOL| An organic reef, generally small but more extensive than a patch reef, with a flat upper surface. { 'plat,fórm ,rēf }

playa |GEOL| **1.** A low, essentially flat part of a basin or other undrained area in an arid region. **2.** A small, generally sandy land area at the mouth of a stream or along the shore of a bay. **3.** A flat, alluvial coastland, as distinguished from a beach. { 'plī·ə }

playa lake |HYD| A shallow temporary sheet of water covering a playa in the wet season. { 'plī·ə ,lāk }

plerotic water |HYD| That part of subsurface ground water that forms the zone of saturation, including underground streams. { plə'räd·ik 'wòd·ər }

pleuropneumonia |VET MED| An infectious disease of cattle producing pleural and lung inflammation, caused by *Mycoplasma* species. { ,plùr·ō·nù'mō·nyə }

plough wind *See* plow wind. { 'plaù ,wind }

plow [AGR] An implement used to cut, lift, turn, and pulverize soil in preparation of a seedbed. { plau̇ }

plowshare [HYD] A wedge-shaped feature developed on a snow surface by further ablation of foam crust. { 'plau̇,sher }

plow wind [METEOROL] A term used in the midwestern United States to describe strong, straight-line winds associated with squall lines and thunderstorms; resulting damage is usually confined to narrow zones like that caused by tornadoes; however, the winds are all in one direction. Also spelled plough wind. Also known as derecho. { 'plau̇ ,wind }

plug reef [GEOL] A small, triangular reef that grows with its apex pointing seaward through openings between linear shelf-edge reefs. { 'pləg ,rēf }

plumbism [MED] Lead poisoning. { 'pləm·biz·əm }

plum blotch [PL PATH] A fungus disease of plums caused by *Phyllosticta congesta* and characterized by minute brown or gray angular leaf spots and brown or gray blotches on the fruit. { 'pləm ,bläch }

plum pocket [PL PATH] A mild fungus disease of plums, caused either by *Taphrina pruni* or *T. communia*, in which the stone of the fruit is aborted. { 'pləm ,päk·ət }

plumule [BOT] The primary bud of a plant embryo. { 'plü·myül }

plunge basin [GEOL] A deep, large hollow or cavity scoured in the bed of a stream at the foot of a waterfall or cataract by the force and eddying effect of the falling water. { 'plənj ,bās·ən }

plunge point [OCEANOGR] The point at which a plunging wave curls over and falls as it moves toward the shore. { 'plənj ,pȯint }

plunge pool [HYD] **1.** The water in a plunge basin. **2.** A deep, circular lake occupying a plunge basin after the waterfall has ceased to exist or the stream has been diverted. Also known as waterfall lake. **3.** A small, deep plunge basin. { 'plənj ,pül }

plunging breaker [OCEANOGR] A breaking wave whose crest curls over and collapses suddenly. Also known as spilling breaker; surging breaker. { 'plənj·iŋ 'brāk·ər }

plutonian *See* plutonic. { plü'tō·nē·ən }

plutonic [GEOL] Pertaining to rocks formed at a great depth. Also known as abyssal; deep-seated; plutonian. { plü'tän·ik }

plutonic water [HYD] Juvenile water in magma, or derived from magma, at a considerable depth, probably several kilometers. { plü'tän·ik 'wȯd·ər }

plutonium oxide [CHEM] PuO_2 A radioactively poisonous pyrophoric oxide of plutonium; particles may be easily airborne. { plü'tō·nē·əm 'äk,sīd }

pluvial [METEOROL] Pertaining to rain, or more broadly, to precipitation, particularly to an abundant amount thereof. { 'plü·vē·əl }

pluvial lake [GEOL] A lake formed during a period of exceptionally heavy rainfall; specifically, a Pleistocene lake formed during a period of glacial advance and now either extinct or only a remnant. { 'plü·vē·əl 'lāk }

pluviilignosa [ECOL] A tropical rain forest. { ¦plü·vē,il·əg'nō·sə }

pluviofluvial [GEOL] Pertaining to the combined action of rainwater and streams. { ¦plü·vē·ō¦flü·vē·əl }

pluviograph *See* recording rain gage. { 'plü·vē·ə,graf }

pluviometer *See* rain gage. { ,plü·vē'äm·əd·ər }

pluviometric coefficient [METEOROL] For any month at a given weather station, the ratio of the monthly normal precipitation to one-twelfth of the annual normal precipitation. Also known as hyetal coefficient. { ¦plü·vē·ə¦me·trik ͵kō·i'fish·ənt }

PNdB *See* perceived noise decibel.

pneumatophore [BOT] **1.** An air bladder in marsh plants. **2.** A submerged or exposed erect root that functions in the respiration of certain marsh plants. { 'nü·məd·ə͵fȯr }

pneumobacillus *See* Klebsiella pneumoniae. { ¦nü·mō·bə'sil·əs }

Pneumocystis carinii pneumonia [MED] A lung infection in humans caused by the protozoan *Pneumocystis carinii*. Also known as interstitial plasma-cell pneumonia. { ¦nü·mə¦sis·təs kə'rin·ē͵ī nü'mō·nyə }

pod [BOT] A dry dehiscent fruit; a legume. { päd }

pod blight [PL PATH] A fungus disease of legumes caused by *Diaporthe* species. { 'päd ͵blīt }

Podosphaera leucotricha [MYCOL] A fungal plant pathogen that causes apple powdery mildew. { ͵päd·ə¦sfī·rə ͵lü·kə'trik·ə }

Podzol [GEOL] A soil group characterized by mats of organic matter in the surface layer and thin horizons of organic minerals overlying gray, leached horizons and dark-brown illuvial horizons; found in coal forests to temperate coniferous or mixed forests. { 'päd ͵zȯl }

podzolic soil *See* Red-Yellow Podzolic soil. { päd¦zäl·ik 'sȯil }

pogonip *See* ice fog. { 'päg·ə͵nip }

poikilotherm [ZOO] An animal, such as reptiles, fishes, and invertebrates, whose body temperature varies with and is usually higher than the temperature of the environment; a cold-blooded animal. { pȯi'kil·ə͵thərm }

point [GEOGR] A tapering piece of land projecting into a body of water; it is generally less prominent than a cape. { pȯint }

point of departure *See* departure. { 'pȯint əv di'pär·chər }

point-placement [AGR] Positioning of fertilizer or some other agricultural chemical within the length of the seed row or in the specific location where the seed is planted. { 'pȯint ͵plās·mənt }

point rainfall [METEOROL] The rainfall during a given time interval (or often one storm) measured in a rain gage, or an estimate of the amount which might have been measured at a given point. { 'pȯint 'rān͵fȯl }

point source [CIV ENG] A municipal or industrial wastewater discharge through a discrete pipe or channel. { 'pȯint ͵sȯrs }

poised stream [HYD] A stream that is neither eroding nor depositing sediment. { 'pȯizd 'strēm }

poison [MED] A substance that in relatively small doses has an action that either destroys life or impairs seriously the functions of organs or tissues. { 'pȯiz·ən }

poison gland [ZOO] Any of various specialized glands in certain fishes and amphibians which secrete poisonous mucuslike substances. { 'pȯiz·ən ͵gland }

poisonous plant [BOT] Any of about 400 species of vascular plants containing principles which initiate pathological conditions in man and animals. { 'pȯiz·ən·əs 'plant }

polar air [METEOROL] A type of air whose characteristics are developed over high latitudes; there are two types: continental polar air and maritime polar air. { 'pō·lər 'er }

polar anticyclone See subpolar high. { 'pō·lər ,ant·i'sī,klōn }

polar automatic weather station [METEOROL] An automatic weather station which measures meteorological elements and transmits them by radio; the station is designed to function primarily in frigid or polar climates in order to fill the need for weather reports from inaccessible regions where manned stations are not practicable. { 'pō·lər ¦öd·ə¦mad·ik 'we<u>th</u>·ər ,stā·shən }

polar cap [HYD] An ice sheet centered at one of the poles of the earth. { 'pō·lər ,kap }

polar-cap ice See polar ice. { 'pō·lər ,kap 'īs }

polar climate [CLIMATOL] The climate of a geographical polar region, most commonly taken to be a climate which is too cold to support the growth of trees. Also known as arctic climate; snow climate. { 'pō·lər 'klī·mət }

polar continental air [METEOROL] Air of an air mass that originates over land or frozen ocean areas in the polar regions; characterized by low temperature, stability, low specific humidity, and shallow vertical extent. { 'pō·lər ,kant·ən'ent·əl 'er }

polar convergence [OCEANOGR] The line of convergence of polar and subpolar water masses in the ocean. { 'pō·lər kən'vər·jəns }

polar cyclone See polar vortex. { 'pō·lər 'sī,klōn }

polar desert [GEOGR] A high-latitude desert where the existing moisture is frozen in ice sheets and is thus unavailable for plant growth. Also known as arctic desert. { 'pō·lər 'dez·ərt }

polar easterlies [METEOROL] The rather shallow and diffuse body of easterly winds located poleward of the subpolar low-pressure belt; in the mean in the Northern Hemisphere, these easterlies exist to an appreciable extent only north of the Aleutian low and Icelandic low. { 'pō·lər 'ēs·tər,lēz }

polar-easterlies index [METEOROL] A measure of the strength of the easterly wind between the latitudes of 55° and 70°N; the index is computed from the average sea-level pressure difference between these latitudes and is expressed as the east to west component of geostrophic wind in meters and tenths of meters per second. { 'pō·lər 'ēs·tər,lēz 'in,deks }

polar firn [HYD] Firn formed at low temperatures with no melting or liquid water present. Also known as dry firn. { 'pō·lər 'fərn }

polar front [METEOROL] The semipermanent, semicontinuous front separating air masses of tropical and polar origin; this is the major front in terms of air mass contrast and susceptibility to cyclonic disturbance. { 'pō·lər 'frənt }

polar-front theory [METEOROL] A theory whereby a polar front, separating air masses of polar and tropical origin, gives rise to cyclonic disturbances which intensify and travel along the front, passing through various phases of a characteristic life history. { 'pō·lər ¦frənt ,thē·ə·rē }

polar glacier [HYD] A glacier whose temperature is below freezing throughout its mass, and on which there is no melting during any season. { 'pō·lər 'glā·shər }

polar high See subpolar high. { 'pō·lər 'hī }

polar ice [OCEANOGR] Sea ice that is more than 1 year old; the thickest form of sea ice. Also known as polar-cap ice. { 'pō·lər 'īs }

polarization isocline [METEOROL] A locus of all points at which the inclination to the vertical of the plane of polarization of the diffuse sky radiation has the same value. { ,pō·lə·rə'zā·shən 'īs·ə,klīn }

polar lake [HYD] A lake whose surface temperature never exceeds 4°C. { 'pō·lər 'lāk }

polar low   *See* polar vortex.   { 'pō·lər 'lō }

polar maritime air   [METEOROL] Air of an air mass that originates in the polar regions and is then modified by passing over a relatively warm ocean surface; characterized by moderately low temperature, moderately high surface specific humidity, and a considerable degree of vertical instability.   { 'pō·lər 'mar·ə,tīm 'er }

polar meteorology   [METEOROL] The application of meteorological principles to a study of atmospheric conditions in the earth's high latitudes or polar-cap regions, northern and southern.   { 'pō·lər ,mē·dē·ə'räl·ə·jē }

polar outbreak   [METEOROL] The movement of a cold air mass from its source region; almost invariably applied to a vigorous equatorward thrust of cold polar air, a rapid equatorward movement of the polar front. Also known as cold-air outbreak.   { 'pō·lər 'aút,brāk }

polar regions   [GEOGR] The regions near the geographic poles; no definite limit for these regions is recognized.   { 'pō·lər ,rē·jənz }

polar trough   [METEOROL] In tropical meteorology, a wave trough in the circumpolar westerlies having sufficient amplitude to reach the tropics in the upper air; at the surface it is reflected as a trough in the tropical easterlies, but at moderate elevations it is characterized by westerly winds.   { 'pō·lər 'trȯf }

polar vortex   [METEOROL] The large-scale cyclonic circulation in the middle and upper troposphere centered generally in the polar regions; specifically, the vortex has two centers in the mean, one near Baffin Island and another over northeastern Siberia; the associated cyclonic wind system comprises the westerlies of middle latitudes. Also known as Antarctic vortex; circumpolar whirl; polar cyclone; polar low.   { 'pō·lər 'vȯr ,teks }

polar westerlies   *See* westerlies.   { 'pō·lər 'wes·tər,lēz }

polder   [CIV ENG] Land reclaimed from the sea or other body of water by the construction of an embankment to restrain the water.   { 'pōl·dər }

pole of inaccessibility   *See* ice pole.   { 'pōl əv ,in·ak,ses·ə'bil·əd·ē }

pole tide   [OCEANOGR] An ocean tide, theoretically 6 millimeters in amplitude, caused by the Chandler wobble of the earth; has a period of 428 days.   { 'pōl ,tīd }

poliomyelitis   [MED] An acute infectious viral disease which in its most serious form involves the central nervous system and, by destruction of motor neurons in the spinal cord, produces flaccid paralysis. Also known as Heine-Medin disease; infantile paralysis.   { ¦pō·lē·ō,mī·ə'līd·əs }

pollen   [BOT] The small male reproductive bodies produced in pollen sacs of the seed plants.   { 'päl·ən }

pollen count   [BOT] The number of grains of pollen that collect on a specified area (often taken as 1 square centimeter) in a specified time.   { 'päl·ən ,kaúnt }

pollen sac   [BOT] In the anther of angiosperms and gymnosperms, a cavity that contains microspores.   { 'päl·ən ,sak }

pollen tube   [BOT] The tube produced by the wall of a pollen grain which enters the embryo sac and provides a passage through which the male nuclei reach the female nuclei.   { 'päl·ən ,tüb }

pollination   [BOT] The transfer of pollen from a stamen to a pistil; fertilization in flowering plants.   { ,päl·ə'nā·shən }

pollinosis   *See* hay fever.   { ,päl·ə'nō·səs }

pollution   [ECOL] Destruction or damage of the natural environment by by-products of human activities such as chemicals, noise, and heat.   { pə'lü·shən }

polonium-210 [PHYS] Radioactive isotope of polonium; mass 210, half-life 140 days, α-radiation; used to calibrate radiation counters, and in oil well logging and atomic batteries. Also known as radium F. { pə'lō·nē·əm ¦tü'ten }

Polychaeta [ZOO] The largest class of the phylum Annelida, distinguished by paired, lateral, fleshy appendages (parapodia) provided with setae, on most segments. { ¸päl·i'kēd·ə }

polychlorinated biphenyl [CHEM] Any one of a number of chlorinated derivatives of biphenyl once used widely in industry due to properties such as nonflammability, chemical stability, and low electrical conductance; due to tendency to accumulate in animal tissues and probable carcinogenicity, further sale of polychlorinated biphenyls for new use was banned by U.S. law in 1979. Abbreviated PCB. { ¦päl·i'klȯr·ə¸nād·əd bī'fen·əl }

polychlorinated dibenzo-*para*-dioxin *See* dioxin. { ¦päl·i'klȯr·ə¸nād·əd dī¦ben·zō ¦par·ə dī'äk·sən }

polyclimax [ECOL] A climax community under the controlling influence of many environmental factors, including soils, topography, fire, and animal interactions. { ¸päl·i'klī¸maks }

Polyctenidae [ZOO] A family of hemipteran insects in the superfamily Cimicoidea; the individuals are bat ectoparasites which resemble bedbugs but lack eyes and have ctenidia and strong claws. { ¸päl·ək'ten·ə¸dē }

polygamous [ZOO] Having both perfect and imperfect flowers on the same plant. { pə'lig·ə·məs }

polymictic [HYD] Pertaining to or characteristic of a lake having no stable thermal stratification. { ¦päl·i¦mik·tik }

polymorphism [GEN] The coexistence of genetically determined distinct forms in the same population, even the rarest of them being too common to be maintained solely by mutation; human blood groups are an example. { ¸päl·i'mȯr¸fiz·əm }

Polypodiophyta [BOT] The ferns, a division of the plant kingdom having well-developed roots, stems, and leaves that contain xylem and phloem and show well-developed alternation of generations. { ¸päl·i¸päd·ē'äf·əd·ē }

polysaprobic [ECOL] Referring to a body of water in which organic matter is decomposing rapidly and free oxygen either is exhausted or is present in very low concentrations. { ¦päl·ə·sə'prō·bik }

polystyrene [CHEM] $(C_6H_5CHCH_2)_x$ A water-white, tough synthetic resin made by polymerization of styrene; soluble in aromatic and chlorinated hydrocarbon solvents; used for injection molding, extrusion or casting for electrical insulation, fabric lamination, and molding of plastic objects. { ¦päl·i'stī¸rēn }

polytropic atmosphere [METEOROL] A model atmosphere in hydrostatic equilibrium with a constant nonzero lapse rate. { ¦päl·i¦träp·ik 'at·mə¸sfir }

polyvinyl acetate [CHEM] $(H_2CCHOOCCH_3)_x$ A thermoplastic polymer; insoluble in water, gasoline, oils, and fats, soluble in ketones, alcohols, benzene, esters, and chlorinated hydrocarbons; used in adhesives, films, lacquers, inks, latex paints, and paper sizes. Abbreviated PVA; PVAc. { ¦päl·i'vīn·əl 'as·ə¸tāt }

polyvinyl chloride [CHEM] $(H_2CCHCL)_x$ Polymer of vinyl chloride; tasteless, odorless; insoluble in most organic solvents; a member of the family of vinyl resins; used in soft flexible films for food packaging and in molded rigid products such as pipes, fibers, upholstery, and bristles. Abbreviated PVC. { ¦päl·i'vīn·əl 'klȯr¸īd }

pompeii worm [ZOO] *Alvinella pompejana.* A polychaetous annelid that lives in sea-floor hydrothermal vent chimneys and may experience extreme thermal gradients between its anterior (80°C; 176°F) and posterior (22°C; 72°F) ends. { päm'pā ¸wərm }

pond |GEOGR| A small natural body of standing fresh water filling a surface depression, usually smaller than a lake. { pänd }

pondage |HYD| Water held in a reservoir for short periods to regulate natural flow, usually for hydroelectric power. { 'pän·dij }

pondage land |GEOL| Land on which water is stored as dead water during flooding, and which does not contribute to the downstream passage of flow. Also known as flood fringe. { 'pän·dij ˌland }

ponded stream |HYD| A stream in which a pond forms due to an interruption of the normal streamflow. { 'pän·dəd ˌstrēm }

ponding |HYD| The natural formation of a pond in a stream by an interruption of the normal streamflow. { 'pänd·iŋ }

pool |HYD| A small deep body of water, often fed by a spring. { pül }

pool stage |HYD| As used along the Ohio and upper Mississippi Rivers of the United States, a low-water condition with the navigation dams up so that the river is a series of shallow pools; when this condition exists, the river is said to be "in pool"; river depth is regulated by the dams so as to be adequate for navigation. { 'pül ˌstāj }

population |BIOL| A group of organisms of a single species occupying a specific geographic area or biome. { ˌpäp·yə'lā·shən }

population density |ECOL| The number of individuals in a population per unit area. { ˌpäp·yə'lā·shən 'den·səd·ē }

population dispersal |BIOL| The process by which groups of living organisms expand the space or range within which they live. { ˌpäp·yə'lā·shən di'spər·səl }

population dispersion |BIOL| The spatial distribution at any particular moment of the individuals of a species of plant or animal. { ˌpäp·yə'lā·shən di'spər·zhən }

population dynamics |BIOL| The aggregate of processes that determine the size and composition of any population. { ˌpäp·yə'lā·shən dī'nam·iks }

population ecology |ECOL| The study of the vital statistics of populations, and the interactions within and between populations that influence survival and reproduction. { ˌpäp·yə,lā·shən ē'käl·ə·jē }

pore ice |HYD| Ice which fills or partially fills pore spaces in permafrost; forms by freezing soil water in place, with no addition of water. { 'pȯr ˌīs }

positive area See positive element. { 'päz·əd·iv 'er·ē·ə }

positive element |GEOGR| A large structural feature of the earth's crust characterized by long-term upward movement (uplift, emergence) or subsidence less rapid than that of adjacent negative elements. Also known as archibole; positive area. { 'päz·əd·iv 'el·ə·mənt }

positive estuary |HYD| An estuary in which there is a measurable dilution of seawater by land drainage. { 'päz·əd·iv 'es·chə,wer·ē }

positive landform |GEOL| An upstanding topographic form, such as a mountain, hill, plateau, or cinder cone. { 'päz·əd·iv 'land,fȯrm }

potable |SCI TECH| Suitable for drinking. { 'pōd·ə·bəl }

potamic |SCI TECH| Pertaining to rivers or river navigation. { pə'tam·ik }

potamology |HYD| The scientific study of rivers. { ˌpäd·ə'mäl·ə·jē }

potamoplankton |BIOL| Plankton found in rivers. { ¦päd·ə·mō¦plaŋk·tən }

potash lake |HYD| An alkali lake whose waters contain a high content of dissolved potassium salts. { 'päd,ash 'lāk }

potassium argentocyanide   *See* silver potassium cyanide.   { pə'tas·ē·əm ¦är·jən·tō'sī·ə ˌnīd }

potassium arsenate   [CHEM] K_3AsO_4 Poisonous, colorless crystals; soluble in water, insoluble in alcohol; used as an insecticide, analytical reagent, and in hide preservation and textile printing. Also known as Macquer's salt.   { pə'tas·ē·əm 'ärs·ən ˌāt }

potassium chloride   [CHEM] KCl Colorless crystals with saline taste; soluble in water, insoluble in alcohol; melts at 776°C; used as a fertilizer and in photography and pharmaceutical preparations. Also known as potassium muriate.   { pə'tas·ē·əm 'klȯr ˌīd }

potassium cyanate   [CHEM] KOCN Colorless, water-soluble crystals; used as an herbicide and for the manufacture of drugs and organic chemicals.   { pə'tas·ē·əm 'sī·ə,nāt }

potassium cyanide   [CHEM] KCN Poisonous, white, deliquescent crystals with bitter almond taste; soluble in water, alcohol, and glycerol; used for metal extraction, for electroplating, for heat-treating steel, and as an analytical reagent and insecticide.   { pə'tas·ē·əm 'sī·ə,nīd }

potassium cyanoargentate   *See* silver potassium cyanide.   { pə'tas·ē·əm ¦sī·ə·nō'är·jən ˌtāt }

potassium fluoride   [CHEM] KF or KF·2H₂O Poisonous, white, deliquescent crystals with saline taste; soluble in water and hydrofluoric acid, insoluble in alcohol; melts at 846°C; used to etch glass and as a preservative and insecticide.   { pə'tas·ē·əm 'flu̇r ˌīd }

potassium-42   [PHYS] Radioactive isotope with mass number of 42; half-life is 12.4 hours, with β- and γ-radiation; radiotoxic; used as radiotracer in medicine.   { pə'tas·ē·əm ¦fȯr·dē'tü }

potassium manganate   [CHEM] K_2MnO_4 Water-soluble dark-green crystals, decomposing at 190°C; used as an analytical reagent, bleach, oxidizing agent, disinfectant, mordant for dyeing wool and in photography, printing, and water purification.   { pə'tas·ē·əm 'maŋ·gə,nāt }

potassium muriate   *See* potassium chloride.   { pə'tas·ē·əm 'myu̇r·ē,āt }

potassium permanganate   [CHEM] KMnO₄ Highly oxidative, water-soluble, purple crystals with sweet taste; decomposes at 240°C; and explodes in contact with oxidizable materials; used as a disinfectant and analytical reagent, in dyes, bleaches, and medicines, and as a chemical intermediate. Also known as purple salt.   { pə'tas·ē·əm pər'man·gə,nāt }

potassium xanthate   [CHEM] KC_2H_5OCSS Water- and alcohol-soluble, yellow crystals; used as an analytical reagent and soil-treatment fungicide.   { pə'tas·ē·əm 'zan,thāt }

potato virus Y   [MICROBIO] A species in the genus *Potyvirus* that is a common pathogen of potato plants. Symptoms of infection can include shortening of the stem internodes, spotting and severe malformation of the upper leaves, defoliation, and early plant death. Abbreviated PVY.   { pə¦tā·dō,vī·rəs 'wī }

potential evaporation   *See* evaporative power.   { pə'ten·chəl i,vap·ə'rā·shən }

potential evapotranspiration   [HYD] Generally, the amount of moisture which, if available, would be removed from a given land area by evapotranspiration; expressed in units of water depth.   { pə'ten·chəl i,vap·ō,tranz·pə'rā·shən }

potentiometric map   [HYD] A map showing the elevation of a potentiometric surface of an aquifer by means of contour lines or other symbols. Also known as pressure-surface map.   { pə¦ten·chē·ə¦me·trik 'map }

potentiometric surface [HYD] An imaginary surface that represents the static head of groundwater (the ratio of its pressure to its weight per unit volume) and is defined by the level to which the water will rise in wells. Also known as isopotential level; piezometric surface; pressure surface. { pə|ten·chē·ə|me·trik 'sər·fəs }

pothole See moulin. { 'pät₁hōl }

powder snow [HYD] A cover of dry snow that has not been compacted in any way. { 'paùd·ər ₁snō }

powdery mildew [MYCOL] A fungus characterized by production of abundant powdery conidia on the host; a member of the family Erysiphaceae or the genus *Oidium*. [PL PATH] A plant disease caused by a powdery mildew fungus. { 'paùd·ə·rē 'mil ₁dü }

powdery scab [PL PATH] A fungus disease of potato tubers caused by *Spongospora subterranea* and characterized by nodular discolored lesions, which burst and expose masses of powdery fungus spores. { 'paùd·ə·rē 'skab }

power barker See barker. { 'paù·ər ₁bärk·ər }

power-law profile [METEOROL] A formula for the variation of wind with height in the surface boundary layer. { 'paù·ər ₁lȯ ₁prō₁fīl }

power plant [ENG] Any unit that converts some form of energy into electrical energy, such as a hydroelectric or steam-generating station, a diesel-electric engine in a locomotive, or a nuclear power plant. Also known as electric power plant. { 'paù·ər ₁plant }

power station See generating station. { 'paù·ər ₁stā·shən }

poxvirus [MICROBIO] A deoxyribonucleic acid-containing animal virus group including the viruses of smallpox, molluscum contagiosum, and various animal pox and fibromas. { 'päks₁vī·rəs }

praecipitatio [METEOROL] Precipitation falling from a cloud and apparently reaching the earth's surface; this supplementary cloud feature is mostly encountered in altostratus, nimbostratus, stratocumulus, stratus, cumulus, and cumulonimbus. { prē|sip·ə|tā·shō }

prairie [GEOGR] An extensive level-to-rolling treeless tract of land in the temperate latitudes of central North America, characterized by deep, fertile soil and a cover of coarse grass and herbaceous plants. { 'prer·ē }

prairie climate See subhumid climate. { 'prer·ē ₁klī·mət }

prairie soil [GEOL] A group of zonal soils having a surface horizon that is dark or grayish brown, which grades through brown soil into lighter-colored parent material; it is 2–5 feet (0.6–1.5 meters) thick and develops under tall grass in a temperate and humid climate. { 'prer·ē ₁sȯil }

preadaptation [GEN] Possession by an organism or group of organisms, specialized to one mode of life, of characters which favor easy adaptation to a new environment. { |prē₁ad·əp'tā·shən }

precancerous [MED] Pertaining to any pathological condition of a tissue which is likely to develop into cancer. { prē'kan·sə·rəs }

precipitable water [METEOROL] The total atmospheric water vapor contained in a vertical column of unit cross-sectional area extending between any two specified levels, commonly expressed in terms of the height to which that water substance would stand if completely condensed and collected in a vessel of the same unit cross section. Also known as precipitable water vapor. { pri'sip·əd·ə·bəl 'wȯd·ər }

precipitable water vapor See precipitable water. { pri'sip·əd·ə·bəl 'wȯd·ər ₁vā·pər }

precipitation |METEOROL| **1.** Any or all of the forms of water particles, whether liquid or solid, that fall from the atmosphere and reach the ground. **2.** The amount, usually expressed in inches of liquid water depth, of the water substance that has fallen at a given point over a specified period of time. { prə₁sip·ə'tā·shən }

precipitation area |METEOROL| **1.** On a synoptic surface chart, an area over which precipitation is falling. **2.** In radar meteorology, the region from which a precipitation echo is received. { prə₁sip·ə'tā·shən ₁er·ē·ə }

precipitation ceiling |METEOROL| After United States weather observing practice, a ceiling classification applied when the ceiling value is the vertical visibility upward into precipitation; this is necessary when precipitation obscures the cloud base and prevents a determination of its height. { prə₁sip·ə'tā·shən ₁sēl·iŋ }

precipitation cell |METEOROL| In radar meteorology, an element of a precipitation area over which the precipitation is more or less continuous. { prə₁sip·ə'tā·shən ₁sel }

precipitation current |METEOROL| The downward transport of charge, from cloud region to earth, that occurs in a fall of electrically charged rain or other hydrometeors. { prə₁sip·ə'tā·shən ₁kə·rənt }

precipitation echo |METEOROL| A type of radar echo (reflected signal) returned by precipitation. { prə₁sip·ə'tā·shən ₁ek·ō }

precipitation effectiveness *See* precipitation-evaporation ratio. { prə₁sip·ə'tā·shən i₁fek·tiv·nəs }

precipitation-evaporation ratio [CLIMATOL] For a given locality and month, an empirical expression devised for the purpose of classifying climates numerically on the basis of precipitation and evaporation. Abbreviated P-E ratio. Also known as precipitation effectiveness. { prə₁sip·ə'tā·shən i₁vap·ə'rā·shən ₁rā·shō }

precipitation excess [HYD] The volume of water from precipitation that is available for direct runoff. { prə₁sip·ə'tā·shən 'ek₁ses }

precipitation gage [ENG] Any device that measures the amount of precipitation; principally, a rain gage or snow gage. { prə₁sip·ə'tā·shən ₁gāj }

precipitation-generating element |METEOROL| In radar meteorology, a relatively small volume of supercooled cloud droplets in which ice crystals form and grow much more rapidly than in a lower, larger cloud mass. { prə₁sip·ə'tā·shən 'jen·ə₁rād·iŋ ₁el·ə·mənt }

precipitation intensity |METEOROL| The rate of precipitation, expressed in inches or millimeters per hour. Also known as rainfall intensity. { prə₁sip·ə'tā·shən in ₁ten·səd·ē }

precipitation inversion |METEOROL| As found in some mountain areas, a decrease of precipitation with increasing elevation of ground above sea level. Also known as rainfall inversion. { prə₁sip·ə'tā·shən in₁vər·zhən }

precipitation physics |METEOROL| The study of the formation and precipitation of liquid and solid hydrometeors from clouds; a branch of cloud physics and of physical meteorology. { prə₁sip·ə'tā·shən ₁fiz·iks }

precipitation station |METEOROL| A weather station at which only precipitation observations are made. { prə₁sip·ə'tā·shən ₁stā·shən }

precipitation trails *See* virga. { prə₁sip·ə'tā·shən ₁trālz }

precision agriculture [AGR] The application of technologies and agronomic principles to manage spatial and temporal variability associated with all aspects of agricultural production for the purpose of improving crop performance and environmental quality. { prə₁sizh·ən 'ag·rə₁kəl·chər }

predation [BIOL] The killing and eating of an individual of one species by an individual of another species. { prə'dā·shən }

predator [ECOL] An animal that preys on other animals as a source of food. { 'pred·əd·ər }

predict See forecast. { pri'dikt }

prediction [METEOROL] **1.** The act of making a weather forecast. **2.** The forecast itself. { prə'dik·shən }

prelogging [FOR] Cutting down and removing small trees before large trees are logged. { 'prē,läg·iŋ }

preserve [ECOL] An area that is maintained for game or fish, especially for sport, and may have limited access requiring a permit for entry. { prə'zərv }

presque isle [GEOGR] A promontory or peninsula extending into a lake, nearly or almost forming an island; its head or end section is connected with the shore by a sag or low gap only slightly above water level or by a strip of lake bottom exposed as a land surface by a drop in lake level. { ¦pres'kīl }

pressure altitude [METEOROL] The height above sea level at which the existing atmospheric pressure would be duplicated in the standard atmosphere; atmospheric pressure expressed as height according to a standard scale. { 'presh·ər ,al·tə,tüd }

pressure-altitude variation [METEOROL] The pressure difference, in feet or meters, between mean sea level and the standard datum plane. { 'presh·ər ¦al·tə,tüd ,ver·ē'ā·shən }

pressure center [METEOROL] **1.** On a synoptic chart (or on a mean chart of atmospheric pressure), a point of local minimum or maximum pressure; the center of a low or high. **2.** A center of cyclonic or anticyclonic circulation. { 'presh·ər ,sen·tər }

pressure-change chart [METEOROL] A chart indicating the change in atmospheric pressure of a constant-height surface over some specified interval of time. Also known as pressure-tendency chart. { 'presh·ər ¦chānj ,chärt }

pressure contour [METEOROL] A line connecting points of equal height of a given barometric pressure; the intersection of a constant pressure surface by a plane parallel to mean sea level. { 'presh·ər ,kän,tür }

pressure depth [OCEANOGR] The depth at which an ocean sample was taken, as inferred from the difference in readings on protected and unprotected thermometers on the sampler; the higher reading is on the unprotected thermometer due to the effect of pressure on the mercury column at the sampling depth. { 'presh·ər ,depth }

pressure field [OCEANOGR] A representation of a pressure gradient as isobar contours, parallel to which ocean currents flow. { 'presh·ər ,fēld }

pressure gradient [METEOROL] The change in atmospheric pressure per unit horizontal distance, usually measured along a line perpendicular to the isobars. { 'presh·ər ,grād·ē·ənt }

pressure ice [OCEANOGR] Ice, especially sea ice, which has been deformed or altered by the lateral stresses of any combination of wind, water currents, tides, waves, and surf; may include ice pressed against the shore, or one piece of ice upon another. { 'presh·ər ,īs }

pressure ice foot [HYD] An ice foot formed along a shore by the freezing together of stranded pressure ice. { 'presh·ər ¦īs ,fút }

pressure pattern [METEOROL] The general geometric characteristics of atmospheric pressure distribution as revealed by isobars on a constant-height chart; usually applied to cyclonic-scale features of a surface chart. { 'presh·ər ,pad·ərn }

pressure ridge [OCEANOGR] A ridge or wall of hummocks where one ice floe has been pressed against another. { 'presh·ər ,rij }

pressure surface See potentiometric surface. { 'presh·ər ,sər·fəs }

pressure-surface map See potentiometric map. { 'presh·ər ¦sər·fəs ˌmap }

pressure system [METEOROL] An individual cyclonic-scale feature of atmospheric circulation, commonly used to denote either a high or a low, less frequently a ridge or a trough. { 'presh·ər ˌsis·təm }

pressure tendency [METEOROL] The character and amount of atmospheric pressure change for a 3-hour or other specified period ending at the time of observation. Also known as barometric tendency. { 'presh·ər ˌten·dən·sē }

pressure-tendency chart See pressure-change chart. { 'presh·ər ¦ten·dən·sē ˌchärt }

pressure topography See height pattern. { 'presh·ər tə,päg·rə·fē }

pressure tube [HYD] A deep, slender, cylindrical hole formed in a glacier by the sinking of an isolated stone that has absorbed more solar radiation than the surrounding ice. { 'presh·ər ˌtüb }

pressure wave [METEOROL] A wave or periodicity which exists in the variation of atmospheric pressure on any time scale, usually excluding normal diurnal or seasonal trends. { 'presh·ər ˌwāv }

prevailing current [OCEANOGR] The ocean current most frequently observed during a given period, such as a month, a season, or a year. { pri'vāl·iŋ ˌkə·rənt }

prevailing visibility [METEOROL] In United States weather observing practice, the greatest horizontal visibility equaled or surpassed throughout half of the horizon circle; in the case of rapidly varying conditions, it is the average of the prevailing visibility while the observation is being taken. { pri'vāl·iŋ ˌviz·ə'bil·əd·ē }

prevailing westerlies [METEOROL] The prevailing westerly winds on the poleward sides of the subtropical high-pressure belts. { pri'vāl·iŋ 'wes·tər·lēz }

prevailing wind See prevailing wind direction. { pri'vāl·iŋ 'wind }

prevailing wind direction [METEOROL] The wind direction most frequently observed during a given period; the periods most often used are the observational day, month, season, and year. Also known as prevailing wind. { pri'vāl·iŋ 'wind di,rek·shən }

prevalence [GEN] The frequency with which a medical condition is found in specific population at a specific time. { 'prev·ə·ləns }

primaquine [MED] $C_{15}H_{21}N_3O$ An ether-soluble viscous liquid, used as the diphosphate salt in medicine to cure malaria. { 'prī·mə,kwēn }

primary [GEOL] A young shoreline whose features are produced chiefly by nonmarine agencies. { 'prī ˌmer·ē }

primary circulation [METEOROL] The prevailing fundamental atmospheric circulation on a planetary scale which must exist in response to radiation differences with latitude, to the rotation of the earth, and to the particular distribution of land and oceans, and which is required from the viewpoint of conservation of energy. { 'prī,mer·ē ˌsər·kyə 'lā·shən }

primary consumer [ECOL] In an ecosystem, an animal that feeds on plants (producers) directly. Also known as a herbivore. { ¦prī,mer·ē kən'sü·mər }

primary cyclone [METEOROL] Any cyclone (or low), especially a frontal cyclone, within whose circulation one or more secondary cyclones have developed. Also known as primary low. { 'prī,mer·ē 'sī,klōn }

primary front [METEOROL] The principal, and usually original, front in any frontal system in which secondary fronts are found. { 'prī,mer·ē 'frənt }

primary host See definitive host. { 'prī,mer·ē 'hōst }

primary hypothermia [MED] A decrease in internal body temperature caused by environmental stress that overwhelms the body's thermoregulation capability. { ¦prī·mərē ‚hī·pōthər·mē·ə }

primary low See primary cyclone. { 'prī‚mer·ē 'lō }

primary pollutant [METEOROL] A pollutant that enters the air directly from a source. { 'prī‚mer·ē pə'lüt·ənt }

primary producer [ECOL] In an ecosystem, an organism (primarily green photosynthetic plants) that utilizes the energy of the sun and inorganic molecules from the environment to synthesize organic molecules. { 'prī‚mer·ē prə'dü·sər }

primary production [ECOL] The total amount of new organic matter produced by photosynthesis. { 'prī‚mer·ē prə'dək·shən }

primary succession See prisere. { 'prī‚mer·ē sək'sesh·ən }

primary treatment [CIV ENG] Removal of floating solids and suspended solids, both fine and coarse, from raw sewage. { 'prī‚mer·ē 'trēt·mənt }

priming of the tides [OCEANOGR] The acceleration in the times of occurrence of high and low tides when the sun's tidal effect comes before that of the moon. { 'prīm·iŋ əv thə 'tīdz }

primitive water [HYD] Water that has been imprisoned in the earth's interior, in either molecular or dissociated form, since the formation of the earth. { 'prim·əd·iv 'wȯd·ər }

prion [BIOL] Any of a group of infectious proteins that cause fatal neurodegenerative diseases in humans and animals, including scrapie and bovine spongiform encephalopathy in animals and Creutzefeldt-Jakob disease and Gerstmann-Straussler-Scheinker disease in humans. { 'prī‚än }

prion diseases [MED] A group of invariably fatal disorders affecting humans and animals that are clinically characterized by neurological and behavioral degeneration caused by the cerebral accumulation of an abnormal prion protein which is resistant to proteolytic enzymes and, in contrast to other infectious agents, does not require nucleic acid for replication. The diseases are transmissible either genetically (for example, Creutzfeldt-Jakob disease) or via infection (new variant Creutzfeldt-Jakob disease and mad cow disease) or can occur spontaneously (classical or sporadic Creutzfeldt-Jakob disease). Also known as spongiform encephalopathies. { 'prī‚än di ‚zēz·əz }

prisere [ECOL] The ecological succession of vegetation that occurs in passing from barren earth or water to a climax community. Also known as primary succession. { 'prī‚sir }

private stream [HYD] Any stream which diverts part or all of the drainage of another stream. { 'prī·vət 'strēm }

probability forecast [METEOROL] A forecast of the probability of occurrence of one or more of a mutually exclusive set of weather contingencies, as distinguished from a series of categorical statements. { ‚präb·ə'bil·əd·ē ‚fȯr‚kast }

probable maximum precipitation [METEOROL] The theoretically greatest depth of precipitation for a given duration that is physically possible over a particular drainage area at a certain time of year; in practice, this is derived over flat terrain by storm transposition and moisture adjustment to observed storm patterns. { 'präb·ə·bəl 'mak·sə·məm pri‚sip·ə'tā·shən }

procaine penicillin G [CHEM] $C_{29}H_{38}N_4O_6S \cdot H_2O$ White crystals or powder, fairly soluble in chloroform; used as an antibiotic in animal feed. { 'prō‚kān ‚pen·ə'sil·ən 'jē }

process lapse rate [METEOROL] The rate of decrease of the temperature of an air parcel as it is lifted, expressed as $-dT/dz$, where z is the altitude, or occasionally dT/dp, where

p is pressure; the concept may be applied to other atmospheric variables, such as the process lapse rate of density. { 'prä,səs 'laps ,rāt }

procumbent [BOT] Having stems that lie flat on the ground but do not root at the nodes. [SCI TECH] **1.** Lying stretched out. **2.** Slanting forward. **3.** Lying face down. { prō 'kəm·bənt }

producer [ECOL] An autotrophic organism of the ecosystem; any of the green plants. { prə'dü·sər }

producer gas [PETR MIN] Fuel gas high in carbon monoxide and hydrogen, produced by burning a solid fuel with a deficiency of air or by passing a mixture of air and steam through a bed of incandescent fuel; used as a cheap, low-Btu industrial fuel. { prə'dü·sər ,gas }

production ecology See ecological energetics. { prə'duk·shən ē'käl·ə·jē }

profile [HYD] A vertical section of a potentiometric surface, such as a water table. { 'prō ,fīl }

profiling snow gage [HYD] A type of radioactive gage for measuring the water equivalent and density/depth distribution of a snowpack, consisting of a radioactive source and a radioactivity detector which move up and down in two adjacent vertical pipes surrounded by snow. Also known as nuclear twin-probe gage. { 'prō,fīl·iŋ 'snō ,gāj }

profundal zone [ECOL] The region occurring below the limnetic zone and extending to the bottom in lakes deep enough to develop temperature stratification. { prō'fənd·əl ,zōn }

prognostic chart [METEOROL] A chart showing, principally, the expected pressure pattern (or height pattern) of a given synoptic chart at a specified future time; usually, positions of fronts are also included, and the forecast values of other meteorological elements may be superimposed. { präg'näs·tik ,chärt }

Prokaryotae [BIOL] Previously used to denote a superkingdom of predominantly unicellular microorganisms lacking a membrane-bound nucleus containing chromosomes and having asexual reproduction by binary fission. Now split into the domains Archaea and Bacteria. { ,prō·kar·ē'ō,dē }

prokaryote [BIOL] **1.** A primitive nucleus, where the deoxyribonucleic acid-containing region lacks a limiting membrane. **2.** Any cell containing such a nucleus, such as the bacteria and the blue-green algae. { prō'kar·ē,ōt }

promontory [GEOL] **1.** A high, prominent projection or point of land, or a rock cliff, jutting out boldly into a body of water. **2.** A cape, either low-lying or of considerable height, with a bold termination. **3.** A bluff or prominent hill overlooking or projecting into a lowland. { 'präm·ən,tór·ē }

prong reef [GEOL] A wall reef that has developed irregular buttresses normal to its axis in both leeward and (to a smaller degree) seaward directions. { präŋ ,rēf }

propagation [BOT] The deliberate, directed reproduction of plants using plant cells, tissues, or organs. { ,präp·ə'gā·shən }

propagule [BOT] **1.** A reproductive structure of brown algae. **2.** A propagable shoot. { 'präp·ə,gyül }

propane [CHEM] $CH_3CH_2CH_3$ A heavy, colorless, gaseous petroleum hydrocarbon gas of the paraffin series; boils at $-44.5°C$; used as a solvent, refrigerant, and chemical intermediate. { 'prō,pān }

1-propanethiol See n-propyl mercaptan. { ¦wən ¦prō,pān'thī,ól }

propham [CHEM] $C_{10}H_{13}NO_2$ A light brown solid with a melting point of 87–88°C; slightly soluble in water; used as a pre- and postemergence herbicide for vegetable crops. Abbreviated IPC (isopropyl-N-phenylcarbamate). { 'prō,fam }

prophylactic vaccination　[MED] Vaccination occurring before exposure to pathogens. { ¦pro·fə¦lak·tic ˌvak·sə'nā·shən }

propionaldehyde　[CHEM] C_2H_5CHO Flammable, water-soluble, water-white liquid, with suffocating aroma; boils at 48.8°C; used to manufacture acetals, plastics, and rubber chemicals, and as a disinfectant and preservative. { ¦prō·pē¦än'al·dəˌhīd }

prop root　[BOT] A root that serves to support or brace the plant. Also known as brace root. { 'präp ˌrüt }

propylene dichloride　[CHEM] $CH_3CHClCH_2Cl$ Water-insoluble, colorless, moderately flammable liquid, with chloroform aroma; boils at 96.3°C; miscible with most common solvents; used as a solvent, dry-cleaning fluid, metal degreaser, and fumigant. { 'prō·pəˌlēn dī'klȯrˌīd }

propylene glycol　[CHEM] $CH_3CHOHCH_2OH$ A viscous, colorless liquid, miscible with water, alcohol, and many solvents; boils at 188°C; used as a chemical intermediate, antifreeze, solvent, lubricant, plasticizer, and bactericide. { 'prō·pəˌlēn 'glīˌkȯl }

***n*-propyl mercaptan**　[CHEM] C_3H_7SH A liquid with an offensive odor and a boiling range of 67–73°C; used as a herbicide. Also known as 1-propanethiol. { ¦en 'prō·pəl mər'kap ˌtan }

propylparaben　[CHEM] $C_{10}H_{12}O_3$ Colorless crystals or white powder with a melting point of 95–98°C; soluble in acetone, ether, and alcohol; used in medicine and as a food preservative and fungicide. { ˌprō·pəl'par·ə·bən }

propylthiopyrophosphate　[CHEM] $C_{12}H_{28}P_2S_2O$ A straw-colored to dark amber liquid with a boiling point of 148°C; used as an insecticide for chinch bugs in lawns and turf. { ¦prō·pəl¦thī·ō,pī·rə'fäˌsfāt }

prospecting seismology　[PETR MIN] The application of seismology to the exploration for natural resources, especially gas and oil. { 'präˌspek·tiŋ sīz'mäl·ə·jē }

protection　[ENG] Any provision to reduce exposure of persons to radiation; for example, protective barriers to reduce external radiation or measures to prevent inhalation of radioactive materials. { prə'tek·shən }

protective action guide　[ENG] The absorbed dose of ionizing radiation in individuals in the general population which would warrant protective action following a contaminating event, such as a nuclear explosion. { prə'tek·tiv 'ak·shən ˌgīd }

protective clothing　[ENG] Special clothing worn by a radiation worker to prevent contamination of the body or personal clothing. { prə'tek·tiv 'klōth·iŋ }

protective coloration　[ZOO] A color pattern that blends with the environment and increases the animal's probability of survival. { prə'tek·tiv ˌkəl·ə'rā·shən }

protective survey　[ENG] An evaluation of the radiation hazards incidental to the production, use, or existence of radioactive materials or other sources of radiation under a specific set of conditions. { prə'tek·tiv 'sərˌvā }

protein coat　*See* capsid. { 'prōˌtēn ˌkōt }

Protista　[BIOL] A proposed kingdom to include all unicellular eukaryotic organisms lacking a definite cellular arrangement, such as algae, diatoms, and fungi. { prə'tis·tə }

protonema　[BOT] A green, filamentous structure that originates from an asexual spore of mosses and some liverworts and that gives rise by budding to a mature plant. { ˌprōt·ən'ē·mə }

prototroph　[MICROBIO] A microorganism that has the ability to synthesize all of its amino acids, nucleic acids, vitamins, and other cellular constituents from inorganic nutrients. { 'prōd·ə,träf }

prototrophic　[MICROBIO] Pertaining to bacteria with the nutritional properties of the wild type, or the strains found in nature. { ¦prōd·ō¦träf·ik }

protoxin [BIOL] A chemical compound that is a precursor to a toxin. { ¦prō¦täk·sən }

Protozoa [ZOO] A diverse phylum of eukaryotic microorganisms; the structure varies from a simple uninucleate protoplast to colonial forms, the body is either naked or covered by a test, locomotion is by means of pseudopodia or cilia or flagella, there is a tendency toward universal symmetry in floating species and radial symmetry in sessile types, and nutrition may be phagotrophic or autotrophic or saprozoic. { ¦prōd·ə¦zō·ə }

province [OCEANOGR] An area composed of a grouping of like bathymetric elements whose features are in obvious contrast with surrounding regions. { 'präv·əns }

prussic acid See hydrocyanic acid. { 'prəs·ik 'as·əd }

pryrrolidine [CHEM] C_4H_9N A colorless to pale yellow liquid with a boiling point of 87°C; soluble in water and alcohol; used in the manufacture of pharmaceuticals, insecticides, and fungicides. { pə'räl·ə,dēn }

psammon [ECOL] **1.** In a body of fresh water, that part of the environment composed of a sandy beach and bottom lakeward from the water line. **2.** Organisms which inhabit the interstitial water in the sands on a lake shore. { 'sa,män }

psammophilic [ECOL] Pertaining to an organism found in sand. { ¦sam·ə¦fil·ik }

psammophyte [ECOL] Thriving (as a plant) on sandy soil. { 'sam·ə,fīt }

psammosere [ECOL] Stages in plant succession which begin in sandy soil. { 'sam·ə ,sir }

psephyte [GEOL] A lake-bottom deposit consisting mainly of coarse, fibrous plant remains. { sē,fīt }

pseudaposematic [ECOL] Pertaining to an imitation in coloration or form by an organism of another organism that possesses dangerous or disagreeable characteristics. { ¦süd·ə,pōz·ə'mad·ik }

pseudoadiabatic chart See Stuve chart. { ¦sü·dō,ad·ē·ə'bad·ik 'chärt }

pseudo cold front See pseudo front. { 'sü·dō'kōld ,frənt }

pseudoequivalent temperature See equivalent temperature. { ¦sü·dō·i'kwiv·ə·lənt 'tem·prə·chər }

pseudo front [METEOROL] A small-scale front, formed in association with organized severe convective activity, between a mass of rain-cooled air from the thunderstorm clouds and the warm surrounding air. Also known as pseudo cold front. { 'sü·dō 'frənt }

Pseudomonas [MICROBIO] A genus of gram-negative, motile, non-spore-forming, rod-shaped bacteria that cause a variety of infectious diseases in animals and humans (such as glanders and melioidosis) and in plants. { ,süd·ə'mōn·əs }

Pseudomonas aeruginosa [MICROBIO] An opportunistic pathogen that is the most significant cause of hospital-acquired infections, particularly in predisposed patients with metabolic, hematologic, and malignant diseases. It produces toxic factors such as lipase, esterase, lecithinase, elastase, and endotoxin, some of which may contribute to its pathogenesis. { ,süd·ə,mōn·əs ,ar·ə·jə'nō·sə }

Pseudomonas mallei [MICROBIO] A mammalian parasite that is the causative agent of glanders, an infectious disease of horses that is occasionally transmitted to humans by direct contact. { ,süd·ə,mōn·əs 'mal·ē,ī }

Pseudomonas pseudomallei [MICROBIO] A bacteria that is the causative agent of melioidosis, an endemic glanders-like disease of humans and animals that occurs most frequently in southeastern Asia and northern Australia. { ,süd·ə,mōn·əs ,süd·ə'mal·ē·ī }

pseudopodium [BOT] A slender, leafless branch of the gametophyte in certain Bryatae. { ‚süd·ə'pōd·ē·əm }

pseudotuberculosis [MED] A bacterial infection in humans and many animals caused by *Pasteurella pseudotuberculosis*; may be severe in humans with septicemia and symptoms resembling typhoid fever. { ¦sü·dō·tə‚bər·kyə'lō·səs }

psilate [BOT] Lacking ornamentation; generally applied to pollen. { 'sī‚lāt }

Psilotophyta [BOT] A division of homosporous (producing only one spore) vascular plants without leaves and roots; for example, whisk ferns. { ‚sī·lō'täf·əd·ə }

psittacosis [MED] Pneumonia and generalized infection of man and of birds caused by *Chlamydia psittaci*, transmitted to humans by psittacine birds (in the parrot family). { ‚sid·ə'kō·səs }

psorosis [PL PATH] A virus disease of tangerine, grapefruit, and sweet orange trees characterized by scaly bark, a gummy exudate, retarded growth, small yellow leaves, and dieback of twigs. Also known as scaly bark. { sə'rō·səs }

psychrophile [BIOL] An organism that thrives at low temperatures. { 'sī·krə‚fīl }

psychrophyte [ECOL] A plant adapted to the climatic conditions of the arctic or alpine regions. { 'sī·krə‚fīt }

psychrosphere [OCEANOGR] The cold deep layer of the ocean, 100–700 meters (330–2300 feet) below the surface, where the water temperature is typically less than 10°C (50°F). { 'sī·krə‚sfir }

public health [MED] **1.** The state of health of a community or of a population. **2.** The art and science dealing with the protection and improvement of community health. { 'pəb·lik 'helth }

Puccinia asparagi [MYCOL] An autoecious fungus of the order Uredinales; the causative agent of asparagus rust. { pü‚sin·ē·ə ə'spär·ə‚gē }

Puccinia graminis [MYCOL] A macrocylic heteroecious fungus of the order Uredinales; the causative agent of black stem rust of cereal grains. { pü‚sin·ē·ə 'gram·ə·nəs }

Puccinia malvacearum [MYCOL] A microcyclic fungus of the order Uredinales; the causative agent of hollyhock rust. { pü‚sin·ē·ə ‚mal·və'sē·ə·rəm }

puelche [METEOROL] An east wind which has crossed the Andes; the Andean foehn of the South American west coast. { 'pwel·chē }

puff ball [BOT] A spherical basidiocarp that retains spores until fully mature and, when disturbed, releases them as puffs of fine dust. { 'pəf ‚bol }

puff of wind [METEOROL] A slight local breeze which causes a patch of ripples on the surface of the sea. { 'pəf əv 'wind }

pullorum disease [VET MED] A highly fatal disease of chickens and other birds caused by *Salmonella pullorum*, characterized by weakness, lassitude, lack of appetite, and whitish or yellowish diarrhea. Also known as bacillary white diarrhea; white diarrhea. { pə'lor·əm di‚zēz }

pulp [BOT] The soft succulent portion of a fruit. [FOR] The cellulosic material produced by reducing wood mechanically or chemically and used in making paper and cellulose products. Also known as wood pulp. { pəlp }

puna [ECOL] An alpine biological community in the central portion of the Andes Mountains of South America characterized by low-growing, widely spaced plants that lack much green color most of the year. { 'pü·nə }

pupa [ZOO] The quiescent, intermediate form assumed by an insect that undergoes complete metamorphosis; it follows the larva and precedes the adult stages and is enclosed in a hardened cuticle or a cocoon. { 'pyü·pə }

pure forest |FOR| A forest in which one species makes up 80% or more of the total number of trees. { 'pyu̇r ¦fär·əst }

purl |HYD| A swirling or eddying stream or rill, moving swiftly around obstructions. { pərl }

purple bacteria |MICROBIO| Any of various photosynthetic bacteria that contain bacteriochlorophyll, distinguished by purplish or reddish-brown pigments. { 'pər·pəl bak'tir·ē·ə }

purple blotch |PL PATH| A fungus disease of onions, garlic, and shallots caused by *Alternaria porri* and characterized by small white spots which become large purplish blotches. { 'pər·pəl ¦bläch }

purple nonsulfur bacteria |MICROBIO| Any of various purple photosynthetic bacteria, especially members of the family Athiorhodaceae, that utilize organic hydrogen donor compounds. { 'pər·pəl ¦nän¦səl·fər bak'tir·ē·ə }

purple salt *See* potassium permanganate. { 'pər·pəl 'sȯlt }

purple sulfur bacteria |MICROBIO| Any of various anaerobic photosynthetic purple bacteria, especially in the family Thiorhodaceae, that utilize H_2S and other inorganic sulfur compounds as a source of hydrogen, while the carbon source can be carbon monoxide. { 'pər·pəl 'səl·fər bak,tir·ē·ə }

purple-top |PL PATH| A virus disease of potato plants characterized by purplish or chlorotic discoloration of the top shoots, swelling of axillary branches, and severe wilting. { 'pər·pəl,täp }

pustule |MED| A small, circumscribed, pus-filled elevation on the skin. |PL PATH| A blisterlike mark on a leaf due to rupture of surface tissues overlying spore masses of a parasitic fungus. { 'pəs·chül }

PVA *See* polyvinyl acetate.

PVAc *See* polyvinyl acetate.

PVY *See* potato virus Y.

pycnocline |OCEANOGR| A region in the ocean where water density increases relatively rapidly with depth. { 'pik·nə,klīn }

pyobacillosis |VET MED| A bacterial infection of sheep, swine, or rarely cattle caused by *Corynebacterium pyogenes*; usually marked by abscess formation, but in sheep takes the form of chronic purulent pneumonia. { ¦pī·ō,bas·ə'lō·səs }

pyramidal iceberg *See* pinnacled iceberg. { pir·ə¦mid·əl 'īs,bərg }

pyramid of numbers |ECOL| The concept that an organism making up the base of a food chain is numerically abundant while each succeeding member of the chain is represented by successively fewer individuals; uses feeding relationship as a basis for the quantitative analysis of an ecological system. { 'pir·ə,mid əv 'nəm·bərz }

pyranometer |ENG| An instrument used to measure the combined intensity of incoming direct solar radiation and diffuse sky radiation; compares heating produced by the radiation on blackened metal strips with that produced by an electric current. Also known as solarimeter. { pir·ə'näm·əd·ər }

pyrethrum |AGR| A toxicant obtained in the form of dried powdered flowers of the plant of the same name; mixed with petroleum distillates, it is used as an insecticide. { pī'rē·thrəm }

pyrheliometer |ENG| An instrument for measuring the total intensity of direct solar radiation received at the earth. { ¦pir,hē·lē'äm·əd·ər }

pyroheliometer |METEOROL| An instrument that measures the sun's radiation output. { ,pī·rō,hē·lē'äm·əd·ər }

pyrometallurgy

pyrometallurgy [ENG] Processes that use chemical reactions at elevated temperatures for the extraction of metals from raw materials such as ores and concentrates, or for the treatment of recycled scrap. { ¦pī·rō'med·əl‚ər·jē }

pyromucic acid *See* furoic acid. { ¦pī·rō'myü·sik 'as·əd }

2-pyrrolidone [CHEM] C_4H_7ON Combustible, light-yellow liquid, boiling at 245°C; soluble in ethyl alcohol, water, chloroform, and carbon disulfide; used as a plasticizer and polymer solvent, in insecticides and specialty inks, and as a nylon-4 precursor. { ¦tü pə'räl·ə‚dōn }

Q

Q fever [MED] An acute, febrile infectious disease of humans, characterized by sudden onset and patchy pneumonitis, and caused by a bacterialike organism, *Coxiella burneti*. { 'kyü ,fē·vər }

quagmire *See* bog. { 'kwäg,mīr }

quality of snow [METEOROL] The amount of ice in a snow sample expressed as a percent of the weight of the sample. Also known as thermal quality of snow. { 'kwäl·əd·ē əv 'snō }

quantitative trait [GEN] A trait that is under the control of many factors, both genetic and environmental, each of which contributes only a small amount to the total variability of the trait. { ,kwänt·ə,tād·iv 'trāt }

quantization [SCI TECH] The restriction of a variable to a discrete number of possible values; thus the age of a person is usually quantized as a whole number of years. { ,kwän·tə'zā·shən }

quasi-stationary front [METEOROL] A front which is stationary or nearly so; conventionally, a front which is moving at a speed less than about 5 knots (0.26 meter per second) is generally considered to be quasi-stationary. Commonly known as stationary front. { ¦kwä·zē 'stā·shə,ner·ē 'frənt }

quicksand [GEOL] A highly mobile mass of fine sand consisting of smooth, rounded grains with little tendency to mutual adherence, usually thoroughly saturated with upward-flowing water; tends to yield under pressure and to readily swallow heavy objects on the surface. Also known as running sand. { 'kwik,sand }

quickwater [HYD] The part of a stream characterized by a strong current. { 'kwik ,wȯd·ər }

quinacrine [MED] $C_{23}H_{30}ClN_3O$ Formerly an important antimalarial drug but now used in the treatment of giardiasis, tapeworm infections, amebiasis, and a variety of other conditions. { 'kwin·ə·krən }

quinine [CHEM] $C_{20}H_{24}N_2O_2 \cdot 3H_2O$ White powder or crystals, soluble in alcohol, ether, carbon disulfide, chloroform, and glycerol; an alkaloid derived from cinchona bark; used as an antimalarial drug and in beverages. { 'kwī,nīn }

quintozene *See* pentachloronitrobenzene. { 'kwin·tə,zēn }

R

R   *See* roentgen.

rabal   [METEOROL] A method of winds-aloft observation, that is, the determination of wind speeds and directions in the atmosphere above a weather station; it is accomplished by recording the elevation and azimuth angles of a radiosonde balloon at specified time intervals while visually tracking the balloon with a theodolite.   { 'rā ,bäl }

rabies   [VET MED] An acute, encephalitic viral infection transmitted to humans by the bite of a rabid animal. Also known as hydrophobia.   { 'rā·bēz }

race   [OCEANOGR] A rapid current, or a constricted channel in which such a current flows; the term is usually used only in connection with a tidal current, which may be called a tide race.   { rās }

raceme   [BOT] An inflorescence on which flowers are borne on stalks of equal length on an unbranched main stalk that continues to grow during flowering.   { rā'sēm }

rad   [PHYS] A special unit of absorbed dose, equal to energy absorption of 100 ergs per gram (0.01 joule per kilogram); equal to 0.01 gray.   { rad }

radappertization   [ENG] The use of radiation for sterilizing foods.   { rā,dap·ərd·ə'zā· shən }

radar-absorbing material   [ENG] A material that is designed to reduce the reflection of electromagnetic radiation by a conducting surface in the frequency range from approximately 100 megahertz to 100 gigahertz.   { 'rā,där əb,sȯrb·iŋ mə,tir·ē·əl }

radar climatology   [CLIMATOL] The statistics in time and space of radar weather echoes (signals).   { 'rā,där ,klī·mə'täl·ə·jē }

radar meteorological observation   [METEOROL] Evaluation of the echoes (signals) appearing on the indicator of a weather radar, in terms of orientation, coverage, intensity, tendency of intensity, height, movement, and unique characteristics of echoes, that may be indicative of certain types of severe storms (such as hurricanes, tornadoes, or thunderstorms) and of anomalous propagation. Also known as radar weather observation.   { 'rā,där ,mēd·ē·ə·rə'läj·ə·kəl ,äb·sər'vā·shən }

radar meteorology   [METEOROL] The study of the scattering of radar waves by all types of atmospheric phenomena and the use of radar for making weather observations and forecasts.   { 'rā,där ,mēd·ē·ə'räl·ə·jē }

radar report   [METEOROL] The encoded and transmitted report of a radar meteorological observation; these reports usually give the azimuth, distance, altitude, intensity, shape and movement, and other characteristics of precipitation echoes observed by the radar. Also known as rain area report. Abbreviated RAREP.   { 'rā,där ri,pȯrt }

radar storm detection   [METEOROL] The detection of certain storms or stormy conditions by means of radar; liquid or frozen water drops within the storm reflect radar echoes.   { 'rā,där 'stȯrm di,tek·shən }

radar weather observation See radar meteorological observation. { 'rā,där 'we<u>th</u>·ər ,äb·zər,vā·shən }

radar wind [METEOROL] Wind of which the movement, speed, and direction is observed or determined by a radar tracking of a balloon carrying a radiosonde, a radio transmitter, or a radar reflector. { 'rā,där ,wind }

radial symmetry [SCI TECH] An arrangement of usually similar parts in a regular pattern around a central axis. { 'rād·ē·əl 'sim·ə·trē }

radiant energy See radiation. { 'rād·ē·ənt 'en·ər·jē }

radiation [PHYS] **1.** The emission and propagation of waves transmitting energy through space or through some medium; for example, the emission and propagation of electromagnetic, sound, or elastic waves. **2.** The energy transmitted by waves through space or some medium; when unqualified, usually refers to electromagnetic radiation. Also known as radiant energy. **3.** A stream of particles, such as electrons, neutrons, protons, α-particles, or high-energy photons, or a mixture of these. { ,rād·ē'ā·shən }

radiation accident [ENG] Any accident resulting in the spread of radioactive materials or in the exposure of individuals to radiation. { ,rād·ē'ā·shən ,ak·sə·dənt }

radiational cooling [METEOROL] The cooling of the earth's surface and adjacent air, accomplished (mainly at night) whenever the earth's surface suffers a net loss of heat due to terrestrial radiation. { ,rād·ē'ā·shən·əl 'kül·iŋ }

radiational inversion [METEOROL] An inversion at the land surface resulting from rapid radiational cooling of lower air; usually occurs on cold winter nights. { ,rad·ē'ā·shən·əl in'vər·zhən }

radiation area [ENG] Any accessible area in which the level of radiation is such that a major portion of an individual's body could receive in any 1 hour a dose in excess of 5 millirem or in any 5 consecutive days a dose in excess of 150 millirem. { ,rād·ē'ā·shən ,er·ē·ə }

radiation biochemistry [BIOL] The study of the response of the constituents of living matter to radiation. { ,rād·ē'ā·shən ¦bī·ō'kem·ə·strē }

radiation biology See radiobiology. { ,rād·ē'ā·shən bī'äl·ə·jē }

radiation biophysics [BIOL] The study of the response of organisms to ionizing radiations and to ultraviolet light. { ,rād·ē'ā·shən ¦bī·ō'fiz·iks }

radiation counter [ENG] An instrument used for detecting or measuring nuclear radiation by counting the resultant ionizing events; examples include Geiger counters and scintillation counters. Also known as counter. { ,rād·ē'ā·shən ,kaúnt·ər }

radiation cytology [BIOL] An aspect of biology that deals with the effects of radiations on living cells. { ,rād·ē'ā·shən sī'täl·ə·jē }

radiation dermatitis See radiodermatitis. { ,rād·ē'ā·shən ,dər·mə'tīd·əs }

radiation detection instrument [ENG] Any device that detects and records the characteristics of ionizing radiation. { ,rād·ē'ā·shən di'tek·shən ,in·strə·mənt }

radiation detector See particle detector. { ,rād·ē'ā·shən di,tek·tər }

radiation dose [PHYS] The total amount of ionizing radiation absorbed by material or tissues, in the sense of absorbed dose (expressed in rads), exposure dose (expressed in roentgens), or dose equivalent (expressed in rems). { ,rād·ē'ā·shən ,dōs }

radiation dosimetry See dosimetry. { ,rād·ē'ā·shən dō'sim·ə·trē }

radiation effects [BIOL] The harmful effects of ionizing radiation on humans and other animals, such as production of cancers, cataracts, and radiation ulcers, loss of hair, reddening of skin, sterilization, nausea, vomiting, mucous or bloody diarrhea, purpura, epilation, and agranulocytic infections. { ,rād·ē'ā·shən i,feks }

radiation fog [METEOROL] A major type of fog, produced over a land area when radiational cooling reduces the air temperature to or below its dew point; thus, strictly, a nighttime occurrence, although the fog may begin to form by evening twilight and often does not dissipate until after sunrise. { ‚rād·ē'ā·shən ‚fäg }

radiation gage [ENG] An instrument for measuring radiation quantity and intensity. { ‚rād·ē'ā·shən ‚gāj }

radiation hazard [MED] Health hazard arising from exposure to ionizing radiation. { ‚rād·ē'ā·shən ‚haz·ərd }

radiation monitoring [ENG] Continuous or periodic determination of the amount of radiation present in a given area. { ‚rād·ē'ā·shən ‚män·ə·triŋ }

radiation preservation [ENG] Exposure of food products to ionizing radiation, such as electrons, x-rays, and γ-rays, in order to destroy microorganisms and thereby aid preservation. { ‚rād·ē'ā·shən ‚prez·ər'vā·shən }

radiation protection [ENG] **1.** Legislation and regulations to protect the public and laboratory or industrial workers against radiation. **2.** Measures to reduce exposure to radiation. { ‚rād·ē'ā·shən prə‚tek·shən }

radiation protection guide [ENG] The officially determined radiation doses which should not be exceeded without careful consideration of the reasons for doing so; these standards, established by the Federal Radiation Council, are equivalent to what was formerly called the maximum permissible dose or maximum permissible exposure. { ‚rād·ē'ā·shən prə'tek·shən ‚gīd }

radiation safety [ENG] Protection of personnel against harmful effects of ionizing radiation by taking steps to ensure that people will not receive excessive doses of radiation and by monitoring all sources of radiation to which they may be exposed. { ‚rād·ē'ā·shən ‚sāf·tē }

radiation shelter *See* fallout shelter. { ‚rād·ē'ā·shən ‚shel·tər }

radiation shield [ENG] A shield or wall of material interposed between a source of radiation and a radiation-sensitive body, such as a person, radiation-detection instrument, or photographic film, to protect the latter. { ‚rād·ē'ā·shən ‚shēld }

radiation sickness [MED] **1.** Illness, usually manifested by nausea and vomiting, resulting from the effects of therapeutic doses of radiation. **2.** Radiation injury following exposure to excessive doses of radiation, such as the explosion of an atomic bomb. { ‚rād·ē'ā·shən ‚sik·nəs }

radiation standards [ENG] Exposure standards, permissible concentrations, rules for safe handling, regulations for transportation, regulations for industrial control of radiation, and control of radiation exposure by legislative means. { ‚rād·ē'ā·shən ‚stan·dərdz }

radiation sterilization [ENG] Exposure of a material, object, or body to ionizing radiation in order to destroy microorganisms. { ‚rād·ē'ā·shən ‚ster·ə·lə'zā·shən }

radiation survey meter [ENG] Portable device to measure the intensity of nuclear radiation in a given region, in such applications as health physics (atomic radiation safety) or supervision of radioactively hot areas. { ‚rād·ē'ā·shən 'sər‚vā ‚mēd·ər }

radiation warning symbol [ENG] A standard symbol used on posters displayed in locations where radiation hazards exist; consists of a magenta trefoil printed on a yellow background. { ‚rād·ē'ā·shən 'wȯrn·iŋ ‚sim·bəl }

radiative diffusivity [METEOROL] A characteristic property of a given layer of the atmosphere which governs the rate at which that layer will warm or cool as a result of the transfer, within it, of infrared radiation; the radiative diffusivity is dependent upon the temperature and water-vapor content of the layer of air and upon the pressure within the layer. { 'rād·ē‚ād·iv ‚di‚fyü'siv·əd·ē }

radiative forcing [METEOROL] The relative effectiveness of greenhouse gases to restrict long-wave radiation from escaping back into space. For a particular greenhouse gas, radiative forcing is measured as the change in average net radiation (in watts per square meter) at the top of the troposphere, and depends on the wavelength at which the gas absorbs the radiation, the strength of absorption per molecule, and the concentration of the gas. { ˌrād·ē·ād·iv 'fȯrs·iŋ }

radiatus [METEOROL] A cloud variety whose elements are arranged in straight parallel bands; owing to the effect of perspective, these bands seem to converge toward a point on the horizon, or, when the bands cross the entire sky, toward two opposite points. { ˌrād·ē'äd·əs }

radicidation [ENG] Destruction by radiation of microorganisms in food that are significant to public health; an example is *Salmonella* species. { ˌrād·ə·sə'dā·shən }

radicle [BOT] The embryonic root of a flowering plant. { 'rad·ə·kəl }

radio- [PHYS] Chemical prefix designating radiation or radioactivity; used to designate radioactive elements (such as radiocarbon) and substances containing them (such as radiochemicals, radiocolloids, or radio compounds). { 'rād·ē·ō }

radioactive age determination *See* radiometric dating. { ¦rād·ē·ō'ak·tiv 'āj di₁tər·mə₁nā·shən }

radioactive carbon dating *See* carbon-14 dating. { ¦rād·ē·ō'ak·tiv ¦kär·bən ˌdād·iŋ }

radioactive dating *See* radiometric dating. { ¦rād·ē·ō'ak·tiv 'dād·iŋ }

radioactive decay [PHYS] The spontaneous transformation of a nuclide into one or more different nuclides, accompanied by either the emission of particles from the nucleus, nuclear capture or ejection of orbital electrons, or fission. Also known as decay; nuclear spontaneous reaction; radioactive disintegration; radioactive transformation; radioactivity. { ¦rād·ē·ō'ak·tiv di'kā }

radioactive disintegration *See* radioactive decay. { ¦rād·ē·ō'ak·tiv di₁sin·tə'grā·shən }

radioactive element [PHYS] A chemical element all of whose isotopes spontaneously transform into one or more different nuclides, giving off various types of radiation; examples include promethium, radium, thorium, and uranium. { ¦rād·ē·ō'ak·tiv 'el·ə·mənt }

radioactive fallout *See* fallout. { ¦rād·ē·ō'ak·tiv 'fȯl₁aut }

radioactive metal [PHYS] A luminous metallic element, such as actinium, radium, or uranium, that spontaneously and continuously emits radiation capable in some degree of penetrating matter impervious to ordinary light. { ¦rād·ē·ō'ak·tiv 'med·əl }

radioactive source [ENG] Any quantity of radioactive material intended for use as a source of ionizing radiation. { ¦rād·ē·ō'ak·tiv 'sȯrs }

radioactive transformation *See* radioactive decay. { ¦rād·ē·ō'ak·tiv ˌtranz·fər'mā·shən }

radioactive waste [ENG] Liquid, solid, or gaseous waste resulting from mining of radioactive ore, production of reactor fuel materials, reactor operation, processing of irradiated reactor fuels, and related operations, and from use of radioactive materials in research, industry, and medicine. { ¦rād·ē·ō'ak·tiv 'wāst }

radioactive-waste disposal [ENG] The disposal of waste radioactive materials and of equipment contaminated by radiation; the two basic disposal methods are concentration for burial underground or in the sea, and dilution for controlled dispersion; reprocessing of reactor fuel is a major source of radioactive waste. { ¦rād·ē·ō'ak·tiv ¦wāst di'spō·zəl }

radioactivity [PHYS] **1.** A particular type of radiation emitted by a radioactive substance, such as alpha radioactivity. **2.** *See* activity; radioactive decay. { ˌrād·ē·ō·ak'tiv·əd·ē }

radioactivity concentration guide [ENG] The concentration of radioactive material in an environment which would result in doses equal, over a period of time, to those in the radiation protection guide; this Federal Radiation Council term replaces the former maximum permissible concentration. { ‚rād·ē·ō·ak'tiv·əd·ē ‚käns·ən'trā·shən ‚gīd }

radio atmometer [ENG] An instrument designed to measure the effect of sunlight upon evaporation from plant foliage; consists of a porous-clay atmometer whose surface has been blackened so that it absorbs radiant energy. { 'rād·ē·ōat'mäm·əd·ər }

radioautography See autoradiography. { ¦rād·ē·ō‚ó'täg·rə·fē }

radiobioassay [BIOL] The analysis of the kind, concentration, and location of radioactive material in the human body by direct measurement (in vivo counting) or the evaluation of materials removed (or excreted). { ‚rād·ē·ō‚bī·ō'as‚ā }

radiobiology [BIOL] Study of the scientific principles, mechanisms, and effects of the interaction of ionizing radiation with living matter. Also known as radiation biology. { 'rād·ē·ō·bī'äl·ə·jē }

radiocarbon See carbon-14. { ¦rad·ē·ō'kär·bən }

radiocarbon dating See carbon-14 dating. { ¦rad·ē·ō'kär·bən 'dād·iŋ }

radiocesium See cesium-137. { ¦rad·ē·ō'sē·zē·əm }

radiochronology [GEOL] An absolute-age dating method based on the existing ratio between radioactive parent elements (such as uranium-238) and their radiogenic daughter isotopes (such as lead-206). { ¦rad·ē·ō·krə'näl·ə·jē }

radio climatology [CLIMATOL] The study of regional and seasonal variations in the manner of propagation of radio energy through the atmosphere. { 'rād·ē·ō 'klī·mə'täl·ə·jē }

radiodermatitis [MED] Degenerative changes in the skin following excessive exposure to ionizing radiation. Also known as radiation dermatitis. { ¦rad·ē·ō‚dər·mə'tīd·əs }

radioecology [ECOL] The interdisciplinary study of organisms, radionuclides, ionizing radiation, and the environment. { ¦rād·ē·ō·ē'käl·ə·jē }

radioelectric meteorology See radio meteorology. { ¦rād·ē·ō·i'lek·trik ‚mēd·ē·ə'räl·ə·jē }

radiogenic age determination See radiometric dating. { ¦rād·ē·ō¦jen·ik 'āj di‚tərm·ə‚nā·shən }

radiogenic dating See radiometric dating. { ¦rād·ē·ō¦jen·ik 'dād·iŋ }

radiogeology [GEOCHEM] The study of the distribution patterns of radioactive elements in the earth's crust and the role of radioactive processes in geologic phenomena. { ¦rād·ē·ō·jē'äl·ə·jē }

radiohydrology [ENG] The study of the hydrologic relationships of extraction, processing, and use (including use in hydrologic investigations) of radioactive materials and the disposal of associated waste products. { ¦rād·ē·ō·hī'dräl·ə·jē }

radiological [PHYS] Pertaining to nuclear radiation, radioactivity, and atomic weapons. { ¦rād·ē·ə¦läj·ə·kəl }

radiological agent [ENG] Any of a family of substances that produce casualties by emitting radiation. { ¦rād·ē·ə¦läj·ə·kəl 'ā·jənt }

radiological dose [ENG] The total amount of ionizing radiation absorbed by an individual exposed to any radiating source. { ¦rād·ē·ə¦läj·ə·kəl 'dōs }

radiological survey [ENG] Determination of the distribution and dose rates of radiation in an area. { ¦rād·ē·ə¦läj·ə·kəl 'sər‚vā }

radio meteorology [METEOROL] That branch of the science of meteorology which embraces the propagation of radio energy through the atmosphere, and the use of radio and radar equipment in meteorology; this is the most general term and includes radar meteorology. Also known as radioelectric meteorology. { 'rād·ē·ō ‚mēd·ē·ə'räl·ə·jē }

radiometric dating [PHYS] A technique for measuring the age of an object or sample of material by determining the ratio of the concentration of a radioisotope to that of a stable isotope in it; for example, the ratio of carbon-14 to carbon-12 reveals the approximate age of bones, pieces of wood, and other archeological specimens. Also known as isotopic age determination; nuclear age determination; radioactive age determination; radioactive dating; radiogenic age determination; radiogenic dating. { ‚rād·ē·ə‚me·trik 'dād·iŋ }

radiomimetic activity [BIOL] The radiationlike effects of certain chemicals, such as nitrogen mustard, urethane, and fluorinated pyrimidines. { ¦rād·ē·ō·mi'med·ik ak'tiv·əd·ē }

radiomimetic substances [CHEM] Chemical substances which cause biological effects similar to those caused by ionizing radiation. { ¦rād·ē·ō·mi'med·ik 'səb·stəns·əz }

radiomutation [GEN] A chromosomal aberration which is the result of exposure of living tissue to ionizing radiation. { ¦rād·ē·ō·myü'tā·shən }

radionecrosis [MED] Destruction of living tissue by radiation. { ¦rād·ē·ō·ne'krō·səs }

radiopasteurization [ENG] Pasteurization by surface treatment with low-energy irradiation. { ¦rād·ē·ō‚pas·chúr·ə'zā·shən }

radioresistance [BIOL] The resistance of organisms or tissues to the harmful effects of various radiations. { ¦rād·ē·ō·ri'zis·təns }

radiosonde [ENG] A balloon-borne instrument for the simultaneous measurement and transmission of meteorological data. { 'rād·ē·ō‚sänd }

radiosonde observation [METEOROL] An evaluation in terms of temperature, relative humidity, and pressure aloft, of radio signals received from a balloon-borne radiosonde; the height of each mandatory and significant pressure level of the observation is computed from these data. Also known as raob. { 'rād·ē·ō‚sänd ‚äb·zər ‚vā·shən }

radiotoxicity [MED] A radioactive compound that is toxic to living cells or tissues, causing radiation sickness. { ‚rād·ē·ō·täk'sis·əd·ē }

radium [CHEM] **1.** A radioactive member of group II, symbol Ra, atomic number 88; the most abundant naturally occurring isotope has mass number 226 and a half-life of 1620 years. **2.** A highly toxic solid that forms water-soluble compounds; decays by emission of α, β, and γ-radiation; melts at 700°C, boils at 1140°C; turns black in air; used in medicine, in industrial radiography, and as a source of neutrons and radon. { 'rād·ē·əm }

radium F See polonium-210. { 'rād·ē·əm 'ef }

raft [HYD] An accumulation or jam of floating logs, driftwood, dislodged trees, or other debris, formed naturally in a stream by caving of the banks. { raft }

rafted ice [OCEANOGR] A form of pressure ice composed of overlying pieces of ice floe. { 'raf·təd 'īs }

rafting [OCEANOGR] The process of forming rafted ice. { 'raft·iŋ }

raft lake [HYD] A relatively short-lived body of water impounded along a stream by a raft. { 'raft ‚lāk }

raft tectonics See plate tectonics. { 'raft tek‚tän·iks }

ragged ceiling See indefinite ceiling. { 'rag·əd'sēl·iŋ }

rain [METEOROL] Precipitation in the form of liquid water drops with diameters greater than 0.5 millimeter, or if widely scattered the drops may be smaller; the only other form of liquid precipitation is drizzle. { rān }

rain and snow mixed [METEOROL] Precipitation consisting of a mixture of rain and wet snow; usually occurs when the temperature of the air layer near the ground is slightly above freezing. { 'rān ən 'snō'mikst }

rain area report *See* radar report. { 'rān ¦er·ē·ə ri,pȯrt }

rain cloud [METEOROL] Any cloud from which rain falls; a popular term having no technical denotation. { 'rān ,klaủd }

rain crust [HYD] A type of snow crust, formed by refreezing after surface snow crystals have been melted and wetted by liquid precipitation; composed of individual ice particles such as firn. { 'rān ,krəst }

rain desert [ECOL] A desert in which rainfall is sufficient to maintain a sparse general vegetation. { 'rān ,dez·ərt }

raindrop [METEOROL] A drop of water of diameter greater than 0.5 millimeter falling through the atmosphere. { 'rān,dräp }

rain factor [HYD] A coefficient designed to measure the combined effect of temperature and moisture on the formation of soil humus; it is obtained by dividing the annual rainfall (in millimeters) by the mean annual temperature (in degrees Celsius). { 'rān ,fak·tər }

rainfall [METEOROL] The amount of precipitation of any type; usually taken as that amount which is measured by means of a rain gage (thus a small, varying amount of direct condensation is included). { 'rān,fȯl }

rainfall frequency [CLIMATOL] The number of times, during a specified period of years, that precipitation of a certain magnitude or greater occurs or will occur at a weather station; numerically, the reciprocal of the frequency is usually given. { 'rān ,fȯl ,frē·kwən·sē }

rainfall intensity *See* precipitation intensity. { 'rān,fȯl in'ten·səd·ē }

rainfall inversion *See* precipitation inversion. { 'rān,fȯl in,vər·zhən }

rainfall penetration [HYD] The depth below the soil surface to which water from a given rainfall has been able to infiltrate. { 'rān,fȯl ,pen·ə,trā·shən }

rainfall regime [CLIMATOL] The character of the seasonal distribution of rainfall at any place; the chief rainfall regimes, as defined by W. G. Kendrew, are equatorial, tropical, monsoonal, oceanic and continental westerlies, and Mediterranean. { 'rān ,fȯl rə,zhēm }

rainforest [ECOL] A forest of broad-leaved, mainly evergreen, trees found in continually moist climates in the tropics, subtropics, and some parts of the temperate zones. { 'rān,fär·əst }

rainforest climate *See* wet climate. { 'rān,fär·əst ,klī·mət }

rain gage [ENG] An instrument designed to collect and measure the amount of rain that has fallen. Also known as ombrometer; pluviometer; udometer. { 'rān ,gāj }

rain gush *See* cloudburst. { 'rān ,gəsh }

rain gust *See* cloudburst. { 'rān ,gəst }

raininess [METEOROL] Generally, the quantitative character of rainfall for a given place. { 'rān·ē·nəs }

rain-intensity gage [ENG] An instrument which measures the instantaneous rate at which rain is falling on a given surface. Also known as rate-of-rainfall gage. { 'rān in 'ten·səd·ē ,gāj }

rainmaking [METEOROL] Popular term applied to all activities designed to increase, through any artificial means, the amount of precipitation released from a cloud. { 'rān,māk·iŋ }

rain shadow [METEOROL] An area of diminished precipitation on the lee side of mountains or other topographic obstacles. { 'rān ,shad·ō }

rainsquall [METEOROL] A squall associated with heavy convective clouds, frequently the cumulonimbus type; usually sets in shortly before the thunderstorm rain, blowing outward from the storm and generally lasting only a short time. Also known as thundersquall. { 'rān,skwȯl }

rainwash [GEOL] **1.** The washing away of loose surface material by rainwater after it has reached the ground but before it has been concentrated into definite streams. **2.** Material transported and accumulated, or washed away, by rainwater. { 'rān,wäsh }

rainwater [HYD] Water that has fallen as rain and is quite soft, as it has not yet collected soluble matter from the soil. { 'rān,wȯd·ər }

rainy climate [CLIMATOL] In W. Koppen's climatic classification, any climate type other than the dry climates; however, it is generally understood that this refers principally to the tree climates and not the polar climates. { 'rān·ē 'klī·mət }

rainy season [CLIMATOL] In certain types of climate, an annually recurring period of one or more months during which precipitation is a maximum for that region. Also known as wet season. { 'rān·ē ,sēz·ən }

raised bog [ECOL] An area of acid, peaty soil, especially that developed from moss, in which the center is relatively higher than the margins. { 'rāzd 'bäg }

ram [HYD] An underwater ledge or projection from an ice wall, ice front, iceberg, or floe, usually caused by the more intensive melting and erosion of the unsubmerged part. Also known as apron; spur. { ram }

ramicolous [BOT] Living on twigs. { rə'mik·ə·ləs }

ramp [HYD] An accumulation of snow forming an inclined plane between land or land ice and sea ice or shelf ice. Also known as drift ice foot. { ramp }

randkluft [HYD] A crevasse at the head of a mountain glacier, separating the moving ice and snow from the surrounding rock wall of the valley, where no ice apron is present. { 'ränt,klúft }

random forecast [METEOROL] A forecast in which one of a set of meteorological contingencies is selected on the basis of chance; it is often used as a standard of comparison in determining the degree of skill of another forecast method. { 'ran·dəm 'fȯr,kast }

random sampling [STAT] A sampling from some population where each entry has an equal chance of being drawn. { 'ran·dəm 'sam·pliŋ }

range [ECOL] The area or region over which a species is distributed. { rānj }

range of tide [OCEANOGR] The difference in height between consecutive high and low tides at a place. { 'rānj əv 'tīd }

raob See radiosonde observation. { 'rā,äb }

rapid [HYD] A portion of a stream in swift, disturbed motion, but without cascade or waterfall; usually used in the plural. { 'rap·əd }

rapid flow [HYD] Water flow whose velocity exceeds the velocity of propagation of a long surface wave in still water. Also known as supercritical flow. { 'rap·əd 'flō }

rapid wasting syndrome [ZOO] A coral reef disease that is characterized by a rapid loss of tissue and destruction of the underlying skeleton. { 'rap·əd 'wāst·iŋ ,sin,drōm }

RAREP See radar report. { 'rer,ep }

rate-of-rainfall gage *See* rain-intensity gage. { 'rāt əv 'rān,fȯl ,gāj }

rate of sedimentation [GEOL] The amount of sediment accumulated in an aquatic environment over a given period of time, usually expressed as thickness of accumulation per unit time. Also known as sedimentation rate. { 'rāt əv ,sed·ə·mən 'tā·shən }

rating curve [HYD] For a given point on a stream, a graph of discharge versus stage. { 'rād·iŋ ,kərv }

rational formula [HYD] The expression of peak discharge as equal to the product of rainfall, drainage area, and a runoff coefficient depending on drainage-basin characteristics. { 'rash·ən·əl 'fȯr·myə·lə }

ratio of rise [OCEANOGR] The ratio of the height of tide at two places. { 'rā·shō əv 'rīz }

ravine [GEOGR] A small and narrow valley with steeply sloping sides. { rə'vēn }

raw humus *See* ectohumus. { 'rȯ 'hyü·məs }

rawin [METEOROL] A method of winds-aloft observation, that is, the determination of wind speeds and directions in the atmosphere above a station; accomplished by tracking a balloon-borne radar target, responder, or radiosonde transmitter with either radar or a radio direction finder. { 'rā,win }

rawinsonde [METEOROL] A method of upper-air observation consisting of an evaluation of the wind speed and direction, temperature, pressure, and relative humidity aloft by means of a balloon-borne radiosonde tracked by a radar or radio direction finder. { 'rā·wən,sänd }

raw material [ENG] A crude, unprocessed or partially processed material used as feedstock for a processing operation; for example, crude petroleum, raw cotton, or steel scrap. Also known as crude material. { 'rȯ mə'tir·ē·əl }

raw sewage [CIV ENG] Untreated waste materials. { 'rȯ 'sü·ij }

raw sludge [CIV ENG] Sewage sludge preliminary to primary and secondary treatment processes. { 'rȯ 'sləj }

reach [HYD] A straight, continuous, or extended part of a river, stream, or restricted waterway. { rēch }

reaction wood [BOT] An abnormal development of a tree and therefore its wood as the result of unusual forces acting on it, such as an atypical gravitational pull. { rē'ak·shən ,wud }

recession [HYD] The gradual upstream retreat of a waterfall due to its erosion of the underlying rock ledge. { ri'sesh·ən }

recession curve [HYD] A hydrograph showing the decrease of the runoff rate after rainfall or the melting of snow. { ri'sesh·ən ,kərv }

recharge [HYD] **1.** The processes involved in the replenishment of water to the zone of saturation. **2.** The amount of water added or absorbed. Also known as groundwater increment; groundwater recharge; groundwater replenishment; increment; intake. { rē'chärj }

recharge area [HYD] An area in which water is absorbed that eventually reaches the zone of saturation in one or more aquifers. Also known as intake area. { 'rē,chärj ,er·ē·ə }

recharge basin [CIV ENG] A basin constructed in sandy material to collect water, as from storm drains, for the purpose of replenishing groundwater supply. { 'rē,chärj ,bās·ən }

recharge well [HYD] A well used as a source of water in the process of artificial recharge. Also known as injection well. { 'rē,chärj ,wel }

reclamation |CIV ENG| **1.** The recovery of land or other natural resource that has been abandoned because of fire, water, or other cause. **2.** Reclaiming dry land by irrigation. { ‚rek·lə'mā·shən }

recording rain gage [ENG] A rain gage which automatically records the amount of precipitation collected, as a function of time. Also known as pluviograph. { ri'kȯrd·iŋ 'rān ‚gāj }

record observation [METEOROL] A type of aviation weather observation; the most complete of all such observations and usually taken at regularly specified and equal intervals (hourly, usually on the hour). Also known as hourly observation. { 'rek·ərd ‚äb·zər‚vā·shən }

recovery [HYD] The rise in static water level in a well, occurring upon the cessation of discharge from that well or a nearby well. { ri'kəv·ə·rē }

recurrence interval [HYD] The average time interval between occurrences of a hydrologic event, such as a flood, of a given or greater magnitude. { ri'kər·əns 'int·ər·vəl }

recurvature [METEOROL] With respect to the motion of severe tropical cyclones (hurricanes and typhoons), the change in direction from westward and poleward to eastward and poleward; such recurvature of the path frequently occurs as the storm moves into middle latitudes. { rē'kər·və·chər }

recurved spit See hook. { rē'kərvd 'spit }

red algae [BOT] The common name for members of the phylum Rhodophyta. { 'red 'al·jē }

Reddish-Brown Lateritic soil [GEOL] One of a zonal, lateritic group of soils developed from a mottled red parent material and characterized by a reddish-brown surface horizon and underlying red clay. { 'red·ish ¦braún ‚lad·ə'rid·ik 'sȯil }

Reddish-Brown soil [GEOL] A group of zonal soils having a reddish, light brown surface horizon overlying a heavier, more reddish horizon and a light-colored lime horizon. { 'red·ish ¦braún 'sȯil }

red earth [GEOL] Leached, red, deep, clayey soil that is characteristic of a tropical climate. Also known as red loam. { 'red ¦ərth }

red loam See red earth. { 'red 'lōm }

red rot [PL PATH] Any of several fungus diseases of plants characterized by red patches on stems or leaves; common in sugarcane, sisal, and various evergreen and deciduous trees. { 'red ‚rät }

red rust [PL PATH] An algal disease of certain subtropical plants, such as tea and citrus, caused by the green alga *Cephaleuros virescens* and characterized by a rusty appearance of the leaves or twigs. { 'red 'rəst }

Red Sea [GEOGR] A body of water that lies between Arabia and northeastern Africa, about 1200 miles (2000 kilometers) long, 180 miles (300 kilometers) wide, and a maximum depth of about 7600 feet (2300 meters). { 'red 'sē }

red snow [HYD] A snow surface of reddish color caused by the presence within it of certain microscopic algae or particles of red dust. { 'red 'snō }

red tide [BIOL] A reddish discoloration of coastal surface waters due to concentrations of certain toxin-producing dinoflagellates. Also known as red water. { 'red 'tīd }

redtop grass [BOT] One of the bent grasses, *Agrostis alba* and its relatives, which grow on a wide variety of soils; it is a perennial, spreads slowly by rootstocks, and has top growth 2–3 feet (60–90 centimeters) tall. { 'red‚täp ‚gras }

reduced pressure [METEOROL] The calculated value of atmospheric pressure at mean sea level or some other specified level, as derived (reduced) from station pressure or actual pressure; thus, sea level pressure is nearly always a reduced pressure.   { ri'düst 'presh·ər }

reducer   *See* decomposer.   { ri'dü·sər }

reduction |GEOL| The lowering of a land surface by erosion.   { ri'dək·shən }

reduction of tidal current |OCEANOGR| The processing of observed tidal current data to obtain mean values of tidal current constants.   { ri'dək·shən əv 'tīd·əl ˌkə·rənt }

reduction of tides |OCEANOGR| The processing of observed tidal data to obtain mean values of tidal constants.   { ri'dək·shən əv 'tīdz }

red water   *See* red tide.|VET MED| **1.** A babesiasis of cattle characterized by hematuria following release of hemoglobin by destruction of erythrocytes.   **2.** A chronic disease of cattle attributed to oxalic acid in the forage; hematuria results from escape of blood from lesions in the bladder.   **3.** An acute febrile septicemia of cattle, and sometimes horses, sheep, and swine, caused by the bacterium *Clostridium hemolyticum* and characterized by hemoglobinuria and sometimes intestinal hemorrhages.   { 'red 'wȯd·ər }

redwood |BOT| *Sequoia sempervirens.* An evergreen tree of the pine family; it is the tallest tree in the Americas, attaining 350 feet (107 meters); its soft heartwood is a valuable building material.   { 'red,wu̇d }

Red-Yellow Podzolic soil |GEOL| Any of a group of acidic, zonal soils having a leached, light-colored surface layer and a subsoil containing clay and oxides of aluminum and iron, varying in color from red to yellowish red to a bright yellowish brown.   { 'red 'yel·ō päd'zäl·ik 'sȯil }

reed |BOT| Any tall grass characterized by a slender jointed stem. |TEXT| A comblike loom attachment that keeps the warp yarns apart and pushes the filling thread against the woven fabric.   { rēd }

reef |GEOL| **1.** A ridge- or moundlike layered sedimentary rock structure built almost exclusively by organisms.   **2.** An offshore chain or range of rock or sand at or near the surface of the water.   { rēf }

reef breccia |GEOL| A rock formed by the consolidation of limestone fragments broken off from a reef by the action of waves and tides.   { 'rēf 'brech·ə }

reef cap |GEOL| A deposit of fossil-reef material overlying or covering an island or mountain.   { 'rēf ˌkap }

reef cluster |GEOL| A group of reefs of wholly or partly contemporaneous growth, found within a circumscribed area or geologic province.   { 'rēf ˌkləs·tər }

reef complex |GEOL| The solid reef core and the heterogeneous and contiguous fragmentary material derived from it by abrasion.   { 'rēf ˌkäm,pleks }

reef conglomerate   *See* reef talus.   { 'rēf kən,gläm·ə·rət }

reef core |GEOL| The rock mass constructed in place, and within the rigid growth lattice formed by reef-building organisms.   { 'rēf ˌkȯr }

reef debris   *See* reef detritus.   { 'rēf də,brē }

reef detritus |GEOL| Fragmental material derived from the erosion of an organic reef. Also known as reef debris.   { 'rēf di,trīd·əs }

reef edge |GEOL| The seaward margin of the reef flat, commonly marked by surge channels.   { 'rēf ˌej }

reef flank |GEOL| The part of the reef that surrounds, interfingers with, and locally overlies the reef core, often indicated by massive or medium beds of reef talus dipping steeply away from the reef core.   { 'rēf ˌflaŋk }

reef flat [GEOL] A flat expanse of dead reef rock which is partly or entirely dry at low tide; shallow pools, potholes, gullies, and patches of coral debris and sand are features of the reef flat. { 'rēf ,flat }

reef front [GEOL] The upper part of the outer or seaward slope of a reef, extending to the reef edge from above the dwindle point of abundant living coral and coralline algae. { 'rēf ,frənt }

reef-front terrace [GEOL] A shelflike or benchlike eroded surface, sometimes veneered with organic growth, sloping seaward to a depth of 8–15 fathoms (15–27 meters). { 'rēf ¦frənt ,ter·əs }

reef knoll [GEOL] **1.** A bioherm or fossil coral reef represented by a small, prominent, rounded hill, up to 330 feet (100 meters) high, consisting of resistant reef material, being either a local exhumation of an original reef feature or a feature produced by later erosion. **2.** A present-day reef in the form of a knoll; a small reef patch developed locally and built upward rather than outward. { 'rēf ,nōl }

reef limestone [GEOL] Limestone composed of the remains of sedentary organisms such as sponges, and of sediment-binding organic constituents such as calcareous algae. Also known as coral rock. { 'rēf 'līm,stōn }

reef patch [GEOL] A single large colony of coral formed independently on a shelf at depths less than 220 feet (70 meters) in the lagoon of a barrier reef or of an atoll. Also known as patch reef. { 'rēf ,pach }

reef pinnacle [GEOL] A small, isolated spire of rock or coral, especially a small reef patch. { 'rēf ,pin·ə·kəl }

reef rock [GEOL] A hard, unstratified rock composed of sand, shale, and the calcareous remains of sedentary organisms, cemented by calcium carbonate. { 'rēf ,räk }

reef segment [GEOL] A part of an organic reef lying between passes, gaps, or channels. { 'rēf ,seg·mənt }

reef slope [GEOL] The face of a reef rising from the sea floor. { 'rēf ,slōp }

reef talus [GEOL] Massive inclined strata composed of reef detritus deposited along the seaward margin of an organic reef. Also known as reef conglomerate. { 'rēf ,tā·ləs }

reef wall [GEOL] A wall-like upgrowth of living coral and the skeletal remains of dead coral and other reef-building organisms, which reaches an intertidal level and acts as a partial barrier between adjacent environments. { 'rēf ,wȯl }

reference level [OCEANOGR] **1.** Level of no motion. **2.** A level for which current is known; allows determination of absolute current from relative current. { 'ref·rəns ,lev·əl }

reference plane *See* datum plane. { 'ref·rəns ,plān }

reference station [OCEANOGR] **1.** A place for which independent daily predictions are given in the tide or current tables, from which corresponding predictions are obtained for other stations by means of differences or factors. **2.** A place for which tidal or tidal current constants have been determined and which is used as a standard for the comparison of simultaneous observations at a second station. Also known as standard station. { 'ref·rəns ,stā·shən }

reforestation [FOR] Establishment of a new forest by seeding or planting seedlings on forest land that fails to restock naturally. { rē,fär·ə'stā·shən }

refraction coefficient [OCEANOGR] The square root of the ratio of the spacing between orthogonals in deep water and in shallow water; it is a measure of the effect of refraction in diminishing wave height by increasing the length of the wave crest. { ri'frak·shən ,kō·i,fish·ənt }

refraction diagram [OCEANOGR] A chart showing the position of the wave crests at a particular time, or the successive positions of a particular wave crest as it moves shoreward. { ri'frak·shən ˌdī·ə,gram }

refrigerant [ENG] A substance that by undergoing a change in phase (liquid to gas, gas to liquid) releases or absorbs a large latent heat in relation to its volume, and thus effects a considerable cooling effect; examples are ammonia, sulfur dioxide, ethyl or methyl chloride (these are no longer widely used), and the fluorocarbons, such as Freon, Ucon, and Genetron. { ri'frij·ə·rənt }

refugium [ECOL] An area that has escaped great changes that have occurred in the region as a whole, often providing conditions in which relic colonies can survive; for example, an area which has escaped the effects of glaciation because it projected above the ice. { rə'fyü·jē·əm }

regelation [HYD] Phenomenon in which ice melts at the bottom of droplets of highly concentrated saline solution that are trapped in ice which has frozen over polar waters, and freezes at the top of these droplets, so that the droplets move downward through the ice, leaving it hard and clear. { ˌrē·jə'lā·shən }

regenerated glacier [HYD] A glacier that becomes active after a period of stagnation. { rē'jen·ə,rād·əd 'glā·shər }

regimen [HYD] **1.** The behavior characteristic of the total amount of water involved in a drainage basin. **2.** Analysis of the total volume of water involved with a lake, including water losses and gains, over a period of a year. **3.** The flow characteristics of a stream with respect to velocity, volume, form and alterations in the channel, capacity to transport sediment, and the amount of material supplied for transportation. { 'rej·ə·mən }

regional forecast See area forecast. { 'rēj·ən·əl 'fȯr,kast }

regional snowline [HYD] The level above which, averaged over a large area, snow accumulation exceeds ablation year after year. { 'rēj·ən·əl 'snō,līn }

region of escape See exosphere. { 'rē·jən əv i'skāp }

regolith [GEOL] The layer rock or blanket of unconsolidated rocky debris of any thickness that overlies bedrock and forms the surface of the land. Also known as mantle rock. { 'reg·ə,lith }

regression [OCEANOGR] Retreat of the sea from land areas, and the consequent evidence of such withdrawal. { ri'gresh·ən }

regressive reef [GEOL] One of a series of nearshore reefs or bioherms superimposed on basinal deposits during the rising of a landmass or the lowering of the sea level, and developed more or less parallel to the shore. { ri'gres·iv 'rēf }

regular [BOT] Having radial symmetry, referring to a flower. { 'reg·yə·lər }

reishi mushroom See Ganoderma lucidum. { rā'ē·shē ,məsh,rüm }

rejected recharge [HYD] Water that infiltrates to the water table but then discharges because the aquifer is full and cannot accept it. { ri'jek·təd 'rē,chärj }

rejuvenate [GEOL] The act of stimulating a stream to renewed erosive activity either by tectonic uplift or a drop in sea level. { ri'jü·və,nāt }

rejuvenated stream [HYD] A mature stream that has reverted to the behavior and forms of a more youthful stage due to rejuvenation, usually as a result of uplift. Also known as revived stream. { ri'jü·və,nād·əd 'strēm }

rejuvenated water [HYD] Water returned to the terrestrial water supply as a result of compaction and metamorphism. { ri'jü·və,nād·əd 'wȯd·ər }

rejuvenation |HYD| **1.** The stimulation of a stream to renew erosive activity. **2.** The renewal of youthful vigor (erosive ability) in a mature stream. { ri¡jü·və'nā·shən }

relapsing fever [MED] An acute infectious disease caused by various species of the spirochete *Borrelia*, characterized by episodes of fever which subside spontaneously and recur over a period of weeks. { ri'laps·iŋ ¡fē·vər }

relative contour *See* thickness line. { 'rel·əd·iv 'kän¡túr }

relative current [OCEANOGR] The current which is a function of the dynamic slope of an isobaric surface and which is determined from an assumed layer of no motion. { 'rel·əd·iv 'kə·rənt }

relative humidity [METEOROL] The (dimensionless) ratio of the actual vapor pressure of the air to the saturation vapor pressure. Abbreviated RH. { 'rel·əd·iv hyü'mid·əd·ē }

relative hypsography *See* thickness pattern. { 'rel·əd·iv hip'säg·rə·fē }

relative isohypse *See* thickness line. { 'rel·əd·iv 'ī·sə¡hips }

relative stability test [CHEM] A color test using methylene blue that indicates when the oxygen present in a sewage plant's effluent or polluted water is exhausted. { 'rel·əd·iv stə'bil·əd·ē ¡test }

relative topography *See* thickness pattern. { 'rel·əd·iv tə'päg·rə·fē }

reliability [STAT] **1.** The amount of credence placed in a result. **2.** The precision of a measurement, as measured by the variance of repeated measurements of the same object. { ri¡lī·ə'bil·əd·ē }

relict glacier |HYD| A remnant of an older and larger glacier. { 'rel·ikt 'glā·shər }

reliction [HYD] The slow and gradual withdrawal or recession of the water in a sea, a lake, or a stream, leaving the former bottom as permanently exposed and uncovered dry land. { rə'lik·shən }

relict lake [HYD] A lake that survives in an area formerly covered by the sea or a larger lake, or a lake that represents a remnant resulting from a partial extinction of the original body of water. { 'rel·ikt 'lāk }

relief well [CIV ENG] A well that drains a pervious stratum, to relieve waterlogging at the surface. { ri'lēf ¡wel }

relogging |FOR| An operation in which small trees are salvaged, often for pulpwood, after the large trees are logged. { rē'läg·iŋ }

rem [PHYS] A unit of ionizing radiation, equal to the amount that produces the same damage to humans as 1 roentgen of high-voltage x-rays. Derived from roentgen equivalent man. { rem }

remotely operated vehicle [OCEANOGR] A crewless submersible vehicle that is tethered to a vessel on the surface by a cable; it has a video camera, lights, thrusters that generally provide three-dimensional maneuverability, depth sensors, and a wide array of manipulative and acoustic devices, as well as special instrumentation to perform a variety of work tasks. Abbreviated ROV. { rə¡mōt·lē ¡äp·ə¡rād·əd 've·ə·kəl }

renewable energy source [ENG] A form of energy that is constantly and rapidly renewed by natural processes such as solar, ocean wave, and wind power. { ri¡nü·ə·bəl 'en·ər·jē ¡sòrs }

reniform [SCI TECH] Bean- or kidney-shaped, as describing the structure of a crystal in which rounded masses occur at the ends of radiating crystals, or certain structures in animals and plants. { 'rēn·ə¡fórm }

reovirus [MICROBIO] A group of ribonucleic acid-containing animal viruses, including agents of encephalitis and phlebotomus fever. { 'rē·ō¡vī·rəs }

repair synthesis   [BIOL] Enzymatic excision and replacement of regions of damaged deoxyribonucleic acid, as in repair of thymine dimers by ultraviolet irradiation.   { ri'per ,sin·thə·səs }

repent   [BOT] Of a stem, creeping along the ground and rooting at the nodes.   { 'rē·pent }

repi   [HYD] A lake, pond, or other standing water body associated with a sink or subsidence of land surface.   { 'rep·ē }

representative sample   [STAT] A sample whose characteristics reflect those of the population from which it is drawn.   { ¦rep·ri¦zen·təd·iv 'sam·pəl }

reproduction   [BIOL] The mechanisms by which organisms give rise to other organisms of the same kind.   { ¦rē·prə¦dək·shən }

reproductive distribution   [ECOL] The range of areas where conditions are favorable to maturation, spawning, and early development of marine animals.   { ¦rē·prə¦dək·tiv ,dis·trə'byü·shən }

reproductive toxicity   [MED] Adverse effects on the male and/or female reproductive systems caused by exposure to a toxic chemical. It may be expressed as alterations in sexual behavior, decreases in fertility, or fetal loss during pregnancy. Developmental toxicity may also be included.   { ,rē·prə,dək·tiv täk'sis·əd·ē }

réseau   [METEOROL] The term adopted by the World Meteorological Organization for the worldwide network of meteorological stations which have been chosen to represent the meteorology of the globe (*réseau mondial*).   { rā'zō }

reservoir   [CIV ENG] A pond or lake built for storage of water, usually by the construction of a dam across a river.   [GEOL] **1.** A subsurface accumulation of crude oil or natural gas under trap conditions.   **2.** An area covered by neve where snow collects to form a glacier.   **3.** A space within the earth that is occupied by magma.   { 'rez·əv,wär }

residual   [GEOL] **1.** Of a mineral deposit, formed by either mechanical or chemical concentration.   **2.** Pertaining to a residue left in place after weathering of rock.   **3.** Of a topographic feature, representing the remains of a formerly great mass or area and rising above the surrounding surface.   { rə'zij·ə·wəl }

resistance factor   *See* R factor.   { ri'zis·təns ,fak·tər }

resonance trough   [METEOROL] A large-scale pressure trough which forms at an appropriate wavelength away from a dominant trough; for example, the mean trough over the Mediterranean in winter is often considered a resonance trough between the two more dynamically active troughs along the east coasts of North America and Asia.   { 'rez·ən·əns ,tróf }

resorbed reef   [GEOL] A reef characterized by embayed margins and by the numerous isolated patches of reef that are closely distributed about the main mass.   { rē'sòrbd 'rēf }

resource   [SCI TECH] A reserve source of supply, such as a material or mineral.   { 'rē ,sórs }

respiration   [BIOL] **1.** The processes by which tissues and organisms exchange gases with their environment.   **2.** The act of breathing with the lungs, consisting of inspiration and expiration.   { ,res·pə'rā·shən }

respirator   [ENG] A device for maintaining artificial respiration to protect the respiratory tract against irritating and poisonous gases, fumes, smoke, and dusts, with or without equipment supplying oxygen or air; some types have a fitting which covers the nose and mouth.   { 'res·pə,rād·ər }

respiratory syncytial virus   [MICROBIO] An enveloped, single-stranded RNA animal virus belonging to the Paramyxoviridae genus *Pneumovirus*; associated with a large proportion of respiratory illnesses in very young children, particularly bronchiolitis and pneumonia.   { 'res·prə,tòr·ē sin'sish·əl 'vī·rəs }

restoration [ECOL] A conservation measure involving the correction of past abuses that have impaired the productivity of the resources base. { ‚res·tə'rā·shən }

restoration ecology [ECOL] The application of ecological principles and field methodologies to the successful restoration of damaged ecosystems. { ‚res·tə‚rā·shən ē'käl·ə·jē }

resultant wind [CLIMATOL] The vectorial average of all wind directions and speeds for a given level at a given place for a certain period, such as a month. { ri'zəlt·ənt 'wind }

resurgence [HYD] The point where an underground stream reappears at the surface to become a surface stream. Also known as emergence; exsurgence; rise. { ri'sər·jəns }

resurrected [HYD] Pertaining to a stream that follows an earlier drainage system after a period of brief submergence has slightly masked the old course by a thin film of sediments. Also known a palingenetic. { ¦rez·ə¦rek·təd }

retained water [HYD] The water remaining in rock or soil after gravity groundwater has been drained out. { ri'tānd 'wȯd·ər }

retardation [OCEANOGR] The amount of time by which corresponding tidal phases grow later day by day, averaging approximately 50 minutes. { ‚rē‚tär'dā·shən }

retrograde wave [METEOROL] An atmospheric wave which moves in a direction opposite to that of the flow in which the wave is embedded; retrogression of a particular wave on daily charts is rarely seen, but is frequently observed on 4-day or monthly mean charts. { 're·trə‚grād 'wāv }

retrograding shoreline [GEOL] A shoreline that is being moved landward by wave erosion. { 're·trə‚grād·iŋ 'shȯr‚līn }

retrogression [METEOROL] The motion of an atmospheric wave or pressure system in a direction opposite to that of the basic flow in which it is embedded. { ‚re·trə'gresh·ən }

Retroviridae [MICROBIO] A family of ribonucleic acid (RNA)–containing animal viruses characterized by spherical enveloped virions containing two single-stranded RNA molecules and reverse transcriptase; includes the subfamilies Oncovirinae, Spumavirinae, and Lentivirinae. { ‚re·trə'vir·ə‚dī }

retrovirus [MICROBIO] A family of ribonucleic acid viruses distinguished by virions which possess reverse transcriptase and which have two proteinaceous structures, a dense core, and an envelope that surrounds the core. { 're·trō‚vī·rəs }

retting [CHEM ENG] Soaking vegetable stalks to decompose the gummy material and release the fibers. { 'red·iŋ }

return flow [HYD] Irrigation water not consumed by evapotranspiration but returned to its source or to another body of ground or surface water. Also known as return water. { ri'tərn ‚flō }

return water See return flow. { ri'tərn ‚wȯd·ər }

reverse cell [METEOROL] A circulating fluid system in which the circulation in a vertical plane is thermally indirect; that is, cooler air rises relative to warmer air. { ri'vərs 'sel }

reversed stream [HYD] A stream whose direction of flow has been reversed, as by glacial action, landsliding, gradual tilting of a region, or capture. { ri'vərst 'strēm }

reversed tide [OCEANOGR] An oceanic tide that is out of phase with the apparent motions of the tide-producing body, so that low tide is directly under the tide-producing body and is accompanied by a low tide on the opposite side of the earth. Also known as inverted tide. { ri'vərst 'tīd }

reverse osmosis [CHEM ENG] A technique used in desalination and waste-water treatment; pressure is applied to the surface of a saline (or waste) solution, forcing pure water to pass from the solution through a membrane (hollow fibers of cellulose acetate or nylon) that will not pass sodium or chloride ions. { ri'vərs äs'mō·səs }

reversing current [OCEANOGR] Any current that changes direction, with a period of slack water at each reversal of direction. { ri'vərs·iŋ ˌkə·rənt }

reversing water bottle *See* Nansen bottle. { ri'vərs·iŋ 'wȯd·ər ˌbäd·əl }

revived stream *See* rejuvenated stream. { ri'vīvd 'strēm }

revolving storm [METEOROL] A cyclonic storm, or one in which the wind revolves about a central low-pressure area. { ri'välv·iŋ 'stȯrm }

Reynolds effect [METEOROL] A process of drop growth in clouds which involves net evaporation from cloud drops warmer than others and net condensation on the cooler drops. { 'ren·əlz iˌfekt }

Reynolds model [OCEANOGR] A laboratory model of ocean currents in which inertial forces and frictional forces predominate, and in which the Reynolds number is used extensively in calculations. { 'ren·əlz ˌmäd·əl }

R factor [GEN] A self-replicating, infectious agent that carries genetic information and transmits drug resistance from bacterium to bacterium by conjugation of cell. Also known as resistance factor. { 'är ˌfak·tər }

RH *See* relative humidity.

rhabdovirus [MICROBIO] A group of ribonucleic acid-containing animal viruses, including rabies virus and certain infective agents of fish and insects. { ¦rab·dō'vī·rəs }

rheophile [ECOL] Living or thriving in running water. { 'rē·əˌfīl }

rheophilous bog [ECOL] A bog which draws its source of water from drainage. { rē'äf·ə·ləs 'bäg }

rheoplankton [ECOL] Plankton found in flowing water. { 'rē·ōˌplaŋk·tən }

rheotaxis [BIOL] Movement of a motile cell or organism in response to the direction of water currents. { ¦rē·ə¦tak·səs }

rheotropism [BIOL] Orientation response of an organism to the stimulus of a flowing fluid, as water. { rē'ä·trəˌpiz·əm }

rheumatic fever [MED] A febrile disease occurring in childhood as a delayed sequel of infection by *Streptococcus hemolyticus*, group A; characterized by arthritis, carditis, nosebleeds, and chorea. { rü'mad·ik 'fē·vər }

rhinocerebral mucormycosis [MED] A mold infection of the sinus that spreads rapidly to the eye and brain. { ˌrīn·ō·sə,rēb·rəl ˌmyü·kō·mī'kō·səs }

rhinoscleroma [MED] A chronic infectious bacterial disease caused by *Klebsiella rhinoscleromatis* and characterized by hard nodules and plaques of inflamed tissue in the nose and adjacent areas. { ˌrīn·ə·sklə'rō·mə }

rhinovirus [MICROBIO] A subgroup of the picornavirus group including small, ribonucleic acid-containing forms which are not inactivated by ether. { ¦rīn·ə'vī·rəs }

rhizanthous [BOT] Producing flowers directly from the root. { rī'zan·thəs }

rhizic water *See* soil water. { 'rīz·ik ˌwȯd·ər }

rhizocarpous [BOT] Pertaining to perennial herbs having perennating underground parts from which stems and foliage arise annually. { ¦rī·zō¦kär·pəs }

rhizoid [BOT] A rootlike structure which helps to hold the plant to a substrate; found on fungi, liverworts, lichens, mosses, and ferns. { 'rī,zȯid }

rhizome [BOT] An underground horizontal stem, often thickened and tuber-shaped, and possessing buds, nodes, and scalelike leaves. { 'rī,zōm }

Rhizophagidae [ZOO] The root-eating beetles, a family of minute coleopteran insects in the superfamily Cucujoidea. { ˌrī·zō'fā·jəˌdē }

rhizotron [BOT] An underground laboratory system designed for examining plant root growth; contains enclosed columns of soil with transparent plastic windows which permit viewing, measuring, and photographing. { 'rīz·ə,trän }

rhodanic acid *See* thiocyanic acid. { rō'dan·ik 'as·əd }

Rhodophyta [BOT] The red algae, a large diverse phylum or division of plants distinguished by having an abundance of the pigment phycoerythrin. { rō'däf·əd·ə }

ria [GEOGR] **1.** Any broad, estuarine river mouth. **2.** A long, narrow coastal inlet, except a fjord, whose depth and width gradually and uniformly diminish inland. { 'rē·ə }

ria coast [GEOGR] A coast with several parallel rias extending far inland and alternating with ridgelike promontories. { 'rē·ə 'kōst }

ria shoreline [GEOGR] A type of coastline developed along a drowning landmass in which numerous long and narrow arms of the sea extend inland parallel with one another and perpendicular to the coastline. { 'rē·ə 'shȯr,līn }

ribbon reef [GEOL] A linear reef within the Great Barrier Reef off the northeast coast of Australia, having inwardly curved extremities, and forming a festoon along the precipitous edge of the continental shelf. { 'rib·ən ,rēf }

rice [BOT] *Oryza sativa*. An annual cereal grass plant of the order Cyperales, cultivated as a source of human food for its carbohydrate-rich grain. { rīs }

ricin [MED] White, poisonous powder derived from pressed castor oil bran. { 'rīs·ən }

Rickettsiales [MICROBIO] An order of prokaryotic microorganisms; gram-negative, obligate, intracellular animal parasites (may be grown in tissue cultures); many cause disease in humans and animals. { ri,ket·sē'ā·lēz }

Rickettsieae [MICROBIO] A tribe of the family Rickettsiaceae; cells are occasionally filamentous; infect arthropods and some vertebrates and are pathogenic for humans, most frequently an incidental host. { ri'ket·sē,ē }

ridge [METEOROL] An elongated area of relatively high atmospheric pressure, almost always associated with, and most clearly identified as, an area of maximum anticyclonic curvature of wind flow. Also known as wedge. { rij }

ridge aloft *See* upper-level ridge. { 'rij ə'lȯft }

ridged ice [OCEANOGR] Sea ice having readily observed surface features in the form of one or more pressure ridges. { 'rijd 'īs }

ridging [OCEANOGR] A form of deformation of floating ice, caused by lateral pressure, whereby ice is forced or piled haphazardly to form ridged ice. { 'rij·iŋ }

riegel [GEOL] A low, traverse ridge of bedrock on the floor of a glacial valley. Also known as rock bar; threshold; verrou. { 'rē·gəl }

riffle [HYD] **1.** A shallows across a stream bed over which water flows swiftly and is broken into waves by submerged obstructions. **2.** Shallow water flowing over a riffle. { 'rif·əl }

rift [GEOL] **1.** A narrow opening in a rock caused by cracking or splitting. **2.** A high, narrow passage in a cave. { rift }

rill [HYD] A small brook or stream. { ril }

rill erosion [GEOL] The formation of numerous, closely spaced rills due to the uneven removal of surface soil by streamlets of running water. { 'ril i'rō·zhən }

rill flow [HYD] Surface runoff flowing in small irregular channels too small to be considered rivulets. { 'ril ,flō }

rime [HYD] A white or milky and opaque granular deposit of ice formed by the rapid freezing of supercooled water drops as they impinge upon an exposed object;

composed essentially of discrete ice granules, and has densities as low as 0.2–0.3 gram per cubic centimeter. { rīm }

rime fog See ice fog. { 'rīm ˌfäg }

rinderpest [VET MED] An acute, contagious, and often fatal virus disease of cattle, sheep, and goats which is characterized by fever and the appearance of ulcers on the mucous membranes of the intestinal tract. { 'rin·dərˌpest }

ring rot [PL PATH] **1.** A fungus disease of the sweet potato root caused by *Rhizopus stolonifer* and marked by rings of dry rot. **2.** A bacterial disease of potatoes caused by *Corynebacterium sepedonicum* and characterized by brown discoloration of the annular vascular tissue. { 'riŋ ˌrät }

ring spot [PL PATH] Any of various virus and fungus diseases of plants characterized by the appearance of a discolored, annular (ring-shaped) lesion. { 'riŋ ˌspät }

ringworm [MED] A fungus infection of skin, hair, or nails producing annular (ring-shaped) lesions with elevated margins. Also known as tinea. { 'riŋˌwərm }

rip [OCEANOGR] A turbulent agitation of water generally caused by the interaction of currents and wind. { rip }

riparian [BIOL] Living or located on a riverbank. { rə'per·ē·ən }

riparian water loss [HYD] Discharge of water through evapo-transpiration along a watercourse, especially water transpired by vegetation growing along the watercourse. { rə'per·ē·ən 'wȯd·ər ˌlȯs }

riparian zone [ECOL] The part of the watershed immediately adjacent to the stream channel. { rī'per·ē·ən ˌzōn }

rip current [OCEANOGR] The return flow of water piled up on shore by incoming waves and wind. { 'rip ˌkə·rənt }

ripe [HYD] Descriptive of snow that is in a condition to discharge meltwater; ripe snow usually has a coarse crystalline structure, a snow density near 0.5, and a temperature near 32°F (0°C). { rīp }

ripple [OCEANOGR] A small curling or undulating wave controlled to a significant degree by both surface tension and gravity. { 'rip·əl }

rips [OCEANOGR] A turbulent agitation of water, generally caused by the interaction of currents and wind; in nearshore regions they may be currents flowing swiftly over an irregular bottom; sometimes referred to erroneously as tide rips. { rips }

rip tide See rip current. { 'rip ˌtīd }

rise See resurgence. { rīz }

rise of tide [OCEANOGR] Vertical distance from the chart datum to a higher water datum. { 'rīz əv 'tīd }

rising limb [HYD] The rising portion of the hydrograph resulting from runoff of rainfall or snowmelt. { 'rīz·iŋ 'lim }

rising tide [OCEANOGR] The portion of the tide cycle between low water and the following high water. { 'rīz·iŋ 'tīd }

river [HYD] A large, natural freshwater surface stream having a permanent or seasonal flow and moving toward a sea, lake, or another river in a definite channel. { 'riv·ər }

river basin [GEOL] The area drained by a river and all of its tributaries. { 'riv·ər ˌbās·ən }

riverbed [GEOL] The channel which contains, or formerly contained, a river. { 'riv·ər ˌbed }

river bottom |GEOL| The low-lying alluvial land along a river. Also known as river flat. { 'riv·ər ,bäd·əm }

river breathing |HYD| Fluctuation of the water level of a river. { 'riv·ər ,brēth·iŋ }

river capture See capture. { 'riv·ər ,kap·chər }

river end |HYD| The lowest point of a river with no outlet to the sea, situated where its water disappears by percolation or evaporation. { 'riv·ər ,end }

river flat See river bottom. { 'riv·ər ,flat }

river forecast |HYD| A forecast of the expected stage or discharge at a specified time, or of the total volume of flow within a specified time interval, at one or more points along a stream. { 'riv·ər ,fȯr,kast }

river ice |HYD| Any ice formed in or carried by a river. { 'riv·ər ,īs }

river morphology |GEOL| The study of the channel pattern and the channel geometry at several points along a river channel, including the network of tributaries within the drainage basin. Also known as channel morphology; fluviomorphology; stream morphology. { 'riv·ər mȯr'fäl·ə·jē }

river piracy See capture. { 'riv·ər ,pī·rə·sē }

river plain See alluvial plain. { 'riv·ər ,plān }

river system |HYD| The aggregate of stream channels draining a river basin. { 'riv·ər ,sis·təm }

river tide |HYD| A tide that occurs in rivers emptying directly into the sea, showing three characteristic modifications of ocean tides: the speed at which the tide travels upstream depends on the depth of the channel, the further upstream the longer the duration of the falling tide and shorter the duration of the rising tide, and the range of the tide decreases with distance upstream. { 'riv·ər ,tīd }

riverwash |GEOL| **1.** Soil material that has been transported and deposited by rivers. **2.** An alluvial deposit in a river bed or flood channel, subject to erosion and deposition during recurring flood periods. { 'riv·ər,wäsh }

river water |HYD| Water having carbonate, sulfate, and calcium as its main dissolved constituents; distinguished from seawater by its chloride and sodium content. { 'riv·ər ,wȯd·ər }

rivulet |HYD| A small stream; a brook. { 'riv·yə·lət }

roaring forties |METEOROL| A popular nautical term for the stormy ocean regions between 40° and 50° latitude; it usually refers to the Southern Hemisphere, where there is an almost completely uninterrupted belt of ocean with strong prevailing westerly winds. { 'rȯr·iŋ 'fȯr·dēz }

robbery See capture. { 'räb·ə·rē }

rock |GEOL| **1.** A consolidated or unconsolidated aggregate of mineral grains consisting of one or more mineral species and having some degree of chemical and mineralogic constancy. **2.** In the popular sense, a hard, compact material with some coherence, derived from the earth. { räk }

rock bar See riegel. { 'räk ,bär }

rock-basin lake See paternoster lake. { 'räk ¦bas·ən ,lāk }

rock cycle |GEOL| The interrelated sequence of events by which rocks are initially formed, altered, destroyed, and reformed as a result of magmatism, erosion, sedimentation, and metamorphism. { 'räk ,sī·kəl }

rockfall |GEOL| **1.** The fastest-moving landslide; free fall of newly detached bedrock segments from a cliff or other steep slope; usually occurs during spring thaw. **2.** The rock material moving in or moved by a rockfall. { 'räk,fȯl }

rock mechanics |GEOPHYS| Application of the principles of mechanics and geology to quantify the response of rock when it is acted upon by environmental forces, particularly when human-induced factors alter the original ambient forces. { 'räk mi‚kan·iks }

rock system |GEOPHYS| In rock mechanics, all natural environmental factors that can influence the behavior of that portion of the earth's crust that will become part of an engineering structure. { 'räk ‚sis·təm }

rock varnish |GEOL| A dark coating on rock surfaces exposed to the atmosphere. It is composed of about 30% manganese and iron oxides, up to 70% clay minerals, and over a dozen trace and rare-earth minerals. Although found in all terrestrial environments, it is mostly developed and best preserved in arid regions. Also know as desert varnish. { 'räk ‚vär·nəsh }

rock wood *See* mountain wood. { 'räk ‚wud }

Rocky Mountain spotted fever |MED| An acute, infectious, typhuslike disease of humans caused by the rickettsial organism *Rickettsia rickettsi* and transmitted by species of hard-shelled ticks; characterized by sudden onset of chills, headache, fever, and an exanthem (skin eruption) on the extremities. Also known as American spotted fever; tick fever; tick typhus. { 'räk·ē 'maunt·ən 'späd·əd 'fē·vər }

rod weeder |AGR| A type of equipment used to prepare the soil during harrowing; it is a power-driven rod, usually square in cross section, which also operates below the surface of loose soil, killing weeds and maintaining the soil in loose mulched condition; adapted to large operations and used in dry areas in the northwestern United States. { 'räd ‚wēd·ər }

roentgen |PHYS| An exposure dose of x- or γ-radiation such that the electrons and positrons liberated by this radiation produce, in air, when stopped completely, ions carrying positive and negative charges of 2.58×10^{-4} coulomb per kilogram of air. Abbreviated R (formerly r). Also spelled röntgen. { 'rent·gən }

ROFOR |METEOROL| An international code word used to indicate a route forecast (along an air route). { 'rō‚for }

ROFOT |METEOROL| An international code word used to indicate a route forecast, with units in the English system. { 'rō‚fät }

roil |HYD| A small section of a stream, characterized by swiftly flowing, turbulent water. { 'roil }

roily water |HYD| **1.** Muddy or sediment-filled water. **2.** Turbulent, agitated, or swirling water. { 'roil·ē 'wod·ər }

roll cloud *See* rotor cloud. { 'rōl ‚klaud }

roller |OCEANOGR| A long, massive wave which usually retains its form without breaking until it reaches the beach or a shoal. { 'rō·lər }

rollers |OCEANOGR| Swells coming from a great distance and forming large breakers on exposed coasts. { 'rō·lərz }

ROMET |METEOROL| An international code word denoting route forecast, with units in the metric system. { 'rō‚met }

röntgen *See* roentgen. { 'rent·gən }

rooster tail |HYD| A plumelike form of water and sometimes spray that occurs at the intersection of two crossing waves. { 'rüs·tər ‚tāl }

root |BOT| The absorbing and anchoring organ of a vascular plant; it bears neither leaves nor flowers and is usually subterranean. { rüt }

root cap |BOT| A thick, protective mass of parenchymal cells covering the meristematic tip of the root. { 'rüt ‚kap }

root hair |BOT| One of the hairlike outgrowths of the root epidermis that function in absorption. { 'rüt ˌher }

root knot |PL PATH| Any of various plant diseases caused by root-knot nematodes which produce gall-like enlargements on the roots. { 'rüt ˌnät }

root nodule |PL PATH| An abnormal nodular growth on a plant root system caused by the establishment of symbiotic nitrogen-fixing bacteria in the host tissue. { 'rüt 'näjˌül }

root rot |PL PATH| Any of various plant diseases characterized by decay of the roots. { 'rüt ˌrät }

rootstock |BOT| A root or part of a root to which a scion (usually a stem or bud) from another plant is attached in grafting. Also known as stock; understock. { 'rütˌstäk }

rootworm |ZOO| **1.** An insect larva that feeds on plant roots. **2.** A nematode that infests the roots of plants. { 'rütˌwərm }

ropak |OCEANOGR| An ice cake standing on edge as a result of excessive pressure. Also known as turret ice. { 'rōˌpak }

Ross Barrier |OCEANOGR| A wall of shelf ice bordering on the Ross Sea. { 'rós 'bar·ē·ər }

Rossby wave |METEOROL| A large, slow-moving, planetary-scale wave generated in the troposphere by ocean-land temperature contrasts and topographic forcing (winds flowing over mountains), and affected by the Coriolis effect due to the earth's rotation. Rossby waves have also been observed in the ocean. Also known as planetary wave. { 'rós·bē ˌwāv }

Rossel Current |OCEANOGR| A seasonal Pacific Ocean current flowing westward and north-westward along both the southern and northeastern coasts of New Guinea, the southern part flowing through Torres Strait and losing its identity in the Arafura Sea, and the northern part curving northeastward to join the equatorial countercurrent of the Pacific Ocean. { 'rós·əl ˌkə·rənt }

Ross Sea |GEOGR| Arm of the South Pacific Ocean off Antarctica. { 'rós 'sē }

rot |PL PATH| Any plant disease characterized by breakdown and decay of plant tissue. { rät }

rotary current |OCEANOGR| A current with the direction of flow rotating through all points of the compass. { 'rōd·ə·rē 'kə·rənt }

rotating models |OCEANOGR| Laboratory models for studying ocean currents, the models being rotated to simulate in part the earth's rotation. { 'rōˌtād·iŋ ˈmäd·əlz }

röteln *See* rubella. { 're,teln }

rotenone |CHEM| $C_{23}H_{22}O_6$ White crystals with a melting point of 163°C; soluble in ether and acetone; used as an insecticide and in flea powders and fly sprays. Also known as tubatoxin. { 'rōt·ənˌōn }

rotor cloud |METEOROL| Turbulent, altocumulus-type cloud formation found in the lee of some large mountain barriers, particularly in the Sierra Nevadas near Bishop, California; the air in the cloud rotates around an axis parallel to the range. Also known as roll cloud. { 'rōd·ər ˌklaůd }

rotten ice |HYD| Any piece, body, or area of ice which is in the process of melting or disintegrating; it is characterized by honeycomb structure, weak bonding between crystals, or the presence of meltwater or sea water between grains. Also known as spring sludge. { 'rät·ən ˈīs }

rough ice |HYD| An expanse of ice having an uneven surface caused by formation of pressure ice or by growlers frozen in place. { 'rəf 'īs }

roughness length *See* dynamic roughness. { 'rəf·nəs ‚leŋkth }

round wind [METEOROL] A wind that gradually changes direction through approximately 180° during the daylight hours. { 'raúnd ‚wind }

route forecast [METEOROL] An aviation weather forecast for one or more specified air routes. { 'rüt ‚fór‚kast }

ROV *See* remotely operated vehicle.

r selection [ECOL] Selection that favors rapid population growth (r represents the intrinsic rate of increase). { 'är si‚lek·shən }

rubber ice [OCEANOGR] Newly formed sea ice which is weak and elastic. { 'rəb·ər ‚īs }

rubble [HYD] Fragments of floating or grounded sea ice in hard, roughly spherical blocks measuring 0.5–1.5 meters (1.5–4.5 feet) in diameter, and resulting from the breakup of larger ice formations. Also known as rubble ice. { 'rəb·əl }

rubble ice *See* rubble. { 'rəb·əl ‚īs }

rubella [MED] An infectious virus disease of humans characterized by coldlike symptoms, fever, and transient, generalized pale-pink rash; its occurrence in early pregnancy is associated with congenital abnormalities. Also known as epidemic roseola; French measles; German measles; röteln. { rü'bel·ə }

rubeola *See* measles. { ‚rü·bē'ō·lə }

ruderal [ECOL] **1.** Growing on rubbish, or waste or disturbed places. **2.** A plant that thrives in such a habitat. { 'rüd·ə·rəl }

rugose mosaic [PL PATH] A virus disease of potatoes marked by dwarfed, wrinkled, and mottled leaves and resulting in premature death. { 'rü‚gōs mō'zā·ik }

runnel [GEOL] A troughlike hollow on a tidal sand beach which carries water drainage off the beach as the tide retreats. { 'rən·əl }

runner [BOT] A horizontally growing, sympodial stem system; adventitious roots form near the apex, and a new runner emerges from the axil of a reduced leaf. Also known as stolon. { 'rən·ər }

running sand *See* quicksand. { 'rən·iŋ ‚sand }

runoff [HYD] **1.** Surface streams that appear after precipitation. **2.** The flow of water in a stream, usually expressed in cubic feet per second; the net effect of storms, accumulation, transpiration, meltage, seepage, evaporation, and percolation. { 'rən ‚óf }

runoff coefficient [HYD] The percentage of precipitation that appears as runoff. { 'rən ‚óf ‚kō·i‚fish·ənt }

runoff cycle [HYD] The part of the hydrologic cycle involving water between the time it reaches the land as precipitation and its subsequent evapotranspiration or runoff. { 'rən‚óf ‚sī·kəl }

runoff desert [ECOL] An arid region in which local rain is insufficient to support any perennial vegetation except in drainage or runoff channels. { 'rən‚óf ‚dez·ərt }

runoff intensity [HYD] The excess of rainfall intensity over infiltration capacity, usually expressed in inches of rainfall per hour. Also known as runoff rate. { 'rən‚óf in ‚ten·səd·ē }

runoff rate *See* runoff intensity. { 'rən‚óf ‚rāt }

runout [HYD] The location where an avalanche slows down or stops, depositing the avalanche debris. { 'rən‚aút }

run-up *See* swash. { 'rən‚əp }

runway observation   [METEOROL] An evaluation of certain meteorological elements observed at a specified point on or near an airport runway; temperature, wind speed and direction, ceiling, and visibility are among the elements frequently observed at such locations, because of the importance of these data to aircraft landing and takeoff operations.   { 'rən,wā ,äb·zər'vā·shən }

rust   [PL PATH] Any plant disease caused by rust fungi (Uredinales) and characterized by reddish-brown lesions on the plant parts.   { rəst }

rust fungi   See Urediniomycetes.   { 'rəst ,fən,jī }

rusting   [GEOL] The formation of red, yellow, or brown iron oxide minerals by oxidation of mineral deposits.   { 'rəst·iŋ }

rusty blotch   [PL PATH] A fungus disease of barley caused by *Helminthosporium californicum* and characterized by brown blotches on the foliage.   { 'rəs·tē 'bläch }

rye   [BOT] *Secale cereale*. A cereal plant of the order Cyperales cultivated for its grain, which contains the most desirable gluten, next to wheat.   { rī }

S

sabadilla [AGR] Ripe seeds of the sabadilla plant (*Schoenocaulon officinale*) that have been dried; used as an insecticide on cattle. Also known as caustic barley; cevedilla. { ,sa·bə'dē·ə }

saccus *See* vesicle. { 'sak·əs }

saddleback [METEOROL] The cloudless air between the "towers" of two cumulus congestus or cumulonimbus clouds and above a lower cloud mass. { 'sad·əl,bak }

saddle point *See* col. { 'sad·əl ,pȯint }

safety button [ENG] A device worn by workers exposed to nuclear radiation to warn of excessive exposure. { 'sāf·tē ,bət·ən }

safrole [CHEM] $C_3H_5C_6H_3O_2CH_2$ A toxic, water-insoluble, colorless oil that boils at 233°C; found in sassafras and camphorwood oils; used in medicine, perfumes, insecticides, and soaps, and as a chemical intermediate. { 'sa,frōl }

sag-and-swell topography [GEOGR] An undulating surface characteristic of till sheets, for example, the landscape of the midwestern United States. { ¦sag ən ¦swel tə'päg·rə·fē }

sahel [METEOROL] A strong dust-bearing desert wind in Morocco. { sə'hel }

Saint John's wort [MED] *Hypericum perforatum*. A herbacious perennial that has been used for millennia for its many medicinal properties, including wound healing and treatment of kidney and lung ailments, insomnia, and depression. { sānt 'jänz ,wȯrt }

Saint Louis encephalitis [MED] A mosquito-borne arbovirus infection of the central nervous system, occurring in the central and western United States and in Florida. { 'sānt 'lü·əs in,sef·ə'līd·əs }

salina [HYD] A body of water containing high concentrations of salt. { sə'lē·nə }

saline-alkali soil [GEOL] A salt-affected soil with a content of exchangeable sodium greater than 15, with much soluble salts, and with a pH value usually less than 9.5. { 'sā,lēn 'al·kə,lī ,sȯil }

saline soil [GEOL] A nonalkali, salt-affected soil with a high content of soluble salts, with exchangeable sodium of less than 15, and with a pH value less than 8.5. { 'sā,lēn ,sȯil }

saline-water reclamation [CHEM ENG] Purification and removal of salts from brine or brackish water by ion exchange, crystallization, distillation, evaporation, and reverse osmosis. { 'sā,lēn 'wȯd·ər ,rek·lə,mā·shən }

salinity [OCEANOGR] The total quantity of dissolved salts in sea water, measured by weight in parts per thousand. { sə'lin·əd·ē }

salinity current [OCEANOGR] A density current in the ocean whose flow is caused, controlled, or maintained by its relatively greater density due to excessive salinity. { sə'lin·əd·ē ,kə·rənt }

salinization [GEOL] In a soil of an arid, poorly drained region, the accumulation of soluble salts by the evaporation of the waters that bore them to the soil zone. { ,sal·ən·ə'zā·shən }

salivation [MED] Mild mercury poisoning suffered by workers in amalgamation plants. { ,sal·ə'vā·shən }

Salmonella [MICROBIO] A genus of gram-negative, facultatively anaerobic bacteria belonging to the family Enterobacteriaceae that cause enteric infections with or without blood invasion. { ,sal·mə'nel·ə }

salt [CHEM] The reaction product when a metal displaces the hydrogen of an acid; for example, $H_2SO_4 + 2NaOH \rightarrow Na_2SO_4$ (a salt) $+ 2H_2O$. { sȯlt }

salt-affected soil [GEOL] A general term for a soil that is not suitable for the growth of crops because of an excess of salts, exchangeable sodium, or both. { 'sȯlt i¦fek·təd 'sȯil }

salt crust [HYD] A salt deposit formed on an ice surface by crystal growth forcing salt out of young sea ice and pushing it upward. { 'sȯlt ,krəst }

salt flowers See ice flowers. { 'sȯlt ,flau̇·ərz }

salt-gradient solar pond See solar pond. { ¦sȯlt ¦grād·ē·ənt ¦sō·lər 'pänd }

salt haze [METEOROL] A haze created by the presence of finely divided particles of sea salt in the air, usually derived from the evaporation of sea spray. { 'sȯlt ,hāz }

salt lake [HYD] A confined inland body of water having a high concentration of salts, principally sodium chloride. { 'sȯlt ,lāk }

saltmarsh [ECOL] A maritime habitat found in temperate regions, but typically associated with tropical and subtropical mangrove swamps, in which excess sodium chloride is the predominant environmental feature. { 'sȯlt,märsh }

saltmarsh plain [ECOL] A salt marsh that has been raised above the level of the highest tide and has become dry land. { 'sȯlt,märsh ,plān }

salt pan [CHEM] A pool used for obtaining salt by the natural evaporation of sea water. { 'sȯlt ,pan }

salt-spray climax [ECOL] A climax community along exposed Atlantic and Gulf seacoasts composed of plants able to tolerate the harmful effects of salt picked up and carried by onshore winds from seawater. { 'sȯlt ¦sprā' klī,maks }

salt water See seawater. { 'sȯlt ¦wȯd·ər }

salt-water front [OCEANOGR] The interface between fresh and salt water in a coastal aquifer or in an estuary. { 'sȯlt ¦wȯd·ər ,frənt }

salt-water intrusion [HYD] Displacement of fresh surface water or groundwater by salt water due to its greater density. { 'sȯlt ¦wȯd·ər in,trü·zhən }

salt-water underrun [OCEANOGR] A type of density current occurring in a tidal estuary, due to the greater salinity of the bottom water. { 'sȯlt ¦wȯd·ər 'ən·də,rən }

salt-water wedge [OCEANOGR] A wedge-shaped intrusion of salty ocean water into a fresh-water estuary or tidal river; it slopes downward in the upstream direction, and salinity increases with depth. { 'sȯlt ¦wȯd·ər ,wej }

salt weathering [GEOL] The granular disintegration or fragmentation of rock material produced by saline solutions or by salt-crystal growth. { 'sȯlt ,weth·ə·riŋ }

samara [BOT] A dry, indehiscent, winged fruit usually containing a single seed, such as sugar maple (*Acer saccharum*). { sə'mar·ə }

sample splitter [ENG] An instrument, generally constructed of acrylic resin, designed to subdivide a total sample of marine plankton while maintaining a quantitatively correct relationship between the various phyla in the sample. { 'sam·pəl ,splid·ər }

sampling [SCI TECH] The obtaining of small representative quantities of materials (gas, liquid, solid) or organisms for the purpose of analysis. { 'sam·pliŋ }

sampling bottle [ENG] A cylindrical container, usually closed at a chosen depth, to trap a water sample and transport it to the surface without introducing contamination. { 'sam·pliŋ ,bäd·əl }

sand [GEOL] Unconsolidated granular material consisting of mineral, rock, or biological fragments between 63 micrometers and 2 millimeters in diameter, usually produced primarily by the chemical or mechanical breakdown of older source rocks, but may also be formed by the direct chemical precipitation of mineral grains or by biological processes. { sand }

sand auger See dust whirl. { 'sand ,ȯg·ər }

sandblow [ECOL] A patch of coarse, sandy soil denuded of vegetation by wind action. { 'san,blō }

sand boil See blowout. { 'san ,bȯil }

sand devil See dust whirl. { 'san ,dev·əl }

sand drain [CIV ENG] A vertical boring through a clay or silty soil filled with sand or gravel to facilitate drainage. { 'san ,drān }

sand dune [GEOL] A mound of loose windblown sand commonly found along low-lying seashores above high-tide level. { 'san ,dün }

sand flood [GEOL] A vast body of sand moving or borne along a desert, as in the Arabian deserts. { 'san ,fləd }

sandfly fever See phlebotomus fever. { 'san,flī ,fē·vər }

sand snow [HYD] Snow that has fallen at very cold temperatures (of the order of −25°C); as a surface cover, it has the consistency of dust or light dry sand. { 'san ,snō }

sandstorm [METEOROL] A strong wind carrying sand through the air, the diameter of most particles ranging from 0.08 to 1 millimeter; in contrast to a duststorm, the sand particles are mostly confined to the lowest 7 feet (2 meters) above ground, rarely rising more than 36 feet (11 meters). { 'san,stȯrm }

sanitary engineering [CIV ENG] A field of civil engineering concerned with works and projects for the protection and promotion of public health. { 'san·ə,ter·ē ,en·jə'nir·iŋ }

sanitary sewer [CIV ENG] A sewer which is restricted to carrying sewage and to which storm and surface waters are not admitted. { 'san·ə,ter·ē 'sü·ər }

sanitation [CIV ENG] The act or process of making healthy environmental conditions. { ,san·ə'tā·shən }

sanitizer [ENG] Disinfectant formulated to clean food-processing equipment and dairy and eating utensils. { 'san·ə,tīz·ər }

San Joaquin Valley fever See coccidioidomycosis. { ¦san wȯ¦kēn ¦val·ē 'fē·vər }

Santa Ana [METEOROL] A hot, dry, foehnlike desert wind, generally from the northeast or east, especially in the pass and river valley of Santa Ana, California, where it is further modified as a mountain-gap wind. { 'san·tə ¦an·ə }

sap [BOT] The fluid part of a plant which circulates through the vascular system and is composed of water, gases, salts, and organic products of metabolism. { sap }

sapling [BOT] A young tree with a trunk less than 4 inches (10 centimeters) in diameter at a point approximately 4 feet (1.2 meters) above the ground. { 'sap·liŋ }

saprobe [ECOL] An organism that lives on decaying organic matter. { 'sa,prōb }

saprobic [BOT] Living on decaying organic matter; applied to plants and microorganisms. { sə'prō·bik }

saprogenous ooze |GEOL| Ooze formed of putrefying organic matter. { sə'präj·ə·nəs 'üz }

saprolite |GEOL| A soft, earthy red or brown, decomposed igneous or metamorphic rock that is rich in clay and formed in place by chemical weathering. Also known as saprolith; sathrolith. { 'sap·rə‚līt }

saprolith See saprolite. { 'sap·rə‚lith }

saprophage [BIOL] An organism that lives on decaying organic matter. { 'sap·rə‚fāj }

saprophyte |BOT| A plant that lives on decaying organic matter. { 'sap·rə‚fīt }

saprozoic |ZOO| Feeding on decaying organic matter; applied to animals. {¦sap·rə¦zō·ik }

sapwood [BOT] The younger, softer, outer layers of a woody stem, between the cambium and heartwood.Also known as alburnum. { 'sap‚wüd }

sarcoma [MED] A malignant tumor arising in connective tissue and composed principally of anaplastic cells that resemble those of supportive tissues. { sär'kō·mə }

sarcosporidiosis [VET MED] A disease of mammals other than humans caused by muscle infestation by sporozoans of the order Sarcosporida. { ‚sär·kō·spə‚rid·ē'ō·səs }

Sargasso Sea |GEOGR| A region of the North Atlantic Ocean; boundaries are defined in the west and north by the Gulf Stream, in the east by longitude 40°W, and in the south by latitude 20°N. { sär'ga·sō 'sē }

SARS See severe acute respiratory syndrome. { särz }

sastruga |HYD| A ridge of snow up to 2 inches (5 centimeters) high formed by wind erosion and aligned parallel to the wind. Also known as skavl; zastruga. { 'zas·trə·gə }

satellite meteorology [METEOROL] That branch of meteorological science that employs sensing elements on meteorological satellites to define the state of the atmosphere. { 'sad·əl‚īt ‚mēd·ē·ə'räl·ə·jē }

sathrolith See saprolite. { 'sath·rə‚lith }

satin ice See acicular ice. { 'sat·ən 'īs }

saturated air [METEOROL] Moist air in a state of equilibrium with a plane surface of pure water or ice at the same temperature and pressure; that is, air whose vapor pressure is the saturation vapor pressure and whose relative humidity is 100. { 'sach·ə‚rād·əd 'er }

saturated surface See water table. { 'sach·ə‚rād·əd 'sər·fəs }

saturated zone See zone of saturation. { 'sach·ə‚rād·əd 'zōn }

saturation [METEOROL] The maximum water vapor per unit volume that a parcel of air can contain at a given temperature. { ‚sach·ə'rā·shən }

saturation-adiabatic process [METEOROL] An adiabatic process in which the air is maintained at saturation by the evaporation or condensation of water substance, the latent heat being supplied by or to the air respectively; the ascent of cloudy air, for example, is often assumed to be such a process. { ‚sach·ə'rā·shən ¦ad·ē·ə¦bad·ik 'prä·səs }

saturation deficit [METEOROL] **1.** The difference between the actual vapor pressure and the saturation vapor pressure at the existing temperature. **2.** The additional amount of water vapor needed to produce saturation at the current temperature and pressure, expressed in grams per cubic meter. Also known as vapor-pressure deficit. { ‚sach·ə'rā·shən 'def·ə·sət }

saturation ratio [METEOROL] The ratio of the actual specific humidity to the specific humidity of saturated air at the same temperature. { ˌsach·ə'rā·shən ˌrā·shō }

sault [HYD] A waterfall or rapids in a stream. { sü }

savane armée See thornbush. { sa'vän är'mā }

savane épineuse See thornbush. { sa'vän ā·pə'nüz }

savanna [ECOL] Any of a variety of physiognomically or environmentally similar vegetation types in tropical and extratropical regions; all contain grasses and one or more species of trees of the families Leguminosae, Bombacaceae, Bignoniaceae, or Dilleniaceae. { sə'van·ə }

savanna climate See tropical savanna climate. { sə'van·ə ˌklī·mət }

savanna-woodland See tropical woodland. { sə'van·ə 'wud·lənd }

saxicolous [ECOL] Living or growing among rocks. { sak'sik·ə·ləs }

saxitoxin [BIOL] A nonprotein toxin produced by the dinoflagellate *Gonyaulax catenella*. { ˌsak·sə'täk·sən }

SBMV See southern bean mosaic virus.

scabland [GEOL] Elevated land that is essentially flat-lying and covered with basalt and has only a thin soil cover, sparse vegetation, and usually deep, dry channels. { 'skab,land }

scale [BOT] The bract of a catkin. { skāl }

scale insect [ZOO] Any of various small, structurally degenerate homopteran insects in the superfamily Coccoidea which resemble scales on the surface of a host plant; serious pests of fruit trees and many other plants. { 'skāl 'in,sekt }

scales of motion [OCEANOGR] A series of increasing characteristic magnitudes of motion, ranging from tiny eddies of turbulence to oceanwide currents, each member of the series interacting with the adjacent members. { 'skālz əv 'mō·shən }

scaly bark See psorosis. { 'skā·lē ˌbärk }

scandent [BOT] Climbing by stem-roots or tendrils. { 'skan·dənt }

scansorial [BOT] Adapted for climbing. { skan'sȯr·ē·əl }

Scarabaeidae [ZOO] The lamellicorn beetles, a large cosmopolitan family of coleopteran insects in the superfamily Scarabaeoidea including the Japanese beetle and other agricultural pests. { ˌskar·ə'bē·ə,dē }

scarf cloud See pileus. { 'skärf ˌklaud }

scarlet fever [MED] An acute, contagious bacterial disease caused by *Streptococcus hemolyticus*; characterized by a papular, or rough, bright-red rash over the body, with fever, sore throat, headache, and vomiting occurring 2–3 days after contact with a carrier. { 'skär·lət 'fē·vər }

scarlet fever streptococcus antitoxin [MED] A sterile aqueous solution of antitoxins obtained from the blood of animals immunized against group A beta hemolytic streptococci toxin; formerly used in the treatment of, and to produce immunity against, scarlet fever. { 'skär·lət 'fē·vər ˌstrep·tə'käk·əs ˌant·i'täk·sən }

scarlet fever streptococcus toxin [MED] Toxic filtrate of cultures of *Streptococcus pyogenes* responsible for the characteristic rash of scarlet fever; the toxin is used in the Dick test. { 'skär·lət 'fē·vər ˌstrep·tə'käk·əs 'täk·sən }

scarp See escarpment. { skärp }

scarp-foot spring [HYD] A spring that flows onto the land surface at or near the foot of an escarpment. { 'skärp ˌfut ˌspriŋ }

scarpland [GEOGR] A region marked by a succession of nearly parallel cuestas separated by lowlands. { 'skärp·lənd }

scattered [METEOROL] Descriptive of a sky cover of 0.1 to 0.5 (5 to 54%), applied only when clouds or obscuring phenomena aloft are present, not applied for surface-based obscuring phenomena. { 'skad·ərd }

scattering layer [OCEANOGR] A layer of organisms in the sea which causes sound to scatter and to return echoes. { 'skad·ə·riŋ ,lā·ər }

scavenger [ECOL] An organism that feeds on carrion, refuse, and similar matter. { 'skav·ən·jər }

scavenger well [HYD] A well located between a good well (or group of wells) and a source of potential contamination, which is pumped (or allowed to flow) as waste to prevent the contaminated water from reaching the good well. { 'skav·ən·jər ,wel }

Scheele's green *See* copper arsenite. { 'shā·ləz 'grēn }

schizocarp [BOT] A dry fruit that separates at maturity into single-seeded indehiscent carpels. { 'skiz·ə,kärp }

scintillation counter [ENG] A device in which the scintillations produced in a fluorescent material by an ionizing radiation are detected and counted by a multiplier phototube and associated circuits; used in medical and nuclear research and in prospecting for radioactive ores. Also known as scintillation detector; scintillometer. { ,sint·əl'ā·shən ,kaunt·ər }

scintillation detector *See* scintillation counter. { ,sint·əl'ā·shən di,tek·tər }

scintillator [ENG] A material that emits optical photons in response to ionizing radiation. { 'sint·əl,ād·ər }

scintillometer *See* scintillation counter. { ,sint·əl'äm·əd·ər }

scion [BOT] A section of a plant, usually a stem or bud, which is attached to the root of a different plant (the stock) in grafting. { 'sī·ən }

sciophilous [ECOL] Capable of thriving in shade. { sī'äf·ə·ləs }

sciophyte [BOT] A plant that thrives at lowered light intensity. { 'sī·ə,fīt }

sclereid [BOT] A thick-walled, lignified plant cell typically found in sclerenchyma. { 'sklir·ē·əd }

sclerenchyma [BOT] A supporting plant tissue composed principally of sclereids whose walls are often mineralized. { sklə'reŋ·kə·mə }

sclerophyllous [BOT] Characterized by thick, hard foliage due to well-developed sclerenchymatous tissue. { ¦skler·ə¦fil·əs }

scorching [PL PATH] Browning of plant tissues caused by heat or parasites; may also be symptomatic of disease. { 'skorch·iŋ }

scotochromogen [MICROBIO] 1. Any microorganism which produces pigment when grown without light as well as with light. 2. A member of group II of the atypical mycobacteria. { ,skäd·ə'krō·mə·jən }

Scott-Darey process [CIV ENG] A chemical precipitation method used for fine solids removal in sewage plants; employs ferric chloride solution made by treating scrap iron with chlorine. { 'skät 'der·ē ,prä·səs }

scrapie [VET MED] A transmissible, usually fatal, virus disease of sheep, characterized by degeneration of the central nervous system. { 'skrā·pē }

screen [ENG] 1. A large sieve of suitably mounted wire cloth, grate bars, or perforated sheet iron used to sort rock, ore, or aggregate according to size. 2. A covering to give physical protection from light, noise, heat, or flying particles. 3. A filter medium for liquid-solid separation. { skrēn }

screw ice [HYD] **1.** Small ice fragments in heaps or ridges, produced by the collision of ice cakes. **2.** A small formation of pressure ice. { 'skrü ˌīs }

scrub [ECOL] A tract of land covered with a generally thick growth of dwarf or stunted trees and shrubs and a poor soil. { skrəb }

scud [METEOROL] Ragged low clouds, usually stratus fractus; most often applied when such clouds are moving rapidly beneath a layer of nimbostratus. { skəd }

scum chamber [CIV ENG] An enclosed compartment in an Imhoff tank, in which gas escapes from the scum which rises to the surface of sludge during sewage digestion. { 'skəm ˌchām·bər }

scythe [ENG] A tool with a long curved blade attached at a more or less right angle to a long handle with grips for both hands; used for cutting grass as well as grain and other crops. { sīth }

SDDC *See* sodium dimethyldithiocarbamate.

sea [OCEANOGR] **1.** A major subdivision of the ocean. **2.** A heavy swell or ocean wave still under the influence of the wind that produced it. { sē }

sea anchor [ENG] An object towed by a usually small vessel to keep the vessel end-on to a heavy sea or surf or to reduce drift; the usual form is a conical canvas bag whose large end is open, and, when towed with the large end in the forward position, the bag offers considerable resistance. { 'sē ˌaŋ·kər }

sea ball [OCEANOGR] A spherical mass of somewhat fibrous material of living or fossil vegetation (especially algae), produced mechanically in shallow waters along a seashore by the compacting effect of wave movement. { 'sē ˌbȯl }

seabed *See* sea floor. { 'sē ˌbed }

sea bottom *See* sea floor. { 'sē ˌbäd·əm }

sea breeze [METEOROL] A coastal, local wind that blows from sea to land, caused by the temperature difference when the sea surface is colder than the adjacent land; it usually blows on relatively calm, sunny summer days, and alternates with the oppositely directed, usually weaker, nighttime land breeze. { 'sē ˌbrēz }

sea breeze of the second kind *See* cold-front-like sea breeze. { 'sē ˌbrēz əv thə 'sek·ənd ˌkīnd }

sea-captured stream [HYD] A stream, flowing parallel to the seashore, that is cut in two as a result of marine erosion and that may enter the sea by way of a waterfall. { 'sē ˌkap·chərd ˌstrēm }

seacoast [GEOGR] The land adjacent to the sea. { 'sē ˌkōst }

sea floor [GEOL] The bottom of the ocean. Also known as seabed; sea bottom. { 'sē ˌflȯr }

sea-floor spreading [GEOL] The hypothesis that the ocean floor is spreading away from the midoceanic ridges and is being conveyed landward by convective cells in the earth's mantle, carrying the continental blocks as passive passengers; the ocean floor moves away from the midoceanic ridge at the rate of 0.4 to 4 inches (1 to 10 centimeters) per year and provides the source of power in the hypothesis of plate tectonics. Also known as ocean-floor spreading; spreading concept; spreading-floor hypothesis. { 'sē ˌflȯr ˌspred·iŋ }

sea fog [METEOROL] A type of advection fog formed over the ocean as a result of any of a variety of processes, as when air that has been lying over a warm water surface is transported over a colder water surface, resulting in a cooling of the lower layer of air below its dew point. { 'sē ˌfäg }

sea front [GEOGR] An area partly bounded by the sea. { 'sē ˌfrənt }

sea gate [GEOGR] A way giving access to the sea such as a gate, channel, or beach. { 'sē ,gāt }

sea glow [OCEANOGR] The luminous, cobalt-blue appearance of very clear water in the open ocean, caused by upward-scattered light from which much of the red has been absorbed. { 'sē ,glō }

sea grass [BOT] Marine plants which are found in shallow brackish or marine waters, are more highly organized than algae, are seed-bearing, and attain lengths of up to 8 feet (2.4 meters). { 'sē ,gras }

sea ice [OCEANOGR] **1.** Ice formed from seawater. **2.** Any ice floating in the sea. { 'sē ,īs }

sea-ice shelf [OCEANOGR] Sea ice floating in the vicinity of its formation and separated from fast ice, of which it may have been a part, by a tide crack or a family of such cracks. { 'sē ¦īs ,shelf }

sea knoll See knoll. { 'sē ,nōl }

sea level [GEOL] The level of the surface of the ocean; especially, the mean level halfway between high and low tide, used as a standard in reckoning land elevation or sea depths. { 'sē ,lev·əl }

sea-level chart See surface chart. { 'sē ¦lev·əl ,chärt }

sea-level pressure [METEOROL] The atmospheric pressure at mean sea level, either directly measured or, most commonly, empirically determined from the observed station pressure. { 'sē ¦lev·əl ,presh·ər }

sea-level pressure chart See surface chart. { 'sē ¦lev·əl ¦presh·ər ,chärt }

sea marsh [ECOL] A salt marsh periodically overflowed or flooded by the sea. Also known as sea meadow. { 'sē ,märsh }

sea meadow [ECOL] See sea marsh. [OCEANOGR] Any of the upper layers of the open ocean that have such an abundance of phytoplankton that they provide food for marine organisms. { 'sē ,med·ō }

sea mist See steam fog. { 'sē ,mist }

sea mud [GEOL] A rich, slimy deposit in a salt marsh or along a seashore, sometimes used as a manure. Also known as sea ooze. { 'sē ,məd }

sea ooze See sea mud. { 'sē ,üz }

sea salt [OCEANOGR] The salt remaining after the evaporation of seawater, containing sodium and magnesium chlorides and magnesium and calcium sulfates. { 'sē ,solt }

sea-salt nucleus [OCEANOGR] A condensation nucleus of a highly hygroscopic nature produced by partial or complete desiccation of particles of sea spray or of seawater droplets derived from breaking bubbles. { 'sē ¦solt ,nü·klē·əs }

seascape [OCEANOGR] The surrounding sea as it appears to an observer. { 'sē,skāp }

seashore [GEOL] **1.** The strip of land that borders a sea or ocean. Also known as seaside; shore. **2.** The ground between the usual tide levels. Also known as seastrand. { 'sē,shȯr }

seashore lake [GEOGR] A lake, containing either fresh or salt water, which lies along a seashore; it is separated from the sea by a river, a delta, or a wall of sediment. { 'sē ,shȯr ,lāk }

seaside See seashore. { 'sē,sīd }

sea smoke See steam fog. { 'sē ,smōk }

season [CLIMATOL] A division of the year according to some regularly recurrent phenomena, usually astronomical or climatic. { 'sēz·ən }

seasonal current [OCEANOGR] An ocean current which has large changes in speed or direction due to seasonal winds. { 'sēz·ən·əl 'kə·rənt }

seasonally frozen ground [GEOL] Ground that is frozen during low temperatures and remains so only during the winter season. Also known as frost zone. { 'sēz·ən·lē ¦frō·zən 'graúnd }

seasonal recovery [HYD] Recharge of groundwater during and after a wet season, with a rise in the level of the water table. { 'sēz·ən·əl ri'kəv·ə·rē }

seasonal stream [HYD] A stream whose flow is not constant because it has water in its course only during certain seasons. { 'sēz·ən·əl 'strēm }

seasonal thermocline [OCEANOGR] A thermocline which develops in the oceans in summer at relatively shallow depths due to surface heating and downward transport of heat caused by mixing of water generated by summer winds. { 'sēz·ən·əl 'thər·mə ‚klīn }

sea state [OCEANOGR] The numerical or written description of ocean-surface roughness. { 'sē ‚stāt }

seastrand *See* seashore. { 'sē‚strand }

sea-surface slope [OCEANOGR] A gradual change in the level of the sea surface with distance, caused by Coriolis and wind forces. { 'sē ¦sər·fəs ‚slōp }

seawater [OCEANOGR] Water of the seas, distinguished by high salinity. Also known as salt water. { 'sē‚wȯd·ər }

seaweed [BOT] A marine plant, especially algae. { 'sē‚wēd }

Secchi disk [ENG] An opaque white disk used to measure the transparency or clarity of seawater by lowering the disk into the water horizontally and noting the greatest depth at which it can be visually detected. { 'sek·ē ‚disk }

secondary cold front [METEOROL] A front which forms behind a frontal cyclone and within a cold air mass, characterized by an appreciable horizontal temperature gradient. { 'sek·ən‚der·ē 'kōld ‚frənt }

secondary consumer [ECOL] In an ecosystem, an animal that feeds on primary consumers. Also known as carnivore. { ¦sek·ən‚der·ē kən'sü·mər }

secondary cyclone [METEOROL] A cyclone which forms near or in association with a primary cyclone. Also known as secondary low. { 'sek·ən‚der·ē 'sī‚klōn }

secondary front [METEOROL] A front which may form within a baroclinic cold air mass which itself is separated from a warm air mass by a primary frontal system; the most common type is the secondary cold front. { 'sek·ən‚der·ē 'frənt }

secondary glacier [HYD] A small valley glacier that joins a larger trunk glacier as a tributary glacier. { 'sek·ən‚der·ē 'glā·shər }

secondary low *See* secondary cyclone. { 'sek·ən‚der·ē 'lō }

secondary pollutant [METEOROL] An air pollutant produced by the reaction of a primary pollutant with some other component in the air. { ¦sek·ən‚der·ē pə'lüt·ənt }

secondary succession [ECOL] Ecological succession that occurs in habitats where the previous community has been destroyed or severely disturbed, such as following forest fire, abandonment of agricultural fields, or epidemic disease or pest attack. { ¦sek·ən‚der·ē sək'sesh·ən }

second-foot [HYD] A contraction of cubic foot per second (cfs), the unit of stream discharge commonly used in the United States. { 'sek·ənd ¦fút }

second-foot day [HYD] The volume of water represented by a flow of 1 cubic foot per second for 24 hours; equal to 86,400 cubic feet (approximately 2446.58 cubic meters);

used extensively as a unit of runoff volume or reservoir capacity, particularly in the eastern United States. { 'sek·ənd ¦fút 'dā }

second growth |FOR| New trees that naturally replace trees which have been removed from a forest by cutting or by fire. { 'sek·ənd 'grōth }

second-order climatological station |CLIMATOL| A station at which observations of atmospheric pressure, temperature, humidity, winds, clouds, and weather are made at least twice daily at fixed hours, and at which the daily maximum and minimum of temperature, the daily amount of precipitation, and the duration of bright sunshine are observed. { 'sek·ənd ¦ör·dər ˌklī·mət·əl'äj·ə·kəl 'stā·shən }

second-order relief |GEOGR| Extensive relief features consisting of major mountain systems and other surface formations of subcontinental extent. { 'sek·ənd ¦ör·dər ri'lēf }

second-order station |METEOROL| After U.S. Weather Bureau practice, a station operated by personnel certified to make aviation weather observations or synoptic weather observations. { 'sek·ənd ¦ör·dər 'stā·shən }

second-year ice |OCEANOGR| Sea ice that has survived only one summer's melt. Also known as two-year ice. { 'sek·ənd ¦yir 'īs }

secretion |BIOL| **1.** The act or process of producing a substance which is specialized to perform a certain function within the organism or is excreted from the body. **2.** The material produced by such a process. { si'krē·shən }

sector |METEOROL| Something resembling the sector of a circle, as a warm sector between the warm and cold fronts of a cyclone. { 'sek·tər }

SED *See* skin erythema dose.

sedentary soil |GEOL| Soil that still lies on the rock from which it was formed. { 'sed·ən,ter·ē 'sȯil }

sediment |GEOL| **1.** A mass of organic or inorganic solid fragmented material, or the solid fragment itself, that comes from weathering of rock and is carried by, suspended in, or dropped by air, water, or ice; or a mass that is accumulated by any other natural agent and that forms in layers on the earth's surface such as sand, gravel, silt, mud, fill, or loess. **2.** A solid material that is not in solution and either is distributed through the liquid or has settled out of the liquid. { 'sed·ə·mənt }

sedimentary cycle *See* cycle of sedimentation. { ¦sed·ə¦men·trē 'sī·kəl }

sedimentary rock |GEOL| A rock formed by consolidated sediment deposited in layers. Also known as derivative rock; neptunic rock; stratified rock. { ¦sed·ə¦men·trē 'räk }

sedimentation |GEOL| **1.** The act or process of accumulating sediment in layers. **2.** The process of deposition of sediment. { ˌsed·ə·mən'tā·shən }

sedimentation rate *See* rate of sedimentation. { ˌsed·ə·mən'tā·shən ˌrāt }

sediment charge |HYD| In a stream, the ratio of the weight or volume of sediment to the weight or volume of water passing a given cross section per unit of time. { 'sed·ə·mənt ˌchärj }

sediment concentration |HYD| The ratio of the dry weight of the sediment in a water-sediment mixture (obtained from a stream or other body of water) to the total weight of the mixture. { 'sed·ə·mənt ˌkän·sən'trā·shən }

sediment discharge |HYD| The amount of sediment moved by a stream in a given time, measured by dry weight or by volume. Also known as sediment-transport rate. { 'sed·ə·mənt 'dis,chärj }

sediment discharge rating |HYD| A relationship between the discharge of sediment and the total discharge of the stream. Also known as silt discharge rating. { 'sed·ə·mənt 'dis,chärj ,rād·iŋ }

sediment load [HYD] The solid material that is transported by a natural agent, especially by a stream. { 'sed·ə·mənt ˌlōd }

sediment station [HYD] A vertical cross-sectional plane of a stream, usually normal to the mean direction of flow, where samples of suspended load are collected on a systematic basis for determining concentration, particle-size distribution, and other characteristics. { 'sed·ə·mənt ˌstā·shən }

sediment-transport rate *See* sediment discharge. { 'sed·ə·mənt 'tranzˌpȯrt ˌrāt }

seed [BOT] A fertilized ovule containing an embryo which forms a new plant upon germination. { sēd }

seed coat [BOT] The envelope which encloses the seed except for a tiny pore, the micropyle. { 'sēd ˌkōt }

seeding [AGR] The planting of seed. { 'sēd·iŋ }

seedling [BOT] **1.** A plant grown from seed. **2.** A tree younger and smaller than a sapling. **3.** A tree grown from a seed. { 'sēd·liŋ }

seep [GEOL] An area, generally small, where water, or another liquid such as oil, percolates slowly to the land surface. { sēp }

seepage [HYD] The slow movement of water through small openings and spaces in the surface of unsaturated soil into or out of a body of surface or subsurface water. { 'sēp·ij }

seepage lake [HYD] **1.** A closed lake that loses water mainly by seepage through the walls and floor of its basin. **2.** A lake that receives its water mainly from seepage. { 'sēp·ij ˌlāk }

segregated ice [HYD] Ice films, seams, lenses, rods, or layers generally 0.04 to 6 inches (1 to 150 millimeters) thick that grow in permafrost by drawing in water as the ground freezes. Also known as Taber ice. { 'seg·rəˌgād·əd 'īs }

seiche [OCEANOGR] A standing-wave oscillation of an enclosed or semienclosed water body, continuing pendulum-fashion after cessation of the originating force, which is usually considered to be strong winds or barometric pressure changes. { sāsh }

seismic area *See* earthquake zone. { 'sīz·mik ˌer·ē·ə }

seismic constant [CIV ENG] In building codes dealing with earthquake hazards, an arbitrarily set quantity of steady acceleration, in units of acceleration of gravity, that a building must withstand. { 'sīz·mik 'kän·stənt }

seismology [GEOPHYS] **1.** The study of earthquakes. **2.** The science of strain-wave propagation in the earth. { sīz'mäl·ə·jē }

sejunction water [HYD] Capillary water bounded by menisci, and in static equilibrium in the soil above the capillary fringe. { sə'jəŋk·shən ˌwȯd·ər }

selection [GEN] Any natural or artificial process which favors the survival and propagation of individuals of a given phenotype in a population. { si'lek·shən }

selective breeding [BIOL] Breeding of animals or plants having desirable characters. { si'lek·tiv 'brēd·iŋ }

selenium [CHEM] A highly toxic, nonmetallic element in group 16, symbol Se, atomic number 34; steel-gray color; soluble in carbon disulfide, insoluble in water and alcohol; melts at 217°C; and boils at 690°C; used in analytical chemistry, metallurgy, and photoelectric cells, and as a lube-oil stabilizer and chemicals intermediate. { sə'lē·nē·əm }

selenosis [MED] Selenium poisoning. { ˌsel·ə'nō·səs }

self-incompatibility [BOT] Pertaining to an individual flower that cannot complete fertilization with its own pollen. { ¦self ˌin·kəmˌpad·ə'bil·əd·ē }

self-pollination [BOT] Transfer of pollen from the anther to the stigma of the same flower or of another flower on the same plant. { ¦self ˌpäl·əˌnā·shən }

selva *See* tropical rainforest. { 'sel·və }

semiarid climate *See* steppe climate. { ¦sem·ē'ar·əd 'klī·mət }

semidesert [ECOL] An area intermediate in character and often located between a desert and a grassland or woodland. { ¦sem·i'dez·ərt }

semidiurnal [METEOROL] Pertaining to a meteorological event that occurs twice a day. { ¦sem·i·dī'ərn·əl }

semidiurnal current [OCEANOGR] A tidal current in which the tidal-day current cycle consists of two flood currents and two ebb currents, separated by slack water, or of two changes in direction of 360° of a rotary current; this is the most common type of tidal current throughout the world. { ¦sem·i·dī'ərn·əl 'kə·rənt }

semidiurnal tide [OCEANOGR] A tide having two high waters and two low waters during a tidal day. { ¦sem·i·dī'ərn·əl 'tīd }

semidormancy [BOT] Decrease in plant growth rate; may be seasonal or associated with unfavorable environmental conditions. { ¦sem·i'dȯr·mən·sē }

semilate [BOT] Pertaining to a plant whose growing season is intermediate between midseason forms and late forms. { ¦sem·i'lāt }

semiochemical [BIOL] Any of a class of substances produced by organisms, especially insects, that participate in regulation of the organisms' behavior in such activities as aggregation of both sexes, sexual stimulation, and trail following. { ¦sem·ē·ə'kem·ə·kəl }

semiparasite *See* hemiparasite. { ¦sem·i'par·əˌsīt }

senescent lake [HYD] A lake that is approaching extinction—for example, from filling by remains of aquatic vegetation. { si'nes·ənt 'lāk }

sensible atmosphere [METEOROL] That part of the atmosphere that offers resistance to a body passing through it. { 'sen·sə·bəl 'at·məˌsfir }

sensible-heat flow [METEOROL] In the atmosphere, the poleward transport of sensible heat (enthalpy) across a given latitude belt by fluid flow. { 'sen·sə·bəl ¦hēt 'flō }

sensible temperature [METEOROL] The temperature at which air with some standard humidity, motion, and radiation would provide the same sensation of human comfort as existing atmospheric conditions. { 'sen·sə·bəl 'tem·prə·chər }

sensitivity [BIOL] The capacity for receiving sensory impressions from the environment. { ˌsen·sə'tiv·əd·ē }

sensitization [MED] The alteration of a body's responsiveness to a foreign antigen, usually an allergen, such that upon subsequent exposures to the allergen there is a heightened immune response. { ˌsen·səd·ə'zā·shən }

sepal [BOT] One of the leaves composing the calyx. { sēp·əl }

separate sewage system [CIV ENG] A drainage system in which sewage and groundwater are carried in separate sewers. { 'sep·rət 'sü·ij ˌsis·təm }

sepsis [MED] **1.** Poisoning by products of putrefaction. **2.** The severe toxic, febrile state resulting from infection with pyogenic (pus-producing) microorganisms, with or without associated septicemia. { 'sep·səs }

septicemia [MED] A clinical syndrome in which infection is disseminated through the body in the bloodstream. Also known as blood poisoning. { ˌsep·tə'sē·mē·ə }

septic tank [CIV ENG] A settling tank in which settled sludge is in immediate contact with sewage flowing through the tank while solids are decomposed by anaerobic bacterial action. { 'sep·tik ˌtaŋk }

sequence of current   [OCEANOGR] The order of occurrence of the tidal current strengths of a day, with special reference to whether the greater flood immediately precedes or follows the greater ebb.   { 'sē·kwəns əv 'kə·rənt }

sequence of tide   [OCEANOGR] The order in which the tides of a day occur, with special reference to whether the higher high water immediately precedes or follows the lower low water.   { 'sē·kwəns əv 'tīd }

Sequoia   [BOT] A genus of conifers having overlapping, scalelike evergreen leaves and vertical grooves in the trunk; the giant sequoia (*Sequoia gigantea*) is the largest and oldest of all living things.   { si'kwói·ə }

serac   [HYD] A sharp ridge or pinnacle of ice among the crevasses of a glacier.   { sə'rak }

sere   [ECOL] A temporary community which occurs during a successional sequence on a given site.   { sir }

serial observation   [OCEANOGR] The procurement of water samples and temperature readings at a number of levels between the surface and the bottom of an ocean.   { 'sir·ē·əl ‚äb·zər'vā·shən }

serial station   [OCEANOGR] An oceanographic station consisting of one or more Nansen casts.   { 'sir·ē·əl ‚stā·shən }

seritinous   [ECOL] Of, pertaining to, or occurring during the latter, drier half of the summer.   { ¦ser·ə'tī·nəs }

serotinous   [BOT] Plants which flower or develop late in a season.   { sə'rät·ən·əs }

serpentine spit   [GEOGR] A spit that is extended in more than one direction due to variable or periodically shifting currents.   { 'sər·pən‚tēn 'spit }

Serpulidae   [ZOO] A family of polychaete annelids belonging to the Sedentaria including many of the feather-duster worms which construct calcareous tubes in the earth, sometimes in such abundance as to clog drains and waterways.   { sər'pyü·lə‚dē }

Serratia marcescens   [MICROBIO] A human pathogen that is intrinsically resistant to many antimicrobials (for example, cephalosporins, polymyxins, and nitrofurans) and occurs predominantly in hospitalized patients.   { sə‚rā·shē·ə mär'ses·əns }

sessile   [BOT] Attached directly to a branch or stem without an intervening stalk.   [ZOO] Permanently attached to the substrate.   { 'ses·əl }

seston   [OCEANOGR] Minute living organisms and particles of nonliving matter which float in water and contribute to turbidity.   { 'se‚stän }

set   [OCEANOGR] The direction toward which an oceanic current flows.   { set }

settleable solids test   [CIV ENG] A test used in examination of sewage to help determine the sludge-producing characteristics of sewage; a measurement of the part of the suspended solids heavy enough to settle is made in an Imhoff cone.   { 'sed·əl·ə·bəl 'säl·ədz ‚test }

settled   [METEOROL] Pertaining to weather, devoid of storms for a considerable period.   { 'sed·əld }

settled snow   [HYD] An old snow that has been strongly metamorphosed (changed structurally) and compacted.   { 'sed·əld 'snō }

settling basin   [ENG] A sedimentation area designed to remove pollutants from factory effluents.   { 'set·liŋ ‚bās·ən }

severe acute respiratory syndrome   [MED] An atypical pneumonia first recognized in February 2003 in Southeast Asia and attributed soon thereafter to a novel coronavirus. Abbreviated SARS.   { si'vir ə'kyüt 'res·pra‚tór·ē ‚sin‚drōm }

severe storm |METEOROL| In general, any destructive storm, but usually applied to a severe local storm, that is, an intense thunderstorm, hail storm, or tornado. { si'vir 'storm }

severe-storm observation |METEOROL| An observation (and report) of the occurrence, location, time, and direction of movement of severe local storms. { si'vir ¦storm ,äb·zər'vā·shən }

severe weather |METEOROL| A more general term for severe storm. { si'vir 'we<u>th</u>·ər }

sewage |CIV ENG| The fluid discharge from medical, domestic, and industrial sanitary appliances. Also known as sewerage. { 'sü·ij }

sewage disposal plant |CIV ENG| The land, building, and apparatus employed in the treatment of sewage by chemical precipitation or filtration, bacterial action, or some other method. { 'sü·ij di¦spōz·əl ,plant }

sewage farm |AGR| A farm in which sewage is used for irrigation and fertilizer. { 'sü·ij ,färm }

sewage sludge |CIV ENG| A semiliquid waste with a solid concentration in excess of 2500 parts per million, obtained from the purification of municipal sewage. Also known as sludge. { 'sü·ij ,sləj }

sewage system |CIV ENG| A drainage system for carrying surface water and sewage for disposal. { 'sü·ij ,sis·təm }

sewage treatment |CIV ENG| A process for the purification of mixtures of human and other domestic wastes; the process can be aerobic or anaerobic. { 'sü·ij ,trēt·mənt }

sewer |CIV ENG| An underground pipe or open channel in a sewage system for carrying water or sewage to a disposal area. { 'sü·ər }

sewerage See sewage. { 'sü·ə·rij }

sewer gas |ENG| The gas evolved from the decomposition of municipal sewage; it has a high content of methane and hydrogen sulfide, and can be used as a fuel gas. { 'sü·ər ,gas }

sex |BIOL| **1.** The characteristics involved with reproduction and raising offspring that also distinguish males, females, and hemaphrodites. **2.** To determine the sex of. { seks }

sex cell See gamete. { 'seks ,sel }

sex ratio |BIOL| The relative proportion of males and females in a population. { 'seks ,rā·shō }

sexually transmitted disease |MED| An infection acquired and transmitted primarily by sexual contact. Abbreviated STD. { 'sek·shə·lē trans'mid·əd di'zēz }

sexual reproduction |BIOL| Reproduction involving the paired union of special cells (gametes) from two individuals. { 'sek·shə·wəl ,rē·prə'dək·shən }

sferics See atmospheric interference. { 'sfir·iks }

sferics fix |METEOROL| The estimated location of a source of atmospheric interference, presumably a lightning discharge. { 'sfir·iks ,fiks }

sferics observation |METEOROL| An evaluation, from one or more sferics receivers, of the location of weather conditions with which lightning is associated; such observations are more commonly obtained from networks of two or three widely spaced stations; simultaneous observations of the azimuth of the discharge are made at all stations, and the location of the storm is determined by triangulation. { 'sfir·iks ,äb·zər,vā·shən }

sferics receiver |METEOROL| An instrument which measures, electronically, the direction of arrival, intensity, and rate of occurrence of atmospheric interference.

In its simplest form, the instrument consists of two orthogonally crossed antennas, whose output signals are connected to an oscillograph so that one loop measures the north-south component while the other measures the east-west component; the signals are combined vertically to give the azimuth. Also known as lightning recorder. { 'sfir·iks ri,sē·vər }

shagbark hickory [FOR] *Carya ovata*. A type of hickory that grows to a height of about 120 ft (36 m) and is found in the eastern half of the United States and adjacent Canada. It is the most important species because of the commercial value of its nuts and of its wood. { ,shag,bärk 'hik·ə·rē }

shale [GEOL] A fine-grained laminated or fissile sedimentary rock made up of silt- or clay-size particles; generally consists of about one-third quartz, one-third clay materials, and one-third miscellaneous minerals, including carbonates, iron oxides, feldspars, and organic matter. { shāl }

shale ice [HYD] A mass of thin and brittle plates of river or lake ice formed when sheets of skim ice break up into small pieces. { 'shāl ,īs }

shale reservoir [GEOL] Underground hydrocarbon reservoir in which the reservoir rock is a brittle, siliceous, fractured shale. { 'shāl 'rez·əv,wär }

shalification [GEOL] The formation of shale. { ,shāl·ə·fə'kā·shən }

shallow fog [METEOROL] In weather-observing terminology, low-lying fog that does not obstruct horizontal visibility at a level 6 feet (1.8 meters) or more above the surface of the earth; this is, almost invariably, a form of radiation fog. { 'shal·ō'fäg }

shallows [HYD] A shallow place or area in a body of water, or an expanse of shallow water. { 'shal·ōz }

shallow water [HYD] Water of such a depth that bottom topography affects surface waves. { 'shal·ō 'wȯd·ər }

shallow-water wave [HYD] A progressive gravity wave in water whose depth is much less than the wavelength. { 'shal·ō ¦wȯd·ər ,wāv }

shallow well [HYD] **1.** A water well, generally dug up by hand or by excavating machinery, or put down by driving or boring, that taps the shallowest aquifer in the vicinity. **2.** A well whose water level is shallow enough to permit use of a suction pump, the practical lift of which is taken as 22 feet (6.7 meters). { 'shal·ō 'wel }

sharp-edged gust [METEOROL] A gust that represents an instantaneous change in wind direction or speed. { 'shap ¦ejd 'gəst }

shear line [METEOROL] A line or narrow zone across which there is an abrupt change in the horizontal wind component parallel to this line; a line of maximum horizontal wind shear. { 'shir ,līn }

shear plane [HYD] A planar surface in a glacier, usually laden with rock debris, attributed to discontinuous shearing or overthrusting. { 'shir ,plān }

sheet *See* sheetflood. { shēt }

sheet composting [AGR] Addition of large amounts of organic residue to a soil; extra nitrogen is usually added for faster decomposition. { 'shēt kəm,pōst·iŋ }

sheet erosion [GEOL] Erosion of thin layers of surface materials by continuous sheets of running water. Also known as sheetflood erosion; sheetwash; surface wash; unconcentrated wash. { 'shēt i,rō·zhən }

sheetflood [HYD] A broad expanse of moving, storm-borne water that spreads as a thin, continuous, relatively uniform film over a large area for a short distance and duration. Also known as sheet; sheetwash. { 'shēt,fläd }

sheetflood erosion *See* sheet erosion. { 'shēd,fläd i,rō·zhən }

sheet flow [HYD] An overland flow or downslope movement of water taking the form of a thin, continuous film over relatively smooth soil or rock surfaces and not concentrated into channels larger than rills. { 'shēt,flō }

sheet frost [HYD] A thick coating of rime formed on windows and other surfaces. { 'shēt ,fròst }

sheet ice [HYD] A smooth, thin layer of ice formed by rapid freezing of the surface layer of a body of water. { 'shēt ,īs }

sheet sand *See* blanket sand. { 'shēt ,sand }

sheetwash [HYD] **1.** A wide, moving expanse of water on an arid plain; the combined result of many streams issuing from the mountains. **2.** *See* sheetflood. { 'shēt, wäsh }

shelf [GEOL] **1.** Solid rock beneath alluvial deposits. **2.** A flat, projecting ledge of rock. **3.** *See* continental shelf. { shelf }

shelf ice [HYD] The ice of an ice shelf. Also known as barrier ice. { 'shelf ,īs }

shelf sea [OCEANOGR] A shallow marginal sea located on the continental shelf, usually less than 150 fathoms (275 meters) in depth; an example is the North Sea. { 'shelf ,sē }

shell ice [HYD] Ice, on a body of water, that remains as an unbroken surface when the water level drops so that a cavity is formed between the water surface and the ice. { 'shel ,īs }

shelterbelt [ECOL] A natural or planned barrier of trees or shrubs to reduce erosion and provide shelter from wind and storm activity. { 'shel·tər,belt }

shelterwood method [FOR] A method for ensuring tree reproduction; older trees are removed by successive cuttings so that the amount of light reaching the seedlings is gradually increased. { 'shel·tər,wúd ,meth·əd }

shield [ENG] The material placed around a nuclear reactor, or other source of radiation, to reduce escaping radiation or particles to a permissible level. Also known as shielding. { shēld }

shielding [ENG] **1.** Reducing the ionizing radiation reaching one region of space from another region by using a shield or other device. **2.** *See* shield. { 'shēld·iŋ }

shielding layer [METEOROL] The layer of air nearest the earth, with reference to the manner in which this layer shields the earth from activity in the free atmosphere above, or vice versa. { 'shēld·iŋ ,lā·ər }

shiitake mushroom *See* Lentinula edodes. { ,shē·ē,tä·kē 'məsh,rüm }

shimmer [METEOROL] To appear tremulous or wavering, due to varying atmospheric refraction in the line of sight. { 'shim·ər }

shingles *See* herpes zoster. { 'shiŋ·gəlz }

ship drift [OCEANOGR] A method of measuring ocean currents; the ship itself is used as a current tracer, its motions being measured by navigating equipment on board. { 'ship ,drift }

shipping fever [VET MED] An acute, occasionally subacute, septicemic disease in cattle and sheep, probably caused by a combination of virus and *Pasteurella multocida* or *P. hemolytica*. { 'ship·iŋ ,fē·vər }

ship report [METEOROL] The encoded and transmitted report of a marine weather observation. { 'ship ri,pòrt }

ship synoptic code [METEOROL] A synoptic code for communicating marine weather observations; it is a modification of the international synoptic code. { 'ship si'näp·tik 'kōd }

shoal [GEOL] A submerged elevation that rises from the bed of a shallow body of water and consists of, or is covered by, unconsolidated material, and may be exposed at low water. { 'shōl }

shoaling [OCEANOGR] The bottom effect which influences the height of waves moving from deep to shallow water. { 'shōl·iŋ }

shoal patches [OCEANOGR] Individual and scattered elevations of the bottom, with depths of 10 fathoms (18 meters) or less, but composed of any material except rock or coral. { 'shōl ,pach·əz }

shoal reef [GEOL] A reef formed in irregular masses amid submerged shoals of calcareous reef detritus. { 'shōl ,rēf }

shoal water [OCEANOGR] Shallow water; over a shoal. { 'shōl ,wȯd·ər }

shock organ [MED] The organ or tissue that exhibits the most marked response to the antigen-antibody interaction in hypersensitivity, as the lungs in allergic asthma or the skin in allergic contact dermatitis. { 'shäk ,ȯr·gən }

shoot [HYD] **1.** A place where a stream flows or descends swiftly. **2.** A natural or artificial channel, passage, or trough through which water is moved to a lower level. **3.** A rush of water down a steep place or a rapids. { shüt }

Shope papilloma [VET MED] A transmissible, virus-induced papilloma occurring naturally on the skin of rabbits. { 'shōp ,pap·ə'lō·mə }

shore [GEOL] **1.** The narrow strip of land immediately bordering a body of water. **2.** See seashore. { shȯr }

shore current [HYD] A water current near a shoreline, often flowing parallel to the shore. { 'shȯr ,kə·rənt }

shore ice [OCEANOGR] Sea ice that has been beached by wind, tides, currents, or ice pressure; it is a type of fast ice, and may sometimes be rafted ice. { 'shȯr ,īs }

shoreline [GEOL] The intersection of a specified plane of water, especially mean high water, with the shore; a limit which changes with the tide or water level. Also known as strandline; waterline. { 'shȯr,līn }

shoreline of emergence [GEOL] A straight or gently curving shoreline formed by the dominant relative emergence of the floor of an ocean or a lake. Also known as emerged shoreline; negative shoreline. { 'shȯr,līn əv i'mər·jəns }

shoreside See onshore. { 'shȯr,sīd }

short-crested wave [OCEANOGR] An ocean wave whose crest is of finite length; that is, the type actually found in nature. { 'shȯrt ¦kres·təd 'wāv }

short-day response [BIOL] A photoperiodic response to decreasing days and increasing nights. { ,shȯrt ,dā ri'späns }

short-range forecast [METEOROL] A weather forecast made for a time period generally not greater than 48 hours in advance. { 'shȯrt ¦rānj 'fȯr,kast }

short wave See deep-water wave. { 'shȯrt ¦wāv }

shortwave radiation [PHYS] A term used loosely to distinguish radiation in the visible and near-visible portions of the electromagnetic spectrum (roughly 0.4 to 1.0 micrometer in wavelength) from long-wave radiation (infrared radiation). { 'shȯrt'wāv ,rād·ē'ā·shən }

Showalter stability index [METEOROL] A measure of the local static stability of the atmosphere, expressed as a numerical index. { 'shō,wȯl·tər stə,bil·əd·ē ,in,deks }

shower [METEOROL] Precipitation from a convective cloud; characterized by the suddenness with which it starts and stops, by the rapid changes of intensity, and usually by rapid changes in the appearance of the sky. { 'shaū·ər }

371

shrub |BOT| A low woody plant with several stems. { shrəb }

shuga |OCEANOGR| A spongy, rather opaque, whitish chunk of ice which forms instead of pancake ice if the freezing takes place in sea water which is considerably agitated. { 'shü·gə }

Siberian anticyclone *See* Siberian high. { sī'bir·ē·ən 'ant·i,sī,klòn }

Siberian high [METEOROL] An area of high pressure which forms over Siberia in winter, and which is particularly apparent on mean charts of sea-level pressure; centered near lake Baikal. Also known as Siberian anticyclone. { sī'bir·ē·ən 'hī }

side stream *See* tributary. { 'sīd,strēm }

sierra |GEOGR| A high range of hills or mountains with irregular peaks that give a sawtooth profile. { sē'er·ə }

sieve tissue *See* phloem. { 'siv ,tish·ü }

sigma-T |OCEANOGR| An abbreviated value of the density of a sea-water sample of temperature T and salinity S: $\sigma T = |\rho(S,T) - 1| \times 10^3$, where $\rho(S,T)$ is the value of the sea-water density in centimeter-gram-second units at standard atmospheric pressure. { 'sig·mə ,tē }

significant wave |OCEANOGR| Statistically, a wave with the average height of the highest third of the waves of a given wave group. { sig'nif·i·kənt ,wāv }

sikussak |OCEANOGR| Very old sea ice trapped in fjords; it resembles glacier ice because snowfall and snow drifts contribute to its formation. { sə'kü,säk }

silage |AGR| Green or mature fodder that is fermented to retard spoilage and produce a succulent winter feed for livestock. { 'sī·lij }

siliceous ooze |GEOL| An ooze composed of siliceous skeletal remains of organisms, such as radiolarians. { sə'lish·əs 'üz }

silicification |GEOL| Introduction of or replacement by silica. Also known as silification. { sə,lis·ə·fə'kā·shən }

silicified wood |GEOL| A material formed by the silicification of wood, generally in the form of opal or chalcedony, in such a manner as to preserve the original form and structure of the wood. Also known as agatized wood; opalized wood; petrified wood; woodstone. { sə'lis·ə,fīd 'wùd }

silicle |BOT| A many-seeded capsule formed from two united carpels, usually of equal length and width, and divided on the inside by a replum. { 'sil·ə·kəl }

silification *See* silicification. { ,sil·ə·fə'kā·shən }

sill depth |OCEANOGR| The maximum depth at which there is horizontal communication between an ocean basin and the open ocean. Also known as threshold depth. { 'sil ,depth }

silt |GEOL| **1.** A rock fragment or a mineral or detrital particle in the soil having a diameter of 0.002–0.05 millimeter that is, smaller than fine sand and larger than coarse clay. **2.** Sediment carried or deposited by water. **3.** Soil containing at least 80% silt and less than 12% clay. { silt }

silt discharge rating *See* sediment discharge rating. { 'silt 'dis,chärj ,rād·iŋ }

silver [CHEM] A white metallic transition element, symbol Ag, with atomic number 47; soluble in acids and alkalies, insoluble in water; melts at 961°C, boils at 2212°C; used in photographic chemicals, alloys, conductors, and plating. { 'sil·vər }

silver frost [METEOROL] A deposit of glaze built up on trees, shrubs, and other exposed objects during a fall of freezing precipitation; the product of an ice storm. Also known as silver thaw. { 'sil·vər 'fròst }

silver potassium cyanide [CHEM] $KAg(CN)_2$ Toxic, white crystals soluble in water and alcohol; used in silver plating and as a bactericide and antiseptic. Also known as potassium argentocyanide; potassium cyanoargentate. { 'sil·vər pə'tas·ē·əm 'sī·ə,nīd }

silver storm See ice storm. { 'sil·vər 'storm }

silver thaw See silver frost. { 'sil·vər 'thȯ }

silviculture [FOR] The theory and practice of controlling the establishment, composition, and growth of stands of trees for any of the goods and benefits that they may be called upon to produce. { 'sil·və,kəl·chər }

simple fruit [BOT] A fruit that has developed from a single carpel or several united carpels. { 'sim·pəl 'früt }

simple leaf [BOT] A leaf having one blade, or a lobed leaf in which the separate parts do not reach down to the midrib. { 'sim·pəl 'lēf }

simple pistil [BOT] A pistil that consists of a single carpel. { 'sim·pəl 'pis·təl }

simple spit [GEOGR] A spit, either straight or recurved, without the development of minor spits at its end or along its inner side. { 'sim·pəl 'spit }

single-station analysis [METEOROL] The analysis or reconstruction of the weather pattern from more or less continuous meteorological observations made at a single geographic location, or the body of techniques employed in such an analysis. { 'siŋ·gəl ¦stā·shən ə'nal·ə·səs }

single-theodolite observation [METEOROL] The usual type of pilot-balloon observation, that is, using one theodolite. { 'siŋ·gəl thē¦äd·əl,īt ,äb·zər'vā·shən }

singular corresponding point [METEOROL] A center of elevation or depression on a constant-pressure chart (or a center of high or low pressure on a constant-height chart) considered as a reappearing characteristic of successive charts. { 'siŋ·gyə·lər ¦kär·ə¦spänd·iŋ 'pȯint }

singularity [METEOROL] A characteristic meteorological condition which tends to occur on or near a specific calendar date more frequently than chance would indicate; an example is the January thaw. { ,siŋ·gyə'lar·əd·ē }

sinkhole [GEOL] Closed surface depressions in regions of karst topography produced by solution of surface limestone or the collapse of cavern roofs. { 'siŋk,hōl }

sinking [OCEANOGR] The downward movement of surface water generally caused by converging currents or when a water mass becomes denser than the surrounding water. Also known as downwelling. { 'siŋk·iŋ }

siphon [BOT] A tubular element in various algae. [GEOL] A passage in a cave system that connects with a water trap. { 'sī·fən }

Siphonaptera [ZOO] The fleas, an order of insects characterized by a small, laterally compressed, oval body armed with spines and setae, three pairs of legs modified for jumping, and sucking mouthparts. { ,sī·fə'näp·trə }

siphonogamous [BOT] In plants, especially seed plants, the accomplishment of fertilization by means of a pollen tube. { ,sī·fə'näg·ə·məs }

sirocco [METEOROL] A warm south or southeast wind in advance of a depression moving eastward across the southern Mediterranean Sea or North Africa. { sə'rä·kō }

sisal [BOT] Agave sisalina. An agave of the family Amaryllidaceae indigenous to Mexico and Central America; a coarse, stiff yellow fiber produced from the leaves is used for making twine and brush bristles. { 'sī·səl }

skauk [HYD] An extensive field of crevasses in a glacier. { skȯk }

skavl See sastruga. { 'skav·əl }

skeleton layer [OCEANOGR] The structure that is formed at the bottom of sea ice while freezing, and consists of vertically oriented platelets of ice separated by layers of brine. { 'skel·ət·ən ,lā·ər }

skill score [METEOROL] In synoptic meteorology, an index of the degree of skill of a set of forecasts, expressed with reference to some standard such as forecasts based upon chance, persistence, or climatology. { 'skil ,skȯr }

skim ice [HYD] First formation of a thin layer of ice on the water surface. { 'skim ,īs }

skimming [HYD] **1.** Diversion of water from a stream or conduit by shallow overflow in order to avoid diverting sand, silt, or other debris carried as bottom load. **2.** Withdrawal of fresh groundwater from a thin body or lens floating on salt water by means of shallow wells or infiltration galleries. { 'skim·iŋ }

skin erythema dose [MED] A unit of radioactive dose resulting from exposure to electromagnetic radiation, equal to the dose that slightly reddens or browns the skin of 80% of all persons within 3 weeks after exposure; it is approximately 1000 roentgens for gamma rays, 600 roentgens for x-rays. Abbreviated SED. { 'skin ,er·ə'thē·mə ,dōs }

sky cover [METEOROL] In surface weather observations, the amount of sky covered but not necessarily concealed by clouds or by obscuring phenomena aloft, the amount of sky concealed by obscuring phenomena that reach the ground, or the amount of sky covered or concealed by a combination of the two phenomena. { 'skī ,kəv·ər }

sky map [METEOROL] A pattern of variable brightness observable on the underside of a cloud layer, and caused by the different reflectivities of material on the earth's surface immediately beneath the clouds; this term is used mainly in polar regions. { 'skī ,map }

slab [HYD] A layer in, or the whole thickness of, a snowpack that is very hard and has the ability to sustain elastic (reversible) deformation under stress. { slab }

slack ice [HYD] Ice fragments on still or slow-moving water. { 'slak 'īs }

slack water [OCEANOGR] The interval when the speed of the tidal current is very weak or zero; usually refers to the period of reversal between ebb and flood currents. { 'slak 'wȯd·ər }

slash [FOR] Debris, such as logs, chunks of wood, bark, and branches, in an open forest tract. { slash }

slate [GEOL] A group name for various very-fine-grained rocks derived from mudstone, siltstone, and other clayey sediment as a result of low-degree regional metamorphism; characterized by perfect fissility or slaty cleavage which is a regular or perfect planar schistosity. { slāt }

sleet [METEOROL] Colloquially in some parts of the United States, precipitation in the form of a mixture of snow and rain. { slēt }

slice method [METEOROL] A method of evaluating the static stability over a limited area at any reference level in the atmosphere; unlike the parcel method, the slice method takes into account continuity of mass by considering both upward and downward motion. { 'slīs ,meth·əd }

slick [OCEANOGR] Area in which capillary waves are absent or suppressed. { slik }

slime disease [PL PATH] Any of several diseases of plants characterized by slimy rot of the parts. { 'slīm di,zēz }

slime fungus See slime mold. { 'slīm ,fəŋ·gəs }

slime mold [MYCOL] The common name for members of the Myxomycetes. Also known as slime fungus. { 'slīm ,mōld }

sliming [OCEANOGR] The formation of films of algae on submerged structures. { 'slīm·iŋ }

slop culture [BOT] A method of growing plants in which surplus nutrient fluid is allowed to run through the sand or other medium in which the plants are growing. { 'släp ‚kǝl·chǝr }

slough [HYD] A minor marshland or tidal waterway which usually connects other tidal areas; often more or less equivalent to a bayou. { slaů }

slough ice [HYD] Slushy ice or snow. { 'slaů ‚īs }

slow sand filter [CIV ENG] A bed of fine sand 20–48 inches (151–122 centimeters) deep through which water, being made suitable for human consumption and other purposes, is passed at a fairly low rate, 2,500,000 to 10,000,000 gallons per acre (23,000 to 94,000 cubic meters per hectare); an underdrain system of graded gravel and perforated pipes carries the water from the filters to the point of discharge. { 'slō 'sand ‚fil·tǝr }

sludge [OCEANOGR] A soft or muddy bottom deposit as on tideland or in a stream bed. { slǝj }

sludge cake [OCEANOGR] An accumulation of sludge hardened into a cake strong enough to bear the weight of a man. { 'slǝj ‚kāk }

sludge floe [OCEANOGR] Sludge that is hardened into a floe strong enough to bear the weight of a person. { 'slǝj ‚flō }

sludge ice *See* sludge. { 'slǝj ‚īs }

sludge lump [OCEANOGR] An irregular mass of sludge formed as a result of strong winds. { 'slǝj ‚lǝmp }

slush [HYD] Snow or ice on the ground that has been reduced to a soft, watery mixture by rain, warm temperature, or chemical treatment. [OCEANOGR] *See* sludge. { slǝsh }

slush ball [HYD] An extremely compact accretion of snow, frazil, and ice particles. { 'slǝsh ‚bȯl }

slush field [HYD] An area of water-saturated snow having a soupy consistency. Also known as snow swamp. { 'slǝsh ‚fēld }

slushflow [HYD] **1.** A mudflow-like outburst of water-saturated snow along a stream course, commonly occurring in the Arctic Zone after intense thawing has produced more meltwater than can drain through the snow, and having a width generally several times greater than that of the stream channel. **2.** A flow of clear slush on a glacier, as in Greenland. { 'shǝsh‚flō }

slush pond [HYD] A pool or lake containing slush, on the ablation surface of a glacier. { 'slǝsh ‚pänd }

small-craft warning [METEOROL] A warning, for marine interests, of impending winds up to 28 knots (32 miles per hour or 52 kilometers per hour). { 'smȯl ¦kraft 'wȯrn·iŋ }

small diurnal range [OCEANOGR] The difference in height between mean lower high water and mean higher low water. { 'smȯl dī¦ǝrn·ǝl 'rānj }

small hail [METEOROL] Frozen precipitation consisting of small, semitransparent, roundish grains, each grain consisting of a snow pellet surrounded by a very thin ice covering, giving it a glazed appearance. { 'smȯl 'hāl }

small ice floe [OCEANOGR] An ice floe of sea ice 30 to 600 feet (9 to 180 meters) across. { 'smȯl 'īs ‚flō }

smallpox [MED] An acute, infectious, viral disease characterized by severe systemic involvement and a single crop of skin lesions which proceeds through macular, papular, vesicular, and pustular stages. Also known as variola. { 'smȯl‚päks }

smallpox vaccine [MED] A vaccine prepared from a glycerinated suspension of the exudate from cowpox vesicles obtained from healthy vaccinated calves or sheep. Also known as antismallpox vaccine; glycerinated vaccine virus; Jennerian vaccine; virus vaccinium. { 'smȯl,päks vak,sēn }

small tropic range [OCEANOGR] The difference in height between tropic lower high water and tropic higher low water. { 'smȯl 'träp·ik 'rānj }

smart structures [ENG] Structures that are capable of sensing and reacting to their environment in a predictable and desired manner, through the integration of various elements, such as sensors, actuators, power sources, signal processors, and communications network. In addition to carrying mechanical loads, smart structures may alleviate vibration, reduce acoustic noise, monitor their own condition and environment, automatically perform precision alignments, or change their shape or mechanical properties on command. { ˌsmärt 'strək·chərz }

smog [METEOROL] Air pollution consisting of smoke and fog. { smäg }

smoke horizon [METEOROL] The top of a smoke layer which is confined by a low-level temperature inversion in such a way as to give the appearance of the horizon when viewed from above against the sky; in such instances the true horizon is usually obscured by the smoke layer. { 'smōk hə,rīz·ən }

smooth [OCEANOGR] Comparatively calm water between heavy seas. { smü<u>th</u> }

smooth sea [OCEANOGR] Sea with waves no higher than ripples or small wavelets. { 'smü<u>th</u> 'sē }

smudge [PL PATH] Any of several fungus diseases of cereals and other plants characterized by dark, sooty discolorations. { sməj }

smudging [ENG] A frost-preventive measure used in orchards; properly, it means the production of heavy smoke, supposed to prevent radiational cooling, but it is generally applied to both heating and smoke production. { 'sməj·iŋ }

smut [PL PATH] Any of various destructive fungus diseases of cereals and other plants characterized by large dusty masses of dark spores on the plant organs. { smət }

smut fungus [MYCOL] The common name for members of the Ustilaginales. { 'smət ˌfəŋ·gəs }

snag [FOR] A standing dead tree. { snag }

snaking stream See meandering stream. { 'snāk·iŋ ˌstrēm }

snap [METEOROL] A brief period of extreme (generally cold) weather setting in suddenly, as in a "cold snap." { snap }

snezhura See snow slush. { 'snezh·ə·rə }

snout [HYD] The protruding lower extremity of a glacier. { snaút }

snow [METEOROL] The most common form of frozen precipitation, usually flakes of starlike crystals, matted ice needles, or combinations, and often rime-coated. { snō }

snow accumulation [METEOROL] The actual depth of snow on the ground at any instant during a storm, or after any single snowstorm or series of snowstorms. { 'snō ə,kyü·myə,lā·shən }

snow avalanche [HYD] An avalanche of relatively pure snow; some rock and earth material may also be carried downward. Also known as snowslide. { 'snō ,av·ə,lanch }

snowbank glacier See nivation glacier. { 'snō,baŋk ,glā·shər }

snow banner [METEOROL] Snow being blown from a mountain crest. Also known as snow plume; snow smoke. { 'snō ,ban·ər }

snow barchan [HYD] A crescentic or horseshoe-shaped snow dune of windblown snow with the ends pointing downwind. Also known as snow medano. { 'snō bär'kän }

snow blink [METEOROL] A bright, white glare on the underside of clouds, produced by the reflection of light from a snow-covered surface; this term is used in polar regions with reference to the sky map. Also known as snow sky. { 'snō ‚bliŋk }

snowbridge [HYD] Snow bridging a crevasse in a glacier. { 'snō‚brij }

snow cap [HYD] **1.** Snow covering a mountain peak when no snow exists at lower elevations. **2.** Snow on the surface of a frozen lake. { 'snō ‚kap }

snow climate *See* polar climate. { 'snō ‚klī·mət }

snow cloud [METEOROL] A popular term for any cloud from which snow falls. { 'snō ‚klaůd }

snow concrete [HYD] Snow that is compacted at low temperatures by heavy objects (as by a vehicle) and that sets into a tough substance of considerably greater strength than uncompressed snow. Also known as snowcrete. { 'snō ‚kän‚krēt }

snow course [HYD] An established line, usually from several hundred feet to as much as a mile long, traversing representative terrain in a mountainous region of appreciable snow accumulation; along this course, measurements of snow cover are made to determine its water equivalent. { 'snō ‚kȯrs }

snow cover [HYD] **1.** All accumulated snow on the ground, including that derived from snowfall, snowslides, and drifting snow. Also known as snow mantle. **2.** The extent, expressed as a percentage, of snow cover in a particular area. { 'snō ‚kəv·ər }

snow-cover chart [METEOROL] A synoptic chart showing areas covered by snow and contour lines of snow depth. { 'snō ‚kəv·ər ‚chärt }

snowcreep [HYD] The slow internal deformation of a snowpack resulting from the stress of its own weight and metamorphism of snow crystals. { 'snō‚krēp }

snowcrete *See* snow concrete. { 'snō‚krēt }

snow crust [HYD] A crisp, firm, outer surface upon snow. { 'snō ‚krəst }

snow crystal [METEOROL] Any of several types of ice crystal found in snow; a snow crystal is a single crystal, in contrast to a snowflake which is usually an aggregate of many single snow crystals. { 'snō ‚krist·əl }

snow cushion [HYD] An accumulation of snow, commonly deep, soft, and unstable, deposited in the lee of a cornice on a steep mountain slope. { 'snō ‚kůsh·ən }

snow density [HYD] The ratio of the volume of meltwater that can be derived from a sample of snow to the original volume of the sample; strictly speaking, this is the specific gravity of the snow sample. { 'snō ‚den·səd·ē }

snowdrift [HYD] Snow deposited on the lee of obstacles, lodged in irregularities of a surface, or collected in heaps by eddies in the wind. { 'snō‚drift }

snowdrift glacier [HYD] A semipermanent mass of firn, formed by drifted snow in depressions in the ground or behind obstructions. Also known as catchment glacier; drift glacier. { 'snō‚drift ‚glā·shər }

snowdrift ice [HYD] Permanent or semipermanent masses of ice, formed by the accumulation of drifted snow in the lee of projections, or in depressions of the ground. Also known as glacieret. { 'snō‚drift ‚īs }

snow dune [HYD] An accumulation of wind-transported snow resembling the forms of sand dunes. { 'snō ‚dün }

snow dust [METEOROL] Fine snow crystals fragmented or driven by the wind. { 'snō ‚dəst }

snow eater |METEOROL| **1.** Any warm wind blowing over a snow surface; usually applied to a foehn wind. **2.** A fog over a snow surface; so called because of the frequently observed rapidity with which a snow cover disappears after a fog sets in. { 'snō ‚ēd·ər }

snowfall |METEOROL| **1.** The rate at which snow falls; in surface weather observations, this is usually expressed as inches of snow depth per 6-hour period. **2.** A snow storm. { 'snō‚fȯl }

snowfield |HYD| **1.** A broad, level, relatively smooth and uniform snow cover on ground or ice at high altitudes or in mountainous regions above the snow line. **2.** The accumulation area of a glacier. **3.** A small glacier or accumulation of perennial ice and snow too small to be designated a glacier. { 'snō‚fēld }

snowflake |METEOROL| An ice crystal or, much more commonly, an aggregation of many crystals which falls from a cloud; simple snowflakes (single crystal) exhibit beautiful variety of form, but the symmetrical shapes reproduced so often in photomicrographs are not actually found frequently in snowfalls; broken single crystals, fragments, or clusters of such elements are much more typical of actual snows. { 'snō‚flāk }

snow flurry |METEOROL| Popular term for snow shower, particularly of a very light and brief nature. { 'snō ‚flər·ē }

snow forest climate |CLIMATOL| A major category in W. Köppen's climatic classification, defined by a coldest-month mean temperature of less than 26.6°F (3°C) and a warmest-month mean temperature of greater than 50°F (10°C). { 'snō¦fär·əst ‚klī·mət }

snow fungus *See* Tremella fuciformis. { 'snō ‚fəŋ·gəs }

snow gage |HYD| An instrument for measuring the amount of water equivalent in a snowpack. Also known as snow sampler. { 'snō ‚gāj }

snow garland |HYD| A rare phenomenon in which snow is festooned from trees, fences, and so on, in the form of a rope of snow, several feet long and several inches in diameter; produced by surface tension acting in thin films of water bonding individual crystals; such garlands form only when the surface temperature is close to the melting point, for only then will the requisite films of slightly supercooled water exist. { 'snō ‚gär·lənd }

snow geyser |METEOROL| Fine, powdery snow blown upward by a snow tremor. { 'snō ‚gī·zər }

snow glide |HYD| The slow slip of a snowpack over the ground surface caused by the stress of its own weight. { 'snō ‚glīd }

snow grains |METEOROL| Precipitation in the form of very small, white opaque particles of ice; the solid equivalent of drizzle; the grains resemble snow pellets in external appearance, but are more flattened and elongated, and generally have diameters of less than 1 millimeter; they neither shatter nor bounce when they hit a hard surface. Also known as granular snow. { 'snō ‚grānz }

snow ice |HYD| Ice crust formed from snow, either by compaction or by the refreezing of partially thawed snow. { 'snō ‚īs }

snow line |GEOGR| **1.** A transient line delineating a snow-covered area or altitude. **2.** An area with more than 50% snow cover. **3.** The altitude or geographic line separating areas in which snow melts in summer from areas having perennial ice and snow. { 'snō ‚līn }

snow mantle *See* snow cover. { 'snō ‚mant·əl }

snow medano *See* snow barchan. { 'snō mə'dä·nō }

snowmelt |HYD| The water resulting from the melting of snow; it may evaporate, seep into the ground, or become a part of runoff. { 'snō‚melt }

snowpack [HYD] The amount of annual accumulation of snow at higher elevations in the western United States, usually expressed in terms of average water equivalent. { 'snō͵pak }

snow pellets [METEOROL] Precipitation consisting of white, opaque, approximately round (sometimes conical) ice particles which have a snowlike structure and are about 2 to 5 millimeters in diameter; snow pellets are crisp and easily crushed, differing in this respect from snow grains, and they rebound when they fall on a hard surface and often break up. Also known as graupel; soft hail; tapioca snow. { 'snō ͵pel·əts }

snow plume *See* snow banner. { 'snō ͵plüm }

snowquake *See* snow tremor. { 'snō͵kwāk }

snow ripple *See* wind ripple. { 'snō ͵rip·əl }

snow roller [HYD] A cylinder mass of snow, rather common in mountainous or hilly regions; it occurs when snow, moist enough to be cohesive, is picked up by wind blowing down a slope and rolled onward and downward until either it becomes too large or the ground levels off too much for the wind to propel it further; snow rollers vary in size from very small cylinders to some as large as 4 feet (1.2 meters) long and 7 feet (2.1 meters) in circumference. { 'snō ͵rō·lər }

snow sampler *See* snow gage. { 'snō ͵sam·plər }

snowshed [HYD] A drainage basin primarily supplied by snowmelt. { 'snō͵shed }

snow sky *See* snow blink. { 'snō ͵skī }

snowslide *See* snow avalanche. { 'snō͵slīd }

snow sludge [OCEANOGR] Sludge formed mainly from snow. { 'snō ͵sləj }

snow slush [HYD] Slush formed from snow that has fallen into water that is at a temperature below that of the snow. Also known as snezhura. { 'snō ͵sləsh }

snow smoke *See* snow banner. { 'snō ͵smōk }

snow stage [METEOROL] The thermodynamic process of sublimation of water vapor into snow in an idealized saturation-adiabatic or pseudoadiabatic expansion (lifting) of moist air; the snow stage begins at the condensation level when it is higher than the freezing level. { 'snō ͵stāj }

snowstorm [METEOROL] A storm in which snow falls. { 'snō͵stȯrm }

snow survey [HYD] The process of determining depth and water content of snow at representative points, for example, along a snow course. { 'snō ͵sər͵vā }

snow swamp *See* slush field. { 'snō ͵swämp }

snow tremor [HYD] A disturbance in a snowfield, caused by the simultaneous settling of a large area of thick snow crust or surface layer. Also known as snowquake. { 'snō ͵trem·ər }

sobole [BOT] An underground creeping stem. { 'sä·bə͵lē }

social insects [ECOL] Insect species in which individuals share resources and reproduce cooperatively. { ¦sō·shəl 'in͵seks }

society [ZOO] An organization of individuals of the same species in which there are divisions of resources and of labor as well as mutual dependence. { sə'sī·əd·ē }

soda lake [HYD] An alkali lake rich in dissolved sodium salts, especially sodium carbonate, sodium chloride, and sodium sulfate. Also known as natron lake. { 'sōd·ə ͵lāk }

sodium aluminum silicofluoride [CHEM] $Na_5Al(SiF_6)_4$ A toxic, white powder, used for mothproofing and in insecticides. { 'sōd·ē·əm ə'lü·mə·nəm ͵sil·ə·kō'flu̇r͵īd }

sodium arsenate [CHEM] $Na_3AsO_4 \cdot 12H_2O$ Water-soluble, poisonous, clear, colorless crystals with a mild alkaline taste; melts at 86°C; used in medicine, insecticides, dry colors, and textiles, and as a germicide and a chemical intermediate. { 'sōd·ē·əm 'ärs·ən,āt }

sodium arsenite [CHEM] $NaAsO_2$ A poisonous, water-soluble, grayish powder; used in antiseptics, dyeing, insecticides, and soaps for taxidermy. { 'sōd·ē·əm 'ärs·ən,īt }

sodium borate [CHEM] $Na_2B_4O_7 \cdot 10H_2O$ A water-soluble, odorless, white powder; melts between 75 and 200°C; used in glass, ceramics, starch and adhesives, detergents, agricultural chemicals, pharmaceuticals, and photography; the impure form is known as borax. Also known as sodium pyroborate; sodium tetraborate. { 'sōd·ē·əm 'bȯ,rāt }

sodium cacodylate [CHEM] $C_2H_6AsNaO_2$ A herbicide used as a harvest aid. Also known as bollseye (trade name). { ,sōd·ē·əm ka'käd·əl,āt }

sodium dehydroacetate [CHEM] $C_8H_7NaO_4 \cdot H_2O$ A tasteless, white powder, soluble in water and propylene glycol; used as a fungicide and plasticizer, in toothpaste, and for pharmaceuticals. { 'sōd·ē·əm dē,hī·drō'as·ə,tāt }

sodium dichloroisocyanate [CHEM] $HC_3N_3O_3NaCl$ A white, crystalline compound, soluble in water; used as a bactericide and algicide in swimming pools. { 'sōd·ē·əm dī,klȯr·ō,ī·sō'sī·ə,nāt }

sodium dichloroisocyanurate [CHEM] $C_3N_3O_3Cl_2Na$ White, crystalline powder; used in dry bleaches, detergents, and cleaning compounds, and for water and sewage treatment. { 'sōd·ē·əm dī,klȯr·ō,ī·sō,sī·ə'nûr,āt }

sodium dimethyldithiocarbamate [CHEM] $(CH_3)_2NCS_2Na$ Amber to light green liquid; used as a fungicide, corrosion inhibitor, and rubber accelerator. Abbreviated SDDC. { 'sōd·ē·əm dī,meth·əl·dī,thī·ō'kär·bə,māt }

sodium dinitro-*ortho*-cresylate [CHEM] $CH_3C_6H_2(NO_2)_2ONa$ A toxic, orange-yellow dye, used as a herbicide and fungicide. { 'sōd·ē·əm dī,nī,trō ,ȯr·thō 'kres·ə,lat }

sodium fluoride [CHEM] NaF A poisonous, water-soluble, white powder, melting at 988°C; used as an insecticide and a wood and adhesive preservative, and in fungicides, vitreous enamels, and dentistry. { 'sōd·ē·əm 'flùr,īd }

sodium fluosilicate [CHEM] Na_2SiF_6 A poisonous, white, amorphous powder; slightly soluble in water; decomposes at red heat; used to fluoridate drinking water and to kill rodents and insects. Also known as sodium silicofluoride. { 'sōd·ē·əm ,flü·ə'sil·ə·kət }

sodium hypochlorite [CHEM] $NaOCl$ Air-unstable, pale-green crystals with sweet aroma; soluble in cold water, decomposes in hot water; used as a bleaching agent for paper pulp and textiles, as a chemical intermediate, and in medicine. { 'sōd·ē·əm ,hī·pō'klȯr,īt }

sodium isopropylxanthate [CHEM] $C_5H_7ONaS_2$ Light yellow, crystalline compound that decomposes at 150°C; soluble in water; used as a postemergence herbicide and as a flotation agent for ores. { 'sōd·ē·əm ,ī·sə,prō·pəl'zan,thāt }

sodium metaborate [CHEM] $NaBO_2$ Water-soluble, white crystals, melting at 966°C; the aqueous solution is alkaline; made by fusing sodium carbonate with borax; used as an herbicide. { 'sōd·ē·əm ,med·ə'bȯr,āt }

sodium *N*-methyldithiocarbamate dihydrate [CHEM] $CH_3NHC(S)SNa \cdot 2H_2O$ A white, water-soluble, crystalline solid; used as a fungicide, insecticide, nematicide, and weed killer. { 'sōd·ē·əm ,en ,meth·əl·dī,thī·ə'kär·bə,māt dī'hī,drāt }

sodium nitrate [CHEM] $NaNO_3$ Fire-hazardous, transparent, colorless crystals with bitter taste; soluble in glycerol and water; melts at 308°C; decomposes when heated; used in manufacture of glass and pottery enamel and as a fertilizer and food preservative. { 'sōd·ē·əm 'nī,trāt }

sodium pentachlorophenate [CHEM] C_6Cl_5ONa A white or tan powder, soluble in water, ethanol, and acetone; used as a fungicide and herbicide. { 'sōd·ē·əm ¦pen·tə ¦klȯr·ə'fe,nāt }

sodium perborate [CHEM] $NaBO_2·H_2O_2·3H_2O$ A white powder with a saline taste; slightly soluble in water, decomposes in moist air; used in deodorants, in dental compositions, and as a germicide. Also known as peroxydol. { 'sōd·ē·əm pər'bȯr,āt }

sodium polysulfide [CHEM] Na_2S_x Yellow-brown granules, used to make dyes and colors, and insecticides, as a petroleum additive, and in electroplating. { 'sōd·ē·əm ¦päl·i'səl,fīd }

sodium propionate [CHEM] CH_3CH_2COONa Deliquescent, transparent crystals; soluble in water, slightly soluble in alcohol; used as a fungicide, and mold preventive. { 'sōd·ē·əm 'prō·pē·ə,nāt }

sodium pyroborate See sodium borate. { 'sōd·ē·əm ¦pī·rō'bȯr,āt }

sodium selenate [CHEM] $Na_2SeO_4·10H_2O$ White, poisonous, water-soluble crystals; used as an insecticide. { 'sōd·ē·əm 'sel·ə,nāt }

sodium silicofluoride See sodium fluosilicate. { 'sōd·ē·əm ¦sil·ə·kō'flur,īd }

sodium TCA See sodium trichloroacetate. { 'sōd·ē·əm ¦tē¦sē'ā }

sodium tetraborate See sodium borate. { 'sōd·ē·əm ¦te·trə'bȯr,āt }

sodium tetrasulfide [CHEM] Na_2S_4 Hygroscopic, yellow or dark-red crystals, melting at 275°C; used for insecticides and fungicides, ore flotation, and dye manufacture, and as a reducing agent. { 'sōd·ē·əm ¦tet·rə'səl,fīd }

sodium trichloroacetate [CHEM] CCl_3COONa A toxic material, used in herbicides and pesticides. Abbreviated sodium TCA. { 'sōd·ē·əm trī¦klȯr·ō'as·ə,tāt }

sodium 2,4,5-trichlorophenate [CHEM] $C_6H_2Cl_3ONa·1\frac{1}{2}H_2O$ Buff to light brown flakes, soluble in water, methanol, and acetone; used as a bactericide and fungicide. { 'sōd·ē·əm ¦tü ¦fȯr ¦fīv trī¦klȯrō'fe,nāt }

soffosian knob See frost mound. { sə'fō·zhən 'näb }

soft hail See snow pellets. { 'sȯft 'hāl }

soft rime [HYD] A white, opaque coating of fine rime deposited chiefly on vertical surfaces, especially on points and edges of objects, generally in supercooled fog. { 'sȯft 'rīm }

soft rot [PL PATH] A mushy, watery, or slimy disintegration of plant parts caused by either fungi or bacteria. { 'sȯft ,rät }

soft water [CHEM ENG] Water that is free of magnesium or calcium salts. { 'sȯft 'wȯd·ər }

soil [GEOL] **1.** Unconsolidated rock material over bedrock. **2.** Freely divided rock-derived material containing an admixture of organic matter and capable of supporting vegetation. { sȯil }

soil air [GEOL] The air and other gases in spaces in the soil; specifically, that which is found within the zone of aeration. Also known as soil atmosphere. { ¦sȯil ¦er }

soil atmosphere See soil air. { ¦sȯil ¦at·mə,sfir }

soil blister See frost mound. { 'sȯil ,blis·tər }

soil chemistry [GEOCHEM] The study and analysis of the inorganic and organic components and the life cycles within soils. { ¦sȯil ¦kem·ə·strē }

soil colloid [GEOL] Colloidal complex of soils composed principally of clay and humus. { ¦sȯil ¦kä,lȯid }

soil complex |GEOL| A mapping unit used in detailed soil surveys; consists of two or more recognized classifications. { ¦sȯil ¦käm‚pleks }

soil conservation [ECOL] Management of soil to prevent or reduce soil erosion and depletion by wind and water. { ¦sȯil ‚kän·sər‚vā·shən }

soil ecology [ECOL] The study of interactions among soil organisms and interactions between biotic and abiotic aspects of the soil environment. { ¦sȯil i‚käl·ə·jē }

soil erosion |GEOL| The detachment and movement of topsoil by the action of wind and flowing water. { 'sȯil i‚rōzh·ən }

soil fertility [AGR] The ability of a soil to supply plant nutrients. { 'sȯil fər‚til·əd·ē }

soil formation See soil genesis. { 'sȯil ‚fȯr·mā·shən }

soil genesis [GEOL] The mode by which soil originates, with particular reference to processes of soil-forming factors responsible for the development of true soil from unconsolidated parent material. Also known as pedogenesis; soil formation. { 'sȯil ‚jen·ə·səs }

soil line See soil pipe. { 'sȯil ‚līn }

soil microbiology [MICROBIO] A study of the microorganisms in soil, their functions, and the effect of their activities on the character of the soil and the growth and health of plant life. { ¦sȯil ¦mī·krə·bī'äl·ə·jē }

soil moisture See soil water. { ¦sȯil ¦mȯis·chər }

soil pipe [CIV ENG] A cast-iron or plastic pipe for carrying discharges from toilet fixtures from a building into the soil drain. Also known as soil line. { 'sȯil ‚pīp }

soil profile [GEOL] A vertical section of a soil, showing horizons and parent material. { ¦sȯil ¦prō‚fīl }

soil remediation [AGR] The removal of harmful contaminants in soil. { 'sȯil rə‚mē·dē‚ā·shən }

soil rot [PL PATH] Plant rot caused by soil microorganisms. { 'sȯil ‚rät }

soil science [GEOL] The study of the formation, properties, and classification of soil; includes mapping. Also known as pedology. { 'sȯil ‚sī·əns }

soil survey [GEOL] The systematic examination of soils, their description and classification, mapping of soil types, and the assessment of soils for various agricultural and engineering uses. { 'sȯil 'sər‚vā }

soil vent See stack vent. { 'sȯil ‚vent }

soil water [HYD] Water in the soil-water zone. Also known as rhizic water; soil moisture. { 'sȯil ‚wȯd·ər }

soil-water belt See belt of soil water. { 'sȯil ¦wȯd·ər ‚belt }

soil-water zone See belt of soil water. { 'sȯil ¦wȯd·ər ‚zōn }

sol-air temperature [METEOROL] The temperature which, under conditions of no direct solar radiation and no air motion, would cause the same heat transfer into a house as that caused by the interplay of all existing atmospheric conditions. { 'säl 'er ‚tem·prə·chər }

solanine [BIOL] A bitter poisonous alkaloid derived from potato sprouts (Solanum tuberosum), tomatoes, and nightshade. { 'sō·lə‚nēn }

solar air mass [METEOROL] The optical air mass penetrated by light from the sun for any given position of the sun. { 'sō·lər 'er ‚mas }

solar array [ENG] An assemblage of individual solar cells into series and parallel circuits to obtain the required working voltage and power. { 'sō·lər ə'rā }

solar battery   [ENG] An array of solar cells, usually connected in parallel and series. { 'sō·lər 'bad·ə·rē }

solar cell   [ENG] A semiconductor device that converts the radiant energy of sunlight directly and efficiently into electrical energy. { 'sō·lər ˌsel }

solar chimney   [ENG] A natural-draft drive device that uses solar radiation to provide upward momentum to a mass of air, thereby converting the thermal energy to kinetic energy, which can be extracted from the air with suitable wind machines. { ˌsō·lər 'chim·nē }

solar climate   [CLIMATOL] The hypothetical climate which would prevail on a uniform solid earth with no atmosphere; thus, it is a climate of temperature alone and is determined only by the amount of solar radiation received. { 'sō·lər 'klī·mət }

solar collector   [ENG] An installation designed to gather and accumulate energy in the form of solar radiation. { 'sō·lər kə'lek·tər }

solar constant   [METEOROL] The rate at which energy from the sun is received just outside the earth's atmosphere on a surface normal to the incident radiation and at the earth's mean distance from the sun; it is approximately 1367 watts per square meter. { 'sō·lər 'kän·stənt }

solar cooking   [ENG] The preparation of food by concentrating solar radiation on a heater plate. { 'sō·lər 'kůk·iŋ }

solar energy   [ENG] Energy emitted from the sun in the form of electromagnetic radiation that can be converted into other forms of energy, suc as heat or electricity. { 'sō·lər 'en·ər·jē }

solar engine   [ENG] An engine which converts solar energy into electrical, mechanical, or refrigeration energy. { 'sō·lər 'en·jən }

solar evaporation   [HYD] The evaporation of water due to the sun's heat. { 'sō·lər iˌvap·ə'rā·shən }

solar generator   [ENG] An electric generator powered by radiation from the sun. { 'sō·lər 'jen·əˌrād·ər }

solar heating   [ENG] The conversion of solar radiation into heat for technological, comfort-heating, and cooking purposes. { 'sō·lər 'hēd·iŋ }

solar heat storage   [ENG] The storage of solar energy for later use; usually accomplished by the heating of water or fusing a salt, although sand and gravel have been used as storage media. { 'sō·lər 'hēt ˌstȯr·ij }

solar house   [ENG] A house with large expanses of glass designed to catch solar radiation for heating. { ¦sō·lər ¦haůs }

solarimeter   [ENG] **1.** A type of pyranometer consisting of a Moll thermopile shielded from the wind by a bell glass. **2.** *See* pyranometer. { ˌsō·lə'rim·əd·ər }

solar pond   [ENG] A type of nonfocusing solar collector consisting of a pool of salt water heated by the sun; used either directly as a source of heat or as a power source for an electric generator. Also known as salt-gradient solar pond. { 'sō·lər 'pänd }

solar power   [ENG] The conversion of the energy of the sun's radiation to useful work. { 'sō·lər 'paů·ər }

solar power satellite   [ENG] A proposed collector of solar energy that would be placed in geostationary orbit where sunlight striking the satellite would be converted to electricity and then to microwaves, which would be beamed to earth. { ˌsō·lər ˌpaů·ər 'sad·əlˌīt }

solar propagation   [BOT] A method of rooting plant cuttings involving the use of a modified hotbed; bottom heat is provided by radiation of stored solar heat from bricks or stones in the bottom of the hotbed frame. { 'sō·lər ˌpräp·ə'gā·shən }

solar radiation |PHYS| The electromagnetic radiation and particles (electrons, protons, and rarer heavy atomic nuclei) emitted by the sun. { 'sō·lər ˌrād·ē'ā·shən }

solar tide |OCEANOGR| The tide caused solely by the tide-producing forces of the sun. { 'sō·lər 'tīd }

solar-topographic theory |CLIMATOL| The theory that the changes of climate through geologic time (the paleoclimates) have been due to changes of land and sea distribution and orography, combined with fluctuations of solar radiation of the order of 10–20% on either side of the mean. { 'sō·lər ˌtäp·əˌgraf·ik 'thē·ə·rē }

solar ultraviolet radiation |PHYS| That portion of the sun's electromagnetic radiation that has wavelengths from about 400 to about 4 nanometers; this radiation may sufficiently ionize the earth's atmosphere so that propagation of radio waves is affected. { 'sō·lər ˌəl·traˌvī·lət ˌrād·ē'ā·shən }

sole |HYD| The basal ice of a glacier, often dirty in appearance due to contained rock fragments. { sōl }

solenoidal index |METEOROL| The difference between the mean virtual temperature from the surface to some specified upper level averaged around the earth at 55° latitude, and the mean virtual temperature for the corresponding layer averaged at 35° latitude. { ˌsäl·əˌnòid·əl 'inˌdeks }

sole plane See sole. { 'sōl ˌplān }

solodize |GEOL| To improve a soil by removing alkalies from it. { 'sō·ləˌdīz }

solstitial tidal currents |OCEANOGR| Tidal currents of especially large tropic diurnal inequality occurring at the time of solstitial tides. { sälz'tish·əl 'tīd·əl ˌkə·rəns }

solstitial tides |OCEANOGR| Tides occurring near the times of the solstices, when the tropic range is especially large. { sälz'tish·əl 'tīdz }

solute compartmentation |BOT| The sequestering of a plant cell's salt in a vacuole so that the salt does not poison the cell. { 'säl·yüt kəmˌpärt·mən'tā·shən }

solvent extraction |ENG| A process for removing uranium fuel residue from used fuel elements of a reactor; it generally involves decay cooling under water for up to 6 months, removal of cladding, dissolution, separation of reusable fuel, decontamination, and disposal of radioactive wastes. Also known as liquid extraction. { 'säl·vənt ikˌstrak·shən }

Somali Current See East Africa Coast Current. { sə'mäl·ē 'kə·rənt }

somatic nervous system |BIOL| The portion of the nervous system concerned with the control of voluntary muscle and relating the organism with its environment. { sō 'mad·ik 'nər·vəs ˌsis·təm }

sone |PHYS| A unit of loudness, equal to the loudness of a simple 1000-hertz tone with a sound pressure level 40 decibels above 0.0002 microbar; a sound that is judged by listeners to be n times as loud as this tone has a loudness of n sones. { sōn }

sonic depth finder |ENG| A sonar-type instrument used to measure ocean depth and to locate underwater objects; a sound pulse is transmitted vertically downward by a piezoelectric or magnetostriction transducer mounted on the hull of the ship; the time required for the pulse to return after reflection is measured electronically. Also known as echo sounder. { 'sän·ik 'depth ˌfīn·dər }

Sonne dysentery |MED| An intestinal bacterial infection caused by *Shigella sonnei*. { 'zòn·ə 'dis·ənˌter·ē }

soot |CHEM| Impure black carbon with oily compounds obtained from the incomplete combustion of resinous materials, oils, wood, or coal. { sùt }

sooty mold |MYCOL| Ascomycetous fungi of the family Capnodiaceae, with dark mycelium and conidia. |PL PATH| A plant disease, common on *Citrus* species, characterized by a dense velvety layer of a sooty mold on exposed parts of the plant. { 'süd·ē 'mōld }

sorbic acid   [CHEM] $CH_3CH=CHCH=CHCOOH$ A white, crystalline compound; soluble in most organic solvents, slightly soluble in water; melts at 135°C; used as a fungicide and food preservative, and in the manufacture of plasticizers and lubricants.   { 'sȯr·bik 'as·əd }

sore shin   [PL PATH] A fungus disease of cowpea, cotton, tobacco, and other plants, beyond the seedling stage, marked by annular (ring-like) growth of the pathogen on the stem at the groundline.   { 'sȯr 'shin }

sorghum   [BOT] Any of a variety of widely cultivated grasses, especially *Sorghum bicolor* in the United States, grown for grain and herbage; growth habit and stem form are similar to Indian corn, but leaf margins are serrate and spikelets occur in pairs on a hairy rachis.   { 'sȯr·gəm }

sorghum downy mildew   [AGR] A serious fungus disease that systematically invades sorghum plants, causing stripped leaves and barren stalks.   { ˌsȯr·gəm ˌdau̇·nē 'mil ˌdü }

sorghum head smut   [AGR] A disease of sorghum plants that completely destroys the normal head and replaces it with masses of smut spores.   { ˌsȯr·gəm 'hed ˌsmət }

sorus   [BOT] **1.** A cluster of sporangia on the lower surface of a fertile fern leaf. **2.** A clump of reproductive bodies or spores in lower plants.   { 'sȯr·əs }

sou'easter   *See* southeaster.   { 'sau̇ˌēs·tər }

souma   [VET MED] A disease caused by *Trypanosoma vivax* in domestic and wild animals; the insect vectors are the tsetse fly and the stable fly.   { 'sü·mə }

sound   [PHYS] An alteration of properties of an elastic medium (most commonly air), such as pressure, particle displacement, or density, that propagates through the medium, or a superposition of such alterations; sound waves having frequencies above the audible (sonic) range are termed ultrasonic waves; those with frequencies below the sonic range are called infrasonic waves. Also known as acoustic wave; sound wave.   { sau̇nd }

sound absorption   [PHYS] A process in which sound energy is reduced when sound waves pass through a medium or strike a surface. Also known as acoustic absorption.   { 'sau̇nd əbˌsȯrp·shən }

sounding   *See* upper-air observation.   { 'sau̇nd·iŋ }

sound level   [PHYS] The sound pressure level (in decibels) at a point in a sound field, averaged over the audible frequency range and over a time interval, with a frequency weighting and the time interval as specified by the American National Standards Association.   { 'sau̇nd ˌlev·əl }

sound-level meter   [ENG] An instrument used to measure noise and sound levels in a specified manner; the meter may be calibrated in decibels or volume units and includes a microphone, an amplifier, an output meter, and frequency-weighting networks.   { 'sau̇nd ˈlev·əl ˌmēd·ər }

sound pressure level   [PHYS] A value in decibels equal to 20 times the logarithm to the base 10 of the ratio of the pressure of the sound under consideration to a reference pressure; reference pressures in common use are 0.0002 microbar and 1 microbar. Abbreviated SPL.   { 'sau̇nd ˈpresh·ər ˌlev·əl }

sound pressure spectrum level   [PHYS] Ten times the logarithm to base 10 of the ratio of the mean square pressure of the portion of sound within a specified frequency band to the mean square pressure of the portion of a reference sound within the same frequency band. Abbreviated SPSL.   { ˈsau̇nd ˌpresh·ər 'spek·trəm ˌlev·əl }

sound wave   *See* sound.   { 'sau̇nd ˌwāv }

source |ENG| A radioactive material packaged so as to produce radiation for experimental or industrial use. { sȯrs }

source data |SCI TECH| Data generated in the course of research. { 'sȯrs ˌdad·ə }

source region |METEOROL| An extensive area of the earth's surface characterized by essentially uniform surface conditions and so situated with respect to the general atmospheric circulation that an air mass may remain over it long enough to acquire its characteristic properties. { 'sȯrs ˌrē·jən }

South African tick-bite fever |MED| An infectious tick-borne rickettsial disease of humans which is similar to fièvre boutonneuse. { 'saůth 'af·ri·kən 'tik ˌbīt ˌfē·vər }

South America |GEOGR| The southernmost of the Western Hemisphere continents, three-fourths of which lies within the tropics. { 'saůth ə'mer·ə·kə }

South American blastomycosis |MED| An infectious, yeastlike fungus disease of humans seen primarily in Brazil; caused by *Blastomyces brasiliensis* and characterized by massive enlargement of the cervical lymph nodes. Also known as paracoccidioidomycosis. { 'saůth ə'mer·ə·kən ˌblas·tō,mī'kō·səs }

South Atlantic Current |OCEANOGR| An eastward-flowing current of the South Atlantic Ocean that is continuous with the northern edge of the West Wind Drift. { 'saůth at'lan·tik 'kə·rənt }

South Australian faunal region |ECOL| A marine littoral region along the southwestern coast of Australia. { 'saůth ȯ'strāl·yən 'fȯn·əl ˌrē·jən }

Southeast Drift Current |OCEANOGR| A North Atlantic Ocean current flowing southeastward and southward from a point west of the Bay of Biscay toward southwestern Europe and the Canary Islands, where it continues as the Canary Current. { saů'thēst 'drift 'kə·rənt }

southeaster |METEOROL| A southeasterly wind, particularly a strong wind or gale; for example, the winter southeast storms of the Bay of San Francisco. Also spelled sou'easter. { saů'thēs·tər }

South Equatorial Current |OCEANOGR| Any of several ocean currents, flowing westward, driven by the southeast trade winds blowing over the tropical oceans of the Southern Hemisphere and extending slightly north of the equator. Also known as Equatorial Current. { 'saůth ˌek·wə'tȯr·ē·əl 'kə·rənt }

souther |METEOROL| A south wind, especially a strong wind or gale. { 'saůth·ər }

southern bean mosaic virus |MICROBIO| The type species of the plant-virus genus *Sobemovirus*. It is transmitted mechanically or via seed or the bean leaf beetle. Symptoms include crinkled leaves expressing a mild mosaic. Abbreviated SBMV. { ¦səth·ərn ˌbēn mō'zā·ik ˌvī·rəs }

Southern Polar Front *See* Antarctic Convergence. { 'səth·ərn 'pō·lər 'frənt }

south foehn |METEOROL| A foehn condition sustained by a strong south-to-north airflow across a transverse mountain barrier; the south foehn of the Alps may well be the most striking foehn in the world. { 'saůth 'fān }

south frigid zone |GEOGR| That part of the earth south of the Antarctic Circle. { 'saůth 'frij·əd ˌzōn }

south geographical pole |GEOGR| The geographical pole in the Southern Hemisphere, at latitude 90°S. Also known as South Pole. { 'saůth ¦jē·ə¦graf·ə·kəl ˌpōl }

South Indian Current |OCEANOGR| An eastward-flowing current of the southern Indian Ocean that is continuous with the northern edge of the West Wind Drift. { 'saůth 'in·dē·ən 'kə·rənt }

South Pacific Current [OCEANOGR] An eastward-flowing current of the South Pacific Ocean that is continuous with the northern edge of the West Wind Drift. { 'sauth pə'sif·ik 'kə·rənt }

South Pole See south geographical pole. { 'sauth 'pōl }

south temperate zone [GEOGR] That part of the earth between the Tropic of Capricorn and the Antarctic Circle. { 'sauth 'tem·prət ,zōn }

southwester [METEOROL] A southwest wind, particularly a strong wind or gale. Also spelled sou'wester. { sauth'wes·tər }

sou'wester See southwester. { sau'wes·tər }

soybean [BOT] *Glycine max.* An erect annual legume native to China and Manchuria and widely cultivated for forage and for its seed. { 'sȯi,bēn }

sparganosis [VET MED] An infection by the plerocercoid larva, or sparganum, of certain species of *Spirometra* tapeworms; occurs primarily in animals, but can be transmitted to humans through contact with or ingestion of infected individuals. { ,spär·gə'nō·səs }

spatial dendrite [METEOROL] A complex ice crystal with fernlike arms that extend in many directions (spatially) from a central nucleus; its form is roughly spherical. Also known as spatial dendritic crystal. { 'spā·shəl 'den,drīt }

spatial dendritic crystal See spatial dendrite. { 'spā·shəl den'drid·ik 'krist·əl }

special observation [METEOROL] A category of aviation weather observation taken to report significant changes in one or more of the observed elements since the last previous record observation. { 'spesh·əl ,äb·zər'vā·shən }

special weather report [METEOROL] The encoded and transmitted weather report of a special observation. { 'spesh·əl 'weth·ər ri,pȯrt }

species [SYST] A taxonomic category ranking immediately below a genus and including closely related, morphologically similar individuals which actually or potentially interbreed. { 'spē·shēz }

species population [ECOL] A group of similar organisms residing in a defined space at a certain time. { 'spē·shēz ,päp·yə'lā·shən }

specific energy [HYD] The energy at any cross section of an open channel, measured above the channel bottom as datum; numerically the specific energy is the sum of the water depth plus the velocity head, $v^2/2g$, where v is the velocity of flow and g the acceleration of gravity. { spə'sif·ik 'en·ər·jē }

specific humidity [METEOROL] In a system of moist air, the (dimensionless) ratio of the mass of water vapor to the total mass of the system. { spə'sif·ik hyü'mid·əd·ē }

specific-volume anomaly [OCEANOGR] The excess of the actual specific volume of the sea water at any point in the ocean over the specific volume of sea water of salinity 35 parts per thousand (‰) and temperature 0°C at the same pressure. Also known as steric anomaly. { spə'sif·ik ¦väl·yəm ə'näm·ə·lē }

specific yield [HYD] The quantity of water which a unit volume of aquifer, after being saturated, will yield by gravity; it is expressed either as a ratio or as a percentage of the volume of the aquifer; specific yield is a measure of the water available to wells. { spə'sif·ik 'yēld }

specimen [SCI TECH] **1.** An item representative of others in the same class or group. **2.** A sample selected for testing, examination, or display. { 'spes·ə·mən }

speck [PL PATH] A fungus or bacterial disease of rice characterized by speckled grains. { spek }

spelean |GEOL| Of or pertaining to a feature in a cave. { spə'lē·ən }

spermatophyte |BOT| Any one of the seed-bearing vascular plants. { spər'mad·ə,fīt }

sperm nucleus |BOT| One of the two nuclei in a pollen grain that function in double fertilization in seed plants. { 'spərm ,nü·klē·əs }

sphagnum bog |ECOL| A bog composed principally of mosses of the genus *Sphagnum* (Sphagnales) but also of other plants, especially acid-tolerant species, which tend to form peat. { 'sfag·nəm 'bäg }

spicule |INVERTEBRATE ZOOLOGY| An empty diatom shell. { 'spik·yül }

spider |AGR| An attachment to a cultivator that pulverizes the soil. |ZOO| The common name for arachnids comprising the order Araneida. { 'spīd·ər }

spike |BOT| An indeterminate inflorescence with sessile flowers. { spīk }

spikelet |BOT| The compound inflorescence of a grass consisting of one or several bracteate spikes. { 'spīk·lət }

spike-tooth harrow |AGR| An implement with steel spikes extending downward from a frame and pulled by a tractor to pulverize and smooth plowed soil. { 'spīk,tüth 'har·ō }

spilling |OCEANOGR| The process by which steep waves break on approaching the shore; white water appears on the crest and the wave top gradually rolls over, without a crash. { 'spil·iŋ }

spilling breaker *See* plunging breaker. { 'spil·iŋ ,brāk·ər }

spillover |METEOROL| That part of orographic precipitation which is carried along by the wind so that it reaches the ground in the nominal rain shadow on the lee side of the barrier. { 'spil,ō·vər }

spill stream *See* overflow stream. { 'spil ,strēm }

spinacene *See* squalene. { 'spin·ə,sēn }

spindle tuber |PL PATH| A virus disease of the potato characterized by spindliness of the tops and tubers. { 'spin·dəl ,tü·bər }

spine |BOT| A rigid sharp-pointed process in plants; many are modified leaves. { spīn }

spinney |ECOL| A small grove of trees or a thicket with undergrowth. { 'spin·ē }

spiral band |METEOROL| Spiral-shaped radar echoes received from precipitation areas within intense tropical cyclones (hurricanes or typhoons); they curve cyclonically in toward the center of the storm and appear to merge to form the wall around the eye of the storm. Also known as hurricane band; hurricane radar band. { 'spī·rəl 'band }

spire |BOT| A narrow, tapering blade or stalk. { spīr }

spiricle |BOT| Any of the coiled threads in certain seed coats which uncoil when moistened. { 'spir·ə·kəl }

Spirochaetales |MICROBIO| An order of bacteria characterized by slender, helically coiled cells sometimes occurring in chains. { ,spī·rə·kē'tā·lēz }

spirochete |MICROBIO| The common name for any member of the order Spirochaetales. { 'spī·rə,kēt }

spit |GEOGR| A small point of land commonly consisting of sand or gravel and which terminates in open water. { spit }

Spitsbergen Current |OCEANOGR| An ocean current flowing northward and westward from a point south of Spitsbergen, and gradually merging with the East Greenland Current in the Greenland Sea; the Spitsbergen Current is the continuation of the northwestern branch of the Norwegian Current. { 'spits,bər·gən 'kə·rənt }

SPL _See_ sound pressure level.

splash erosion [GEOL] Erosion resulting from the impact of falling raindrops. { 'splash i,rōzh·ən }

splenic fever _See_ anthrax. { 'splen·ik 'fē·vər }

spongiform encephalopathies _See_ prion diseases. { ,spən·jə,fȯrm in,sef·ə'läp·ə·thēz }

spontaneous generation _See_ abiogenesis. { spän'tā·nē·əs ,jen·ə'rā·shən }

sporangiophore [BOT] A stalk or filament on which sporangia are borne. { spə 'ran·jē·ə,fȯr }

sporangiospore [BOT] A spore that forms in a sporangium. { spə'ran·jē·ə,spȯr }

sporangium [BOT] A case in which asexual spores are formed and borne. Plural, sporangia. { spə'ran·jē·əm }

spore [BIOL] A uni- or multicellular, asexual, reproductive or resting body that is resistant to unfavorable environmental conditions and produces a new vegetative individual when the environment is favorable. { spȯr }

sporocarp [BOT] Any multicellular structure in or on which spores are formed. { 'spȯr·ə,kärp }

sporocyst [BOT] A unicellular resting body from which asexual spores arise. { 'spȯr·ə ,sist }

sporophyll [BOT] A modified leaf that develops sporangia. { 'spȯr·ə,fil }

sporophyte [BOT] **1.** An individual of the spore-bearing generation in plants exhibiting alternation of generation. **2.** The spore-producing generation. **3.** The diploid in a plant life cycle. { 'spȯr·ə,fīt }

Sporozoa [ZOO] A subphylum of parasitic Protozoa, typically producing spores during the asexual stages of the life cycle. { ,spȯr·ə'zō·ə }

sporozoite [ZOO] A motile, infective stage of certain sporozoans, which is the result of sexual reproduction and which gives rise to an asexual cycle in the new host. { ,spȯr·ə'zō,īt }

spot blotch [PL PATH] A fungus disease of barley caused by _Helminthosporium sativum_ and characterized by the appearance of dark, elongated spots on the foliage. { 'spät ,bläch }

spotted wilt [PL PATH] A virus disease of various crop and wild plants, especially tomato, characterized by bronzing and downward curling of the leaves. { 'späd·əd 'wilt }

spot wind [METEOROL] In air navigation, wind direction and speed, either observed or forecast if so specified, at a designated altitude over a fixed location. { 'spät ,wind }

spray oil [AGR] A low-viscosity petroleum oil similar to lubricating oil; used to combat pests that attack trees and shrubbery. { 'sprā ,ȯil }

spreading concept _See_ sea-floor spreading. { 'spred·iŋ ,kän,sept }

spreading-floor hypothesis _See_ sea-floor spreading. { 'spred·iŋ ¦flȯr hī,päth·ə·səs }

spring [HYD] A general name for any flow of groundwater to the surface, usually occurring where the water table intersects the surface. { spriŋ }

spring crust [HYD] A type of snow crust, formed when loose firn is recemented by a decrease in temperature; it is most common in late winter and spring. { 'spriŋ 'krəst }

spring high water _See_ mean high-water springs. { 'spriŋ 'hī ,wȯd·ər }

spring low water _See_ mean low-water springs. { 'spriŋ 'lō,wȯd·ər }

spring rise [OCEANOGR] The height of mean high-water springs above the chart datum. { 'spriŋ 'rīz }

spring seepage [HYD] A spring of small discharge. Also known as weeping spring. { 'spriŋ ˌsēp·ij }

spring sludge *See* rotten ice. { 'spriŋ 'sləj }

spring snow [HYD] A coarse, granular snow formed during spring by alternate freezing and thawing. Also known as corn snow. { 'spriŋ 'snō }

spring tidal currents [OCEANOGR] Tidal currents of increased speed occurring at the time of spring tides. { 'spriŋ 'tīd·əl ˌkə·rəns }

spring tide [OCEANOGR] Tide of increased range which occurs about every 2 weeks when the moon is new or full. { 'spriŋ 'tīd }

spring-tooth harrow [AGR] An implement pulled over plowed soil that has long curved teeth of spring steel, used to break clods, level the surface, and destroy weeds. { ˌspriŋ ˌtüth 'har·ō }

spring velocity [OCEANOGR] The average speed of the maximum flood and maximum ebb of a tidal current at the time of spring tides. { 'spriŋ və'läs·əd·ē }

springwood [BOT] The portion of an annual ring that is formed principally during the growing season; it is softer, more porous, and lighter than summerwood because of its higher proportion of large, thin-walled cells. { 'spriŋˌwud }

sprinkle [METEOROL] A very light shower of rain. { 'spriŋ·kəl }

sprinkler irrigation [AGR] A method of providing water to plants by pipelines which carry water under pressure from a pump or elevated source to lines, with sprinkler heads spaced at appropriate intervals. { ¦spriŋ·klər ˌir·i'gā·shən }

sprite [METEOROL] A transient illumination that can appear over a laterally extensive thunderstorm, with a red body about 20 kilometers (12 miles) in diameter, extending up to an altitude of 85–90 kilometers (51–54 miles), and blue tendrils extending down to an altitude of about 45 kilometers (27 miles). { sprīt }

spruce budworm [ZOO] The larva of a common moth, *Choristoneura fumiferana*, that is a destructive pest primarily of spruce and balsam fir. { ˌsprüs 'bəd,wərm }

SPSL *See* sound pressure spectrum level.

spur *See* ram. { spər }

spur blight [PL PATH] A fungus disease of raspberries and blackberries caused by *Didymella applanata* which kills the fruit spurs and causes dark spotting of the cane. { 'spər ˌblīt }

squalene [BIOL] $C_{30}H_{50}$A liquid triterpene which is found in large quantities in shark liver oil, and which appears to play a role in the biosynthesis of sterols and polycyclic terpenes; used as a bactericide and as an intermediate in the synthesis of pharmaceuticals. Also known as spinacene. { 'skwā,lēn }

squall [METEOROL] A strong wind with sudden onset and more gradual decline, lasting for several minutes; in the United States observational practice, a squall is reported only if a wind speed of 16 knots (8.23 meters per second) or higher is sustained for at least 2 minutes. { skwȯl }

squall cloud [METEOROL] A small eddy cloud sometimes formed below the leading edge of a thunderstorm cloud, between the upward and downward currents. { 'skwȯl ˌklaud }

squall line [METEOROL] A line of thunderstorms near whose advancing edge squalls occur along an extensive front; the region of thunderstorms is typically 12 to 30 miles

(20 to 50 kilometers) wide and a few hundred to 1200 miles (2000 kilometers) long. { 'skwȯl ‚līn }

stability chart [METEOROL] A synoptic chart that shows the distribution of a stability index. { stə'bil·əd·ē ‚chärt }

stability index [METEOROL] An indication of the local static stability of a layer of air. { stə'bil·əd·ē ‚in‚deks }

stabilizer [AGR] Any powdered or liquid additive used as an agent in soil stabilization. [ENG] Any substance that tends to maintain the physical and chemical properties of a material. { 'stā·bə‚līz·ər }

stable equilibrium [SCI TECH] Equilibrium in which any departure from the equilibrium state gives rise to forces or influences which tend to return the system to equilibrium. { 'stā·bəl ‚ē·kwə'lib·rē·əm }

stack [ENG] **1.** Any structure or part thereof that contains a flue or flues for the discharge of gases. **2.** Tall, vertical conduit (such as smokestack, flue) for venting of combustion or evaporation products or gaseous process wastes. { stak }

stack pollutants [ENG] Smokestack emissions subject to Environmental Protection Agency standards regulations, including sulfur oxides, particulates, nitrogen oxides, hydrocarbons, carbon monoxide, and photochemical oxidants. { 'stak pə‚lüt·əns }

stack vent [ENG] An extension to the atmosphere of a waste stack or a soil stack above the highest horizontal branch drain or fixture branch that is connected to the stack. Also known as soil vent; waste vent. { 'stak ‚vent }

stage [HYD] The elevation of the water surface in a stream as measured by a river gage with reference to some arbitrarily selected zero datum. Also known as stream stage. { stāj }

stagnant glacier [HYD] A glacier which has ceased to move. { 'stag·nənt 'glā·shər }

stagnant water [HYD] Motionless water, not flowing in a stream or current. Also known as standing water. { 'stag·nənt 'wȯd·ər }

stagnation [HYD] **1.** The condition of a body of water unstirred by a current or wave. **2.** The condition of a glacier that has stopped flowing. { stag'nā·shən }

stagnum [HYD] A pool of water with no outlet. { 'stag·nəm }

stamen [BOT] The male reproductive structure of a flower, consisting of an anther and a filament. { 'stā·mən }

staminate flower [BOT] A flower having stamens but lacking functional carpels. { 'stam·ə·nət 'flau̇·ər }

staminode [BOT] A stamen with no functional anther. { 'stā·mə‚nōd }

stamukha [OCEANOGR] An individual piece of stranded ice. { ‚sta‚mü·kə }

stand [ECOL] A group of plants, distinguishable from adjacent vegetation, which is generally uniform in species composition, age, and condition. [OCEANOGR] The interval at high or low water when there is no appreciable change in the height of the side. Also known as tidal stand. { stand }

standard atmosphere [METEOROL] A hypothetical vertical distribution of atmospheric temperature, pressure, and density which is taken to be representative of the atmosphere for purposes of pressure altimeter calibrations, aircraft performance calculations, aircraft and missile design, and ballistic tables; the air is assumed to obey the perfect gas law and hydrostatic equation, which, taken together, relate temperature, pressure, and density variations in the vertical; it is further assumed that the air contains no water vapor, and that the acceleration of gravity does not change with height. { 'stan·dərd 'at·mə‚sfir }

standard pressure [METEOROL] The arbitrarily selected atmospheric pressure of 1000 millibars to which adiabatic processes are referred for definitions of potential temperature, equivalent potential temperature, and so on. { 'stan·dərd 'presh·ər }

standard project flood [HYD] The volume of streamflow expected to result from the most severe combination of meteorological and hydrologic conditions which are reasonably characteristic of the geographic region involved, excluding extremely rare combinations. { 'stan·dərd 'präj₁ekt ‚fləd }

standard seawater *See* normal water. { 'stan·dərd 'sē‚wȯd·ər }

standard station *See* reference station. { 'stan·dərd 'stā·shən }

stand fire [FOR] A forest fire igniting in the trunks of trees. { 'stand ‚fīr }

standing cloud [METEOROL] Any stationary cloud maintaining its position with respect to a mountain peak or ridge. { 'stand·iŋ 'klaůd }

standing crop [ECOL] The number of individuals or total biomass present in a community at one particular time. { 'stand·iŋ 'kräp }

standing water *See* stagnant water. { ¦stand·iŋ 'wȯd·ər }

stand method [FOR] The practice of successively cutting trees of different ages so that ultimately the trees in the stand are new growth of a uniform age. { 'stand ‚meth·əd }

staphylotoxin [BIOL] Any of the various toxins elaborated by strains of *Staphylococcus aureus* bacteria, including hemolysins, enterotoxins, and leukocidin. { ‚staf·ə·lō 'täk·sən }

state of the sea [OCEANOGR] A description of the properties of the wind-generated waves on the surface of the sea. { 'stāt əv <u>th</u>ə 'sē }

state of the sky [METEOROL] The aspect of the sky in reference to the cloud cover; the state of the sky is fully described when the amounts, kinds, directions of movement, and heights of all clouds are given. { 'stāt əv <u>th</u>ə 'skī }

static level [HYD] The height to which water will rise in an artesian well; the static level of a flowing well is above the ground surface. { 'stad·ik 'lev·əl }

static oceanography [OCEANOGR] Branch of oceanography that deals with the physical and chemical nature of water in the ocean and with the shape and composition of the ocean bottom. { 'stad·ik ‚ō·shə'näg·rə·fē }

static stability [METEOROL] The stability of an atmosphere in hydrostatic equilibrium with respect to vertical displacements, usually considered by the parcel method. Also known as convectional stability; convection stability; hydrostatic stability; vertical stability. { 'stad·ik stə'bil·əd·ē }

stationary front *See* quasi-stationary front. { 'stā·shə‚ner·ē 'frənt }

stationary population [ECOL] A population in which the proportion of individuals in each group stays the same from one generation to the next. { 'stā·shə‚ner·ē ‚päp·yə'lā·shən }

station continuity chart [METEOROL] A chart or graph on which time is one coordinate, and one or more of the observed meteorological elements at that station is the other coordinate. { 'stā·shən ‚kan·tə'nü·əd·ē ‚chärt }

station elevation [METEOROL] The vertical distance above mean sea level that is adopted as the reference datum level for all current measurements of atmospheric pressure at the station. { 'stā·shən ‚el·ə‚vā·shən }

station model [METEOROL] A specified pattern for entering, on a weather map, the meteorological symbols that represent the state of the weather at a particular observation station. { 'stā·shən ‚mäd·əl }

station pressure [METEOROL] The atmospheric pressure computed for the level of the station elevation. { 'stā·shən ¸presh·ər }

statistical analysis [STAT] The body of techniques used in statistical inference concerning a population. { stə'tis·tə·kəl ə'nal·ə·səs }

statistical forecast [METEOROL] A weather forecast based upon a systematic statistical examination of the past behavior of the atmosphere, as distinguished from a forecast based upon thermodynamic and hydrodynamic considerations. { stə'tis·tə·kəl 'fȯr ¸kast }

statistical inference [STAT] The process of reaching conclusions concerning a population upon the basis of random samplings. { stə'tis·tə·kəl 'in·frəns }

statistical map [GEOGR] A special type of map in which the variation in quantity of a factor such as rainfall, population, or crops in a geographic area is indicated; a dot map is one type. { stə'tis·tə·kəl 'map }

statocyst [BOT] A cell containing statoliths in a fluid medium. Also known as statocyte. { 'stad·ə¸sist }

statocyte *See* statocyst. { 'stad·ə¸sīt }

statolith [BOT] A sand grain or other solid inclusion which moves readily in the fluid contents of a statocyst, comes to rest on the lower surface of the cell, and is believed to function in gravity perception. { 'stad·ə¸lith }

STD *See* sexually transmitted disease.

steadiness *See* persistence. { 'sted·ē·nəs }

steam fog [METEOROL] Fog formed when water vapor is added to air which is much colder than the vapor's source; most commonly, when very cold air drifts across relatively warm water. Also known as frost smoke; sea mist; sea smoke; steam mist; water smoke. { 'stēm ¸fäg }

steam mist *See* steam fog. { 'stēm ¸mist }

steering [METEOROL] Loosely used for any influence upon the direction of movement of an atmospheric disturbance exerted by another aspect of the state of the atmosphere; for example, a surface pressure system tends to be steered by isotherms, contour lines, or streamlines aloft, or by warm-sector isobars or the orientation of a warm front. { 'stir·iŋ }

steering level [METEOROL] A hypothetical level, in the atmosphere, where the velocity of the basic flow bears a direct relationship to the velocity of movement of an atmospheric disturbance embedded in the flow. { 'stir·iŋ ¸lev·əl }

Stefan's formula [OCEANOGR] A formula for the growth of thickness h of an ice cover on the ocean at various freezing temperatures, expressed as

$$h \approx \sqrt{\left(\frac{2l}{\lambda_i \rho_i} \right) \Psi}$$

where l is the coefficient of thermal conductivity, λ_i is the latent heat of fusion, ρ_i is the density of ice, and ψ is the cold sum (in degree days below 0°C). { 'shte¸fänz ¸fȯr·myə·lə }

stele [BOT] The part of a plant stem including all tissues and regions of plants from the cortex inward, including the pericycle, phloem, cambium, xylem, and pith. { 'stēl }

stem [BOT] The organ of vascular plants that usually develops branches and bears leaves and flowers. { stem }

stem blight [PL PATH] Any of various fungus blights that affect the plant stem. { 'stem ,blīt }

stem break *See* browning. { 'stem ,brāk }

stem rust [PL PATH] Any of several fungus diseases, especially of grasses, affecting the stem and marked by black or reddish-brown lesions. { 'stem ,rəst }

stenohaline [ECOL] In marine organisms, indicating the ability to tolerate only a narrow range of salinities. { ¦sten·ə¦ha¦līn }

stenoplastic [BIOL] Relating to an organism which exhibits a limited capacity for modification or adaptation to a new environment. { ¦sten·ə¦plas·tik }

stenotherm [BIOL] An organism able to tolerate only a small variation of temperature in the environment. { 'sten·ə,thərm }

stenothermic [BIOL] Indicating the ability to tolerate only a limited range of temperatures. { ¦sten·ə¦thər·mik }

stenotopic [ECOL] Referring to an organism with a restricted distribution. { ¦sten·ə¦täp·ik }

step aeration [CIV ENG] An activated sludge process in which the settled sewage is introduced into the aeration tank at more than one point. { 'step e,rā·shən }

step-down photophobic response [BIOL] A photophobic response elicited by a sudden decrease in light intensity. { ¦step ¦daún ¦fōd·ə,fō·bik ri'späns }

step lake *See* paternoster lake. { 'step ,lāk }

steppe [GEOGR] An extensive grassland in the semiarid climates of southeastern Europe and Asia; it is similar to but more arid than the prairie of the United States. { step }

steppe climate [CLIMATOL] The type of climate in which precipitation though very slight, is sufficient for growth of short, sparse grass; typical of the steppe regions of south-central Eurasia. Also known as semiarid climate. { 'step ,klī·mət }

stepped leader [GEOPHYS] The initial streamer of a lightning discharge; an intermittently advancing column of high ion density which established the channel for subsequent return streamers and dart leaders. { 'stept 'lēd·ər }

step-up photophobic response [BIOL] A photophobic response elicited by a sudden increase in light intensity. { ¦step ¦əp ¦fōd·ə,fō·bik ri'späns }

stereotaxis [BIOL] An orientation movement in response to stimulation by contact with a solid body. Also known as thigmotaxis. { ,ster·ē·ə'tak·səs }

stereotropism [BIOL] Growth or orientation of a sessile organism or part of an organism in response to the stimulus of a solid body. Also known as thigmotropism. { ,ster·ē'ä·trə,piz·əm }

steric anomaly *See* specific-volume anomaly. { 'ster·ik ə'näm·ə·lē }

sterile distribution [ECOL] A range of areas in which marine animals may live and spawn, but in which eggs do not hatch and larvae do not survive. { 'ster·əl ,dis·trə'byü·shən }

sterilization [MICROBIO] An act or process of destroying all forms of microbial life on and in an object. { ,ster·ə·lə'zā·shən }

stigma [BOT] The rough or sticky apical surface of the pistil for reception of the pollen. { 'stig·mə }

stillage [ENG] The residue grain from the manufacture of alcohol from grain; used as a feed supplement. { 'stil·ij }

stilling basin [ENG] A depressed area in a channel or reservoir that is deep enough to reduce the velocity of the flow. Also known as stilling box. { 'stil·iŋ ,bas·ən }

stilling box *See* stilling basin. { 'stil·iŋ ,bäks }

still water [HYD] A portion of a stream having a very slight gradient and no visible current. { 'stil 'wȯd·ər }

still-water level [OCEANOGR] The level that the sea surface would assume in the absence of wind waves. { 'stil ¦wȯd·ər ,lev·əl }

stilt root [BOT] A prop root of a mangrove tree. { 'stilt ,rüt }

stimulation deafness [MED] Deafness induced by noise; involves changes in the chemical interchange between the canals of the cochlea, as well as nerve destruction. { ,stim·yə'lā·shən 'def·nəs }

stinger [ZOO] A sharp piercing organ, as of a bee, stingray, or wasp, usually connected with a poison gland. { 'stiŋ·ər }

stipe [BOT] **1.** The petiole of a fern frond. **2.** The stemlike portion of the thallus in certain algae. [MYCOL] The short stalk or stem of the fruit body of a fungus, such as a mushroom. { stīp }

stipule [BOT] Either of a pair of appendages that are often present at the base of the petiole of a leaf. { 'stip·yül }

stock *See* pipe. { stäk }

stolon *See* runner. [MYCOL] A hypha produced above the surface and connecting a group of conidiophores. { 'stō·lən }

stoma [BOT] One of the minute openings in the epidermis of higher plants which are regulated by guard cells and through which gases and water vapor are exchanged between internal spaces and the external atmosphere. { 'stō·mə }

stone [GEOL] A small fragment of rock or mineral. { stōn }

stone fruit *See* drupe. { 'stōn ,früt }

stone ice *See* ground ice. { 'stōn ,īs }

stooping [METEOROL] An atmospheric refraction phenomenon; a special case of sinking in which the curvature of light rays due to atmospheric refraction decreases with elevation so that the visual image of a distant object is foreshortened in the vertical. { 'stüp·iŋ }

storage equation [HYD] An equation applied to unsteady fluid flow that states that the fluid inflow to a given space during an interval of time minus the outflow during the same interval is equal to the change in storage; it is applied in hydrology to the routing of floods through a reservoir or a reach of a stream. { 'stȯr·ij i,kwā·zhən }

storage routing *See* flood routing. { 'stȯr·ij ,rüd·iŋ }

storm [METEOROL] An atmospheric disturbance involving perturbations of the prevailing pressure and wind fields on scales ranging from tornadoes (0.6 mile or 1 kilometer across) to extratropical cyclones (up to 1800 miles or 3000 kilometers across); also the associated weather (rain storm or blizzard) and the like. { stȯrm }

storm center [METEOROL] The area of lowest atmospheric pressure of a cyclone; this is a more general expression than eye of the storm, which refers only to the center of a well-developed tropical cyclone, in which there is a tendency of the skies to clear. { 'stȯrm ,sen·tər }

storm detection [METEOROL] Any of the methods and techniques used to ascertain the formation of storms, including procedures for locating, tracking, and forecasting; special tools adapted to this purpose are radar and satellites to supplement meteorological charts and visual observations. { 'stȯrm di,tek·shən }

storm drain [CIV ENG] A drain which conducts storm surface, or wash water, or drainage after a heavy rain from a building to a storm or a combined sewer. Also known as storm sewer. { 'stȯrm ˌdrān }

storm ice foot [OCEANOGR] An ice foot produced by the breaking of a heavy sea or the freezing of wind-driven spray. { 'stȯrm 'īs ˌfu̇t }

storm model [METEOROL] A physical, three-dimensional representation of the inflow, outflow, and vertical motion of air and water vapor in a storm. { 'stȯrm ˌmäd·əl }

storm sewage [CIV ENG] Refuse liquids and waste carried by sewers during or following a period of heavy rainfall. { 'stȯrm ˌsü·ij }

storm sewer *See* storm drain. { 'stȯrm ˌsü·ər }

storm surge [OCEANOGR] A rise above normal water level on the open coast due only to the action of wind stress on the water surface; includes the rise in level due to atmospheric pressure reduction as well as that due to wind stress. Also known as storm wave; surge. { 'stȯrm ˌsərj }

storm tide [OCEANOGR] Height of a storm surge or hurricane wave above the astronomically predicted sea level. { 'stȯrm ˌtīd }

storm track [METEOROL] The path followed by a center of low atmospheric pressure. { 'stȯrm ˌtrak }

storm warning [METEOROL] A specially worded forecast of severe weather conditions, designed to alert the public to impending dangers; usually, this refers to a warning of potentially dangerous wind conditions for marine interests. { 'stȯrm ˌwȯrn·iŋ }

storm-warning signal [METEOROL] An arrangement of flags or pennants (by day) and lanterns (by night) displayed on a coastal storm-warning tower. { 'stȯrm ¦wȯrn·iŋ ˌsig·nəl }

storm-warning tower [METEOROL] A tower, generally constructed of steel, for displaying coastal storm-warning signals. { 'stȯrm ¦wȯrn·iŋ ˌtau̇·ər }

storm wave *See* storm surge. { 'stȯrm ˌwāv }

storm wind [METEOROL] In the Beaufort wind scale, a wind whose speed is from 56 to 63 knots (64 to 72 miles per hour or 104 to 117 kilometers per hour). { 'stȯrm ˌwind }

strain [BIOL] A population of cells derived either from a primary culture or from a cell line by the selection or cloning of cells having specific properties or markers. { strān }

strait [GEOGR] **1.** A neck of land. **2.** A narrow waterway connecting two larger bodies of water. { strāt }

stranded-floe ice foot *See* stranded ice foot. { 'stran·dəd ¦flō'īs ˌfu̇t }

stranded ice [OCEANOGR] Ice held in place by virtue of being grounded. Also known as grounded ice. { 'stran·dəd 'īs }

stranded ice foot [OCEANOGR] An ice foot formed by the stranding of floes or small icebergs along a shore; it may be built up by freezing spray or breaking seas. Also known as stranded-floe ice foot. { 'stran·dəd 'īs ˌfu̇t }

strandline [GEOL] **1.** A beach raised above the present sea level. **2.** The level at which a body of standing water meets the land. **3.** *See* shoreline. { 'strand,līn }

strategy [ECOL] A group of related traits that evolved under the influence of natural selection and solve particular problems encountered by organisms. { 'strad·ə·jē }

stratification [HYD] **1.** The arrangement of a body of water, as a lake, into two or more horizontal layers of differing characteristics, especially densities. **2.** The formation of layers in a mass of snow, ice, or firn. { ˌstrad·ə·fə'kā·shən }

stratified ocean   [OCEANOGR] An ocean where there is a vertical gradient of density. { 'strad·ə‚fīd 'ō·shən }

stratified rock   *See* sedimentary rock. { 'strad·ə‚fīd 'räk }

stratiform   [METEOROL] Description of clouds of extensive horizontal development, as contrasted to the vertically developed cumuliform types.   { 'strad·ə‚fórm }

stratiformis   [METEOROL] A cloud species consisting of a very extensive horizontal layer or layers which need not be continuous; this species is the most common form of the genera altocumulus and stratocumulus and is occasionally found in cirrocumulus. { ¦strad·ə¦fór·məs }

stratocumulus   [METEOROL] A principal cloud type predominantly stratiform, in the form of a gray or whitish layer of patch, which nearly always has dark parts.   { ¦strad·ō 'kyü·myə‚ləs }

stratopause   [METEOROL] The boundary or zone of transition separating the strato-sphere and the mesosphere; it marks a reversal of temperature change with altitude. { 'strad·ə‚póz }

stratosphere   [METEOROL] The atmospheric shell above the troposphere and below the mesosphere; it extends, therefore, from the tropopause to about 33 miles (55 kilometers), where the temperature begins again to increase with altitude. { 'strad·ə‚sfir }

stratospheric coupling   [METEOROL] The interaction between disturbances in the stratosphere and those in the troposphere.   { ¦strad·ə¦sfir·ik 'kəp·liŋ }

stratospheric ozone   [METEOROL] Atmospheric ozone that is relatively concentrated in the lower stratosphere in a layer between 9 and 18 miles (15 and 30 kilometers) above the earth's surface, and plays a critical role for the biosphere by absorbing the damaging ultraviolet radiation with wavelengths 320 nanometers and lower. Also known as ozone layer.   { ‚strad·ə‚sfir·ik 'ō‚zōn }

stratospheric steering   [METEOROL] The steering of lower-level atmospheric distur-bances along the contour lines of the tropopause, which lines are presumably roughly parallel to the direction of the wind at the tropopause level.   { ¦strad·ə¦sfir·ik 'stir·iŋ }

stratum   [GEOL] A mass of homogeneous or gradational sedimentary material, either consolidated rock or unconsolidated soil, occurring in a distinct layer and visually separable from other layers above and below.   [SCI TECH] One in a sequence of distinct layers.   { 'strad·əm }

stratus   [METEOROL] A principal cloud type in the form of a gray layer with a rather uniform base; a stratus does not usually produce precipitation, but when it does occur it is in the form of minute particles, such as drizzle, ice crystals, or snow grains. { 'strad·əs }

stratus fractus   [METEOROL] Irregularly fragmented stratus clouds that appear as if they had been shred or torn.   { 'strad·əs 'frak·təs }

strays   *See* atmospheric interference.   { strāz }

stream   [HYD] A body of running water moving under the influence of gravity to lower levels in a narrow, clearly defined natural channel.   { strēm }

stream capacity   [GEOL] The ability of a stream to carry detritus, measured at a given point per unit of time.   { 'strēm kə‚pas·əd·ē }

stream capture   *See* capture.   { 'strēm ‚kap·chər }

stream current   [OCEANOGR] A deep, narrow, well-defined fast-moving ocean current. { 'strēm ‚kə·rənt }

stream erosion   [GEOL] The progressive removal of exposed matter from the surface of a stream channel by a stream.   { 'strēm i‚rō·zhən }

streamflow |HYD| A type of channel flow, applied to surface runoff moving in a stream. { 'strēm‚flō }

streamflow routing See flood routing. { 'strēm‚flō'rüd·iŋ }

stream-length ratio |HYD| Ratio of the mean length of a stream of a given order to the mean length of the next lower order stream in the same basin. { 'strēm ¦leŋkth ‚rā·shō }

stream load |GEOL| Solid material transported by a stream. { 'strēm ‚lōd }

stream morphology See river morphology. { 'strēm mȯr'fäl·ə·jē }

stream order |HYD| The designation by a dimensionless integer series (1, 2, 3, . . .) of the relative position of stream segments in the network of a drainage basin. Also known as channel order. { 'strēm ¦ȯr·dər }

stream piracy See capture. { 'strēm ‚pī·rə·sē }

stream profile |HYD| The longitudinal profile of a stream. { 'strēm ‚prō‚fīl }

stream robbery See capture. { 'strēm ‚räb·ə·rē }

stream segment |HYD| The part of a stream extending between designated tributary junctions. Also known as channel segment. { 'strēm ‚seg·mənt }

stream stage See stage. { 'strēm ‚stāj }

strength of current |OCEANOGR| **1.** The phase of a tidal current at which the speed is a maximum. **2.** The velocity of the current at this time. { 'streŋkth əv 'kə·rənt }

strength of ebb |OCEANOGR| **1.** The ebb current at the time of maximum speed. **2.** The speed of the current at this time. { 'streŋkth əv 'eb }

strength-of-ebb interval |OCEANOGR| The time interval between the transit (upper or lower) of the moon and the next maximum ebb current at a place. { 'streŋkth əv 'eb 'in·tər·vəl }

strength of flood |OCEANOGR| **1.** The flood current at the time of maximum speed. **2.** The speed of the current at this time. { 'streŋkth əv 'fləd }

strength-of-flood interval |OCEANOGR| The time interval between the transit (upper or lower) of the moon and the next maximum flood current at a place. { 'streŋkth əv 'fləd 'in·tər·vəl }

streptobacillary fever See Haverhill fever. { ¦strep·tō·bə'sil·ə·rē 'fē·vər }

streptomycin |MICROBIO| $C_{21}H_{39}O_{12}N_7$ A water-soluble antibiotic obtained from Streptomyces griseus that is used principally in the treatment of tuberculosis. { ‚strep·tə'mīs·ən }

stress |BIOL| A stimulus or succession of stimuli of such magnitude as to tend to disrupt the homeostasis of the organism. { stres }

stretch See reach. { strech }

strike |GEOL| The direction taken by a structural surface, such as a fault plane, as it intersects the horizontal. Also known as line of strike. { strīk }

strike stream See subsequent stream. { 'strīk ‚strēm }

strip-cropping |AGR| Growing separate crops in adjacent strips that follow the contour of the land as a method of reducing soil erosion. { 'strip ‚kräp·iŋ }

strip method |FOR| A lumbering method in which timbers are cleared from a forest in strips; new growth in the strip results from seeds sown in the adjoining forest. { 'strip ‚meth·əd }

strip survey |FOR| A survey of the value of a strip of forest; used to estimate the value of a larger area of the forest. { 'strip ¦sər‚vā }

strobilus [BOT] **1.** A conelike structure made up of sporophylls, or spore-bearing leaves, as in Equisetales. **2.** The cone of members of the Pinophyta. { 'sträb·ə·ləs }

strong breeze [METEOROL] In the Beaufort wind scale, a wind whose speed is from 22 to 27 knots (25 to 31 miles per hour or 41 to 50 kilometers per hour). { 'strȯŋ 'brēz }

strong gale [METEOROL] In the Beaufort wind scale, a wind whose speed is from 41 to 47 knots (47 to 54 miles per hour or 76 to 87 kilometers per hour). { 'strȯŋ 'gāl }

strong liquor *See* filtrate. { 'strȯŋ 'lik·ər }

strontium-90 [PHYS] A poisonous, radioactive isotope of strontium; 28-year half life with β radiation; derived from reactor-fuel fission products; used in thickness gages, medical treatment, phosphor activation, and atomic batteries. { 'strän·tē·əm 'nīn·tē }

stubborn disease [PL PATH] A virus disease of citrus trees characterized by short internodes resulting in stiff brushy growth and chlorotic leaves. { 'stəb·ərn di'zēz }

Stuve chart [METEOROL] A thermodynamic diagram with atmospheric temperature as the x axis and atmospheric pressure to the power 0.286 as the y ordinate, increasing downward; named after G. Stuve. Also known as adiabatic chart; pseudoadiabatic chart. { 'stüv·ə ‚chärt }

style [BOT] The portion of a pistil connecting the stigma and ovary. { stīl }

subalkaline [GEOCHEM] Pertaining to a soil in which the pH is 8.0 to 8.5, usually in a limestone or salt-marsh region. { ¦səb'al·kə‚līn }

subalpine *See* alpestrine. { ¦səb'al‚pīn }

Subantarctic Intermediate Water [OCEANOGR] A layer of water above the deep-water layer in the South Atlantic. { ¦səb·ant'ärd·ik ‚in·tər‚mēd·ē·ət 'wȯd·ər }

subaqueous [HYD] Pertaining to conditions and processes occurring in, under, or beneath the surface of water, especially fresh water. { ¦səb'ā·kwē·əs }

subarctic [GEOGR] Pertaining to regions adjacent to the Arctic Circle or having characteristics somewhat similar to those of these regions. { ¦səb'ärd·ik }

subarctic climate *See* taiga climate. { ¦səb'ärd·ik 'klī·mət }

subarid [CLIMATOL] Pertaining to regions that are moderately or slightly arid. { ¦səb'ar·əd }

subartesian well [HYD] A well that requires artificial pumping to raise water to the surface because confining pressure forces the water only part of the distance up the well shaft. { ¦səb·är'tē·zhən 'wel }

subboreal [ECOL] A biogeographic zone whose climatic condition approaches that of the boreal. { ¦səb'bȯr·ē·əl }

subclimax [ECOL] A community immediately preceding a climax in an ecological succession. { ¦səb'klī‚maks }

subcontinent [GEOGR] **1.** A landmass such as Greenland that is large but not as large as the generally recognized continents. **2.** A large subdivision of a continent (for example, the Indian subcontinent) distinguished geologically or geomorphically from the rest of the continent. { ¦səb'känt·ən·ənt }

subdominant [ECOL] A species which may appear more abundant at particular times of the year than the true dominant in a climax; for example, in a savannah trees and shrubs are more conspicuous than the grasses, which are the true dominants. { ¦səb'däm·ə·nənt }

suberose [BOT] Having a texture like cork due to or resembling that due to suberization. { 'sü·bə‚rōs }

subgeostrophic wind |METEOROL| Any wind of lower speed than the geostrophic wind required by the existing pressure gradient. { ¦səb¦jē·ō¦sträf·ik 'wind }

subgradient wind |METEOROL| A wind of lower speed than the gradient wind required by the existing pressure gradient and centrifugal force. { ¦səb'grād·ē·ənt 'wind }

subhumid climate |CLIMATOL| A humidity province based on its typical vegetation. Also known as grassland climate; prairie climate. { ¦səb'hyü·məd 'klī·mat }

subirrigation *See* subsurface irrigation. { ¦səb‚ir·ə'gā·shən }

sublimation |METEOROL| The process by which solids are transformed directly to the vapor state or vica versa without passing through the liquid phase; for example, the creation of ice crystals directly from water vapor. { ‚səb·lə'mā·shən }

sublimation nucleus |METEOROL| Any particle upon which an ice crystal may grow by the process of sublimation. { ‚səb·lə'mā·shən ¦nü·klē·əs }

sublittoral zone |OCEANOGR| The benthic region extending from mean low water (2–3 fathoms or 40–60 meters, according to some authorities) to a depth of about 110 fathoms (200 meters), or the edge of a continental shelf, beyond which most abundant attached plants do not grow. { ¦səb'lid·ə·rəl ‚zōn }

submarine |OCEANOGR| Being or functioning in the sea. { ¦səb·mə'rēn }

submarine geology *See* geological oceanography. { ¦səb·mə'rēn jē'äl·ə·jē }

submarine plain *See* plain. { ¦səb·mə'rēn 'plān }

submarine spring |HYD| A spring of water issuing from the bottom of the sea. { ¦səb·mə'rēn 'spriŋ }

submarine station |OCEANOGR| **1.** One of the places for which tide or tidal current predictions are determined by applying a correction to the predictions of a reference station. **2.** A tide or tidal current station at which a short series of observations have been made; these observations are reduced by comparison with simultaneous observations at a reference station. { ¦səb·mə'rēn 'stā·shən }

submarine trench *See* trench. { ¦səb·mə'rēn 'trench }

submarine valley *See* valley. { ¦səb·mə'rēn 'val·ē }

submerged breakwater |OCEANOGR| A breakwater with its top below the still water level; when struck by a wave, part of the wave energy is reflected seaward and the remaining energy is largely dissipated in a breaker, transmitted shoreward as a multiple crest system, or transmitted shoreward as a simple wave system. { səb'mərjd 'brāk‚wȯd·ər }

submerged coastal plain |GEOL| The continental shelf as the seaward extension of a coastal plain on the land. Also known as coast shelf. { səb'mərjd 'kōst·əl 'plān }

submerged lands |GEOL| Lands covered by water at any stage of the tide, as distinguished from tidelands which are attached to the mainland or an island and are covered or uncovered with the tide; tidelands presuppose a high-water line as the upper boundary, submerged lands do not. { səb'mərjd 'lanz }

submergence |GEOL| A change in the relative levels of water and land either from a sinking of the land or a rise of the water level. { səb'mər·jəns }

subpolar anticyclone *See* subpolar high. { ¦səb'pō·lər ¦ant·i'sī‚klōn }

subpolar glacier |HYD| A polar glacier with 30 to 60 feet (10 to 20 meters) of firn in the accumulation area where some melting occurs. { ¦səb'pō·lər 'glā·shər }

subpolar high |METEOROL| A high that forms over the cold continental surfaces of subpolar latitudes, principally in Northern Hemisphere winters; these highs typically migrate eastward and southward. Also known as polar anticyclone; polar high; subpolar anticyclone. { ¦səb'pō·lər 'hī }

subpolar low-pressure belt [METEOROL] A belt of low pressure located, in the mean, between 50 and 70° latitude; in the Northern Hemisphere, this belt consists of the Aleutian low and the Icelandic low; in the Southern Hemisphere, it is supposed to exist around the periphery of the Antarctic continent. { ¦səb'pō·lər ¦lō'presh·ər ‚belt }

subpolar westerlies *See* westerlies. { ¦səb'pō·lər 'wes·tər‚lēz }

subsequent drainage [HYD] Drainage by a stream developed subsequent to the system of which it is a part; drainage follows belts of weak rocks. { 'səb·sə·kwənt 'drā·nij }

subsequent stream [HYD] A stream that flows in the general direction of the strike of the underlying strata and develops subsequent to the formation of the consequent stream of which it is a tributary. Also known as longitudinal stream; strike stream. { 'səb·sə·kwənt 'strēm }

subsere [ECOL] A secondary community that succeeds an interrupted climax. { 'səb ‚sir }

subsidence [PETR MIN] A sinking down of a part of the earth's crust due to underground excavations. { səb'sīd·əns }

subsidence inversion [METEOROL] A temperature inversion produced by the adiabatic warming of a layer of subsiding air; this inversion is enhanced by vertical mixing in the air layer below the inversion. { səb'sīd·əns in‚vər·zhən }

subsoil ice *See* ground ice. { 'səb‚sȯil 'īs }

subspecies [SYST] A geographically defined grouping of local populations which differs taxonomically from similar subdivisions of species. { ¦səb'spē·shēz }

substrate [ECOL] The foundation to which a sessile organism is attached. { 'səb ‚strāt }

substratosphere [METEOROL] A region of indefinite lower limit just below the stratosphere. { ¦səb'strad·ə‚sfir }

subsurface current [OCEANOGR] An underwater current which is not present at the surface or whose core (region of maximum velocity) is below the surface. { ¦səb'sər·fəs 'kə·rənt }

subsurface irrigation [AGR] A method of providing water to plants by raising the water table to the root zone of the crop or by carrying moisture to the root zone by perforated underground pipe. Also known as subirrigation. { səb‚sər·fəs ‚ir·ə'gā·shən }

subsurface tillage [AGR] A method of stirring the soil with blades that leaves stubble on or just below the surface. { ¦səb'sər·fəs 'til·ij }

subsurface waste disposal [ENG] Disposal of manufacturing wastes in porous underground rock formations. { ¦səb'sər·fəs 'wāst di‚spōz·əl }

subterranean ice *See* ground ice. { ¦səb·tə'rā·nē·ən 'īs }

subterranean stream [HYD] A subsurface stream that flows through a cave or a group of communicating caves. { ¦səb·tə'rā·nē·ən 'strēm }

subtropic [METEOROL] An indefinite belt in each hemisphere between the tropic and temperate regions; the polar boundaries are considered to be roughly 35–40° northern and southern latitudes, but vary greatly according to continental influence, being farther poleward on the western coasts of continents and farther equatorward on the eastern coasts. { ¦səb'träp·ik }

subtropical anticyclone *See* subtropical high. { ‚səb'träp·ə·kəl ¦ant·i'sī‚klōn }

Subtropical Convergence [OCEANOGR] The zone of converging currents, generally located in midlatitudes. { ‚səb'träp·ə·kəl kən'vər·jəns }

subtropical cyclone [METEOROL] The low-level (surface chart) manifestation of a cutoff low. { ‚səb'träp·ə·kəl 'sī‚klōn }

subtropical easterlies   *See* tropical easterlies.   { ˌsəb'träp·ə·kəl 'ēs·tər‚lēz }

subtropical easterlies index   [METEOROL] A measure of the strength of the easterly wind between the latitudes of 20° and 35°N; the index is computed from the average sea-level pressure difference between these latitudes and is expressed as the east to west component of the corresponding geostrophic wind in meters and tenths of meters per second.   { ˌsəb'träp·ə·kəl 'ēs·tər‚lēz ‚in‚deks }

subtropical forest   *See* temperate rainforest.   { ˌsəb'träp·ə·kəl 'fär·əst }

subtropical high   [METEOROL] One of the semipermanent highs of the subtropical high-pressure belt; these highs appear as centers of action on mean charts of surface pressure; they lie over oceans and are best developed in the summer season. Also known as oceanic anticyclone; oceanic high; subtropical anticyclone.   { ˌsəb'träp·ə·kəl 'hī }

subtropical high-pressure belt   [METEOROL] One of the two belts of high atmospheric pressure that are centered, in the mean, near 30°N and 30°S latitudes; these belts are formed by the subtropical highs.   { ˌsəb'träp·ə·kəl ‚hī 'presh·ər ‚belt }

subtropical westerlies   *See* westerlies.   { ˌsəb'träp·ə·kəl 'wes·tər‚lēz }

succession   *See* ecological succession.   { sək'sesh·ən }

succulent   [BOT] Describing a plant having juicy, fleshy tissue.   { 'sək·yə·lənt }

sucker   [BOT] A shoot that develops rapidly from the lower portion of a plant, and usually at the expense of the plant.   [ZOO] A disk-shaped organ in various animals for adhering to or holding onto an individual, usually of another species.   { 'sək·ər }

suffrutescent   [BOT] Of or pertaining to a stem intermediate between herbaceous and shrubby, becoming partly woody and perennial at the base.   { ‚sə‚frü'tes·ənt }

suffruticose   [BOT] Low stems which are woody, grading into herbaceous at the top.   { sə'früd·ə‚kōs }

sugar berg   [OCEANOGR] An iceberg of porous glacier ice.   { 'shủg·ər ‚bərg }

sugarcane gummosis   *See* Cobb's disease.   { 'shủg·ər‚kān gə'mō·səs }

sugarloaf sea   [OCEANOGR] A sea characterized by waves that rise into sugarloaf shapes, with little wind, possibly resulting from intersecting waves.   { 'shủg·ər ‚lōf ‚sē }

sugar snow   *See* depth hoar.   { 'shủg·ər ‚snō }

sulfallate   [CHEM] $C_8H_{14}NS_2Cl$ An oily liquid, used as a preemergence herbicide for vegetable crops and ornamentals. Also known as 2-chloroallyl diethyldithiocarbamate (CDEC).   { səl'fa‚lāt }

sulfate   [CHEM] **1.** A compound containing the $-SO_4$ group, as in sodium sulfate, Na_2SO_4.   **2.** A salt of sulfuric acid.   { 'səl‚fāt }

sulfide   [CHEM] Any compound with one or more sulfur atoms in which the sulfur is connected directly to a carbon, metal, or other nonoxygen atom; for example, sodium sulfide, Na_2S.   { 'səl‚fīd }

sulfidogen   [MICROBIO] A strict anaerobe that reduces sulfur to hydrogen sulfide.   { səl'fīd·ə‚jen }

sulfocyanic acid   *See* thiocyanic acid.   { ‚səl·fō·sī‚an·ik 'as·əd }

sulfofication   [GEOCHEM] Oxidation of sulfur and sulfur compounds into sulfates, occurring in soils by the agency of bacteria.   { ‚səl·fə·fə'kā·shən }

sulfonyl chloride   *See* sulfuryl chloride.   { 'səl·fə‚nil 'klȯr‚īd }

sulfur [CHEM] A nonmetallic element in group 16, symbol S, atomic number 16, atomic weight 32.06, existing in a crystalline or amorphous form and in four stable isotopes; used as a chemical intermediate and fungicide, and in rubber vulcanization. { 'səl·fər }

sulfur bacteria [MICROBIO] Any of various bacteria having the ability to oxidize sulfur compounds. { 'səl·fər bak,tir·ē·ə }

sulfur bichloride See sulfur dichloride. { 'səl·fər bī'klȯr,īd }

sulfur chloride [CHEM] S_2Cl_2 A combustible, water-soluble, oily, fuming, amber to yellow-red liquid with an irritating effect on the eyes and lungs, boils at 138°C; used to make military gas and insecticides, in rubber substitutes and cements, to purify sugar juices, and as a chemical intermediate. Also known as sulfur subchloride. { 'səl·fər 'klȯr,īd }

sulfur dichloride [CHEM] SCl_2 A red-brown liquid boiling (when heated rapidly) at 60°C, decomposes in water; used to make insecticides, for rubber vulcanization, and as a chemical intermediate and a solvent. Also known as sulfur bichloride. { 'səl·fər dī'klȯr,īd }

sulfur dioxide [CHEM] SO_2 A toxic, irritating, colorless gas soluble in water, alcohol, and ether; boils at −10°C; used as a chemical intermediate, in artificial ice, paper pulping, and ore refining, and as a solvent. Also known as sulfurous acid anhydride. { 'səl·fər dī'äk,sīd }

sulfuric acid [CHEM] H_2SO_4 A toxic, corrosive, strongly acid, colorless liquid that is miscible with water and dissolves most metals, and melts at 10°C; used in industry in the manufacture of chemicals, fertilizers, and explosives, and in petroleum refining. Also known as dipping acid; oil of vitriol; vitriolic acid. { ¦səl¦fyu̇r·ik 'as·əd }

sulfuric chloride See sulfuryl chloride. { ¦səl¦fyu̇r·ik 'klȯr,īd }

sulfurous acid anhydride See sulfur dioxide. { 'səl·fə·rəs 'as·əd an'hī,drīd }

sulfur spring [HYD] A spring containing sulfur compounds such as hydrogen sulfide. { 'səl·fər ,sprin }

sulfur subchloride See sulfur chloride. { 'səl·fər ¦səb'klȯr,īd }

sulfur-35 [PHYS] Radioactive sulfur with mass number 35; radiotoxic, with 87.1-day half-life, β radiation; derived from pile irradiation; used as a tracer to study chemical reactions, engine wear, and protein metabolism. { 'səl·fər ,thərd·ē'fīv }

sulfuryl chloride [CHEM] SO_2Cl_2 A colorless liquid with a pungent aroma, boils at 69°C, decomposed by hot water and alkalies; used as a chlorinating agent and solvent and for pharmaceuticals, dyestuffs, rayon, and poison gas. Also known as sulfonyl chloride; sulfuric chloride. { 'səl·fə,ril 'klȯr,īd }

sulfuryl fluoride [CHEM] SO_2F_2 A colorless gas with a melting point of −136.7°C and a boiling point of 55.4°C; used as an insecticide and fumigant. { 'səl·fə,ril 'flu̇r,īd }

sullage [CIV ENG] Drainage or wastewater from a building, farmyard, or street. { 'səl·ij }

sultriness [METEOROL] An oppressively uncomfortable state of the weather which results from the simultaneous occurrence of high temperature and high humidity, and often enhanced by calm air and cloudiness. { 'səl·trē·nəs }

summation principle [METEOROL] In United States weather observing practice, the rule which governs the assignment of sky cover amount to any layer of cloud or obscuring phenomenon, and to the total sky cover; in essence, this principle states that the sky cover at any level is equal to the summation of the sky cover of the lowest layer plus the additional sky cover provided at all successively higher layers up to and including the layer in question; thus, no layer can be assigned a sky cover less than a lower layer, and no sky cover can be greater than 1.0 (10/10). { sə'mā·shən ,prin·sə·pəl }

summerwood |BOT| The less porous, usually harder portion of an annual ring that forms in the latter part of the growing season. { 'səm·ər‚wüd }

sun cross |METEOROL| A rare halo phenomenon in which bands of white light intersect over the sun at right angles. { 'sən ‚krós }

sun crust |HYD| A type of snow crust, formed by refreezing of surface snow crystals after having been melted by the sun. { 'sən ‚krəst }

sun drawing water |METEOROL| Popular designation for a phenomenon of the sun showing through scattered openings in a layer of clouds into a layer of turbid air that is hazy or dusty; bright bands are seen where the several beams of sunlight pass down through the subcloud layer; sailors called the phenomenon the backstays of the sun. { 'sən 'dró·iŋ 'wód·ər }

sun pillar |METEOROL| A luminous streak of light, white or slightly reddened, extending above and below the sun, most frequently observed near sunrise or sunset; it may extend to about 20° above the sun, and generally ends in a point. Also known as light pillar. { 'sən ‚pil·ər }

sunscald |PL PATH| An injury to woody plants which results in local death of the plant tissues; in summer it is caused by excessive action of the sun's rays, in winter, by the great variation of temperature on the side of trees that is exposed to the sun in cold weather. { 'sən‚skóld }

supercell |METEOROL| A thunderstorm with a persistent rotating updraft. While rare, it produces the most severe weather such as tornadoes, strong winds, and hail. { 'sü·pər‚sel }

supercooled cloud |METEOROL| A cloud composed of supercooled liquid waterdrops. { ¦sü·pər'küld 'klaůd }

supercritical flow *See* rapid flow. { ¦sü·pər'krid·ə·kəl 'flō }

supergeostrophic wind |METEOROL| Any wind of greater speed than the geostrophic wind required by the pressure gradient. { ¦sü·pər¦jē·ö¦sträf·ik 'wind }

superglacial |HYD| Of or pertaining to the upper surface of a glacier or ice sheet. { ¦sü·pər'glā·shəl }

supergradient wind |METEOROL| A wind of greater speed than the gradient wind required by the existing pressure gradient and centrifugal force. { ¦sü·pər'grād·ē·ənt 'wind }

superimposed drainage |HYD| A naturally evolved drainage system that became established on a preexisting surface, now eroded, and whose course is unrelated to the present underlying geological structure. { ¦sü·pər·im'pōzd 'drā·nij }

superimposed stream |HYD| A stream, started on a new surface, that kept its course through the different preexisting lithologies and structures encountered as it eroded downward into the underlying rock. Also known as superinduced stream. { ¦sü·pər·im'pōzd 'strēm }

superinduced stream *See* superimposed stream. { ¦sü·pər·in'düst 'strēm }

superior air |METEOROL| An exceptionally dry mass of air formed by subsidence and usually found aloft but occasionally reaching the earth's surface during extreme subsidence processes. { sə'pir·ē·ər 'er }

superior tide |OCEANOGR| The tide in the hemisphere in which the moon is above the horizon. { sə'pir·ē·ər 'tīd }

superjacent waters |OCEANOGR| The waters above the continental shelf. { ¦sü·pər'jā·sənt 'wód·ərz }

superresolution [OCEANOGR] Separation of tides into components of different frequencies, without taking measurements for the full extent of the longest-period component. { ¦sü·pər‚rez·ə'lü·shən }

supersaturation [METEOROL] The condition existing in a given portion of the atmosphere when the relative humidity is greater than 100%, in respect to a plane surface of pure water or pure ice. { ¦sü·pər‚sach·ə'rā·shən }

supersolubility *See* supersaturation. { ¦sü·pər‚säl·yə'bil·əd·ē }

supragelisol *See* suprapermafrost layer. { ¦sü·prə'jel·ə‚sòl }

supralateral tangent arcs [METEOROL] Two oblique luminous arcs, concave to the sun and tangent to the halo of 46° at points above the altitude of the sun. { ¦sü·prə'lad·ə·rəl 'tan·jənt 'ärks }

suprapermafrost layer [HYD] The layer of ground above permafrost; it includes the active layer and possibly occurrences of talik and perelotok. Also known as supragelisol. { ¦sü·prə'pər·mə‚fròst ‚lā·ər }

surf [OCEANOGR] Wave activity in the area between the shoreline and the outermost limit of breakers, that is, in the surf zone. { sərf }

surface boundary layer [METEOROL] That thin layer of air adjacent to the earth's surface, extending up to the so-called anemometer level (the base of the Ekman layer); within this layer the wind distribution is determined largely by the vertical temperature gradient and the nature and contours of the underlying surface, and shearing stresses are approximately constant. Also known as atmospheric boundary layer; friction layer; ground layer; surface layer. { 'sər·fəs 'baùn·drē ‚lā·ər }

surface chart [METEOROL] An analyzed synoptic chart of surface weather observations; essentially, a surface chart shows the distribution of sea-level pressure (therefore, the positions of highs, lows, ridges, and troughs) and the location and nature of fronts and air masses, plus the symbols of occurring weather phenomena, analysis of pressure tendency (isallobars), and indications of the movement of pressure systems and fronts. Also known as sea-level chart; sea-level-pressure chart; surface map. { 'sər·fəs ‚chärt }

surface current [OCEANOGR] **1.** Water movement which extends to depths of 3–10 feet (1–3 meters) below the surface in nearshore areas, and to about 33 feet (10 meters) in deep-ocean areas. **2.** Any current whose maximum velocity core is at or near the surface. { 'sər·fəs ‚kə·rənt }

surface detention [HYD] Water in temporary storage as a thin sheet over the soil surface during the occurrence of overland flow. { 'sər·fəs di‚ten·chən }

surface drainage [HYD] Natural or artificial removal of excess groundwater. { 'sər·fəs ‚drā·nij }

surface fire [FOR] A forest fire in which only surface litter and undergrowth burn. { 'sər·fəs ‚fīr }

surface flow *See* overland flow. { 'sər·fəs ‚flō }

surface hoar [HYD] **1.** Fernlike ice crystals formed directly on a snow surface by sublimation; a type of hoarfrost. **2.** Hoarfrost that has grown primarily in two dimensions, as on a window or other smooth surface. { 'sər·fəs ‚hòr }

surface inversion [METEOROL] A temperature inversion based at the earth's surface; that is, an increase of temperature with height beginning at ground level. Also known as ground inversion. { 'sər·fəs in‚vər·zhən }

surface irrigation [AGR] Application of water to the soil by means of pipes or furrows along the surface. { 'sər·fəs ‚ir·ə‚gā·shən }

surface layer *See* surface boundary layer. { 'sər·fəs ‚lā·ər }

surface map *See* surface chart. { 'sər·fəs ‚map }

surface of discontinuity [METEOROL] An interface, applied to the atmosphere; for example, an atmospheric front is represented ideally by a surface of discontinuity of velocity, density, temperature, and pressure gradient. { ¦sər·fəs əv ‚dis‚känt·ən'ü·əd·ē }

surface pressure [METEOROL] The atmospheric pressure at a given location on the earth's surface; the expression is applied loosely and about equally to the more specific terms: station pressure and sea-level pressure. { 'sər·fəs ‚presh·ər }

surface retention See surface storage. { 'sər·fəs ri‚ten·chən }

surface runoff [HYD] Runoff that moves over the soil surface to the nearest surface stream. { 'sər·fəs 'rən‚óf }

surface soil [GEOL] The soil extending 5 to 8 inches (13 to 20 centimeters) below the surface. { 'sər·fəs ‚sóil }

surface storage [HYD] The part of precipitation retained temporarily at the ground surface as interception or depression storage so that it does not appear as infiltration or surface runoff either during the rainfall period or shortly thereafter. Also known as initial detention; surface retention. { 'sər·fəs ‚stór·ij }

surface temperature [OCEANOGR] Temperature of the layer of seawater nearest the atmosphere. { 'sər·fəs ‚tem·prə·chər }

surface visibility [METEOROL] The visibility determined from a point on the ground, as opposed to control-tower visibility. { 'sər·fəs ‚viz·ə'bil·əd·ē }

surface wash See sheet erosion. { 'sər·fəs ‚wäsh }

surface water See mixed layer. { 'sər·fəs ‚wód·ər }

surface weather observation [METEOROL] An evaluation of the state of the atmosphere as observed from a point at the surface of the earth, as opposed to an upper-air observation, and applied mainly to observations which are taken for the primary purpose of preparing surface synoptic charts. { 'sər·fəs 'weth·ər ‚äb·zər‚vā·shən }

surface wind [METEOROL] The wind measured at a surface observing station; customarily, it is measured at some distance above the ground itself to minimize the distorting effects of local obstacles and terrain. { 'sər·fəs ‚wind }

surf beat [OCEANOGR] Oscillations of water level near shore, associated with groups of high breakers. { 'sərf ‚bēt }

surf zone [OCEANOGR] The area between the landward limit of wave uprush and the farthest seaward breaker. { 'sərf ‚zōn }

surge [OCEANOGR] **1.** Wave motion of low height and short period, from about ½ to 60 minutes. **2.** See storm surge. { sərj }

surge line [METEOROL] A line along which a discontinuity in the wind speed occurs. { 'sərj ‚līn }

surging breaker See plunging breaker. { 'sərj·iŋ 'brā·kər }

surging glacier [HYD] A glacier that alternates periodically between surges (brief periods of rapid flow) and stagnation. { 'sərj·iŋ 'glā·shər }

survey instrument [ENG] A portable instrument used to detect and measure radiation. Also known as survey meter. { 'sər‚vā ‚in·strə·mənt }

survey meter See survey instrument. { 'sər‚vā ‚mēd·ər }

survival curve [ENG] The curve obtained by plotting the number or percentage of organisms surviving at a given time against the dose of radiation, or the number surviving at different intervals after a particular dose of radiation. { sər'vī·vəl ‚kərv }

survival ratio [BIOL] The number of organisms surviving irradiation by ionizing radiation divided by the number of organisms before irradiation. { sər'vī·vəl ‚rā·shō }

suscept [PL PATH] A plant that is susceptible to disease caused by either parasitic or nonparasitic plant pathogens. { sə'sept }

suspended water *See* vadose water. { sə'spen·dəd 'wȯd·ər }

suspension current *See* turbidity current. { sə'spen·shən ‚kə·rənt }

sustainable agriculture [AGR] An integration of traditional, conservation-minded farming techniques with modern scientific advances that maximizes use of on-farm renewable resources instead of imported and nonrenewable resources, while earning a return that is large enough for the farmer to continue in an ecologically harmless, regenerative way. { sə‚stān·ə·bəl 'ag·rə‚kəl·chər }

sustainable development [ENG] Development of industrial and natural resources that meets the energy needs of the present without compromising the ability of future generations to meet their needs in a similar manner. { sə‚stān·ə·bəl di'vel·əp·mənt }

sustainable forest management [FOR] Integrated management of the full range of environmental, social, and economic values of the forest to ensure future health and usefulness of the forest. { sə‚stān·ə·bəl 'fär·əst ‚man·ij·mənt }

sustained yield [BIOL] In a biological resource such as timber or grain, the replacement of a harvest yield by growth or reproduction before another harvest occurs. { sə'stānd 'yēld }

swamp [ECOL] A waterlogged land supporting a natural vegetation predominantly of shrubs and trees. { swämp }

swash [OCEANOGR] The rush of water up onto the beach following the breaking of a wave. Also known as run-up; uprush. { swäsh }

sweepstakes route [ECOL] A means that allows chance migration across a sea on natural rafts, so that oceanic islands can be colonized. { 'swēp‚stāks ‚rüt }

swell [OCEANOGR] Ocean waves which have traveled away from their generating area; these waves are of relatively long length and period, and regular in character. { swel }

swell-and-swale topography [GEOGR] A low-relief, undulating landscape characterized by gentle slopes and rounded hills interspersed with shallow depressions. { ¦swel ən ¦swāl tə'päg·rə·fē }

swell direction [OCEANOGR] The direction from which swell is moving. { 'swel di‚rek·shən }

swelled ground [GEOL] A soil or rock that expands when wetted. { 'sweld 'grau̇nd }

swell forecast [OCEANOGR] Prediction of the frequency and height of swell waves in a remote area from the characteristics of the waves at their origin. { 'swel ‚fȯr‚kast }

swep [CHEM] $C_8H_7Cl_2NO_2$ A white, crystalline compound with a melting point of 112–114°C; insoluble in water; used as a pre- and postemergence herbicide for rice, carrots, potatoes, and cotton. Also known as methyl-N-(3,4-dichlorophenyl)carbamate. { swep }

swine influenza [VET MED] A disease of swine caused by the associated effects of a filterable virus and *Hemophilus suis*, characterized by inflammation of the upper respiratory tract. { 'swin ‚in·flü'en·zə }

swine plague [VET MED] Hemorrhagic septicemia of swine caused by *Pasteurella suiseptica*, characterized by pleuropneumonia. { 'swin ¦plāg }

syconium [BOT] A fleshy fruit, as a fig, with an enlarged pulpy receptacle internally lined with minute flowers. { sī'kō·nē·əm }

sylvatic plague [VET MED] Plague occurring in rodents; may be transmitted to humans. Also known as endemic rural plague. { sil'vad·ik 'plāg }

symbiont [ECOL] A member of a symbiotic pair. { 'sim·bē‚änt }

symbiosis [ECOL] **1.** An interrelationship between two different species. **2.** An interrelationship between two different organisms in which the effects of that relationship is expressed as being harmful or beneficial. Also known as consortism. { ˌsim·bē'ō·səs }

symclosene *See* trichloroisocyanuric acid. { 'sim·klə͵zēn }

sympatric [ECOL] Of a species, occupying the same range as another species but maintaining identity by not interbreeding. { sim'pa·trik }

symphile [ECOL] An organism, usually a beetle, living as a guest in the nest of a social insect, such as an ant, where it is reared and bred in exchange for its exudates. { 'sim͵fīl }

sympodium [BOT] A branching system in trees in which the main axis is composed of successive secondary branches, each representing the dominant fork of a dichotomy. { sim'pōd·ē·əm }

synangium [BOT] A compound sorus made up of united sporangia. { sə'nan·jē·əm }

syncarp [BOT] A compound fleshy fruit. { 'sin͵kärp }

syncarpous [BOT] Descriptive of a gynoecium having the carpels united in a compound ovary. { sin'kär·pəs }

synchorology [ECOL] A study which involves the distribution ranges of plant communities, phytosociological zones, vegetation and geographical complexes, dissemination spectra, and current plant migration patterns. { ˌsin·kə'räl·ə·jē }

synecology [ECOL] The study of environmental relations of groups of organisms, such as communities. { ¦sin·i'käl·ə·jē }

synergism [ECOL] An ecological association in which the physiological processes or behavior of an individual are enhanced by the nearby presence of another organism. { 'sin·ər͵jiz·əm }

synfuel *See* synthetic fuel. { 'sin͵fyül }

synoptic [METEOROL] Refers to the use of meteorological data obtained simultaneously over a wide area for the purpose of presenting a comprehensive and nearly instantaneous picture of the state of the atmosphere. { sə'näp·tik }

synoptic chart [METEOROL] Any chart or map on which data and analyses are presented that describe the state of the atmosphere over a large area at a given moment in time. { sə'näp·tik 'chärt }

synoptic climatology [CLIMATOL] The study and analysis of climate in terms of synoptic weather information, principally in the form of synoptic charts; the information thus obtained gives the climate (that is, average weather) of a given locality in a given synoptic situation rather than the usual climatic parameters which represent averages over all synoptic conditions. { sə'näp·tik ͵klī·mə'täl·ə·jē }

synoptic code [METEOROL] In general, any code by which synoptic weather observations are communicated; among the synoptic codes in use are the international synoptic code, ship synoptic code, U.S. Airways code, and RECCO code. { sə'näp·tik 'kōd }

synoptic meteorology [METEOROL] The study and analysis of synoptic weather information. { sə'näp·tik ͵mēd·ē·ə'räl·ə·jē }

synoptic model [METEOROL] Any model specifying a space distribution of some meteorological elements; the distribution of clouds, precipitation, wind, temperature, and pressure in the vicinity of a front is an example of a synoptic model. { sə'näp·tik 'mäd·əl }

synoptic oceanography   [OCEANOGR] The study of the physical spatial parameters of the ocean through analysis of simultaneous observations from many stations. { sə'näp·tik ˌō·shə'näg·rə·fē }

synoptic report   [METEOROL] An encoded and transmitted synoptic weather observation.   { sə'näp·tik ri'pórt }

synoptic scale   *See* cyclonic scale.   { sə'näp·tik 'skāl }

synoptic wave chart   [OCEANOGR] A chart of an ocean area on which is plotted synoptic wave reports from vessels, along with computed wave heights for areas where reports are lacking; atmospheric fronts, highs, and lows are also shown; isolines of wave height and the boundaries of areas having the same dominant wave direction are drawn.   { sə'näp·tik 'wāv ˌchärt }

synoptic weather observation   [METEOROL] A surface weather observation, made at periodic times (usually at 3- and 6-hourly intervals specified by the World Meteorological Organization), of sky cover, state of the sky, cloud height, atmospheric pressure reduced to sea level, temperature, dew point, wind speed and direction, amount of precipitation, hydrometeors and lithometeors, and special phenomena that prevail at the time of the observation or have been observed since the previous specified observation.   { sə'näp·tik 'weth·ər ˌäb·zər,vā·shən }

synphylogeny   [ECOL] The study of the trends and changes in plant communities through historical and evolutionary perspectives.   { ¦sin·fə'läj·ə·nē }

synphysiology   [ECOL] The study of the metabolic processes of plant communities or species which constantly compete with each other, by investigating water needs, transpiration, assimilation and production or organic matter, physiological effects of light, temperature, root exudates, and various other ecological factors.   { ¦sin ˌfiz·ē'äl·ə·jē }

synthetic fuel   [PETR MIN] A fuel that is artificially formulated and manufactured; frequently derived from fossil fuels that are less convenient or environmentally undesirable for direct use. Also known as synfuel.   { sin'thed·ik 'fyül }

synusia   [ECOL] A structural unit of a community characterized by uniformity of life-form or of height.   { sə'nü·zhə }

syphilis   [MED] An infectious disease caused by the spirochete *Treponema pallidum*, transmitted principally by sexual intercourse.   { 'sif·ə·ləs }

syphilitic meningoencephalitis   *See* general paresis.   { ¦sif·ə¦lid·ik mə¦niŋ·gō·in ˌsef·ə'līd·əs }

systematics   [BIOL] The science of animal and plant classification.   { ˌsis·tə'mad·iks }

systems ecology   [ECOL] The combined approaches of systems analysis and the ecology of whole ecosystems and subsystems.   { 'sis·təmz i'käl·ə·jē }

T

2,4,5-T *See* 2,4,5-trichlorophenoxyacetic acid.

2,4,6-T *See* trichlorophenol.

Taber ice *See* segregated ice. { 'tā·bər ˌīs }

table iceberg *See* tabular iceberg **2.** { 'tā·bəl 'īs,bərg }

table knoll [GEOGR] A knoll with a comparatively smooth, flat top. { 'tā·bəl ,nōl }

tableland [GEOGR] A broad, elevated, nearly level, and extensive region of land that has been deeply cut at intervals by valleys or broken by escarpments. Also known as continental plateau. { 'tā·bəl,and }

table mountain [GEOGR] A flat-topped mountain. { 'tā·bəl ,maunt·ən }

table reef [GEOL] A small, isolated organic reef which has a flat top and does not enclose a lagoon. { 'tā·bəl ,rēf }

tabular berg *See* tabular iceberg. { 'tab·yə·lər 'bərg }

tabular iceberg [OCEANOGR] An iceberg with clifflike sides and a flat top; usually arises by detachment from an ice shelf. Also known as table iceberg; tabular berg. { 'tab·yə·lər 'īs,bərg }

taele *See* frozen ground. { 'tā·lə }

taiga [ECOL] A zone of forest vegetation encircling the Northern Hemisphere between the arctic-subarctic tundras in the north and the steppes, hardwood forests, and prairies in the south. Also known as boreal forest. { 'tī·gə }

taiga climate [CLIMATOL] In general, a climate which produces taiga vegetation, that is, too cold for prolific tree growth but milder than the tundra climate and moist enough to promote appreciable vegetation. Also known as subarctic climate. { 'tī·gə ,klī·mət }

tailwater ditch [AGR] A channel made along the lower end of a field to carry surface runoff from irrigation furrows off the field. { 'tāl,wod·ər ,dich }

tailwind [METEOROL] A wind which assists the intended progress of an exposed, moving object, for example, rendering an airborne object's ground speed greater than its airspeed; the opposite of a headwind. Also known as following wind. { 'tāl,wind }

tall fescue toxicosis [VET MED] A group of several animal disorders caused by grazing on tall fescue infected with the endophytic symbiotic fungus *Acremonium coenophialum.* { ,tȯl ,fes·kyü ,tak·sə'kō·səs }

tan rot [PL PATH] A fungus disease of strawberries caused by *Pezizella lythri* and characterized by the appearance of tan depressions on the fruit. { 'tan ,rät }

tantalum [CHEM] A metallic transition element, symbol Ta, atomic number 73, atomic weight 180.9479; black powder or steel-blue solid soluble in fused alkalies, insoluble in acids (except hydrofluoric and fuming sulfuric); melts about 3000°C. { 'tant·əl·əm }

tape grass [BOT] *Vallisnerida spiralis.* An aquatic flowering plant belonging to the family Hydrocharitaceae. Also known as eel grass. { 'tāp ,gras }

tapeworm |ZOO| Any member of the class Cestoidea; all are vertebrate endoparasites, characterized by a ribbonlike body divided into proglottids, and the anterior end modified into a holdfast organ. { 'tāp‚wərm }

Taphrina caerulescens |MYCOL| A fungal pathogen that is the cause of leaf blister of oaks. { ta‚frī·nə ‚kī·rə'les·ənz }

Taphrina deformans [MYCOL] A fungal pathogen that is the cause of leaf curl of peach and almond trees. { ta‚frī·nə di'fȯr·mənz }

tapioca snow *See* snow pellets. { ‚tap·ē'ō·kə 'snō }

taproot |BOT| A root system in which the primary root forms a dominant central axis that penetrates vertically and rather deeply into the soil; it is generally larger in diameter than its branches. { 'tap‚rüt }

tar acid |AGR| A mixture of phenols (phenols, cresols, and xylenols) found in tars and tar distillates; toxic, combustible, and soluble in alcohol and coal-tar hydrocarbons; used as a wood preservative and an insecticide for farm animals and also to make disinfectants. { 'tär ‚as·əd }

tar camphor *See* naphthalene. { 'tär ‚kam·fər }

target spot |PL PATH| Any plant disease characterized by lesions in the form of concentric markings. { 'tär·gət ‚spät }

tarn |GEOGR| A landlocked pool or small lake that may occur in a marsh or swamp, or that may occupy a basin amid mountain ranges. { tärn }

tassel |BOT| The male inflorescence of corn and certain other plants. { 'tas·əl }

taxon |SYST| A group of organisms within a given taxonomic category tht share characteristics differentiating them from other groups; for example, all the organisms in a given species. Plural, taxa. Also known as taxonomic group; taxonomic unit. { 'tak ‚sän }

taxonomic category |SYST| One of a hierarchy of levels in the biological classification of organisms; the eight major categories are domain, kingdom, phylum, class, order, family, genus, species. { ¦tak·sə¦näm·ik 'kad·ə‚gȯr·ē }

taxonomic group *See* taxon. { ¦tak·sə¦näm·ik 'grüp }

taxonomic unit *See* taxon. { ¦tak·sə¦näm·ik 'yü·nət }

TBH *See* 1,2,3,4,5,6-hexachlorocyclohexane.

TCA *See* trichloroacetic acid.

TDE *See* 2,2-bis(*para*-chlorophenyl)-1,1-dichloroethane.

technosphere |ECOL| The part of the physical environment affected through building or modification by humans. { 'tek·nə‚sfir }

tectogenesis *See* orogeny. { ¦tek·tə'jen·ə·səs }

tectonoeustatism |OCEANOGR| Fluctuations of sea level due to changes in the capacities of the ocean basins resulting from earth movements. { ¦tek·tə·nō'yü·stə ‚tiz·əm }

tectoquinone |CHEM| $C_{15}H_{10}O_2$ A white compound with needlelike crystals; sublimes at 177°C; insoluble in water; used as an insecticide to treat wood. Also known as 2-methyl anthraquinone. { ¦tek·tō·kwə'nōn }

TEG *See* triethylene glycol.

T-E index *See* temperature-efficiency index. { ¦tē'ē ‚in‚deks }

TEL *See* tetraethyllead.

teleceptor [BIOL] A sense receptor which transmits information about portions of the external environment which are not necessarily in direct contact with the organism, such as the receptors of the ear, eye, and nose. { 'tel·ə,sep·tər }

telemeteorometry [METEOROL] The study of making meteorological observations at a distance. { ¦tel·ə,mēd·ē·ə'räm·ə·trē }

teleutospore See teliospore. { tə'lüd·ə,spór }

teliospore [MYCOL] A thick-walled spore of the rust and smut fungi that germinates to form a basidium. Also known as teleutospore. { 'tē·lē·ə,spór }

telvar [CHEM] The common name for the herbicide 3-(*para*-chlorophenyl)-1,1-dimethylurea; used as a soil sterilant. { 'tel,vär }

TEM See triethylenemelamine.

temperate and cold savannah [ECOL] A regional vegetation zone, very extensively represented in North America and in Eurasia at high altitudes; consists of scattered or clumped trees (very often conifers and mostly needle-leaved evergreens) and a shrub layer of varying coverage; mosses and, even more abundantly, lichens form an almost continuous carpet. { ¦tem·prət ən ¦kōld sə'van·ə }

temperate and cold scrub [ECOL] Regional vegetation zone whose density and periodicity vary a good deal; requires a considerable amount of moisture in the soil, whether from mist, seasonal downpour, or snowmelt; shrubs may be evergreen or deciduous; and undergrowth of ferns and other large-leaved herbs are quite frequent, especially at subalpine level; wind shearing and very cold winters prevent tree growth. Also known as bosque; fourré; heath. { ¦tem·prət ən ¦kōld 'skrəb }

temperate belt [CLIMATOL] A belt around the earth within which the annual mean temperature is less than 20°C (68°F) and the mean temperature of the warmest month is higher than 10°C (50°F). { 'tem·prət ,belt }

temperate climate [CLIMATOL] The climate of the middle latitudes; the climate between the extremes of tropical climate and polar climate. { 'tem·prət 'klī·mət }

temperate glacier [HYD] A glacier which, at the end of the melting season, is composed of firn and ice at the melting point. { 'tem·prət 'glā·shər }

temperate mixed forest [ECOL] A forest of the North Temperate Zone containing a high proportion of conifers with a few broad-leafed species. { 'tem·prət 'mikst ¦fär·əst }

temperate rainforest [ECOL] A vegetation class in temperate areas of high and evenly distributed rainfall characterized by comparatively few species with large populations of each species; evergreens are somewhat short with small leaves, and there is an abundance of large tree ferns. Also known as cloud forest; laurel forest; laurisilva; moss forest; subtropical forest. { 'tem·prət 'rān,fär·əst }

temperate rainy climate [CLIMATOL] One of the major categories in W. Kippen's climatic classification; the coldest-month mean temperature is less than 64.4°F (18°C) and greater than 26.6°F (−3°C), and the warmest-month mean temperature is more than 50°F (10°C). { 'tem·prət 'rān·ē ,klī·mət }

temperate westerlies See westerlies. { 'tem·prət 'wes·tər·lēz }

temperate-westerlies index [METEOROL] A measure of the strength of the westerly wind between latitudes 35°N and 55°N; the index is computed from the average sea-level pressure difference between these latitudes and is expressed as the west to east component of geostrophic wind in meters and tenths of meters per second. { 'tem·prət ¦wes·tər·lez 'in,deks }

temperate woodland [ECOL] A vegetation class similar to tropical woodland in spacing, height, and stratification, but it can be either deciduous or evergreen, broad-leaved or needle-leaved. Also known as parkland; woodland. { 'tem·prət 'wúd·lənd }

Temperate Zone |CLIMATOL| Either of the two latitudinal zones on the earth's surface which lie between 23°27' and 66°32' N and S (the North Temperate Zone and South Temperate Zone, respectively). { 'tem·prət ˌzōn }

temperature belt |METEOROL| The belt which may be drawn on a thermograph trace or other temperature graph by connecting the daily maxima with one line and the daily minima with another. { 'tem·prə·chər ˌbelt }

temperature-efficiency index |ECOL| For a given location, a measure of the long-range effectiveness of temperature (thermal efficiency) in promoting plant growth. Abbreviated T-E index. Also known as thermal-efficiency index. { 'tem·prə·chər i'fish·ən·sē ˌin,deks }

temperature-efficiency ratio |ECOL| For a given location and month, a measure of thermal efficiency; it is equal to the departure, in degrees Fahrenheit, of the normal monthly temperature above 32°F (0°C) divided by 4: (T − 32)/4. Abbreviated T-E ratio. Also known as thermal-efficiency ratio. { 'tem·prə·chər i'fish·ən·sē ˌrā·shō }

temperature-humidity index |METEOROL| An index which gives a numerical value, in the general range of 70-80, reflecting outdoor atmospheric conditions of temperature and humidity as a measure of comfort (or discomfort) during the warm season of the year; equal to 15 plus 0.4 times the sum of the dry-bulb and wet-bulb temperatures in degrees Fahrenheit. Also known as comfort index; discomfort index. Abbreviated CI; DI; THI. { 'tem·prə·chər hyü'mid·ə·dē ˌin,deks }

temperature inversion |METEOROL| A layer in the atmosphere in which temperature incrreases with altitude; the principal characteristic of an inversion layer is its marked stability, so that very little turbulent exchange can occur within it; strong wind shears often occur across inversion layers, and abrupt changes in concentrations of atmospheric particulates and atmospheric water vapor may be encountered on ascending through the inversion layer. Also known as thermal inversion. |OCEANOGR| A layer of a large body of water in which temperature increases with depth. { 'tem·prə·chər in,vər·zhən }

temperature province |CLIMATOL| A major division of C.W. Thornthwaite's schemes of climatic classification, determined as a function of the temperature-efficiency index or the potential evapotranspiration. { 'tem·prə·chər ˌpräv·əns }

temperature-salinity diagram |OCEANOGR| The plot of temperature versus salinity data of a water column; the resulting diagram identifies the water masses within the column, the column's stability, indicates the σ_T value via lines of constant σ_T printed on paper, and allows an estimate of the accuracy of the temperature and salinity measurements. Also known as T-S curve; T-S diagram; T-S relation. { 'tem·prə·chər sə'lin·əd·ē ˌdī·ə,gram }

temperature zone |CLIMATOL| A portion of the earth's surface defined by relatively uniform temperature characteristics, and usually bounded by selected values of some measure of temperature or temperature effect. { 'tem·prə·chər ˌzōn }

temporal |SCI TECH| Pertaining to or limited by time. { 'tem·prəl }

temporale |METEOROL| A rainy wind from the southwest to west resulting from a deflection of the southeast trades of the eastern South Pacific onto the Pacific coast of Central America. { ˌtem·pȯ'rä·lē }

temporary plankton See meroplankton. { 'tem·pəˌrer·ē 'plaŋk·tən }

tendency |METEOROL| The local rate of change of a vector or scalar quality with time at a given point in space. { 'ten·dən·sē }

tendency chart See change chart. { 'ten·dən·sē ˌchärt }

tendency equation |METEOROL| An equation for the local change of pressure at any point in the atmosphere, derived by combining the equation of continuity with an integrated form of the hydrostatic equation. { 'ten·dən·sē iˌkwā·zhən }

tendency interval [METEOROL] The finite increment of time over which a change of the value of a meteorological element is measured in order to estimate its tendency; the most familiar example is the three-hour time interval over which local pressure differences are measured in determining pressure tendency. { 'ten·dən·sē ,in·tər·vəl }

tender plant [BOT] A plant that is incapable of resisting cold. { ¦ten·dər ¦plant }

tendril [BOT] A stem modification in the form of a slender coiling structure capable of twining about a support to which the plant is then attached. { 'ten·drəl }

Tenebrionidae [ZOO] The darkling beetles, a large cosmopolitan family of coleopteran insects in the superfamily Tenebrionoidea; members are common pests of grains, dried fruits, beans, and other food products. { tə,neb·rē'än·ə,dē }

tension wood [BOT] In some hardwood trees, wood characterized by the presence of gelatinous fibers and excessive longitudinal shrinkage; causes trees to lean. { 'ten·chən ,wud }

tented ice [OCEANOGR] A type of pressure ice formed when ice is pushed up vertically, producing a flat-sided arch with a cavity between the raised ice and the water beneath. { 'ten·təd 'īs }

Tenthredinidae [ZOO] A family of hymenopteran insects in the superfamily Tenthredinoidea including economically important species whose larvae are plant pests. { ,ten·thrə'din·ə,dē }

tenting [OCEANOGR] The vertical displacement upward of ice under pressure to form a flat-sided arch with a cavity beneath. { 'tent·iŋ }

tentoxin [PL PATH] A species-selective pathotoxin produced by the fungus *Alternaria alternata* that causes variegated chlorosis in cucumber, cotton, lettuce, and many other sensitive plants. { 'ten,täk·sən }

TEP   *See* triethyl phosphate.

tepee butte [GEOGR] A tepeelike hill or knoll, especially one comprising soft material capped by more resistant rock. { 'tē·pē ,byüt }

tephigram [METEOROL] A thermodynamic diagram designed by Napier Shaw with temperature and logarithm of potential temperature as coordinates. { 'tef·ə,gram }

T-E ratio   *See* temperature-efficiency ratio. { ¦tē'ē 'rā·shō }

terbacil [CHEM] $C_9H_{13}ClN_2O_2$ A colorless, crystalline compound with a melting point of 175–177°C; used as an herbicide to control weeds in sugarcane, apples, peaches, citrus, and mints. { 'tər·bə,sil }

terbutol [CHEM] The common name for the herbicide 2,6-di-*tert*-butyl-*p*-tolylmethylcarbamate; used as a selective preemergence crabgrass herbicide for turf. { 'tər·byə,tòl }

terbutryn [CHEM] $C_{13}H_{19}N_5S$ A colorless powder with a melting point of 104–105°C; used for weed control for wheat, barley, and grain sorghum. { tər'byü·trən }

terbutylhylazine [CHEM] $C_9H_{16}N_5Cl$ A white solid with a melting point of 177–179°C; used as a preemergence herbicide. { ,tər,byüd·əl'hī·lə,zēn }

terdiurnal [METEOROL] Pertaining to a meteorological event that occurs three times a day. { ,tər·dī'ərn·əl }

terminal bud [BOT] A bud that develops at the apex of a stem.Also known as apical bud. { 'tər·mən·əl ,bəd }

termite [ZOO] A soft-bodied insect of the order Isoptera; individuals feed on cellulose and live in colonies with a caste system comprising three types of functional individuals: sterile workers and soldiers, and the reproductives. Also known as white ant. { 'tər,mīt }

termiticole |ECOL| An organism that lives in a termites' nest. { tər'mīd·ə,kōl }

termitophile |ECOL| An organism that lives in a termites' nest in a symbiotic association with the termites. { tər'mīd·ə,fīl }

terpene |CHEM| $C_{10}H_{16}$ A moderately toxic, flammable, unsaturated hydrocarbon liquid found in essential oils and plant oleoresins; used as an intermediate for camphor, menthol, and terpineol. { 'tər,pēn }

terrace |GEOL| **1.** A horizontal or gently sloping embankment of earth along the contours of a slope to reduce erosion, control runoff, or conserve moisture. **2.** A narrow coastal strip sloping gently toward the water. **3.** A long, narrow, nearly level surface bounded by a steeper descending slope on one side and by a steeper ascending slope on the other side. **4.** A benchlike structure bordering an undersea feature. { 'ter·əs }

terraced pool |GEOGR| A shallow, rimmed pool on the surface of a reef. { 'ter·əst 'pül }

terrachlor See pentachloronitrobenzene. { 'ter·ə,klȯr }

terracing See contour plowing. { 'ter·əs·iŋ }

terrain sensing |ENG| The gathering and recording of information about terrain surfaces without actual contact with the object or area being investigated; in particular, the use of photography, radar, and infrared sensing in airplanes and artificial satellites. { tə'rān ,sens·iŋ }

terra miraculosa See bole. { 'ter·ə mi,rak·yə'lō·sə }

terrestrial |SCI TECH| Of or pertaining to the earth. { tə'res·trē·əl }

terrestrial coordinates See geographical coordinates. { tə'res·trē·əl kō'ȯrd·ən·əts }

terrestrial ecosystem |ECOL| A community of organisms and their environment that occurs on the landmasses of continents and islands. { tə,res·trē·əl 'ek·ō,sis·təm }

terrestrial environment |GEOGR| The earth's land area, including its human-made and natural surface and subsurface features, and its interfaces and interactions with the atmosphere and the oceans. { tə'res·trē·əl in'vī·rən·mənt }

terrestrial frozen water |HYD| Seasonally or perennially frozen waters of the earth, exclusive of the atmosphere. { tə'res·trē·əl 'frō·zən 'wȯd·ər }

tertiary circulation |METEOROL| The generally small, localized atmospheric circulations, represented by such phenomena as local winds, thunderstorms, and tornadoes. { 'tər·shē,er·ē ,sər·kyə'lā·shən }

tertiary sewage treatment |CIV ENG| A process for purification of wastewater in which nitrates and phosphates, as well as fine particles, are removed; the process follows removal of raw sludge and biological treatment. Also known as advanced sewage treatment. { ¦tər·shē,er·ē ¦sü·ij ,trēt·mənt }

testa |BOT| A seed coat.Also known as episperm. { 'tes·tə }

tetanospasmin |BIOL| A neurotoxin elaborated by the bacterium *Clostridium tetani* and which is responsible for the manifestations of tetanus. { ,tet·ən·ō'spaz·mən }

tetanus |MED| An infectious disease of humans and animals caused by the toxin of *Clostridium tetani* and characterized by convulsive tonic contractions of voluntary muscles; infection commonly follows dirt contamination of deep wounds or other injured tissue. Also known as lockjaw. { 'tet·ən·əs }

tetanus antitoxin |MED| A serum containing antibodies that neutralize tetanus toxin. { 'tet·ən·əs 'ant·i,täk·sən }

tetanus toxoid |MED| Detoxified tetanus toxin used to produce active immunity against tetanus. { 'tet·ən·əs 'täk,sȯid }

tetrachlorobenzene [CHEM] $C_6H_2Cl_4$ Water-insoluble, combustible white crystals that appear in two forms: 1,2,3,4-tetrachlorobenzene which melts at 47°C and is used in chemical synthesis and in dielectric fluids; and 1,2,4,5-tetrachlorobenzene which melts at 138°C and is used to make herbicides, defoliants, and electrical insulation. { ¦te·tra¦klȯr·ə'ben,zēn }

sym-tetrachloroethane [CHEM] $CHCl_2CHCl_2$ A colorless, corrosive, toxic liquid with a chloroform scent, soluble in alcohol and ether, slightly soluble in water, boils at 147°C; used as a solvent, metal cleaner, paint remover, and weed killer. { ¦sim ¦te·tra¦klȯr·ō 'eth,ān }

tetrachlorophenol [CHEM] C_6HCl_4OH Either of two toxic compounds: 2,3,4,6-tetrachlorophenol comprises brown flakes, soluble in common solvents, melting at 70°C, and is used as a fungicide; 2,4,5,6-tetrachlorophenol is a brown solid, insoluble in water, soluble in sodium hydroxide, has a phenol scent, melts at about 50°C, and is used as a fungicide and for wood preservatives. { ¦te·tra¦klȯr·ə'fē,nȯl }

tetracycline [MICROBIO] **1.** Any of a group of broad-spectrum antibiotics produced biosynthetically by fermentation with a strain of *Streptomyces aureofaciens* and certain other species or chemically by hydrogenolysis of chlortetracycline. **2.** $C_{22}H_{24}O_8N_2$ A broad-spectrum antibiotic belonging to the tetracycline group of antibiotics; useful because of broad antimicrobial action, with low toxicity, in the therapy of infections caused by gram-positive and gram-negative bacteria as well as rickettsiae and large viruses such as psittacosis-lymphogranuloma viruses. { ,te·tra'sī,klēn }

tetraethyllead [CHEM] $Pb(C_2H_5)_4$ A highly toxic lead compound that, when added in small proportions to gasoline, increases the fuel's antiknock quality. Abbreviated TEL. { ¦te·tre¦eth¦əl'led }

tetraethylpyrophosphate [CHEM] $C_8H_{20}O_7P_2$ A hygroscopic corrosive liquid miscible with although decomposed by water, and miscible with many organic solvents; inhibits the enzyme acetylcholinesterase; used as an insecticide in place of nicotine sulfate. { ¦te·tra¦eth·əl¦pī·rō'fä,sfāt }

tetrahydro-3,5-dimethyl-2H-1,3,5-thiadiazine-6-thione *See* dazomet. { ¦te·tra¦hī·drō ¦thrē ¦fīv dī,meth·əl ¦tu ¦āch ¦wən ¦thrē ¦fīv thī·ə'dī·ə,zēn ¦sicks 'thī·ōn }

tetraiodoethylene [CHEM] $I_2C:CI_2$ Light yellow crystals with a melting point of 187°C; soluble in organic solvents; used in surgical dusting powder and antiseptic ointments, and as a fungicide. Also known as iodoethylene. { ¦te·tra¦ī·ə,dō'eth·ə,lēn }

TGA *See* nitrilotriacetic acid.

thalassic [OCEANOGR] Of or pertaining to the smaller seas. { thə'las·ik }

thalassophile element [GEOCHEM] An element that is relatively more abundant in sea water than in normal continental waters, such as sodium and chlorine. { thə'las·ə,fīl ,el·ə·mənt }

thallium sulfate [CHEM] Tl_2SO_4 Toxic, water-soluble, colorless crystals melting at 632°C; used as an analytical reagent and in medicine, rodenticides, and pesticides. Also known as thallous sulfate. { 'thal·ē·əm 'səl,fāt }

Thallobionta [BOT] One of the two subkingdoms of plants, characterized by the absence of specialized tissues or organs and multicellular sex organs. { ¦thal·ō·bī'änt·ə }

Thallophyta [BOT] The equivalent name for Thallobionta. { thə'läf·əd·ə }

thallospore [BOT] A spore that develops by budding of hyphal cells. { 'thal·ə,spȯr }

thallous sulfate *See* thallium sulfate. { 'thal·əs 'səl,fāt }

thallus [BOT] A plant body that is not differentiated into special tissue systems or organs and may vary from a single cell to a complex, branching multicellular structure. { 'thal·əs }

thalweg [HYD] Water seeping through the ground below the surface in the same direction as a surface stream course. { 'täl,veg }

thaw [CLIMATOL] A warm spell during which ice and snow melt, as a January thaw. { thȯ }

theobromine [CHEM] $C_7H_8N_4O_2$ A toxic alkaloid found in cocoa, chocolate products, tea, and cola nuts; closely related to caffeine. { ‚thē·ə'brō‚mēn }

theodolite [PHYS] An optical instrument used in surveying which consists of a sighting telescope mounted so that it is free to rotate around horizontal and vertical axes, and graduated scales so that the angles of rotation may be measured; the telescope is usually fitted with a right-angle prism so that the observer continues to look horizontally into the eyepiece, whatever the variation of the elevation angle; in meteorology, it is used principally to observe the motion of a pilot balloon. { thē'äd·əl ‚īt }

theoretical community ecology [ECOL] A branch of theoretical ecology that is concerned with factors determining the species composition and functional organization of communities, with a particular emphasis on interspecific interactions such as competition, predation, and mutualism. { ‚thē·ə¦red·i·kəl kə‚myün·əd·ē ē'käl·ə·jē }

theoretical ecology [ECOL] The use of mathematical models and verbal reasoning to provide a conceptual framework for the analysis of ecological systems. { ‚thē·ə ‚red·ə·kəl ē'käl·ə·jē }

thermal [METEOROL] A relatively small-scale, rising current of air produced when the atmosphere is heated enough locally by the earth's surface to produce absolute instability in its lower layers. { 'thər·məl }

thermal belt [ECOL] Any one of several possible horizontal belts of a vegetation type found in mountainous terrain, resulting primarily from vertical temperature variation. Also known as thermal zone. { 'thər·məl ‚belt }

thermal climate [CLIMATOL] Climate as defined by temperature, and divided regionally into temperature zones. { 'thər·məl 'klī·mət }

thermal convection [METEOROL] A net upward transport of heat resulting from the vertical ascent of relatively hot (lighter) material in some regions and the vertical descent of relatively cold (heavier) material in other regions; commonly produced by solar heating of the ground; the cause of convective (cumulus) clouds. Also known as free convection; gravitational convection. { 'thər·məl kən'vek·shən }

thermal ecology [ECOL] Study of the independent and interactive biotic and abiotic components of naturally heated environments. { 'thər·məl i'käl·ə·jē }

thermal-efficiency index See temperature-efficiency index. { 'thər·məl i¦fish·ən·sē ‚in ‚deks }

thermal-efficiency ratio See temperature-efficiency ratio. { 'thər·məl i¦fish·ən·sē ‚rā· shō }

thermal equator See heat equator. { 'thər·məl i'kwād·ər }

thermal high [METEOROL] A high resulting from the cooling of air by a cold underlying surface, and remaining relatively stationary over the cold surface. { 'thər·məl 'hī }

thermal inversion See temperature inversion. { 'thər·məl in'vər·zhən }

thermal jet [METEOROL] A region in the atmosphere where isotherms or thickness lines are closely packed; therefore, a region of very strong thermal wind. { 'thər·məl 'jet }

thermal low [METEOROL] An area of low atmospheric pressure due to high temperatures caused by intensive heating at the earth's surface; common to the continental subtropics in summer, thermal lows remain stationary over the area that produces them, their cyclonic circulation is generally weak and diffuse, and they are nonfrontal. Also known as heat low. { 'thər·məl 'lō }

thermal pollution [ECOL] The discharge of heated effluent into natural waters that causes a rise in temperature sufficient to upset the ecological balance of the waterway. { 'thər·məl pə'lü·shən }

thermal quality of snow *See* quality of snow. { 'thər·məl 'kwäl·əd·ē əv 'snō }

thermal radiation *See* heat radiation. { 'thər·məl ˌrād·ē'ā·shən }

thermal spring [HYD] A spring whose water temperature is higher than the local mean annual temperature of the atmosphere. { 'thər·məl 'spriŋ }

thermal steering [METEOROL] The steering of an atmospheric disturbance in the direction of the thermal wind in its vicinity; equivalent to steering along thickness lines. { 'thər·məl 'stir·iŋ }

thermal stratification [HYD] Horizontal layers of differing densities produced in a lake by temperature changes at different depths. { 'thər·məl ˌstrad·ə·fə'kā·shən }

thermal tide [METEOROL] A variation in atmospheric pressure due to the diurnal differential heating of the atmosphere by the sun; so-called in analogy to the conventional gravitational tide. { 'thər·məl 'tīd }

thermal vorticity [METEOROL] The vorticity of a thermal wind. { 'thər·məl vȯr'tis·əd·ē }

thermal wind [METEOROL] The mean wind-shear vector in geostrophic balance with the gradient of mean temperature of a layer bounded by two isobaric surfaces. { 'thər·məl 'wind }

thermal zone *See* thermal belt. { 'thər·məl 'zōn }

thermoacidophile [BIOL] An organism that grows under extremely acidic conditions and at very high temperatures. { ¦thər·mō·a'sid·ə,fīl }

thermocline [GEOPHYS] **1.** A temperature gradient as in a layer of sea water, in which the temperature decrease with depth is greater than that of the overlying and underlying water. Also known as metalimnion. **2.** A layer in a thermally stratified body of water in which such a gradient occurs. { 'thər·mə,klīn }

thermocyclogenesis [METEOROL] A theory of cyclogenesis by G. Stüve, in which the disturbance is initiated in the stratosphere and is reflected in the development of a disturbance in the lower troposphere. { ¦thər·mō,sī·klə'jen·ə·səs }

thermoduric bacteria [MICROBIO] Bacteria which survive pasteurization, but do not grow at temperatures used in a pasteurizing process. { ¦thər·mȯ'dur·ik bak'tir·ē·ə }

thermoelectric solar cell [ENG] A solar cell in which the sun's energy is first converted into heat by a sheet of metal, and the heat is converted into electricity by a semiconductor material sandwiched between the first metal sheet and a metal collector sheet. { ¦thər·mō·i'lek·trik 'sō·lər 'sel }

thermohaline [OCEANOGR] Pertaining to the joint activity of salinity and temperature in the oceans. { ¦thər·mō'hā,līn }

thermohaline convection [OCEANOGR] Vertical water movement observed when sea water, due to conditions of decreasing temperature or increasing salinity, becomes heavier than the water beneath it. { ¦thər·mō'hā,līn kən'vek·shən }

thermoisopleth [CLIMATOL] An isopleth of temperature; specifically, a line on a climatic graph showing the variation of temperature in relation to two coordinates. { ¦thər·mō 'īs·ə,pleth }

thermometric depth [OCEANOGR] The ocean depth, in meters, deduced from the difference between the paired protected and unprotected reversing thermometer readings; the unprotected reversing thermometer indicates higher temperature due to pressure effects on the instrument. { ¦thər·mə¦me·trik 'depth }

thermoperiodicity |BOT| The totality of responses of a plant to appropriately fluctuating temperatures. { ¦thər·mō,pir·ē·ə'dis·əd·ē }

thermophile [BIOL] An organism that thrives at high temperatures. { 'thər·mə,fzīl }

thermoplastic |CHEM ENG| A polymeric material with a linear macromolecular structure that will repeatedly soften when heated and harden when cooled; for example, styrene, acrylics, polyethylenes, vinyls, nylons, and fluorocarbons. Also known as thermoplastic resin. { 'thər·mə,plas·tik }

thermoplastic resin See thermoplastic. { ¦thər·mə¦plas·tik 'rez·ən }

thermoreception [BIOL] The process by which environmental temperature affects specialized sense organs (thermoreceptors). { ¦thər·mō·ri'sep·shən }

thermoregulation [BIOL] A mechanism by which mammals and birds attempt to balance heat gain and heat loss in order to maintain a constant body temperature when exposed to variations in cooling power of the external medium. { ¦thər·mō ,reg·yə'lā·shən }

thermosphere [METEOROL] The atmospheric shell extending from the top of the mesosphere to outer space; it is a region of more or less steadily increasing temperature with height, starting at 40 to 50 miles (70 to 80 kilometers); the thermosphere includes, therefore, the exosphere and most or all of the ionosphere. { 'thər·mə,sfir }

thermotaxis [BIOL] Orientation movement of a motile organism in response to the stimulus of a temperature gradient. { ¦thər·mō¦tak·səs }

thermotropic model [METEOROL] A model atmosphere used in numerical forecasting, in which the parameters are the height of one constant-pressure surface (usually 500 millibars) and one temperature (usually the mean temperature between 100 and 500 millibars). { ¦thər·mō¦träp·ik 'mad·əl }

therophyte [ECOL] An annual plant whose seed is the only overwintering structure. { 'ther·ə,fīt }

THI See temperature-humidity index.

thicket See tropical scrub. { 'thik·ət }

thickness [METEOROL] The vertical depth, measured in geometric or geopotential units, of a layer in the atmosphere bounded by surfaces of two different values of the same physical quantity, usually constant-pressure surfaces. { 'thik·nəs }

thickness chart [METEOROL] A type of synoptic chart showing the thickness of a certain physically defined layer in the atmosphere; it almost always refers to an isobaric thickness chart, that is, a chart of vertical distance between two constant-pressure surfaces. { 'thik·nəs ,chärt }

thickness line [METEOROL] A line drawn through all geographic points at which the thickness of a given atmospheric layer is the same. Also known as relative contour; relative isohypse. { 'thik·nəs ,līn }

thickness pattern [METEOROL] The general geometric distribution of thickness lines on a thickness chart. Also known as relative hypsography; relative topography. { 'thik·nəs ,pad·ərn }

Thiessen polygon method [METEOROL] A method of assigning areal significance to point rainfall values: perpendicular bisectors are constructed to the lines joining each measuring station with those immediately surrounding it; the bisectors form a series of polygons, each polygon containing one station; the value of precipitation measured at a station is assigned to the whole area covered by the enclosing polygon. { 'tē·sən 'päl·i,gän ,meth·əd }

thigmotaxis See stereotaxis. { ¦thig·mə¦tak·səs }

420

thigmotropism   *See* stereotropism.   { thig'mä·trə,piz·əm }

thin   [METEOROL] In aviation weather observations, the description of a sky cover that is predominantly transparent.   { thin }

thin-film solar cell   [ENG] A solar cell in which a thin film of gallium arsenide, cadmium sulfide, or other semiconductor material is evaporated on a thin, flexible metal or plastic substrate; advantages include flexibility, light weight, and relatively low cost of construction.   { 'thin ¦film 'sō·lər 'sel }

thiocyanic acid   [CHEM] HSC:N A colorless, water-soluble liquid decomposing at 200°C; used to inhibit paper deterioration due to the action of light, and (in the form of organic esters) as an insecticide. Also known as rhodanic acid; sulfocyanic acid.   { ¦thī-ō-sī ¦an·ik 'as·əd }

thiram   [CHEM] $(CH_3)_2$N-CS-S-S-CS-N$(CH_3)_2$ Tetramethylthioperoxydicarbonic dia-mide; a fungicide, bacteriostat (in soap), antimicrobial agent (chemotherapeutic for plants), seed disinfectant, and vulcanizing agent.   { 'thī,ram }

third-order climatological station   [CLIMATOL] As defined by the World Meteorological Organization, a station, other than a precipitation station, at which the observations are of the same kind as those at a second-order climatological station, but are not so comprehensive, are made once a day only, and are made at other than the specified hours.   { 'thərd ¦òr·dər ,klī·mə·tə'läj·ə·kəl 'stā·shən }

third-order relief   [GEOGR] Specific landform complexes that are smaller in extent and size than formations of subcontinental extent.   { 'thərd ¦òr·dər ri'lēf }

thirty-day forecast   [METEOROL] A weather forecast for a period of 30 days; as issued by the U.S. Weather Bureau, the forecast concerns expected departures of temperature and precipitation from normal.   { ¦thər·dē ¦dā'fòr,kast }

thistle   [BOT] Any of the various prickly plants comprising the family Compositae.   { 'this·əl }

thorium   [CHEM] An element of the actinium series, symbol Th, atomic number 90, atomic weight 232; soft, radioactive, insoluble in water and alkalies, soluble in acids, melts at 1750°C, boils at 4500°C.   [ENG] A heavy malleable metal that changes from silvery-white to dark gray or black in air; potential source of nuclear energy; used in manufacture of sunlamps.   { 'thòr·ē·əm }

thorn   [BOT] A short, sharp, rigid, leafless branch on a plant.   { thòrn }

thornbush   [ECOL] A vegetation class that is dominated by tall succulents and profusely branching smooth-barked deciduous hardwoods which vary in density from mesquite bush in the Caribbean to the open spurge thicket in Central Africa; the climate is that of a warm desert, except for a rather short intense rainy season. Also known as Dorngeholz; Dorngestrauch; dornveld; savane armée; savane épineuse; thorn scrub.   { 'thòrn,bùsh }

thorn forest   [ECOL] A type of forest formation, mostly tropical and subtropical, intermediate between desert and steppe; dominated by small trees and shrubs, many armed with thorns and spines; leaves are absent, succulent, or deciduous during long dry periods, which may also be cool; an example is the caatinga of northeastern Brazil.   { 'thòrn 'fär·əst }

thorn scrub   *See* thornbush.   { 'thòrn 'skrəb }

thread blight   [PL PATH] A fungus disease of a number of tropical and semitropical woody plants, including cocoa and tea, caused by species of *Pellicularia* and *Marasmius* which form filamentous mycelia on the surface of twigs and leaves.   { 'thred ,blīt }

three-day fever   *See* phlebotomus fever.   { 'thrē ¦dā 'fē·vər }

threshold   [BIOL] The minimum level of a stimulus that will evoke a response in an irritable tissue.   { 'thresh,hōld }

threshold depth   *See* sill depth.   { 'thresh,hōld ,depth }

threshold limit value   [MED] The average concentration of toxic gas to which the normal person can be exposed without injury for 8 hours per day, 5 days per week for an unlimited period; differs slightly from maximum allowable concentration in that threshold limit value is an average concentration. Abbreviated TLV.   { 'thresh,hōld 'lim·ət ,val·yü }

through glacier   [HYD] A two-ended glacier, consisting of two valley glaciers in a depression, flowing in opposite directions.   { 'thrü ,glā·shər }

thrush   [MED] A form of candidiasis due to infection by *Candida albicans* and characterized by small whitish spots on the tip and sides of the tongue and the mucous membranes of the buccal cavity. Also known as mycotic stomatitis; parasitic stomatitis.   [VET MED] A disease of the frog of a horse's foot (the triangle-shaped pad on the bottom of the foot) accompanied by a fetid discharge.   { thrəsh }

thundercloud   [METEOROL] A convenient and often used term for the cloud mass of a thunderstorm, that is, a cumulonimbus.   { 'thən·dər,klaüd }

thunderhead   *See* incus.   { 'thən·dər,hed }

thundersquall   *See* rainsquall.   { 'thən·dər,skwól }

thunderstorm   [METEOROL] A convective storm accompanied by lightning and thunder and rain, rarely snow showers but often hail, and gusty squall winds at the onset of precipitation; the characteristic cloud is the cumulonimbus.   { 'thən·dər,stórm }

thunderstorm cell   [METEOROL] The convection cell of a cumulonimbus cloud.   { 'thən·dər,stórm ,sel }

thylakoid   [BIOL] An internal membrane system which occupies the main body of a plastid; particularly well developed in chloroplasts.   { 'thī·lə,kóid }

tick   [ZOO] Any arachnid comprising Ixodoidea; a bloodsucking parasite and important vector of various infectious diseases of humans and lower animals.   { tik }

tick-bite paralysis   [VET MED] A flaccid paralysis in animals, and occasionally in humans, caused by a feeding tick attached to the body.   { 'tik ¦bīt pə'ral·ə·səs }

tick-borne typhus fever of Africa   [MED] Any of several infections caused by *Rickettsia conori*, transmitted by ixodid ticks, and occurring in Africa and adjacent areas; includes boutonneuse fever, Marseilles fever, Kenya tick typhus fever, and South African tick-bite fever.   { 'tik ¦bórn 'tī·fəs ,fē·vər əv 'af·ri·kə }

tick fever   *See* Rocky Mountain spotted fever.   { 'tik ,fē·vər }

tick typhus   *See* Rocky Mountain spotted fever.   { 'tik ,ti·fəs }

tidal bore   *See* bore.   { 'tīd·əl 'bór }

tidal channel   [OCEANOGR] A major channel followed by tidal currents, extending from the ocean into a tidal marsh or tidal flat.   { 'tīd·əl 'chan·əl }

tidal component   *See* partial tide.   { 'tīd·əl kəm'pō·nənt }

tidal constants   [OCEANOGR] Tidal relations that remain essentially constant for any particular locality.   { 'tīd·əl 'kän·stəns }

tidal constituent   *See* partial tide.   { 'tīd·əl kən'stich·ə·wənt }

tidal current   [OCEANOGR] The alternating horizontal movement of water associated with the rise and fall of the tide caused by the astronomical tide-producing forces.   { 'tīd·əl 'kə·rənt }

tidal-current chart   [OCEANOGR] A chart showing by arrows and numbers the average direction and speed of tidal currents at a particular part of the current cycle.   { 'tīd·əl ¦kə·rənt ,chärt }

tidal-current tables [OCEANOGR] Tables issued annually which give daily predictions of the times of slack water and the times and velocities of the strength of flood and ebb currents for a number of reference stations, together with differences and constants for obtaining predictions at subordinate stations. { 'tīd·əl ¦kə·rənt ˌtā·bəlz }

tidal cycle *See* tide cycle. { 'tīd·əl ˌsī·kəl }

tidal datum [OCEANOGR] A level of the sea, defined by some phase of the tide, from which water depths and heights of tide are reckoned. Also known as tidal datum plane. { 'tīd·əl 'dad·əm }

tidal datum plane *See* tidal datum. { 'tīd·əl ¦dad·əm ˌplān }

tidal day [OCEANOGR] The interval between two consecutive high waters of the tide at a given place, averaging 24 hours 51 minutes. { 'tīd·əl 'dā }

tidal difference [OCEANOGR] The difference in time or height of a high or low water at a subordinate station and at a reference station for which predictions are given in the tide tables; the difference applied to the prediction at the reference station gives the corresponding time or height for the subordinate station. { 'tīd·əl 'dif·rəns }

tidal energy [OCEANOGR] The energy in a tide flowing from a basin into an open sea. { 'tīd·əl 'en·ər·jē }

tidal epoch *See* phase lag. { 'tīd·əl 'ep·ik }

tidal excursion [OCEANOGR] The net horizontal distance over which a water particle moves during one tidal cycle of flood and ebb; the distances traversed during ebb and flood are rarely equal in nature, since there is usually a layered circulation in an estuary, with a net surface flow in one direction compensated by an opposite flow at depth. { 'tīd·əl ik'skər·zhən }

tidal flat [GEOL] A marshy, sandy, or muddy nearly horizontal coastal flatland which is alternately covered and exposed as the tide rises and falls. { 'tīd·əl 'flat }

tidal frequency [OCEANOGR] The rate of travel, in degrees per day, of a component of a tide, the component being created by a particular juxtaposition of forces in the sun-earth-moon system. { 'tīd·əl 'frē·kwən·sē }

tidal friction [OCEANOGR] The frictional effect of the tidal wave particularly in shallow waters that lengthens the tidal epoch and tends to slow the rotational velocity of the earth, thus increasing very slowly the length of the day. { 'tīd·əl 'frik·shən }

tidal glacier *See* tidewater glacier. { 'tīd·əl 'glā·shər }

tidal marsh [GEOGR] Any marsh whose surface is covered and uncovered by tidal flow. { 'tīd·əl 'märsh }

tidal platform ice foot [OCEANOGR] An ice foot between high and low water levels, produced by the the rise and fall of the tide. { 'tīd·əl ¦plat,fòrm 'īs ˌfùt }

tidal pool [OCEANOGR] An accumulation of sea water remaining in a depression on a beach or reef after the tide recedes. { 'tīd·əl 'pül }

tidal potential [OCEANOGR] Tidal forces expressed as components of a vector field. { 'tīd·əl pə,ten·chəl }

tidal prism [OCEANOGR] The difference between the mean high-water volume and the mean low-water volume of an estuary. { 'tīd·əl 'priz·əm }

tidal range *See* tide range. { 'tīd·əl 'rānj }

tidal stand *See* stand. { 'tīd·əl 'stand }

tidal water [OCEANOGR] Any water whose level changes periodically due to tidal action. { 'tīd·əl ˌwòd·ər }

tidal wave [OCEANOGR] **1.** Any unusually high and generally destructive sea wave or water level along a shore. **2.** *See* tide wave. { 'tīd·əl ,wāv }

tidal wind [METEOROL] A very light breeze which occurs in calm weather in inlets where the tide sets strongly; it blows onshore with rising tide and offshore with ebbing tide. { 'tīd·əl 'wind }

tide [OCEANOGR] The periodic rising and falling of the oceans resulting from lunar and solar tide-producing forces acting upon the rotating earth. { tīd }

tide amplitude [OCEANOGR] One-half of the difference in height between consecutive high water and low water; half the tide range. { 'tīd 'am·plə,tüd }

tide bulge *See* tide wave. { 'tīd ,bəlj }

tide crack [OCEANOGR] A crack in sea ice, parallel to the shore, caused by the vertical movement of the water due to tides; several such cracks often appear as a family. { 'tīd ,krak }

tide curve [OCEANOGR] Any graphic representation of the rise and fall of the tide; time is generally represented by the abscissas, and the height of the tide by the ordinates; for normal tides the curve so produced approximates a sine curve. { 'tīd ,kərv }

tide cycle [OCEANOGR] A period which includes a complete set of tide conditions or characteristics, such as a tidal day or a lunar month. Also known as tidal cycle. { 'tīd ,sī·kəl }

tidehead [OCEANOGR] The inland limit of water affected by a tide. { 'tīd,hed }

tide hole [OCEANOGR] A hole made in ice to observe the height of the tide. { 'tīd ,hōl }

tideland [GEOGR] Land which is under water at high tide and uncovered at low tide. { 'tīd·lənd }

tidemark [OCEANOGR] **1.** A high-water mark left by tidal water. **2.** The highest point reached by a high tide. { 'tīd,märk }

tide notes [OCEANOGR] Notes included on nautical charts which give information on the mean range or the diurnal range of the tide, mean tide level, and extreme low water at key places on the chart. { 'tīd ,nōts }

tide prediction [OCEANOGR] The mathematical process by which the times and heights of the tide are determined in advance from the harmonic constituents at a place. { 'tīd pri,dik·shən }

tide race [OCEANOGR] A strong tidal current or a channel in which such a current flows. { 'tīd ,rās }

tide range [OCEANOGR] The difference in height between consecutive high and low waters. Also known as tidal range. { 'tīd ,rānj }

tide rips *See* rips. { 'tīd ,rips }

tide station [OCEANOGR] A place where observations of the tides are obtained. { 'tīd ,stā·shən }

tide table [OCEANOGR] A table giving daily predictions, usually a year in advance, of the times and heights of the tide for a number of reference stations. { 'tīd ,tā·bəl }

tidewater [OCEANOGR] **1.** A body of water, such as a river, affected by tides. **2.** Water inundating land at flood tide. { 'tīd,wȯd·ər }

tidewater glacier [HYD] A glacier that descends into the sea and usually has a terminal ice cliff. Also known as tidal glacier. { 'tīd,wȯd·ər 'glā·shər }

tide wave [OCEANOGR] A long-period wave associated with the tide-producing forces of the moon and sun, and identified with the rising and falling of the tide. Also known as tidal wave; tide bulge. { 'tīd ,wāv }

tideway [OCEANOGR] A channel through which a tidal current runs. { 'tīd,wā }

till [GEOL] Unsorted and unstratified drift consisting of a heterogeneous mixture of clay, sand, gravel, and boulders which is deposited by and underneath a glacier. Also known as boulder clay; glacial till; ice-laid drift. { til }

tillage [AGR] The operation or practice of cultivating soil in order to improve it for agricultural purposes. { 'til·ij }

tilt [METEOROL] The inclination to the vertical of a significant feature of the circulation (or pressure) pattern or of the field of temperature or moisture; for example, troughs in the westerlies usually display a westward tilt with altitude in the lower and middle troposphere. { tilt }

tilted iceberg [OCEANOGR] A tabular iceberg that has become unbalanced, so that the flat, level top is inclined. { 'til·təd 'īs,bərg }

timberline *See* tree line. { 'tim·bər,līn }

time-weighted average [SCI TECH] The average exposure to a contaminant or condition (such as noise) to which workers may be exposed without adverse effect over a period such as in an 8-hour day or 40-hour week. Abbreviated TWA. { ¦tīm ,wād·əd 'av·rij }

tinea [MED] Group of skin diseases caused by various fungi, for example, tinea pedis (athlete's foot) and tinea capitis (ringworm infection of the scalp). { ,tin·ē·ə }

tipburn [PL PATH] A disease of certain cultivated plants, such as potato and lettuce, characterized by browning of the leaf margins due to excessive loss of water. { 'tip ,bərn }

tipping-bucket rain gage [ENG] A type of recording rain gage; the precipitation collected by the receiver empties into one side of a chamber which is partitioned transversely at its center and is balanced bistably upon a horizontal axis; when a predetermined amount of water has been collected, the chamber tips, spilling out the water and placing the other half of the chamber under the receiver; each tip of the bucket is recorded on a chronograph, and the record obtained indicates the amount and rate of rainfall. { 'tip·iŋ ,bək·ət 'rān ,gāj }

tjaele *See* frozen ground. { 'chā·lē }

TLV *See* threshold limit value.

TMA *See* trimethylamine.

TMV *See* tobacco mosaic virus.

TNT *See* 2,4,6-trinitrotoluene.

toad [ZOO] Any of several species of the amphibian order Anura, especially in the family Bufonidae; glandular structures in the skin secrete acrid, irritating substances of varying toxicity. { tōd }

toadstool [MYCOL] Any of various fleshy, poisonous or inedible fungi with a large umbrella-shaped fruiting body. { 'tōd,stül }

tobacco [BOT] **1.** Any plant of the genus *Nicotinia* cultivated for its leaves, which contain 1–3% of the alkaloid nicotine. **2.** The dried leaves of the plant. { tə'bak·ō }

tobacco mosaic [PL PATH] Any of a complex of virus diseases of tobacco and other solanaceous plants (in the family Solanaceae) in which the leaves are mottled with light- and dark-green patches, sometimes interspersed with yellow. { tə'bak·ō mō ,zā·ik }

tobacco mosaic virus [MICROBIO] The type species of the genus *Tobamovirus*, it infects tobacco, tomato, and other solanaceous plants (in the family Solanaceae), causing defoliation and/or mosaic symptoms on leaves, stems, and fruit. Abbreviated TMV. { tə¦bak·ō mō′zā·ik ‚vī·rəs }

tobacco rattle virus [MICROBIO] The type species of the genus *Tobravirus*, it infects a wide range of plants, usually via a nematode vector. Abbreviated TRV. { tə¦bak·ō ′rad·əl ‚vī·rəs }

tobacco ring spot virus [MICROBIO] The type species of the genus *Nepovirus*, it has a wide host range and is transmitted via seeds and nematodes. Abbreviated TRSV. { tə ¦bak·ō′riŋ ‚spät ‚vī·rəs }

Togaviridae [MICROBIO] A family of positive-strand ribonucleic acid (RNA)-containing viruses characterized by spherical enveloped particles with an icosahedral nucleocapsid containing linear single-stranded RNA, it contains the genera *Alphavirus* (arbovirus A; prototype Sindbis virus), *Flavivirus* (arbovirus B; prototype yellow fever), Rubivirus (rubella virus), and Pestivirus (mucosal disease virus). { ‚tō·gə′vir·ə‚dī }

tolerance [MED] **1.** The ability of enduring or being less responsive to the influence of a drug or poison, particularly when acquired by continued use of the substance. **2.** The allowable deviation from a standard, as the range of variation permitted for the content of a drug in one of its dosage forms. { ′täl·ə·rəns }

tolnaftate [CHEM] $C_{19}H_{17}NOS$ An agricultural fungicide; it is also used medically as an antifungal agent. { tōl′naf‚tāt }

***ortho*-toluic acid** [CHEM] $C_6H_4CH_3COOH$ White, combustible crystals soluble in alcohol and chloroform, slightly soluble in water, melts at 104°C; used as a bacteriostat. Also known as *ortho*-toluylic acid. { ¦òr·thō tə′lü·ik ′as·əd }

***ortho*-toluylic acid** See *ortho*-toluic acid. { ¦òr·thō ¦täl·yə¦wil·ik ′as·əd }

tomatine [CHEM] $C_{50}H_{83}NO_{21}$ A glycosidal alkaloid obtained from the leaves and stems from the tomato plant; the crude extract is known as tomatin: white, toxic crystals; used as a plant fungicide and as a precipitating agent for cholesterol. { ′täm·ə‚tēn }

tongue [OCEANOGR] **1.** A protrusion of water into a region of different temperature, or salinity, or dissolved oxygen concentrating. **2.** A protrusion of one water mass into a region occupied by a different water mass. { təŋ }

tonoplast [BOT] The membrane surrounding a plant-cell vacuole. { ′tän·ə‚plast }

top See overburden. { täp }

topocline [ECOL] A graded series of characters exhibited by a species or other closely related organisms along a geographical axis. { ′täp·ə‚klīn }

topographic climax [ECOL] A climax plant community under a uniform macroclimate over which minor topographic features such as hills, rivers, valleys, or undrained depressions exert a controlling influence. { ¦täp·ə¦graf·ik ′klī‚maks }

topographic curl effect [OCEANOGR] A term in Ekman's differential equation for the effects of variable wind stress, variable depth, variable friction, and variable latitude on the deep current; tends to make the curl G (velocity of deep current) positive when the current flows over increasing depth and negative when the depth decreases in the direction of the current. { ¦täp·ə¦graf·ik ′kərl i‚fəkt }

topographic passage [OCEANOGR] A pass or gap through a sea-floor feature that possesses high topography, such as a ridge or a plateau. { ¦täp·ə‚graf·ik ′pas·ij }

topographic unconformity [GEOGR] A lack of harmony or conformity between two parts of a landscape or two kinds of topography. { ¦täp·ə¦graf·ik ‚ən·kən′fòr·məd·ē }

topography [GEOGR] **1.** The general configuration of a surface, including its relief; may be a land or water-bottom surface. **2.** The natural surface features of a region, treated collectively as to form. { tə′päg·rə·fē }

topotaxis See tropism. { ¦täp·ə¦tak·səs }

topsoil [GEOL] **1.** Soil presumed to be fertile and used to cover areas of special planting. **2.** Surface soil, usually corresponding with the A horizon, as distinguished from subsoil. { 'täp,sȯil }

topwork [BOT] A procedure employed to propagate seedless varieties of fruit and hybrids, to change the variety of fruit, and to correct pollination problems, using any of three methods: root grafting, crown grafting, and top grafting. { 'täp,wərk }

tor [GEOGR] An isolated, rough pinnacle or rocky peak. { tȯr }

tornado [METEOROL] An intense rotary storm of small diameter, the most violent of weather phenomena; tornadoes always extend downward from the base of a convective-type cloud, generally in the vicinity of a severe thunderstorm. { tȯr 'nād·ō }

tornado belt [METEOROL] The district of the United States in which tornadoes are most frequent; it encompasses the great lowland areas of the central and upper Mississippi, the Ohio, and lower Missouri River valleys. { tȯr'nād·ō,belt }

tornado cloud *See* tuba. { tȯr'nād·ō,klaȯd }

tornado echo [METEOROL] A type of radar precipitation echo which has been observed in connection with a number of tornadoes; it frequently appears, on plan-position-indicator scopes, in the form of the figure 6 in the southwest sector of the storm; this echo has not been noted with all radar-observed tornadoes. { tȯr'nād·ō ,ek·ō }

torpor [BIOL] The condition in hibernating poikilotherms during winter when body temperature drops in a parallel relation to ambient environmental temperatures. { 'tȯr·pər }

Torrert [GEOL] A suborder of the soil order Vertisol; it is the driest soil of the order and forms cracks that tend to remain open; occurs in arid regions. { 'tȯr·ərt }

Torrid Zone [CLIMATOL] The zone of the earth's surface which lies between the Tropics of Cancer and Capricorn. { 'tär·əd ,zōn }

Torrox [GEOL] A suborder of the soil order Oxisol that is low in organic matter, well drained, and dry most of the year; believed to have been formed under rainier climates of past eras. { 'tȯr,äks }

torulosis *See* cryptococcosis. { ,tȯr·ə'lō·səs }

total allowable catch [OCEANOGR] A fishery management approach to assign an annual quota that, if exceeded, will terminate the fishery for that year; the total allowable catch is set at a level to prevent a catch so large that the stock will be overfished. { ¦tōd·əl ə,laȯ·ə·bəl 'kach }

total evaporation *See* evapotranspiration. { 'tōd·əl i,vap·ə'rā·shən }

towering cumulus [METEOROL] A descriptive term, used mostly in weather observing, for the cloud type cumulus congestus. { 'taȯ·ə·riŋ 'kyü·myə·ləs }

toxemia [MED] A condition in which the blood contains toxic substances, either of microbial origin or as by-products of abnormal protein metabolism. { täk'sē·mē·ə }

toxic [MED] Relating to a harmful effect by a poisonous substance on the human body by physical contact, ingestion, or inhalation. { 'täk·sik }

toxic amaurosis [MED] Blindness following the introduction of toxic substances into the body, such as ethyl and methyl alcohol, tobacco, lead, and metabolites of uremia and diabetes. { 'täk·sik ,a,mȯ'rō·səs }

toxic hepatitis [MED] Inflammation of the liver caused by chemical agents ingested or inhaled into the body, such as chlorinated hydrocarbons and some alkaloids. { 'täk·sik ,hep·ə'tīd·əs }

toxicity [MED] **1.** The quality of being toxic. **2.** The kind and amount of poison or toxin produced by a microorganism, or possessed by a chemical substance not of biological origin. { täk'sis·əd·ē }

toxicological study [MED] The study of how much poison must be present to produce an effect on animals or plant systems, may also include what type of effect is produced and how it is detected. { ˌtäk·sə·kəˌläj·ə·kəl 'stəd·ē }

toxicology [MED] The study of poisons, including their nature, effects, and detection, and methods of treatment. { ˌtak·sə'käl·ə·jē }

toxic psychosis [MED] A brain disorder due to a toxic agent such as lead or alcohol. { 'täk·sik sī'kō·səs }

toxigenicity [MICROBIO] A microorganism's capability for producing toxic substances. { ˌtäk·sə·jə'nis·əd·ē }

toxin [BIOL] Any of various poisonous substances produced by certain plant and animal cells, including bacterial toxins, phytotoxins, and zootoxins. { 'täk·sən }

toxoid [MED] Detoxified toxin, but with antigenic properties intact; toxoids of tetanus and diphtheria are used for immunization. { 'täkˌsȯid }

trace [METEOROL] A precipitation of less than 0.005 inch (0.127 millimeter). { trās }

trace element [BIOL] A chemical element that is needed in minute quantities for the proper growth, development, and physiology of the organism. Also known as micronutrient. [GEOCHEM] An element found in small quantities (usually less than 1.0%) in a mineral. Also known as accessory element; guest element. { 'trās ˌel·ə·mənt }

Tracheophyta [BOT] A large group of plants characterized by the presence of specialized conducting tissues (xylem and phloem) in the roots, stems, and leaves. { ˌtrā·kē'äf·əd·ə }

tracheophyte See vascular plant. { 'trā·kē·əˌfīt }

trachoma [MED] An infectious disease of the conjunctiva and cornea caused by *Chlamydia trachomatis* producing photophobia, pain, and excessive lacrimation. { trə'kō·mə }

trade cumulus See trade-wind cumulus. { 'trād ¦kyü·myə·ləs }

trade-wind [METEOROL] The wind system, occupying most of the tropics, which blows from the subtropical highs toward the equatorial trough; a major component of the general circulation of the atmosphere; the winds are northeasterly in the Northern Hemisphere and southeasterly in the Southern Hemisphere; hence they are known as the northeast trades and southeast trades, respectively. { 'trād ¦wind }

trade-wind cumulus [METEOROL] The characteristic cumulus cloud of the trade winds over the oceans in average, undisturbed weather conditions; the individual cloud usually exhibits a blocklike appearance since its vertical growth ends abruptly in the lower stratum of the trade-wind inversion; a group of fully grown clouds shows considerable uniformity in size and shape. Also known as trade cumulus. { 'trād ¦wind ˌkyü·myə·ləs }

trade-wind desert [CLIMATOL] **1.** An area of very little rainfall and high temperature which occurs where the trade winds or their equivalent blow over land; the best examples are the Sahara and Kalahari deserts. **2.** The arid cold-water coasts on the western shores of North and South America and Africa. { 'trād ¦wind ˌdez·ərt }

trade-wind inversion [METEOROL] A characteristic temperature inversion usually present in the trade-wind streams over the eastern portions of the tropical oceans; it is formed by broad-scale subsidence of air from high altitudes in the eastern extremities of the subtropical highs; while descending, the current meets the opposition of the low-level maritime air flowing equatorward; the inversion forms at the meeting point of these two strata which flow horizontally in the same direction. { 'trād ¦wind in ¸vər·zhən }

trail pheromone [BIOL] A type of pheromone used by social insects and some lepidopterans to recruit others of its species to a food source. { 'trāl ¸fer·ə¸mōn }

Trametes versicolor [MYCOL] A brightly colored mushroom that appears to have antitumor properties, is a common inhabitant of the woods worldwide. Also known as Coriolus versicolor; turkey tail mushroom. { trə¸mēd¸ēz 'vər·sə¸kəl·ər }

transect [SCI TECH] To cut across, or to cut transversely. { tran'sekt }

transgression [OCEANOGR] Extension of the sea over land areas. { tranz'gresh·ən }

translocation [BOT] Movement of water, mineral salts, and organic substances from one part of a plant to another. { ¦tranz·lō'kā·shən }

translucidus [METEOROL] A cloud variety occurring in a layer, patch, or extensive sheet, the greater part of which is sufficiently translucent to reveal the position of the sun, or through which higher clouds may be discerned; this variety is found in the general altocumulus, altostratus, stratocumulus, and stratus. { tran'slüs·əd·əs }

transparent sky cover [METEOROL] In United States weather-observing practice, that portion of sky cover through which higher clouds and blue sky may be observed; opposed to opaque sky cover. { tranz'par·ənt 'skī ¸kəv·ər }

transpiration [BIOL] The passage of a gas or liquid (in the form of vapor) through the skin, a membrane, or other tissue. [BOT] The process whereby water vapor is lost from plants (primarily from the leaves). { ¸tranz·pə'rā·shən }

trap [CIV ENG] A bend or dip in a soil drain which is always full of water, providing a water seal to prevent odors from entering the building. { trap }

treating [CHEM ENG] Usually, the contacting of a fluid stream (for example, water, sewage, petroleum products, or mixed gases) with chemicals to improve the fluid properties by removing, sequestering, or converting undesirable impurities. { 'trēd·iŋ }

tree [BOT] A perennial woody plant at least 20 feet (6 meters) in height at maturity, having an erect stem or trunk and a well-developed crown or leaf canopy. { trē }

tree climate [CLIMATOL] Any type of climate which supports the growth of trees, including the tropical rainy climates, temperate rainy climates, and snow-forest climates. { 'trē ¸klī·mət }

tree fern [BOT] The common name for plants belonging to the families Cyatheaceae and Dicksoniaceae; all are ferns that exhibit an arborescent habit. { 'trē ¸fərn }

tree line [ECOL] The altitudinal or latitudinal limit beyond which conditions do not permit the growth of trees. Also known as timberline. { 'trē ¸līn }

tree-ring hydrology See dendrohydrology. { ¦trē ¦riŋ hī'dräl·ə·jē }

trellis drainage [HYD] A drainage pattern characterized by parallel main streams and secondary tributaries intersected at right angles by tributaries. Also known as espalier drainage; grapevine drainage. { 'trel·əs ¸drā·nij }

trembling ill See louping ill. { 'trem·bliŋ ¦il }

Tremella fuciformis [MYCOL] A mushroom that grows on deciduous trees in the southern United States and in warm climates worldwide. Once primarily grown for its medicinal properties (it boosts immunological function and stimulates leukocyte activity), it is now used mostly for food. Also known as snow fungus. { trə¸mel·ə ¸fyü·sə'fór·məs }

trench [GEOGR] **1.** A narrow, straight, elongate, U-shaped valley between two mountain ranges. **2.** A narrow stream-eroded canyon, gulley, or depression with steep sides. [GEOL] A long, narrow, deep depression of the sea floor, with relatively steep sides. Also known as submarine trench. { trench }

trench fever [MED] A louse-borne infection that is caused by *Rickettsia quintana* and is characterized by headache, chills, rash, pain in the legs and back, and often by a relapsing fever. { 'trench ˌfē·vər }

Treponema pallidum [MICROBIO] A pathogenic spirochete that causes the sexually transmitted disease syphilis. { ˌtrep·ə‚nē·mə 'pal·ə·dəm }

tretamine *See* triethylenemelamine. { 'tred·ə‚mēn }

tributary [HYD] A stream that feeds or flows into or joins a larger stream or a lake. Also known as contributory; feeder; side stream; tributary stream. { 'trib·yə‚ter·ē }

tributary stream *See* tributary. { 'trib·yə‚ter·ē 'strēm }

tributary waterway [HYD] Any body of water that flows into a larger body, that is, a creek in relation to a river, a river in relation to a bay, and a bay in relation to the open sea. { 'trib·yə‚ter·ē 'wȯd·ər‚wā }

tributyltin chloride [CHEM] $(C_4H_9)_3SnCl$ A colorless liquid with a boiling point of 145–147°C; soluble in alcohol, benzene, and other organic solvents; used as a rodenticide. { trī'byüd·əl·tən 'klȯr‚īd }

trichloroacetic acid [CHEM] CCl_3COOH Toxic, deliquescent, colorless crystals with a pungent aroma; soluble in water, alcohol, and ether; boils at 198°C; used as a chemical intermediate and laboratory reagent, and in medicine, pharmacy, and herbicides. Abbreviated TCA. { trī‚klȯr·ō·ə‚sed·ik 'as·əd }

trichlorobenzene [CHEM] $C_6H_3Cl_3$ Either of two toxic compounds: 1,2,3-trichlorobenzene forms white crystals, soluble in ether, insoluble in water, boiling at 221°C, and is used as a chemical intermediate; 1,2,4-trichlorobenzene is a combustible, colorless liquid, soluble in most organic solvents and oils, insoluble in water, boiling at 213°C, and is used as a solvent and in dielectric fluids, synthetic transformer oils, lubricants, and insecticides. { trī‚klȯr·ō'ben‚zēn }

trichloroethane [CHEM] $C_2H_3Cl_3$ Either of two nonflammable, irritating liquid isomeric compounds: 1,1,1-trichloroethane (CH_3CCl_3) is toxic, soluble in alcohol and ether, insoluble in water, and boils at 75°C; it is used as a solvent, aerosol propellant, and pesticide and for metal degreasing, and is also known as methyl chloroform; 1,1,2-trichloroethane ($CHCl_2CH_2Cl$) is clear and colorless, is soluble in alcohols, ethers, esters, and ketones, insoluble in water, has a sweet aroma, and boils at 114°C; it is used as a chemical intermediate and solvent, and is also known as vinyl trichloride. { trī‚klȯr·ō'eth‚ān }

trichloroethylene [CHEM] $CHCl:CCl_2$ A heavy, stable, toxic liquid with a chloroform aroma; slightly soluble in water, soluble with greases and common organic solvents; boils at 87°C; used for metal degreasing, solvent extraction, and dry cleaning and as a fumigant and chemical intermediate. { trī‚klȯr·ō'eth·ə‚lēn }

trichlorofluoromethane [CHEM] CCl_3F A toxic, noncombustible, colorless liquid boiling at 24°C; used as a chemical intermediate, solvent, refrigerant, aerosol prepellant, and blowing agent (plastic foams) and in fire extinguishers. Also known as fluorocarbon-11; fluorotrichloromethane. { trī‚klȯr·ō‚flür·ō'meth‚ān }

trichloroiminocyanuric acid *See* trichloroisocyanuric acid. { trī‚klȯr·ō‚im·ə·nō‚sī·ə‚nùr·ik 'as·əd }

trichloroisocyanuric acid [CHEM] $C_3Cl_3N_3O_3$ A crystalline substance that releases hypochlorous acid on contact with water; melting point is 246–247°C; soluble in chlorinated and highly polar solvents; used as a chlorinating agent, disinfectant, and industrial deodorant. Also known as symclosene; trichloroiminocyanuric acid. { trī ¦klȯr·ō¦i·sō¦sī·ə¦nür·ik 'as·əd }

trichloromethane   See chloroform.   { trī¦klȯr·ō'meth‚ān }

trichloronate   See ortho-ethyl(O-2,4,5-trichlorophenyl)ethylphosphonothioate.   { trī 'klȯr·ə‚nāt }

trichloronitromethane   See chloropicrin.   { trī¦klȯr·ō¦nī·trō'meth‚ān }

trichlorophenol   [CHEM] $C_6H_2Cl_3OH$ Either of two toxic nonflammable compounds with a phenol aroma: 2,4,5-trichlorophenol is a gray solid, is soluble in alcohol, acetone, and ether, melts at 69°C, and is used as a fungicide and bactericide; 2,4,6-trichlorophenol forms yellow flakes, is soluble in alcohol, acetone, and ether, boils at 248°C, and is used as a fungicide, defoliant, and herbicide; it is also known as 2,4,6-T. { trī¦klȯr·ō'fē‚nȯl }

2,4,5-trichlorophenoxyacetic acid   [CHEM] $C_6H_2Cl_3OCH_2CO_2H$ A toxic, light-tan solid; soluble in alcohol, insoluble in water; melts at 152°C; used as a defoliant, plant hormone, and herbicide. Also known as 2,4,5-T.   { ¦tü ¦fȯr ¦fīv trī¦klȯr·ō·fə¦näk·sē·ə ¦sēd·ik 'as·əd }

1,1,2-trichloro-1,2,2-trifluoroethane   [CHEM] CCl_2FCClF_2 A colorless, volatile liquid with a boiling point of 47.6°C; used as a solvent for dry cleaning, as a refrigerant, and in fire extinguishers. Also known as trifluorotrichloroethane.   { ¦wən ¦wən ¦tü trī ¦klȯr·ō ¦wən ¦tü ¦tü trī¦flür·ō'eth‚ān }

trickle drain   [CIV ENG]   A drain that is set vertically in water, such as a pond, with its top open and level with the normal water surface in order to carry off excess water. { 'trik·əl ‚drān }

trickling filter   [CIV ENG]   A bed of broken rock or other coarse aggregate onto which sewage or industrial waste is sprayed intermittently and allowed to trickle through, leaving organic matter on the surface of the rocks, where it is oxidized and removed by biological growths.   { 'trik·liŋ ‚fil·tər }

tricyclic dibenzopyran   See xanthene.   { trī'sī·klik dī¦ben·zō'pī·rən }

triethanolamine   [CHEM] $(HOCH_2CH_2)_3N$ A viscous, hygroscopic liquid with an ammonia aroma, soluble in chloroform, water, and alcohol, and boiling at 335°C; used in dry-cleaning soaps, cosmetics, household detergents, and textile processing, for wool scouring, and as a corrosion inhibitor.   { trī‚eth·ə'näl·ə‚mēn }

triethylamine   [CHEM] $(C_2H_5)_3N$ A colorless, toxic, flammable liquid with an ammonia aroma; soluble in water and alcohol; boils at 90°C; used as a solvent, rubber-accelerator activator, corrosion inhibitor, and propellant, and in penetrating and waterproofing agents.   { trī¦eth·ə·lə¦mēn }

triethylene glycol   [CHEM] $HO(C_2H_4O)_3H$ A colorless, combustible, hygroscopic, water-soluble liquid; boils at 287°C; used as a chemical intermediate, solvent, bactericide, humectant (a substance that absorbs or retains moisture), and fungicide. Abbreviated TEG.   { trī'eth·ə‚lēn 'glī‚kȯl }

triethylenemelamine   [CHEM] $NC[N(CH_2)_2]NC[N(CH_2)_2]NC[N(CH_2)_2]$ White crystals, soluble in water, alcohol, acetone, chloroform, and methanol; polymerizes at 160°C; used in medicine and insecticides and as a chemosterilant. Abbreviated TEM. Also known as tretamine.   { trī¦eth·ə‚lēn'mel·ə‚mēn }

triethyl phosphate   [CHEM] $(C_2H_5)_3PO_4$ A toxic, colorless liquid that acts as a cholinesterase inhibitor; boiling point is 216°C; soluble in organic solvents; used as a solvent and plasticizer and for pesticides manufacture. Abbreviated TEP.   { trī'eth·əl 'fä‚sfāt }

trifluorotrichloroethane   See 1,1,2-trichloro-1,2,2-trifluoroethane.   { trī¦flür·ō·trī¦klȯr·ō 'eth‚ān }

trimethylamine [CHEM] $(CH_3)_3N$ A colorless, liquefied gas with a fishy odor and a boiling point of $-4°C$; soluble in water, ether, and alcohol; used as a warning agent for natural gas, a flotation agent, and insect attractant. Abbreviated TMA. { trī¦meth·ə·lə ¦mēn }

2,4,6-trinitrotoluene [CHEM] $CH_3C_6H_2(NO_2)_3$ Toxic, flammable, explosive, yellow crystals; soluble in alcohol and ether, insoluble in water; melts at $81°C$; used as an explosive and chemical intermediate and in photographic chemicals. Abbreviated TNT. { ¦tü ¦fór ¦siks trī¦nī·trō'täl·yə,wēn }

triphenyltinacetate See fentinacetate. { trī¦fen·əl·tə'nas·ə,tāt }

tris[2-(2,4-dichlorophenoxy)ethyl]phosphite [CHEM] $C_{24}H_{21}Cl_6O_6P$ A dark liquid that boils above $200°C$; used as a preemergence herbicide for corn, peanuts, and strawberries. Abbreviated 2,4-DEP. { ¦tris¦tü ¦tü ¦fór dī¦klór·ō·fə¦näk·sē¦eth·əl'fä,sfīt }

tristeza [PL PATH] A viral disease spread by three species of aphids that causes rapid decline or death of trees of sweet orange, grapefruit, and tangerine propagated on certain susceptible rootstock varieties. { tris'tā·zə }

TRMM See Tropical Rainfall Measuring Mission.

Tropept [GEOL] A suborder of the order Inceptisol, characterized by moderately dark A horizons with modest additions of organic matter, B horizons with brown or reddish colors, and slightly pale C horizons; restricted to tropical regions with moderate or high rainfall. { 'trä,pept }

trophallaxis [ECOL] Exchange of food between organisms, not only of the same species but between different species, especially among social insects. { ,träf·ə'lak·səs }

trophic [BIOL] Pertaining to or functioning in nutrition. { 'träf·ik }

trophic ecology [ECOL] The study of the feeding relationships of organisms in communities and ecosystems. { ,träf·ik ē'käl·ə·jē }

trophic level [ECOL] Any of the feeding levels through which the passage of energy through an ecosystem proceeds; examples are photosynthetic plants, herbivorous animals, and microorganisms of decay. { 'träf·ik ,lev·əl }

trophobiosis [ECOL] A nutritional relationship associated only with certain species of ants in which alien insects supply food to the ants and are milked by the ants for their secretions. { ,träf·ō,bī'ō·səs }

trophogenic [ZOO] Originating from nutritional differences rather than resulting from genetic determinants, usually referring to various castes of social insects. { ¦träf·ə ¦jen·ik }

tropical air [METEOROL] A type of air whose characteristics are developed over low latitudes. { 'träp·ə·kəl 'er }

tropical climate [CLIMATOL] A climate which is typical of equatorial and tropical regions, that is, one with continually high temperatures and with considerable precipitation, at least during part of the year. { 'träp·ə·kəl 'klī·mət }

tropical cyclone [METEOROL] The general term for a cyclone that originates over tropical oceans; at maturity, the tropical cyclone is one of the most intense storms of the world; winds exceeding 175 knots (324 kilometers per hour) have been measured, and the rain is torrential. { 'träp·ə·kəl 'sī,klōn }

tropical disturbance [METEOROL] A cyclonic wind system of the tropics, of lesser intensity than a tropical cyclone. { 'träp·ə·kəl di'stər·bəns }

tropical easterlies [METEOROL] The trade winds when shallow and exhibiting a strong vertical shear; at about 500 feet (152 meters) the easterlies give way to the upper westerlies, which are sufficiently strong and deep to govern the course of cloudiness and weather. Also known as subtropical easterlies. { 'träp·ə·kəl 'ēs·tər·lēz }

tropical front   See intertropical front.   { 'träp·ə·kəl 'frənt }

tropical life zone   [ECOL] A zone comprising the climate and biotic communities of the extreme southern edge of the United States, the Mexican lowlands, and Central America.   { 'träp·ə·kəl 'līf ,zōn }

tropical meteorology   [METEOROL] The study of the tropical atmosphere; the dividing lines, in each hemisphere, between the tropical easterlies and the mid-latitude westerlies in the middle troposphere roughly define the poleward boundaries of this region.   { 'träp·ə·kəl ,mēd·ē·ə'räl·ə·jē }

tropical monsoon climate   [CLIMATOL] One of the tropical rainy climates; it is sufficiently warm and rainy to produce tropical rainforest vegetation, but it does exhibit the monsoon climate influences in that it has a winter dry season.   { 'träp·ə·kəl män'sün ,klī·mət }

Tropical Rainfall Measuring Mission   [METEOROL] A meteorological satellite used for mapping tropical precipitation in order to better understand the earth's climate system and to verify climate models. Abbreviated TRMM.   { ¦träp·ə·kəl 'rān,fól ,mezh·ər·iŋ ,mish·ən }

tropical rainforest   [ECOL] A vegetation class consisting of tall, close-growing trees, their columnar trunks more or less unbranched in the lower two-thirds, and forming a spreading and frequently flat crown; occurs in areas of high temperature and high rainfall. Also known as hylaea; selva.   { 'träp·ə·kəl 'rān,fär·əst }

tropical rainforest climate   [CLIMATOL] In general, the climate which produces tropical rainforest vegetation, that is, a climate of unbroken warmth, high humidity, and heavy annual precipitation. Also known as tropical wet climate.   { 'träp·ə·kəl 'rān,fär·əst ,klī·mət }

tropical rainy climate   [CLIMATOL] A major category in W. Köppen's climatic classification, characterized by a mean temperature of the coldest month of 64.4°F (18°C) or higher, and by a mean annual precipitation, in inches, greater than $0.44(t − a)$, where t is the mean annual temperature in degrees Fahrenheit, and a equals 32 for precipitation chiefly in winter, 19.4 for evenly distributed precipitation, and 6.8 for precipitation chiefly in summer.   { 'träp·ə·kəl ¦rän·ē ,klī·mət }

tropical savanna   See tropical woodland.   { 'träp·ə·kəlsə'van·ə }

tropical savanna climate   [CLIMATOL] In general, the type of climate which produces the vegetation of the tropical and subtropical savanna; thus, a climate with a winter dry season, a relatively short but heavy rainy summer season, and high year-round temperatures. Also known as savanna climate; tropical wet and dry climate.   { 'träp·ə·kəl sə'van·ə ,klī·mət }

tropical scrub   [ECOL] A class of vegetation composed of low woody plants (shrubs), sometimes growing quite close together, but more often separated by large patches of bare ground, with clumps of herbs scattered throughout; an example is the Ghanaian evergreen coastal thicket. Also known as brush; bush; fourré; mallee; thicket.   { 'träp·ə·kəl 'skrəb }

tropical wet and dry climate   See tropical savanna climate.   { 'träp·ə·kəl ¦wet ən ¦drī ,klī·mət }

tropical wet climate   See tropical rainforest climate.   { 'träp·ə·kəl 'wet ,klī·mət }

tropical woodland   [ECOL] A vegetation class similar to a forest but with wider spacing between trees and sparse lower strata characterized by evergreen shrubs and seasonal graminoids; the climate is warm and moist. Also known as parkland; savanna-woodland; tropical savanna.   { 'träp·ə·kəl 'wúd·lənd }

tropic high-water inequality   [OCEANOGR] The average difference between the heights of the two high waters of the tidal day at the time of tropic tides.   { 'träp·ik 'hī ,wód·ər ,in·i'kwäl·əd·ē }

tropic low-water inequality [OCEANOGR] The average difference between the heights of the two low waters of the tidal day at the time of tropic tides. { 'träp·ik 'lō‚wȯd·ər ‚in·i'kwäl·əd·ē }

tropics [CLIMATOL] Any portion of the earth characterized by a tropical climate. { 'träp·iks }

tropic tidal currents [OCEANOGR] Tidal currents of increased diurnal inequality occurring at the time of tropic tides. { 'träp·ik 'tīd·əl 'kə·rəns }

tropic tide [OCEANOGR] A tide occurring when the moon is near maximum declination; the diurnal inequality is then at a maximum. { 'träp·ik 'tīd }

tropic velocity [OCEANOGR] The speed of the greater flood or greater ebb at the time of tropic currents. { 'träp·ik və'läs·əd·ē }

tropism [BIOL] Orientation movement of a sessile organism in response to a stimulus. Also known as topotaxis. { 'trō‚piz·əm }

tropopause [METEOROL] The boundary between the troposphere and stratosphere, usually characterized by an abrupt change of lapse rate; the change is in the direction of increased atmospheric stability from regions below to regions above the tropopause; its height varies from 9 to 12 miles (15 to 20 kilometers) in the tropics to about 6 miles (10 kilometers) in polar regions. { 'trōp·ə‚pȯz }

tropopause chart [METEOROL] A synoptic chart showing the contour lines of the tropopause and tropopause break lines. { 'trōp·ə‚pȯz ‚chärt }

tropopause fold [METEOROL] A phenomenon occurring in the stratosphere in which a tapering cone of dry, ozone-rich air intrudes into the troposphere. { 'trōp·ə‚pȯz 'fōld }

tropopause inversion [METEOROL] The decrease in the lapse rate of temperature encountered at the level of the tropopause. Also known as upper inversion. { 'trōp·ə‚pȯz in'vər·zhən }

tropophytia [BOT] Plants that thrive in a climate that undergoes marked periodic changes. { ‚träp·ə'fī·shə }

troposphere [METEOROL] That portion of the atmosphere from the earth's surface to the tropopause, that is, the lowest 10 to 20 kilometers of the atmosphere. { 'trōp·ə ‚sfir }

trough [METEOROL] An elongated area of relatively low atmospheric pressure; the opposite of a ridge. { trȯf }

trough aloft *See* upper-level trough. { 'trȯf ə'lȯft }

TRSV *See* tobacco ring spot virus.

true air temperature [METEOROL] Basic air temperature corrected for heat of compression error due to high-speed motion of the thermometer through the air, as on an aircraft. { 'trü 'er ‚tem·prə·chər }

true mean temperature [METEOROL] As adopted by the International Meteorological Organization, a monthly or annual mean air temperature based upon hourly observations at a given place, or on some combination of less frequent observations designed to represent this mean as nearly as possible. { 'trü 'mēn 'tem·prə·chər }

true wind [METEOROL] Wind relative to a fixed point on the earth. { 'trü 'wind }

true wind direction [METEOROL] The direction, with respect to true north, from which the wind is blowing. { 'trü 'wind də‚rek·shən }

truncated landform [GEOGR] A landform which has been cut off by erosion, creating a steep side or cliff. { 'trəŋ‚kād·əd 'land‚fȯrm }

trunk [BOT] The main stem of a tree. { trəŋk }

trunk sewer [CIV ENG] A sewer receiving sewage from many tributaries serving a large territory. { 'trəŋk ˌsü·ər }

trunk stream *See* main stream. { 'trəŋk ˌstrēm }

TRV *See* tobacco rattle virus.

trypanosome [ZOO] A flagellated protozoan of the genus *Trypanosoma*. { trə'pan·ə,sōm }

trypanosomiasis [MED] Any of many diseases of humans and animals caused by infection with species of *Trypanosoma* and transmitted by tsetse flies and other insects. { trə,pan·ə·sō'mī·ə·səs }

T-S curve *See* temperature-salinity diagram. { ¦tē¦es ˌkərv }

T-S diagram *See* temperature-salinity diagram. { ¦tē¦es ˌdī·ə,gram }

T-S relation *See* temperature-salinity diagram. { ¦tē¦es ri,lā·shən }

tsunami [OCEANOGR] An ocean wave or series of waves generated by any large, abrupt disturbance of the sea-surface by an earthquake in marine and coastal regions, as well as by a suboceanic landslide, volcanic eruption, or asteroid impact. { tsü'nä·mē }

Tsushima Current [OCEANOGR] That part of the Kuroshio Current flowing northeastward through the Korea Strait and along the Japanese coast in the Sea of Japan. { 'tsü·shē,mä 'kə·rənt }

tuba [METEOROL] A cloud column or inverted cloud cone, pendant from a cloud base; this supplementary feature occurs mostly with cumulus and cumulonimbus; when it reaches the earth's surface it constitutes the cloudy manifestation of an intense vortex, namely, a tornado or waterspout. Also known as pendant cloud; tornado cloud. { 'tü·bə }

tubatoxin *See* rotenone. { 'tü·bə,täk·sən }

tuber [BOT] The enlarged end of a rhizome in which food accumulates, as in the potato. { 'tü·bər }

tuberculosis [MED] A chronic infectious disease of humans and animals primarily involving the lungs caused by the tubercle bacillus, *Mycobacterium tuberculosis*, or by M. *bovis*. Also known as consumption. { tə,bər·kyə'lō·səs }

tularemia [VET MED] A bacterial infection of wild rodents caused by *Pasteurella tularensis*; it may be generalized, or it may be localized in the eyes, skin, or lymph nodes, or in the respiratory tract or gastrointestinal tract; may be transmitted to humans and to some domesticated animals. { ˌtü·lə'rē·mē·ə }

tumbleweed [BOT] Any of various plants that break loose from their roots in autumn and are driven by the wind in rolling masses over the ground. { 'təm·bəl,wēd }

tundra [ECOL] An area supporting some vegetation between the northern upper limit of trees and the lower limit of perennial snow on mountains, and on the fringes of the Antarctic continent and its neighboring islands. Also known as cold desert. { 'tən·drə }

tundra climate [CLIMATOL] The climate which produces tundra vegetation; it is too cold for the growth of trees but does not have a permanent snow-ice cover. { 'tən·drə ˌklī·mət }

turbidity [METEOROL] Any condition of the atmosphere which reduces its transparency to radiation, especially to visible radiation. { tər'bid·əd·ē }

turbidity current [OCEANOGR] A highly turbid, relatively dense current carrying large quantities of clay, silt, and sand in suspension which flows down a submarine slope through less dense sea water. Also known as density current; suspension current. { tər'bid·əd·ē ,kə·rənt }

turbinate [BOT] Shaped like an inverted cone. { 'tər·bə·nət }

turbosphere [METEOROL] The region of the atmosphere in which turbulence frequently exists. { 'tər·bə‚sfir }

turbulent heat conduction [OCEANOGR] Conduction of heat in water by lateral and vertical eddy diffusion, with currents. { 'tər·byə·lənt 'hēt kən‚dək·shən }

turgor [BOT] Distension of a plant cell wall and membrane by the fluid contents. { 'tər·gər }

turgor movement [BOT] A reversible change in the position of plant parts due to a change in turgor pressure in certain specialized cells; movement of *Mimosa* leaves when touched is an example. { 'tər·gər ‚müv·mənt }

turgor pressure [BOT] The actual pressure developed by the fluid content of a turgid plant cell. { 'tər·gər ‚presh·ər }

turion [BOT] A scaly shoot, such as asparagus, developed from an underground bud. { 'tür·ē‚än }

turkey tail mushroom See Trametes versicolor. { 'tər·kē ‚tāl ‚məsh‚rüm }

turnip yellow mosaic virus [MICROBIO] The type species of the genus *Tymovirus*, it is transmitted mechanically and via beetles, causing chloroplast clumping. Abbreviated TYMV. { ¦tər·nəp ‚yel·ōmō'zā·ik ‚vī·rəs }

turn of the tide See change of tide. { 'tərn əv thə 'tīd }

turret ice See ropak. { 'tə·rət ‚īs }

tussock [ECOL] A small hummock of generally solid ground in a bog or marsh, usually covered with and bound together by the roots of low vegetation such as grasses, sedges, or ericaceous shrubs. { 'təs·ək }

TWA See time-weighted average.

twiner [BOT] A climbing stem that winds about its support, as pole beans or many tropical lianas. { 'twī·nər }

twister [METEOROL] In the United States, a colloquial term for tornado. { 'twis·tər }

two-layer ocean [OCEANOGR] An idealized ocean in which a layer of uniform density near the surface overlays a deep layer of uniform but distinctly higher-density water. { 'tü ¦lā·ər 'ō·shən }

two-year ice See second-year ice. { 'tü ¦yir 'īs }

TYMV See turnip yellow mosaic virus.

Tyndall flowers [HYD] Small water-filled cavities, often of basically hexagonal shape, which appear in the interior of ice masses upon which light is falling. { 'tind·əl ‚flau·erz }

typhoid fever [MED] A highly infectious, septicemic (blood-borne) disease of humans caused by *Salmonella typhi* which enters the body by the oral route through ingestion of food or water contaminated by contact with fecal matter. { 'tī‚fȯid 'fē·vər }

typhoon [METEOROL] A severe tropical cyclone in the western Pacific. { tī'fün }

typhoon wind See hurricane wind. { tī'fün ‚wind }

ubac [METEOROL] The shady (usually north) side of an Alpine mountain, characterized by a lower timberline and snow line than the sunny side. { 'ü,bäk }

udometer *See* rain gage. { yü'däm·əd·ər }

Ultisol [GEOL] A soil order characterized by typically moist soils, with horizons of clay accumulation and a low supply of bases. { 'əl·tə,sȯl }

ultrasonic transmitter [ENG] A device used to track seals, fish, and other aquatic animals: the device is fastened to the outside of the animal or fed to it, and has a loudspeaker which is made to vibrate at an ultrasonic frequency, propagating ultrasonic waves through the water to a special microphone or hydrophone. { ¦əl·trə'sän·ik tranz'mid·ər }

ultraviolet [PHYS] Pertaining to ultraviolet radiation. Abbreviated UV. { ¦əl·trə'vī·lət }

ultraviolet light *See* ultraviolet radiation. { ¦əl·trə'vī·lət 'līt }

ultraviolet radiation [PHYS] Electromagnetic radiation in the wavelength range 4–400 nanometers; this range begins at the short-wavelength limit of visible light and overlaps the wavelengths of long x-rays (some scientists place the lower limit at higher values, up to 40 nanometers). Also known as ultraviolet light. { ¦əl·trə'vī·lət ,rād·ē'ā·shən }

umbel [BOT] An indeterminate inflorescence with the pedicels all arising at the top of the peduncle and radiating like umbrella ribs; there are two types, simple and compound. { 'əm·bəl }

Umbrept [GEOL] A suborder of the Inceptisol soil order; has dark A horizon more than 10 inches (25 centimeters) thick, brown B horizons, and slightly paler C horizons; soil is strongly acid, and clay minerals are crystalline; occurs in cool or temperate climates. { 'əm,brept }

uncertainty [SCI TECH] The estimated amount by which an observed or calculated value may depart from the true value. { ¦ən'sərt·ən·tē }

unconcentrated wash *See* sheet erosion. { ¦ən'käns·ən,trād·əd 'wäsh }

unconformity iceberg [OCEANOGR] An iceberg consisting of more than one kind of ice, such as blue water-formed ice and névé; such an iceberg often contains many crevasses and silt bands. { ¦ən·kən'fȯr·məd·ē 'īs,bərg }

undercurrent [OCEANOGR] A water current flowing beneath a surface current at a different speed or in a different direction. { 'ən·dər,kə·rənt }

underdrain [CIV ENG] A subsurface drain with holes into which water flows when the water table reaches the drain level. { 'ən·dər,drān }

underfit stream [HYD] A misfit stream that appears to be too small to have eroded the valley in which it flows. { 'ən·dər,fit 'strēm }

underflow *See* bottom flow. { 'ən·dər,flō }

underground ice *See* ground ice. { ¦ən·dər¦graủnd 'īs }

underground stem [BOT] Any of the stems that grow underground and are often mistaken for roots; principal kinds are rhizomes, tubers, corms, bulbs, and rhizomorphic droppers. { ¦ən·dər¦graủnd 'stem }

underground stream [HYD] A subsurface body of water flowing in a definite current in a distinct channel. { ¦ən·dər¦graủnd 'strēm }

undermelting [HYD] The melting from below of any floating ice. { ¦ən·dər¦melt·iŋ }

understock *See* rootstock. { 'ən·dər,stäk }

understory [FOR] A foliage layer occurring beneath and shaded by the main canopy of a forest. { 'ən·dər,stȯr·ē }

undertow [OCEANOGR] A subsurface seaward movement by gravity flow of water carried up on a sloping beach by waves or breakers. { 'ən·dər,tō }

underwater vehicle [OCEANOGR] A submersible work platform designed to be operated either remotely or directly. { ,ən·dər,wȯd·ər 'vē·ə·kəl }

undulant fever *See* brucellosis. { 'ən·jə·lənt 'fē·vər }

undulatus *See* billow cloud. { ən·jə'läd·əs }

United States airways code [METEOROL] A synoptic code for communicating aviation weather observations. Also known as airways code. { yə'nīd·əd 'stāts 'er,wāz ,kōd }

unlimited ceiling [METEOROL] A ceiling that exists when the total sky cover is less than 0.6%, or when the total transparent sky cover is 0.5% or more, or when surface-based obscuring phenomena are classed as partial obscuration (that is, they obscure 0.9% or less of the sky) and no layer aloft is reported as broken or overcast. { ¦ən'lim·əd·əd 'sē·liŋ }

unrestricted visibility [METEOROL] The visibility when no obstruction to vision exists in sufficient quantity to reduce the visibility to less than 7 miles (11.3 kilometers). { ¦ən·ri'strik·təd ,viz·ə'bil·əd·ē }

unsaturated zone *See* zone of aeration. { ¦ən'sach·ə,rād·əd 'zōn }

unsettled [METEOROL] Pertaining to fair weather which may at any time become rainy, cloudy, or stormy. { ¦ən'sed·əld }

unwater [ENG] To remove or draw off water; to drain. { ¦ən'wȯd·ər }

updrift [OCEANOGR] The direction which is opposite that of the prevailing movement of littoral material. { 'əp,drift }

upland [GEOGR] **1.** An extensive region of high land. **2.** The higher ground of a region, in contrast to a valley, plain, or other low-lying land. **3.** The elevated land above the low areas along a stream or between hills. { 'əp·lənd }

upper air [METEOROL] The region of the atmosphere which is above the lower troposphere; although no distinct lower limit is set, the term is generally applied to levels above that at which the pressure is 850 millibars. { 'əp·ər 'er }

upper-air chart *See* upper-level chart. { ¦əp·ər ¦er 'chärt }

upper-air disturbance [METEOROL] A disturbance of the flow pattern in the upper air, particularly one which is more strongly developed aloft than near the ground. Also known as upper-level disturbance. { ¦əp·ər ¦er di,stər·bəns }

upper-air observation [METEOROL] A measurement of atmospheric conditions aloft, above the effective range of a surface weather observation. Also known as sounding; upper-air sounding. { ¦əp·ər ¦er ,äb·zər,vā·shən }

upper-air sounding *See* upper-air observation. { ¦əp·ər ¦er ,saủnd·iŋ }

upper anticyclone   See upper-level anticyclone.   { 'əp·ər ¦ant·i'sī·klōn }

upper atmosphere   [METEOROL] The general term applied to the atmosphere above the troposphere.   { 'əp·ər 'at·mə‚sfir }

upper-atmosphere dynamics   [METEOROL] Motion of the atmosphere above 300 miles (500 kilometers); predominant dynamical phenomena are internal gravity waves, tides, sound waves, turbulence, and large-scale circulation.   { ¦əp·ər ¦at·mə‚sfir dī'nam·iks }

upper cyclone   See upper-level cyclone.   { 'əp·ər 'sī‚klōn }

upper front   [METEOROL] A front which is present in the upper air but does not extend to the ground.   { 'əp·ər 'frənt }

upper high   See upper-level anticyclone.   { 'əp·ər 'hī }

upper inversion   See tropopause inversion.   { 'əp·ər in¦vər·zhən }

upper-level anticyclone   [METEOROL] An anticyclonic circulation existing in the upper air; this often refers to such anticyclones only when they are much more pronounced at upper levels than at and near the earth's surface. Also known as high aloft; high-level anticyclone; upper anticyclone; upper high; upper-level high.   { ¦əp·ər ¦lev·əl 'ant·i'sī ‚klōn }

upper-level chart   [METEOROL] A synoptic chart of meteorological conditions in the upper air, almost invariably referring to a standard constant-pressure chart. Also known as upper-air chart.   { ¦əp·ər ¦lev·əl 'chärt }

upper-level cyclone   [METEOROL] A cyclonic circulation existing in the upper air, and specifically, as seen on an upper-level constant-pressure chart; often restricted to describe cyclones associated with relatively little cyclonic circulation in the lower atmosphere. Also known as high-level cyclone; low aloft; upper cyclone; upper-level low; upper low.   { ¦əp·ər ¦lev·əl 'sī‚klōn }

upper-level disturbance   See upper-air disturbance.   { ¦əp·ər ¦lev·əl di'stər·bəns }

upper-level high   See upper-level anticyclone.   { ¦əp·ər ¦lev·əl 'hī }

upper-level low   See upper-level cyclone.   { ¦əp·ər ¦lev·əl 'lō }

upper-level ridge   [METEOROL] A pressure ridge existing in the upper air, especially one that is stronger aloft than near the earth's surface. Also known as high-level ridge; ridge aloft; upper ridge.   { ¦əp·ər ¦lev·əl 'rij }

upper-level trough   [METEOROL] A pressure trough existing in the upper air, but sometimes restricted to the troughs that are much more pronounced aloft than near the earth's surface. Also known as high-level trough; trough aloft; upper trough.   { ¦əp·ər ¦lev·əl 'trȯf }

upper-level winds   See winds aloft.   { ¦əp·ər ¦lev·əl 'winz }

upper low   See upper-level cyclone.   { 'əp·ər 'lō }

upper mixing layer   [METEOROL] The region of the upper mesosphere between about 30 and 50 miles (50 and 80 kilometers; that is, immediately above the mesopeak) through which there is a rapid decrease of temperature with height and where there appears to be considerable turbulence.   { 'əp·ər 'miks·iŋ ‚lā·ər }

upper ridge   See upper-level ridge.   { 'əp·ər 'rij }

Upper Sonoran life zone   [ECOL] A life zone characterized by semiarid climate, moderately high altitude, and pinyon-juniper forests.   { 'əp·ər sə'nȯr·ən 'līf ‚zōn }

upper trough   See upper-level trough.   { 'əp·ər 'trȯf }

upper winds   See winds aloft.   { 'əp·ər 'winz }

uprush   See swash.   { 'əp‚rəsh }

upslope fog [METEOROL] A type of fog formed when air flows upward over rising terrain and, consequently, is adiabatically cooled to or below its dew point. { 'əp₁slōp 'fäg }

upstream [HYD] Toward the source of a stream. { 'əp₁strēm }

upwelling [OCEANOGR] The process by which water rises from a deeper to a shallower depth, usually as a result of divergence of offshore currents. { ¦əp¦wel·iŋ }

upwind [METEOROL] In the direction from which the wind is flowing. { 'əp₁wind }

upwind effect [METEOROL] The effect of an orographic barrier in producing orographic precipitation windward of the base of the barrier, because the airflow is forced upward before the barrier slope is actually reached. { 'əp¦wind i₁fekt }

urania See uranium dioxide. { yə'rā·nē·ə }

uranic oxide See uranium dioxide. { yü'ran·ik 'äk₁sīd }

uranium [CHEM] A metallic element in the actinide series, symbol U, atomic number 92, atomic weight 238.03; highly toxic and radioactive; ignites spontaneously in air and reacts with nearly all nonmetals; melts at 1132°C, boils at 3818°C; used in nuclear fuel and as the source of ^{235}U and plutonium. { yə'rā·nē·əm }

uranium dioxide [CHEM] UO_2 Black, highly toxic, spontaneously flammable, radioactive crystals; insoluble in water, soluble in nitric and sulfuric acids; melts at approximately 3000°C; used to pack nuclear fuel rods and in ceramics, pigments, and photographic chemicals. Also known as urania; uranic oxide; uranium oxide. { yə'rā·nē·əm dī'äk₁sīd }

uranium oxide See uranium dioxide. { yə'rā·nē·əm 'äk₁sīd }

urban forestry [FOR] The management of tree resources in and around cities and towns. { ₁ər·bən 'fär·ə·strē }

urban geography [GEOGR] The study of the site, evolution, morphology, spatial patterns, and classification of densely populated areas. { ¦ər·bən jē'äg·rə·fē }

urban geology [GEOL] The study of geological aspects of planning and managing high-density population centers and their surroundings. { ¦ər·bən jē'äl·ə·jē }

urban heat island [METEOROL] Increased urban temperatures of 1–2°C higher for daily maxima and 1–9°C for daily minima compared to rural environs resulting from changes in moisture balance due to impermeable surfaces, decreased humidity, or alteration in heat balance. { 'ər·bən 'hēt ₁ī·lənd }

urbanization [CIV ENG] The state of being or becoming a community with urban characteristics. { ₁ər·bə·nə'zā·shən }

urban renewal [CIV ENG] Redevelopment and revitalization of a deteriorated urban community. { 'ər·bən ri'nü·əl }

urea [CHEM] $CO(HN_2)_2$ A natural product of protein metabolism found in urine; synthesized as white crystals or powder with a melting point of 132.7°C; soluble in water, alcohol, and benzene; used as a fertilizer, in plastics, adhesives, and flameproofing agents, and in medicine. Also known as carbamide. { yü'rē·ə }

Uredinales [MYCOL] An order of parasitic fungi of the subclass Heterobasidiomycetidae characterized by the teleutospore, a spore with one or more cells, each of which is a modified hypobasidium; members cause plant diseases known as rusts. { yə₁red·ən'ā·lēz }

Urediniomycetes [MYCOL] A class of fungi in the subdivision Basidiomycotina that causes plant rust diseases, members have thick-walled teliospores, produced in the terminal state in pustules. Also known as rust fungi. { ₁yür·ə₁din·ē·ō·mī'sē₁dēz }

urediniospore [MYCOL] A spore produced by a uredinium. Also known as urediospore; uredospore. { ₁yür·ə'din·ē·ə₁spȯr }

urediospore   *See* urediniospore.   { yü'red·ē·ə,spȯr }

uredospore   *See* urediniospore.   { yü'red·ə,spȯr }

ureotelic   [BIOL] Referring to animals that produce urea as their main nitrogenous excretion.   { yə¦rē·ə¦tel·ik }

urethane   [CHEM] $CO(NH_2)OC_2H_5$ A combustible, toxic, colorless powder; soluble in water and alcohol; melts at 49°C; used as a solvent and chemical intermediate and in biochemical research and veterinary medicine. Also known as ethyl carbamate; ethyl urethane.   { 'yür·ə,thān }

uricotelism   [BIOL] An adaptation of terrestrial reptiles and birds which effectively provides for detoxification of ammonia and also for efficient conservation of water due to a relatively low rate of glomerular filtration and active secretion of uric acid by the tubules to form a urine practically saturated with urate.   { ,yür·ə'käd·əl,iz·əm }

Ustilaginales   [MYCOL] An order of the subclass Heterobasidiomycetidae comprising the smut fungi which parasitize plants and cause diseases known as smut or bunt.   { ,əs·tə,laj·ə'nā·lēz }

Uukuvirus   [MICROBIO] A genus of the viral family Bunyaviridae that is transmitted via ticks to a wide range of vertebrate hosts.   { yü'yük·ə,vī·rəs }

UV   *See* ultraviolet.

UV-A   [METEOROL] Ultraviolet radiation produced by the sun, ranging in wavelength from 320 to 400 nanometers; biologically, it is the least damaging of the sun's rays.

uvala   [GEOGR] Broad-bottomed lowlands.   { 'ü·və·lə }

UV-B   [METEOROL] Ultraviolet radiation produced by the sun, ranging in wavelength from 280 to 320 nanometers; it is biologically damaging. Stratospheric ozone absorbs much of it.

UV-C   [METEOROL] Ultraviolet radiation produced by the sun, ranging in wavelength from 200 to 280 nanometers, biologically, it is the most damaging of the sun's rays. Stratospheric ozone strongly absorbs it and, as a result, the solar spectrum at the earth's surface contains only the UV-A and UV-B radiation.

V

vaccinia [VET MED] A contagious disease of cows which is characterized by vesicopustular lesions of the skin that are prone to appear on the teats and udder, and which is transmissible to humans by handling infected cows and by vaccination; confers immunity against smallpox. Also known as cowpox. { vak'sin·ē·ə }

vacuole *See* vesicle. { 'vak·yə,wōl }

vadose water [HYD] Water in the zone of aeration. Also known as kremastic water; suspended water; wandering water. { 'vā,dōs ,wȯd·ər }

vadose zone *See* zone of aeration. { 'vā,dōs ,zōn }

vagility [ECOL] The ability of organisms in a given species to disperse geographically. { və'jil·əd·ē }

valerian oil [CHEM ENG] A combustible, yellow to brown liquid with a penetrating aroma; soluble in alcohol, acetone, and other organic solvents; derived from the roots and rhizome of the garden heliotrope (*Valeriana officinalis*), the main components being pinene, camphene, borneol, and esters, used in medicine, flavors, and industrial odorants and to perfume tobacco. { və'lir·ē·ən ,ȯil }

valley [GEOGR] A generally broad area of flat, low-lying land bordered by higher ground. [GEOL] A relatively shallow, wide depression of the sea floor with gentle slopes. Also known as submarine valley. { 'val·ē }

valley breeze [METEOROL] A gentle wind blowing up a valley or mountain slope in the absence of cyclonic or anticyclonic winds, caused by the warming of the mountainside and valley floor by the sun. { 'val·ē ,brēz }

valley glacier [HYD] A glacier that flows down the walls of a mountain valley. { 'val·ē ,glā·shər }

valley iceberg [OCEANOGR] An iceberg weathered in such a manner that a large U-shaped slot extends through the iceberg. Also known as dry-dock iceberg. { 'val·ē 'īs,bərg }

valley line *See* thalweg. { 'val·ē ,līn }

valley wind [METEOROL] A wind which ascends a mountain valley (up-valley wind) during the day; the daytime component of a mountain and valley wind system. { 'val·ē ,wind }

vamidothion [CHEM] $C_7H_{16}NO_4PS_2$ A white wax with a melting point of 40°C; very soluble in water; used to control pests in orchards, vineyards, rice, cotton, and ornamentals. { ¦vam·əd·ō'thī,än }

vanadic acid anhydride *See* vanadium pentoxide. { və'nād·ik 'as·əd an'hī,drīd }

vanadium pentoxide [CHEM] V_2O_5 A toxic, yellow to red powder, soluble in alkalies and acids, slightly soluble in water; melts at 690°C; used in medicine, as a catalyst, as a ceramics coloring, for ultraviolet-resistant glass, photographic developers, textiles dyeing, and nuclear reactors. Also known as vanadic acid anhydride. { və'nād·ē·əm ¦pen'täk₁sīd }

vancomycin [MICROBIO] A complex antibiotic substance produced by *Streptomyces orientalis*; useful for treatment of severe staphylococcic infections. { ₁vaŋ·kə'mīs·ən }

vanishing tide [OCEANOGR] When a high water and low water "melt" together into a period of several hours with a nearly constant water level. { 'van·ish·iŋ 'tīd }

vapor pressure [METEOROL] The partial pressure of water vapor in the atmosphere. { 'vā·pər ₁presh·ər }

vapor-pressure deficit *See* saturation deficit. { 'vā·pər ¦presh·ər 'def·ə·sət }

vapor trail *See* condensation trail. { 'vā·pər ₁trāl }

variable ceiling [METEOROL] After United States weather-observing practice, a condition in which the ceiling rapidly increases and decreases while the ceiling observation is being made; the average of the observed values is used as the reported ceiling, and it is reported only for ceilings of less than 3000 feet (914 meters). { 'ver·ē·ə·bəl 'sēl·iŋ }

variable visibility [METEOROL] After United States weather observing practice, a condition in which the prevailing visibility fluctuates rapidly while the observation is being made; the average of the observed values is used as the reported visibility, and it is reported only for visibilities of less than 3 miles (4.8 kilometers). { 'ver·ē·ə·bəl ₁viz·ə'bil·əd·ē }

Varicella *See* chickenpox. { ₁var·ə'sel·ə }

variola *See* smallpox. { ₁ver·ē'ō·lə }

vascular bundle [BOT] A strandlike part of the plant vascular system containing xylem and phloem. { 'vas·kyə·lər 'bənd·əl }

vascular cambium [BOT] The lateral meristem which produces secondary xylem and phloem. { 'vas·kyə·lər 'kam·bē·əm }

vascular plant [BOT] A plant charcterized by the presence of specialized conducting tissues (xylem and phloem) in the roots, stems, and leaves. Also known as tracheophyte. { 'vas·kyə·lər 'plant }

vascular ray [BOT] A ray derived from cambium and found in the stele of some vascular plants, often separating vascular bundles. { 'vas·kyə·lər 'rā }

vascular tissue [BOT] The conducting tissue found in higher plants, consisting principally of xylem and phloem. { 'vas·kyə·lər 'tish·ü }

vectopluviometer [ENG] A rain gage or array of rain gages designed to measure the inclination and direction of falling rain; vectopluviometers may be constructed in the fashion of a wind vane so that the receiver always faces the wind, or they may consist of four or more receivers arranged to point in cardinal directions. { ¦vek·tō ₁plü·vē'äm·əd·ər }

vector [MED] An agent, such as an insect, capable of mechanically or biologically transferring a pathogen from one organism to another. { 'vek·tər }

veering [METEOROL] **1.** In international usage, a change in wind direction in a clockwise sense (for example, south to southwest to west) in either hemisphere of the earth. **2.** According to widespread usage among United States meteorologists, a change in wind direction in a clockwise sense in the Northern Hemisphere, counterclockwise in the Southern Hemisphere. { 'vir·iŋ }

vegetable [AGR] The edible portion of a usually herbaceous plant; customarily served with the main course of a meal. [BOT] Resembling or relating to plants. { 'vej·tə·bəl }

vegetation [BOT] The total mass of plant life that occupies a given area. { ˌvej·ə'tā·shən }

vegetational plant geography [ECOL] A field of study concerned with the mapping of vegetation regions and the interpretation of these in terms of environmental or ecological influences. { ˌvej·ə'tā·shən·əl 'plant jē,äg·rə·fē }

vegetation and ecosystem mapping [BOT] The drawing of maps which locate different kinds of plant cover in a geographic area. { ˌvej·ə'tā·shən ən 'ek·ō,sis·təm 'map·iŋ }

vegetation management [ECOL] The art and practice of manipulating vegetation such as timber, forage, crops, or wild life, so as to produce a desired part or aspect of that material in higher quantity or quality. { ˌvej·ə'tā·shən ˌman·ij·mənt }

vegetation zone [ECOL] **1.** An extensive, even transcontinental, band of physiognomically similar vegetation on the earth's surface. **2.** Plant communities assembled into regional patterns by the area's physiography, geological parent material, and history. { ˌvej·ə'tā·shən ˌzōn }

vegetative [BIOL] Having nutritive or growth functions, as opposed to reproductive. { 'vej·ə,tād·iv }

vegetative propagation [BOT] Production of a new plant from a nonreproductive portion of another plant, such as a stem or branch. { 'vej·ə,tād·iv ˌpräp·ə'gā·shən }

veil [METEOROL] A very thin cloud through which objects are visible. { vāl }

Veillonellaceae [MICROBIO] The single family of gram-negative, anaerobic cocci; characteristically occur in pairs with adjacent sides flattened; parasites of homotherms, including humans, rodents, and pigs. { ˌvā·yō·nə'lās·ē,ē }

vein [BOT] One of the vascular bundles in a leaf. [GEOL] A mineral deposit in tabular or shell-like form filling a fracture in a host rock. { vān }

veld See veldt. { velt }

veldt [ECOL] Grasslands of eastern and southern Africa that are usually level and mixed with trees and shrubs. Also spelled veld. { velt }

velocity ratio [OCEANOGR] The ratio of the speed of tidal current at a subordinate station to the speed of the corresponding current at the reference station. { və'läs·əd·ē ˌrā·shō }

velum [METEOROL] An accessory cloud veil of great horizontal extent draped over or penetrated by cumuliform clouds; velum occurs with cumulus and cumulonimbus. { 'vē·ləm }

venation [BOT] The system or pattern of veins in the tissues of a leaf. { ve'nā·shən }

venom [BIOL] Any of various poisonous materials secreted by certain animals, such as snakes or bees. { 'ven·əm }

ventilation [METEOROL] The process of causing representative air to be in contact with the sensing elements of observing instruments; especially applied to producing a flow of air past the bulb of a wet-bulb thermometer. { ˌvent·əl'ā·shən }

Venturia inaequalis [MYCOL] A fungal pathogen that causes apple scab disease. { venˌtür·ē·ə ˌin·ē'kwäl·əs }

verdant zone See frostless zone. { 'vərd·ənt ˌzōn }

verglas See glaze. { vər'glä }

vernalization [BOT] The induction in plants of the competence or ripeness to flower by the influence of cold, that is, at temperatures below the optimal temperature for growth. { ˌvərn·əl·ə'zā·shən }

verrou   See riegel.   { və'rü }

vertebratus   [METEOROL] A cloud variety (applied mainly to the genus cirrus), the elements of which are arranged in a manner suggestive of vertebrae, ribs, or a fish skeleton.   { ‚vərd·ə'bräd·əs }

vertical anemometer   [METEOROL] An instrument which records the vertical component of the wind speed.   { 'vərd·ə·kəl ‚an·ə'mäm·əd·ər }

vertical differential chart   [METEOROL] A synoptic chart showing the difference in value of a meteorological element between two levels in the atmosphere; a common example is the thickness chart.   { 'vərd·ə·kəl ‚dif·ə'ren·chəl ‚chärt }

vertical jet   See uprush.   { 'vərd·ə·kəl 'jet }

vertical stability   See static stability.   { 'vərd·ə·kəl stə'bil·əd·ē }

vertical stretching   [METEOROL] A process in which ascending vertical motion of air increases with altitude, or descending motion decreases with (increasing) altitude. { 'vərd·ə·kəl 'strech·iŋ }

vertical visibility   [METEOROL] According to United States weather observing practice, the distance that an observer can see vertically into a surface-based obscuring phenomenon, such as fog, rain, or snow.   { 'vərd·ə·kəl ‚vis·ə'bil·əd·ē }

Vertisol   [GEOL] A soil order formed in regoliths high in clay; subject to marked shrinking and swelling with changes in water content; low in organic content and high in bases. { 'vərd·ə‚sȯl }

very close pack ice   [OCEANOGR] Sea ice so concentrated that there is little if any open water.   { ¦ver·ē ¦klōs 'pak ‚īs }

very open pack ice   [OCEANOGR] Sea ice whose concentration ranges between one-tenth and three-tenths of the sea surface.   { ¦ver·ē ¦ō·pən 'pak ‚īs }

vesicle   [BIOL] A small, thin-walled bladderlike cavity, usually filled with fluid.   [GEOL] A cavity in lava formed by entrapment of a gas bubble during solidification. Also known as air sac; bladder; saccus; vacuole; wing.   { 'ves·ə·kəl }

vesicular-arbuscular mycorrhizal fungi   [MYCOL] Mycorrhizal fungi that grow into the root cortex of the host plant and penetrate root cells to form two kinds of specialized structures, arbuscules and vesicles. Also known as arbuscular mycorrhizae.   { və ¦sik·yə·lər är¦bəs·kyə·lər ‚mī·kə‚rīz·əl 'fən‚jī }

vesicular stomatitis   [VET MED] A viral disease, most often of horses, cattle, and pigs, characterized by fever and by vesicular and erosive lesions on the tongue, gums, lips, feet, and teats.   { ve‚sik·yə·lər ‚stō·mə'tīd·əs }

vesicular stomatitis virus   [MICROBIO] A virus in the genus *Vesiculovirus*, family Rhabdoviridae, the causative agent of vesicular stomatitis.   { və¦sik·yə·lər ‚stō·mə'tīd·əs ‚vī·rəs }

vessel   [BOT] A water-conducting tube or duct in the xylem.   { 'ves·əl }

viable   [BIOL] Able to live and develop normally.   { 'vī·ə·bəl }

vibriosis   [VET MED] An infectious bacterial disease, primarily of cattle, sheep, and goats, caused by *Vibrio fetus* and characterized by abortion, retained placenta, and metritis.   { ‚vib·rē'ō·səs }

vicariants   [ECOL] Two or more closely related taxa, presumably derived from one another or from a common immediate ancestor, that inhabit geographically distinct areas.   { vī'kar·ē·əns }

vinblastine   [MED] $C_{46}H_{58}O_9N_4$ An alkaloid obtained from the periwinkle plant (*Vinca rosea*) and used, as the sulfate salt, as an antineoplastic drug.   { vin'bla‚stēn }

vincristine [MED] $C_{46}H_{56}O_{10}N_4$ An alkaloid extracted from the periwinkle plant (*Vinca rosea*) and used, as the sulfate salt, as an antineoplastic drug. Also known as leurocristine. { vin′kri‚stēn }

vine [BOT] A plant having a stem that is too flexible or weak to support itself. { vīn }

vinyl chloride [CHEM] CH_2:$CHCl$ A flammable, explosive gas with an ethereal aroma; soluble in alcohol and ether, slightly soluble in water; boils at $-14°C$; an important monomer for polyvinyl chloride and its copolymers; used in organic synthesis and in adhesives. Also known as chloroethene; chloroethylene. { ′vīn·əl ′klȯr‚īd }

vinyl ether [CHEM] CH_2:$CHOCH$:CH_2 A colorless, light-sensitive, flammable, explosive liquid; soluble in alcohol, acetone, ether, and chloroform, slightly soluble in water; boils at $39°C$; used as an anesthetic and a comonomer in polyvinyl chloride polymers. Also known as divinyl ether; divinyl oxide. { ′vīn·əl ′ē·thər }

vinylpyridine [CHEM] C_5H_4NCH:CH_2 A toxic, combustible liquid; soluble in water, alcohol, hydrocarbons, esters, ketones, and dilute acids; used to manufacture elastomers and pharmaceuticals. { ‚vīn·əl′pir·ə‚dēn }

vinyltoluene [CHEM] CH_2:$CHC_6H_4CH_3$ A colorless, flammable, moderately toxic liquid; soluble in ether and methanol, slightly soluble in water; boils at $170°C$; used as a chemical intermediate and solvent. Also known as methyl styrene. { ‚vīn·əl′täl·yə ‚wēn }

viral hepatitis [MED] A type of hepatitis caused by two distinct viruses, A and B; type A is also known as infectious hepatitis, type B as serum hepatitis. { ′vī·rəl ‚hep·ə′tīd·əs }

virga [METEOROL] Wisps or streaks of water or ice particles falling out of a cloud but evaporating before reaching the earth's surface as precipitation. Also known as fall streaks; Fallstreifen; precipitation trails. { ′vər·gə }

virion [MICROBIO] The complete, mature virus particle. { ′vir·ē‚än }

viroid [MICROBIO] The smallest known agents of infectious disease, characterized by the absence of encapsidated proteins. { ′vī‚rȯid }

virology [MICROBIO] The study of viruses. { vī′räl·ə·jē }

virotoxin [BIOL] One of a group of toxins present in the mushroom *Amanita virosa*. { ‚vī·rə‚täk·sən }

virtual gravity [METEOROL] The force of gravity on a parcel of air, reduced by centrifugal force due to the motion of the parcel relative to the earth. { ′vər·chə·wəl ′grav·əd·ē }

virtual pressure [METEOROL] The pressure of a parcel of moist air when it has the same density as a parcel of dry air at the same temperature. { ′vər·chə·wəl ′presh·ər }

virtual temperature [METEOROL] In a system of moist air, the temperature of dry air having the same density and pressure as the moist air. { ′vər·chə·wəl ′tem·prə·chər }

virulence [MICROBIO] The disease-producing power of a microorganism; infectiousness. { ′vir·ə·ləns }

virus [MICROBIO] A large group of infectious agents ranging from 10 to 250 nanometers in diameter, composed of a protein sheath surrounding a nucleic acid core and capable of infecting all animals, plants, and bacteria; characterized by total dependence on living cells for reproduction and by lack of independent metabolism. { ′vī·rəs }

virus hepatitis *See* infectious hepatitis. { ′vī·rəs ‚hep·ə′tīd·əs }

virus vaccinium *See* smallpox vaccine. { ′vī·rəs vak′sin·ē·əm }

visceral leishmaniasis [MED] A severe, generalized, and often fatal infection, caused by any of three pathogenic hemoflagellates of the genus *Leishmania*, affecting organs rich in endothelial cells; accompanied by fever, spleen and liver enlargement, anemia, leukopenia (decrease in white blood cell count), skin pigmentation, and changes in plasma protein. { 'vis·ə·rəl ‚lēsh·mə'nī·ə·səs }

visibility [METEOROL] In weather observing practice, the greatest distance in a given direction at which it is just possible to see and identify with the unaided eye, in the daytime, a prominent dark object against the sky at the horizon and, at nighttime, a known, preferably unfocused, moderately intense light source. { ‚viz·ə'bil·əd·ē }

visible radiation See light. { 'viz·ə·bəl ‚rād·ē'ā·shən }

visual range [METEOROL] The distance, under daylight conditions, at which the apparent contrast between a specified type of target and its background becomes just equal to the threshold contrast of an observer. { 'vizh·ə·wəl 'rānj }

vitriolic acid See sulfuric acid. { 'vi·trē‚äl·ik 'əs·ed }

VOC See volatile organic compounds.

void ratio [SCI TECH] The ratio of the volume of void space to the volume of solid substance in any material consisting of void space and solid material, such as a soil sample, a sediment, or a powder. { 'vȯid ‚rā·shō }

volatile organic compounds [ENG] Organic chemicals that produce vapors readily at room temperature and normal atmospheric pressure, including gasoline and solvents such as toluene, xylene, and tetrachloroethylene. They form photochemical oxidants (including ground-level ozone) that affect health, damage materials, and cause crop and forest losses. Many are also hazardous air pollutants. Abbreviated VOC. { ¦väl·ə·təl ȯr‚gan·ik 'käm‚paůnz }

volcano [GEOL] **1.** A mountain or hill, generally with steep sides, formed by the accumulation of magma erupted through openings or volcanic vents. **2.** The vent itself. { väl'kā·nō }

volume transport [OCEANOGR] The volume of moving water measured between two points of reference and expressed in cubic meters per second. { 'väl·yəm ‚tranz‚pȯrt }

wake stream theory [OCEANOGR] The theory that, in a stratified ocean, a compensation current must develop on the right side of a wake stream, flowing in the same direction, and a countercurrent in the opposite direction must appear to the left. { 'wāk ¦strēm ‚thē·ə·rē }

wall reef [GEOL] A linear, steep-sided coral reef constructed on a reef wall. { 'wȯl ¦rēf }

wall-sided glacier [HYD] A glacier unconfined by a marked ravine or valley. { 'wȯl ¦sīd·əd 'glā·shər }

wandering water *See* vadose water. { 'wän·də·riŋ 'wȯd·ər }

warm-air drop *See* warm pool. { 'wȯrm ¦er 'dräp }

warm air mass [METEOROL] An air mass that is warmer than the surrounding air; an implication that the air mass is warmer than the surface over which it is moving. { 'wȯrm ¦er 'mas }

warm anticyclone *See* warm high. { 'wȯrm ¦ant·i'sī‚klōn }

warm-blooded *See* homoiothermal. { 'wȯrm ¦bləd·əd }

warm-core anticyclone *See* warm high. { 'wȯrm ¦kȯr ¦ant·i'sī‚klōn }

warm-core cyclone *See* warm low. { 'wȯrm ¦kȯr 'sī‚klōn }

warm-core high *See* warm high. { 'wȯrm ¦kȯr 'hī }

warm-core low *See* warm low. { 'wȯrm ¦kȯr 'lō }

warm cyclone *See* warm low. { 'wȯrm 'sī‚klōn }

warm drop *See* warm pool. { 'wȯrm 'dräp }

warm front [METEOROL] Any nonoccluded front, or portion thereof, which moves in such a way that warmer air replaces colder air. { 'wȯrm ‚frənt }

warm high [METEOROL] At a given level in the atmosphere, any high that is warmer at its center than at its periphery. Also known as warm anticyclone; warm-core anticyclone; warm-core high. { 'wȯrm 'hī }

warm low [METEOROL] At a given level in the atmosphere, any low that is warmer at its center than at its periphery; the opposite of a cold low. Also known as warm-core cyclone; warm-core low; warm cyclone. { 'wȯrm 'lō }

warm pool [METEOROL] A region, or pool, of relatively warm air surrounded by colder air; the opposite of a cold pool; commonly applied to warm air of appreciable vertical extent isolated in high latitudes when a cutoff high is formed. Also known as warm-air drop; warm drop. { 'wȯrm 'pül }

warm sector [METEOROL] The area of warm air, within the circulation of a wave cyclone, which lies between the cold front and warm front of a storm. { 'wȯrm ‚sek·tər }

warm tongue [METEOROL] A pronounced poleward extension or protrusion of warm air. { 'wȯrm 'təŋ }

warm-tongue steering [METEOROL] The steering influence apparently exerted upon a tropical cyclone by an upper-level warm tongue which often extends a considerable distance into regions adjacent to the cyclone. { 'wȯrm ˌtəŋ 'stir·iŋ }

warm wave *See* heat wave. { 'wȯrm ˌwāv }

warning stage [HYD] The stage, on a fixed river gage, at which it is necessary to begin issuing warnings or river forecasts if adequate precautionary measures are to be taken before flood stage is reached. { 'wȯrn·iŋ ˌstāj }

wash-and-strain ice foot [OCEANOGR] An ice foot formed from ice casts and slush and attached to a shelving beach, between the high and low waterlines; high waves and spray may cause it to build up above the high waterline. { ¦wäsh ən ¦strān 'īs ˌfu̇t }

wasp [ZOO] The common name for members of 67 families of the order Hymenoptera; all are important as parasites or predators of injurious pests. { wäsp }

waste pipe [CIV ENG] A pipe to carry waste water from a basin, bath, or sink in a building. { 'wāst ˌpīp }

waste plain *See* alluvial plain. { 'wāst ˌplān }

waste vent *See* stack vent. { 'wāst ˌvent }

water atmosphere [METEOROL] The concept of a separate atmosphere composed only of water vapor. { 'wȯd·ər 'at·mə,sfir }

water-bearing strata [GEOL] Ground layers below the standing water level. { 'wȯd·ər ¦ber·iŋ 'strad·ə }

water-borne [SCI TECH] Floating on or transported by water. { 'wȯd·ər ˌbȯrn }

water budget *See* hydrologic accounting. { 'wȯd·ər ˌbəj·ət }

water cloud [METEOROL] Any cloud composed entirely of liquid water drops; to be distinguished from an ice-crystal cloud and from a mixed cloud. { 'wȯd·ər ˌklau̇d }

water conservation [ECOL] The protection, development, and efficient management of water resources for beneficial purposes. { 'wȯd·ər ˌkän·sər'vā·shən }

water content [HYD] The liquid water present within a sample of snow (or soil) usually expressed in percent by weight; the water content in percent of water equivalent is 100 minus the quality of snow. Also known as free-water content; liquid-water content. { 'wȯd·ər ˌkän,tent }

watercourse [HYD] **1.** A stream of water. **2.** A natural channel through which water may run or does run. { 'wȯd·ər,kȯrs }

water cycle *See* hydrologic cycle. { 'wȯd·ər ˌsī·kəl }

water equivalent [METEOROL] The depth of water that would result from the melting of the snowpack or of a snow sample; thus, the water equivalent of a new snowfall is the same as the amount of precipitation represented by that snowfall. { 'wȯd·ər i'kwiv·ə·lənt }

water exchange [OCEANOGR] The volume and rate of water exchange between air and a body of water in a specific location, or between several bodies of water, controlled by such factors as tides, winds, river discharge, and currents. { 'wȯd·ər iks,chānj }

waterfall [HYD] A perpendicular or nearly perpendicular descent of water in a stream. { 'wȯd·ər,fȯl }

waterfall lake *See* plunge pool. { 'wȯd·ər,fȯl ˌlāk }

waterflooding *See* flooding. { 'wȯd·ər¦fläd·iŋ }

water front [GEOGR] An area partly bounded by water. { 'wȯd·ər ˌfrənt }

water gap [GEOL] A deep and narrow pass that cuts to the base of a mountain ridge, and through which a stream flows; the Delaware Water Gap is an example. { 'wȯd·ər ,gap }

waterless zone [HYD] The lowest hydrologic zone, generally beginning several miles beneath the land surface and characterized by the absence of water in the pore spaces due to the great pressure and density of the rock. { 'wȯd·ər·ləs 'zōn }

waterline *See* water table. { 'wȯd·ər,līn }

water loss *See* evapotranspiration. { 'wȯd·ər ,lȯs }

water mass [OCEANOGR] A body of water identified by its temperature-salinity curve or chemical composition, and normally consisting of a mixture of two or more water types. { 'wȯd·ər ,mas }

water opening [OCEANOGR] A break in sea ice, revealing the sea surface. { 'wȯd·ər ,ō·pə·niŋ }

water pollution [ECOL] Contamination of water by materials such as sewage effluent, chemicals, detergents, and fertilizer runoff. { 'wȯd·ər pə,lü·shən }

water requirement [HYD] The total quantity of water required to mature a specified crop under field conditions; includes applied irrigation, water precipitation, and groundwater available to the crop. { 'wȯd·ər ri,kwīr·mənt }

water retting [MICROBIO] A type of retting process in which the stalks of fiber plants are immersed in cold or warm, slowly renewed water, for 4 days to several weeks. The active organism is *Clostridium felsineum* and related types, which break down the pectin to a mixture of organic acids (chiefly acetic and butyric), alcohols (butanol, ethanol, and methanol), carbon dioxide (CO_2), and hydrogen (H2). { 'wȯd·ər ,red·iŋ }

watershed [HYD] The drainage area of a stream. { 'wȯd·ər,shed }

water sky [METEOROL] The dark appearance of the underside of a cloud layer when it is over a surface of open water. { 'wȯd·ər ,skī }

water smoke *See* steam fog. { 'wȯd·ər ,smōk }

water snow [HYD] Snow that, when melted, yields a more than average amount of water; thus, any snow with a high water content. { 'wȯd·ər ,snō }

waterspout [METEOROL] A tornado occurring over water; rarely, a lesser whirlwind over water, comparable in intensity to a dust devil over land. { 'wȯd·ər,spaut }

water table [HYD] The planar surface between the zone of saturation and the zone of aeration. Also known as free-water elevation; free-water surface; groundwater level; groundwater surface; groundwater table; level of saturation; phreatic surface; plane of saturation; saturated surface; water level; waterline. { 'wȯd·ər ,tā·bəl }

water type [OCEANOGR] Ocean water of a specified temperature and salinity. { 'wȯd·ər ,tīp }

water-vapor absorption [METEOROL] The absorption of certain wavelengths of infrared radiation by atmospheric water vapor; a process of fundamental importance in the energy budget of the earth's atmosphere. { 'wȯd·ər ¦vā·pər əb,sȯrp·shən }

waterwheel [ENG] A vertical wheel on a horizontal shaft that is made to revolve by the action or weight of water on or in containers attached to the rim. { 'wȯd·ər,wēl }

water year [HYD] Any 12-month period, usually selected to begin and end during a relatively dry season, used as a basis for processing streamflow and other hydrologic data; the period from October 1 to September 30 is most widely used in the United States. { 'wȯd·ər 'yir }

wave base [HYD] The depth at which sediments are not stirred by wave action, usually about 33 feet (10 meters). Also known as wave depth. { 'wāv ,bās }

wave basin [GEOGR] A basin close to the inner entrance of a harbor in which the waves from the outer entrance are absorbed, thus reducing the size of the waves entering the inner harbor. { 'wāv ,bās·ən }

wave cyclone [METEOROL] A cyclone which forms and moves along a front; the circulation about the cyclone center tends to produce a wavelike deformation of the front. Also known as wave depression. { 'wāv ,sī,klōn }

wave depression *See* wave cyclone. { 'wāv di,presh·ən }

wave depth *See* wave base. { 'wāv ,depth }

wave disturbance [METEOROL] In synoptic meteorology, the same as wave cyclone, but usually denoting an early state in the development of a wave cyclone, or a poorly developed one. { 'wāv di,stər·bəns }

wave erosion *See* marine abrasion. { 'wāv i,rō·zhən }

wave forecasting [OCEANOGR] The theoretical determination of future wave characteristics based on observed or forecasted meteorological phenomena. { 'wāv 'fȯr ,kast·iŋ }

wave height [OCEANOGR] The height of a water-surface wave; generally taken as the height difference between the wave crest and the preceding trough. { 'wāv ,hīt }

wave setdown [OCEANOGR] A decrease in the mean water level in the region in which breakers form near the seashore, caused by the presence of a pressure field. { 'wāv 'set,daún }

wave setup [OCEANOGR] An increase in the mean water level shoreward of the region in which breakers form at the seashore, caused by the onshore flux of momentum against the beach. { 'wāv 'sed,əp }

wave system [OCEANOGR] In ocean wave studies, a group of waves which have the same height, length, and direction of movement. { 'wāv ,sis·təm }

wave theory of cyclones [METEOROL] A theory of cyclone development based upon the principle of wave formation on an interface between two fluids; in the atmosphere, a front is taken as such an interface. { 'wāv 'thē·ə·rē əv 'sī,klōnz }

weak acid [CHEM] An acid that does not ionize greatly; for example, acetic acid or carbonic acid. { 'wēk 'as·əd }

weather [METEOROL] **1.** The state of the atmosphere, mainly with respect to its effects upon life and human activities; as distinguished from climate, weather consists of the short-term (minutes to months) variations of the atmosphere. **2.** As used in the making of surface weather observations, a category of individual and combined atmospheric phenomena which must be drawn upon to describe the local atmospheric activity at the time of observation. { 'weth·ər }

weather central [METEOROL] An organization which collects, collates, evaluates, and disseminates meteorological information in such a manner that it becomes a principal source of such information for a given area. { 'weth·ər 'sen·trəl }

weathered iceberg [OCEANOGR] An iceberg which is irregular in shape, due to an advanced stage of ablation; it may have overturned. { 'weth·ərd 'īs,bərg }

weather forecast [METEOROL] A forecast of the future state of the atmosphere with specific reference to one or more associated weather elements. { 'weth·ər ,fȯr,kast }

weathering [GEOL] Physical disintegration and chemical decomposition of earthy and rocky materials on exposure to atmospheric agents, producing an in-place mantle of waste. Also known as clastation; demorphism. { 'weth·ə,riŋ }

weather map [METEOROL] A chart portraying the state of the atmospheric circulation and weather at a particular time over a wide area; it is derived from a careful analysis of simultaneous weather observations made at many observing points in the area. { 'we<u>th</u>·ər ,map }

weather-map type   *See* weather type.   { 'we<u>th</u>·ər ,map ,tīp }

weather minimum   [METEOROL.] The worst weather conditions under which aviation operations may be conducted under either visual or instrument flight rules; usually prescribed by directives and standing operating procedures in terms of minimum ceiling, visibility, or specific hazards to flight.   { 'we<u>th</u>·ər ,min·ə·məm }

weather modification   [METEOROL] The changing of natural weather phenomena by technical means; so far, only on the microscale of condensation and freezing nuclei has it been possible to exert modifying influences.   { 'we<u>th</u>·ər ,mäd·ə·fə'kā·shən }

weather observation   [METEOROL] An evaluation of one or more meteorological elements that describe the state of the atmosphere either at the earth's surface or aloft.   { 'we<u>th</u>·ər ,äb·zər,vā·shən }

weather shore   [METEOROL] As observed from a vessel, the shore lying in the direction from which the wind is blowing.   { 'we<u>th</u>·ər ,shȯr }

weather side   [METEOROL] The side of a ship exposed to the wind or weather.   { 'we<u>th</u>·ər ,sīd }

weather signal   [METEOROL] A visual signal displayed to indicate a weather forecast. { 'we<u>th</u>·ər ,sig·nəl }

weather station   [METEOROL] A place and facility for the observation, measurement, and recording and transmission of data of the variable elements of weather; one of the most effective network facilities is that of the U.S. Weather Bureau.   { 'we<u>th</u>·ər ,stā·shən }

weather type   [METEOROL] A series of generalized synoptic situations, usually presented in chart form; weather types are selected to represent typical pressure patterns, and were originally devised as a method for lengthening the effective time-range of forecasts. Also known as weather-map type.   { 'we<u>th</u>·ər ,tīp }

Weddell Current   [OCEANOGR] A surface current which flows in an easterly direction from the Weddell Sea outside the limit of the West Wind Drift.   { we'del 'kə·rənt }

Wedener-Bergeron process   *See* Bergeron-Findeisen theory.   { 'väd·ən·ər ¦ber·zhə¦rōn ,prä·səs }

wedge   *See* ridge.   { wej }

weed   [BOT] A plant that is useless or of low economic value, especially one growing on cultivated land to the detriment of the crop.   { wēd }

weeping spring   *See* spring seepage.   { 'wēp·iŋ 'spriŋ }

weevil   [ZOO] Any of various snout beetles whose larvae destroy crops by eating the interior of the fruit or grain, or bore through the bark into the pith of many trees. { 'wē·vəl }

weighing rain gage   [ENG] A type of recording rain gage, consisting of a receiver in the shape of a funnel which empties into a bucket mounted upon a weighing mechanism; the weight of the catch is recorded, on a clock-driven chart, as inches of precipitation; used at climatological stations.   { 'wā·iŋ 'rān gāj }

wellhead   [HYD] The place where a stream emerges from the ground.   { 'wel,hed }

West Australia Current   [OCEANOGR] The complex current flowing northward along the west coast of Australia; it is strongest from November to January, and weakest and variable from May to July; it curves toward the west to join the South Equatorial Current.   { 'west ȯ'strāl·yə 'kə·rənt }

westerlies [METEOROL] The dominant west-to-east motion of the atmosphere, centered over the middle latitudes of both hemispheres; at the earth's surface, the westerly belt (or west-wind belt) extends, on the average, from about 35 to 65° latitude. Also known as circumpolar westerlies; middle-latitude westerlies; mid-latitude westerlies; polar westerlies; subpolar westerlies; subtropical westerlies; temperate westerlies; zonal westerlies; zonal winds. { 'wes·tər·lēz }

westerly wave [METEOROL] An atmospheric wave disturbance embedded in the mid-latitude westerlies. { 'wes·tər·lē 'wāv }

Western Equatorial Countercurrent [OCEANOGR] Weak, narrow bands of eastward-flowing water observed in some winter months in the western Atlantic near the equator. { 'wes·tərn ‚ek·wə'tȯr·ē·əl kaunt·ər‚kə·rənt }

West Greenland Current [OCEANOGR] The current flowing northward along the west coast of Greenland into the Davis Strait; part of this current joins the Labrador Current, while the other part continues into Baffin Bay. { 'west ¦grēn·lənd 'kə·rənt }

westward intensification [OCEANOGR] The intensification of ocean currents to the west, derived from a mathematical model that includes the effects of zonal wind stress at the sea surface and internal friction. { 'west·wərd in‚ten·sə·fə'kā·shən }

West Wind Drift *See* Antarctic Circumpolar Current. { 'west ¦wind 'drift }

wet-bulb depression [METEOROL] The difference in degrees between the dry-bulb temperature and the wet-bulb temperature. { 'wet ¦bəlb di'presh·ən }

wet-bulb temperature [METEOROL] **1.** Isobaric wet-bulb temperature, that is, the temperature an air parcel would have if cooled adiabatically to saturation at constant pressure by evaporation of water into it, all latent heat being supplied by the parcel. **2.** The temperature read from the wet-bulb thermometer; for practical purposes, the temperature so obtained is identified with the isobaric wet-bulb temperature. { 'wet ¦bəlb 'tem·prə·chər }

wet climate [CLIMATOL] A climate whose vegetation is of the rainforest type. Also known as rainforest climate. { 'wet 'klī·mət }

wet gas [PETR MIN] Natural gas produced along with crude petroleum in oil fields or from gas-condensate fields; in addition to methane, it contains ethane, propane, butanes, and some higher hydrocarbons, such as pentane and hexane. { 'wet 'gas }

wetlands [ECOL] An area characterized by a high content of soil moisture, such as a swamp or bog. { 'wet ‚lanz }

wet season *See* rainy season. { 'wet ‚sēz·ən }

wet snow [METEOROL] Deposited snow that contains a great deal of liquid water. { 'wet 'snō }

wheat [BOT] A food grain crop of the genus *Triticum*; plants are self-pollinating; the inflorescence is a spike bearing sessile spikelets arranged alternately on a zigzag rachis. { wēt }

whippoorwill storm *See* frog storm. { ¦wip·ər¦wil ‚stȯrm }

whirlpool [OCEANOGR] Water in rapid rotary motion. { 'wərl‚pül }

whirly [METEOROL] A small violent storm, a few yards (or meters) to 100 yards (91 meters) or more in diameter, frequent in Antarctica near the time of the equinoxes. { 'wər·lē }

white ant *See* termite. { 'wīt 'ant }

white band disease [ZOO] A coral reef disease that is typified by a loss of tissue that is visible as a band of bare white skeleton. { ‚wīt 'band diz‚ēz }

whitecap [OCEANOGR] A cloud of bubbles at the sea surface caused by a breaking wave. { 'wīt‚kap }

white copperas *See* zinc sulfate. { 'wīt 'käp·rəs }

white damp [PETR MIN] In mining, carbon monoxide (CO); a gas that may be present in the afterdamp of a gas or coal-dust explosion, or in the gases given off by a mine fire; it is an important constituent of illuminating gas, supports combustion, and is very poisonous. { 'wīt ‚damp }

white diarrhea *See* pullorum disease. { 'wīt ‚dī·ə'rē·ə }

white frost *See* hoarfrost. { 'wīt 'frȯst }

white lead [CHEM] Basic lead carbonate of variable composition, the oldest and most important lead paint pigment; also used in putty and ceramics. { 'wīt 'led }

white light [PHYS] Any radiation producing the same color sensation as average noon sunlight. { 'wīt 'līt }

whiteout [METEOROL] An atmospheric optical phenomenon of the polar regions in which the observer appears to be engulfed in a uniformly white glow: shadows, horizon, and clouds are not discernible; sense of depth and orientation are lost; dark objects in the field of view appear to float at an indeterminable distance. Also known as milky weather. { 'wīd‚aȯt }

white phosphorus [CHEM] The element phosphorus in its allotropic form, a soft, waxy, poisonous solid melting at 44.5°C; soluble in carbon disulfide, insoluble in water and alcohol; self-igniting in air. Also known as yellow phosphorus. { 'wīt 'fä·sfə·rəs }

white squall [METEOROL] A sudden squall in tropical or subtropical waters, which lacks the usual squall cloud and whose approach is signaled only by the whiteness of a line of broken water or whitecaps. { 'wīt 'skwȯl }

white vitriol *See* zinc sulfate. { 'wīt 'vi·trē‚ȯl }

white water [OCEANOGR] Frothy water, as in whitecaps or breakers. { 'wīt ‚wȯd·ər }

whiting [OCEANOGR] A patch of seawater that contains a substantial amount of calcium carbonate and therefore appears white relative to surrounding water. { 'wīd·iŋ }

whole-body counter [ENG] A radiation counter that directly measures radioactivity in the entire human body. { 'hōl ‚bäd·ē 'kaȯnt·ər }

whole gale [METEOROL] **1.** In storm-warning terminology, a wind of 48 to 63 knots (55 to 72 miles, or 89 to 133 kilometers, per hour). **2.** In the Beaufort wind scale, a wind whose speed is from 48 to 55 knots (55 to 63 miles, or 89 to 102 kilometers, per hour). { 'hōl 'gāl }

whooping cough *See* pertussis. { 'huṗ·iŋ ‚kȯf }

whorl [BOT] An arrangement of several identical anatomical parts, such as petals, in a circle around the same point. { wərl }

Wiesen *See* meadow. { 'vēz·ən }

Wild fence [METEOROL] A wooden enclosure about 16 feet 4.8 meters square and 8 feet 2.4 meters high with a precipitation gage in its center; the function of the fence is to minimize eddies around the gage, and thus ensure a catch which will be representative of the actual rainfall or snowfall. { 'wīld ‚fens }

wildfire [FOR] An uncontrolled fire that burns surface vegetation (grass, weeds, grainfields, brush, chaparral, tundra, and forest and woodland). [PL PATH] A bacterial disease of tobacco caused by *Pseudomonas tabaci* and characterized by the appearance of brown spots surrounded by yellow rings, which turn dark, rot, and fall out. { 'wīl‚fīr }

wild snow [METEOROL] Newly deposited snow which is very fluffy and unstable; in general, it falls only during a dead calm at very low air temperatures. { 'wīld 'snō }

wilt [PL PATH] Any of various plant diseases characterized by drooping and shriveling, following loss of turgidity. { wilt }

wilting point [BOT] A condition in which a plant begins to use water from its own tissues for transpiration because soil water has been exhausted. { 'wilt·iŋ ,pȯint }

wind [METEOROL] The motion of air relative to the earth's surface; usually means horizontal air motion, as distinguished from vertical motion, and air motion averaged over the response period of the particular anemometer. { wind }

windburn [BOT] Injury to plant foliage, caused by strong, hot, dry winds. { 'win,bərn }

wind chill [METEOROL] That part of the total cooling of a body caused by air motion. { 'win ,chil }

wind-chill index [METEOROL] The cooling effect of any combination of temperature and wind, expressed as the loss of body heat in kilogram calories per hour per square meter of skin surface; it is only an approximation because of individual body variations in shape, size, and metabolic rate. { 'win ¦chil ,in,deks }

wind crust [HYD] A type of snow crust, formed by the packing action of wind on previously deposited snow; wind crust may break locally but, unlike wind slab, does not constitute an avalanche hazard. { 'win 'krəst }

wind current [METEOROL] Generally, any of the quasi-permanent, large-scale wind systems of the atmosphere, for example, the westerlies, trade winds, equatorial easterlies, or polar easterlies. { 'win 'kə,rənt }

wind direction [METEOROL] The direction from which wind blows. { 'win də,rek·shən }

wind divide [METEOROL] A semipermanent feature of the atmospheric circulation (usually a high-pressure ridge) on opposite sides of which the prevailing wind directions differ greatly. { 'win də,vīd }

wind drift See drift current. { 'win ,drift }

wind-driven current See drift current. { 'win ¦driv·ən 'kə·rənt }

wind erosion [GEOL] Detachment, transportation, and deposition of loose topsoil or sand by the action of wind. { 'wind i,rō·zhən }

wind measurement [METEOROL] The determination of three parameters: the size of an air sample, its speed, and its direction of motion. { 'win ,mezh·ər·mənt }

window [HYD] The unfrozen part of a river surrounded by river ice during the winter. { 'win·dō }

window frost [HYD] A thin deposit of hoarfrost often found on interior surfaces of windows in winter, and frequently exhibiting beautiful fernlike patterns. { 'win·dō ,frȯst }

window ice [HYD] A thin deposit of ice which forms by the freezing of many tiny drops of water that have condensed on the indoors side of a cold window surface. { 'win·dō ,īs }

wind power [ENG] The extraction of kinetic energy from the wind and conversion of it into a useful type of energy: thermal, mechanical, or electrical. { 'win ,paů·ər }

wind ripple [METEOROL] One of a series of wavelike formations on a snow surface, an inch or so in height, at right angles to the direction of wind. Also known as snow ripple. { 'win ,drip·əl }

wind rose [METEOROL] A diagram in which statistical information concerning direction and speed of the wind at a location may be summarized; a line segment is drawn in each of perhaps eight compass directions from a common origin; the length of a particular segment is proportional to the frequency with which winds blow from that direction; thicknesses of a segment indicate frequencies of occurrence of various classes of wind speed. { 'win ‚drōz }

windrow [GEOL] Any accumulation of material formed by wind or tide action. { 'win ‚drō }

winds aloft [METEOROL] Generally, the wind speeds and directions at various levels in the atmosphere above the domain of surface weather observations, as determined by any method of winds-aloft observation. Also known as upper-level winds; upper winds. { 'winz ə'lȯft }

winds-aloft observation [METEOROL] The measurement and computation of wind speeds and directions at various levels above the surface of the earth. { 'winz ə'lȯft ‚äb·zər'vā·shən }

wind scoop [METEOROL] A saucerlike depression in the snow near obstructions such as trees, houses, and rocks, caused by the eddying action of the deflected wind. { 'win ‚sküp }

wind shear [METEOROL] The local variation of the wind vector or any of its components in a given direction. { 'win ‚shir }

wind-shift line [METEOROL] A line or narrow zone along which there is an abrupt change of wind direction. { 'win ¦shift ‚līn }

wind slab [HYD] A type of snow crust; a patch of hard-packed snow, which is packed as it is deposited in favored spots by the wind, in contrast to wind crust. { 'wind ‚slab }

wind speed [METEOROL] The rate of motion of air. { 'win ‚spēd }

windstorm [METEOROL] A storm in which strong wind is the most prominent characteristic. { 'win‚stȯrm }

wind stress [METEOROL] The drag or tangential force per unit area exerted on the surface of the earth by the adjacent layer of moving air. { 'win ‚stres }

wind tide [OCEANOGR] **1.** The vertical rise in the still-water level on the leeward side of a body of water, particularly the ocean or other large body, caused by wind stresses on the surface of the water. **2.** The difference in still-water level between the windward and leeward sides of such a body caused by wind stresses. { 'win ‚tīd }

wind turbine [ENG] Machine that converts wind power to electricity; as moving air flows past the rotors of the turbine, the rotors spin and drive the shaft of an electric generator. { 'win 'ter‚bīn }

wind velocity [METEOROL] The speed and direction of wind. { 'win və‚läs·əd·ē }

windward [METEOROL] In the general direction from which the wind blows. { 'win·wərd }

wind wave [OCEANOGR] A wave resulting from the action of wind on a water surface. { 'win ‚wāv }

wing See vesicle. { wiŋ }

winged headland [GEOGR] A seacliff with two bays or spits, one on either side. { 'wiŋd 'hed·lənd }

winter ice [OCEANOGR] Level sea ice more than 8 inches (20 centimeters) thick, and less than 1 year old; the stage which follows young ice. { 'win·tər 'īs }

Witte-Margules equation [OCEANOGR] A formula expressing the slope of the boundary layer between two water masses of different densities and velocities, taking into account the rotation of the earth. Also known as Margules equation. { 'vid·ə 'mär·gyə·lēz i‚kwā·zhən }

wood [ECOL] A dense growth of trees, more extensive than a grove and smaller than a forest. { wùd }

woodland *See* forest; temperate woodland. { 'wùd·lənd }

wood preservative [ENG] A material used to coat wood to kill insects and fungi, but not usually classed as an insecticide; coal tar creosote and its derivatives are the most widely used wood preservatives. { 'wùd pri,zər·vəd·iv }

wood pulp *See* pulp. { 'wùd ,pəlp }

woodstone *See* silicified wood. { 'wùd,stōn }

wool-sorter's disease *See* anthrax. { 'wùl ¦sȯrd·ərz di,zēz }

worker [ZOO] One of the neuter, usually sterile individuals making up a caste of social insects, such as ants, termites, or bees, which labor for the colony. { 'wər·kər }

xanthan gum [BIOL] A polysaccharide produced by the bacterium *Xanthomonas campestris* that is used as a thickener and stabilizer in foods; also used in oil recovery to help improve water flooding and oil displacement. { 'zan·thən ˌgəm }

xanthene [CHEM] $CH_2(C_6H_4)_2O$ Yellowish crystals that are soluble in ether, slightly soluble in water and alcohol; melts at 100°C; used as a fungicide and chemical intermediate. Also known as tricyclic dibenzopyran. { 'zanˌthēn }

9-xanthenone *See* genicide. { ¦nīn 'zan·thə·nōn }

Xanthomonas citri [MICROBIO] The bacterial pathogen that causes citrus canker. { ˌzan·thəˌmō·nəs 'siˌtrē }

xanthone [CHEM] $CO(C_6H_4)_2O$ White needle crystals that are found in some plant pigments; insoluble in water, soluble in alcohol, chloroform, and benzene; melts at 173°C, sublimes at 350°C; used as a larvicide, as a dye intermediate, and in perfumes and pharmaceuticals. { 'zanˌthōn }

xenogamy [BOT] Cross-fertilization between flowers on different plants. { zə 'näg·ə·mē }

xenon [CHEM] An element, symbol Xe, member of the noble gas family, group 0, atomic number 54, atomic weight 131.291; colorless, boiling point −108°C (1 atmosphere, or 101,325 pascals), noncombustible, nontoxic, and nonreactive; used in photographic flash lamps, luminescent tubes, and lasers, and as an anesthetic. { 'zēˌnän }

xerarch succession [ECOL] A type of succession that originates in a dry habitat. { 'zer ˌärk səkˌsesh·ən }

xeric [ECOL] **1.** Of or pertaining to a habitat having a low or inadequate supply of moisture. **2.** Of or pertaining to an organism living in such an environment. { 'zer·ik }

xeromorphic [ECOL] Referring to a plant that is able to survive in dry environments. { ˌzir·ə'mȯr·fik }

xerophyte [ECOL] A plant adapted to life in areas where the water supply is limited. { 'zir·əˌfīt }

xerosere [ECOL] A temporary community in an ecological succession on dry, sterile ground such as rock, sand, or clay. { 'zir·əˌsir }

xerothermic [CLIMATOL] Characterized by dryness and heat. { ¦zir·ə¦thər·mik }

xerotolerance [BIOL] The ability to grow in extremely dry habitats. { 'zer·əˌtäl·ə·rəns }

xylem [BOT] The principal water-conducting tissue and the chief supporting tissue of higher plants; composed of tracheids, vessel members, fibers, and parenchyma. { 'zī·ləm }

xylene [CHEM] $C_6H_4(CH_3)_2$ Any one of the family of isomeric, colorless aromatic hydrocarbon liquids, produced by the destructive distillation of coal or by the catalytic reforming of petroleum naphthenic fractions; used for high-octane and aviation gasolines, solvents, chemical intermediates, and the manufacture of polyester resins. Also known as dimethylbenzene; xylol.   { 'zī,lēn }

ortho-xylene   [CHEM] 1,2-$C_6H_4(CH_3)_2$ A flammable, moderately toxic liquid; insoluble in water, soluble in alcohol and ether; boils at 144°C; used to make phthalic anhydride, vitamins, pharmaceuticals, and dyes, and in insecticides and motor fuels.   { ¦ȯr·thō 'zī,lēn }

xylenol   [CHEM] $(CH_3)_2C_6H_3OH$ Highly toxic, combustible crystals; slightly soluble in water, soluble in most organic solvents; melts at 20–76°C; used as a chemical intermediate, disinfectant, solvent, and fungicide, and for pharmaceuticals and dyestuffs.   { 'zī·lə,nȯl }

xylol   *See* xylene.   { 'zī,lȯl }

xylophagous   [BIOL] Referring to an organism which feeds on wood.   { zī'läf·ə·gəs }

Y

yaws [MED] An infectious tropical disease of humans caused by the spirochete *Treponema pertenue*; manifested by a primary cutaneous lesion followed by a granulomatous skin eruption. { 'yȯz }

yeast [MYCOL] A collective name for those fungi which possess, under normal conditions of growth, a vegetative body (thallus) consisting, at least in part, of simple, individual cells. { yēst }

yellow dwarf [PL PATH] Any of several plant viral diseases characterized by yellowing of the foliage and stunting of the plant. { 'yel·ō 'dwȯrf }

yellow-green algae [BOT] The common name for members of the class Xanthophyceae. { 'yel·ō ¦grēn 'al·jē }

yellow phosphorus *See* white phosphorus. { 'yel·ō 'fä·sfə·rəs }

yellows [PL PATH] Any of various fungus diseases of plants characterized by yellowing of the leaves which later turn brown, become brittle, and die; affects cabbage, lettuce, cauliflower, peach, sugarbeet, and other plants. { 'yel·ōz }

Yellow Sea [GEOGR] An inlet of the Pacific Ocean between northeastern China and Korea. { 'yel·ō 'sē }

yellow snow [HYD] Snow with a golden or yellow appearance because of the presence of pine or cypress pollen. { 'yel·ō 'snō }

young ice [HYD] Newly formed ice in the transitional stage of development from ice crust to winter ice. { 'yəŋ ¦īs }

Yucatán Current [OCEANOGR] A rapid northward flowing current along the western side of the Yucatán Strait; generally loops to the north and exits as the Florida Current. { ¦yü·kə¦tän 'kə·rənt }

Z

zastruga *See* sastruga. { 'zas·trə·gə }

Zeitgeber |BIOL| A periodic environmental condition or event that acts to set or reset an innate biological rhythm of an organism. { 'tsīt₁gā·bər }

zenithal rain |METEOROL| In the tropics or subtropics, the rainy season which recurs annually or semiannually at about the time that the sun is most nearly overhead (at zenith). { 'zē·nə·thəl 'rān }

zephyr |METEOROL| Any soft, gentle breeze. { 'zef·ər }

zero layer |OCEANOGR| A reference level in the ocean, at which horizontal motion is at a minimum. { 'zir·ō₁lā·ər }

zero population growth |ECOL| Situation in which each organism in a population is replaced by only one offspring, so there is no increase in population size. { 'zir·ō 'päp·yə₁lā·shən ₁grōth }

zinc arsenite |CHEM| $Zn(AsO_2)_2$ A toxic white powder that is insoluble in water, soluble in alkalies; used as an insecticide and timber preservative. Also known as zinc meta-arsenite. { 'ziŋk 'ärs·ən₁īt }

zinc chloride |CHEM| $ZnCl_2$ Water- and alcohol-soluble, white, fire-hazardous crystals that melt at 290°C, and are irritating to the skin; used as a catalyst and in electroplating, wood preservation, textile processing, petroleum refining, medicine, and feed additives. { 'ziŋk 'klȯr₁īd }

zinc metaarsenite *See* zinc arsenite. { 'ziŋk ¦med·ə'ärs·ən₁āt }

zinc naphthenate |CHEM| $Zn(C_6H_5COO)_2$ A combustible, viscous, acetone-soluble solid; used in paints, varnishes, and resins, and as a drier and wetting agent, insecticide, fungicide, and mildewstat. { 'ziŋk 'naf·thə₁nāt }

zinc orthoarsenate |CHEM| $Zn_3(AsO_4)_2$ A toxic white powder that is insoluble in water, soluble in alkalies; used as an insecticide and wood preservative. { 'ziŋk¦ȯr·thō'ärs·ən ₁āt }

zinc sulfate |CHEM| $ZnSO_4 \cdot 7H_2O$ Efflorescent, water-soluble, colorless crystals with an astringent taste; used to preserve skins and wood and as a paper bleach, analytical reagent, feed additive, and fungicide. Also known as white copperas; white vitriol; zinc vitriol. { 'ziŋk 'səl₁fāt }

zinc sulfide |CHEM| ZnS A yellowish powder that is insoluble in water, soluble in acids; exists in two crystalline forms (alpha, or wurtzite, and beta, or sphalerite); beta becomes alpha at 1020°C, and sublimes at 1180°C; used as a pigment for paints and linoleum, in opaque glass, rubber, and plastics, for hydrosulfite dyeing process, as x-ray and television screen phosphor, and as a fungicide. { 'ziŋk 'səl₁fīd }

zinc vitriol *See* zinc sulfate. { 'ziŋk 'vi·trē₁ȯl }

zonal |METEOROL| Latitudinal, easterly or westerly, opposed to meridional. { 'zōn·əl }

zonal circulation   See zonal flow.   { 'zōn·əl ˌsər·kyə'lā·shən }

zonal flow   [METEOROL] The flow of air along a latitude circle; more specifically, the latitudinal (east or west) component of existing flow. Also known as zonal circulation. { 'zōn·əl 'flō }

zonal index   [METEOROL] A measure of strength of the middle-latitude westerlies, expressed as the horizontal pressure difference between 35° and 55° latitude, or as the corresponding geostrophic wind.   { 'zōn·əl 'in‚deks }

zonal kinetic energy   [METEOROL] The kinetic energy of the mean zonal wind, obtained by averaging the zonal component of the wind along a fixed latitude circle.   { 'zōn·əl ki'ned·ik 'en·ər·jē }

zonal soil   [GEOL] In early classification systems in the United States, a soil order including soils with well-developed characteristics that reflect the influence of agents of soil genesis. Also known as mature soil.   { 'zōn·əl 'sȯil }

zonal westerlies   See westerlies.   { 'zōn·əl 'wes·tər·lēz }

zonal wind   [METEOROL] The wind, or wind component, along the local parallel of latitude, as distinguished from the meridional wind.   { 'zōn·əl 'wind }

zonal winds   See westerlies.   { 'zōn·əl 'winz }

zonal wind-speed profile   [METEOROL] A diagram in which the speed of the zonal flow is one coordinate and latitude the other.   { 'zōn·əl 'win ‚spēd ‚prō‚fīl }

zonation   [ECOL] Arrangement of organisms in biogeographic zones.   { zō'nā·shən }

zone   [GEOGR] An area or region of latitudinal character.   [GEOL] A belt, layer, band, or strip of earth material such as rock or soil.   { zōn }

zone of accumulation   See B horizon.   { 'zōn əv ə‚kyü·mə'lā·shən }

zone of aeration   [GEOL] The subsurface sediment above the water table containing air and water. Also known as unsaturated zone; vadose zone; zone of suspended water. { 'zōn əv e'rā·shən }

zone of illuviation   See B horizon.   { 'zōn əv i‚lü·vē'ā·shən }

zone of maximum precipitation   [METEOROL] In a mountain region, the belt of elevation at which the annual precipitation is greatest.   { 'zōn əv 'mak·sə·məm pri ‚sip·ə'tā·shən }

zone of saturation   [HYD] A subsurface zone in which water fills the interstices in soil and is under pressure greater than atmospheric pressure; the top of the zone of saturation marks the water table for the area. Also known as phreatic zone; saturated zone.   { 'zōn əv ‚sach·ə'rā·shən }

zone of soil water   See belt of soil water.   { 'zōn əv 'sȯil ‚wȯd·ər }

zone of suspended water   See zone of aeration.   { 'zōn əv sə‚spen·dəd ‚wȯd·ər }

zoocecidium   [PL PATH] A plant gall usually caused by an insect.   { ‚zō·ə·sə'sid·ē·əm }

zoochlorellae   [BIOL] Unicellular green algae which live as symbionts in the cytoplasm of certain protozoans, sponges, and other invertebrates.   { ‚zō·ə·klə'rel‚ē }

zoochory   [BOT] Dispersal of plant disseminules by animals.   { 'zō·ə‚klȯr·ē }

zoogeographic region   [ECOL] A major unit of the earth's surface characterized by faunal homogeneity.   { ‚zō·ə‚jē·ə‚graf·ik 'rē·jən }

zoogeography   [BIOL] The science that attempts to describe and explain the distribution of animals in space and time.   { ‚zō·ə‚jē'äg·rə·fē }

zoology   [BIOL] The science that deals with knowledge of animal life.   { zō'äl·ə·jē }

zoonoses [BIOL] Diseases which are biologically adapted to and normally found in lower animals but which under some conditions also infect humans. { ˌzō·ə'nō·sēz }

zooplankton [ECOL] Microscopic animals which move passively in aquatic ecosystems. { ¦zō·ə'plaŋk·tən }

zoosphere [ECOL] The world community of animals. { 'zō·ə,sfir }

zoosporangium [BOT] A spore case bearing zoospores. { ¦zō·ə·spə'ran·jē·əm }

zootoxin [BIOL] A toxic substance or poison produced by an animal, for example, snake venom. { 'zō·ə'täk·sən }

zooxanthellae [BIOL] Microscopic yellow-green algae which live symbiotically in certain radiolarians and marine invertebrates. { ˌzō·ə·zan'thē,lē }

zoster *See* herpes zoster. { 'zäs·tər }

zygospore [BOT] A thick-walled cell or resting spore that results from the fusion of similar reproductive cells, especially in organisms that reproduce by conjugation. { 'zī·gə,spȯr }

zymosis *See* fermentation. { zī'mō·səs }

Zythiaceae [MYCOL] A family of fungi of the order Sphaeropsidales which contains many plant and insect pathogens. { ˌzith·ē·'ās·ē,ē }

Appendix

Base units of the International System

Quantity	Name of unit	Unit symbol
Length	meter	m
Mass	kilogram	kg
Time	second	s
Electric current	ampere	A
Temperature	kelvin	K
Luminous intensity	candela	cd
Amount of substance	mole	mol

Appendix

Derived units of the International System*

Quantity	Name of unit	Unit symbol, or unit expressed in terms of other SI units	Unit expressed in terms of SI base units
Plane angle	radian	rad	$m/m = 1$
Solid angle	steradian	sr	$m^2/m^2 = 1$
Area	square meter		m^2
Volume	cubic meter		m^3
Frequency	hertz	Hz	s^{-1}
Density	kilogram per cubic meter		kg/m^3
Velocity	meter per second		m/s
Angular velocity	radian per second	rad/s	$m/(m \cdot s) = s^{-1}$
Acceleration	meter per second squared		m/s^2
Angular acceleration	radian per second squared	rad/s^2	$m/(m \cdot s^2) = s^{-2}$
Volumetric flow rate	cubic meter per second		m^3/s
Force	newton	N	$kg \cdot m/s^2$
Surface tension	newton per meter, joule per square meter	N/m, J/m^2	kg/s^2
Pressure	pascal, newton per square meter	Pa, N/m^2	$kg/(m \cdot s^2)$
Viscosity, dynamic	pascal-second, newton-second per square meter	$Pa \cdot s$, $N \cdot s/m^2$	$kg/(m \cdot s)$
Viscosity, kinematic	meter squared per second		m^2/s
Work, torque, energy, quantity of heat	joule, newton-meter, watt-second	J, $N \cdot m$, $W \cdot s$	$kg \cdot m^2/s^2$
Power, heat flux	watt, joule per second	W, J/s	$kg \cdot m^2/s^3$
Heat flux density	watt per square meter	W/m^2	kg/s^3
Volumetric heat release rate	watt per cubic meter	W/m^3	$kg/(m \cdot s^3)$
Heat transfer coefficient	watt per square meter kelvin	$W/(m^2 \cdot K)$	$kg/(s^3 \cdot K)$

Heat capacity (specific)	joule per kilogram kelvin	$J/kg \cdot K$	$m^2/(s^2 \cdot K)$
Capacity rate	watt per kelvin	W/K	$kg \cdot m^2/(s^3 \cdot K)$
Thermal conductivity	watt per meter kelvin	$W/(m \cdot K),\ \dfrac{J \cdot m}{s \cdot m^2 \cdot K}$	$kg \cdot m/(s^3 \cdot K)$
Quantity of electricity	coulomb	C	$A \cdot s$
Electromotive force	volt	$V,\ W/A$	$kg \cdot m^2/(A \cdot s^3)$
Electric field strength	volt per meter	V/m	$kg \cdot m/(A \cdot s^3)$
Electric resistance	ohm	$\Omega,\ V/A$	$kg \cdot m^2/(A^2 \cdot s^3)$
Electric conductance	siemens	$S,\ A/V$	$A^2 \cdot s^3/(kg \cdot m^2)$
Electric conductivity	ampere per volt meter	$A/(V \cdot m)$	$A^2 \cdot s^3/(kg \cdot m^3)$
Electric capacitance	farad	$F,\ A \cdot s/V$	$A^2 \cdot s^4/(kg \cdot m^2)$
Magnetic flux	weber	$Wb,\ V \cdot s$	$kg \cdot m^2/(A \cdot s^2)$
Inductance	henry	$H,\ V \cdot s/A$	$kg \cdot m^2/(A^2 \cdot s^2)$
Magnetic permeability	henry per meter	H/m	$kg \cdot m/(A^2 \cdot s^2)$
Magnetic flux density	tesla, weber per square meter	$T,\ Wb/m^2$	$kg/(A \cdot s^2)$
Magnetic field strength	ampere per meter	A/m	A/m
Magnetomotive force	ampere	A	A
Luminous flux	lumen	$lm,\ cd \cdot sr$	$cd \cdot m^2/m^2 = cd$
Luminance	candela per square meter		cd/m^2
Illumination	lux, lumen per square meter	$lx,\ lm/m^2,\ cd \cdot sr/m^2$	$cd \cdot m^2/m^4 = cd/m^2$
Activity (of radionuclides)	becquerel	Bq	s^{-1}
Absorbed dose	gray	$Gy,\ J/kg$	m^2/s^2
Dose equivalent	sievert	$sv,\ J/kg$	m^2/s^2
Catalytic activity	katal	kat	mol/s

The degree Celsius (°C) is also a derived unit of the International System.

Appendix

Prefixes for units in the International System

Prefix	Symbol	Power	Example	Prefix	Symbol	Power	Example
Yotta	Y	10^{24}		Deci	d	10^{-1}	
Zetta	Z	10^{21}		Centi	c	10^{-2}	centimeter (cm)
Exa	E	10^{18}		Milli	m	10^{-3}	milligram (mg)
Peta	P	10^{15}		Micro	μ	10^{-6}	microgram (μg)
Tera	T	10^{12}	terawatt (TW)	Nano	n	10^{-9}	nanosecond (ns)
Giga	G	10^{9}	gigawatt (GW)	Pico	p	10^{-16}	picofarad (pF)
Mega	M	10^{6}	megahertz (MHz)	Femto	f	10^{-15}	femtosecond (fs)
Kilo	k	10^{3}	kilometer (km)	Atto	a	10^{-18}	
Hecto	h	10^{2}		Zepto	z	10^{-21}	
Deka	da	10^{1}		Yocto	y	10^{-24}	

Some common units defined in terms of SI units

Quantity	Name of unit	Unit symbol	Definition of unit
Length	inch	in.	2.54×10^{-2} m
Mass	pound (avoirdupois)	lb	0.45359237 kg
Force	kilogram-force	kgf	9.80665 N
Pressure	atmosphere	atm	101325 Pa
Pressure	torr	torr	(101325/760) Pa
Pressure	conventional millimeter of mercury*	mmHg	$13.5951 \times 980.685 \times 10^{-2}$ Pa
Energy	kilowatt-hour	kWh	3.6×10^{5} J
Energy	thermochemical calorie	cal	4.184 J
Energy	international steam table calorie	cal_{IT}	4.1868 J
Thermodynamic temperature (T)	degree Rankine	°R	(5/9) K
Customary temperature (t)	degree Celsius	°C	$t(°C) = T(K) - 273.15$
Customary temperature (t)	degree Fahrenheit	°F	$t(°F) = [1.8 \times t(°C)] + 32 = T(°R) - 459.67$
Radioactivity	curie	Ci	3.7×10^{10} Bq
Energy[†]	electronvolt	eV	$eV = 1.60218 \times 10^{-19}$ J
Mass[†]	unified atomic mass unit	u	$u = 1.66054 \times 10^{-27}$ kg

*The conventional millimeter of mercury, symbol mmHg (not mm Hg!), is the pressure exerted by a column exactly 1 mm high of a fluid of density exactly 13.5951 g cm^{-3} in a place where the gravitational acceleration is exactly 980.665 cm · s^{-2}. The mmHg differs from the torr by less then 2×10^{-7} torr.

[†]These units defined in terms of the best available experimental values of certain physical constants may be converted to SI units. The factors for conversion of these units are subject to change in the light of new experimental measurements of the constants involved.

Appendix

Equivalents of commonly used units for the U.S. Customary System and the metric system

1 inch = 2.5 centimeters (25 millimeters)
1 foot = 0.3 meter (30 centimeters)
1 yard = 0.9 meter
1 mile = 1.6 kilometers

1 acre = 0.4 hectare
1 acre = 4047 square meters

1 gallon = 3.8 liters
1 fluid ounce = 29.6 milliliters
32 fluid ounces = 946.4 milliliters

1 quart = 0.95 liter
1 ounce = 28.35 grams
1 pound = 0.45 kilogram
1 ton = 907.18 kilograms

°F = (1.8 × °C) + 32

1 centimeter = 0.4 inch
1 meter = 3.3 feet
1 meter = 1.1 yards
1 kilometer = 0.62 mile

1 hectare = 2.47 acres
1 square meter = 0.00025 acre

1 liter = 1.06 quarts = 0.26 gallon
1 milliliter = 0.034 fluid ounce

1 gram = 0.035 ounce
1 kilogram = 2.2 pounds
1 kilogram = 1.1×10^{-3} ton

°C = (°F − 32) ÷ 1.8

1 inch = 0.083 foot
1 foot = 0.33 yard (12 inches)
1 yard = 3 feet (36 inches)
1 mile = 5280 feet (1760 yards)

1 quart = 0.25 gallon (32 ounces; 2 pints)
1 pint = 0.125 gallon (16 ounces)
1 gallon = 4 quarts (8 pints)

1 ounce = 0.0625 pound
1 pound = 16 ounces
1 ton = 2000 pounds

Appendix

Conversion factors for the U.S. Customary System, metric system, and International System

A. Units of length

Units		cm	m	in.	ft	yd	mi
1 cm	= 1		0.01^*	0.3937008	0.0328840	0.01093613	6.213712×10^{-6}
1 m	= 100.		1	39.37008	3.280840	1.093613	6.213712×10^{-4}
1 in.	$= 2.54^*$		0.0254	1	$0.08333333\ldots$	$0.02777777\ldots$	1.578283×10^{-5}
1 ft	= 30.48		0.3048	$12.^*$	1	$0.3333333\ldots$	$1.893939\ldots \times 10^{-4}$
1 yd	= 91.44		0.9144	36.	$3.^*$	1	$5.681818\ldots \times 10^{-4}$
1 mi	$= 1.609344 \times 10^5$		1.609344×10^3	6.336×10^4	$5280.^*$	1760.	1

B. Units of area

Units		cm^2	m^2	$in.^2$	ft^2	yd^2	mi^2
1 cm²	= 1		10^{-4*}	0.1550003	1.076391×10^{-3}	1.195990×10^{-4}	3.861022×10^{-11}
1 m²	$= 10^4$		1	1550.003	10.76391	1.195990	3.861022×10^{-7}
1 in.²	$= 6.4516^*$		6.4516×10^{-4}	1	$6.944444\ldots \times 10^{-3}$	7.716049×10^{-4}	2.490977×10^{-10}
1 ft²	= 929.0304		0.09290304	$144.^*$	1	$0.7777777\ldots$	3.587007×10^{-8}
1 yd²	= 8361.273		0.8361273	1296.	$9.^*$	1	3.228306×10^{-7}
1 mi²	$= 2.589988 \times 10^{10}$		2.589988×10^6	4.014490×10^9	$2.78784 \times 10^{7*}$	3.0976×10^6	1

Appendix

Conversion factors for the U.S. Customary System, metric system, and International System (cont.)

C. Units of volume

Units	m^3	cm^3	liter	$in.^3$	ft^3	qt	gal
1 m^3 = 1	1	10^6	10^3	6.102374×10^4	35.31467×10^{-3}	1.056688	264.1721
1 cm^3 = 10^{-6}	10^{-6}	1	10^{-3}	0.06102374	3.531467×10^{-5}	1.056688×10^{-3}	2.641721×10^{-4}
1 liter = 10^{-3}	10^{-3}	1000.*	1	61.02374	0.03531467	1.056688	0.2641721
1 $in.^3$ = 1.638706×10^{-5}	1.638706×10^{-5}	16.38706*	0.01638706	1	5.787037×10^{-4}	0.01731602	4.329004×10^{-3}
1 ft^3 = 2.831685×10^{-2}	2.831685×10^{-2}	28316.85	28.31685	1728.*	1	2.992208	7.480520
1 qt = 9.46353×10^{-4}	9.46353×10^{-4}	946.353	0.946353	57.75	0.0342014	1	0.25
1 gal (U.S.) = 3.785412×10^{-3}	3.785412×10^{-3}	3785.412	3.785412	231.*	0.1336806	4.*	1

D. Units of mass

Units	g	kg	oz	lb	metric ton	ton
1 g = 1	1	10^{-3}	0.03527396	2.204623×10^{-3}	10^{-6}	1.102311×10^{-6}
1 kg = 1000.	1000.	1	35.27396	2.204623	10^{-3}	1.102311×10^{-3}
1 oz (avdp) = 28.34952	28.34952	0.02834952	1	0.0625	2.834952×10^{-5}	3.125×10^{-5}
1 lb (avdp) = 453.5924	453.5924	0.4535924	16.*	1	4.535924×10^{-4}	$5. \times 10^{-4}$
1 metric ton = 10^8	10^8	1000.*	35273.96	2204.623	1	1.102311
1 ton = 907184.7	907184.7	907.1847	32000.	2000.*	0.9071847	1

E. Units of density

Units	$g \cdot cm^{-3}$	$g \cdot L^{-1}, kg \cdot m^{-3}$	$oz \cdot in.^{-3}$	$lb \cdot in.^{-3}$	$lb \cdot ft^{-3}$	$lb \cdot gal^{-1}$
$1\ g \cdot cm^{-3}$ $= 1$		1000.	0.5780365	0.03612728	62.42795	8.345403
$1\ g \cdot L^{-1}, kg \cdot m^{-3}$ $= 10^{-3}$	1	5.780365×10^{-4}	3.612728×10^{-5}	0.06242795	8.345403×10^{-3}
$1\ oz \cdot in.^{-3}$ $= 1.729994$	1729.994	1	0.0625	108.	14.4375
$1\ lb \cdot in.^{-3}$ $= 27.67991$	27679.91	16.	1	1728.	231.
$1\ lb \cdot ft^{-3}$ $= 0.01601847$	16.01847	9.259259×10^{-3}	5.7870370×10^{-4}	1	0.1336806
$1\ lb \cdot gal^{-1}$ $= 0.1198264$	119.8264	4.749536×10^{-3}	4.3290043×10^{-3}	7.480519	1

F. Units of pressure

Units	$Pa,\ N\ m^{-2}$	$dyn \cdot cm^{-2}$	bar	atm	$kgf \cdot cm^{-2}$	$mmHg\ (torr)$	$in.\ Hg$	$lbf \cdot in.^{-2}$
$1\ Pa,\ 1\ N\ m^{-2}$ $= 1$	10	10^{-5}	9.869233×10^{-6}	1.019716×10^{-5}	7.500617×10^{-3}	2.952999×10^{-4}	1.450377×10^{-4}
$1\ dyn \cdot cm^{-2}$ $= 0.1$	1	10^{-6}	9.869233×10^{-7}	1.019716×10^{-6}	7.500617×10^{-4}	2.952999×10^{-5}	1.450377×10^{-5}
$1\ bar$ $= 10^{5*}$	10^6	1	0.9869233	1.019716	750.0617	29.52999	14.50377
$1\ atm$ $= 101325.0^*$	1013250	1.013250	1	1.033227	760.	29.92126	14.69595
$1\ kgf \cdot cm^{-2}$ $= 98066.5$	980665	0.980665	0.9678411	1	735.5592	28.95903	14.22334
$1\ mmHg\ (torr)$ $= 133.3224$	1333.224	1.333224×10^{-3}	1.3157895×10^{-3}	2.78784×10^{-3}	1	0.03937008	0.01933678
$1\ in.\ Hg$ $= 3386.388$	33863.88	0.03386388	0.03342105	0.03453155	25.4	1	0.4911541
$1\ lbf \cdot in.^{-2}$ $= 6894.757$	68947.57	0.06894757	0.06804596	0.07030696	51.71493	2.036021	1

Conversion factors for the U.S. Customary System, metric system, and International System (*cont.*)

G. Units of energy

Units	g mass (energy equiv)	J	eV	cal	cal$_{IT}$	Btu$_{IT}$	kWh	hp-h	ft-lbf	ft³·lbf·in^{-2}	liter-atm
1 g mass (energy equiv)	= 1	8.987552×10^{13}	5.609589×10^{32}	2.148076×10^{13}	2.146640×10^{13}	8.518555×10^{10}	2.499542×10^{7}	3.347918×10^{7}	6.628878×10^{13}	4.603388×10^{11}	8.870024×10^{11}
1 J	$= 1.112750 \times 10^{-14}$	1	6.241510×10^{18}	0.2390057	0.2388459	9.478172×10^{-4}	$2.777777... \times 10^{-7}$	3.725062×10^{-7}	0.7375622	5.121960×10^{-3}	9.869233×10^{-3}
1 eV	$= 1.782662 \times 10^{-33}$	1.602176×10^{-19}	1	3.829293×10^{-20}	3.826733×10^{-20}	1.518570×10^{-22}	4.450490×10^{-26}	5.968206×10^{-26}	1.181705×10^{-19}	8.206283×10^{-22}	1.581225×10^{-21}
1 cal	$= 4.655328 \times 10^{-14}$	4.184*	2.611448×10^{19}	1	0.9993312	3.965667×10^{-3}	$1.1622222... \times 10^{-6}$	1.558562×10^{-6}	3.085960	2.143028×10^{-2}	0.04129287
1 cal$_{IT}$	$= 4.658443 \times 10^{-14}$	4.1868*	2.613195×10^{19}	1.000669	1	3.968321×10^{-3}	1.163000×10^{-6}	1.559609×10^{-6}	3.088025	2.144462×10^{-2}	0.04132050
1 Btu$_{IT}$	$= 1.173908 \times 10^{-11}$	1055.056	6.585141×10^{21}	252.1644	251.9958	1	2.930711×10^{-4}	3.930148×10^{-4}	778.1693	5.403953	10.41259
1 kWh	$= 4.005540 \times 10^{-8}$	3600000.*	2.246944×10^{25}	860420.7	859845.2	3412.142	1	1.341022	2655224.	18349.06	35529.24
1 hp-h	$= 2.986931 \times 10^{-8}$	2384519.	1.675545×10^{25}	641615.6	641186.5	2544.33	0.7456998	1	1980000.*	13750.	26494.15
1 ft-lbf	$= 1.508551 \times 10^{-14}$	1.355818	8.462351×10^{18}	0.3240483	0.3238315	1.285067×10^{-3}	3.766161×10^{-7}	$5.050505... \times 10^{-7}$	1	$6.944444... \times 10^{-3}$	0.01338088
1 ft³ lbf·in^{-2}	$= 2.172313 \times 10^{-12}$	195.2378	1.218579×10^{21}	46.66295.	46.63174	0.1850497	5.423272×10^{-5}	$7.272727... \times 10^{-5}$	144.*	1	1.926847
1 liter-atm	$= 1.127393 \times 10^{-12}$	101.3250	6.324210×10^{20}	24.21726	24.20106	0.09603757	2.814583×10^{-5}	3.774419×10^{-5}	74.73349	0.5189825	1

*Numbers followed by an asterisk are definitions of the relation between the two units.

The chemical elements

Name	Symbol	At. no.	Name	Symbol	At. no.
Actinium	Ac	89	Manganese	Mn	25
Aluminum	Al	13	Meitnerium	Mt	109
Americium	Am	95	Mendelevium	Md	101
Antimony	Sb	51	Mercury	Hg	80
Argon	Ar	18	Molybdenum	Mo	42
Arsenic	As	33	Neodymium	Nd	60
Astatine	At	85	Neon	Ne	10
Barium	Ba	56	Neptunium	Np	93
Berkelium	Bk	97	Nickel	Ni	28
Beryllium	Be	4	Niobium	Nb	41
Bismuth	Bi	83	Nitrogen	N	7
Bohrium	Bh	107	Nobelium	No	102
Boron	B	5	Osmium	Os	76
Bromine	Br	35	Oxygen	O	8
Cadmium	Cd	48	Palladium	Pd	46
Calcium	Ca	20	Phosphorus	P	15
Californium	Cf	98	Platinum	Pt	78
Carbon	C	6	Plutonium	Pu	94
Cerium	Ce	58	Polonium	Po	84
Cesium	Cs	55	Potassium	K	19
Chlorine	Cl	17	Praseodymium	Pr	59
Chromium	Cr	24	Promethium	Pm	61
Cobalt	Co	27	Protactinium	Pa	91
Copper	Cu	29	Radium	Ra	88
Curium	Cm	96	Radon	Rn	86
Darmstadtium	Ds	110	Rhenium	Re	75
Dubnium	Db	105	Rhodium	Rh	45
Dysprosium	Dy	66	Rubidium	Rb	37
Einsteinium	Es	99	Ruthenium	Ru	44
Element 111*		111	Rutherfordium	Rf	104
Element 112*		112	Samarium	Sm	62
Erbium	Er	68	Scandium	Sc	21
Europium	Eu	63	Seaborgium	Sg	106
Fermium	Fm	100	Selenium	Se	34
Fluorine	F	9	Silicon	Si	14
Francium	Fr	87	Silver	Ag	47
Gadolinium	Gd	64	Sodium	Na	11
Gallium	Ga	31	Strontium	Sr	38
Germanium	Ge	32	Sulfur	S	16
Gold	Au	79	Tantalum	Ta	73
Hafnium	Hf	72	Technetium	Tc	43
Hassium	Hs	108	Tellurium	Te	52
Helium	He	2	Terbium	Tb	65
Holmium	Ho	67	Thallium	Tl	81
Hydrogen	H	1	Thorium	Th	90
Indium	In	49	Thulium	Tm	69
Iodine	I	53	Tin	Sn	50
Iridium	Ir	77	Titanium	Ti	22
Iron	Fe	26	Tungsten	W	74
Krypton	Kr	36	Uranium	U	92
Lanthanum	La	57	Vanadium	V	23
Lawrencium	Lr	103	Xenon	Xe	54
Lead	Pb	82	Ytterbium	Yb	70
Lithium	Li	3	Yttrium	Y	39
Lutetium	Lu	71	Zinc	Zn	30
Magnesium	Mg	12	Zirconium	Zr	40

*This element does not have an official name or symbol.

Appendix

Periodic table

(The atomic numbers are listed above the symbols identifying the elements. The heavy line separates metals from nonmetals.)

s			d										p				s
1	2	3	4	5	6	7	8	9	10	11	12	13	14	15	16	17	18
3 Li Lithium	4 Be Beryllium											5 B Boron	6 C Carbon	7 N Nitrogen	8 O Oxygen	9 F Fluorine	2 He Helium / 1 H Hydrogen
11 Na Sodium	12 Mg Magnesium											13 Al Aluminum	14 Si Silicon	15 P Phosphorus	16 S Sulfur	17 Cl Chlorine	10 Ne Neon
19 K Potassium	20 Ca Calcium	21 Sc Scandium	22 Ti Titanium	23 V Vanadium	24 Cr Chromium	25 Mn Manganese	26 Fe Iron	27 Co Cobalt	28 Ni Nickel	29 Cu Copper	30 Zn Zinc	31 Ga Gallium	32 Ge Germanium	33 As Arsenic	34 Se Selenium	35 Br Bromine	18 Ar Argon
37 Rb Rubidium	38 Sr Strontium	39 Y Yttrium	40 Zr Zirconium	41 Nb Niobium	42 Mo Molybdenum	43 Tc Technetium	44 Ru Ruthenium	45 Rh Rhodium	46 Pd Palladium	47 Ag Silver	48 Cd Cadmium	49 In Indium	50 Sn Tin	51 Sb Antimony	52 Te Tellurium	53 I Iodine	36 Kr Krypton
55 Cs Cesium	56 Ba Barium	71 Lu Lutetium	72 Hf Hafnium	73 Ta Tantalum	74 W Tungsten	75 Re Rhenium	76 Os Osmium	77 Ir Iridium	78 Pt Platinum	79 Au Gold	80 Hg Mercury	81 Tl Thallium	82 Pb Lead	83 Bi Bismuth	84 Po Polonium	85 At Astatine	54 Xe Xenon
87 Fr Francium	88 Ra Radium	103 Lr Lawrencium	104 Rf Rutherfordium	105 Db Dubnium	106 Sg Seaborgium	107 Bh Bohrium	108 Hs Hassium	109 Mt Meitnerium	110 Ds Darmstadtium	111	112	113	114	115	116	117	86 Rn Radon
																	118

f

57 La Lanthanum	58 Ce Cerium	59 Pr Praseodymium	60 Nd Neodymium	61 Pm Promethium	62 Sm Samarium	63 Eu Europium	64 Gd Gadolinium	65 Tb Terbium	66 Dy Dysprosium	67 Ho Holmium	68 Er Erbium	69 Tm Thulium	70 Yb Ytterbium
89 Ac Actinium	90 Th Thorium	91 Pa Protactinium	92 U Uranium	93 Np Neptunium	94 Pu Plutonium	95 Am Americium	96 Cm Curium	97 Bk Berkelium	98 Cf Californium	99 Es Einsteinium	100 Fm Fermium	101 Md Mendelevium	102 No Nobelium

480

Classification of living organisms

Domain Archaea[a]
- Phylum Crenarchaeota
  - Class Thermoprotei
    - Order Thermoproteales
    - Order Desulfurococcales
    - Order Sulfolobales
- Phylum Euryarchaeota
  - Class Methanobacteria
    - Order Methanobacteriales
  - Class Methanoccoci
    - Order Methanococcales
    - Order Methanomicrobiales
    - Order Methanosarcinales
  - Class Halobacteria
    - Order Halobacteriales
  - Class Thermoplasmata
    - Order Thermoplasmatales
  - Class Thermococci
    - Order Thermococcales
  - Class Archaeoglobi
  - Class Methanopyrl
    - Order Methanopyrales

Domain Bacteria
- Phylium Aquificae
  - Class Aquificae
    - Order Aquificales
- Phylum Thermotogae
  - Class Thermotogae
    - Order Thermotogales
- Phylum Thermodesulfobacteria
  - Class Thermodesulfobacteria
    - Order Thermodesulfobacteriales
- Phylum Deinococcus-Thermus
  - Class Deinococci
    - Order Deinococcales
    - Order Thermales
- Phylum Chryslogenetes
  - Class Chrysiogenetes
    - Order Chrysiogenales
- Phylum Chloroflexi
  - Class Chloroflexi
    - Order Chloroflexales
    - Order Herpetosiphonales
- Phylum Thermomicrobia
  - Class Thermomicrobia
    - Order Thermomicrobiales
- Phylum Nitrospira
  - Class Nitrospira
    - Order Nitrospirales

- Phylum Deferribacteres
  - Class Deferribacteres
    - Order Deferribacterales
- Phylum Cyanobacteria
  - Class Cyanobacteria
- Phylum Chlorobi
  - Class Chlorobia
    - Order Chlorobiales
- Phylum Proteobacteria
  - Class Alphaproteobacteria
    - Order Rhodospirillales
    - Order Rickettsiales
    - Order Rhodobacterales
    - Order Sphingomonadales
    - Order Caulobacterales
    - Order Rhizobiales
  - Class Betaproteobacteria
    - Order Burkholderiales
    - Order Hydrogenophilales
    - Order Methylophilales
    - Order Neisseriales
    - Order Nitrosomonadales
    - Order Rhodocyclales
  - Class Cammaproteobacteria
    - Order Chromatiales
    - Order Acidithiobacillales
    - Order Xanthomonadales
    - Order Cardiobacteriales
    - Order Thiotrichales
    - Order Legionellals
    - Order Methylococcales
    - Order Oceanospirillales
    - Order Pseudomonadales
    - Order Alteromonadales
    - Order Vibrionales
    - Order Aeromonadales
    - Order Enterobacteriales
    - Order Pasteurellales
  - Class Deltaproteobacteria
    - Order Desulfurellales
    - Order Desulfovibrionales
    - Order Desulfobacterales
    - Order Desulfuromonadales
    - Order Syntrophobacterales
    - Order Bdellovibrionales
    - Order Myxococcales
  - Class Epsilonproteobacteria
    - Order Campylobacterales
- Phylum Firmicutes

- Class Clostridia
  - Order Clostridiales
  - Order Thermoanaerobacteriales
  - Order Haloanaerobiales
- Class Mollicutes
  - Order Mycoplasmatales
  - Order Entomoplasmatales
  - Order Acholeplasmatales
  - Order Anaeroplasmatales
- Class Bacilli
  - Order Bacillales
  - Order Lactobacillales
- Phylum Actinobacteria
  - Class Actinobacteria
    - Subclass Acidimicrobidae
      - Order Acidimicrobiales
      - Suborder Acidimicrobineae
    - Subclass Rubrobacteridae
      - Order Rubrobacterles
      - Suborder Rubrobacterineae
    - Subclass Coriobacteridae
      - Order Coriobacteriales
      - Suborder Cariobacterineae
    - Subclass Sphaerobacteridae
      - Order Sphaeriobacteriales
      - Suborder Sphaerobacterineae
    - Subclass Actinobacteridae
      - Order Actinomyietales
      - Suborder Actiomycineae
      - Suborder Micrococcineae
      - Suborder Corynebacterineae
      - Suborder Micromonosporineae
      - Suborder Propionibacterineae
      - Suborder Pseudonocardineae
      - Suborder Streptomycineae
      - Suborder Streptosporangineae
      - Suborder Frankineae
      - Suborder Glycomycineae
      - Order Bifidobacteriales
- Phylum Planctomycetes

Appendix

Classification of living organisms (*cont.*)

Class Planctomycetacia
  Order Planctomycetales
Phylum Chlamydiae
  Class Chlamydiae
    Order Chlamydiales
Phylum Spirochaetes
  Class Spirochaetes
    Order Spirochaetales
Phylum Fibrobacteres
  Class Fibrobacteres
    Order Fibrobacterales
Phylum Acidobacteria
  Class Acidobacteria
    Order Acidobacteriales
Phylum Bacteroidetes
  Class Bacteroidetes
    Order Bacteroidales
  Class Flavobacteria
    Order Flavobacteriales
  Class Sphingobacteria
    Order Sphingobacteriales
Phylum Fusobacteria
  Class Fusobacteria
    Order Fusobacteriales
Phylum Verrucomicrobia
  Class Verrucomicrobiae
    Order Verrucomicrobiales
Phylum Dictyoglomus
  Class Dictyoglomi
    Order Dictyoglomales

Domain Eukarya[b]

Kingdom Protista
  Phylum Metamonada
  Phylum Trichozoa
    Subphylum Parabasala
      Class Trichomonadea
      Class Hypermastigotea

Subkingdom Neozoa
  Phylum Choanozoa
  Phylum Amoebozoa
    Subphylum Lobosa
    Subphylum Conosa
      Class Archamoebae
      Class Mycetozoa
  Phylum Foraminifera
  Phylum Percolozoa
  Phylum Euglenozoa
    Class Euglenoidea

Class Saccostomae
Phylum Sporozoa
  Subphylum Gregarinae
  Subphylum Coccidiomorpha
  Subphylum Perkinsida
  Subphylum Manubrispora
Phylum Ciliophora
Phylum Radiozoa
Phylum Heliozoa
Phylum Rhodophyta
  Class Rhodophyceae
    Subclass Banglophycidae
      Order Bangiales
      Order Compsopogonales
      Order Porphyridiales
      Order Rhodochaetales
    Subclass Florideophycidae
      Order Acrochaetiales
      Order Ahnfeltiales
      Order Balbianiales
      Order Balliales
      Order Batrachospermales
      Order Bonnemaisoniales
      Order Ceramiales
      Order Colaconematales
      Order Corallinales
      Order Gelidiales
      Order Gigartinales
      Order Gracilariales
      Order Halymeniales
      Order Hildenbrandiales
      Order Nemaliales
      Order Palmariales
      Order Plocamiales
      Order Rhodogorgonales
      Order Rhodymeniales
      Order Thoreales
Phylum Chrysophyta
  Class Bacillariophyceae
    Subclass Bacillariophycidae
      Order Achnanthales
      Order Bacillariales
      Order Cymbellales
      Order Dictyoneidales
      Order Lyrellales
      Order Mastogloiales
      Order Naviculales
      Order Rhopalodiales
      Order Surirellales
      Order Thallassiophysales
    Subclass Biddulphiophycidae

Order Anaulales
Order Biddulphiales
Order Hemlaulales
Order Triceratiales
Subclass Chaetocerotophycidae
  Order Chaetocerotales
  Order Leptocylindrales
Subclass Corethrophycidae
  Order Cymatosirales
Subclass Coscinodiscophycidae
  Order Arachnoidiscales
  Order Asterolamprales
  Order Aulacoseirales
  Order Chrysaanthemodiscales
  Order Coscinodiscales
  Order Ethmodiscales
  Order Melosirales
  Order Orthoseirales
  Order Parallales
  Order Stictocyclales
  Order Stictodiscales
Subclass Cymatosirophycidae
  Order Cymatosirales
Subclass Eunotiophycidae
  Order Eunotiales
Subclass Fragilariophycidae
  Order Ardissoneales
  Order Cyclophorales
  Order Climacospheniales
  Order Fragllariales
  Order Licmorphorales
  Order Protoraphidales
  Order Rhabdonematales
  Order Rhaphoneidales
  Order Striatellales
  Order Tabellariales
  Order Thalassionematales
  Order Toxariales
Subclass Lithodesmiophycidae
  Order Lithodesmialescidae
Subclass Rhizosoleniophycidae
  Order Rhizosoleniales
Subclass Thalassiosirophycidae
  Order Thalassiosirales
Class Bolidophyceae
  Order Bolidomonadales
Class Chrysomerophyceae
  Order Chrysomeridales
    nom. nud.
Class Chrysophyceae
  Order Chromulinales

Classification of living organisms (*cont.*)

Order Hibberdiales
Class Dictyochophyceae
  Order Dictyochales
  Order Pedinellales
  Order Rhizochromulinales
Class Eustigmatophyceae
  Order Eustigmatales
Class Pelagophyceae
  Order Pelagomonadales
  Order Sarcinochrysidales
Class Phaeophyceae
  Order Ascoseirales
  Order Chordariales
  Order Cutleriales
  Order Desmarestiales
  Order Dictysiphonales
  Order Dictyotales
  Order Durvillaeales
  Order Ectocarpales
  Order Fucales
  Order Laminariales
  Order Scytosiphonales
  Order Sphacelariales
  Order Sporochnales
  Order Tilopteridiales
Class Phaeothamniophyceae
  Order Phaeothamniales
  Order Pleurochloridellales
Class Pinguiophyceae
  Order Pinguiochrysidales
Class Raphidophyceae
  Order Rhaphidomonadales
Class Synurophyceae
  Order Synurales
Class Xanthophyceae
  (=Tribophyceae)
  Order Botrydiales
  Order Chloramoebales
  Order Heterogloeales
  Order Mischococcales
  Order Rhizochloridales
  Order Tribonematales
  Order Vaucheriales
Phylum Cryptophyta
Class Cryptophyceae
  Order Cryptomonadales
  Order Cryptococcales
Phylum Glaucocystophyta
Class Glaucocystophyceae
  Order Cyanophorales
  Order Glaucocystales

Order Gloeochaetales
Phylum Prymnesiophyta
  (=Haptophyta)
Class Pavlovophyceae
  Order Pavlovales
Class Prymnesiophyceae
  Order Coccolithales
  Order Isochrysidales
  Order Phaeocystales
  Order Prymneslales
Phylum Dinophyta
Class Dinophyceae
  Order Actiniscales
  Order Blastodiniales
  Order Chytriodiniales
  Order Desmocapsales
  Order Desmomonadales
  Order Dinophysales
  Order Gonyaulacales
  Order Gymnodiniales
  Order Kokwitziellaless
  Order Nannoceratopslales
  Order Noctilucales
  Order Oxyrthinales
  Order Peridiniales
  Order Phytodiniales
  Order Prorocentrales
  Order Ptychodiscales
  Order Pyrocysales
  Order Suessiales
  Order Syndiniales
  Order Thoracosphaerales
Phylum Chlorophyta
Class Charophyceae
  Order Charales
  Order Chlorokybaees
  Order Coleochaetales
  Order Klebsormidiales
  Order Zygnematales
Class Chlorophyceae
  Order Chaetophorales
  Order Chlorococcales
  Order Cladophorales
  Order Odeogoniales
  Order Sphaeropleales
  Order Volvocales
  Order Pleurastrales
Class Prasinophyceae
  Order Chlorodendrales
  Order Mamiellales
  Order Pseudoscourfeldiales

Order Pyramimonidales
Class Trebouxiophyceae
  Order Trebouxiales
Class Ulvophyceae
  Order Bryopsidales
  Order Caulerpales
  Order Codiolales
  Order Dasycladales
  Order Halimedales
  Order Prasioeales
  Order Siphonocladales
  Order Trentepohliales
  Order Ulotrichales
  Order Ulvales
Phylum Euglenophyta
Class Euglenophyceae
  Order Euglenales
  Order Euglenamorphales
  Order Eutreptiales
  Order Heteronematales
  Order Rhabdomonadales
  Order Sphenomonadales
Phylum Acrasiomycota
Class Acrasiomycetes
  Order Acrasiales
Phylum Dictyosteliomycota
Class Dictyosteliomycetes
  Order Dictyosteliales
Phylum Myxomycota
Class Myxomycetes
  Order Liceales
  Order Echinosteliales
  Order Trichiales
  Order Physarales
  Order Stemonitales
  Order Ceratiomyxales
Class Protosteliomycetes
  Order Protosteliales
Phylum Plasmodiophoromycota
Class Plasmodiophoromy-
  cetes
  Order Plasmodiophorales
Phylum Oomycota
Class Oomycetes
  Order Saprolegniales
  Order Salilagenidiales
  Order Leptomitales
  Order Myzocytiopsidales
  Order Rhipidiales
  Order Pythiales
  Order Peronosporales

Appendix

Classification of living organisms (*cont.*)

Phylum Hyphochytriomycota
Class Hyphochytriomycetes
Order Hyphochytriales
Phylum Labyrinthulomycota
Class Labyrinthulomycetes
Order Labyrinthulales
Phylum Chytridiomycota
Class Chytridiomycetes
Order Blastocladiales
Order Chytridiales
Order Monoblepharidales
Order Neocallimastigales
Order Spizellomycetales
Phylum Zygomycota
Class Trichomycetes
Order Amoebidiales
Order Asellariales
Order Eccrinales
Order Harpellales
Class Zygomycetes
Order Mucorales
Order Dimargaritales
Order Kickxellales
Order Endogonales
Order Glomales
Order Entomophthorales
Order Zoopagales
Phylum Ascomycota
Class Archiascomycetes
Order Taphrinales
Order Schizosaccharomyce-
tales
Class Saccharomycetes
Order Saccharomycetales
Class Plectomycetes
Order Eurotiales
Order Ascosphaerales
Order Onygenales
Class Laboulbeniomycetes
Order Laboulbeniales
Order Spathulosporales
Class Pyrenomycetes
Order Hypocreales
Order Melanosporales
Order Microascales
Order Phylachorales
Order Ophiostomatales
Order Diaporthales
Order Calosphaeriales
Order Xylariales
Order Sordariales

Order Meliolales
Order Halosphaeriales
Class Discomycetes
Order Medeolarlales
Order Rhytismatales
Order Ostropales
Order Cyttariales.
Order Helotiales
Order Neolectales
Order Gyalectales
Order Lecanorales
Order Lichinales
Order Peltigerales
Order Pertusariales
Order Teloschistales
Order Caliciales
Order Pezizales
Class Loculoascomycetes
Order Coryneliales
Order Dothideales
Order Myriangiales
Order Arthoniales
Order Pyrenulales
Order Asterinales
Order Capnodiales
Order Chaetothyriales
Order Patellariales
Order Pleosporales
Order Melanommatales
Order Trichotheliales
Order Verrucariales
Phylum Basidiomycota
Class Basidiomycetes
Subclass Heterobasidiomy-
cetes
Order Agricostibales
Order Atractiellales
Order Auriculariales
Order Heterogastridiales
Order Tremellales
Subclass Homobasidiomycetes
Order Agaricales
Order Boletales
Order Bondarzewiales
Order Cantharellales
Order Ceratobasidiales
Order Cortinariales
Order Dacrymycetales
Order Fistulinales
Order Ganodermatales
Order Gautieriales

Order Gomphales
Order Hericiales
Order Hymenoghaetales
Order Hymenogastrales
Order Lachnocladiales
Order Lycoperdales
Order Melanogastrales
Order Nidulariales
Order Phallales
Order Poriales
Order Russulales
Order Schizophyllales
Order Sclerodermatales
Order Stereales
Order Thelephorales
Order Tulasnellales
Order Tulostomatales
Class Ustomycetes
Order Cryptobasidiales
Order Cryptomycocola-
cales
Order Exobasidiales
Order Graphiolales
Order Platyglocales
Order Sporidiales
Order Ustilaginales
Class Tellomycetes
Order Septobasidiales
Order Uredinales
Phylum Deuteromycetes
(Asexual Ascomycetes
and Basidiomycetes)
Class Hyphomycetes
Order Hyphomycetales
Order Stibeilales
Order Tuberculariales
Class Agonomycetes
Order Agonomycetales
Class Coelomycetes
Order Melanconiales
Order Sphaeropsidales
Order Pycnothyriales

Kingdom Plantae

Subkingdom Embryobionta
Division Hepaticophyta
Class Junermanniopsida
Order Calobryales
Order Jungermanniales
Order Metzgeriales

Classification of living organisms (*cont.*)

Class Marchantiopsida
  Order Sphaerocarpales
  Order Monocleales
  Order Marchantiales
Division Anthocerotophyta
  Class Anthocerotopsida
    Order Anthocerotales
Division Bryophyta
  Class Sphagnicopsida
    Order Sphagnicales
  Class Andreaeopsida
    Order Andreaeles
  Class Bryopsida
    Order Archidiales
    Order Bryales
    Order Buxbaumiales
    Order Dicranales
    Order Encalyptales
    Order Fissidentales
    Order Funariales
    Order Grimmiales
    Order Hookeriales
    Order Hypnobryales
    Order Isobryales
    Order Orthotrichales
    Order Pottiales
    Order Orthotrichales
    Order Seligerales
    Order Splachnales
Division Lycophyta
  Class Lycopsida
    Order Isoetales
    Order Lycopodiales
    Order Selaginellales
Division Polypodiophyta
  Class Polypodopsida
    Order Equisetales
    Order Marattiales
    Order Ophioglossales
    Order Polypodiales
    Order Psilotales
Division Pinophyta
  Class Ginkgopsida
    Order Ginkgoales
  Class Cycadopsida
    Order Cycadales
  Class Pinopsida
    Order Pinales
    Order Podocarpales
    Order Gnetales
Division Magnoliophyta

|unplaced orders|
  Order Ceratophyllales
  Order Chloranthales
Class Amborellopsida
  Order Amborellales
Class Austrobaileyales
  Order Austrobaileyales
Class Liliopsida
  Order Acorales
  Order Alismatales
  Order Arecales
  Order Asparagales
  Order Commelinales
  Order Dioscoreales
  Order Liliales
  Order Pandanales
  Order Poales
  Order Zingiberales
Class Magnoliopsida
  Order Magnoliales
  Order Laurales
  Order Piperales
  Order Canellales
Class Nymphaeopsida
  Order Nymphaeales
Class Rosopsida
  |unplaced orders|
  Order Berberidopsidales
  Order Buxales
  Order Gunnerales
  Order Proteales
  Order Saxifragales
  Order Santalales
  Order Trochodendrales
  Subclass Caryophyllidae
    Order Caryophyliales
    Order Dilleniales
  Subclass Ranunculidae
    Order Ranunculales
  Subclass Rosidae
    |unplaced orders|
    Order Crossosomatales
    Order Geraniales
    Order Myrtales
    Order Vitales
    Superorder Rosanae
    Order Celastrales
    Order Cucurbitales
    Order Fabales
    Order Fagales
    Order Malpighiales

    Order Oxalidales
    Order Rosales
    Order Zygophyllales
    Superorder Malvanae
    Order Brassicales
    Order Malvales
    Order Sapindales
  Subclass Asteridae
    |unplaced order|
    Order Boraginales
    Superorder Cornanae
    Order Cornales
    Superorder Ericanae
    Order Ericles
    Superorder Lamianae
    Order Garryales
    Order Gentianales
    Order Lamiales
    Order Solanales
    Superorder Asteranae
    Order Apiales
    Order Aquifoliales
    Order Asterales
    Order Dipsacales

Kingdom Animalia

Subkingdom Parazoa
  Phylum Porifera
    Subphylum Cellularia
      Class Demosponglae
      Class Calcarea
    Subphylum Symplasma
      Class Hexactinellida
  Phylum Placozoa

Subkingdom Eumetazoa
  Phylum Cnidaria
    (=Coelenterata)
    Class Scyphozoa
      Order Stauromedusae
      Order Coronatae
      Order Semaeostomeae
      Order Rhizostomeae
    Class Cubozoa
      Order Cubomedusae
    Class Hydrozoa
      Order Hydroida
      Order Milleporina
      Order Stylasterina
      Order Trachylina

Appendix

Classification of living organisms (*cont.*)

Order Siphonophora
Order Chondrophora
Order Actinulida
Class Anthozoa
Subclass Alcyonaria
(=Octocorallia)
Order Stolonifera
Order Gorgonacea
Order Alcyonacea
Order Pennatulacea
Subclass Zoantharia
(=Hexacorallia)
Order Actinaria
Order Corallimorpharia
Order Scleractinia
Order Zoanthinaria
(=Zoanthidea)
Order Ceriantharia
Order Ptychodactiaria
Order Antipatharia
Phylum Ctenophora
Class Tentaculata
Order Cydippida
Order Platyctenida
Order Lobata
Order Cestida
Order Ganeshida
Order Thalassocalycida
Class Nuda
Order Beroida
Phylum Platyhelminthes
Class Turbellaria
Order Acoela
Order Rhabdocoela
Order Catenullda
Order Macrostomida
Order Nemertodermatida
Order Lecithoepitheliata
Order Polycladida
Order Prolecithophora
(=Holocoela)
Order Proseriata
Order Tricladida
Order Neorhabdocoela
Class Cestoda
Subclass Cestodaria
Subclass Eucestoda
Order Caryophyllidea
Order Spathebothriidea
Order Trypanorhyncha
Order Pseudophyllidea

Order Tetraphyllidea
Order Cyclophyllidea
Class Monogenea
Class Trematoda
Subclass Digenea
Order Strigeidida
Order Azygiida
Order Echinostomida
Order Plagiorchiida
Order Opisthorchiida
Subclass Aspidogastrea
(=Aspidobothrea)
Phylum Mesozoa
Class Orthonectida
Class Rhombozoa
Order Dicyemida
Order Heterocyemida
Phylum Myxozoa
(=Myxospora)
Phylum Nemertea
(=Rhynchocoela,
Nemertinea)
Class Anopia
Order Palaeonemertea
(=Palaeonemertini)
Order Heteronemertea
Class Enopia
Order Hoplonemertea
(=Hoplonemertini)
Order Bdelionemertea
Phylum Gnathostomuilda
Order Filospermoidea
Order Bursovaginoidea
Phylum Gastrotricha
Order Chaetonotida
Order Macrodasyida
Phylum Cycliophora
Phylum Rotifera
Class Monogononta
Order Ploima
Order Flosculariaceae
Order Collothecaceae
Class Bdelloidea
Class Seisonidea
Phylum Acanthocephaia
Class Archiacanthocephaia
Class Eoacanthocephaia
Class Palaeacanthocephala
Phylum Nematoda (=Nemata)
Class Adenophorea
Subclass Enoplia

Order Enoplida
Order Dorylaimida
Order Trichocephalida
Order Mermithida
Subclass Chromadoria
Class Secernentea
Subclass Rhabditia
Order Rhabditida
Order Ascaridida
Order Strongylida
Subclass Spiruria
Order Spirurida
Order Camallanida
Subclass Diplogasteria
Phylum Nematomorpha
Class Nectonematoida
Class Gordioida
Phylum Priapulida
Phylum Kinorhyncha
(=Echinoderida)
Class Cyclorhagida
Class Homalorhagida
Phylum Loricifera
Phylum Mollusca
Subphylum Aculifera
Class Polyplacophora
Class Aplacophora
Subclass Neomeniophora
(=Solenogastres)
Subclass Chaetodermomor-
pha
(=Caudofoveata)
Subphylum Conchifera
Class Monoplacophora
Class Gastropoda
Subclass Prosobranchia
Order Archaeogastropoda
Order Mesogastropoda
(=Taenioglossa)
Order Neogastropoda
Subclass Opisthobranchia
Order Cephalaspidea
Order Runcinoidea
Order Acochlidioidea
Order Sacoglossa
(=Ascoglossa)
Order Anaspidea
(=Aplysiacea)
Order Notaspidea
Order Thecosomata
Order Gymnosomata

Classification of living organisms (*cont.*)

Order Nudibranchia
Subclass Pulmonata
Order Archaeopulmonata
Order Basommatophora
Order Stylommatophora
Order Systellommato-
phora
Class Bivalvia
(=Pelecypoda)
Subclass Protobranchia
(=Palaeotaxodonta,
Cryptodonta)
Subclass Pteriomorphia
Subclass Paleoheterodonta
Subclass Heterodonta
Subclass Anomalodesmata
Class Scaphopoda
Class Cephalopoda
Subclass Nautiloidea
Subclass Coleoidea
(=Dibranchiata)
Order Sepioidea
Order Teuthoidea
(=Decapoda)
Order Vampyromorpha
Order Octopoda
Phylum Annelida
Class Polychaeta
Order Phyllodocida
Order Spintherida
Order Eunicida
Order Spionida
Order Chaetopterida
Order Magelonida
Order Psammodrilida
Order Cirratulida
Order Flabelligerida
Order Ophelida
Order Capitellida
Order Owenilda
Order Terebellida
Order Sabellida
Order Protodrilida
Order Myzostomida
Class Clitellata
Subclass Oligochaeta
Order Lumbriculida
Order Haplotaxida
Subclass Hirudinea
Order Rhynchobdeilae
Order Arhynchobdellae

Order Branchiobdellida
Order Acanthobdellida
Class Pogonophora
(=Siboglinidae)
Subclass Perviata
(=Frenulata)
Subclass Obturata
(=Vestimentifera)
Class Echiura
Order Echiura
Order Xenopneusta
Order Heteromyota
Phylum Sipuncula
Phylum Arthropoda
Subphylum Chelicerata
Class Merostomata
Order Xiphosura
Class Arachnida
Order Scorpiones
Order Uropygi
Order Amblypygi
Order Araneae
Order Ricinulei
Order Pseudoscorpiones
Order Solifugae
(=Solpugida)
Order Opiliones
Order Acari
Class Pycnogonida
(=Pantopoda)
Subphylum Mandibulata
Class Myriapoda
Order Chilopoda
Order Dipiopoda
Order Symphyia
Order Pauropoda
Class insecta (=Hexapoda)
Subclass Apterygota
Order Thysanura
Order Collembola
Subclass Pterygota
Superorder Hemime-
tabola
Order Ephemeroptera
Order Odonata
Order Blattaria
Order Mantodea
Order isoptera
Order Grylioblattaria
Order Orthoptera
Order Phasmida

(=Phasmatoptera)
Order Dermaptera
Order Embiidina
Order Plecoptera
Order Psocoptera
Order Anoplura
Order Mallophaga
Order Thysanoptera
Order Hemiptera
Order Homoptera
Superorder Holometabola
Order Neuroptera
Order Coleoptera
Order Strepsiptera
Order Mecoptera
Order Siphonaptera
Order Diptera
Order Trichoptera
Order Lepidoptera
Order Hymenoptera
Class Crustacea
Subclass Cephalocarida
Subclass Malacostraca
Superorder Syncarida
Superorder Hoplocarida
Order Stomatopoda
Superorder Peracarida
Order Thermosbaenacea
Order Mysidacea
Order Cumacea
Order Tanaidacea
Order Isopoda
Order Amphipoda
Superorder Eucarida
Order Euphausiacea
Order Decapoda
Subclass Branchiopoda
Order Notostraca
Order Cladocera
Order Conchostraca
Order Anostraca
Subclass Ostracoda
Order Myodocopa
Order Podocopa
Subclass Mystacocarida
Subclass Copepoda
Order Calanoida
Order Harpacticoida
Order Cyclopoida
Order Monstrilloida
Order Siphonostomatoida

Appendix

Classification of living organisms (*cont.*)

Order Poecilostomatoida
Subclass Branchiura
Subclass Pentastomida
Order Cephalobaenida
Order Porocephalida
Subclass Tantulocarida
Subclass Remipedia
Subclass Cirripedia
Order Acrothoracica
Order Ascothoracica
Order Thoracica
Order Rhizocephala
Phylum Tardigrada
Class Heterotardigrada
Class Mesotardigrada
Class Eutardigrada
Order Parachela
Order Apochela
Phylum Onychophora
Phylum Phoronida
Phylum Brachiopoda
Class inarticulata
Order Lingulida
Order Acrotretida
Class Articulata
Order Rhynchonellida
Order Terebratulida
Phylum Bryozoa
(=Ectoprocta, polyzoa)
Class Phylactolaemata
Class Stenolaemata
Class Gymnolaemata
Order Ctenostomata
Order Chellostomata
Phylum Entoprocta
(=Kamptozoa)
Phylum Chaetognatha
Class Sagittoidea
Order Phragmophora
Order Aphragmophora
Phylum Echinodermata
Subphylum Crinozoa
Class Crinoidea
Order Millericrinida
Order Cyrtocrinida
Order Bourgueticrinida
Order Isocrinida
Order Comatulida
Subphylum Asterozoa
Class Stelleroidea
Subclass Somasteroidea

Subclass Ophiuroidea
Order Phrynophiurida
Order Ophiurida
Subclass Asteroidea
Order Platyasterida
Order Paxillosida
Order Valvatida
Order Spinulosida
Order Forcipulata
Order Brisingida
Class Concentricycloidea
Subphylum Echinozoa
Class Echinoidea
Order Cidaroida
Order Echinothuroida
Order Diadematoida
Order Arbacioida
Order Temnopleuroida
Order Echinoida
Order Holectypoida
Order Clypeasteroida
Order Spatangoida
Class Holothuroidea
Order Dendrochirotida
Order Aspidochirotida
Order Elasipodida
Order Apodida
Order Molpadiida
Phylum Hemichordata
Class Enteropneusta
Class Pterobranchia
Phylum Chordata
Subphylum Urochordata
(=Tunicata)
Class Ascidiacea
Order Aspiousobranchia
Order Phlebobranchia
Order Stolidobranchia
Class Larvacea
(=Appendicularia)
Class Thaliacea
Order Pyrosomida
Order Doliolida
Order Salpida
Subphylum Cephalochordata
(=Acrania)
Phylum Chordata[c]
Subphylum Vertebrata
Superclass Agnatha
Class Myxini
Order Myxiniformes

Class Cephalaspidomorphi
Order Petromyzontiformes
Superclass Gnathostomata
Class Chondrichthyes
Subclass Holocephali
Order Chimaeriformes
Subclass Elasmobranchii
Order Hexanchiformes
Order Squaliformes
Order Pristiophoriformes
Order Squatiniformes
Order Pristiformes
Order Rhinobatiformes
Order Torpediniformes
Order Myliobatiformes
Order Heterodontiformes
Order Orectolobiformes
Order Lamniformes
Order Carchiniformes
Class Sarcopterygii
Subclass Coelacanthimor-
morpha
Order Coelacanthiformes
Subclass Porolepimorpha
and Dipnol
Order Ceratodontiformes
Order Lepidosireniformes
Class Actinopterygii
Subclass Chondrostei
Order Polypteriformes
Order Acipenseriformes
Subclass Neopterygli
Order Semionotiformes
Order Amiiformes
Division Teiestei
Subdivision Osteoglosso-
morpha
Order Osteoglossiformes
Subdivision Elopomorpha
Order Elopiformes
Order Albuliformes
Order Anguilliformes
Order Saccopharyngi-
formes
Subdivision Clupeomorpha
Order Clupeiformes
Subdivision Euteleostei
Superorder Ostariophysi
Order Gonorthynchiformes
Order Cypriniformes
Order Characiformes

Classification of living organisms (*cont.*)

Order Siluriformes
Order Gymnotiformes
Superorder Protacanthop-
terygii
Order Esociformes
Order Osmeriformes
Order Salmoniformes
Superorder Stenopterygii
Order Stomiformes
Order Ateleopodiformes
Superorder Cyclosqua-
mata
Order Aulopiformes
Superoder Scopelo-
morpha
Order Myctophiformes
Superorder Lampridio-
morpha
Order Lampridiformes
Superorder Polymixio-
morpha
Order Polymixiiformes
Superorder Paracanthop-
terygii
Order Percopsiformes
Order Ophidiiformes
Order Gadiformes
Order Batrachoidiformes
Order Lophiiformes
Superorder Acanthop-
terygil
Order Mugiliformes
Order Atherinomorpha
Order Beloniformes
Order Cyprinodontiformes
Order Stephanoberyci-
formes
Order Beryciformes
Order Zeiformes

Order Gasterosteiformes
Order Synbranchiformes
Order Scorpeaniformes
Order Perciformes
Order Pleurnectiformes
Order Tetraodontiformes
Class Amphibia
Subclass Lissamphibia
Order Gymnophiona
Order Caudata (Urodela)
Order Anura–frogs and
toads
Class Reptilia
Subclass Anapsida
Order Testudines
Subclass Diapsida
Order Sphenodonta
Order Squamata
Suborder Lacertilia
Suborder Serpentes
Order Crocodylia
Infraclass Eoaves
Order Struthioniformes
Order Tinamiformes
Infraclass Neoaves
Order Craciformes
Order Galliformes
Order Anseriformes
Order Turniciformes
Order Piciformes
Order Galbuliformes
Order Bucerotiformes
Order Upupiformes
Order Trogoniformes
Order Coraciiformes
Order Coliiformes
Order Cuculiformes
Order Psittaciformes
Order Apodiformes

Order Trochiliformes
Order Musophagiformes
Order Strigiformes
Order Columbiformes
Order Grulformes
Order Ciconliformes
Suborder Charadrii
Suborder Ciconii
Order Passeriformes
Class Mammalia (Synapsida)
Order Monotremata
Order Didelophimorphia
Order Paucituberculata
Order Microbiotheria
Order Dasyuromorphia
Order Peramelemorphia
Order Notoryctemorphia
Order Diprotodontia
Order Xenarthra
Order insectivora
Order Scandentia
Order Dermoptera
Order Chiroptera
Order Primates
Order Carnivora
Order Cetacea
Order Sirenia
Order Proboscidea
Order Perissodactyla
Order Hyracoidea
Order Tubulidentata
Order Artiodactyla
Order Pholidota
Order Rodentia
Suborder Sciurognathi
Suborder Hystricognathi
Order Lagomorpha
Order Macroscelidea

[a]Derived from G. M. Garrity et. al., *Taxonomic Outline of the Prokaryotes* Release 2, January 2002, Springer-Verlag. New York. http://dx.doi.org/10.1007/bergeysouthline. Readers interested in determning taxonomic composition of lower taxa may obtain this document, free of charge.

[b]Condensed from Jan A. Pechnik, *Biology of the Inverlebrates*, 4th ed., McGraw-Hill, 2000.

[c]Condensed from Donald Linzey. *Vertebrate Biology*. Appendix 1: Classification of Living Vertebrates, McGraw-Hill, 2001.

Note: The contributions of the following to the updating of this classification scheme are gratefully acknowledged: Dr. Craig Balley: Dr. Mark Chase: Dr. George M. Garrity: Dr. S.C. long; Dr. Robert Knowlton; Dr. Donald Linzey.

Appendix

Soil orders

Order	Formative element in name	General nature
Alfisols	alf	Soils with gray to brown surface horizons, medium to high base supply, with horizons of clay accumulation; usually moist, but may be dry during summer
Aridisols	id	Soils with pedogenic horizons, low in organic matter, and usually dry
Entisols	ent	Soils without pedogenic horizons
Histosols	ist	Organic soils (peats and mucks)
Inceptisols	ept	Soils that are usually moist, with pedogenic horizons of alteration of parent materials but not of illuviation
Mollisols	oll	Soils with nearly black, organic-rich surface horizons and high base supply
Oxisols	ox	Soils with residual accumulations of inactive clays, free oxides, kaolin, and quartz; mostly tropical
Spodosols	od	Soils with accumulation of amorphous materials in subsurface horizons
Ultisols	ult	Soils that are usually moist, with horizons of clay accumulation and a low supply of bases
Vertisols	ert	Soils with high content of swelling clays and wide deep cracks during some seasons

Carbon cycle

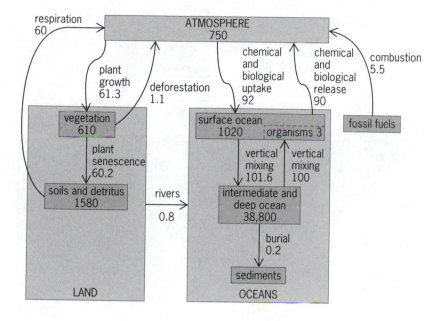

The storage of carbon in the atmosphere and terrestrial and oceanic ecosystems is in picograms ($1\ Pg = 10^{15}$ g), and fluxes of carbon between boxes are in Pg y^{-1}.

Appendix

Nitrogen cycle

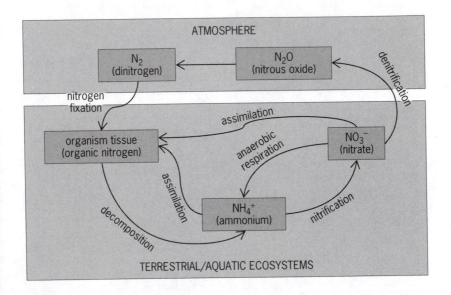

Structure of the atmosphere

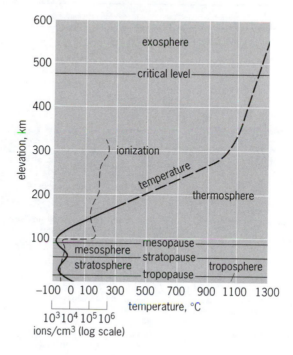

The log scale applies only to the ionization curve. 1 km=0.6 mi; °F=(°C × 1.8) + 32.

Appendix

Major sources and types of indoor air pollutants	
Sources	*Pollutants*
Combustion with appliances using fossil fuels or wood	Particulate matter Nitrogen oxides Carbon monoxide Carbon dioxide Lead and trace metals Hydrocarbons Volatile organic compounds
Tobacco smoking	Particulate matter Carbon monoxide Carbon dioxide Nitrogen oxides Hydrocarbons Volatile organic compounds Radon progeny
Building and furnishing materials	Hydrocarbons (especially aldehydes) Volatile organic compounds Particulate matter Radon progeny Molds and other allergens
Water reservoirs (fixtures for air conditioning, cleaning, or treating)	Molds Bacilli and other bacteria
Consumer products	Halogenated hydrocarbons Volatile organic compounds Trace metals
Animals (pets and opportunistic dwellers) and plants	Allergens Carbon dioxide
Infiltration	Particulate matter Nitrogen oxides Sulfur oxides Pollen Molds

Major categories of water pollutants*

Category	Examples	Sources
A. Causes health problems		
1. Infectious agents	Bacteria, viruses, parasites	Human and animal excreta
2. Organic chemicals	Pesticides, plastics, detergents, oil, and gasoline	Industrial, household, and farm use
3. Inorganic chemicals	Acids, caustics, salts, metals	Industrial effluents, household cleansers, surface runoff
4. Radioactive materials	Uranium, thorium, cesium, iodine, radon	Mining and processing of ores, power plants, weapons production, natural sources
B. Causes ecosystem disruption		
1. Sediment	Soil, silt	Land erosion
2. Plant nutrients	Nitrates, phosphates, ammonium	Agricultural and urban fertilizers, sewage, manure
3. Oxygen-demanding wastes	Animal manure and plant residues	Sewage, agricultural runoff, paper mills, food processing
4. Thermal	Heat	Power plants, industrial cooling

*Reproduced with permission from W. P. Cunningham et al., *Environmental Science. A Global Concern*, 7th ed., McGraw-Hill, 2003.

Appendix

Top fifteen hazardous substances, 2001

Substance	Source	Toxic effects
Arsenic	From elevated levels in soil or water	Multiple organ systems affected. Heart and blood vessel abnormalities, liver and kidney damage, impaired nervous system function.
Lead	Lead-based paint	Neurological damage. Affects brain development in children.
	Lead additives in gasoline	Large doses affect brain and kidneys in adults and children.
Metallic mercury	Air or water at contaminated sites	Permanent damage to brain, kidneys, developing fetus.
Vinyl chloride	Plastics manufacturing	Acute effects: dizziness, headache, unconsciousness, death.
	Air or water at contaminated sites	Chronic effects: liver, lung, and circulatory damage.
Polychlorinated biphenyls (PCBs)	Eating contaminated fish	Probable carcinogens. Acne and skin lesions.
Benzene	Industrial exposure	Acute effects: drowsiness, headache, death at high levels.
	Industrial exposure	Chronic effects: damages blood-forming tissues and immune system; also carcinogenic.
	Glues, cleaning products, gasoline	Probable carcinogen, kidney damage, lung damage, high blood pressure.
Cadmium	Released during combustion	
	Living near a smelter or power plant	
	Picked up in food	
Benzo[a]pyrene	Product of combustion of gasoline or other fuels	Probable carcinogen, possible birth defects.
	In smoke and soot	
Polycyclic aromatic hydrocarbons	Tobacco smoke and charbroiled meats	Probable carcinogens. Difficulty reproducing and possible birth defects.
Benzo[b]fluoranthene	Product of combustion of gasoline and other fuels	Probable carcinogen.
	Inhaled in smoke	
Chloroform	Contaminated air and water	Affects central nervous system, liver, and kidneys; probable carcinogen.
	Many kinds of industrial settings	
DDT	From food with low levels of contamination	Probable carcinogen; possible long-term effect on liver; possible reproductive problems.
	Still used as pesticide in parts of world	
Aroclor 1254 (a mixture of PCBs)	From food and air	Probable carcinogens
		Acne and skin lesions.
Aroclor 1260 (a mixture of PCBs)	From food and air	Probable carcinogens
		Acne and skin lesions.
Trichloroethylene	Used as a degreaser, evaporates into air	Dizziness, numbness, unconsciousness, death.

SOURCE: Data from Agency for Toxic Substances and Disease Registry. (Adapted from E. D. Enger and B. F. Smith, *Environmental Science: A Study of Interrelationships*, 8th ed., McGraw-Hill, 2002)